Giovanni Ciccotti, Mauro Ferrario and Christof Schuette (Eds.)

Molecular Dynamics Simulation

MDPI

This book is a reprint of the special issue that appeared in the online open access journal *Entropy* (ISSN 1099-4300) from 2013–2014 (available at: http://www.mdpi.com/journal/entropy/special_issues/mol_dyn_sim).

Guest Editors
Prof. Dr. Giovanni Ciccotti
Department of Physics, University of Roma "La Sapienza"
Roma, Italy

Prof. Dr. Mauro Ferrario
Department of Physics, University of Modena and Reggio Emilia
Modena, Italy

Prof. Dr. Christof Schuette
Freie Universitaet Berlin, Institute of Mathematics
Berlin, Germany

Editorial Office
MDPI
St. Alban-Anlage 66
4052 Basel, Switzerland

Publisher
Shu-Kun Lin

Managing Editor
Jely He

1. Edition 2014

MDPI • Basel • Beijing • Wuhan

ISBN 978-3-906980-65-2 (Hbk)
ISBN 978-3-906980-66-9 (PDF)

Table of Contents

Preface

Condensed matter systems, ranging from simple fluids and solids to complex multi-component materials and even biological matter, are governed by well understood laws of physics, within the formal theoretical framework of quantum theory and statistical mechanics. On the relevant scales of length and time, the appropriate 'first-principles' description needs only the Schroedinger equation together with Gibbs averaging over the relevant statistical ensemble. However, this program cannot be carried out straightforwardly—dealing with electron correlations is still a challenge for the methods of quantum chemistry. Similarly, standard statistical mechanics makes precise explicit statements only on the properties of systems for which the many-body problem can be effectively reduced to one of independent particles or quasi-particles.

As the interactions among so many degrees of freedom introduce nontrivial correlations between them, only computer simulation provides us with a methodic route to make accurate explicit predictions for the static and dynamic properties of many-body physical systems starting from first principles. The molecular dynamics simulation method (MD) was introduced in the 1950s, shortly after the 'companion' Monte Carlo method. Since then, the scope of both has been rapidly expanding. Despite the fact that suitable computing facilities were scarce, very slow, and with very small storage capacities compared to present-day facilities, immediately important and, at the time, rather surprising discoveries were made—notably that hard spheres crystallize at a density long before close packing has been achieved and that dynamic correlations in fluids exhibit long time tails. These have been the starting point of a great variety of methodological developments, with many exciting technical extensions still under development, providing broad applications and opportunities for important discoveries.

Nowadays, with pervasive high-speed networking and powerful massively-parallel computers at the hands of every scientist, advances in simulation methods are progressing at a breathtaking speed. Molecular dynamics computer simulation offers the advantage that connections can be established between the models of condensed matter on different scales and the hierarchy, from the sub-Angstrom scale—where one deals with effects due to the electrons, up to the mesoscopic and macroscopic scales relevant for living matter. Applications cut across extremely diverse fields, from fundamental problems in solid state physics to the rich world of phenomena exhibited by complex fluids and biological systems—elucidating the electronic properties of materials as well as the major non-equilibrium processes that take place in the living cell. The goal is to develop a simulation approach for complex materials and biological matter that successfully bridges the gap from the small scales of electronic structure calculations to the mesoscopic scales of pattern formation in soft matter (where one uses coarse-grained techniques such as dissipative particle dynamics and multiscale collision dynamics). This is a goal that will remain an exciting challenge for many years to come.

The contributions collected in this book move from the quantum-statistical description to the validity of classical modeling; they present some perspectives in the algorithmic and in the enhanced sampling approaches, tackling some longstanding challenges to simulation in the area of non-equilibrium, rare events, mesoscale and quantum-classical simulation. Initially, the book deals with the validity of molecular dynamics modeling, starting from the adiabatic hypothesis for the electronic ground state; the first contribution explores different descriptions of the potential energy surfaces one can use in a molecular dynamics simulation; the second analyzes in detail the Born-Oppenheimer schemes for *ab initio* MD within Kohn–Sham density functional theory, while the third one tackles the problem from the alternative perspective of a quantum Monte Carlo approach. The next contribution dwells on how to improve the statistical ensemble properties of time integrators for Langevin dynamics by including an acceptance–rejection scheme. The subject of free energy calculations by molecular dynamics is illustrated in the next two contributions, first with a presentation of alternative dynamical approaches for performing enhanced sampling by force biasing and temperature acceleration, then using non-equilibrium path sampling within the framework of Jarzynski identity and Crooks fluctuation theorem. The general ideas behind non-equilibrium molecular dynamics are the focus of the next two contributions, regarding calculation of dynamical responses and the application of Malliavin weight sampling to dynamical trajectories. Many of the same ideas are at the core of the study of rare, reactive, events by molecular dynamics as discussed in the next two contributions, more in general in the first and then with specific reference to the Markov state models approach. The last four invited contributions are dedicated to the problem of dealing with well separated space and time scales. First, the general philosophy of multiscale approaches and the related computational strategies within molecular dynamics are discussed in a concept paper, while the other three deals with specific non-adiabatic dynamical approaches for systems with a mixed quantum-classical description, based upon alternative approaches borrowing either from the Wigner transform representation or from the Bohmian formulation of quantum dynamics. The book is completed by the contributed papers to the molecular dynamics special issue.

The reader will find answers to a number of questions, a few of which we can briefly recall here:

- How to generate averages in statistical mechanics ensembles, other than the microcanonical one, or, in other words, how to couple the system to temperature, pressure or particle baths.
- How to deal with the simultaneous occurrence of slow and fast degrees of freedom that makes straightforward implementations of MD very inefficient, with a great waste of computer resources.
- How to evolve in time a quantum subsystem immersed in a classical environment, using a consistent description based on the Wigner formulation of quantum statistical mechanics, allowing the study of transport phenomena in such mixed quantum-classical systems.
- How to combine *ab initio* MD with classical MD using hybrid approaches in the environment of the reactive groups, by suitable "quantum mechanical/molecular mechanical (QM/MM)" partitioning.
- How to extend the standard quantum Monte Carlo approach to obtain a description of electronic structure that provides an interesting alternative to the density functional based methods.
- How to efficiently sample rare events, e.g., a nucleation process where a huge free energy barrier needs to be crossed to form a critical nucleus of the new stable phase on the background of a metastable phase, and develop sampling schemes for computing the relevant properties and studying the mechanisms of transitions between metastable states.
- How to eliminate or treat in a simplified way, by coarse-graining, some small-scale degrees of freedom, which are considered less relevant to the considered questions.

This is what you will find in the present book but many more questions, some certainly yet to be posed, will certainly find their answers in the forthcoming developments of molecular dynamics simulation.

We wish to acknowledge the collaboration of the many people who have made possible this special issue. First of all, the authors, whose rigor, good work and speed have, of course, been instrumental. Also, we are very grateful to the many anonymous referees for the invaluable work of guaranteeing the quality and soundness of the contributions. Thanks, finally, to Jely He: She and the entire MDPI staff of the Editorial Office of Entropy have generously given invaluable help and good professional skill to bring this adventure to a successful conclusion.

Giovanni Ciccotti, Mauro Ferrario and Christof Schuette
Guest Editors

Reprinted from *Entropy*. Cite as: Ballone, P. Modeling Potential Energy Surfaces: From First-Principle Approaches to Empirical Force Fields. *Entropy* **2014**, *16*, 322–349.

Article

Modeling Potential Energy Surfaces: From First-Principle Approaches to Empirical Force Fields

Pietro Ballone

Department of Physics, Università di Roma "La Sapienza", Roma 00185, Italy;
E-Mail: pballone58@gmail.com; Tel.: +39-06-4991-4248; Fax: +39-06-4991-7697

Received: 17 September 2013; in revised form: 15 October 2013 / Accepted: 18 October 2013 / Published: 30 December 2013

Abstract: Explicit or implicit expressions of potential energy surfaces (PES) represent the basis of our ability to simulate condensed matter systems, possibly understanding and sometimes predicting their properties by purely computational methods. The paper provides an outline of the major approaches currently used to approximate and represent PESs and contains a brief discussion of what still needs to be achieved. The paper also analyses the relative role of empirical and *ab initio* methods, which represents a crucial issue affecting the future of modeling in chemical physics and materials science.

Keywords: atomistic modeling; bond-order potentials; *ab initio* methods

1. Introduction

Most, if not all, of computer simulations using particles require the specification of the system potential energy as a function of particles' coordinates [1]. The most *ab initio* methods, such as those discussed in [2], represent systems as made of electrons and atomic nuclei, and Coulomb's law is sufficient to account for every interaction. In all other cases, *particles* represent composite objects, such as atoms or atomic nuclei, dressed by core electrons, possibly embedded into a sea of valence electrons described at some approximate level of a many-body theory. Then, all the relevant interactions need to be worked out on a case by case basis, and the effort required to determine inter-particle forces may represent a sizeable fraction of the work to be done to investigate condensed matter systems [3].

The sections that follow contain an overview of modeling approaches and a discussion of their relative merits and limitations. Needless to say, the variety of systems and methods, together with the shear size of the knowledge accumulated over decades, impose strict limits to the scope of this presentation. First of all, the focus is on atomistic models, *i.e.*, models in which the number and geometry of interaction centers follows the distribution of atoms closely. A second major branch of modeling, concerning coarse graining approaches, is the subject of a separate contribution (see [4]).

Moreover, again, for limitations of space, the discussion that follows mainly concerns the most restrictive picture of interatomic interactions, based on the assumption that the potential energy of a system of N atoms can be expressed as a single-valued function of their $3N$ coordinates $\{\mathbf{R}_i, i = 1, ..., N\}$, which represents the so-called potential energy surface (PES) of the system. This assumption relies, first of all, on the so-called Born-Oppenheimer approximation [5], whose validity is loosely attributed to the ~ 3–4 orders of magnitude difference in the mass of electrons and atomic nuclei, giving rise to a clear separation of the characteristic energy and time scales for the motion of electrons and atomic nuclei. Then, for any given instantaneous configuration of the atomic cores, electrons will be able to reach their electronic ground state, justifying the single-value assumption for the system potential energy. Experience shows that this "adiabatic assumption" is fairly well justified for a wide variety of systems and thermodynamic conditions. To be precise, it turns out that some cases are left out of this picture and often represent systems and phenomena of great interest. Methods suitable to deal with these cases are discussed in [6].

Computational science and simulation, in particular, always have a practical and an algorithmic aspect to them, and a central theme of research is the development of efficient ways to approximate and represent PESs. The availability of simple and computationally-convenient models of inter-particle interactions, for instance, has been instrumental in the dawning of computer simulation. Since then, the two complementary stages of determining the relevant interactions and of working out their structural, thermodynamic and dynamical consequences have cross fertilized each other, so much that the terms, *modeling* and *simulation*, often appear together in the title of books, papers, conferences, workshops and funding proposals.

Nowadays, the general perception of atomistic modeling is that of an overwhelmingly important and successful field, steadily expanding its reach towards more complex systems, which in this context means systems combining a wider variety of chemical bonds. In this respect, it is clear that much remains to be done, for instance, to bring under the cover of simulation heterogeneous systems and interfaces at which organic, semiconducting and metal phases meet each other or to model systems in which chemical transformations take place.

During the last few decades, *ab initio* simulation methods have progressively come to play the role of the elephant in the (modeling) room. Methods, such as density functional theory [7,8] and *ab initio* molecular dynamics [9], could, in principle, replace all other approaches, reducing the variety of modeling problems to just one, concerning the effective and accurate representation of the energy of valence electrons in the field of atomic nuclei or ionic cores.

Up to now, this replacement has not been pervasive, mainly because of the size and time limitations of *ab initio* methods running on present day computers and partly because the approximations that make *ab initio* computations feasible still somewhat limit their accuracy on the energy scale of thermal motion, especially for molecular systems whose properties are determined by weak interactions among closed shell molecules. *Ab initio* modeling, however, is progressing and extending its reach. For what concerns atomistic simulation, therefore, empirical and semi-empirical models might eventually be squeezed out by the combination of *ab initio* methods and coarse-grained approaches. Simple models of atom-atom interactions, however, are likely to retain their appeal, because of their unique ability to represent and rationalize the microscopic forces underlying the properties and behaviors of condensed matter systems.

2. The Potential Energy Surface (PES) of a Many-Atom System

From a physicist point of view, ordinary matter consists of an assembly of electrons and atomic nuclei, evolving according to the laws of quantum mechanics. The non-relativistic limit is adequate for many of the systems and properties of interest for the present discussion, and unless differently specified, we shall restrict ourselves to this case.

Let us therefore consider a system made of N electrons and K nuclei, and let $\{\mathbf{r_i}, i = 1, ..., N\}$ and $\{\mathbf{R_\alpha}, \alpha = 1, ..., K\}$ be the coordinates of electrons and nuclei, respectively. The corresponding linear momenta are denoted by $\{\mathbf{p_i}\}$ and $\{\mathbf{P_\alpha}\}$. In the absence of external fields, the system Hamiltonian is:

$$\hat{H}_0 = \sum_{\alpha=1}^{K} \frac{\mathbf{P_\alpha^2}}{2M_\alpha} + \sum_{i=1}^{N} \frac{\mathbf{p_i^2}}{2m} + \frac{1}{2} \sum_{\alpha \neq \beta} \frac{Z_\alpha Z_\beta e^2}{|\mathbf{R_\alpha} - \mathbf{R_\beta}|} - \sum_{i,\alpha} \frac{Z_\alpha e^2}{|\mathbf{r_i} - \mathbf{R_\alpha}|} + \frac{1}{2} \sum_{i \neq j} \frac{e^2}{|\mathbf{r_i} - \mathbf{r_j}|} \tag{1}$$

that, for the sake of simplicity, we re-write as:

$$\hat{H}_0 = T_{ion} + T_{ele} + V_{ion-ion} + V_{ion-ele} + V_{ele-ele} \tag{2}$$

with an obvious correspondence between Equations (1) and (2). The Hamiltonian does not depend on the spin of electrons and nuclei, since we restrict ourselves to the non-relativistic limit, and we do not include any spin-orbit interaction into our Hamiltonian. Unless differently specified, Hartree atomic units ($\hbar = e^2 = m = 1$) are used in this section.

Let us assume that the system is described by a many-body wave function, $\Psi(\mathbf{r_1}, ..., \mathbf{r_N}; \mathbf{R_1}, ..., \mathbf{R_k}; t)$, whose time evolution is determined by the time-dependent Schrodinger equation:

$$i\hbar \frac{\partial \Psi(\{\mathbf{r_i}\}; \{\mathbf{R_\alpha}\}; t)}{\partial t} = \hat{H}_0 \Psi(\{\mathbf{r_i}\}; \{\mathbf{R_\alpha}\}; t) \tag{3}$$

with appropriate boundary conditions in space and in time. Since the Hamiltonian is time independent, let us turn to the equivalent version of this same problem, concerned with the stationary states, $\Psi_k(\{\mathbf{r_i}\}; \{\mathbf{R_\alpha}\})$ of $\hat{H}_0$.

The first important step towards the definition of a potential energy surface for the atomic nuclei is provided by the Born-Oppenheimer approximation (BO), which, under suitable and often verified conditions, opens the way to a separate description of the time evolution of electrons and nuclei [5]. The intuitive justification of BO is the observation that the motion of electrons and nuclei takes place over different time scales, since M_α/m is at least $M_n/m \sim 1,800$, and usually approaches $2Z_\alpha M_n/m$, where M_n is the mass of a nucleon (proton or neutron). Moreover, the ratio of vibrational and rotational excitations is again $\sim \sqrt{M_\alpha/m}$. Experimental data confirm that, indeed, typical electronic excitations are of the order of a few eV; vibrational energies reach up to a few hundred meV, and even for small molecules, the separation of rotational levels is of the order of 1 meV. The conclusion is that the excitation of electrons, because of vibrational or rotational motion, is very unlikely. We can therefore represent the motion of electrons as taking place in the slowly varying field of the nuclei. Consistently with these qualitative arguments, the BO approximation breaks down whenever the energy of relevant electronic excitations becomes comparable to typical vibrational energies (or, much less likely, comparable to rotational energies). In those cases, vibrational and electronic excitations need to be considered on the same footing.

The core of the so-called adiabatic approximation can be given a semi-rigorous mathematical formulation in the following way [5]. Let us re-write $\hat{H}_0$ as:

$$\hat{H}_0 = \hat{T}_{ion} + \hat{H}_{ele} \tag{4}$$

where $\hat{H}_{ele} = \hat{T}_{ele} + V_{ion-ion} + V_{ion-ele} + V_{ele-ele}$. The energy term, $V_{ion-ion}$, commutes with all other terms in $\hat{H}_{ele}$, and its inclusion in the electronic part is just a matter of convenience.

For every choice of the nuclear coordinates, $\{\mathbf{R}_\alpha, \alpha = 1, ..., K\}$, the eigenvalue problem:

$$\hat{H}_{ele}\psi_j(\{\mathbf{r_i}\} \mid \{\mathbf{R}_\alpha\}) = E_j(\{\mathbf{R}_\alpha\})\psi_j(\{\mathbf{r_i}\} \mid \{\mathbf{R}_\alpha\}) \tag{5}$$

is well defined and provides a sequence of eigenvalues, $E_j(\{\mathbf{R}_\alpha\})$, and eigenfunctions $\psi_j(\{\mathbf{r_i}\} \mid \{\mathbf{R}_\alpha\})$. At this stage, nuclei are "clamped", i.e., they are no longer treated as particles embodied with a mass and a momentum, but only as sources of the potential acting on the electrons. The notation, $(\mathbf{r_i} \mid \mathbf{R}_\alpha)$, means that ψ_j is an explicit function of $\mathbf{r_i}$ and depends parametrically on the nuclear coordinates, $\{\mathbf{R}_\alpha\}$.

The functions, ψ_j, are a basis for the Hilbert space spanned by the electron coordinates, and we can represent Ψ_k as follows:

$$\Psi_k(\{\mathbf{r_i}\}, \{\mathbf{R}_\alpha\}) = \sum_j \psi_j(\{\mathbf{r_i}\} \mid \{\mathbf{R}_\alpha\})\chi_j^{(k)}(\mathbf{R}_\alpha) \tag{6}$$

where, at this stage, $\chi_j^{(k)}(\mathbf{R}_\alpha)$ is simply the coefficient expressing the projection of Ψ_k on ψ_j:

$$\chi_j^{(k)}(\{\mathbf{R}_\alpha\}) = \int \psi_j^*(\{\mathbf{r_i}\} \mid \{\mathbf{R}_\alpha\})\Psi_k(\{\mathbf{r_i}\}, \{\mathbf{R}_\alpha\})\Pi_{i=1}^N d\mathbf{r_i} \tag{7}$$

The equation for Ψ_k becomes:

$$\hat{H}_0\Psi_k(\{\mathbf{r_i}\}, \{\mathbf{R}_\alpha\}) = (\hat{T}_{ion} + \hat{H}_{ele})\Psi_k(\{\mathbf{r_i}\}, \{\mathbf{R}_\alpha\}) \tag{8}$$

$$= \sum_j \chi_j^{(k)}(\{\mathbf{R}_\alpha\})E_j(\mathbf{R}_\alpha)\psi_j(\{\mathbf{r_i}\} \mid \{\mathbf{R}_\alpha\}) + \psi_j(\{\mathbf{r_i}\} \mid \{\mathbf{R}_\alpha\})\hat{T}_{ion}\chi_j^{(k)}(\{\mathbf{R}_\alpha\})$$

$$+ \chi_j^{(k)}(\{\mathbf{R}_\alpha\})\hat{T}_{ion}\psi_j(\{\mathbf{r_i}\} \mid \{\mathbf{R}_\alpha\}) = \mathcal{E}_k\Psi_k(\{\mathbf{r_i}\}, \{\mathbf{R}_\alpha\})$$

Let us now multiply on the left by $\psi_m^*(\{\mathbf{r_i}\} \mid \{\mathbf{R}_\alpha\})$ and integrate over the electron coordinates. One obtains in this way a set of coupled partial differential equations for the $\chi_m^{(k)}(\{\mathbf{R}_\alpha\})$ functions:

$$E_m(\{\mathbf{R}_\alpha\})\chi_m^{(k)}(\{\mathbf{R}_\alpha\}) + \hat{T}_{ion}\chi_m^{(k)}(\{\mathbf{R}_\alpha\}) + \sum_j \chi_j^{(k)}(\{\mathbf{R}_\alpha\})\langle\psi_m \mid \hat{T}_{ion} \mid \psi_j\rangle = \mathcal{E}_k\chi_m^{(k)}(\{\mathbf{R}_\alpha\}) \quad (9)$$

where $\mathcal{E}_k$ is the eigenvalue of the full, *i.e.*, electrons and ions Hamiltonian $\hat{H}_0$, and the relation, $\langle\psi_m \mid \psi_j\rangle = \delta_{mj}$, has been used. The coupling among the equations is due to the non-diagonal part of $\langle\psi_m \mid \hat{T}_{ion} \mid \psi_j\rangle$:

$$\langle\psi_m \mid \hat{T}_{ion} \mid \psi_j\rangle = \sum_\alpha \frac{1}{M_\alpha} \int \left[-i\frac{\partial\psi_m(\{\mathbf{r_i}\} \mid \{\mathbf{R}_\alpha\})}{\partial\mathbf{R}_\alpha}\right]^* \left[-i\frac{\partial\psi_j(\{\mathbf{r_i}\} \mid \{\mathbf{R}_\alpha\})}{\partial\mathbf{R}_\alpha}\right] \Pi_{i=1}^N d\mathbf{r_i} \quad (10)$$

whose computation requires the parametric dependence of $\chi_m(\mathbf{R}_\alpha)$ on the $\{\mathbf{R}_\alpha\}$ coordinates to be continuous and differentiable.

Neglecting these non-diagonal terms, the equations for the electronic and ionic coordinates are decoupled, and the picture emerging from this manipulation of Equation (6) is that of nuclei evolving on the potential energy surfaces $U_j[\{\mathbf{R}_\alpha\}] = E_j(\{\mathbf{R}_\alpha\}) + \langle\psi_j \mid \hat{T}_{ion} \mid \psi_j\rangle$. This last expression, corresponding to the so-called Born-Huang approximation [10], represents, in fact, an upper bound for the system's potential energy. A lower bound, instead, is given by the original BO approximation, *i.e.*, $U_j[\{\mathbf{R}_\alpha\}] = E_j(\{\mathbf{R}_\alpha\})$.

The nuclear motion in general is quantum mechanical, and, depending on initial conditions, it might occur on any of the U_j potential energy surfaces (PESs). More precisely, since the equations for different j's are separated, it will take place on a single surface of index j, provided the starting point is consistent with this choice. This condition, that we identify with *adiabatic motion*, underlies most of the simulations that are routinely carried out in computational-condensed matter physics. Moreover, again, in most cases, but with noticeable exceptions, the relevant PES corresponds to the electronic ground state, and the scale of times and energies of interest allows the usage of classical dynamics instead of quantum mechanics [6].

The following sections are devoted to the discussion of the general properties of PESs, and of computationally tractable approaches to approximate them. Before doing that, it might be interesting to consider briefly when the BO approximation and the conditions for adiabatic motion are no longer valid.

An estimate of the $\langle\psi_m \mid \hat{T}_{ion} \mid \psi_j\rangle$ terms can be obtained by perturbation theory, showing that the strength of the non-diagonal coupling is proportional to:

$$\langle\psi_m \mid \hat{T}_{ion} \mid \psi_j\rangle \propto \frac{1}{E_m - E_j}\langle\psi_m \mid [\mathbf{P}_\alpha, \hat{H}_{ele}] \mid \psi_j\rangle \quad (11)$$

Moreover, the matrix element of the commutator can be shown to depend primarily on the properties of individual atoms and to be only moderately dependent on the $\{\mathbf{R}_\alpha\}$ coordinates. Then, the major factor determining the coupling strength among different adiabatic surfaces is the energy gap separating different PESs. Whenever $(E_m - E_j)$ becomes comparable to the typical energies of the atomic motion, the BO decoupling is no longer valid, the electronic and ionic motion are intimately intertwined and both need to be treated quantum mechanically. The range of quantum mechanical features that become relevant in the non-BO case go beyond delocalization and diffraction, but includes the appearance of geometric (Berry-Pancharatnam) phases [11].

Far from being the exception, violations of the BO approximation are pervasive. They occur often, but not exclusively, at the so-called conical intersections [11], playing a major role in chemical reactions and, for instance, challenging our ability to model catalysis [12]. Apparent non-BO effects are routinely highlighted by clever experiments [13,14].

Metals, whose occupied states are immediately contiguous in energy to the empty states, may appear as the most obvious candidates for large deviations from the BO picture. In the vicinity of the Fermi surface, however, single particle excitations are the only relevant excitations, but the coupling of each of these excitations to the nuclear motion (through Equation (11)) is vanishingly small. Collective electron excitations, such as plasmons, couple to the atomic motion, but their energies are of the order of several eV and, thus, are comparable to, if not higher than, those of closed shell atoms and molecules. As a result, vibrational properties of metals are generally well described by adiabatic dynamics. Exceptions are represented by Kohn anomalies, resulting from the nesting of reciprocal lattice vectors with the Fermi surface. Metals also provide the setting for a type of BO violation qualitatively different from those considered until now, represented by superconductors, in which the coupling of the electron and nuclear motion changes the symmetry of the ground state.

The isolated system picture underlying the BO decoupling has been generalized in [15–17] to the case of electrons and nuclei evolving in an external time-dependent potential. It was shown, in particular, that the full wave function can be factorized exactly into an electronic and a nuclear wave function, again opening the way to the definition of a time-dependent PES. The picture is less simple than in the static case, since it involves the introduction of a Berry vector potential and of Berry-Pancharatnam geometric phases [18,19] into the problem. This approach has already provided the basis for the real-time simulation of molecular systems in strong (laser) external fields. For completeness, I mention that some details of the formal framework might still need to be worked out for a fully rigorous treatment [20].

3. Properties of Potential Energy Surfaces

Basic features of the PES can be anticipated even without an explicit solution of the standard electronic problem in Equation (5). A surprisingly realistic intuition of what a PES looks like was outlined in elegant Latin prose long before quantum mechanics [21], based on an atomistic hypothesis and on the assumption that the still undiscovered atoms felt each other mainly at short distances.

The modern interpretation confirms this picture and adds a wealth of microscopic detail. The direct Coulomb repulsion among nuclei, unscreened by electrons at short distances, prevents the close contact of atoms and their eventual collapse. The kinetic energy of the electrons tightly bound to the nuclei will provide an additional repulsive contribution, resulting from the need to preserve the Pauli principle. On the other hand, the formation of chemical bonds gives rise to attractive potentials, binding atoms together. Even in the case of inert species, subtle quantum mechanical effects give rise to dispersion forces, which provide a weak, but pervasive, attraction.

Arguably, the simplest and most intuitive picture of atomic interactions is provided by pair potential models, in which the system energy is written as:

$$U[\{\mathbf{R}_\alpha\}] = \frac{1}{2} \sum_{\alpha,\beta} \phi_{\alpha\beta}(|\mathbf{R}_\alpha - \mathbf{R}_\beta|) \tag{12}$$

where the α, β label on $\phi_{\alpha,\beta}$ indicates that the interaction depends on the chemical identity of particles α and β. A spherically symmetric potential has been assumed for the sake of simplicity.

Computations and comparison with experiments have shown that an expression of this kind is suitable for rare gases [22] and for simple ionic compounds [23]. Systems and models of this kind have been instrumental in establishing computer simulation as a quantitative research tool in condensed matter and in chemical physics.

Needless to say, the scope of pair potentials is very narrow, and limitations of this model were already apparent well before the dawn of computer simulation, based on the results of lattice dynamics models in metals and semiconductors.

One could think of the pair potential expression as being only the lowest order approximation of the PES into an n-body expansion of the form:

$$U[\{\mathbf{R}_\alpha\}] = \frac{1}{2!} \sum_{\alpha,\beta} V_2(\mathbf{R}_\alpha, \mathbf{R}_\beta) + \frac{1}{3!} \sum_{\alpha,\beta\gamma} V_3(\mathbf{R}_\alpha, \mathbf{R}_\beta, \mathbf{R}_\gamma) + \dots \tag{13}$$

For a system made of a finite and constant number of particles, such an expression can always be written down. For instance, one could define V_2 as the interaction energy of two isolated atoms, V_3 as the corresponding energy of trimers, minus the symmetrized combination of V_2 contributions, *etc.* Such an expansion, however, is useful only if it converges within a few terms, at least because the cost of evaluating successive n body terms grows rapidly with increasing n. Moreover, it contributes to the physical understanding of the system behavior only when its convergence is absolute, *i.e.*, it does not require the cancellation of contributions of alternating sign, whose amplitude is constant or even increasing with increasing order. Model computations based on a tight binding Hamiltonian [24], however, show that even for simple systems, the expansion in Equation (13) is not well behaved and, thus, is seldom useful for practical computations.

More fruitful than the systematic expansion of Equation (13) has been the introduction of the *cluster potential* idea [25,26], loosely and sometimes more closely based on the bond-order concept introduced by Pauling [27]. In this approach, a fixed and low number of terms is retained; the expression looses its character of a systematic series to become an asymptotic expansion. Each of the few terms that are retained describe low-order potentials whose strength depends on the local environment. Approaches of this kind have given origin to the most popular family of potentials used to simulate metals and metallic alloys and also to some important approaches to approximate the PES of semi-conductors, which are discussed in the following sections.

4. Many-Body Interactions: Metals and Metal Alloys

Metals and their alloys posed an early challenge to the pair or few-body potential picture, since their basic properties manifest essential many-body interactions [28].

The successful and physically-motivated incorporation of these effects into tractable models in the early eighties of the last century has spawned a vast simulation activity, aiming, at first, at reproducing phase diagrams, then at analyzing in detail surfaces and interfaces and further progressing towards the prediction of mechanical properties through multi-scale approaches. Physical metallurgy is currently one of the most active and productive subfields of atomistic simulation [29,30].

Many-body interactions in metals were first identified by the analysis of their elastic properties. For instance, the elastic constants of cubic materials consisting of atoms interacting via spherically symmetric pair potentials have to satisfy the so-called Cauchy relations, stating, for instance, that $C_{12} = C_{44}$. The violation of this relation, known in the solid state literature as a Cauchy anomaly, is the rule more than the exception in metals, unambiguously pointing to a deviation from the pair potential picture.

These features were first rationalized by considering the basic representation of a metal, as made of ions embedded into a sea of valence electrons. Since the major ingredient, *i.e.*, the homogeneous electron gas could be solved analytically, and, at least for sp metals, the electron-ion interaction is weak, the full problem could be attacked by perturbation theory [28,31]. Carried up to the second order, this approach provides an expression for the system total energy that consists of a large volume (or, equivalently, density) term and a pair potential contribution. The volume term is able to account for the Cauchy anomaly. In simple metals, such as the alkalis, the pair potential is relatively soft at short distances and oscillates at large distances, reflecting Friedel oscillations. These features explain the bccstructure of these systems at normal conditions and provide a clue to understand more complex structures adopted by the lighter alkali metals at very low temperature or found in slightly more complex systems, such as alloys, or heavier sp metals, such as gallium, indium or tin.

Approaches of this kind are now mainly of historical interest, since most of the cases relevant for applications involve transition metals, and in those systems, the valence electron-ion interaction is by no means weak; the perturbation expansion cannot be limited to the second order and becomes rapidly untreatable beyond that point [32]. Besides these fundamental problems, other practical difficulties concern the definition and the zero-order solution of an electron gas problem suitable for inhomogeneous systems and for alloys. Electron gas perturbative approaches, therefore, could not solve problems, such as the inward relaxation of crystal surfaces, the quantitative description of stacking faults or the overestimation by pair potentials of the vacancy formation energy in metals.

To overcome these problems, new models have been proposed in [33–35], conforming to the cluster-potential idea [26], and representing low-order approximations to a bond-order potential. The embedded atom model (EAM) of [33,34], loosely based on density functional theory, has the broadest appeal, and for this reason, it is used here as a representative of a wider class of models.

According to EAM, each metal ion, i, at position $\mathbf{R_i}$ gains an energy, $E[\rho_e(\mathbf{R_i})]$, upon being immersed into the valence electron distribution at density $\rho_e(\mathbf{R_i})$ and interacts with neighboring ions by a short range repulsive pair potential, $V_2(R)$. The energy of N metal atoms, therefore, is:

$$U[\{\mathbf{R_i}\}] = \frac{1}{2}\sum_{\substack{i \neq j}}^{N} V_2(|\,\mathbf{R_i} - \mathbf{R_j}\,|) + \sum_{i=1}^{N} E[\rho_e(\mathbf{R_i})] \tag{14}$$

The picture is completed by a prescription to compute the electron density, ρ_e, at the position, $\mathbf{R_i}$, of each atomic core. EAM represents such a density as the sum of contributions from every other atom:

$$\rho_e(\mathbf{R_i}) = \sum_{j \neq i} t_j(|\mathbf{R_i} - \mathbf{R_j}|) \tag{15}$$

where the $t_j(R)$ are again relatively short-range functions, mimicking the tail of the electron distribution around an isolated atom. Since it introduces a *local* embedding density, this prescription overcomes most of the limitations of the free electron models, which instead rely on a global definition of the valence electron density.

Parameters and auxiliary functions, such as $t(R)$, $E[\rho_e]$ and $V_2(R)$, could be computed from first principles [36], but this approach has been only moderately successful. Far more effective has been the strategy of adopting the EAM potential energy expression as a general framework, relying on fitting experimental quantities to tune a few parameters distributed into the functional form.

The success of EAM has been due to its ability to overcome the limitations of simpler models, easily accounting for the Cauchy anomaly, the reduced value of the vacancy formation energy, the inward relaxation of compact metal surfaces and the reconstruction of more open ones. Its broad acceptance relies also on the many and physically appealing properties of the model, discussed in a number of publications, such as the ease of extending EAM to alloys or the close relation with pair potentials in the case of homogeneous systems at constant volume.

From the computational point of view, the efficiency of EAM is due to the pair potential form of both the repulsive contribution, V_2, and the embedding density expression in Equation (15). The time required to carry out a simulation based on EAM is expected to be twice that of a pair potential model, since a pass on all atom pairs is required to compute the repulsive potentials and the embedding density, while a second pass is needed to compute forces on atoms arising from the embedding energy. With suitable lists of neighbors, and depending on the range of $V_2(R)$ and of $t(R)$, EAM can be used to carry out MDsimulations for systems of 10^4 atoms over several nanoseconds using laptops or inexpensive PCs. Supercomputers extend these ranges to several million atoms, and μs time scales.

Needless to say, an empirical and approximate approach, such as EAM, cannot provide the final answer to the problem of modeling metals, and transition metals, in particular. A comprehensive discussion of inaccuracies and limitations identified during thirty years of applications is beyond the scope of this short review, and only two examples are briefly mentioned here. Phonons in transition metal crystals, a property routinely measured by inelastic neutron scattering, are not well reproduced by EAM. The elastic constants usually enter the fitting of the potential, and thus, the low-frequency acoustic phonons close to the Γ-point of the first Brillouin zone are usually well reproduced. Higher frequency modes at the zone boundary, however, turn out to be too soft with respect to the experimental data (see Figure 1). Transition metal clusters from a few to several thousand atoms are important for catalysis and represent a basic ingredient of nanotechnology. EAM neglects the details of the electronic structure of the atoms, leaving out quantum mechanical effects, such as Jahn-Teller. Thus, EAM is unable to quantitatively reproduce the structure and cohesive properties of the very small aggregates as provided by density functional computations. Beyond ~ 100 atoms, cluster properties are expected to evolve more continuously with size, approaching those of bulk phases beyond 10^4 atoms. EAM has been used extensively to investigate clusters across this range, but a quantitative validation of the model is still lacking and difficult to achieve, since more *ab initio* computations become too expensive to carry out, and experiments find it difficult to probe this range of cluster sizes.

A step beyond EAM, needed to quantitatively model the fine details of the structure, thermodynamics and dynamics of transition metal systems, requires the introduction of explicit angular terms into the potential energy expression. This can be achieved through a conceptually simple extension of EAM, known as modified EAM (MEAM) [34], or resorting to a chemically accurate bond-order potential model, including the directionality of d and f electron orbitals, as well as the distinction of σ, π, δ, ..., bonding, anti-bonding and non-bonding orbitals [37].

The MEAM is somewhat more complex to use than EAM, and probably for this reason, it has been less extensively applied. Moreover, its ability to quantitatively overcome the limitations of the simpler model is not always so apparent. The other approaches, more closely based on the bond order approach, appear to be cumbersome to use in simulations, and the number of applications based on these models has been limited.

Figure 1. Phonon frequencies of fccpalladium from experiments (symbols, see [38]) and from the embedded atom model (EAM) model of [33].

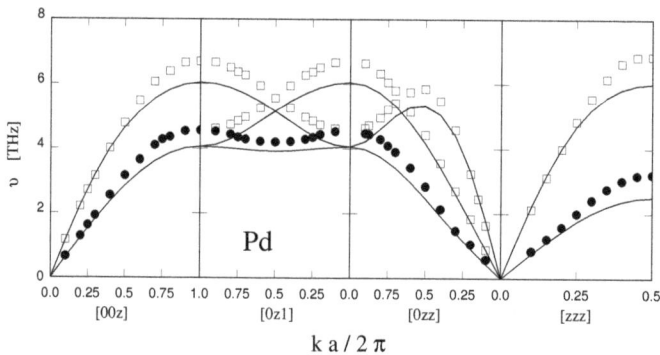

Because of the inclusion of angularly dependent forces, the scope of MEAM could, in principle, cover semiconductors. Successful applications have been published [34], but more specific models, described in the following section, have received broader attention in this subfield.

5. Semiconductors and Insulators

Semiconductor materials, exemplified by silicon, germanium, gallium arsenite, *etc.*, are characterized by fairly open and complex structures of relatively low coordination, stabilized by sizeable angular forces, arising from the directionality of covalent bonds. Apart from elemental systems, most inorganic semiconductors are characterized, in fact, by a combination of covalent and ionic bonding. Several of these systems, most notably silicon and germanium, turn into metals upon melting.

Despite the difficulty of reproducing these properties by few-body potentials, the urgency of investigating the elements and compounds that fueled the electronic revolution stimulated the first bold attempts. The two- and three-body potential for silicon proposed by Stillinger and Weber [39] arguably has been the most representative example of this first generation of models.

Despite their interest, approaches of this kind have been only moderately successful, and once again, the bond-order concept [27] proved more fruitful. Its application to semiconductors was first discussed by Abell [25] before being used in a more empirical setting by Tersoff [40,41] and extended by Brenner [42] to a wider class of systems and problems.

According to these models, the potential energy of an assembly of N atoms of coordinates $\{\mathbf{R_i}\}$ is written as:

$$E_N = \sum_{i \neq j} [A \exp(-\lambda_1 R_{ij}) - B_{ij} \exp(-\lambda_2 R_{ij})] \tag{16}$$

where $R_{ij} = |\mathbf{R_i} - \mathbf{R_j}|$. The first term, representing the short-range repulsion, is a genuine pair potential. The second term contains many-body contributions via the dependence of B_{ij} on the local environment around the interacting pair, ij.

This form has obvious analogies with the EAM case. The difference is that B_{ij} not only counts neighbors, as the embedding density does, but takes into account also the angular correlation among their mutual positions. This addition is required to enforce the dominance of tetrahedral sp^3 coordination, but also to carve a secondary role for other structures, from the sp^2 bonding of graphite, to the octahedral coordination of liquid silicon and germanium [40,41].

Parallel to the EAM case for metals, potentials of this type replaced previous models and established a new standard in modeling semiconducting systems. Success, however, has been somewhat less pervasive than in the case of EAM, for reasons that are relatively easy to identify. First of all, interactions in semiconductors are more complex and propagate at a longer range, since screening is not as effective as in metals. Moreover, semiconducting alloys and compounds give rise to partially Coulombic interactions, whose combination with covalent bonding has seldom been modeled, even by bond-order potentials.

Furthermore, in this case, the systematic improvement beyond the semi-empirical Tersoff and Brenner potentials has to rely on the analytical development of chemically accurate bond-order models [43]. Work along these lines is underway and has shown promising developments, but current models still appear fairly difficult to implement in molecular dynamics or Monte Carlo packages.

An important development of Brenner's scheme has been the introduction of *reactive* force fields, able to describe chemical transformations in the system under consideration. The majority of the parameterizations and applications published until now concern organic systems, but potentials of this kind are mentioned here for their similarity with models first introduced for semiconductor systems. Prototypical examples of a reactive force field are the so-called ReaxFF [44] and the REBOpotential [45]. Both models require a massive parametrization effort, and for this reason, they appear to be fairly *ad hoc* and system specific.

A different line of attack to modeling semiconducting systems is suggested by the observation that in many cases, force fields of the form currently used to model organic systems and consisting on stretching, bending and torsion might indeed provide a good representation of structural and dynamical properties of semiconductors and of network insulators, such as silica. Models of this kind, in fact, were developed well before the age of computer simulation, and extensively used in lattice dynamics studies of semiconductors and insulators [46]. The problem of these models is that, mainly because of the established tradition, the topology of bonds is kept fixed, bonds are harmonic and can neither form nor break. These models, therefore, describe only low amplitude oscillations around a pre-assigned minimum of the potential energy surface. Removing these inessential constraints by introducing rules to break, form and interchange bonds results in a far more realistic picture. It was shown, for instance, that such a reactive force field model of silica undergoes melting at approximately the right conditions [47] (see Figure 2), and the same model has been used to provide an intriguing view of the amorphous silica surface at length and time scales unachievable by other methods [48].

Figure 2. Average potential energy per atom $\langle U(T)\rangle/K_B$ of SiO_2 computed by the force field of [47]. k_B is the Boltzmann constant, introduced to express energies in temperature units (K). Solid dots: heating a β-cristobalite sample. Solid line: cooling the same sample from high temperature. The potential energy contribution, C_p, to the constant pressure specific heat computed on heating the full model is shown in the inset. The peak in C_p and the anomaly in $\langle U(T)\rangle$ are around the same temperature point to a melting transition at $T_M \sim 2150$ K.

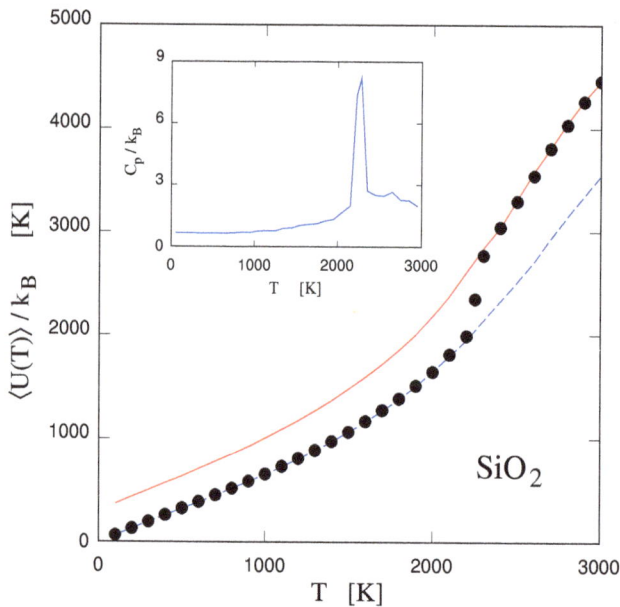

Progressively increasing the electronegativity difference in compound semiconductors enhances the charge transfer among atoms, widening the band gap and turning the system into an ionic insulator. In the limit of strongly ionic materials, of course, pair potentials are adequate, but only a few compounds belong to this class, such as, for instance, alkali-halides or the oxides and chlorides of Group IIA and Group IIB metals. In between ionic insulators and polar semiconductors, there is a vast number of systems, including technologically relevant compounds, such as ceramics, transition metal oxides, ferroelectric and ferroid materials, minerals and bio-minerals, in particular, for which no current model is fully satisfactory. One of the major issues for these systems is the inclusion of polarizability into ionic and polar models [49]. Unfortunately, simulation approaches using polarizable models require either the minimization at every step of a polarization energy functional or the inclusion into the model of charged shells [50]. These last represent electronic degrees of freedom and react to electric fields on a time scale much faster than that of ionic vibrations [51]. Both methods are significantly at a disadvantage with respect to cases in which the potential energy is an explicit function of the atomic coordinates, and the simulation of systems bound by a combination of covalent and ionic forces appears to be split between oversimplified pair potential models and *ab initio* approaches.

6. Force Fields for Molecular Systems

Although every material ultimately consists of atoms, many systems are more easily understood as being made of molecules.

Modeling the PES of small and relatively unreactive species, such as N_2, O_2, CO, CO_2, but, also, PF_6, BF_4, BH_4, *etc.*, requires only a slight extension of the pair-potential picture. Each molecule is represented by a small number of interaction centers, which may or may not coincide with atoms in number and position. The intra-molecular configuration is enforced by constraints representing rigid bonds or, less often, by harmonic springs, while centers on different molecules interact pair-wise. Because of their simplicity, models for small inorganic molecules have been used since the early days of computer simulation. Perhaps the most remarkable observation concerning these systems is that the quantitative details of their PES are still under investigations and require surprisingly sophisticated models to be reproduced [52,53].

Conspicuously absent in the list of small unreactive and supposedly simple molecules is water, whose peculiar properties and special role have motivated an extraordinary modeling effort, which is discussed separately in Section 7.

A specialized subfield of modeling simple species concerns systems in which a weakly bound molecular fluid is physisorbed on an inert solid surface, such as MgO, mica, graphite and flat or stepped transition metal surfaces. In this case, the effect of the solid substrate on the molecular fluid often is represented as an external field. In the case of crystal surfaces, the in-plane dependence of the field strength can be expanded in plane waves, whose wave vectors reflect the periodicity and symmetry of the surface lattice [54].

6.1. Organic Molecular Systems

In many respects, organic molecular systems are not so different from any other molecular systems, but the range and impact of their applications together with the explosive expansion of simulation in bio-physics and bio-chemistry amply justify a separate discussion. Systems of interest in this context include polymers, hydrocarbons, sugars, cellulose, *etc.*, but also the endless variety of biological molecules, from phospholipids to proteins and nucleic acids. Other molecular organic systems of biological interest include drugs, simple nutrients, signal molecules, such as hormones, metabolic species, such as ATP, GTP, NADP, coenzymes, including vitamins, and prosthetic groups.

The modeling and simulation of systems of this kind arguably is the computational condensed matter activity with the largest economic relevance, both directly via the commercialization of packages and force fields and indirectly through the impact it has on applied research.

Despite the complexity of the structures they form, the PESs of organic systems turns out to be approximated fairly well by simple analytic expressions. First of all, the organic and biological species of interest are made primarily of light elements, forming strong covalent bonds through their s and p orbitals, giving origin to closed shell molecules. Systems of this kind, therefore, can be thought of as consisting of atoms connected by a fixed topology of bonds, with inter-molecular, *i.e.*, non-bonded, interactions consisting of pair-wise Coulomb and dispersion forces. Because of their sp character, intra-molecular angular forces are relatively simple. Whenever d electron metals are involved, as in metal centers and in prosthetic groups, modeling becomes far more challenging.

In the standard cases, the PES of organic and biological systems is written as the sum of contributions from bonded (U_b) and non-bonded (U_{nb}) interactions:

$$U = U_b + U_{nb} \tag{17}$$

The bonded energy, in turn, is given by the sum of two-, three- and four-body terms from atoms joined by one ($\{ij\}$), two ($\{ijk\}$) and three ($\{ijkl\}$) consecutive covalent bonds:

$$U_b = \frac{1}{2}\sum_{\{ij\}} K_{ij}^s[R_{ij} - \bar{R}_{ij}]^2 + \frac{1}{2}\sum_{\{ijk\}} K_{ijk}^b[\theta_{ijk} - \bar{\theta}_{ijk}]^2 + \frac{1}{2}\sum_{\{ijkl\}} K_{ijkl}^\tau \left[1 + \cos\left(n\phi_{ijkl} - \bar{\phi}_{ijkl}\right)\right] \tag{18}$$

K_{ij}^s, K_{ijk}^b and K_{ijkl}^τ are suitable force constants; $\bar{R}_{ij}$, $\bar{\theta}_{ijk}$, $\bar{\phi}_{ijkl}$ and n reflect the length, bending and dihedral angles of unstrained bonds. The sub-indices, ij, *etc.*, indicate that each of these parameters depends on the chemical identity of the atoms involved. The form for the dihedral contribution in Equation (18) is just one of a few different expressions used in popular force fields, while the choice for stretching and bending terms is more uniform.

Non-bonded interactions are written as.

$$U_{nb} = \frac{1}{4\pi\epsilon_0}\sum_{i\neq j}' \frac{q_i q_j}{R_{ij}} + \sum_{i\neq j}' 4\epsilon_{ij}\left[\left(\frac{\sigma_{ij}}{R_{ij}}\right)^{12} - \left(\frac{\sigma_{ij}}{R_{ij}}\right)^6\right] \tag{19}$$

where the $\{q_i\}$ are atomic charges, Coulomb forces are assumed to be acting in vacuum and σ_{ij} and ϵ_{ij} are suitable coefficients for the dispersion interaction. The prime on each sum indicates that pairs of atoms separated by one and two consecutive bonds are excluded, and the contribution from pairs separated by three consecutive bonds might be reduced.

The remarkable and, to same extent, unique property of the PES of organic and biological systems is that the bonds, whose properties are described in Equation (18), are fairly transferable, meaning that the equilibrium length, stiffness, *etc.*, of a given organic bond is nearly the same in a large number of homologous compounds. Highlighting these similarities and exploiting them to endow the model with broad transferability is the most challenging and most rewarding part of modeling organic molecular systems.

The parametrization and, especially, validation of these potentials may require sizeable computations and are the playground of large collaborations, since it requires the convergence of several types of complementary expertise. Any single system might be analyzed by *ab initio* computations to derive intra-molecular force constants and atomic charges. These need to be complemented by suitable coefficients for the dispersion part, which are usually obtained by fitting measured properties, such as the equilibrium density and enthalpy per molecule or the molecular diffusion constant.

Generic potentials covering large classes of compounds and widely used by the community include Amber [55], CHARMM [56], OPLS [57] and Gromos [58]. More specialized parameterizations, tuned on the properties of specific families of compounds, are too many to be listed.

In many respects, the most uncertain part of the parametrization is the choice of coefficients for the non-bonded interactions. The definition of atomic charges is not unique, and different methods provide fairly different results. The most popular approach [59] attributes charges by fitting the electrostatic potential outside gas-phase molecules, as provided by *ab initio* computations. The method is physically sound, but the fit becomes ill conditioned whenever the molecular size exceeds $\sim$ 15–20 atoms or when the geometry is compact, thus reducing the number of multipolar momenta whose modulus is significantly different from zero. Constraints and minimum conditions on the size of individual charges do improve the fit [60], but the choice of these parameters remains fairly uncertain. For each individual system, the error introduced by the choice of the charge may be compensated for by the selection of the dispersion coefficients. In fact, it has been observed many times that it was possible to accurately reproduce the target properties of condensed phases such as the density or the molecular diffusion even starting from the fairly different charges provided by different methods. Unfortunately, this cancellation of errors limits the transferability of the potential, since an equivalent compensation might not occur when a given organic molecule is transferred into a different environment.

Especially for large biological systems, computational cost considerations have motivated approximations and shortcuts that might reduce the size of the simulated system. One obvious saving is obtained by representing CH_2 and CH_3 groups in aliphatic chains by a single particle. This *united-atoms* approximation is fairly well justified, since these groups are small and and the non-bonded potential arising from them is fairly spherical. Moreover, the motion of hydrogen in each of these groups is frozen by quantum effects up to fairly high temperature.

16

A second more drastic approximation concerns systems in solution. Since, especially in biochemistry, one is interested in the properties of the solute, implicit solvent models [61] have been developed to replace the effect of the solvent by suitable modifications of the solute force field. In many respects, implicit solvent models are a special case of coarse graining and, as such, are left out of our discussion.

In summary, the force field modeling of organic and biological systems is a largely successful enterprise, validated by a vast number of applications and supporting the research of a large portion of the simulation community. Furthermore, in this case, and almost needless to say, the vast simulation activity has highlighted many cases of inaccuracies or outright failures. The general feeling, however, is that the scale of most of these simulations is too large to allow, at present, the usage of significantly more sophisticated and more expensive approaches. Polarizability is likely to be the single most relevant missing ingredient, but the available methods to include it into simulations are still fairly expensive, and for this reason, explicitly polarizable models have been used only for a limited number of large-scale studies.

At present, a very active research field is the development of force fields for organo-metallic complexes, which represent prosthetic groups in proteins or active groups in a variety of organic opto-electronic devices and are important also for homogeneous catalysis. Peculiar difficulties are represented by the variety of coordination numbers, sometimes corresponding to different spin states, thus pointing to multiple PESs fairly close in energy. Moreover, the structure of organo-metallic complexes is characterized by the importance of quantum mechanical effects, such as Jahn-Teller, or by the so-called *trans influence*, defined as the "tendency of a ligand to selectively weaken the bond *trans* to itself" [62]. Models to include these effects in empirical PES models might turn out to be too complex to be used in practice. A more promising alternative is provided by QM/MMapproaches, using classical force fields for most of the system and resorting to *ab initio* methods for the challenging portion around the metal center.

An intriguing subset of mainly, but not exclusively, organic compounds is represented by the so-called room temperature ionic liquids [63], defined as molecular ionic systems whose melting temperature is below 100°. Prototypical systems are made by an alkane substituted imidazolium cation, joined to an organic or inorganic anion. Systems of this kind are relevant here, not only because of the intense simulation activity that concerns them, but mainly because they provide a bridge between different classes of bonding and, thus, pose special modeling problems.

The bulk of the extensive simulation work carried out at present relies on Amber-like force fields, with specialized parameterizations (see, for instance, [64,65]). Models of this kind are fairly successful, but issues concerning polarizability and the attribution of partial charges to atoms become particularly important for these systems. Despite these difficulties, a number of simulations have successfully addressed the properties of very complex systems, consisting of room temperature ionic liquids in combination with a variety of solvents and neutral organic compounds, including bio-molecular species (see Figure 3).

A few carbon systems, such as fullerenes, carbon nanotubes and graphene, lie at the boundary between inorganic and organic species and even blur the distinction between covalent and metal character. Not surprisingly, systems of this kind have been represented by a variety of models, from Tersoff-Brenner to a molecular force field, such as those described in this section.

Figure 3. Snapshot from a molecular dynamics simulation of a room temperature ionic liquid/water solution at 0.5 M concentration in contact with a POPCphospholipid bilayer [66]. Green balls: [Cl]$^-$; gray-silver molecules: [bmim]$^+$. wireframe molecules: POPC. Water has been removed to highlight the incorporation of [bmim]$^+$ cations into the phospholipid bilayer.

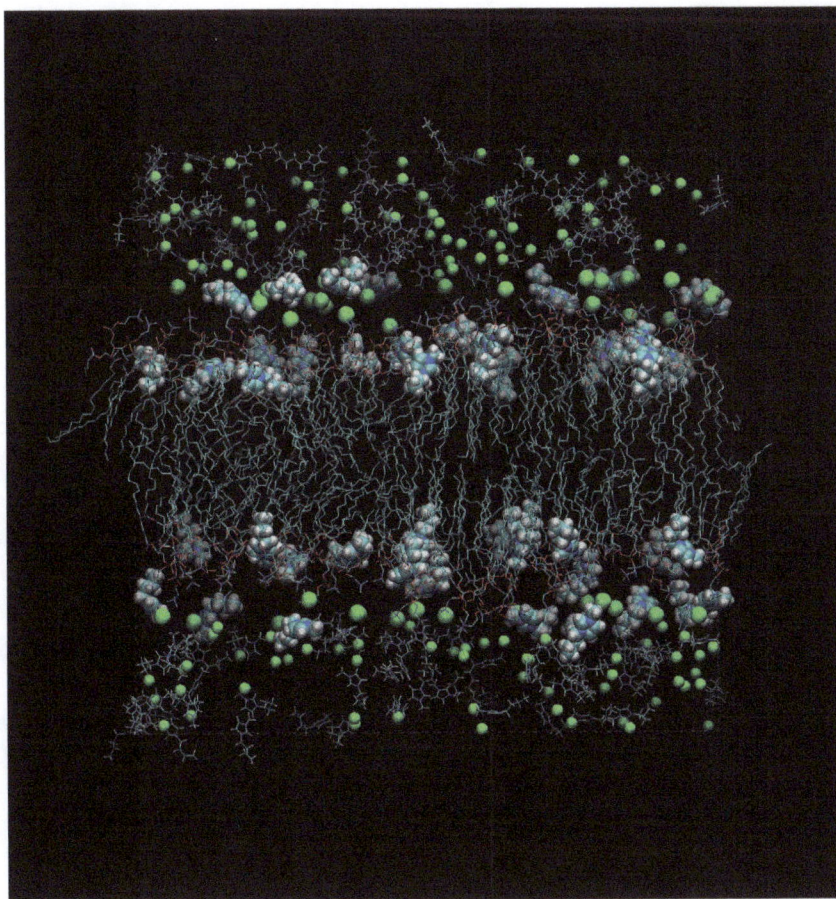

7. Water

Because of its fundamental role in life and of its widespread and generally benign presence in nature, water has always been the object of interest and fascination. In this respect, computational physicists and chemists are no exception, although the reasons for their interest are somewhat different from those of the rest of humankind. A number of measurements have highlighted a wide variety of peculiarities, if not anomalies, in the properties of water [67]. These include the surprising expansion of water upon freezing, the density anomaly observed at 4 °C at ambient pressure, and, more in general, the non-monotonic variation of several physics-chemical properties in the vicinity of this remarkable density maximum. Other peculiar features consist in the wide temperature range of super-cooling, the high liquid-vapor critical temperature and the large value of the latent heat of the liquid water-ice transition.

To a large extent, these anomalous behaviors are embodied into the PES of water systems and arise from the strength and directionality of the hydrogen bond network that provides the bulk of water cohesion. In part, however, they are due to the light mass of the water molecule, causing non-negligible quantum effects that influence the properties of hydrogen bonds. Heavy water, for instance, is already somewhat different from ordinary water, so much that D_2O is known to have peculiar and generally adverse biological effects. This duality of potential energy *versus* quantum mechanical effects poses apparent and significant problems to modeling [68]. Potentials tuned on the *exact* PES of water do not reproduce its properties when used in a classical simulation. On the other hand, potentials tuned on experimental properties of water do not necessarily reflect the details of the exact PES.

Work to provide a quantitative and comprehensive description of water properties is still in progress [69,70]. In the meantime, a vast number of simulations in which water is the unique or an essential component are being carried out with a variety of simple potentials, reflecting the basic atomistic and electronic structure of the water molecule. Two major families are in use: TIPnP [71–73], with $n = 3$, 4 and 5, and SPC [74–77], both based on fixed charges (rigid ions) and centers of short range interactions, joined by rigid or harmonic bonds.

Models of this kind allow the routine simulation by MD of systems of 50×10^3 water molecules solvating whole proteins, covering times well in excess of 100 ns. Results are generally good, and a large number of successful applications clearly validate these models, at least up to the accuracy needed for these large-scale applications. However, it is fair to say that no single model of the rigid ion type is able to provide a uniformly satisfactory account of water properties over a wide range of regimes and thermodynamic conditions. Several of these models, in particular, do not display the experimental density maximum of water or place it at (P, T)conditions far from the experimental ones [69,70]. The liquid-vapor coexistence curve is also poorly predicted by rigid ion models, unless the potential parameters are explicitly adjusted for this purpose. In such a case, however, the accurate description of some other quantity might need to be sacrificed. The description of critical properties, that are accurately known from measurements, are only moderately well reproduced [78].

Water clusters and droplets are another, distinct subfield of water research. Thermodynamic and spectroscopic data are available from experiments, but are not sufficiently detailed to provide a full description of structural and dynamical properties. In this case, state-of-the-art quantum chemistry computations supplement the experimental information [79]. Once again, it turns out that rigid ion models are only moderately successful in predicting their properties and usually fail to reproduce the reduced binding of very small clusters. The oxygen-oxygen equilibrium distance in the water dimer, for instance, is greatly underestimated by popular models, and its cohesive energy is correspondingly overestimated. These discrepancies decrease in importance with increasing cluster size, but the convergence to the bulk cohesive properties, reliably described by current DFTmodels of water, is fairly slow (See Table 1). In these small systems, the rigid-ion assumption, or, in other terms, the lack of polarizability, again seems to be the major problem. The molecular dipole moment of water, for instance, changes from $\mu = 1.855$ D in the gas phase molecule, to nearly $\mu = 3$ D in ice and in liquid water, but rigid ion models cannot reproduce this change. Moreover, within rigid-ion models, hydrogen bonds have only a Coulombic origin, contradicting the results of experiments and quantum chemistry computations showing that both Coulomb and covalent contributions are important [80] and change in slightly different ways upon changing the aggregation state of water.

Table 1. Cohesive energy (kJ/mol per water molecule) of $(H_2O)_2$, of cyclic water clusters $(H_2O)_n$, $n = 3, 4, 5, 6$, and of the cubic D_{2d} form of $(H_2O)_8$ computed by an SPC, rigid ion model (SPC/Fw, [77]). Deviations from dispersion-corrected [81] DFT [82] results are given in parentheses. Data are from [83].

n	2	3	4	5	6	8
PBE+vdW	12.06	25.95	34.48	36.09	36.82	45.28
SPC/Fw	14.35	26.66	33.44	35.10	35.67	40.69
	[18.99%]	[2.7%]	[−3.0%]	[−2.7%]	[−3.1%]	[−10.1%]

Somewhat surprisingly, the inclusion of polarizability into simple models has not resulted yet into the systematic improvement of the description of the properties for extended water systems [84], while it has been more successful for clusters.

All these difficulties have stimulated a large number of new attempts. It might be worth mentioning the representation of electron polarizability via classical [85] and quantum [86] Drude oscillators, the application to water [87] of the empirical valence band (EVB) theory [88] and the usage of polarizable Thole models [89].

Ab initio modeling, discussed in more detailed below, will eventually provide the method of choice to study water [90]. Until now, however, approaches of this kind using standard approximations for the exchange-correlation energy (see next section) have given rather mixed results [91].

8. The *Ab initio* Route

Over the last twenty years, the art of representing PES as a function of atomic coordinates has seen its role increasingly challenged by the explosive growth of *ab initio* simulation methods.

As discussed in Section 2, the exact PES of a system made by N electrons evolving in the field of K nuclei can be determined point by point by computing the energy eigenvalues of the $\hat{H}_{ele}$ Hamiltonian:

$$E_k(\{\mathbf{R}_\alpha\}) = \frac{\hat{H}_{ele}\psi_k(\{\mathbf{r}_i\} \mid \{\mathbf{R}_\alpha\})}{\psi_k(\{\mathbf{r}_i\} \mid \{\mathbf{R}_\alpha\})} \tag{20}$$

For any single choice of the $\{\mathbf{R}_\alpha\}$ coordinates, a fairly extended array of quantum chemistry *ab initio* methods, such as configuration interaction, Møller-Plesset perturbation theory or coupled clusters, are available to find all or a few of the lowest energy eigenvalues and eigenvectors of this so-called *standard problem* in electronic structure computations.

For what concerns the direct application of *ab initio* methods to simulation, however, progress came primarily through the advent of density functional theory, whose recognized theoretical and practical foundation is provided by the Hohenberg-Kohn (HK) theorem [92] and by the seminal paper by Kohn and Sham (KS) [93]. In a very schematic way, density functional theory in the popular Kohn-Sham formulation represents the ground state electron density, $\rho(\mathbf{r})$, in terms of an auxiliary set of non-interacting electron orbitals $\{\phi_i(\mathbf{r}), i = 1, ..., K\}$, generally known as the Kohn-Sham orbitals:

$$\rho(\mathbf{r}) = \sum_{i=1}^{K} \mid \phi_i(\mathbf{r}) \mid^2 \tag{21}$$

To reproduce the exact density, the (unspecified) potential acting on the non-interacting electrons has to be different from the one acting on their interacting counterpart. The properties of such a potential and, in particular, its local, *i.e.*, multiplicative nature are a corollary of the HKtheorem.

Then, according to KS, the system ground state energy is the minimum of the unique and universal functional:

$$E_{KS}[\rho \mid \{\mathbf{R}_\alpha\}] = -\frac{1}{2} \sum_{i=1}^{K} \langle \phi_i \mid \nabla^2 \mid \phi_i \rangle + \frac{1}{2} \int \frac{\rho(\mathbf{r})\rho(\mathbf{r}')}{\mid \mathbf{r} - \mathbf{r}' \mid} d\mathbf{r} d\mathbf{r}' - \sum_{\alpha=1}^{K} Z_\alpha \int \frac{\rho(\mathbf{r}) d\mathbf{r}}{\mid \mathbf{r} - \mathbf{R}_\alpha \mid} + U_{XC}[\rho]$$
$$\tag{22}$$

where $U_{XC}[\rho]$ is the so-called exchange correlation energy, a functional of the electron density, $\rho(\mathbf{r})$, which also contains a small fraction of the kinetic energy of the interacting electrons. Minimization of Equation (22) under the constraint of ortho-normality for the Kohn-Sham orbitals results in a set of coupled partial differential equations for $\{\phi_i\}$.

Methods to solve this problem have been developed and discussed in a vast numbers of papers and textbooks [7,8]. The accuracy of the solution depends on the functional used to approximate $U_{XC}[\rho]$, and on the choice of the basis used to represent the orbitals. Popular choices for the exchange-correlation energy are generalized gradient corrections, such as PBE [82], or hybrid functionals, such as B3LYP [94]. Basis sets range from atomic orbitals to wavelets, but plane waves [95,96] and Gaussian functions [97] are probably the most widely used choice for implementations tuned on molecular dynamics applications.

The solution of the standard problem in Equation (5) obtained through Equation (22) is restricted to the ground state PES. Even within this limited scope, the PES itself can only be determined point by point. Nevertheless, the KS energy expression can be used to evolve the atomic positions in time, thus opening the way to MD, provided one can: (i) minimize Equation (22) fast enough; and (ii) evaluate forces on the atoms through:

$$\mathbf{F}_\beta = -\nabla_{\mathbf{R}_\beta} E_{KS}[\rho \mid \{\mathbf{R}_\alpha\}] \tag{23}$$

Towards this goal, the work of Car and Parrinello [9] has truly represented the single most important breakthrough, whose major innovation consisted of the introduction of direct minimization approaches for Equation (22), exploiting the close similarity of the electronic configuration at two successive steps of MD. Evaluation of forces, moreover, was greatly eased by the choice of plane waves as the basis set to represent KS orbitals, whose unbiased coverage of the entire space allows the application of the Hellmann-Feynman theorem in its simplest form to compute gradients of the ground state energy [95,98].

Atoms evolve on the adiabatic PES implicitly defined by Equation (22) classically or quantum mechanically. The validity of a classical time evolution for the atoms according to Newton's equations relies on conditions discussed in detail in Chapter [6]. Outside these conditions, one could resort to a path integral approach, as done, for instance, in [99].

The method can be extended to simulate the atomic dynamics on the single PES of an electronically excited state [100], provided the different symmetry of the ground and excited state allows a meaningful definition of both PESs by density functional methods. As apparent from the discussion of the Born-Oppenheimer approximation, multiple PESs close in energy make it impossible to disentangle the ionic and electron dynamics, and in these cases, resorting to semiclassical or to more accurate quantum mechanical approaches [6] is mandatory.

Somewhat simplified versions of the density-functional-based MD, resorting to localized bases and relying on a self-consistent tight-binding approach have been developed [101,102] and provide a cheaper and popular alternative to unrestricted DFT methods. The price to be paid is a slight limitation in the quality of the solution, as well as occasional failures of the method.

The amazing success of density-functional-based simulation methods is due to the fact that they represent the only method endowed with truly predictive power, which can be used for systems of several hundred atoms, with up to a few thousand valence electrons. *ab initio* simulation, therefore, is the method of choice whenever we cannot guess a suitable representation of the PES or when we need an accuracy that cannot be provided by the empirical models that are available. *Ab initio* simulation is also strictly required for systems whose structure is affected by electronic effects, such as Jahn-Teller, and also enjoys a clear advantage in describing spin-polarization effects or systems undergoing chemical transformations and non-stoichiometric compounds exhibiting different valence states.

Well known drawbacks are represented by the computational cost that limits the size and especially the time scale of *ab initio* simulations, even though the reach of the method is constantly expanding. At present, large computations running on state-of-the-art facilities may involve $\sim 1,000$ atoms and $\sim 4,000$–$5,000$ valence electrons. Early problems with metals have been progressively eased by approaches relying on the accurate step-by-step minimization of the KS energy functional. Problems, however, remain with transition and, especially, rare-earth metals, for which standard exchange-correlation approximations give unsatisfactory results, and quantum chemistry hybrid methods fail fairly spectacularly [103]. Progress is being achieved with methods incorporating strong correlation at some approximate level, such as LSD+U [104].

Difficulties remain also in the limit of weakly interacting molecular systems. Furthermore, in this case, early methods lacked essential components, such as the dispersion interaction, which in molecular systems provide a good portion of cohesion. Dispersion interactions are now increasingly included in *ab initio* simulations [81], especially for molecular systems and for water, in particular.

Results are encouraging, although not yet in full quantitative agreement with experiments. However, the accuracy, reliability and computational efficiency of these methods are improving rapidly.

The major problem in current MD applications of *ab initio* methods arguably is that achieving accurate results for *difficult* systems, such as transition metals and oxides or molecular systems, still require an extensive preliminary calibration stage and system-specific exchange correlation approximations [105], effectively spoiling the *ab initio* character of these methods. Perhaps more importantly, these adjustments of the model decrease their reliability for systems exhibiting different bonding types, since the improvement on one type might worsen the description of the other type.

Most of the cost of KS-DFT computations is due to the representation of the density in terms of KS orbitals. Approaches relying on genuine density functional formalism, such as a refined Thomas-Fermi method, could enjoy a huge computational advantage, but no successful scheme has emerged during the years, and only very idealized Gordon-Kim approaches [106] have been used with some success.

9. Conclusions

Explicit or implicit expressions of the PES of condensed matter systems represent the basis of our ability to simulate them, possibly understanding and sometimes predicting their properties by purely computational methods. For this reason, the development of approximations and efficient representations of PES is the focus of an intense research effort, involving a sizable portion of the computational community.

Such a modeling activity is an art as much as a science. It is a science in the systematic derivation of interatomic forces from more fundamental interactions. It is an art in the invention of effective ways to incorporate new ideas in physically transparent and computationally efficient mathematical expressions. Like many other forms of art, it relies on a big deal of craftsmanship, required in the stage of parameterizing force fields, validating them and incorporating them into widely used computer packages, using sophisticated programming techniques, tuned on state-of-the-art computational hardware.

It should be apparent from the discussion of the previous sections that the last thirty years have seen an amazing enhancement of our ability to model a wide variety of systems at the atomistic level, fueling the explosive growth of simulation studies, while, at the same time, being driven by it. Equally amazing, however, is the extent of what we are still unable to model satisfactorily. Interfaces between different materials, for instance, are intrinsically difficult to describe by simple approaches. Excluding *ab initio*, no reliable, general and widely accepted model is available to simulate water and electrolyte solutions in contact with neutral or charged electrodes, organic and biological molecules on solid surfaces or the junction of metal and semiconducting phases. Even homogeneous phases, such as non-stoichiometric oxides, still represent a formidable challenge for models suitable for simulating 10^4 atoms over 100 ns or more. Systems undergoing chemical transformations are another sore point, even though methods, such as ReaxFF and REBO, are achieving progress in this direction.

At this stage, strategic decisions on the directions and aims of the modeling effort have to take into account the rapid growth of *ab initio* methods, which easily account for the intermixing of different bonding categories, cover electrostatic polarizability, provide information on excited state PES and may include magnetic interactions and spin effects through their approximate description of exchange.

The rapid progress of methods and computational equipment implies that the foreseeable future spans at most ten to fifteen years from now. Over this time, empirical models of PES will continue to play an important and useful role in the atomistic simulation of large systems ($N \gtrsim 10^4$ atoms) over times in excess of 100 ns. Most biochemistry and biophysics simulations fall into this class.

On the longer run, however, the general picture of modeling might indeed change. First of all, the domain proper to atomistic modeling concerns the investigation of the microscopic details underlying larger-scale phenomena. In this context, the scales of interest rarely exceed $\sim 10^4$ atoms and correspondingly short times of less than ~ 10 ns. Beyond this range, simulation may become the exclusive domain of coarse graining and multi-scale approaches, provided refined versions of these methods are developed over the next few years.

Ab initio methods already represent the method of choice for systems for which we do not have reliable approximations of their PES, for phenomena that can be represented by 100 to 1,000 atoms and that take place within a 50–100 ps time span. Mixed QM/MM approaches extend this reach and represent the most appealing method to treat systems, such as protein reaction centers, organometallic catalysts, *etc.*, in which a small portion of a large system needs to be represented in full chemical detail.

The parallel development of *ab initio* and of refined coarse graining and multi-scale methods, therefore, could greatly shrink the role of empirical PES approximations in atomistic simulation. Even these likely developments, however, might not mark the end of atomistic potential models, since simple and transparent representations of PES will continue to provide the conceptual basis to rationalize the properties of condensed matter systems in terms of atoms, of molecules and of their microscopic interactions.

Acknowledgments

I thank Carlo Pierleoni for useful discussions and for a careful reading of the manuscript.

Conflicts of Interest

The authors declare no conflict of interest.

References

1. Frenkel, D.; Smit, B. *Understanding Molecular Simulation*, 2nd ed.; Academic Press: San Diego, CA, USA, 2002.
2. Morales, M.A.; Clay, R.; Pierleoni, C.; Ceperley, D.M. First-principle methods: A perspective from quantum Monte Carlo. *Entropy* **2014**, *16*, 287–321.
3. Finnis, M. *Interatomic Forces in Condensed Matter*; Oxford Series on Materials Modeling: Oxford, UK, 2010.

4. Delle Site, L. What is a multiscale problem in molecular dynamics? *Entropy* **2014**, *16*, 23–40.

5. Ziman, J.M. *Electrons and Phonons*; Oxford University Press: Oxford, UK, 1960; Chapter 5.

6. De Carvalho, F.F.; Bouduban, M.E.F.; Curchod, B.F.E.; Tavernelli, I. Nonadiabatic molecular dynamics based on trajectories. *Entropy* **2014**, *16*, 62–85.

7. Martin, R.M. *Electronic Structure, Basic Theory and Practical Methods*; Cambridge University Press: Cambridge, UK, 2004.

8. Kaxiras, E. *Atomic and Electronic Structure of Solids*; Cambridge University Press: Cambridge, UK, 2003.

9. Car, R.; Parrinello, M. Unified approach for molecular dynamics and density-functional theory. *Phys. Rev. Lett.* **1985**, *55*, 2471–2474.

10. Born, M.; Huang, K. *Dynamical Theory of Crystal Lattices*; Oxford University Press: Oxford, UK, 1954.

11. Yarkony, D.R. Diabolical conical intersections. *Rev. Mod. Phys.* **1996**, *68*, 985–1013.

12. Kroes, G.J.; Gross, A.; Baerends, E.J.; Scheffler, M.; McCormack, D.A. Quantum theory of dissociative chemisorption on metal surfaces. *Acc. Chem. Res.* **2002**, *35*, 193–200.

13. Bowman, J.M. Beyond born-oppenheimer. *Science* **2008**, *319*, 40.

14. White, J.D.; Chen, J.; Matsiev, D.; Auerbach, D.J.; Wodtke, A.M. Conversion of large-amplitude vibration to electron excitation at a metal surface. *Nature* **2005**, *433*, 503–505.

15. Abedi, A.; Maitra, N.T.; Gross, E.K.U. Exact factorization of the time-dependent electron-nuclear wave function, *Phys. Rev. Lett.* **2010**, *105*, 123002.

16. Abedi, A.; Maitra, N.T.; Gross, E.K.U. Correlated electron-nuclear dynamics: Exact factorization of the molecular wavefunction, *J. Chem. Phys.* **2012**, *137*, 22A530.

17. Hunter, G. Conditional probability amplitudes in wave mechanics, *Int. J. Quantum Chem.* **1975**, *9*, 237–242.

18. Berry, M.V. Quantal phase factors accompanying adiabatic changes, *Proc. R. Soc. Lond. Ser. A* **1984**, *392*, 45–57.

19. Pancharatnam, S. Generalized theory of interference, and its applications. *Proc. Indian Acad. Sci. A* **1956**, *44*, 247–262.

20. Sutcliffe, B. Is there an exact potential energy surface? *Theor. Chem. Acc.* **2012**, *131*, doi:10.1007/s00214-012-1215-x.

21. Boscovich, R.J. *Theoria Philosophiae Naturalis*; Venice, Italy, 1758.

22. Lennard-Jones, J.E. On the determination of molecular fields. *Proc. R. Soc. Lond. A* **1924**, *106*, 463–477.

23. Fumi, F.G.; Tosi, M.P. Ionic sizes and Born repulsive parameters in the NaCl-type alkali halides. *J. Phys. Chem. Solids* **1964**, *25*, 31–43.

24. Carlsson, A.E.; Ashcroft, N.W. Pair potentials from band theory: Application to vacancy-formation energies. *Phys. Rev. B* **1983**, *27*, 2101–2110.

25. Abell, G.C. Empirical chemical pseudopotential theory of molecular and metallic bonding. *Phys. Rev. B* **1985**, *31*, 6184–6196.

26. Carlsson, A.E. Beyond Pair Potentials in Elemental Transition Metals and Semiconductors. In *Solid State Physics: Advances in Research and Applications*; Ehrenreich, H., Turnbull, D., Eds.; Academic Press: Boston, MA, USA, 1990; Volume 43, pp. 1–91.

27. Pauling, L. *The Nature of the Chemical Bond*, 3rd ed.; Cornell University Press: Ithaca, NY, USA, 1960.

28. Hafner, J. *From Hamiltonians to Phase Diagrams*; Springer: Berlin, Germany, 1987.

29. *Handbook of Materials Modeling*; Yip, S., Ed.; Springer: Berlin, Germany, 2005.

30. *Comprehensive Nuclear Materials*; Basic Aspects of Radiation Effects in Solids/Basic Aspects of Multi-Scale Modeling, Konings, R.J.M., Ed.; Elsevier: Amsterdam, The Netherlands, 2012; Volume 1.

31. Dagens, L.; Rasolt, M.; Taylor, R. Charge densities and interionic potentials in simple metals. *Phys. Rev. B* **1975**, *11*, 2726–2734.

32. Moriarty, J.A. First-principles interatomic potentials in transition metals. *Phys. Rev. Lett.* **1985**, *55*, 1502–1505.

33. Daw, M.S.; Baskes, M.I. Semiempirical, quantum mechanical calculation of hydrogen embrittlement in metals. *Phys. Rev. Lett.* **1983**, *50*, 1285–1288.

34. Baskes, M.I. Application of the embedded-atom method to covalent materials: A semiempirical potential for silicon. *Phys. Rev. Lett.* **1987**, *59*, 2666–2669.

35. Rosato, V.; Guillope, M.; Legrand, B. Thermodynamical and structural properties of f.c.c. transition metals using a simple tight-binding model. *Philos. Mag. A* **1989**, *59*, 321–336.

36. Jacobsen, K.W.; Nørskov, J.K.; Puska, M.J. Interatomic interactions in the effective-medium theory. *Phys. Rev. B* **1987**, *35*, 7423–7442.

37. Drautz, R.; Pettifor, D.G. Valence-dependent analytic bond-order potential for transition metals. *Phys. Rev. B* **2006**, *74*, 174117.

38. Miiller, A.P.; Brockhouse, B.N. Crystal dynamics and electronic specific heats of palladium and copper. *Can. J. Phys.* **1971**, *49*, 704–723.

39. Stillinger, F.; Weber, T. Computer simulation of local order in condensed phases of silicon. *Phys. Rev. B* **1985**, *31*, 5262–5271.

40. Tersoff, J. New empirical model for the structural properties of silicon. *Phys. Rev. Lett.* **1986**, *56*, 632–635.

41. Tersoff, J. New empirical approach for the structure and energy of covalent systems. *Phys. Rev. B* **1988**, *37*, 6991–7000.

42. Brenner, D.W. Empirical potential for hydrocarbons for use in simulating the chemical vapor deposition of diamond films. *Phys. Rev. B* **1990**, *42*, 9458–9471.

43. Pettifor, D.G.; Oleinik, I.I. Analytic bond-order potentials beyond Tersof-Brenner. I. Theory. *Phys. Rev. B* **1999**, *59*, 8487–8499.

44. Van Duin, A.C.T.; Dasgupta, S.; Lorant, F.; Goddard, W.A., III. ReaxFF: A reactive force field for hydrocarbons. *J. Phys. Chem. A* **2001**, *105*, 9396–9409.

45. Brenner, D.W.; Shenderova, O.A.; Harrison, J.A.; Stuart, S.J.; Ni, B.; Sinnott, S.B. A second-generation reactive empirical bond order (REBO) potential energy expression for hydrocarbons. *J. Phys.: Condens. Matter* **2002**, *14*, 783–802.

46. Keating, P.N. Effect of invariance requirements on the elastic strain energy of crystals with application to the diamond structure. *Phys. Rev.* **1966**, *145*, 637–645.

47. Cabriolu, R.; Del Popolo, M.G.; Ballone, P. Melting of a tetrahedral network model of silica. *Chem. Phys. Phys. Chem.* **2009**, *11*, 10820–10823.

48. Cabriolu, R.; Ballone, P. Thermodynamic properties and atomistic structure of the dry amorphous silica surface from a reactive force field model. *Phys. Rev. B* **2010**, *81*, 155432.

49. Ponder, J.W.; Wu, C.; Ren, P.; Pande, V.S.; Chodera, J.D.; Schnieders, M.J.; Haque, I.; Mobley, D.L.; Lambrecht, D.S.; DiStasio, R.A.; *et al.* Current status of the AMOEBA polarizable force field. *J. Phys. Chem. B* **2010**, *114*, 2549–2564.

50. Dick, B.G.; Overhauser, A.W. Theory of the dielectric constants of alkali halide crystals. *Phys. Rev.* **1958**, *112*, 90–103.

51. Mitchell, P.J.; Fincham, D. Shell model simulations by adiabatic dynamics. *J. Phys.: Condens. Matter* **1993**, *5*, 1031–1038.

52. Rezaei, M.; Sheybani-Deloui, S.; Moazzen-Ahmadi, N.; Michaelian, K.H.; McKellar, A.R.W. Spectroscopic evidence for a planar cyclic CO trimer. *J. Chem. Phys.* **2013**, *138*, 071102.

53. Visser, G.W.M.; Hesselmann, A.; Jansen, G.; Wormer, P.E.S.; van der Avoird, A. New CO-CO interaction potential tested by rovibrational calculations. *J. Chem. Phys.* **2005**, *122*, 054306.

54. Righi, M.C.; Ferrario, M. Potential energy surface for rare gases adsorbed on Cu(111): Parameterization of the gas/metal interaction potential. *J. Phys.: Condens. Matter* **2007**, *19* 305008.

55. Cornell, W.D.; Cieplak, P.; Bayly, C.I.; Gould, I.R.; Merz, K.M., Jr; Ferguson, D.M.; Spellmeyer, D.C.; Fox, T.; Caldwell, J.W.; Kollman, P.A. A second generation force field for the simulation of proteins, nucleic acids, and organic molecules. *J. Am. Chem. Soc.* **1995**, *117* 5179–5197.

56. Patel, S.; Brooks, C.L., III. CHARMM fluctuating charge force field for proteins: I parameterization and application to bulk organic liquid simulations. *J. Comput. Chem.* **2004**, *25*, 1–16.

57. Jorgensen, W.L.; Maxwell, D.S.; Tirado-Rives, J. Development and testing of the OPLS all-atom force field on conformational energetics and properties of organic liquids. *J. Am. Chem. Soc.* **1996**, *118*, 11225–11236.

58. Schuler, L.D.; Daura, X.; van Gunsteren, W.F. An improved GROMOS96 force field for aliphatic hydrocarbons in the condensed phase. *J. Comp. Chem.* **2001**, *22*, 1205–1218.

59. Chandra Singh, U.; Kollman, P.A. An approach to computing electrostatic charges for molecules. *J. Comput. Chem.* **1984**, *5*, 129–145.

60. Cornell, W.D.; Cieplak, P.; Bayly, C.I.; Kollman, P.A. A well-behaved electrostatic potential based method using charge restraints for deriving atomic charge. *J. Phys. Chem.* **1993**, *97*, 9620–9631.

61. Roux, B; Simonson, T. Implicit solvent models. *Biophys Chem.* **1999**, *78*, 1–20.

62. Pidcock, A.; Richards, R.E.; Venanzi, L.M. [195]Pt–[31]P nuclear spin coupling constants and the nature of the trans-effect in platinum complexes. *J. Chem. Soc. A* **1966**, 1707–1710.

63. Welton, T. Room-temperature ionic liquids. Solvents for synthesis and catalysis. *Chem. Rev.* **1999**, *99*, 2071–2084.

64. Canongia Lopes, J.N.; Deschamps, J.; Padua, A.A.H. Modeling ionic liquids using a systematic all-atom force field. *J. Phys. Chem. B* **2004**, *108*, 2038–2047.

65. Canongia Lopes, J.N.; Deschamps, J.; Padua, A.A.H. Modeling ionic liquids using a systematic all-atom force field. *J. Phys. Chem. B* **2004**, *108*, 11250.

66. Bingham, R.; Ballone, P. Computational study of room-temperature ionic liquids interacting with a POPC phospholipid bilayer. *J. Phys. Chem. B* **2012**, *116*, 11205–11216.

67. Brovchenko, I.; Oleinikova, A. Multiple phases of liquid water. *ChemPhysChem* **2008**, *9*, 2660–2675.

68. Billeter, S.R.; King, P.M.; van Gunsteren, W.F. Can the density maximum of water be found by computer-simulation. *J. Chem. Phys.* **1994**, *100*, 6692–6699.

69. Guillot, B.; Guissani, Y. How to build a better pair potential for water. *J. Chem. Phys.* **2001**, *114*, 6720–6733.

70. Guillot, B. A reappraisal of what we have learnt during three decades of computer simulations on water. *J. Mol. Liquids* **2002**, *101*, 219–260.

71. Jorgensen, W.L.; Chandrasekhar, J.; Madura, J.D.; Impey, R.W.; Klein, M.L. Comparison of simple potential functions for simulating liquid water. *J. Chem. Phys.* **1983**, *79*, 926–935.

72. Mahoney, M.W.; Jorgensen, W.L. A five-site model for liquid water and the reproduction of the density anomaly by rigid, nonpolarizable potential functions. *J. Chem. Phys.* **2000**, *112*, 8910–8922.

73. Abascal, J.L.F.; Sanz, E.; Garcia Fernandez, R.; Vega, C. A potential model for the study of ices and amorphous water: TIP4P/Ice. *J. Chem. Phys.* **2005**, *122*, 234511.

74. Berendsen, H.J.C.; Postma, J.P.M.; von Gunsteren, W.F.; Hermans, J. *Intermolecular Forces*; Pullman, P., Ed.; Reidel: Dordrecht, Holland, 1981.

75. Berendsen, H.J.C.; Grigera, J.R.; Straatsma, T.P. The missing term in effective pair potentials. *J. Phys. Chem.* **1987**, *91*, 6269–6271.

76. Dang, L.X.; Pettitt, B.M. Simple intermolecular model potentials for water. *J. Phys. Chem.* **1987**, *91*, 3349–3354.

77. Wu, Y.; Tepper, H.L.; Voth, G.A. Flexible simple point-charge water model with improved liquid-state properties. *J. Chem. Phys.* **2006**, *124*, 024503.

78. Chialvo, A.A.; Cummings, P.T. Molecular-based modeling of water and aqueous solutions at supercritical conditions. *Adv. Chem. Phys.* **1999**, *109*, 115–205.

79. Temelso, B.; Archer, K.A.; Shields, G.C. Benchmark structures and binding energies of small water clusters with anharmonicity corrections. *J. Phys. Chem. A* **2011**, *115*, 12034–12046.

80. Khaliullin, R.Z.; Bell, A.T.; Head-Gordon, M. Electron donation in the water-water hydrogen bond. *Chem.-Eur. J.* **2009**, *15*, 851–855.

81. Grimme, S. Semiempirical GGA-type density functional constructed with a long-range dispersion correction. *J. Comput. Chem.* **2006**, *27*, 1787–1799.

82. Perdew, J.P.; Burke, K.; Ernzerhof, M. Generalized gradient approximation made simple. *Phys. Rev. Lett.* **1996**, *77*, 3865–3868.

83. Bingham, R.J.; Ballone, P. Energy, structure and vibrational modes of small water clusters by a simple many-body potential mimicking polarization effects. *Mol. Phys.* **2013**, doi:10.1080/00268976.2013.830788.

84. Chen, B.; Potoff, J.J.; Siepmann, I.J.I. Adiabatic nuclear and electronic sampling Monte Carlo simulations in the Gibbs ensemble: Application to polarizable force fields for water. *J. Phys. Chem. B* **2000**, *104*, 2391–2401.

85. Lamoureux, G.; Harder, E.; Vorobyov, I.V.; Roux, B.; MacKerell, A.D., Jr. A polarizable model of water for molecular dynamics simulations of biomolecules. *Chem. Phys. Lett.* **2006**, *418*, 245–249.

86. Jones, A.; Cipcigan, F.; Sokhan, V.P.; Crain, J. Martyna, G.J. Electronically coarse-grained model for water. *Phys. Rev. Lett.* **2013**, *110*, 227801.

87. Day, T.J.F.; Soudackov, A.V.; Cuma, M.; Schmitt, U.W.; Voth, G.A. A second generation multistate empirical valence bond model for proton transport in aqueous systems. *J. Chem. Phys.* **2002**, *117*, 5839–5849.

88. Aqvist, J.; Warshel, A. Simulation of enzyme-reactions using valence-bond force-fields and other hybrid quantum-classical approaches. *Chem. Rev.* **1993**, *93*, 2523–2544.

89. Thole, B.T. Molecular polarizabilities calculated with a modified dipole interaction. *Chem. Phys.* **1981**, *59*, 341–350.

90. Zhang, C.; Wu, J.; Galli, G.; Gygi, F. Structural and vibrational properties of liquid water from van der Waals density functionals. *J. Chem. Theory Comput.* **2011**, *7*, 3054–3061.

91. Yoo, S.; Zeng, X.C.; Xantheas, S.S. On the phase diagram of water with density functional theory potentials: The melting temperature of ice Ih with the PBE and BLYP functionals. *J. Chem. Phys.* **2009**, *130*, 221102.

92. Hohenberg, P.; Kohn, W. Inhomogeneous electron gas. *Phys. Rev.* **1964**, *136*, B864–B871.

93. Kohn, W.; Sham, L.J. Self-consistent equations including exchange and correlation effects. *Phys. Rev.* **1965**, *140*, A1133–A1138.

94. Becke, A.D. Density-functional thermochemistry. III. The role of exact exchange. *J. Chem. Phys.* **1993**, *98*, 5648–5652

95. Marx, D.; Hutter, J. *Modern Methods and Algorithms of Quantum Chemistry*; Grotendorst, J., Ed.; John von Neumann Institute for Computing: Jülich, Germany, 2000; NIC Series, Volume 1, pp. 301–449.

96. Quantum Espresso is an Open Source Distribution. Available online: http://www.quantum-espresso.org (accessed on 15 October 2013).

97. Cp2k is an Open Source Distribution. Available online: http://www.cp2k.org (accessed on 15 October 2013).

98. Srivastava, G.P.; Weaire, D. The theory of the cohesive energies of solids. *Adv. Phys.* **1997**, *36*, 463–517.

99. Morrone, J.A.; Car, R. Nuclear quantum effects in water. *Phys. Rev. Lett.* **2008** *101*, 017801.

100. Frank, I.; Hutter, J.; Marx, D.; Parrinello, M. Molecular dynamics in low-spin excited states. *J. Chem. Phys.* **1998**, *108*, 4060–4069.

101. Soler, J.M.; Artacho, E.; Gale, J.D; García, A.; Junquera, J.; Ordejón, P.; Sánchez-Portal, D. The SIESTA method for ab initio order-N materials simulation. *J. Phys.: Condens. Matter* **2002**, *14*, 2745–2779.

102. Porezag, D.; Frauenheim, Th.; Köhler, Th.; Seifert, G.; Kaschner, R. Construction of tight-binding-like potentials on the basis of density-functional theory: Application to carbon. *Phys. Rev. B* **1995**, *51*, 12947–12957.

103. Paier, J.; Marsman, M.; Kresse, G. Why does the B3LYP hybrid functional fail for metals? *J. Chem. Phys.* **2007**, *127*, 024103.

104. Anisimov, V.I.; Zaanen, J.; Andersen, O.K. Band theory and Mott insulators: Hubbard U instead of Stoner I. *Phys. Rev. B* **1991**, *44*, 943–954.

105. Tonigold, K.; Gross, A. Dispersive interactions in water bilayers at metallic surfaces: A comparison of the PBE and RPBE functional including semiempirical dispersion corrections. *J. Comp. Chem.* **2012**, *33*, 695–701.

106. Tabacchi, G.; Hutter, J.; Mundy, C.J A density-functional approach to polarizable models: A Kim-Gordon response density interaction potential for molecular simulations. *J. Chem. Phys.* **2005**, *123*, 074108.

Reprinted from *Entropy*. Cite as: Lin, L.; Lu, J.; Shao, S. Analysis of Time Reversible Born-Oppenheimer Molecular Dynamics. *Entropy* **2014**, *16*, 110–137.

Article

Analysis of Time Reversible Born-Oppenheimer Molecular Dynamics [†]

Lin Lin [1], Jianfeng Lu [2] and Sihong Shao [3],*

[1] Computational Research Division, Lawrence Berkeley National Laboratory, Berkeley, CA 94720, USA; E-Mail: linlin@lbl.gov

[2] Department of Mathematics and Department of Physics, Duke University, Box 90320, Durham, NC 27708, USA; E-Mail: jianfeng@math.duke.edu

[3] LMAM and School of Mathematical Sciences, Peking University, Beijing 100871, China

[†] Dedicated to the celebration of the 100th anniversary of the establishment of modern education system in mathematical sciences at Peking University.

* Author to whom correspondence should be addressed; E-Mail: sihong@math.pku.edu.cn; Tel.: +86-10-6275-3433; Fax: +86-10-6275-1801.

Received:13 June 2013; in revised form: 10 July 2013 / Accepted: 9 September 2013 / Published: 27 December 2013

Abstract: We analyze the time reversible Born-Oppenheimer molecular dynamics (TRBOMD) scheme, which preserves the time reversibility of the Born-Oppenheimer molecular dynamics even with non-convergent self-consistent field iteration. In the linear response regime, we derive the stability condition, as well as the accuracy of TRBOMD for computing physical properties, such as the phonon frequency obtained from the molecular dynamics simulation. We connect and compare TRBOMD with Car-Parrinello molecular dynamics in terms of accuracy and stability. We further discuss the accuracy of TRBOMD beyond the linear response regime for non-equilibrium dynamics of nuclei. Our results are demonstrated through numerical experiments using a simplified one-dimensional model for Kohn-Sham density functional theory.

Keywords: ab initio molecular dynamics; self-consistent field iteration; time reversibility; stability

Classification: PACS 31.15.xv; 71.15.Pd

1. Introduction

Ab initio molecular dynamics (AIMD) [1–6] has been greatly developed in the past few decades, so that nowadays, it is able to quantitatively predict the equilibrium and non-equilibrium properties for a vast range of systems. AIMD has become widely used in chemistry, biology, materials science, *etc*. A coherent and comprehensive presentation of AIMD with both the basic theory and advanced methods can be found in [7]. Most AIMD methods treat the nuclei as classical particles following Newtonian dynamics (known as the time-dependent Born-Oppenheimer approximation), and the interactive force among nuclei is provided directly from electronic structure theory, such as the Kohn-Sham density functional theory [8,9] (KSDFT), without the need of using empirical atomic potentials. KSDFT consists of a set of nonlinear equations that are solved at each molecular dynamics time step *self-consistently* via the self-consistent field (SCF) iteration. In Born-Oppenheimer molecular dynamics (BOMD), KSDFT is solved until full self-consistency for each atomic configuration per time step. Since many iterations are usually needed to reach full self-consistency and each iteration takes a considerable amount of time, until recently, this procedure was still found to be prohibitively expensive for producing meaningful dynamical information. On the other hand, if the self-consistent iterations are truncated before convergence is reached, it is often the case that the energy of the system is no longer conservative, even for an NVE system. The error in SCF iteration acts as a sink or source, gradually draining or adding energy to the atomic system within a short period of molecular dynamics simulation [10]. This is one of the main challenges for accelerating Born-Oppenheimer molecular dynamics.

AIMD was made practical by the ground-breaking work of Car-Parrinello molecular dynamics (CPMD) [11]. CPMD introduces an extended Lagrangian, including the degrees of freedom of both nuclei and electrons without the necessity of a convergent SCF iteration. The dynamics of electronic orbitals can be loosely viewed as a special way for performing the SCF iteration at each molecular dynamics (MD) step. Thanks to the Hamiltonian structure, numerical simulation for CPMD is stable, and the energy is conservative over a much longer time period compared to that for BOMD with non-convergent SCF iteration. When the system has a spectral gap, the accuracy of CPMD is controlled by a single parameter, the fictitious electron mass, μ. The result of CPMD approaches that of BOMD as μ goes to zero [12,13]. However, it has also been shown that CPMD does not work as well for systems with a vanishing gap, for example, for metallic systems [12].

To reduce the cost of BOMD, in particular, the number of SCF iterations needed per MD time step, a new type of AIMD method, the time reversible Born-Oppenheimer molecular dynamics (TRBOMD) method has been recently proposed by Niklasson, Tymczak and Challacombe in [14]. The method has been further developed in [15–18]. The idea of TRBOMD can be summarized as follows: TRBOMD assumes that the SCF iteration is a *deterministic* procedure, with the outcome determined only by the initial guess of the variable to be determined self-consistently. For instance, this variable can be the electron density, and the SCF iteration procedure can be simple mixing with a fixed number of iteration steps without reaching full self-consistency. Then, a fictitious dynamics governed by a second order ordinary differential equation (ODE) is introduced on this initial guess variable. The resulting coupled dynamics is then time-reversible and supposed to be more stable, since it has been found that time-reversible numerical schemes are more stable for long time simulation [19,20]. Besides TRBOMD, alternative ideas based on time-reversible predictor-corrector methods [21] and Langevin dynamics [22,23] can also relax the requirement on the accuracy of the force for AIMD simulation. For these methods, we refer the readers to a recent review paper [24] for more information.

Although TRBOMD has been found to be effective and significantly reduces the number of SCF iterations needed in practice, to the extent of our knowledge, there has been so far no detailed analysis of TRBOMD, other than the numerical stability condition of the Verlet or generalized Verlet scheme for time discretization [17]. Accuracy, stability, as well as the applicability range of TRBOMD remain unclear. In particular, it is not known how the choice of SCF iteration scheme affects TRBOMD. These are crucial issues for guiding the practical use of TRBOMD. The full TRBOMD method for general systems is highly nonlinear and is difficult to analyze. In this work, we first focus on the linear response regime, i.e., we assume that each atom oscillates around their equilibrium position and the electron density stays around the "true" electron density. Under such assumptions, we analyze the accuracy and stability of TRBOMD. We then extend the results to the regime where the atom position is not near equilibrium using the averaging principle.

The rest of the paper is organized as follows. We illustrate the idea of TRBOMD and its analysis in the linear response regime using a simple model in Section 2 and introduce TRBOMD for AIMD in Section 3. We analyze TRBOMD in the linear response regime and compare TRBOMD with CPMD in Section 4. The numerical results for TRBOMD in the linear response regime are given in Section 5. We present the analysis of TRBOMD beyond the linear response regime, such as the non-equilibrium dynamics in Section 6, and conclude with a few remarks in Section 7.

2. An Illustrative Model

To start, let us illustrate the main idea for a simple model problem, which provides the essence of TRBOMD in a much simplified setting. Consider the following nonlinear ODE:

$$\ddot{x}(t) = f(x(t)) \tag{1}$$

where we assume that the right-hand side $f(x)$ is difficult to compute, and it can be approximated by an iterative procedure. Starting from an initial guess, $s \approx f(x)$, the final approximation via the iterative procedure is denoted by $g(x, s)$. We assume the approximation, $g(x, s)$, is consistent, i.e.,:

$$g(x, f(x)) = f(x) \tag{2}$$

To numerically solve the ODE Equation (1), we discretize it by some numerical scheme; then, it remains to decide the initial guess, s, at each time step. A natural choice of s would be $g(x, s)$ from the previous step, as x does not change much in successive steps. For instance, if the Verlet algorithm is used and $t_k = k\Delta t$ with Δt being the time step, the discretized ODE becomes:

$$x_{k+1} = 2x_k - x_{k-1} + (\Delta t)^2 g(x_k, s_k)$$
$$s_{k+1} = g(x_k, s_k) \tag{3}$$

We immediately observe that the discretization scheme Equation (3) breaks the time reversibility of the original ODE Equation (1). In other words, for the original ODE Equation (1), we propagate the system forward in time from $(x(t_0), \dot{x}(t_0))$ to $(x(t_1), \dot{x}(t_1))$. Then, if we use $(x(t_1), \dot{x}(t_1))$ as the initial data at $t = t_1$ and propagate the system backward in time to time $t = t_0$, we will be at the state, $(x(t_0), \dot{x}(t_0))$. The loss of the time reversible structure can introduce large error in long time numerical simulation [20]. This is the main reason why BOMD with non-convergent SCF iteration fails for long time simulations [14]. To overcome this obstacle, the idea of TRBOMD is to introduce a fictitious dynamics for the initial guess, s. Namely, we consider the time reversible coupled system:

$$\ddot{x}(t) = g(x(t), s(t))$$
$$\ddot{s}(t) = \omega^2(g(x(t), s(t)) - s(t))$$

(4)

where ω is an artificial frequency. We analyze, now, the accuracy and stability of Equation (4) in the linear response regime by assuming that the trajectory, $x(t)$, oscillates around an equilibrium position, x^*. We denote by $\tilde{x}(t) = x(t) - x^*$ the deviation from the equilibrium position and $\tilde{s}(t) = s(t) - f(x(t))$, the deviation of the initial guess from the exact force term. Consequently, the equation of motion (4) can be rewritten as (for simplicity we suppress the t-dependence in the notation for the rest of the section):

$$\ddot{\tilde{x}} = g(x, s)$$
$$\ddot{\tilde{s}} = \omega^2(g(x, s) - s) - f''(x)(\dot{x})^2 - f'(x)\ddot{x}$$

(5)

where the term, $-f''(x)(\dot{x})^2 - f'(x)\ddot{x}$, comes from the term, $f(x)$ in $\tilde{s}$, by the chain rule.

In the linear response regime, we assume the linear approximation of force for x around x^*:

$$f(x) \approx -\Omega^2(x - x^*) = -\Omega^2\tilde{x}$$

(6)

where Ω is the oscillation frequency of x in the linear response regime. We also linearize g with respect to $\tilde{s}$ and $\tilde{x}$ and dropping all higher order terms as:

$$g(x, s) = g(x, f(x) + \tilde{s})$$
$$\approx g(x, f(x)) + g_s(x, f(x))\tilde{s}$$
$$\approx -\Omega^2\tilde{x} + g_s(x^*, f(x^*))\tilde{s}$$

(7)

where g_s denotes the partial derivative of g with respect to s, and the consistency condition (2) is applied. We then have:

$$g(x, s) - s = (g(x, f(x) + \tilde{s}) - f(x)) - (s - f(x))$$
$$\approx (g_s(x, f(x)) - 1)\tilde{s}$$
$$\approx (g_s(x^*, f(x^*)) - 1)\tilde{s}$$

(8)

In accord with notations used in later discussions, let us denote:

$$\mathcal{L} = g_s(x^*, f(x^*)), \quad \mathcal{K} = 1 - g_s(x^*, f(x^*))$$

(9)

with which the linearized system of Equation (5) becomes:

$$\frac{d^2}{dt^2}\begin{pmatrix}\widetilde{x}\\\widetilde{s}\end{pmatrix} = \begin{pmatrix}-\Omega^2 & \mathcal{L}\\f'(x^*)\Omega^2 & -f'(x^*)\mathcal{L} - \omega^2\mathcal{K}\end{pmatrix}\begin{pmatrix}\widetilde{x}\\\widetilde{s}\end{pmatrix} := A\begin{pmatrix}\widetilde{x}\\\widetilde{s}\end{pmatrix} \tag{10}$$

Note that when the force is computed accurately, *i.e.*,

$$g(x, s) = f(x), \quad \forall s \tag{11}$$

we have:

$$\mathcal{L} = 0, \quad \mathcal{K} = 1 \tag{12}$$

meaning that the motion of $\widetilde{x}$ is decoupled from that of $\widetilde{s}$, and $\widetilde{x}$ follows the exact harmonic motion in the linear response regime with the accurate frequency, Ω. When the force is computed inaccurately, $\widetilde{x}$ is coupled with $\widetilde{s}$ in Equation (10). Actually, we can solve (10) analytically, and the eigenvalues of A are:

$$\begin{pmatrix}\lambda_{\widetilde{\Omega}}\\\lambda_{\widetilde{\omega}}\end{pmatrix} = \begin{pmatrix}\frac{1}{2}\left(\sqrt{(\mathcal{L}f'(x^*) + \mathcal{K}\omega^2 + \Omega^2)^2 - 4\mathcal{K}\omega^2\Omega^2} - \mathcal{L}f'(x^*) - \mathcal{K}\omega^2 - \Omega^2\right)\\\frac{1}{2}\left(-\sqrt{(\mathcal{L}f'(x^*) + \mathcal{K}\omega^2 + \Omega^2)^2 - 4\mathcal{K}\omega^2\Omega^2} - \mathcal{L}f'(x^*) - \mathcal{K}\omega^2 - \Omega^2\right)\end{pmatrix} \tag{13}$$

Then, the frequencies of the normal modes of the ODE are $\widetilde{\Omega} = \sqrt{-\lambda_{\widetilde{\Omega}}}$ and $\widetilde{\omega} = \sqrt{-\lambda_{\widetilde{\omega}}}$, respectively. Assume $\omega^2 \gg \Omega^2$ and expand the solution to the order of $\mathcal{O}(1/\omega^2)$; we have:

$$\widetilde{\Omega} = \Omega\left(1 - \frac{f'(x^*)}{2\omega^2}\mathcal{L}\mathcal{K}^{-1}\right) + \mathcal{O}(1/\omega^4) \tag{14}$$

Similarly, the frequency for the other normal mode, which is dominated by the motion of $\widetilde{s}$, is:

$$\widetilde{\omega} = \sqrt{\mathcal{K}}\omega\left(1 + \frac{f'(x^*)}{2\omega^2}\mathcal{L}\mathcal{K}^{-1}\right) + \mathcal{O}(1/\omega^3) \tag{15}$$

It is found that one of the normal modes of Equation (10) has frequency $\widetilde{\Omega} \approx \Omega$. We can therefore measure the accuracy of Equation (4) using the relative error between $\widetilde{\Omega}$ and Ω. Furthermore, if the dynamics (4) is stable in the linear response regime, it is necessary to have $\mathcal{K} > 0$.

From Equation (14), we conclude that if the time reversible numerical scheme (4) is used for simulating the ODE Equation (1) and if we neglect the error due to the Verlet scheme, the error introduced in computing the frequency, Ω, is proportional to ω^{-2}. This seems to indicate that very large ω (*i.e.*, very small time step Δt) might be needed to obtain accurate results. Fortunately, the ω^{-2} term in Equation (14) has the prefactor, $f'(x^*)\mathcal{L}\mathcal{K}^{-1}$. Equation (6) shows that $f'(x^*) \approx -\Omega^2$, which is small compared to ω^2. If $g_s(x^*, f(x^*))$ is small, then $\mathcal{K} \approx 1$, and the accuracy of $\widetilde{\Omega}$ is determined by $\mathcal{L}$ or $g_s(x^*, f(x^*))$, which indicates the sensitivity of the computed force with respect to the initial guess, or the accuracy of the iterative procedure for computing the force. If a "good" iterative procedure is used, $g_s(x^*, f(x^*))$ will be small. Therefore, the presence of the term, $\mathcal{L}$, allows one to obtain relatively accurate approximation to the frequency, Ω, without using a large ω. The same behavior can be observed when using TRBOMD to approximate BOMD (*vide post*).

Finally, we remark that even though Equation (1) is a much simplified system, it will be seen below that for BOMD with M atoms and N interacting electrons, the analysis in the linear response regime follows the same line, and the result for the frequency is similar to Equation (14).

3. Time Reversible Born-Oppenheimer Molecular Dynamics

Consider a system with M atoms and N electrons. The position of the atoms at time t is denoted by $\mathbf{R}(t) = (R_1(t), \ldots, R_M(t))^T$. In BOMD, the motion of atoms follows Newton's law:

$$m\ddot{R}_I(t) = f_I(\mathbf{R}(t)) = -\frac{\partial E(\mathbf{R}(t))}{\partial R_I} \tag{16}$$

where $E(\mathbf{R}(t))$ is the total energy of the system at the atomic configuration, $\mathbf{R}(t)$. In KSDFT, the total energy is expressed as a functional of a set of Kohn-Sham orbitals, $\{\psi_i(x)\}_{i=1}^N$. To illustrate the idea with minimal technicality, let us consider for the moment a system of N electrons at zero temperature. The energy functional in KSDFT takes the form:

$$E(\{\psi_i(x)\}_{i=1}^N; \mathbf{R}) = \frac{1}{2}\sum_{i=1}^N \int |\nabla\psi_i(x)|^2 \, dx + \int \rho(x) V_{\text{ion}}(x; \mathbf{R}) \, dx + E_{\text{hxc}}[\rho]$$

$$\rho(x) = \sum_{i=1}^N |\psi_i(x)|^2 \tag{17}$$

The first term in the energy functional is the kinetic energy of the electrons. The second term contains the electron-ion interaction energy. The ion-ion interaction energy usually takes the form $\sum_{I<J} \frac{Z_I Z_J}{|R_I - R_J|}$, where Z_I is the charge for the nucleus, I. The ion-ion interaction energy does not depend on the electron density, ρ. To simplify the notation, we include the ion-ion interaction energy in the V_{ion} term as a constant shift that is independent of the x variable. The third term does not explicitly depend on the atomic configuration, $\mathbf{R}$, and is a nonlinear functional of the electron density, ρ. It represents the Hartree part of electron-electron interaction energy (h) and the exchange-correlation energy (xc) characterizing many body effects. The energy, $E(\mathbf{R})$, as a function of atomic positions is given by the following minimization problem:

$$E(\mathbf{R}) = \min_{\{\psi_i(x)\}_{i=1}^N} E(\{\psi_i(x)\}_{i=1}^N; \mathbf{R})$$

$$\text{s.t.} \quad \int \psi_i^\dagger(x)\psi_j(x) \, dx = \delta_{ij}, \quad i, j = 1, \ldots, N \tag{18}$$

We denote by $\{\psi_i(x; \mathbf{R})\}_{i=1}^N$ the (local) minimizer and $\rho^*(x; \mathbf{R}) = \sum_{i=1}^N |\psi_i(x; \mathbf{R})|^2$, the converged electron density corresponding to the minimizer (here, we assume that the minimizing electron density is unique). Then, the force acting on the atom I is:

$$f_I(\mathbf{R}; \rho^*(x; \mathbf{R})) = -\frac{\partial E(\mathbf{R})}{\partial R_I} = -\int \rho^*(x; \mathbf{R})\frac{\partial V_{\text{ion}}(x; \mathbf{R})}{\partial R_I} \, dx \tag{19}$$

In the physics literature, the force formula in Equation (19) is referred to as the Hellmann-Feynman force. The validity of the Hellmann-Feynman formula relies on the electron density, $\rho^*(x; \mathbf{R})$, corresponding to the minimizers of the Kohn-Sham energy functional. Since $E_{\text{hxc}}[\rho]$ is a nonlinear functional of ρ, the electron density, ρ, is usually determined through the self-consistent field (SCF) iteration as follows.

Starting from an inaccurate input electron density, ρ^{in}, one first computes the output electron density by solving the lowest N eigenfunctions of the problem:

$$\left(-\frac{1}{2}\Delta_x + \mathcal{V}(x; \mathbf{R}, \rho^{\text{in}})\right)\psi_i = \varepsilon_i\psi_i \qquad (20)$$

with:

$$\mathcal{V}(x; \mathbf{R}, \rho) = V_{\text{ion}}(x; \mathbf{R}) + \frac{\delta E_{\text{hxc}}[\rho]}{\delta\rho}(x) \qquad (21)$$

and the output electron density, ρ^{out}, is defined by:

$$\rho^{\text{out}}(x) := F[\rho^{\text{in}}](x) = \sum_{i=1}^{N}|\psi_i(x)|^2 \qquad (22)$$

Here, the operator, F, is called the Kohn-Sham map. ρ^{out} can be used directly as the input electron density, ρ^{in}, in the next iteration. This is called the *fixed point iteration*. Unfortunately, in most electronic structure calculations, the fixed point iteration does not converge, even when ρ^{in} is very close to the true electron density, ρ^*. The fixed point iteration can be improved by the simple mixing method, which takes the linear combination of the electron density:

$$\alpha\rho^{\text{out}} + (1-\alpha)\rho^{\text{in}} \qquad (23)$$

as the input density for the next iteration with $0 < \alpha \leq 1$. Simple mixing can greatly improve the convergence properties of the SCF iteration over the fixed point iteration, but the convergence rate can still be slow in practice. There are more complicated SCF iteration schemes, such as the Anderson mixing scheme [25], the Pulay mixing scheme [26] and the Broyden mixing scheme [27]. Furthermore, preconditioners can be applied to the SCF iteration to enhance convergence properties, such as the Kerker preconditioner [28]. More detailed discussion on the convergence properties of these SCF schemes can be found in [29]. In the following discussions, we denote by $\rho_{\text{SCF}}(x; \mathbf{R}, \rho)$ the final electron density after the SCF iteration starting from an initial guess, ρ. We assume that ρ_{SCF} satisfies the consistency condition:

$$\rho_{\text{SCF}}(x; \mathbf{R}, \rho^*(\cdot; \mathbf{R})) = \rho^*(x; \mathbf{R}) \qquad (24)$$

If a non-convergent SCF iteration procedure is used, $\rho_{\text{SCF}}(x; \mathbf{R}, \rho)$ might deviate from $\rho^*(x; \mathbf{R})$. Such deviation introduces error in the force, and the error can accumulate in the long time molecular dynamics simulation and lead to inaccurate results in computing the statistical and dynamical properties of the systems.

The map, ρ_{SCF}, is usually highly nonlinear, which makes it difficult to correct the error in the force. The TRBOMD scheme avoids the direct correction for the inaccurate ρ_{SCF}, but allows the initial guess to dynamically evolve together with the motion of the atoms. We denote by $\rho(x, t)$ the initial guess for the SCF iteration at time t. When $\rho(\cdot, t)$ is used as an argument, we also write $\rho_{\text{SCF}}(x; \mathbf{R}(t), \rho(t)) := \rho_{\text{SCF}}(x; \mathbf{R}(t), \rho(\cdot, t))$. The Hellmann-Feynman formula (19) is used to compute the force at the electron density, $\rho_{\mathrm{SCF}}(x; \mathbf{R}(t), \rho(t))$, even though $\rho^*(x; \mathbf{R}(t))$ is not available. Thus, the equation of motion in TRBOMD reads:

$$m\ddot{R}_I(t) = f_I(\mathbf{R}(t); \rho_{\mathrm{SCF}}(x; \mathbf{R}(t), \rho(t))) = -\int \rho_{\mathrm{SCF}}(x; \mathbf{R}(t), \rho(t)) \frac{\partial V_{\mathrm{ion}}(x; \mathbf{R}(t))}{\partial R_I} \, dx \tag{25}$$

$$\ddot{\rho}(x, t) = \omega^2(\rho_{\mathrm{SCF}}(x; \mathbf{R}(t), \rho(t)) - \rho(x, t))$$

It is clear that TRBOMD is time reversible. The discretized TRBOMD is still time reversible if the numerical scheme is time reversible. For instance, if the Verlet scheme is used, the discretized equation of motion becomes:

$$R_I(t_{k+1}) = 2R_I(t_k) - R_I(t_{k-1}) - \frac{\Delta t^2}{m} f_I(\mathbf{R}(t_k); \rho_{\mathrm{SCF}}(x; \mathbf{R}(t_k), \rho(t_k)) \tag{26}$$

$$\rho(x, t_{k+1}) = 2\rho(x, t_k) - \rho(x, t_{k-1}) + \Delta t^2 \omega^2(\rho_{\mathrm{SCF}}(x; \mathbf{R}(t_k), \rho(t_k)) - \rho(x, t_k))$$

which is evidently time reversible. The artificial frequency, ω, controls the frequency of the fictitious dynamics of $\rho(x, t)$ and is generally chosen to be larger than the frequency of the motion of the atoms. The numerical stability of the Verlet algorithm requires that the dimensionless quantity, $\kappa := (\omega \Delta t)^2$, be small [30]. When κ is fixed, ω controls the stiffness or, equivalently, the time step $\Delta t = \frac{\sqrt{\kappa}}{\omega}$ for the equation of motion (26).

Let us mention that TRBOMD is closely related to CPMD. In CPMD, the equation of motion is given by:

$$m\ddot{R}_I(t) = f_I(\mathbf{R}(t), \rho(t)) = -\int \rho(t) \frac{\partial V_{\mathrm{ion}}(x; \mathbf{R}(t))}{\partial R_I} \, dx \tag{27}$$

$$\mu\ddot{\psi}_i(t) = -\frac{\delta E(\mathbf{R}(t), \{\psi_i(t)\})}{\delta \psi_i^\dagger} + \sum_j \psi_j(t)\Lambda_{ji}(t)$$

where μ is the fictitious electron mass for the fake electron dynamics in CPMD and Λ's are the Lagrange multipliers determined so that $\{\psi_i(t)\}$ is an orthonormal set of functions for any time. The CPMD scheme (27) can be viewed as the equation of motion with an extended Lagrangian:

$$\mathcal{L}_{\mathrm{CP}}(\mathbf{R}, \dot{\mathbf{R}}, \{\psi_i\}, \{\dot{\psi}_i\}) = \sum_I \frac{m}{2}|\dot{R}_I|^2 + \sum_i \frac{\mu}{2} \int |\dot{\psi}_i|^2 - E(\mathbf{R}, \{\psi_i\}) \tag{28}$$

which contains both ionic and electronic degrees of freedom. Therefore, CPMD is a Hamiltonian dynamics and, thus, time reversible.

Note that the frequency of the evolution equation for $\{\psi_i\}$ in CPMD is adjusted by the fictitious mass parameter, μ. Comparing with TRBOMD, the parameter, μ, plays a similar role as ω^{-2}, which controls the frequency of the fictitious dynamics of the initial density guess in SCF iteration. This connection will be made more explicit in the sequel.

We remark that the papers, [16,17], took a further step in viewing TRBOMD by an extended Lagrangian approach in a vanishing mass limit. This was also interpreted differently in [24] by starting from a Lagrangian and, then, using inaccurate forces in the equation of motions. However, unless a very specific and restrictive form of the error due to non-convergent SCF iterations is assumed, the equation of motion in TRBOMD does not have an associated Lagrangian in general. The connection to Lagrangian dynamics remains formal, and hence, we will not further explore it here.

4. Analysis of TRBOMD in the Linear Response Regime

In this section, we consider Equation (25) in the linear response regime, in which each atom, I, oscillates around its equilibrium position, R_I^*. The displacement of the atomic configuration, $\mathbf{R}$, from the equilibrium position is denoted by $\widetilde{\mathbf{R}}(t) := \mathbf{R}(t) - \mathbf{R}^*$, and the deviation of the electron density from the converged density is denoted by $\widetilde{\rho}(x,t) := \rho(x,t) - \rho^*(x;\mathbf{R}(t))$. Both $\widetilde{\mathbf{R}}(t)$ and $\widetilde{\rho}(x,t)$ are small quantities in the linear response regime and contain the same information as $\mathbf{R}(t)$ and $\rho(x,t)$. Using $\widetilde{\mathbf{R}}(t)$ and $\widetilde{\rho}(x,t)$ as the new variables and noting the chain rule due to the $\mathbf{R}$-dependence in $\rho^*(x;\mathbf{R}(t))$, the equation of motion in TRBOMD becomes:

$$m\ddot{\widetilde{R}}_I(t) = -\int \rho_{\mathrm{SCF}}(x;\mathbf{R}(t),\rho(t))\frac{\partial V_{\mathrm{ion}}(x;\mathbf{R}(t))}{\partial R_I}\,\mathrm{d}x$$

$$\ddot{\widetilde{\rho}}(x,t) = \omega^2(\rho_{\mathrm{SCF}}(x;\mathbf{R}(t),\rho(t)) - \rho(x,t)) - \sum_{I=1}^{M}\frac{\partial \rho^*(x;\mathbf{R}(t))}{\partial R_I}\ddot{\widetilde{R}}_I(t) \qquad (29)$$

$$-\sum_{I,J=1}^{M}\dot{\widetilde{R}}_I(t)\dot{\widetilde{R}}_J(t)\frac{\partial^2 \rho^*(x;\mathbf{R}(t))}{\partial R_I \partial R_J}$$

To simplify notation, from now on, we suppress the t-dependence in all variables, and Equation (29) becomes:

$$m\ddot{\widetilde{R}}_I = -\int \rho_{\mathrm{SCF}}(x;\mathbf{R},\rho)\frac{\partial V_{\mathrm{ion}}(x;\mathbf{R})}{\partial R_I}\,\mathrm{d}x \qquad (30a)$$

$$\ddot{\widetilde{\rho}}(x) = \omega^2(\rho_{\mathrm{SCF}}(x;\mathbf{R},\rho) - \rho(x)) - \sum_{I=1}^{M}\frac{\partial \rho^*}{\partial R_I}(x;\mathbf{R})\ddot{\widetilde{R}}_I - \sum_{I,J=1}^{M}\dot{\widetilde{R}}_I\dot{\widetilde{R}}_J\frac{\partial^2 \rho^*}{\partial R_I \partial R_J}(x;\mathbf{R}) \qquad (30b)$$

In the linear response regime, we expand Equation (30) and only keep terms that are linear with respect to $\widetilde{\mathbf{R}}$ and $\widetilde{\rho}$. All the higher order terms, including all the cross products of $\widetilde{R}_I$, $\dot{\widetilde{R}}_I$ and $\widetilde{\rho}$, will be dropped. First, we linearize the force on atom I with respect to $\widetilde{\rho}$ as:

$$f_I(\mathbf{R};\rho_{\mathrm{SCF}}(x;\mathbf{R},\rho))$$

$$= -\int \rho_{\mathrm{SCF}}(x;\mathbf{R},\rho)\frac{\partial V_{\mathrm{ion}}(x;\mathbf{R})}{\partial R_I}\,\mathrm{d}x$$

$$= -\int \rho^*(x;\mathbf{R})\frac{\partial V_{\mathrm{ion}}(x;\mathbf{R})}{\partial R_I}\,\mathrm{d}x - \int (\rho_{\mathrm{SCF}}(x;\mathbf{R},\rho^*(\mathbf{R})+\widetilde{\rho}) - \rho^*(x;\mathbf{R}))\frac{\partial V_{\mathrm{ion}}(x;\mathbf{R})}{\partial R_I}\,\mathrm{d}x \qquad (31)$$

$$\approx -\int \rho^*(x;\mathbf{R})\frac{\partial V_{\mathrm{ion}}(x;\mathbf{R})}{\partial R_I}\,\mathrm{d}x - \int \frac{\delta \rho_{\mathrm{SCF}}}{\delta \rho}(x,y;\mathbf{R})\widetilde{\rho}(y)\frac{\partial V_{\mathrm{ion}}(x;\mathbf{R})}{\partial R_I}\,\mathrm{d}x\,\mathrm{d}y$$

Next, we linearize with respect to $\widetilde{\mathbf{R}}$; we have:

$$\int \rho^*(x;\mathbf{R})\frac{\partial V_{\mathrm{ion}}(x;\mathbf{R})}{\partial R_I}\,\mathrm{d}x \approx -m\sum_{I,J=1}^{M} D_{IJ}\widetilde{R}_J \qquad (32)$$

Here, the matrix, $\{\mathcal{D}_{IJ}\}$, is the dynamical matrix for the atoms. For the last term in Equation (31), we have:

$$
\begin{aligned}
&\int \frac{\delta\rho_{\text{SCF}}}{\delta\rho}(x,y;\mathbf{R})\widetilde{\rho}(y)\frac{\partial V_{\text{ion}}(x;\mathbf{R})}{\partial R_I}\,dx\,dy \\
&\approx \int \frac{\delta\rho_{\text{SCF}}}{\delta\rho}(x,y;\mathbf{R}^*)\widetilde{\rho}(y)\frac{\partial V_{\text{ion}}(x;\mathbf{R}^*)}{\partial R_I}\,dx\,dy \\
&:= -m\mathcal{L}_I[\widetilde{\rho}]
\end{aligned}
\tag{33}
$$

The last equation in Equation (33) defines a linear functional, $\mathcal{L}_I$, with $\frac{\delta\rho_{\text{SCF}}}{\delta\rho}(x,y;\mathbf{R}^*)$ and $\frac{\partial V_{\text{ion}}(x;\mathbf{R}^*)}{\partial R_I}$ evaluated at the fixed equilibrium point, $\mathbf{R}^*$.

In the linear response regime, the operator, $\frac{\delta\rho_{\text{SCF}}}{\delta\rho}(x,y;\mathbf{R}^*)$, carries all the information of the SCF iteration scheme. Let us now derive the explicit form of $\frac{\delta\rho_{\text{SCF}}}{\delta\rho}(x,y;\mathbf{R}^*)$ for the k-step simple mixing scheme with mixing parameter (step length) α ($0 < \alpha \leq 1$). If $k = 1$, the simple mixing scheme reads:

$$
\rho_{\text{SCF}}(x;\mathbf{R},\rho^*(\mathbf{R})+\widetilde{\rho}) = \alpha F[\rho^*(\mathbf{R})+\widetilde{\rho}] + (1-\alpha)(\rho^*(\mathbf{R})+\widetilde{\rho})
\tag{34}
$$

so:

$$
\frac{\delta\rho_{\text{SCF}}}{\delta\rho}(x,y;\mathbf{R}^*) = \delta(x-y) - \alpha\left(\delta(x-y) - \frac{\delta F}{\delta\rho}(x,y)\right)
\tag{35}
$$

Here, $\delta(x)$ is the Dirac δ-function, and the operator, $\left(\delta(x-y) - \frac{\delta F}{\delta\rho}(x,y)\right) := \varepsilon(x,y)$, is usually refereed to as the *dielectric operator* [31,32]. To simplify the notation, we would not distinguish the kernel of an integral operator from the integral operator itself. For example, $\varepsilon(x,y)$ is denoted by ε. Neither will we distinguish integral operators defined on continuous space from the corresponding finite dimensional matrices obtained from certain numerical discretization. This slight abuse of notation allows us to simply denote $f(x) = \int A(x,y)g(y)\,dy$ by $f = Ag$ as a matrix-vector multiplication and to denote the composition of kernels of integral operators $C(x,y) = \int dz\,A(x,z)B(z,y)$ by $C = AB$ as a matrix-matrix multiplication. Using such notations, Equation (35) can be written in a more compact form:

$$
\frac{\delta\rho_{\text{SCF}}}{\delta\rho} = I - \alpha\varepsilon
\tag{36}
$$

Similarly, for the k-step simple mixing method, we have:

$$
\frac{\delta\rho_{\text{SCF}}}{\delta\rho} = (1-\alpha\varepsilon)^k
\tag{37}
$$

In general, the dielectric operator is diagonalizable, and all eigenvalues of ε are real. Therefore, the linear response operator, $\frac{\delta\rho_{\text{SCF}}}{\delta\rho}$, for the k-th step simple mixing method is also diagonalizable with real eigenvalues.

From Equation (30b), we have:

$$
\begin{aligned}
&\rho_{\text{SCF}}(x; \mathbf{R}, \rho) - \rho(x) \\
&= (\rho_{\text{SCF}}(x; \mathbf{R}, \widetilde{\rho} + \rho^*(\mathbf{R})) - \rho^*(x; \mathbf{R})) - (\rho(x) - \rho^*(x; \mathbf{R})) \\
&\approx \int \frac{\delta \rho_{\text{SCF}}}{\delta \rho}(x, y; \mathbf{R})\widetilde{\rho}(y)\, \mathrm{d}y - \widetilde{\rho}(x) \\
&\approx \int \frac{\delta \rho_{\text{SCF}}}{\delta \rho}(x, y; \mathbf{R}^*)\widetilde{\rho}(y)\, \mathrm{d}y - \widetilde{\rho}(x) \\
&:= -\int \mathcal{K}(x, y)\widetilde{\rho}(y)\, \mathrm{d}y
\end{aligned}
\tag{38}
$$

Here, we have used consistency condition (24). The last line of Equation (38) defines a kernel:

$$
\mathcal{K}(x, y) = \delta(x - y) - \frac{\delta \rho_{\text{SCF}}}{\delta \rho}(x, y; \mathbf{R}^*)
\tag{39}
$$

which is an important quantity for the stability of TRBOMD, as will be seen later. Using Equations (33) and (38), the equation of motion, (30), can be written in the linear response regime as:

$$
\ddot{\widetilde{R}}_I = -\sum_{J=1}^{M} \mathcal{D}_{IJ}\widetilde{R}_J + \mathcal{L}_I[\widetilde{\rho}]
$$

$$
\ddot{\widetilde{\rho}}(x) = -\omega^2 \int \mathcal{K}(x, y)\widetilde{\rho}(y)\, \mathrm{d}y - \sum_{I=1}^{M} \frac{\partial \rho^*}{\partial R_I}(x; \mathbf{R}^*)\left(-\sum_{J=1}^{M}\mathcal{D}_{IJ}\widetilde{R}_J + \mathcal{L}_I[\widetilde{\rho}]\right)
\tag{40}
$$

Define:

$$
\mathcal{L} = (\mathcal{L}_1, \cdots, \mathcal{L}_M)^T
\tag{41}
$$

then Equation (40) can be rewritten in a more compact form as:

$$
\ddot{\widetilde{\mathbf{R}}} = -\mathcal{D}\widetilde{\mathbf{R}} + \mathcal{L}[\widetilde{\rho}],
\tag{42a}
$$

$$
\ddot{\widetilde{\rho}}(x) = -\omega^2 \int \mathcal{K}(x, y)\widetilde{\rho}(y)\, \mathrm{d}y - \left(\frac{\partial \rho^*}{\partial \mathbf{R}}(x; \mathbf{R}^*)\right)^T\left(-\mathcal{D}\widetilde{\mathbf{R}} + \mathcal{L}[\widetilde{\rho}]\right)
\tag{42b}
$$

Now, if the self-consistent iteration is performed accurately regardless of the initial guess, *i.e.*,

$$
\rho_{\text{SCF}}(x; \mathbf{R}, \rho) = \rho^*(x; \mathbf{R}), \quad \forall \rho
\tag{43}
$$

which implies:

$$
\frac{\delta \rho_{\text{SCF}}}{\delta \rho}(x, y; \mathbf{R}^*) = 0, \quad \mathcal{L} = 0, \quad \mathcal{K}(x, y) = \delta(x - y)
\tag{44}
$$

The linearized equation of motion (42) becomes:

$$
\ddot{\widetilde{\mathbf{R}}} = -\mathcal{D}\widetilde{\mathbf{R}},
\tag{45a}
$$

$$
\ddot{\widetilde{\rho}}(x) = -\omega^2\widetilde{\rho}(x) + \left(\frac{\partial \rho^*}{\partial \mathbf{R}}(x; \mathbf{R}^*)\right)^T \mathcal{D}\widetilde{\mathbf{R}}
\tag{45b}
$$

Therefore, in the case of accurate SCF iteration, according to Equation (45a), the equation of the motion of atoms follows the accurate linearized equation and is decoupled from the fictitious dynamics of $\widetilde{\rho}$. The normal modes of the equation of motion of atoms can be obtained by diagonalizing the dynamical matrix, $\mathcal{D}$, as:

$$\mathcal{D}\mathbf{v}_l = \Omega_l^2 \mathbf{v}_l, \quad l = 1, \ldots, M \tag{46}$$

The frequencies, $\{\Omega_l\}$ ($\Omega_l > 0$), are known as *phonon frequencies*. When the SCF iterations are performed inaccurately, it is meaningless to assess the accuracy of the approximate dynamics (42) by direct investigation of the trajectories, $\widetilde{R}(t)$, since small difference in the phonon frequency can cause large error in the phase of the periodic motion, $\widetilde{R}(t)$, over a long time. However, it is possible to compute the approximate phonon frequencies, $\{\widetilde{\Omega}_l\}$, from Equation (42) and measure the accuracy of TRBOMD in the linearized regime from the relative error:

$$\text{err}_l = \frac{\widetilde{\Omega}_l - \Omega_l}{\Omega_l} \tag{47}$$

The operator, $\mathcal{K}(x, y)$, in Equation (39) is directly related to the stability of the dynamics. Equation (42b) also suggests that in the linear response regime, the spectrum of $\mathcal{K}(x, y)$ must be on the real line, which requires that the matrix, $\frac{\delta\rho_{\text{SCF}}}{\delta\rho}(x, y; \mathbf{R}^*)$, be diagonalizable with real eigenvalues. This has been shown for the simple mixing scheme. However, we remark that the condition that all eigenvalues of $\mathcal{K}(x, y)$ are real may not hold for general preconditioners or for more complicated SCF iterations (for instance, Anderson mixing). This is one important restriction of the linear response analysis. Of course, this may not be a restriction for practical TRBOMD simulation for real systems. We will leave further understanding of this to future works.

Let us now assume that all eigenvalues of $\mathcal{K}$ are real. The lower bound of the spectrum of $\mathcal{K}$, denoted by $\lambda_{\min}(\mathcal{K})$, should satisfy:

$$\lambda_{\min}(\mathcal{K}) > 0 \tag{48}$$

Equation (48) is a necessary condition for TRBOMD to be stable, which will be referred to as the *stability condition* in the following. Furthermore, ω should be chosen large enough in order to avoid resonance between the motion of $\widetilde{\mathbf{R}}$ and $\widetilde{\rho}$. Therefore, the *adiabatic condition*:

$$\omega^2 \gg \frac{\lambda_{\max}(\mathcal{D})}{\lambda_{\min}(\mathcal{K})} = \frac{\max_l \Omega_l^2}{\lambda_{\min}(\mathcal{K})} \tag{49}$$

should also be satisfied. Due to Equation (49), we may assume $\epsilon = 1/\omega^2$ is a small number and expand Ω_l in the perturbation series of ϵ to quantify the error in the linear response regime. Following the derivation in the appendix, we have:

$$\widetilde{\Omega}_l = \Omega_l \left(1 - \frac{1}{2\omega^2} \mathbf{v}_l^T \mathcal{L} \left[\mathcal{K}^{-1} \left[\left(\frac{\partial \rho^*}{\partial \mathbf{R}} \right)^T \mathbf{v}_l \right] \right] \right) + \mathcal{O}(1/\omega^4) \tag{50}$$

where $\mathcal{K}^{-1}$ is the inverse operator of $\mathcal{K}$ ($\mathcal{K}$ is invertible, due to the stability condition). Since $\omega = \sqrt{\kappa}/\Delta t$, Equation (50) suggests that the accuracy of TRBOMD in the linear response regime is $(\Delta t)^2$, with the pre-constant mainly determined by $\mathcal{L}$, *i.e.*, the accuracy of the SCF iteration.

Let us compare TRBOMD with CPMD. It is well known that CPMD accurately approximates the results of BOMD, provided that the electronic and ionic degrees of freedom remain adiabatically separated, as well as the electrons stay close to the Born-Oppenheimer surface [12,13]. More specifically, the fictitious electron mass should be chosen, so that the lowest electronic frequency is well above ionic frequencies:

$$\mu \ll \frac{E_{\text{gap}}}{\max_l \Omega_l^2} \tag{51}$$

where E_{gap} is the spectral gap (between the highest occupied and the lowest unoccupied states) of the system, and recall that Ω_l is the vibration frequency of the lattice phonon. For CPMD, a similar analysis in the linear response regime as above (we omit the derivation here) shows that:

$$\widetilde{\Omega}_l = \Omega_l(1 + \mathcal{O}(\mu)) \tag{52}$$

under assumption (51). The adiabaticity (51), as well as the role of the fictitious electron mass on physical quantities have been investigated extensively in [33–35]. The linear relationship (52) between the fictitious electron mass and the dynamical frequencies of CPMD was also presented in [34].

Note that condition (51) implies that CPMD no longer works if the system has a small gap or is even metallic. The usual work-around for this is to add a heat bath for the electronic degrees of freedom in CPMD [33], so that it maintains a fictitious temperature for the electronic degree of freedom. Nonetheless, the adiabaticity is lost for metallic systems, and CPMD is no longer accurate over long time simulation. In contrast, as we have discussed previously, TRBOMD may work for both insulating and metallic systems without any modification, provided that the SCF iteration is accurate and no resonance occurs. This is an important advantage of TRBOMD, which we will illustrate using numerical examples in the next section.

When the system has a gap, we can take μ sufficiently small to satisfy the adiabatic separation condition (51). Compare Equation (52) with Equation (50); we see that μ in CPMD plays a similar role as ω^{-2} in TRBOMD. The accuracy (in the linear regime) for CPMD and TRBOMD is the first order in μ and ω^{-2}, respectively. At the same time, as taking a small μ or large ω increases the stiffness of the equation, the computational cost is proportional to μ^{-1} and ω^2, respectively.

Let us remark that the above analysis is done in the linear response regime. As shown in [12,13], the accuracy of CPMD, in general, is only $\mathcal{O}(\mu^{1/2})$ instead of $\mathcal{O}(\mu)$ for the linear regime. Due to the close connection between these two parameters, we do not expect $\mathcal{O}(\omega^{-2})$ accuracy for TRBOMD in general, either. Actually, as will be discussed in Section 6, if the deviation of atom positions from equilibrium is not so small that we cannot linearize the nuclei motion, the error of TRBOMD in general will be $\mathcal{O}(\omega^{-1})$.

5. Numerical Results in the Linear Response Regime

In this section, we present numerical results for TRBOMD in the linear response regime using a one-dimensional (1D) model for KSDFT without the exchange correlation functional. The model problem can be tuned to exhibit both metallic and insulating features. Such a model was used before in mathematical analysis of ionization conjecture [36].

The total energy functional in our 1D density functional theory (DFT) model is given by:

$$E(\{\psi_i(x)\}_{i=1}^N; \mathbf{R}) = \frac{1}{2}\sum_{i=1}^N \int \left|\frac{d}{dx}\psi_i(x)\right|^2 dx + \frac{1}{2}\int K(x,y)(\rho(x)+m(x;\mathbf{R}))(\rho(y)+m(y;\mathbf{R}))\,dx\,dy \tag{53}$$

with $\rho(x) = \sum_{i=1}^N |\psi_i(x)|^2$. The associated Hamiltonian is given by:

$$H(\mathbf{R}) = -\frac{1}{2}\frac{d^2}{dx^2} + \int K(x,y)(\rho(y)+m(y;\mathbf{R}))\,dy \tag{54}$$

Here, $m(x;\mathbf{R}) = \sum_{I=1}^M m_I(x-R_I)$, with the position of the I-th nucleus denoted by R_I. Each function, $m_I(x)$, takes the form:

$$m_I(x) = -\frac{Z_I}{\sqrt{2\pi\sigma_I^2}}e^{-\frac{x^2}{2\sigma_I^2}} \tag{55}$$

where Z_I is an integer representing the charge of the i-th nucleus. This can be understood as a local pseudopotential approximation to represent the electron-ion interaction. The second term on the right-hand side of Equation (53) represents the electron-ion, electron-electron and ion-ion interaction energy. The parameter, σ_I, represents the width of the nuclei in the pseudopotential theory. Clearly, as $\sigma_I \to 0$, $m_I(x) \to -Z_I\delta(x)$, which is the charge density for an ideal nucleus. In our numerical simulation, we set σ_I to a finite value. The corresponding $m_I(x)$ is called a *pseudo charge density* for the I-th nucleus. We refer to the function, $m(x)$, as the total pseudo-charge density of the nuclei. The system satisfies the charge neutrality condition, *i.e.*,

$$\int \rho(x) + m(x;\mathbf{R})\,dx = 0 \tag{56}$$

Since $\int m_I(x)\,dx = -Z_I$, the charge neutrality condition (56) implies:

$$\int \rho(x)\,dx = \sum_{I=1}^M Z_I = N \tag{57}$$

where N is the total number of electrons in the system. To simplify discussion, we omit the spin degeneracy here. The Hellmann-Feynman force is given by:

$$f_I = -\int K(x,y)(\rho(y)+m(y;\mathbf{R}))\frac{\partial m(x;\mathbf{R})}{\partial R_I}\,dx\,dy \tag{58}$$

Instead of using a bare Coulomb interaction, which diverges in 1D, we adopt a Yukawa kernel:

$$K(x,y) = \frac{2\pi e^{-\kappa|x-y|}}{\kappa\epsilon_0} \tag{59}$$

which satisfies the equation:

$$-\frac{d^2}{dx^2}K(x,y) + \kappa^2 K(x,y) = \frac{4\pi}{\epsilon_0}\delta(x-y) \tag{60}$$

As $\kappa \to 0$, the Yukawa kernel approaches the bare Coulomb interaction given by the Poisson equation. The parameter, ϵ_0, is used to make the magnitude of the electron static contribution comparable to that of the kinetic energy.

The parameters used in the 1D DFT model are chosen as follows. Atomic units are used throughout the discussion unless otherwise mentioned. The Yukawa parameter, $\kappa = 0.01$, is small enough so that the range of the electrostatic interaction is sufficiently long, and ϵ_0 is set to 10.00. The nuclear charge, Z_I, is set to one for all atoms. Since spin is neglected, $Z_I = 1$ implies that each atom contributes to one occupied state. The Hamiltonian operator is represented in a planewave basis set. All the examples presented in this section consists of 32 atoms. Initially, the atoms are at their equilibrium positions, and the distance between each atom and its nearest neighbor is set to 10 au. Starting from the equilibrium position, each ion is given a finite velocity, so that the velocity on the centroid of mass is zero. In the numerical experiments below, the system contains only one single phonon, which is obtained by assigning an initial velocity, $v_0 \propto (1, -1, 1, -1, \cdots)$, to the atoms. We denote by Ω^{Ref} the corresponding phonon frequency. We choose v_0, so that $\frac{1}{2}mv_0^2 = k_B T_{\text{ion}}$, where k_B is the Boltzmann constant and T_{ion} is 10 K, to make sure that the system is in the linear response regime. In the atomic unit, the mass of the electron is one, and the mass of each nuclei is set to $42,000$. By adjusting the parameters, $\{\sigma_I\}$, the 1D DFT model model can be tuned to resemble an insulating (with $\sigma_I = 2.0$) or a metallic system (with $\sigma_I = 6.0$) throughout the MD simulation. Figure 1 shows the spectrum of the insulating and the metallic system after running $1,000$ BOMD steps with converged SCF iteration.

Figure 1. Spectrum for the insulator and metal with 32 atoms after $1,000$ Born-Oppenheimer molecular dynamics (BOMD) steps with converged self-consistent field (SCF) iteration. (**a**) Insulator; (**b**) metal.

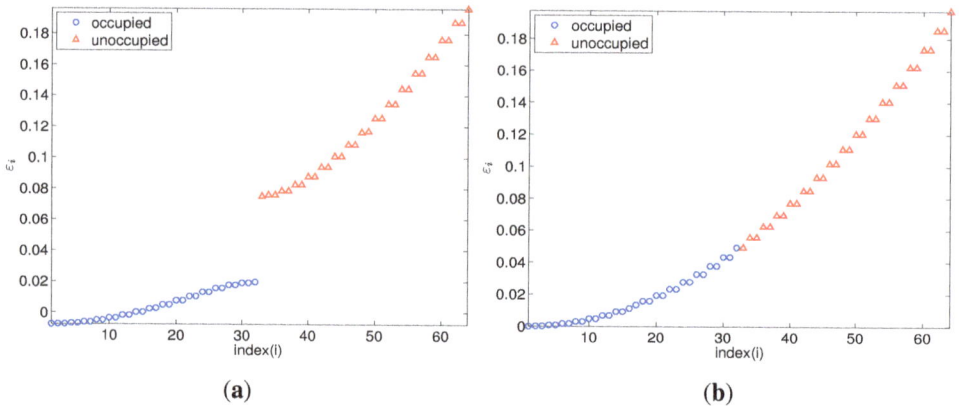

(**a**)

(**b**)

In the linear response regime, we measure the error of the phonon frequency calculated from TRBOMD. This can be done in two ways. The first is given by Equation (50), namely, all quantities in the big parentheses in Equation (50) can be directly obtained by using the finite difference method at the equilibrium position, $\mathbf{R}^*$. The second is to explore the fact that in the linear response regime, there is a linear relation between the force and the atomic position, as in Equation (32), *i.e.*, Hooke's law:

$$f_I(t_l) \approx -m \sum_J \mathcal{D}_{IJ} \widetilde{R}_J(t_l) \qquad (61)$$

holds approximately at each time step. Here, $\{f_I(t_l)\}$ and $\{\widetilde{R}_I(t_l)\}$ are obtained from the trajectory of the TRBOMD simulation directly. To numerically compute $\mathcal{D}_{IJ}$, we solve the least square problem:

$$\min_{\mathcal{D}} \sum_{l,I} \left\| f_I(t_l) + m \sum_J \mathcal{D}_{IJ}\widetilde{R}_J(t_l) \right\|^2 \tag{62}$$

which yields:

$$\mathcal{D} = -\frac{1}{m}S^{fR}\left(S^{RR}\right)^{-1} \tag{63}$$

where:

$$S^{fR}_{IJ} = \sum_l f_I(t_l)\widetilde{R}_J(t_l), \quad S^{RR}_{IJ} = \sum_l \widetilde{R}_I(t_l)\widetilde{R}_J(t_l) \tag{64}$$

The frequencies, $\{\widetilde{\Omega}_l\}$, can be obtained by diagonalizing the matrix, $\mathcal{D}$. Similarly, one can perform the calculation for the accurate BOMD simulation and obtain the exact value of the frequencies, $\{\Omega_l\}$.

In order to compare the performance among BOMD, TRBOMD and CPMD, we define the following relative errors:

$$\text{err}_{\Omega}^{\text{Hooke}} = \frac{\widetilde{\Omega}^{\text{Hooke}} - \Omega^{\text{Ref}}}{\Omega^{\text{Ref}}} \tag{65}$$

$$\text{err}_{\Omega}^{\text{LR}} = \frac{\widetilde{\Omega}^{\text{LR}} - \Omega^{\text{Ref}}}{\Omega^{\text{Ref}}} \tag{66}$$

$$\text{err}_{\overline{E}} = \frac{\overline{E} - \overline{E}^{\text{Ref}}}{\overline{E}^{\text{Ref}}} \tag{67}$$

$$\text{err}_R^{L^2} = \frac{\|R_1(t) - R_1^{\text{Ref}}(t)\|_{L^2}}{\|R_1^{\text{Ref}}(t)\|_{L^2}} \tag{68}$$

$$\text{err}_R^{L^\infty} = \frac{\|R_1(t) - R_1^{\text{Ref}}(t)\|_{L^\infty}}{\|R_1^{\text{Ref}}(t)\|_{L^\infty}} \tag{69}$$

where the results from BOMD with convergent SCF iteration are taken to be corresponding reference values, $\overline{E}$ is the average total energy over time, the frequencies, $\widetilde{\Omega}^{\text{Hooke}}$ and Ω^{Ref}, are obtained via solving the least square problem (62), the frequency, $\widetilde{\Omega}^{\text{LR}}$, is measured by Equation (50) with finite difference methods and $R_1(t)$ is the trajectory of the left-most atom.

5.1. Numerical Comparison between BOMD and TRBOMD

The first run is to validate the performance of TRBOMD. We set the time step $\Delta t = 250$, the artificial frequency $\omega = \frac{1}{\Delta t} = 4.00E\text{-}03$, the final time $T = 2.50E\text{+}06$ and employ the simple mixing with step length $\alpha = 0.3$ and the Kerker preconditioner in SCF cycles. Figure 2 plots the energy drift for BOMD with the converged SCF iteration (denoted by BOMD(c)) where the tolerance is $1.00E\text{-}08$; BOMD with five SCF iterations per time step (denoted by BOMD(5)) and TRBOMD with five SCF iterations per time step (denoted by TRBOMD(5)). We see clearly there that BOMD(5) produces large drift for both insulator and metal, but TRBOMD(5) does not. Actually, from Table 1, the relative error in the average total energy over time between TRBOMD(5) and BOMD(c) is under $1.30E\text{-}05$, but BOMD(c) needs about an average of 45 SCF iterations per time step to reach the tolerance

1.00E-08. Figure 3 plots corresponding trajectory of the left-most atom during about the first 25 periods and shows that the trajectory from TRBOMD (five) almost coincides with that from BOMD (c), which is also confirmed by the data of $\mathrm{err}_R^{L^2}$ and $\mathrm{err}_R^{L^\infty}$ in Table 1. However, for BOMD(5), the atom will cease oscillation after a while. A similar phenomena occurs for other atoms. In Table 1, we present more results for TRBOMD(n) with $n = 3, 5, 7$. We observe there that TRBOMD(n) gives more accurate results with larger n, and $\mathrm{err}_\Omega^{\mathrm{Hooke}}$ has a similar behavior as n increases to $\mathrm{err}_\Omega^{\mathrm{LR}}$, which is in accord with our previous linear response analysis in Section 4.

Figure 2. The energy fluctuations around the starting energy, $E(t = 0)$, as a function of time. The time step is $\Delta t = 250$. The final time is 2.50E+06 and $\omega = 1/\Delta t = 4.00\text{E-}03$. The simple mixing with the Kerker preconditioner is applied in SCF cycles. BOMD (c) denotes the BOMD simulation with converged SCF iteration, and BOMD (n) (resp. TRBOMD(n)) represents the BOMD (resp. TRBOMD) simulation with n SCF iterations per time step. It shows clearly that BOMD (five) produces large drift for both the insulator (a) and the metal (b), but TRBOMD (five) does not.

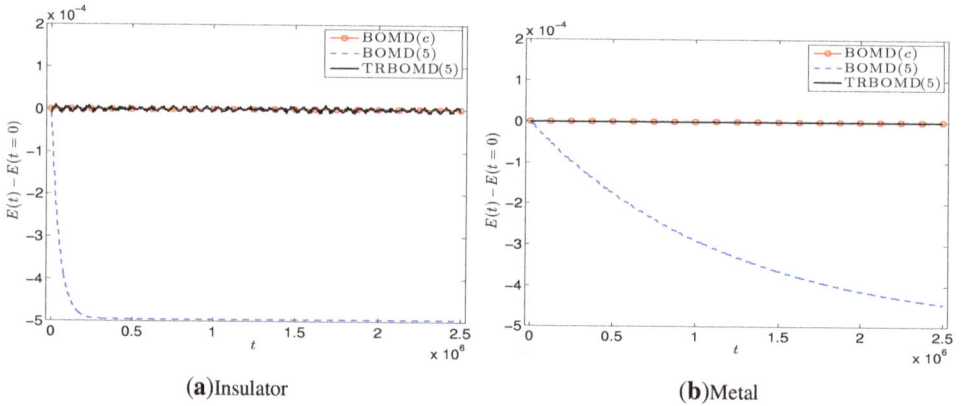

(**a**)Insulator   (**b**)Metal

Table 1. The errors for time reversible Born-Oppenheimer molecular dynamics (TRBOMD) (n). The settings are the same as those in Figure 2, except for the number of SCF iterations.

		Insulator: $\Omega^{\mathrm{Ref}} = 2.51\text{E-}04$, $\overline{E}^{\mathrm{Ref}} = 8.66\text{E-}01$		
n	$\mathrm{err}_\Omega^{\mathrm{LR}}$	$\mathrm{err}_\Omega^{\mathrm{Hooke}}$	$\mathrm{err}_{\overline{E}}$	$\mathrm{err}_R^{L^2}$	$\mathrm{err}_R^{L^\infty}$
3	$-6.53\text{E-}03$	$-1.63\text{E-}02$	$-7.63\text{E-}05$	$2.26\text{E-}02$	$4.25\text{E-}02$
5	$-1.08\text{E-}03$	$-2.38\text{E-}03$	$-1.30\text{E-}05$	$1.27\text{E-}02$	$2.92\text{E-}02$
7	$-2.76\text{E-}04$	$-5.41\text{E-}04$	$-3.32\text{E-}06$	$3.02\text{E-}03$	$7.22\text{E-}03$
		Metal: $\Omega^{\mathrm{Ref}} = 1.06\text{E-}04$, $\overline{E}^{\mathrm{Ref}} = 5.28\text{E-}01$		
3	$-2.65\text{E-}04$	$-6.92\text{E-}04$	$-4.36\text{E-}06$	$3.86\text{E-}03$	$8.95\text{E-}03$
5	$-3.65\text{E-}05$	$-7.31\text{E-}05$	$-4.44\text{E-}07$	$4.14\text{E-}04$	$9.60\text{E-}04$
7	$-5.24\text{E-}06$	$2.93\text{E-}06$	$-1.10\text{E-}07$	$1.63\text{E-}05$	$3.78\text{E-}05$

Figure 3. The position of the left-most atom as a function of time. The settings are the same as those in Figure 2. It shows clearly that the trajectory from TRBOMD (five) almost coincides with that from BOMD (c). However, for BOMD (five), the atom will cease oscillation after a while. (**a**) Insulator; (**b**) metal.

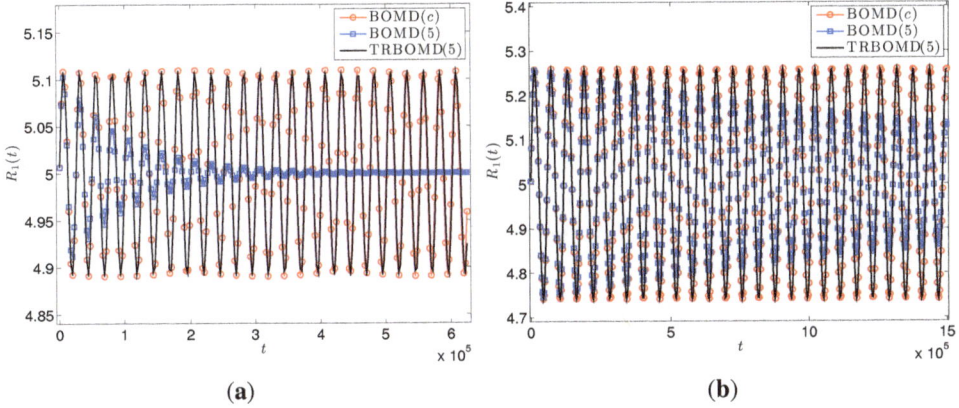

Figure 4. The absolute value of the error for TRBOMD (three) as a function of $1/\omega^2$ in logarithmic scales. The time step is $\Delta t = 20$, and the final time is 6.00E+05. For the readers' reference, within each plot, the red straight line denotes corresponding linear dependence, while the red solid point on the x axis represents the critical value of $\lambda_{\min}(\mathcal{K})/\lambda_{\max}(\mathcal{D})$. (**a**) Insulator; (**b**) metal.

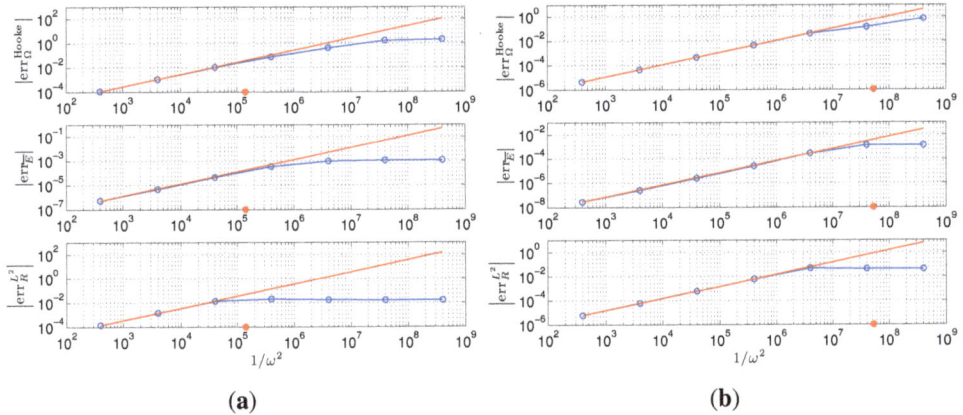

According to Equation (50), we have that $\mathrm{err}_\Omega^{\mathrm{LR}}$ is proportional to $1/\omega^2$ for large ω. We verify this behavior using TRBOMD(3) as an example. In this example, a smaller time step, $\Delta t = 20$, is set to allow bigger artificial frequency ω. The final time is $T = 6.00\mathrm{E}{+}05$, and the simple mixing with $\alpha = 0.3$ and the Kerker preconditioner is applied in SCF iterations. For TRBOMD (three) under these settings, we have $\lambda_{\min}(\mathcal{K}) \simeq 8.81\mathrm{E}{-}03$ for the insulator and $\lambda_{\min}(\mathcal{K}) \simeq 5.92\mathrm{E}{-}01$

for the metal, and thus, the critical values of $(\Omega^{\text{Ref}})^2/\lambda_{\min}(\mathcal{K})$ in Equation (49) are about 7.12E-06 and 1.90E-08, respectively. We choose $\omega^2 = $ 2.50E-03, 2.50E-04, 2.50E-05, 2.50E-06, 2.50E-07, 2.50E-08, 2.50E-09, and plot in Figure 4 the absolute values of $\text{err}_{\Omega}^{\text{Hooke}}$, $\text{err}_{\overline{E}}$, $\text{err}_R^{L^2}$ for TRBOMD (three) as a function of $1/\omega^2$ in logarithmic scales. When $1/\omega^2 \ll \lambda_{\min}(\mathcal{K})/(\Omega^{\text{Ref}})^2$, Figure 4 shows clearly that all of $|\text{err}_{\Omega}^{\text{Hooke}}|$, $|\text{err}_{\overline{E}}|$, $|\text{err}_R^{L^2}|$ depend linearly on $1/\omega^2$. The error, $\text{err}_R^{L^\infty}$, has a similar behavior to $\text{err}_R^{L^2}$ and is skipped here for saving space.

The last example illustrates the possible unstable behavior of TRBOMD when the stability condition $\lambda_{\min}(\mathcal{K}) > 0$ in Equation (48) is violated. Here, we take the insulator as an example and set the time step $\Delta t = 250$, the final time to 2.50E+05 and the artificial frequency $\omega = \frac{1}{\Delta t} = 4.00\text{E-}03$. The simple mixing with $\alpha = 0.3$ is now applied in SCF iterations. Under these setting, we have $\lambda_{\min}(\mathcal{K}) < 0$, e.g., $\lambda_{\min}(\mathcal{K}) = -2.42\text{E+}03$ for TRBOMD (three). Figure 5a plots the energy drift for TRBOMD (n) with $n = 3, 5, 7, 45$. We see clearly there that TRBOMD is unstable even using 45 SCF iterations per time step (recall that BOMD (c) in the first run needs about average 45 SCF iterations per time step). Figure 5b plots the corresponding trajectory of the left-most atom and shows that the atom is driven wildly by the non-convergent SCF iteration.

Figure 5. The unstable behavior of TRBOMD with the simple mixing for the insulator. The time step is $\Delta t = 250$. The final time is 2.50E+05 and $\omega = 1/\Delta t = 4.00\text{E-}03$. **(a)** The energy drift; **(b)** the trajectory of the left-most atom.

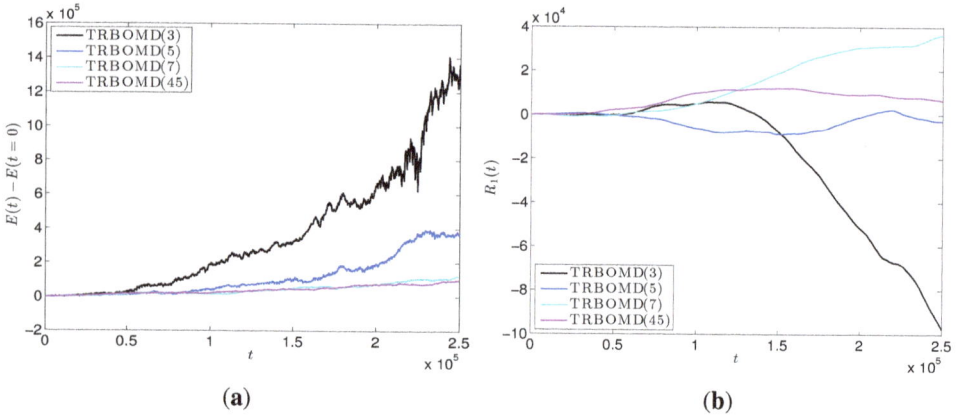

(a)

(b)

5.2. Numerical Comparison between TRBOMD and CPMD

We now present some numerical examples for CPMD illustrating the difference between CPMD and TRBOMD. As we have discussed, TRBOMD is applicable to both metallic and insulting systems, while CPMD becomes inaccurate when the gap vanishes. To make this statement more concrete, we apply CPMD to the same atom chain system. We implement CPMD using a standard velocity Verlet scheme combined with RATTLEfor the orthonormality constraints [37–39].

We present in Figure 6 the error of CPMD simulation for different choices of fictitious electron mass μ. We study the relative error of the phonon frequency, $\text{err}_{\Omega}^{\text{Hooke}}$, the relative error of the position of the left-most atom measured in L^2 norm, *i.e.*, $\text{err}_R^{L^2}$. We observe in Figure 6a linear convergence of CPMD to the BOMD result as the parameter, μ, decreases. This is consistent with our analysis. Recall that in CPMD, μ plays a similar role as ω^{-2} in TRBOMD. For the metallic example, the behavior is quite different; actually, Figure 6b shows a systematic error as μ decreases. For metallic system, as the spectral gap vanishes, the adiabatic separation between ionic and electronic degrees of freedom cannot be achieved no matter how small μ is. The adiabatic separation for TRBOMD, on the other hand, relies on the choice of an effective ρ_{SCF}, and hence, TRBOMD also works for a metallic system, as Figure 4 indicates.

Figure 6. The absolute value of the error for Car-Parrinello molecular dynamics (CPMD) as a function of μ in logarithmic scales. The time step is $\Delta t = 20$, and the final time is 6.00E+05. (**a**) Insulator; (**b**) metal.

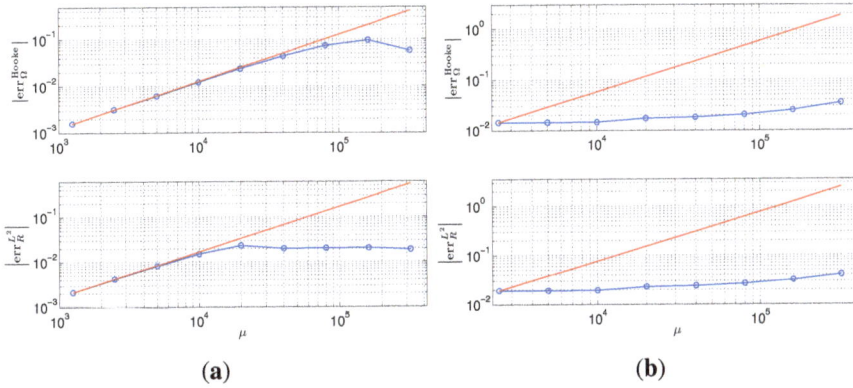

(**a**)   (**b**)

Figure 7. The trajectory of the position of the left-most atom. The dashed line is the result from BOMD with converged SCF iteration. Colored solid lines are the results from CPMD with fictitious electron mass $\mu = 2,500, 5,000, 10,000$ and $20,000$. The time step is $\Delta t = 20$; the trajectory plotted is within the time interval, $[2.00\text{E}+05, 4.00\text{E}+05]$. (**a**) Insulator; (**b**) metal.

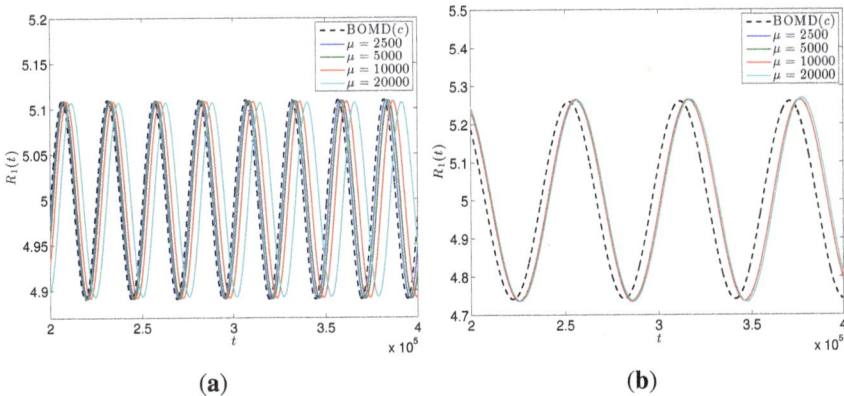

(**a**)   (**b**)

The different behavior of CPMD for insulating and metallic systems is further illustrated by Figure 7, which shows the trajectory of the position of the left-most atom during the simulation. The phase error is apparent from the two subfigures. While the phase error decreases so that the trajectory approaches that of BOMD for the insulator in Figure 7a, the result in Figure 7b shows a systematic error for a metallic system.

6. Beyond the Linear Response Regime: Non-Equilibrium Dynamics

The discussion so far has been limited to the linear response regime so that we can make linear approximations for the degrees of freedom of both nuclei and electrons. In this case, as the system becomes linear, explicit error analysis has been given. For practical applications, we will be also interested in non-equilibrium nuclei dynamics, so that the deviation of atom positions is no longer small. In this section, we will investigate the non-equilibrium case using the averaging principle (see e.g., [40,41] for a general introduction on the averaging principle).

Figure 8. Comparison of the trajectories of the first three atoms from the left for a non-equilibrium system. Different atoms are distinguished by color (blue for the initially left-most atom; green for the initially second left-most atom; red for the initially third left-most atom). Solid lines are the results from BOMD (c); circled lines are the results from TRBOMD (seven); dashed lines are the results from BOMD (seven). It is evident that while the results from BOMD with a non-convergent SCF iteration have a huge deviation, the results from TRBOMD are hardly distinguishable from the "true" results from BOMD.

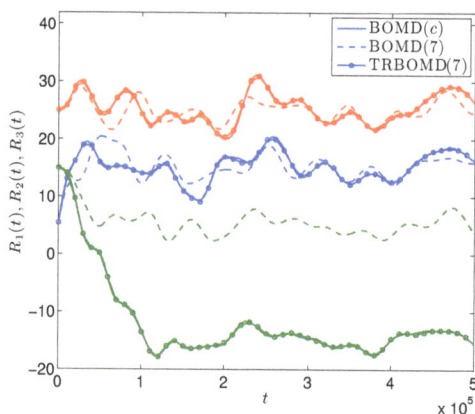

Let us first show numerically a non-equilibrium situation for the atom chain example discussed before. Initially, the 32 atoms stay at their equilibrium position. We set the initial velocity so that the left-most atom has a large velocity towards the right and other atoms have equal velocity towards the left. The mean velocity is equal to zero; so, the center of mass does not move. Figure 8 shows the trajectory of the positions of the first three atoms from the left. We observe that the results from

TRBOMD agree very well with the BOMD results with convergent SCF iterations. Let us note that in the simulation, the left-most atom crosses over the second left-most atom. This happens since, in our model, we have taken a $1D$ analog of Coulomb interaction, the nuclei background charges are smeared out and, hence, the interaction is "soft" without hard-core repulsion. In Figure 9, we plot the difference between ρ_{SCF} and the converged electron density of the SCF iteration (denoted by ρ_{KS}) along the TRBOMD simulation. We see that the electron density used in TRBOMD stays close to the ground state electron density corresponding to the atom configuration.

To understand the performance of TRBOMD, recall that the equations of motion are given by:

$$m\ddot{R}_I(t) = -\int \rho_{\mathrm{SCF}}(x; \mathbf{R}(t), \rho(t)) \frac{\partial V_{\mathrm{ion}}(x; \mathbf{R}(t))}{\partial R_I} \, dx$$

$$\ddot{\rho}(x, t) = \omega^2(\rho_{\mathrm{SCF}}(x; \mathbf{R}(t), \rho(t)) - \rho(x, t))$$

To satisfy the adiabatic condition (49) from the linear analysis, ω here is a large parameter. As a result, the time scales of the motions of the nuclei and of the electrons are quite different: The electronic degrees of freedom move much faster than the nuclear degrees of freedom.

Let us consider the limit, $\omega \to \infty$. In this case, we may freeze the $\mathbf{R}$ degree of freedom in the equation of motion for ρ, as ρ changes on a much faster time scale. To capture the two time scale behavior, we introduce a heuristic two-scale asymptotic expansion with faster time variable given by $\tau = \omega t$ (with some abuse of notation):

$$R(t) = R(t) \quad \text{and} \quad \rho(x, t) = \rho(x, t, \tau) \tag{70}$$

and hence:

$$\ddot{\rho}(x, t) = \omega^2 \partial_\tau^2 \rho(x, t, \tau) + 2\omega \partial_\tau \partial_t \rho(x, t, \tau) + \partial_t^2 \rho(x, t, \tau) \tag{71}$$

Therefore, to the leading order, after neglecting the terms of $\mathcal{O}(\omega^{-1})$, we obtain:

$$m\ddot{R}_I(t) = -\int \rho_{\mathrm{SCF}}(x; \mathbf{R}(t), \rho(t, \tau)) \frac{\partial V_{\mathrm{ion}}(x; \mathbf{R}(t))}{\partial R_I} \, dx \tag{72}$$

$$\partial_\tau^2 \rho(x, t, \tau) = \rho_{\mathrm{SCF}}(x; \mathbf{R}(t), \rho(t, \tau)) - \rho(x, t, \tau) \tag{73}$$

For the equation of motion for ρ, note that as $\mathbf{R}$ only depends on t, the nuclear positions are fixed parameters in Equation (73).

To proceed, we consider the scenario that $\rho(t, \tau)$ is close to the ground state electron density corresponding to the current atom configuration, $\rho^*(\mathbf{R}(t))$. We have seen from numerical examples (Figure 9) that this is indeed the case for a good choice of SCF iteration, while we do not have a proof of this in the general case. Hence, we linearize the map: ρ_{SCF}.

$$\rho_{\mathrm{SCF}}(x; \mathbf{R}, \rho) = \rho^*(x; \mathbf{R}) + \int \frac{\delta \rho_{\mathrm{SCF}}}{\delta \rho}(x, y; \mathbf{R}, \rho^*(\mathbf{R}))(\rho(y) - \rho^*(y; \mathbf{R})) \, dy \tag{74}$$

and Equation (73) becomes:

$$\partial_\tau^2 \rho(x, t, \tau) = -\mathcal{K}(\mathbf{R})(\rho(x, t, \tau) - \rho^*(x; \mathbf{R}(t))) \tag{75}$$

where $\mathcal{K}(\mathbf{R})$ is the same as in Equation (39), except it is now defined for each atom configuration, $\mathbf{R}$. Let us emphasize that here we have only taken the linear approximation for the electronic degrees of freedom, while keeping the possibly nonlinear dynamics of $\mathbf{R}$. This is different from the linear response regime considered before, where the nuclei motion is also linearized.

Figure 9. The difference of ρ_{SCF} with the converged electron density of SCF iteration (denoted by ρ_{KS}) measured in L^1 norm along the TRBOMD simulation for a non-equilibrium system.

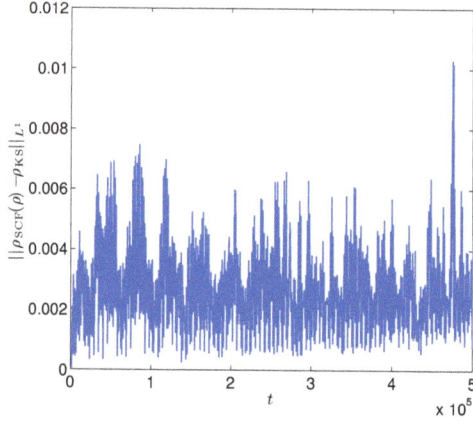

Under the stability condition (48), it is easy to see that for $\rho(t, \tau)$ satisfying Equation (75), the limit of the time average:

$$\bar{\rho}(x; \mathbf{R}(t)) = \lim_{T \to \infty} \frac{1}{T} \int_0^T \rho_{\text{SCF}}(x; \mathbf{R}(t), \rho(t, \tau)) \, d\tau$$

$$\approx \rho^*(x; \mathbf{R}(t)) + \int \frac{\delta \rho_{\text{SCF}}}{\delta \rho}(x, y; \mathbf{R}, \rho^*(\mathbf{R})) \left(\lim_{T \to \infty} \frac{1}{T} \int_0^T \rho(y; t, \tau) - \rho^*(y; \mathbf{R}(t)) \, d\tau \right) dy$$

$$= \rho^*(x; \mathbf{R}(t))$$

(76)

Take the average of Equation (72) in τ, we have:

$$m\ddot{R}_I(t) = - \int \bar{\rho}(x; \mathbf{R}(t)) \frac{\partial V_{\text{ion}}(x; \mathbf{R}(t))}{\partial R_I} \, dx$$

(77)

Because of Equation (76), the above dynamics is given by:

$$m\ddot{R}_I(t) = - \int \rho^*(x; \mathbf{R}(t)) \frac{\partial V_{\text{ion}}(x; \mathbf{R}(t))}{\partial R_I} \, dx$$

(78)

which agrees with the equation of the motion of atoms in BOMD. As we have neglected $\mathcal{O}(\omega^{-1})$ terms in the averaging, the difference in the trajectory of BOMD and TRBOMD is on the order of $\mathcal{O}(\omega^{-1})$ for finite ω.

Remark. If we do not make the linear approximation for the electronic degree of freedom, as the map, ρ_{SCF}, is quite nonlinear and complicated, the analysis of the long time (in τ) behavior of Equation (73) is not as straightforward. In particular, it is not clear to us whether the limit:

$$\bar{\rho}(x; \mathbf{R}(t)) = \lim_{T\to\infty} \frac{1}{T} \int_0^T \rho_{SCF}(x; \mathbf{R}(t), \rho(t,\tau)) \, d\tau \tag{79}$$

exists or how close the limit is to $\rho^*(x; \mathbf{R}(t))$ in a fully nonlinear regime. One particular difficulty lies in the fact that unlike BOMD or CPMD, we do not have a conserved Lagrangian for the TRBOMD. Actually, it is easy to construct a much simplified analog of Equation (73), the average of which is different from ρ^*. For example, if we consider the following analog, which only has one degree of freedom, ξ:

$$\ddot{\xi} = (\xi/2 + a\xi^2) - \xi \tag{80}$$

where $(\xi/2 + a\xi^2)$ is the analog of ρ_{SCF}, here, and $a > 0$ is a small parameter, which characterizes the nonlinearity of the map. Note that:

$$\ddot{\xi} = -\xi/2 + a\xi^2 = -\partial_\xi(\xi^2/4 - a\xi^3/3) \tag{81}$$

The motion of ξ is equivalent to the motion of a particle in an anharmonic potential. It is clear that if, initially, $\xi(0) \neq 0$, the long time average of ξ will not be zero. Furthermore, if, initially, $\xi(0)$ is too large, the orbit is not closed (ξ escapes the well around $\xi = 0$). If phenomena similar to this occur for a general ρ_{SCF}, then even in the limit, $\omega \to \infty$, there will be a systematic uncontrolled bias between BOMD and TRBOMD. This is in contrast with Car-Parrinello molecular dynamics, which agrees with BOMD in the limit fictitious mass going to zero ($\mu \to 0$) if the adiabatic condition holds.

As a result of this discussion, in practice, when we apply TRBOMD to a particular system, we need to be cautious whether the electronic degree of freedom remains around the converged Kohn-Sham electron density, which is not necessarily guaranteed (in contrast to CPMD for systems with gaps).

7. Conclusions

The recently developed time reversible Born-Oppenheimer molecular dynamics (TRBOMD) scheme provides a promising way for reducing the number of self-consistent field (SCF) iterations in molecular dynamics simulation. By introducing auxiliary dynamics to the initial guess of the SCF iteration, TRBOMD preserves the time-reversibility of the NVE dynamics, both at the continuous and at the discrete level, and exhibits improved long time stability over the Born-Oppenheimer molecular dynamics with the same accuracy. In this paper we analyze, for the first time, the accuracy and the stability of the TRBOMD scheme, and our analysis is verified through numerical experiments using a one-dimensional density functional theory (DFT) model without exchange correlation potential. The validity of the stability condition in TRBOMD is directly associated with the quality of the SCF iteration procedure. In particular, we demonstrate in the case in which the SCF iteration procedure is not very accurate, the stability condition can be violated, and TRBOMD becomes unstable. We also compare TRBOMD with the Car-Parrinello molecular dynamics (CPMD) scheme. CPMD relies on the adiabatic evolution of the occupied electron states, and therefore, CPMD works better for insulators than for metals. However, TRBOMD may be effective for both insulating and metallic systems. The present study is restricted to the NVE system and to simplified DFT models. Moreover, the analysis in the present work is mainly focused on the accuracy of trajectories and harmonic frequencies in the perturbation regime. However, in practice, the more important question is how the introduced artificial dynamics influence static properties, like distribution functions, and the most critical capability is to reproduce the correct distribution functions. The performance of TRBOMD for the NVT system and for realistic DFT systems with emphasis on the accuracy of static properties will be our future work.

Acknowledgments

This work was partially supported by the Laboratory Directed Research and Development Program of Lawrence Berkeley National Laboratory under the US Department of Energy contract number DE-AC02-05CH11231 and the Scientific Discovery through Advanced Computing (SciDAC) program funded by the US Department of Energy, Office of Science, Advanced Scientific Computing Research and Basic Energy Sciences (L.L.), the Alfred P. Sloan Foundation and the National Science Foundation (J.L.), the National Natural Science Foundation of China under the Grant Nos. 11101011 and 91330110 and the Specialized Research Fund for the Doctoral Program of Higher Education under the Grant No. 20110001120112 (S.S.). The authors would also like to thank the referees for many useful suggestions.

Conflicts of Interest

The authors declare no conflict of interest.

Appendix

Here, we derive the perturbation analysis result in Equation (50). When deriving the perturbation analysis below, we use linear algebra notation and do not distinguish matrices from operators. We use the linear algebra notation, replace all the integrals by matrix-vector multiplication and drop all the dependencies of the electron degrees of freedom, x and y. For instance, $\mathcal{K}\tilde{\rho}$ should be understood as $\int \mathcal{K}(x,y)\tilde{\rho}(y)\,\mathrm{d}y$. We also denote $\frac{\partial \rho^*}{\partial \mathbf{R}}(x; \mathbf{R}^*)$ simply by $\frac{\partial \rho^*}{\partial \mathbf{R}}$; then, Equation (42) can be rewritten as:

$$\begin{pmatrix} \ddot{\tilde{\mathbf{R}}} \\ \ddot{\tilde{\rho}} \end{pmatrix} = A \begin{pmatrix} \tilde{\mathbf{R}} \\ \tilde{\rho} \end{pmatrix} = \left(A_0 + \frac{1}{\epsilon} A_1 \right) \begin{pmatrix} \tilde{\mathbf{R}} \\ \tilde{\rho} \end{pmatrix} \tag{82}$$

Here:

$$A_1 = \begin{pmatrix} 0 & 0 \\ 0 & -\mathcal{K} \end{pmatrix} \tag{83}$$

is a block diagonal matrix, and:

$$A_0 = \begin{pmatrix} -\mathcal{D} & \mathcal{L} \\ \left(\frac{\partial\rho^*}{\partial\mathbf{R}}\right)^T \mathcal{D} & -\left(\frac{\partial\rho^*}{\partial\mathbf{R}}\right)^T \mathcal{L} \end{pmatrix} = \begin{pmatrix} \mathcal{I} \\ -\left(\frac{\partial\rho^*}{\partial\mathbf{R}}\right)^T \end{pmatrix} \begin{pmatrix} -\mathcal{D} & \mathcal{L} \end{pmatrix} \tag{84}$$

is a rank-M matrix. $\mathcal{I}$ is a $M \times M$ identity matrix. Now, assume the eigenvalues and eigenvectors of A follow the expansion:

$$\lambda = \lambda_0 + \epsilon\lambda_1 + \cdots, \quad v = v_0 + \epsilon v_1 + \cdots \tag{85}$$

Match the equation up to $\mathcal{O}(\epsilon)$, and:

$$A_1 v_0 = 0 \tag{86a}$$
$$A_0 v_0 + A_1 v_1 = \lambda_0 v_0 \tag{86b}$$
$$A_0 v_1 + A_1 v_2 = \lambda_0 v_1 + \lambda_1 v_0 \tag{86c}$$

Equation (86a) implies that $v_0 \in \mathrm{Ker}A_1$. Apply the projection operator, $P_{\mathrm{Ker}A_1}$, to both sides of Equation (86b), and use $v_0 = P_{\mathrm{Ker}A_1} v_0$; we have:

$$P_{\mathrm{Ker}A_1} A_0 P_{\mathrm{Ker}A_1} v_0 = \lambda_0 P_{\mathrm{Ker}A_1} v_0 \tag{87}$$

or:

$$\begin{pmatrix} -\mathcal{D} & 0 \\ 0 & 0 \end{pmatrix} v_0 = \lambda_0 v_0 \tag{88}$$

From the eigen-decomposition of $\mathcal{D}$ in Equation (46), we have $\lambda_0 = -\Omega_l^2$ for some $l = 1, \ldots, M$. For a fixed l, the corresponding eigenvector to the 0-th order is:

$$v_0 = (\mathbf{v}_l, \mathbf{0})^T \tag{89}$$

From Equation (86b), we also have:

$$A_1 v_1 = \lambda_0 v_0 - A_0 v_0 = \begin{pmatrix} \mathbf{0} \\ -\Omega_l^2 \left(\frac{\partial\rho^*}{\partial\mathbf{R}}\right)^T \mathbf{v}_l \end{pmatrix} \tag{90}$$

and therefore:

$$v_1 = \Omega_l^2 \left(\mathbf{0}, \mathcal{K}^{-1} \left[\left(\frac{\partial\rho^*}{\partial\mathbf{R}}\right)^T \mathbf{v}_l \right] \right)^T \tag{91}$$

Finally, we apply v_0 to both sides of Equation (86c); we have:

$$\lambda_1 = (v_0, A_0 v_1) - (v_0, \lambda_0 v_1) = \Omega_l^2 \mathbf{v}_l^T \mathcal{L} \left[\mathcal{K}^{-1} \left[\left(\frac{\partial\rho^*}{\partial\mathbf{R}}\right)^T \mathbf{v}_l \right] \right] \tag{92}$$

Therefore:

$$\lambda = -\Omega_l^2 + \epsilon\Omega_l^2 \mathbf{v}_l^T \mathcal{L} \left[\mathcal{K}^{-1} \left[\left(\frac{\partial\rho^*}{\partial\mathbf{R}}\right)^T \mathbf{v}_l \right] \right] + \mathcal{O}(\epsilon^2) \tag{93}$$

56

In other words, the phonon frequency, $\widetilde{\Omega}_l = \sqrt{-\lambda}$, up to the leading order is:

$$\widetilde{\Omega}_l = \Omega_l \left(1 - \frac{1}{2\omega^2} \mathbf{v}_l^T \mathcal{L} \left[\mathcal{K}^{-1} \left[\left(\frac{\partial \rho^*}{\partial \mathbf{R}} \right)^T \mathbf{v}_l \right] \right] \right) + \mathcal{O}(1/\omega^4) \tag{94}$$

which is Equation (50).

References

1. Marx, D.; Hutter, J. Ab initio molecular dynamics: Theory and implementation. *Mod. Methods Algorithms Quantum Chem.* **2000**, *1*, 301–449.
2. Kirchner, B.; di Dio, P.J.; Hutter, J. Real-world predictions from ab initio molecular dynamics simulations. *Top. Curr. Chem.* **2012**, *307*, 109–153.
3. Payne, M.C.; Teter, M.P.; Allen, D.C.; Arias, T.A.; Joannopoulos, J.D. Iterative minimization techniques for *ab initio* total energy calculation: Molecular dynamics and conjugate gradients. *Rev. Mod. Phys.* **1992**, *64*, 1045–1097.
4. Deumens, E.; Diz, A.; Longo, R.; Öhrn, Y. Time-dependent theoretical treatments of the dynamics of electrons and nuclei in molecular systems. *Rev. Mod. Phys.* **1994**, *66*, 917–983.
5. Tuckerman, M.E.; Ungar, P.J.; von Rosenvinge, T.; Klein, M.L. Ab initio molecular dynamics simulations. *J. Phys. Chem.* **1996**, *100*, 12878–12887.
6. Parrinello, M. From silicon to RNA: The coming of age of ab initio molecular dynamics. *Solid State Commun.* **1997**, *102*, 107–120.
7. Marx, D.; Hutter, J. *Ab initio Molecular Dynamics: Basic Theory and Advanced Methods*; Cambridge University Press: Cambridge, UK, 2009.
8. Hohenberg, P.; Kohn, W. Inhomogeneous electron gas. *Phys. Rev.* **1964**, *136*, B864–B871.
9. Kohn, W.; Sham, L. Self-consistent equations including exchange and correlation effects. *Phys. Rev.* **1965**, *140*, A1133–A1138.
10. Remler, D.K.; Madden, P.A. Molecular dynamics without effective potentials via the Car-Parrinello approach. *Mol. Phys.* **1990**, *70*, 921–966.
11. Car, R.; Parrinello, M. Unified approach for molecular dynamics and density-functional theory. *Phys. Rev. Lett.* **1985**, *55*, 2471–2474.
12. Pastore, G.; Smargiassi, E.; Buda, F. Theory of ab initio molecular-dynamics calculations. *Phys. Rev. A* **1991**, *44*, 6334–6347.
13. Bornemann, F.A.; Schütte, C. A mathematical investigation of the Car-Parrinello method. *Numer. Math.* **1998**, *78*, 359–376.
14. Niklasson, A.M.N.; Tymczak, C.J.; Challacombe, M. Time-reversible Born-Oppenheimer molecular dynamics. *Phys. Rev. Lett.* **2006**, *97*, 123001:1–123001:4.
15. Niklasson, A.M.N.; Tymczak, C.J.; Challacombe, M. Time-reversible ab initio molecular dynamics. *J. Chem. Phys.* **2007**, *126*, 144103:1–144103:9.
16. Niklasson, A.M.N. Extended Born-Oppenheimer molecular dynamics. *Phys. Rev. Lett.* **2008**, *100*, 123004:1–123004:4.

17. Niklasson, A.M.N.; Steneteg, P.; Odell, A.; Bock, N.; Challacombe, M.; Tymczak, C.J.; Holmström, E.; Zheng, G.; Weber, V. Extended Lagrangian Born-Oppenheimer molecular dynamics with dissipation. *J. Chem. Phys.* **2009**, *130*, 214109, doi:10.1063/1.3148075.

18. Niklasson, A.M.N.; Cawkwell, M.J. Fast method for quantum mechanical molecular dynamics. *Phys. Rev. B* **2012**, *86*, 174308:1–174308:12.

19. Hairer, E.; Lubich, C.; Wanner, G. *Geometric Numerical Integration*, 2nd ed.; Springer: Berlin/Heidelberg, Germany, 2006.

20. McLachlan, R.I.; Perlmutter, M. Energy drift in reversible time integration. *J. Phys. A: Math. Gen.* **2004**, *37*, L593–L598.

21. Kolafa, J. Time-reversible always stable predictor-corrector method for molecular dynamics of polarizable molecules. *J. Comput. Chem.* **2004**, *25*, 335–342.

22. Kühne, T.D.; Krack, M.; Mohamed, F.R.; Parrinello, M. Efficient and accurate Car-Parrinello-like approach to Born-Oppenheimer molecular dynamics. *Phys. Rev. Lett.* **2007**, *98*, 066401:1–066401:4.

23. Dai, J.; Yuan, J. Large-scale efficient Langevin dynamics, and why it works. *EPL* **2009**, *88*, 20001, doi:10.1209/0295-5075/88/20001.

24. Hutter, J. Car-Parrinello molecular dynamics. *WIREs Comput. Mol. Sci.* **2012**, *2*, 604–612.

25. Anderson, D.G. Iterative procedures for nonlinear integral equations. *J. Assoc. Comput. Mach.* **1965**, *12*, 547–560.

26. Pulay, P. Convergence acceleration of iterative sequences: The case of SCF iteration. *Chem. Phys. Lett.* **1980**, *73*, 393–398.

27. Johnson, D.D. Modified Broyden's method for accelerating convergence in self-consistent calculations. *Phys. Rev. B* **1988**, *38*, 12807–12813.

28. Kerker, G.P. Efficient iteration scheme for self-consistent pseudopotential calculations. *Phys. Rev. B* **1981**, *23*, 3082–3084.

29. Lin, L.; Yang, C. Elliptic preconditioner for accelerating self consistent field iteration in Kohn-Sham density functional theory. *SIAM J. Sci. Comput.* **2013**, *35*, S277–S298.

30. McLachlan, R.I.; Atela, P. The accuracy of symplectic integrators. *Nonlinearity* **1992**, *5*, 541–562.

31. Adler, S.L. Quantum theory of the dielectric constant in real solids. *Phys. Rev.* **1962**, *126*, 413–420.

32. Wiser, N. Dielectric constant with local field effects included. *Phys. Rev.* **1963**, *129*, 62–69.

33. Blöchl, P.E.; Parrinello, M. Adiabaticity in first-principles molecular dynamics. *Phys. Rev. B* **1992**, *45*, 9413–9416.

34. Tangney, P.; Scandolo, S. How well do Car-Parrinello simulations reproduce the Born-Oppenheimer surface? Theory and examples. *J. Chem. Phys.* **2002**, *116*, 14–24.

35. Tangney, P. On the theory underlying the Car-Parrinello method and the role of the fictitious mass parameter. *J. Chem. Phys.* **2006**, *124*, 044111:1–044111:14.

36. Solovej, J.P. Proof of the ionization conjecture in a reduced Hartree-Fock model. *Invent. Math.* **1991**, *104*, 291–311.

37. Ryckaert, J.P.; Ciccotti, G.; Berendsen, H.J.C. Numerical integration of the cartesian equations of motion of a system with constraints: Molecular dynamics of n-alkanes. *J. Comput. Phys.* **1977**, *23*, 327–341.

38. Ciccotti, G.; Ferrario, M.; Ryckaert, J.P. Molecular dynamics of rigid systems in cartesian coordinates: A general formulation. *Mol. Phys.* **1982**, *47*, 1253–1264.

39. Andersen, H.C. Rattle: A "velocity" version of the Shake algorithm for molecular dynmiacs calculations. *J. Comput. Phys.* **1983**, *52*, 24–34.

40. E, W. *Principles of Multiscale Modeling*; Cambridge University Press: Cambridge, UK, 2011.

41. Pavliotis, G.; Stuart, A. *Multiscale Methods: Averaging and Homogenization*; Springer: Berlin/Heidelberg, Germany, 2008.

Reprinted from *Entropy*. Cite as: Morales, M.A.; Clay, R.; Pierleoni, C.; Ceperley, D.M. First Principles Methods: A Perspective from Quantum Monte Carlo. *Entropy* **2014**, *16*, 287–321.

Article

First Principles Methods: A Perspective from Quantum Monte Carlo

Miguel A. Morales [1], **Raymond Clay** [2], **Carlo Pierleoni** [3,4,*] and **David M. Ceperley** [2]

[1] Lawrence Livermore National Laboratory, 7000 East Ave., Livermore, CA 94550, USA;
  E-Mail: moralessilva2@llnl.gov
[2] Department of Physics, University of Illinois at Urbana-Champaign, 1110 West Green Street
  Urbana, IL 61801-3080, USA; E-Mails: rcclay2@illinois.edu (R.C.);
  ceperley@uiuc.edu (D.M.C.)
[3] Dipartimento di Scienze Fisiche e Chimiche, Università de L'Aquila, Via Vetoio 10,
  L'Aquila 67100, Italy
[4] Dipartimento di Fisica, Sapienza Università di Roma, P.le A. moro 2, Rome 00185, Italy

* Author to whom correspondence should be addressed; E-Mail: carlo.pierleoni@aquila.infn.it;
  Tel.: +39-0862433056.

Received: 22 September 2013; in revised form: 27 November 2013 / Accepted: 28 November 2013 / Published: 30 December 2013

Abstract: Quantum Monte Carlo methods are among the most accurate algorithms for predicting properties of general quantum systems. We briefly introduce ground state, path integral at finite temperature and coupled electron-ion Monte Carlo methods, their merits and limitations. We then discuss recent calculations using these methods for dense liquid hydrogen as it undergoes a molecular/atomic (metal/insulator) transition. We then discuss a procedure that can be used to assess electronic density functionals, which in turn can be used on a larger scale for first principles calculations and apply this technique to dense hydrogen and liquid water.

Keywords: quantum Monte Carlo; first-principles simulations; hydrogen; Coupled Electron-Ion Monte Carlo; high pressure

1. Introduction

With the increasing computational power and the greater access to large clusters seen during the last decade, simulation methods have become an increasingly useful tool for many fields of science, including chemistry, materials science, condensed matter physics, and biophysics. In this article we explore some of the future impact of Quantum Monte Carlo in the field of first principles simulation (FPS). By this we mean reliable simulation methods that can be performed on condensed matter systems in the absence of detailed experimental information on those systems. Starting with the general Hamiltonian in Equation (1), and taking as input only the chemical compositions, masses, density, temperature *etc*, currently there is a hierarchy of methods that are used to perform such a simulation. In this introduction we focus on three classes of methods: the use of semi-empirical interatomic potentials together with Monte Carlo (MC) or molecular dynamics (MD) simulations, Density Functional Theory-based simulation methods, and Quantum Monte Carlo simulations.

The first member of the hierarchy uses semi-empirical interatomic potentials among effective atoms considered as point particles, the best known of which is the Lennard-Jones potential. Such potentials are routinely used in the vast majority of simulations (soft condensed matter, biophysics, materials science) and are reviewed in a different contribution to this issue [1]. The first question is how do we construct such a potential? The typical approach is to use available experimental data. However, it is well known that those potentials are not very accurate in the vast majority of systems, even if they match experimental data. Hence, though they can be used to say something about generic properties of systems, quantitative predictions for defect energies, energy barriers, melting temperatures, cannot be trusted. (If the potential has been adjusted to reproduce experimental measurements, then the method is no longer first principles, and the question becomes whether the potential is transferrable, *i.e.*, reliable for properties that are not fitted for.) Another fundamental limitation of this approach is that it becomes difficult to construct reliable interatomic potentials for complex systems containing several types of atoms, for example a solvent with various solutes, or systems under extreme conditions, since it becomes difficult to get enough reliable experimental data to constrain all of the parameters. For these reasons, it is highly desirable to have methods that can provide reliable predictions without input from experimental measurements.

Density Functional Theory (DFT) in the Kohn-Sham formulation maps the problem of many interacting electrons in the external field of the nuclei onto a system of non-interacting electrons in external field, a one body problem, and adds electronic correlation through an exchange-correlation functional. A breakthrough in the usefulness and popularity of simulations occurred with the development of the first-principle molecular dynamics (FPMD) approach by Car and Parrinello [2], where they combined molecular dynamics and DFT to perform simulations of complex chemical systems. Due to its favorable ratio between accuracy and computational cost, DFT has become the workhorse as electronic solver in the field of first-principles simulations. In fact, the recent explosion in the popularity of first-principles methods is, to a large part, due to the success of DFT in providing a fairly accurate description of the electronic structure of materials at a reasonable computational cost. DFT also gives access to a large range of observables. While DFT has been very successful in the description of many types of materials, e.g., metals and weakly correlated systems, many of the currently available exchange-correlation functionals in DFT possess well-known limitations [3], including the failure to properly describe strongly correlated materials, self-interaction errors, *etc*. It is recognized that even for such a fundamental system as water, the FPMD procedure is not accurate enough, giving large errors in many basic properties including the melting temperature, the diffusion constant, the compressibility, among others [4].

In the past decade there has been an explosion of new DFT exchange-correlation functionals with various characteristics. The reason is the difficulty of making systematic improvements to the functional or judging the accuracy of a functional. If the DFT functional is considered as "variable" then how does the user, in the absence of experimental data, decide on the functional? In the case of finite molecular systems, the availability of high-level quantum chemistry methods, like Coupled-Cluster theory offers a possible path towards the improvement of approximated functionals in DFT, for example by minimizing errors in a training set between DFT and Coupled Cluster theory results at various level of accuracy (with Single, Double or Triple excitations). In fact, many exchange-correlation functionals contain optimizable parameters that are obtained from calculations on finite molecular systems (exceptions to this include LDA, PBE, among others), where results of quantum chemistry methods are routinely used as a references. In solids, accurate calculations using many-body methods are computationally expensive, which has limited their use in the development of density functionals. While there has also been considerable developments in other correlated approaches for bulk systems, such as the many-body Green's function methods (GW approximation and Bethe-Salpeter equation), and Dynamical Mean Field Theory (DMFT), they are more expensive and still leave questions of accuracy. For reasons of space, we do not discuss these approaches further.

The third approach in our hierarchy is the use of Quantum Monte Carlo (QMC) methods, which are generalizations of the classical Monte Carlo techniques to quantum statistical physics and fundamentally based on imaginary-time path integrals. For a class of systems (bosons and systems in one dimension) such techniques provide an exact computational method. For general problems, though not exact, they are highly accurate *and* systematically improvable. Although there are a variety of QMC methods (ground state, variational, path integral, auxiliary field...) fundamentally they are closely related. QMC are the most accurate general methods but are less developed and require much more computational facilities than DFT methods (although the scaling of computer time versus system size is similar) limiting the systems on which such simulations can and have been performed. The largest impact to date of QMC has been in the development and improvement of DFT methods; specifically we mention the correlation energy of the electron gas [5], a fundamental component in almost all exchange-correlation functionals used in DFT. Recent calculations [6] give the corresponding correlation energies at finite electronic temperature.

Later in this paper we give an example of work in progress in this direction where QMC is used to directly rank various DFT functionals. We suggest that this benchmark quality data could be used to improve directly the best functionals. One can then envision using the highest ranked functional to develop intermolecular potentials that would then be of higher quality. Ercolessi *et al.* [7] have developed the force-matching procedure to find the optimal effective potential reproducing the forces appearing in an FPMD simulation. Such an approach is now feasible using QMC calculated forces and energies.

First principles simulation methods entirely based on QMC have also been developed in the last decade. These are the Coupled Electron-Ion Monte Carlo method [8] and the QMC-Molecular Dynamics [9], and have been recently reviewed in [10]. However their application to condensed phases has been limited so far to high pressure hydrogen, and hydrogen-helium mixtures because of their considerable computation cost. In this paper we will illustrate their use to investigate the dissociation of liquid molecular hydrogen under pressure, a problem which is still unsolved by DFT methods.

The article is organized as follows. We first describe in Section 2 the various QMC methods. Section 3 is devoted to few applications of QMC. In Section 3.1 we present a QMC study of high pressure phases of hydrogen. This is followed in Section 3.2 by a description of the use of these methods to provide quantitative information on the accuracy of various DFT functionals. Finally we close with a discussion in Section 4.

2. Computational Methods

In this section, we review some of the Quantum Monte Carlo methods used in the first principles modeling of condensed matter systems. Under normal conditions of temperature and pressure, such systems are described to a high degree of accuracy by the non-relativistic Hamiltonian for a collection of electrons and ions. We will use atomic units throughout the paper, where Planck's constant $\hbar = m_e = k_B = e = 4\pi\epsilon_0 = 1$ with k_B being Boltzmann's constant, and the energy is measured in Hartrees $E_h = 315,775$ K $= 27.2114$ eV. Note that, in these units, the energy of a hydrogen atom is $0.5E_h$, the binding energy of a hydrogen molecule is $0.17E_h$, the unit of length is the Bohr Radius $a_0 = 0.0529$ nm, and the molecular equilibrium bond length is $1.4a_0$. The Hamiltonian of the systems reads

$$\hat{H} = \hat{T}_n + \hat{H}_{el} = \hat{T}_n + \hat{T}_e + \hat{V}, \tag{1}$$

$$\hat{T}_n = -\sum_{I=1}^{N_n} \lambda_I \hat{\nabla}_I^2, \quad \hat{T}_e = -\lambda_e \sum_{i=1}^{N_e} \hat{\nabla}_i^2, \tag{2}$$

$$\hat{V} = \sum_{I<J} \frac{z_I z_J}{|\vec{R}_I - \vec{R}_J|} + \sum_{i<j} \frac{1}{|\vec{r}_i - \vec{r}_j|} - \sum_{i,I} \frac{z_I}{|\vec{r}_i - \vec{R}_I|}, \tag{3}$$

where N_n and N_e are the number of ions and electrons, respectively, in atomic units $\lambda_e = 1/2$, $\lambda_I = 1/(2M_I)$, and M_I and z_I are the mass and charge (in units of the electron mass m_e and charge e) of the nucleus I. The system occupies a volume Ω. Note that $\vec{r}$ with lower case indexes $(i, j, ...)$ is used to denote the position of electrons and $\vec{R}$ with upper case indexes $(I, J, ...)$ is used for the nuclei. When no indices are used, $\vec{r}$ and $\vec{R}$ represent the full $3N_e$ and $3N_n$ dimensional vectors, respectively. The electronic Hamiltonian $\hat{H}_{el}$ corresponds to the solution of the problem in the clamped-nuclei approximation, where the ions produce a fixed external potential for the electrons. Another quantity that will be of interest is the electron number-density given by $\rho = N_e/\Omega$, and parameterized with $r_s = a/a_0$, where $4\pi a^3/3 = \rho^{-1}$. Given Equation (1), we only need to add the temperature, particle statistics and boundary conditions to completely specify the physical and numerical problem to be solved.

Finding the eigenvalues and eigenfunctions of the Hamiltonian in Equation (1) is a formidable task, impossible to do analytically except for a few simple systems such as the single hydrogen atom. In practice, numerical or approximate theoretical methods must be used. Two of the most widely applicable methods are based either on imaginary-time path integrals or density functional theory (DFT), as discussed in the following subsections.

2.1. Ground State Methods

The following ground state methods seek to evaluate expectation values of physical observables taken over the ground state wavefunction $\phi_0(R)$:

$$\langle \hat{O} \rangle = \frac{\int dR\, \phi_0^*(R)\hat{O}\phi_0(R)}{\int dR\, |\phi_0(R)|^2} \tag{4}$$

Two problems are evident from this formula. The first is that we almost never know $\phi_0(R)$ exactly. The second is that even if we did, Equation (4) is a high dimensional integral. The following methods address both these problems. For sake of notation simplicity, throughout the Sections 2.1–2.3 we will indicate by R the set of all coordinates of the quantum degrees of freedom without distinction between electrons and nuclei.

2.1.1. Variational Monte Carlo

Variational Monte Carlo (VMC) is conceptually the simplest of the ground-state QMC methods. It works by approximating the true ground-state wavefunction $\phi_0(R)$ with some trial wavefunction $\Psi_T(R)$. Integrals like Equation (4) are then performed using Metropolis Monte Carlo sampling, with $\Psi_T(R)$ in place of $\phi_0(R)$ [11]. The accuracy of this method depends strongly on how closely $\Psi_T(R)$ approximates $\phi_0(R)$. Fortunately, the variational principle of quantum mechanics gives us a metric by which to improve the quality of trial wavefunctions. Consider the expectation value of the Hamiltonian and its variance:

$$E[\Psi_T] = \frac{\int dR\, \Psi_T^*(R)\hat{H}\Psi_T(R)}{\int |\Psi_T(R)|^2 dR} = \frac{\int dR\, |\Psi_T(R)|^2 E_L(R)}{\int dR\, |\Psi_T(R)|^2} \tag{5}$$

$$\sigma_E^2[\Psi_T] = \frac{\int dR\, \Psi_T^*(R)(\hat{H} - E[\Psi_T])^2 \Psi_T(R)}{\int dR\, |\Psi_T(R)|^2} \tag{6}$$

$$\tag{7}$$

where $E_L(R) = [\hat{H}\Psi_T(R)]/\Psi_T(R)$ in Equation (5) is called *local energy*. The variational theorem states that:

$$E[\Psi_T] \geq E[\phi_0] \tag{8}$$

$$\sigma_E^2[\Psi_T] \geq \sigma_E^2[\phi_0] = 0 \tag{9}$$

64

Based on this, improvements to the wavefunction can be quickly gauged by whether they lower the energy and variance.

A popular approach for fermionic problems is to assume a Slater-Jastrow wavefunction. This type of wavefunction possesses the correct fermionic antisymmetry, and symbolically is given by $\Psi_T(R) = \det(M(R))e^{J(R)}$. Here, $M(R)_{ij} = \phi_j(r_i)$ is a Slater determinant of single-particle orbitals. The single-particle orbitals $\phi_j(r)$ are typically taken from other quantum-chemistry methods (Hartree-Fock, DFT, *etc.*). $J(R)$ is called a "Jastrow" factor, and is constructed to be symmetric under particle exchange [12,13]. The Jastrow factor is typically chosen to be a sum of species dependent one-body, two-body, and sometimes three-body functions, which are designed to capture bosonic correlations. The form of these functions can vary from analytically derived forms with few to no free parameters, like the RPA jastrow [14,15], to functions with a large number of variational parameters, like b-splines. The interested reader is encouraged to look at the references for more information on Slater-Jastrow wavefunctions [12,16]. One can also go beyond the Slater-Jastrow form; other possible choices include multi-Slater determinant expansions [17], geminals [18], *etc.*

VMC can be improved if we consider classes of trial wavefunctions $\Psi_T(R, \alpha)$ parameterized by $\alpha = (\alpha_1, ..., \alpha_m)$ free parameters. We then minimize the energy and/or variance with respect to these parameters. Recent improvements to optimization algorithms allow the optimization of thousands of variational parameters [19,20]. Traditionally, only the Jastrow functions have been parameterized, although work has been done using parameterized single particle orbitals and multi-Slater determinantal expansions.

VMC has some advantages that keep it in use. First, it is usually computationally cheaper than more accurate QMC methods (to be discussed later). VMC can also include several different types of electron correlations (various forms of electronic wave functions). Lastly, it doesn't suffer from a sign problem. However, it is at heart an approximate method, and does depend on the choice of trial wavefunction.

2.1.2. Projector Methods

2.1.2.1. Formalism

Projector methods attempt to stochastically project out the exact many-body ground state, allowing us to sample this distribution for Monte Carlo integration. The "projector", or imaginary-time Green's function $G(R', R, \beta' - \beta)$, is the operator solution to the imaginary-time Schrödinger equation:

$$\frac{\partial \Psi}{\partial \beta} = -\hat{H}\Psi(R, \beta) \tag{10}$$

subject to the boundary condition that $\lim_{\beta' \to \beta} G(R', R, \beta' - \beta) = \delta(R' - R)$. One can verify that the formal solution is $\hat{G} = \exp(-\beta\hat{H})$. Now consider an arbitrary wavefunction $\Psi(R, \beta = 0)$ that is not orthogonal to the ground state $\phi_0(R)$ (in general this is an optimized trial function Ψ_T). Expanding this function in terms of the eigenfunctions of the Hamiltonian, and applying the projector to this, we find:

$$\Psi(R,\beta) = \sum_i a_i \phi_i(R) e^{-\beta \epsilon_i}$$

$$\propto a_0 \phi_0(R) + \sum_i a_i \phi_i(R) e^{-\beta(\epsilon_i - \epsilon_0)} \tag{11}$$

This implies that as $\beta \to \infty$, we are left with just the ground state wavefunction.

For efficiency reasons, it is better to use the "importance-sampled" Schrödinger's equation [12,21,22]. We obtain this by writing the original equation in terms of $f(R,\beta) = \Psi_T(R)\Psi(R,\beta)$. After some algebra [12], we find that

$$\frac{\partial f(R,\beta)}{\partial \beta} = \hat{L}f(R,\beta) \tag{12}$$

$$= \lambda \nabla \cdot [\nabla - F(R)] f(R,\beta) + [E_T - E_L(R)] f(R,\beta)$$

$F(R)$ is the quantum force defined by $F(R) = \nabla \ln |\Psi_T(R)|^2$ and $E_L(R)$ is the local energy defined above. E_T, the trial energy, is an arbitrary energy shift, unessential for the physics, but important for the numerical algorithm. If $f(R,\beta) \geq 0$ everywhere, then we can interpret f as a probability distribution. This amounts to demanding a bosonic many-body ground state (fermions will be covered in a later section). Equation (12) can then be interpreted as a generalized Smoluchowski equation for a drift-diffusion process with sources and sinks. The first term represents a drift-diffusion process, whereas the second term represents an exponential growth/decay process. When we get around to simulating this equation, we will use the mapping between a Smoluchowski equation governing probability distributions, and Langevin-like equations, governing the diffusion and growth of *particles*.

The solution of Equation (12) satisfy the following integral equation

$$f(R,\beta) = \int dR' \tilde{G}(R,R',\beta) f(R',\beta) \tag{13}$$

where the Green's function for this equation is formally $\tilde{G}(R',R,\beta) = \langle R'| \exp(\beta \hat{L})|R\rangle$, and it is easy to show that this is related to the original projector by the transformation $\tilde{G}(R',R,\beta) = \Psi_T(R')G(R',R,\beta)\Psi_T(R)^{-1}$. In the short-time approximation ($\tau\lambda \ll 1$), we can decouple the drift-diffusion and growth operators by the Trotter formula. The result (for the symmetric decomposition) is:

$$\tilde{G}(R',R,\tau) \simeq G_{DD}(R',R,\tau)G_B(R',R,\tau) \tag{14}$$

$$G_{DD} = \exp\left(-\frac{(R'-R-2\lambda\tau F(R))^2}{4\lambda\tau}\right) \tag{15}$$

$$G_B(R',R,\tau) = \exp(-\frac{\tau}{2}[E_L(R') + E_L(R) - 2E_T]) \tag{16}$$

where λ indicates either λ_e or λ_I as defined after Equation(1). The short-time approximation allows us to deal with the full propagator as a product of short-time propagators, $\hat{G}(\beta) = (\hat{G}(\tau = \beta/M))^M$. The cost is that we have now incurred in a time-step error that we must take into account.

2.1.2.2. Diffusion Monte Carlo

In diffusion Monte Carlo (DMC) [22–24], we represent the distribution function $f(R, \beta)$ as an ensemble of 3N-dimensional samples $\{R_1, ..., R_M\}$, which are known as "walkers". The average density of walkers at position R in configurational space is proportional to the distribution function $f(R)$.

As in classical diffusion, we would then simulate Equations (13) and (14) by a Langevin-like process acting on the walkers. Assuming that the time step $\tau = \beta/M$ is sufficiently small, we advance from $f(R, \beta) \rightarrow f(R, \beta + \tau)$ by first proposing to move each walker R_i to R_i' by a drift-diffusion step, prescribed by Equation (15). Then we accumulate a weight associated with walker i, given by $w_i(\beta + \tau) = w_i(\beta)G_B(R_i', R_i, \tau)$. To calculate the expectation value of an operator $\hat{O}$ over $f(R, \beta) = \Psi_T(R)\Psi(R, \beta)$, we average over the ensemble of walkers, including the appropriate weights:

$$\langle \hat{O} \rangle = \frac{\sum_{i=1}^{M} w_i(\beta)\mathcal{O}(R_i)}{\sum_{i=1}^{M} w_i(\beta)} \tag{17}$$

If we stopped here, this would be the basis of pure-diffusion Monte Carlo [25]. Because these weights are exponential factors, the variance associated with Equation (17) will increase exponentially as the simulation progresses: the weights of a few walkers will exponentially grow, whereas the rest will exponentially tend to zero.

Branching diffusion Monte Carlo [23], by far the most used form of DMC, fixes this problem by using the weights to either replicate or kill off walkers. After each drift-diffusion step, the number of walkers associated with the single walker R_i to advance to the next time-step, M_{next}^i is chosen to be $M_{next}^i = \text{INT}(w_i(\beta + \tau) + \xi)$, where ξ is a random number between $[0, 1]$. The weights of the replicated walkers are all adjusted to conserve the total weight of walker i as much as possible. Modern methods are typically hybrids, where the weights of walkers are carried until they exceed certain established bounds, at which point they are branched [26].

The simulation is run by initializing the starting ensemble according to $f(R, 0) = |\Psi_T(R)|^2$. Assuming β is the projection time required to reach the ground-state, the simulation is incremented $M = \beta/\tau$ steps, at which point our ensemble is distributed according to $f_0(R) = \Psi_T(R)\phi_0(R)$. Samples can then be accumulated, and the simulation is run for a long enough time to achieve the desired statistical error bars.

It is important to note that since we are sampling $f_0(R)$, this corresponds to the following type of expectation value, known as a "mixed-estimate":

$$\langle \hat{O} \rangle_{DMC} = \frac{\langle \Psi_T | \hat{O} | \phi_0 \rangle}{\langle \Psi_T | \phi_0 \rangle} \tag{18}$$

For observables that commute with the Hamiltonian, this gives us exact, unbiased estimates over the true many-body ground state wavefunction. For those that don't, the estimators will be biased by the quality of the trial wavefunction. This bias is less than that encountered by VMC, but still present. This can be alleviated somewhat by the use of "extrapolated estimators", and by the "forward-walking" method [27].

2.1.2.3. Reptation Monte Carlo

Reptation Monte Carlo (RMC) is based on the path-integral representation of the projector. Assuming that β is large enough to guarantee sufficient convergence to the ground state, we begin by partitioning the full projector into M segments of time-interval $\tau = \beta/M$, called "time slices". Inserting a resolution of the identity between each short-time projector, we find the following path-integral expression for the mixed distribution $\langle \Psi_T | \phi_0 \rangle$:

$$\langle \Psi_T | \phi_0 \rangle = \int dR_0 \ldots dR_M \Psi_T(R_0) G(R_0, R_1, \tau) \ldots G(R_{M-1}, R_M, \tau) \Psi_T(R_M) \qquad (19)$$

Using the short-time approximate Green's function at the beginning of this section, we can recast this expectation value in a more traditional path-integral form:

$$\langle \Psi_T | \phi_0 \rangle \;=\; \mathcal{Z} = \int \mathcal{D} X e^{S[X]} \qquad (20)$$

$$S[X] \;=\; \ln \Psi_T(R_0) + \ln \Psi_T(R_M) - \sum_{i=0}^{M-1} L_s(R_i, R_{i+1}) \qquad (21)$$

$$L_s(R', R) \;=\; \frac{(R' - R)^2}{4\lambda\tau} + \frac{1}{2}(R' - R) \cdot (F' - F) \qquad (22)$$

$$+ \frac{\tau}{2}\left[E_L(R') + E_L(R) + \lambda(F^2(R') + F^2(R)) \right] \qquad (23)$$

Here, X is shorthand for the directed path $X = R_0, \ldots, R_M$. Equation (20) plays the role of a partition function in statistical mechanics, where the $\Pi[X] = e^{S[X]}/\mathcal{Z}$ is the probability of a given path X, $-S[X]$ is the path action, which includes the trial wavefunctions at the ends of the path, as well as a sum over "link-actions" $L_s(R', R)$, (see Equations (22) and (23)). The form we used for the link-action comes from imposing symmetry of the normal Green's function under the exchange of two end-points, and writing it in terms of the importance-sampled Green's functions [28].

The versatility of reptation Monte Carlo comes from how $\Pi[X]$ is sampled. In the original method [29], one takes a given path X and chooses a growth direction at random. One then proposes a new path X^* by adding δ time slices to the "head" and removing δ slices from the "tail". Acceptance or rejection of this move is based on the usual Metropolis acceptance step. This type of move is called "reptation", reminiscent of a "reptile", from which the method derives its name. The proposed head move is done by a sequence of drift-diffusion moves, as in DMC, and rigorously preserves detailed balance.

Most practical implementations use what's known as the "bounce algorithm" [28]. Rather than choosing the growth direction randomly, it is set at the beginning of the simulation and is changed only after a rejection step, hence the name "bounce". This method does not satisfy detailed balance, but does satisfy the more general stationarity condition required for Markov chain Monte Carlo. This dramatically decreases the autocorrelation time of the method, and also tames ergodicity problems that have been observed to crop up in the method.

RMC is appealing for two reasons. It gives us the same level of accuracy for the energy as DMC but correlated sampling between different configurations can be done without approximation. This is particularly useful in methods like the Coupled Electron-Ion Monte Carlo. RMC also gives us the ability to sample expectation values over the pure distribution, as seen below:

$$\langle \hat{O} \rangle_{pure} = \frac{\langle \Psi_T | e^{-\frac{\beta}{2}\hat{H}} \hat{O} e^{-\frac{\beta}{2}\hat{H}} | \Psi_T \rangle}{\langle \Psi_T | e^{-\beta \hat{H}} | \Psi_T \rangle} \tag{24}$$

$$= \frac{1}{Z} \int \mathcal{D}X e^{-S[X]} \mathcal{O}(R_{\beta/2}) \tag{25}$$

This shows that the center time slice of the reptile is distributed according to $|\phi_0(R)|^2$, whereas the ends are distributed according to the mixed distribution $f(R)$. This easy access to the pure distribution makes RMC ideal for calculations of unbiased observables and correlation functions, doing so in a more efficient manner than "forward-walking" in DMC. Estimation of observables over the pure distribution works whenever we can write a meaningful estimator in terms of position space coordinates. Diagonal position space observables, like the average potential energy and pair-correlation function, can be measured directly from the sampled pure distribution. Observables that aren't diagonal in position space, like off-diagonal density matrix elements and the momentum distribution, can be measured from the pure distribution with suitable additions to the basic algorithm. This procedure does not work for all estimators however; one can show that evaluating the local kinetic energy over the pure distribution does not yield a correct estimate of the average ground-state kinetic energy.

2.1.2.4. The Fixed-Node Approximation

The previous projector methods we mentioned are in principle *exact* for bosonic systems, since the mapping to a diffusion process is valid when $\phi_0(R) \geq 0$ everywhere. However, since the wavefunction for a fermion systems must be antisymmetric under exchange, the ground state wavefunction will have as many negative configurations as positive ones (in many cases the wavefunction can be made real). We can restore the probabilistic interpretation of the wavefunction $\Psi(R, \beta)$ if we factor its sign into the weight of the walker, or into the observable itself. It turns out that in doing so, we will have large and almost equal contributions to the expectation value of opposite signs. This leads to an exponentially decaying signal to noise ratio, implying that the computational effort required to treat the fermion problem directly scales exponentially. This is the well known "fermion sign problem".

By far, the most common means of alleviating the sign-problem in both DMC and RMC is applying the "fixed-node" approximation [23,24]. We assume that the nodes of $\phi_0(R)$ are the same as the nodes for $\Psi_T(R)$. We then propagate our ensemble of walkers or our reptile strictly within restricted space where $\Psi_T(R)$ doesn't change sign. This can be implemented by rejecting moves that carry walkers across a node, or bouncing a reptile whenever a head move is proposed across a nodal surface. Though this is an uncontrolled approximation, it turns out to be an extremely good one in most cases. Fixed-node energies are proved to be upper bounds of the exact energy [16], which allows us to optimize the nodal surfaces and to compare fixed-node DMC and fixed-node RMC energies with other methods. It turns out that both of these methods are among the most accurate computational methods known for electronic systems.

2.2. Scaling of QMC Methods

Like DFT, fermionic QMC typically has scaling between $\mathcal{O}(N^3)$ and $\mathcal{O}(N^4)$ depending on the property computed and the trial function. Here N is the number of particles. In contrast, popular quantum chemistry methods like Moller-Plesset Perturbation Theory, coupled-cluster, or configuration interaction, scale at least like $O(N^7)$. This makes QMC one of the few accurate many-body theories that is able to treat bulk systems.

Unlike DFT, whose scaling prefactor is governed by the solution of a generalized eigenvalue problem, Monte Carlo methods, in general, have statistical error bars which reduce as the inverse of the square root of the sampled configurations as a consequence of the central limit theorem. This makes quantum Monte Carlo significantly more expensive than DFT to reach chemical accuracy, though it has a smaller uncontrolled bias. The necessity for a much smaller time step in projector monte carlo than in VMC can make projector monte carlo about an order of magnitude more expensive for the same statistical uncertainty.

The cost of a single N-particle monte carlo step in VMC and projector monte carlo methods are determined by the evaluation of the trial wavefunction. For bosonic trial wavefunctions with pair-wise correlations, these calculations scale like $\mathcal{O}(N^2)$ per N-particle step. If these correlations are short-ranged, linear scaling can be achieved.

For fermionic trial wavefunctions, the computational cost is determined by the evaluation of single-particle orbitals and by the evaluation of a Slater determinant. The scaling of orbital evaluations depends on whether the electrons are localized since evaluating localized orbitals can be done in constant time. For plane waves basis sets, the cost scales like $\mathcal{O}(N)$. If we seek to include the effects of backflow, this can increase the computational cost by an additional factor of N. The remaining bottleneck is then the evaluation of the Slater determinant, which scales like $\mathcal{O}(N^3)$ per N-particle step. In theory, the cost of the determinant evaluation could be brought down by almost a factor of N if the Slater determinant is sparse, however, the crossover point is prohibitive (greater than 3000 particles for a model system) [30]. This causes VMC and projector monte carlo to realistically scale like $\mathcal{O}(N^{3-4})$ depending on whether one uses backflow or not.

2.3. Finite-Temperature Methods

Next, we summarize path integral methods. These methods are similar to DMC but can treat systems at non-zero temperature: a many-body density matrix replaces the trial wave function. Concerning first principles simulations the path integral method can be used either to simulate the properties of thermal electrons or to simulate the zero point effects of light nuclei or both. For electronic simulations there are two major problems. First, the energy scale of electrons is 1 Hartree or above, thus to reach ambient temperature requires very long paths. Second, since electrons are fermions, antisymmetrization and hence the sign problem is inevitable. For a more complete overview of the method and its application to fermion systems, see [31,32] respectively.

70

2.3.1. Path Integrals

To begin, we define the many particle density matrix for a system in equilibrium with an external reservoir at inverse temperature $\beta = 1/k_B T$ (canonical ensemble)

$$\rho(R, R'; \beta) = \langle R \mid e^{-\beta \hat{H}} \mid R' \rangle \tag{26}$$

where $R \equiv (r^{(1)}, \ldots, r^{(N)})$ with $r^{(i)}$ specifying the spacial coordinates of the ith of N particles. The partition function is defined as the trace of the density matrix,

$$Z(\beta) = Tr(\rho) = \int dR \langle R \mid e^{-\beta \hat{H}} \mid R \rangle = \int dR \rho(R, R; \beta) \tag{27}$$

The expectation value of any observable may be computed from this definition as

$$\langle \hat{O} \rangle = Tr(\hat{O}\rho)/Z = Tr(\hat{O}\rho)/Tr(\rho) \tag{28}$$

Using the product property of the density matrix M times, such that $\beta = M\tau$, we write the partition function (or the diagonal density matrix) as an integral over a discrete path:

$$Z(\beta) = \int \left[\prod_{i=0}^{M-1} dR_i \right] \rho(R_0, R_1; \tau)\rho(R_1, R_2; \tau) \ldots \rho(R_{M-1}, R_0; \tau) \tag{29}$$

We have reduced the problem of sampling a low temperature density matrix to one of finding a high temperature density matrix and integrating over the path. The action, defined as

$$S(R_i, R_j; \tau) \equiv -ln[\rho(R_i, R_j; \tau)] \tag{30}$$

can be broken into kinetic and potential parts, using Trotter's formula. The integration over all of the path variables is done using a specialized form of either Metropolis Monte Carlo or Molecular Dynamics, generating the Path Integral Monte Carlo (PIMC) or Path Integral Molecular Dynamics (PIMD) methods.

Finally, in order to account for the particle statistics of the simulated system, we must sum over permutations $\mathcal{P}$, giving

$$Z(\beta) = \frac{1}{N!} \sum_{\mathcal{P}} (\pm 1)^{\mathcal{P}} \int_{R \to \mathcal{P}R} dR_t e^{-S[R_t]} \tag{31}$$

where R_t represents the generic path starting at R and ending at $\mathcal{P}R$ while t varies from 0 to β.

2.3.2. Restricted Paths

For fermions, negative terms enter in this sum, leading to a sign problem. As was done in the previous discussion of DMC, one way to circumvent this issue is to impose a nodal constraint [33]. We define the *nodal surface* $\Upsilon_{R_*\beta}$ for a given point R_* and inverse temperature β to be

$$\Upsilon_{R_*\beta} = \{R \mid \rho(R, R_*; \beta) = 0\} \tag{32}$$

which is a $(dN-1)$-dimensional manifold in dN-dimensional configuration space (d is the space dimensionality). Here, $R_\star$ is dubbed the *reference point*, as it is needed to define the nodal surfaces. Inside a nodal cell, by definition the sign of the density matrix is uniform. Using Dirichlet boundary conditions, we may solve the Bloch equation within each nodal cell. We define the *reach* $\Gamma_\beta(R_\star)$ as the set of all continuous paths R_t, for which $\rho(R_t, R_\star, \beta) \neq 0$ for all intermediate t ($0 < t \leq \beta$), *i.e.*, node-avoiding paths

$$\Gamma_\beta(R_\star) = \{\gamma : R_\star \to R_t \mid \rho(R_\star, R_t; \beta) \neq 0\} \tag{33}$$

Since paths are continuous Brownian objects, all paths contributing to the Bloch equation solution must belong to this reach. For all diagonal contributions, odd permutations must cross a node an odd number of times and thus are not allowed by this constraint and are exactly cancelled by all paths of node-crossing even permutations. This leaves us with the following expression for the density matrix,

$$\rho(R, R; \beta) = \frac{1}{N!} \sum_{\mathcal{P}, even} \int_{\gamma : R \to \mathcal{P}R}^{\gamma \in \Gamma_\beta(R)} \mathcal{D}R_t e^{-S[R_t]/\hbar} \tag{34}$$

We have thus turned the sign-full expression for the density matrix into one which includes only terms of a single sign, allowing efficient computation. However, because ρ appears on both sides of Equation (34) (in the r.h.s. it appears into the definition of the reach), this requires a priori knowledge of the density matrix nodal structure, which is generally unknown. To escape this self-consistency issue, an ansatz density matrix that approximates the actual nodal structure, is introduced. This will give an exact sampling of the Fermi density matrix if its nodes are correct. This method is called *restricted* PIMC (RPIMC). The density matrix for non-interacting fermions is a Slater determinant of single-particle distinguishable density matrices, $\rho(R, R_\star; \beta) = \frac{1}{N!} \det \rho_{ij_\star}$ where

$$\rho_{ij_\star} = (4\pi\lambda\beta)^{-d/2} \exp(-\frac{(r_i - r_{j\star})^2}{4\lambda\beta}) \tag{35}$$

It is a good approximation to use the free particle density matrix at high temperatures (say for temperatures greater than the Fermi energy) and when correlation effects are weak. Furthermore, due to the constraint of translational invariance, free particle nodes are quite reasonable for homogeneous systems.

The nodal error, arising from using an approximate restriction is problematic since it is uncontrollable. The finite temperature variational principle is through the free energy, as opposed to the internal energy in the ground state. Thus one possible solution is to parameterize the nodal ansatz, and then minimize the free energy by varying the parameters. This will require a thermodynamic integration, in general. Systems analyzed to date suggest that the nodal error arising from the free-particle ansatz is small since the correlation from the interacting potential is fully taken into account.

2.3.3. Path Integrals for Nuclei

Even when quantum particles can be considered distinguishable, as for instance light nuclei in condensed phases, there could be substantial physical effects arising from their quantum behavior, *i.e.*, resulting from the $\hat{T}_n$ in Equation (1). For example in bulk hydrogen and in water, the zero point motion of the protons must be taken into account for an accurate description. Furthermore, in the crystalline phase the frequently used harmonic approximation is often inadequate since non-harmonic effects can be as significant as harmonic effects. In contrast to the situation with electrons, our ability to simulate the nuclei with current algorithms and hardware is well controlled; because the nuclei are thousands of time heavier, they are much closer to the classical limit, so that fewer path steps are needed. For hydrogen-containing compounds at room temperature, one can often get away with about few tens of imaginary time slices. A second consequence is that particle statistics (either Fermi or Bose) can typically be ignored; a notable exception is the difference between para- and ortho-hydrogen, important for modeling the low-temperature low-pressure crystals of molecular hydrogen and deuterium.

A frequent use of path integrals for nuclei occurs when DFT is used to integrate out the electronic degrees of freedom. However, one wants to use the DFT energy surface for the properties of the quantum nuclei in equilibrium, using the path integral method. To perform the path integration, it is advantageous to use molecular dynamics instead of Monte Carlo since that will allow the electronic wave functions to evolve smoothly in time, and thus reduce the time to convergence in solving the DFT self-consistency conditions. M. Ceriotti, *et al.* [34] have devised an ingenious noise filtering scheme to reduce the number of needed path integral steps. Assuming the density functional description of the electrons is accurate, thermodynamic (static) properties of the simulated system will be accurate. Conversely the dynamical properties are not to be trusted. In general a reliable method for quantum time correlation functions or, even worse, quantum dynamics is still missing.

2.4. Coupled Electron-Ion Monte Carlo

The QMC methods described so far, when applied to an ion-electron system, treat all particles on the same footing, either both in the ground state [35–37] or both at the same finite temperature [38–40]. However the large nucleon-electron mass ratio implies a wide separation of time and energy scales and it is a common practice to adopt the adiabatic, or Born-Oppenheimer (BO), approximation. Ignoring such an approximation in QMC causes difficulties. The imaginary time step of the path integral representation (both in DMC/RMC and PIMC) is imposed by the light electron mass. In DMC this means that nuclear "dynamics" (the speed of sampling configuration space) is much slower than electron "dynamics" requiring very long (and time consuming) trajectories. In PIMC the separation of time scales presents itself as a separation in the regions where thermal effects are relevant: in high pressure hydrogen for instance nuclear quantum effects becomes relevant below $\sim$2000 K where electrons are, to a very good approximation, in their ground state. Performing PIMC in this region of temperatures requires very long electronic paths causing a slowing down of the exploration of configuration space and effectively limiting the ability of PIMC to perform accurate calculations at low temperatures.

The Coupled Electron-Ion Monte Carlo method (CEIMC) is a QMC method based on the BO approximation [8]. In CEIMC a Monte Carlo calculation for finite temperature nuclei (either classical or quantum represented by path integrals) is performed using the Metropolis method with the BO energy obtained by a separate QMC calculation for ground state electrons. CEIMC has been extensively reviewed in [8,10]. Here, we only briefly report the main features of the method.

2.4.1. Penalty Method

In CEIMC the difference of BO energies of two nearby nuclear configurations in a MC attempted step, as obtained by an electronic QMC run, is affected by statistical noise which, if ignored, results in a biased nuclear sampling. To cope with this situation either the statistical noise needs to be reduced to a negligible value by long electronic calculations (very inefficient), or the Metropolis acceptance/rejection scheme needs modifications to cope with noisy energy differences. The latter strategy is implemented in the Penalty Method [41] which enforces detailed balance to hold on average over the noise distribution. The presence of statistical noise causes an extra rejection for a single nuclear move with respect to the noiseless situation. An extra "penalty" defined as the variance of the energy difference over the square of the physical temperature is added to the energy differences. Therefore running at lower temperatures requires a reduced variance to keep an acceptable efficiency of the nuclear sampling. Small variances can be obtained if correlated sampling is used to compute the energy of the two competing nuclear configurations. In an attempted nuclear MC step, a single ground state electronic run is performed with a trial wave function which is a linear combination of the wave functions of the two nuclear configurations considered. The BO energy of the two nuclear configurations is obtained by a reweighting procedure which provides energy differences with a much reduced variance with respect to performing two independent electronic runs if the "distance" between the two nuclear configurations is limited (*i.e.*, the overlap between the trial wave functions of the two configurations is large) [42]. This strategy allows an efficient sampling of nuclear configuration space for high pressure hydrogen and helium down to temperature as low as $\sim$200 K.

2.4.2. Nuclear PIMC

When nuclear quantum effects are included using a path integral representation (see §2.3), the relevant inverse temperature in the penalty method is the imaginary time discretization step τ, so that no loss of efficiency is experienced when lowering the temperature (*i.e.*, taking longer paths). For quantum protons in high pressure hydrogen, CEIMC can be used to efficiently study systems at temperatures as low as $\sim$200 K. In the present implementation of nuclear quantum effects in CEIMC, we introduce an effective pair potential between nuclei and use the pair density matrix corresponding to the effective potential to factorize the imaginary time propagator. The residual difference between the energy of the effective system and the BO energy of the original system is considered at the primitive approximation level of the Trotter break-up of the proton propagator [8]. In high pressure hydrogen ($r_s = 1.40$) it is found that with this strategy, an inverse time step of $\tau^{-1} \simeq 4800$ K is enough to reach convergence of the thermodynamics properties, which allows to study systems at low temperature with a limited number of time slices ($\leq$50).

In CEIMC many-body nuclear moves are preferred to single-body moves. The reason is that even if only few nuclei are moved the entire electronic calculation must be repeated, by far the most expensive part of the method. For this reason we sample nuclear configuration by a smart Monte Carlo method [43] in the normal mode space of the path [44] with forces from the effective two body potential. This strategy allows us to simulate systems of ~ 100 protons (for hydrogen) at temperature as low as 200 K with an acceptable efficiency.

2.4.3. VMC *vs.* RMC

The main ingredient of CEIMC is the electronic QMC engine used to compute the BO energy. As mentioned a very important aspect for the efficiency of CEIMC is the noise level which is related to the variance of the local energy. In ground state QMC (see §2.1) the "zero variance principle" applies: if the trial wave function is an eigenfunction of the Hamiltonian, the local energy is no longer a function of the electronic coordinates and a single calculation provides the exact corresponding eigenvalues. Therefore by improving the trial wave function and approaching the exact ground state, the variance of the local energy decreases to zero. In connection with CEIMC, this is important not only for the accuracy of the BO energy but also for the efficiency of the nuclear sampling since the extra rejection due to the noise is reduced for a more accurate trial wave function.

To go beyond VMC accuracy in CEIMC we have implemented Reptation QMC method (RMC) [8,29]. RMC is superior to DMC in the CEIMC context since it uses an explicit representation of the statistical weight of each path and therefore the reweighting procedure needed for estimating energy differences is easily applied. Going from VMC to RMC accuracy in CEIMC requires at least one order of magnitude more computer time. This is because it is in general more difficult to properly sample the configuration space of a 3N-dimensional path than of a 3N-dimensional point. It is analogous to the difficulty of sampling the configuration space of a long polymers with respect to point particles. For any proposed nuclear move one has to relax the electronic path to the new equilibrium state and perform long enough sampling of the electronic configuration space to compute the energy difference with the required noise level.

In order to improve the efficiency of CEIMC while keeping the RMC accuracy, we have recently developed a method, based on a peculiar thermodynamic integration, to estimate the free energy of the system with RMC based BO energy from the knowledge of the free energy of the system with VMC based BO energy [45]. This allows to extensively use VMC rather than RMC, performing RMC on selected thermodynamic states only.

2.4.4. Hydrogen Trial Wave Function

For high pressure hydrogen we have developed a quite accurate trial function of the Slater-Jastrow, single determinant, form. The Jastrow part has an electron-proton and electron-electron Random Phase Approximation (RPA) term plus two-body and three-body empirical terms depending on few variational parameters. The Slater determinants (one for each spin state) are built with single electron orbitals obtained by a self-consistent DFT solution. We have recently integrated the PWSCF-DFT solver [46] into our CEIMC code to ensure a faster and uniform convergence of the single electron orbitals in different physical conditions. Further, the argument of the orbitals are not the bare electron positions but rather the quasiparticle positions defined by the backflow transformation [47,48]. We combined both the RPA analytical form and the Gaussian-like empirical terms depending on variational parameters. Our trial wave function has a total of 13 variational parameters to be optimized [42,48].

Figure 1. Variational energy of four different crystalline molecular structures versus r_s: C_2/c upper-left panel, Cmca-12 upper-right panel, P63m lower-left panel and Pbcn lower-right panel. Energies from wave functions with different orbitals relatives to the energy with LDA orbitals: PBE orbitals (red triangles), HSE orbitals (green closed circles) and vdW-DF2 orbitals (blue closed squares).

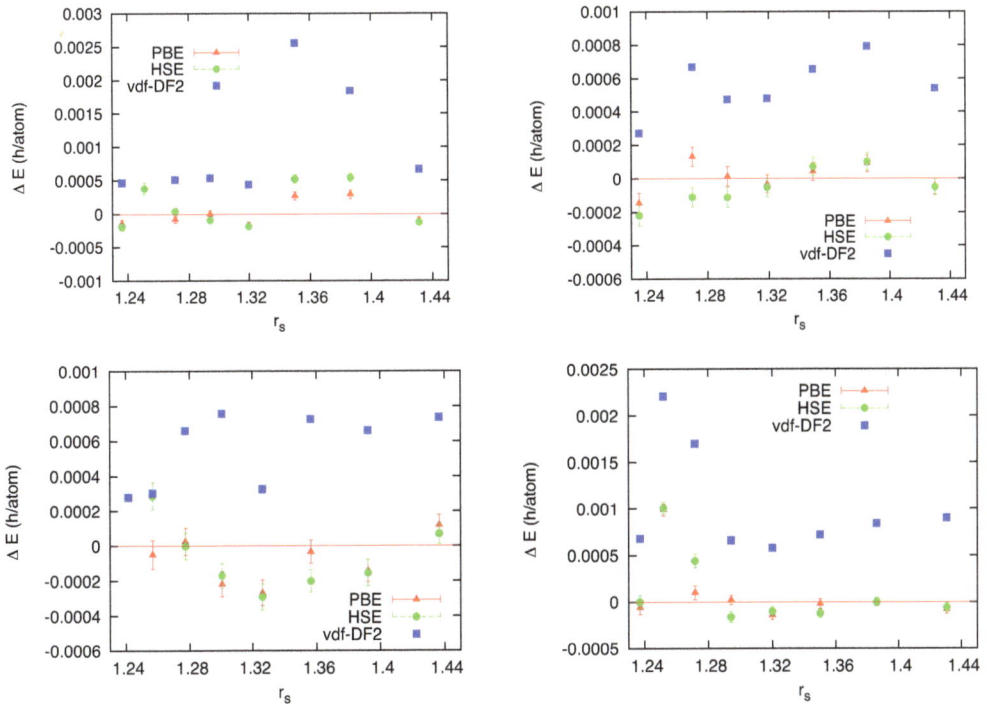

In view of the large variability of DFT results from different exchange-correlation approximations in the dissociation region of high pressure hydrogen (see next section), one interesting question is about the sensitivity of the trial wave function to the particular form of the adopted Kohn-Sham orbitals in the Slater determinant. This is particularly relevant since the form of the orbitals determine the nodal surface of the trial wave function, the ultimate limit in the accuracy of fermionic QMC. On the one hand one could hope to further improve the quality of the trial wave function by varying the type of orbitals, on the other hand a large sensitivity to the form of the Kohn-Sham orbitals will signal a too constrained form of the wave function, probably with a large room for improvements. The recent technical advance of the CEIMC code, namely the integration of PWSCF, allowed us to test several different types of orbitals: standard local (LDA) and semilocal (GGA-PBE) approximation, a non-local functional devised to reduce the self-interaction error and improve the description of the electronic correlation in DFT (HSE [49]) and a functional devised to improve the description of the dispersion interactions which are absent in a self-consistent mean-field theory (vdW-DF2 [50–52]). In the range of coupling parameter $1.22 \leq r_s \leq 1.44$ which corresponds approximatively to the range of pressure between 200 GPa and 550 GPa according to DFT, we have considered four recently proposed candidate structures for the molecular crystal [53], namely C2/c, Cmca-12, Pbcn and P63m. For each structure we have performed parameter optimizations for the four mentioned forms of the orbitals and at eight different densities. Supercells of 96 atoms were considered for C2/c, Cmca-12 and Pbcn structures, while a supercell of 128 atoms was studied for the P63m structure. Moreover for a single structure, Pbcn, at a single value of $r_s = 1.35$ we have performed a complete RMC study. In Figure 1 we report for all densities investigated the variational energies from the different orbitals relative to the energy of the trial function with LDA orbitals.

Figure 2. Pbcn structure of molecular hydrogen at $r_s = 1.35$. Left panel: energy per atom versus projection time in RMC from different kind of orbitals: LDA (closed red squares), PBE (green closed circles), HSE (upward blue triangles), vdW-DF2 (downward purple triangles). Also results from the old LDA implementation (cyan open circles) are reported. Right panel: Energy per atom versus variance in RMC from different kind of orbitals.

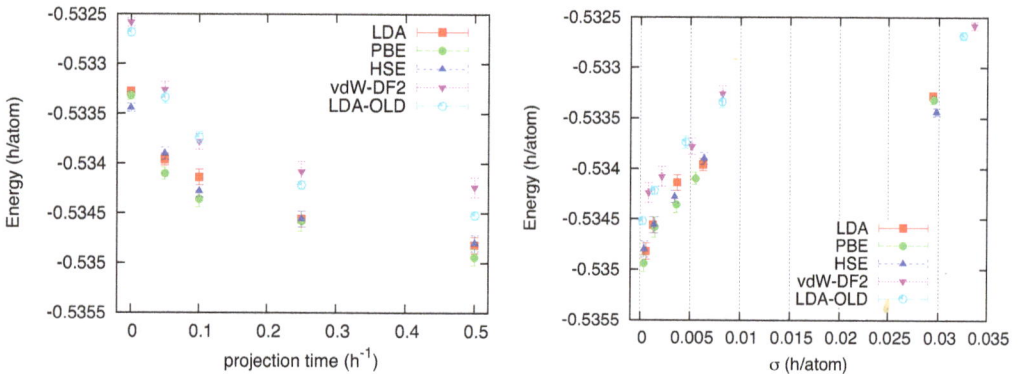

We note that for all structures and at all densities LDA, PBE and HSE orbitals provides trial functions of the same quality (differences are of the order of 0.2 mH/atoms = 90K). Instead the trial function with orbitals from vdW-DF2 functional provides higher energies, by roughly 0.4 mH/at with values up to 1.4 mH/atom ($\simeq$630 K). This first result is quite indicative that our trial function is flexible and general enough to be very little sensitive to the form of the orbitals. In order to check whether the observed differences from vdW-DF2 orbitals could be due to optimization problems only, we performed a complete RMC study for a single case, namely the Pbcn structure at $r_s = 1.35$. A time step of $\tau = 0.005$ h^{-1} was used, which is fairly typical in this sort of calculation. No further time step error extrapolation study has been performed. In Figure 2 the energy versus projection time is reported for all kind of orbitals. We also added results from our old DFT solver with LDA orbitals plagued by the truncation error. For all kinds of trial function we observe a very similar relaxation with projection time meaning that the quality of the trial function is similar in all cases. The differences observed at the variational level among different trial functions essentially remain along the projection and therefore in the extrapolated value for the total energy. A quantitative way to estimate the extrapolated ($\beta \to \infty$) value of the total energy is to plot energy versus its variance (pure estimate) and use a linear extrapolation at small values of σ^2. This plot for all studied cases is shown in the right panel of Figure 2. We see that the three kinds of orbitals, LDA, PBE and HSE all provides extrapolated energies within error bars ($E_0 = -0.5350(2)$), while the vdW-DF2 orbitals provides a higher value ($E_0 = -0.5342(2)$). The fact that the RMC projection is not able to remove the difference observed at the VMC level means that the nodes from the vdW-DF2 are less accurate than for the other kind of orbitals, which instead, despite their differences, provide essentially the same nodal structure. Finally we note that our old implementation of LDA orbitals provides a less accurate determination of the energy with correspondingly larger variance.

3. Applications

3.1. High-Pressure Hydrogen

Hydrogen is the simplest element of the periodic table and also the most abundant element in the Universe. Because of its simple electronic structure, it has been instrumental in the development of quantum mechanics and remains important for developing ideas and theoretical methods. In the next section we explore its use in developing DFT functionals. Its phase diagram at high pressure has received considerable attention from the first-principles simulation community due to its critical importance in many fields like planetary science, high pressure physics, astrophysics, inertial confinement fusion, among many others [10,54,55]. The phase diagram of hydrogen at high pressure contains many interesting features including: a maximum in the melting line with a subsequent negative slope [56,57], a predicted liquid-liquid transition between an insulating molecular and a conducting atomic phase [58,59], exotic molecular phases at low temperature, and a predicted metal-insulator transition in the solid phase [10,55].

The ground state structure of crystalline hydrogen across the pressure-induced molecular dissociation has been studied by DMC [35–37] which predicted molecular dissociation at density corresponding to $r_s \simeq 1.3$. RPIMC has been applied to investigate the Warm Dense Matter regime, namely the regime of high pressure and density where thermal and pressure molecular dissociation and ionization occur simultaneously [38,39,60]. Particularly relevant for our current understanding of the phase diagram and the Equation of State (EOS) of compressed hydrogen has been the determination of the primary and secondary Hugoniots lines of deuterium which could be directly compared with experimental data [40,61]. RPIMC predictions for the principal Hugoniot of deuterium were first in disagreement with pulsed laser-produced shock compression experiments [62–64], but were later confirmed by magnetically generated shock compression experiments at the Z-pinch machine [65–70] and by converging explosive-driven shock waves techniques [71,72]. Also relevant for the development and fine tuning of simulation methods for

Warm Dense Matter has been the comparison with the less demanding, but also less fundamental methods based on Density Functional Theory (either Kohn-Sham or Orbital-Free flavours). A general agreement between RPIMC and FPMD predictions for the Hugoniot lines was observed [10] except at the lowest temperatures that could be reached by RPIMC ($\sim$10,000 K). More recently the synergetic use of Born-Oppenheimer molecular dynamics (BOMD) and RPIMC has allowed to produce first-principle based EOS's in a wide range of physical conditions for hydrogen, helium and hydrogen-helium mixtures [73,74] instrumental in planetary modeling and crucial ingredients for the hydrodynamic codes used in the large facilities for extreme conditions experiments.

Temperatures lower than $\sim$10,000 K cannot be easily reached by RPIMC without reducing the level of accuracy. However, most of the interesting phenomena in high pressure hydrogen, like molecular dissociation under pressure, metallization, solid-fluid transition, a possible liquid-liquid phase transition and its interplay with melting, the various crystalline phases and the transition to the atomic phases [10], occur at lower temperature out of the reach of RPIMC. Investigating this regime by QMC methods has been the main motivation in developing CEIMC. The other motivation, as mentioned above, is the benchmark of the much more developed (and less demanding) alternative theoretical method, namely FPMD based on DFT. Indeed the numerical implementation of DFT is based on approximations (the exchange-correlation functional) the accuracy of which can only be established against experiments or, better, against more accurate theories. As mentioned earlier, QMC energy is an upper bound and therefore has an internal measure of accuracy.

CEIMC has been applied to investigate the WDM regime of hydrogen and helium and benchmark FPMD [48,75,76]. In [76] an investigation of the fully ionized state of hydrogen in a region of pressure and temperature relevant for Jovian planets found that FPMD based on the GGA-PBE exchange-correlation functional and CEIMC are in very good agreement but both deviates from a widely accepted phenomenological EOS. The agreement between the simulation methods becomes less good when approaching the molecular dissociation regime at slightly lower temperature and pressure. Both CEIMC and FPMD with different approximated functionals has been applied to investigate the Liquid-Liquid phase transition (LLPT) region in hydrogen [45,59,77]. The emerging picture is that a weak first-order phase transition occurs in hydrogen between a molecular-insulating fluid and a metallic-mostly monoatomic fluid. At higher temperature, molecular dissociation and metallization occur continuously. However the precise location of the transition line and the critical point are still matter of debate since several levels of the theory provide different locations. Within FPMD-DFT the location of the transition line depends strongly on the exchange-correlation functional employed and on whether classical or quantum protons are considered [77]. Transition lines from the PBE and vdW-DF2 approximations differ by roughly 200–250 GPa, the PBE one being located at lower pressure. The PBE melting line with quantum protons is not in agreement with experiments, which highlights the failure of the PBE approximation when employed together with the quantum description of the nuclei. On the other hand, optical properties for the vdW-DF2 approximation are in agreement with experiments supporting the use of this functional for hydrogen in the WDM regime. The LLPT line from CEIMC lies in between the lines from PBE and vdW-DF2 functionals [45,59]. However, those results were plagued by a truncation error in the calculations of the single electron orbitals which showed up only around the metallization and which resulted in biased estimates. We have now changed the DFT solver in our CEIMC code and checked the convergence. We find a roughly uniform shift of the transition line of $\sim$50 GPa to higher pressure and we are performing new calculations with quantum nuclei. Preliminary results, based on VMC electronic energies, suggests that, similarly to the DFT scenario, nuclear quantum effects favor molecular dissociation and become increasingly important at lower temperatures. We estimate that the transition pressure is decreased, because of nuclear quantum effects, by $\sim$60 GPa at 600 K and by $\sim$150 GPa at 300 K (from $\sim$430 GPa for classical nuclei to $\sim$290 GPa for quantum nuclei). RQMC corrections to the transition lines was previously found to be small and we expect an even smaller effect with the new CEIMC implementation since the VMC variance is roughly half of what it was in the previous code [45].

The last estimate however is for a metastable liquid state obtained by an instantaneous quenching of the fluid at higher temperature, while it is expected that the equilibrium state at 300 K and $\sim$290 GPa be crystalline (of unknown structure) [10]. Those results are preliminary since the calculation is performed for a small system of 54 protons (we employ Twist Averaged boundary conditions to reduce size effects on the single-electron properties with a $4 \times 4 \times 4$ twist angle grid) and we are presently estimating size effects, both by direct size extrapolation and by the analytic treatment of size effects [78,79]. In Figure 3 we report CEIMC proton-proton g(r) at various densities along the T = 600 K isotherm to illustrate the relevance of nuclear quantum effects on the pressure dissociation. The preliminary CEIMC results suggest that, despite the good performance observed on band gap calculations in the crystalline phases [80], the vdW-DF2 exchange-correlation functional has a tendency to over-stabilize molecules.

Although our results demonstrate the power of CEIMC in predicting the physical properties of hydrogen, its use is still quite demanding in terms of computer time, a fact that limits its applicability. This is particularly true when a much larger exploration of external conditions is needed to clarify the physics. For example, to study the crystalline state of the molecular system and clarify the molecular-atomic transition mechanism in the solid state, it is necessary to consider a large number of candidate structures, some of which have very large unit cells (the recently proposed Pc structure for phase IV of molecular hydrogen [81] contains 192 proton, more than three times larger than the system considered in the LLPT). Moreover, in studying those structure at finite temperature it is important to apply a constant stress algorithm allowing the simulation box to deform and release the excess internal stress that otherwise would produce metastable states. While larger systems (>250 particles) and constant pressure algorithms are routinely applied in FP methods based on DFT, their use in conjunction with CEIMC is still problematic. Therefore, it is important to apply CEIMC and other QMC methods to validate DFT predictions and determine the most accurate functional for a given system. The same considerations apply to systems more complex than hydrogen. In the next section we will describe our effort to benchmark functionals for high pressure hydrogen and for water in condensed phase.

Figure 3. Proton-proton radial distribution function at various densities along the isotherm T = 600 K. Comparison between classical nuclei (red continuous line) and quantum nuclei (blue dashed line) for hydrogen nuclear mass. It is evident the molecular dissociation with increasing density.

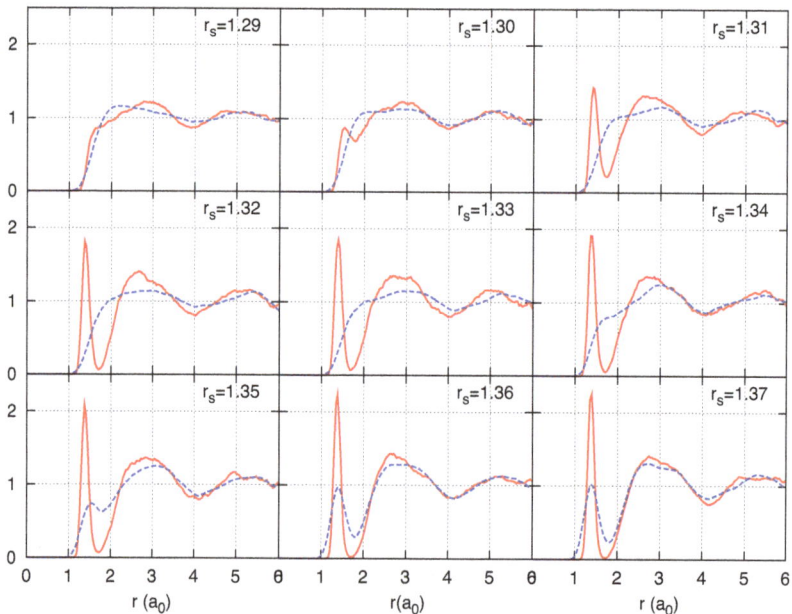

3.2. QMC Benchmarks of DFT

Within the Born-Oppenheimer approximation at low temperatures, the only interaction between ions and electrons comes through the potential energy surface $E_0(\mathbf{R})$, defined as the solution of the electronic Hamiltonian for a fixed set of ionic coordinates. $E_0(\mathbf{R})$ is typically approximated by $E_{DFT}(\mathbf{R})$ in first-principles calculations, and obtained from a density functional theory (DFT) calculation. Over the last several years, many-body methods for solids have been developed to the point that the prospect of developing density functionals from accurate reference calculations is now a possibility. In this section, we show how quantum Monte Carlo calculations can be used to benchmark the accuracy of DFT in the description of the potential energy surface. The quality of $E_{DFT}(\mathbf{R})$ defines the predictive capabilities of the resulting first-principles simulation. We use large sets of representative configuration from PIMD simulations, and compare the mean absolute error between accurate QMC calculations and various DFT functionals. We present preliminary calculations on high pressure hydrogen and liquid water at ambient conditions, two materials that are particularly challenging to DFT due to the subtle competition between dispersion interactions, nuclear quantum effects, hydrogen bonding, and anisotropic interactions.

3.2.1. Hydrogen

The phase diagram of hydrogen at high pressure has been extensively explored using first-principles simulations with DFT [58,59,82–85]. In spite of the large number of studies, most of the work so far has employed either the local density (LDA) [86] approximation to the exchange-correlation potential or the Perdew-Burke-Ehrzenhof (PBE) [87] generalized gradient approximation. These are two of the simplest functionals currently available in DFT. In fact, both of them suffer from self-interaction errors and lack a proper treatment of dispersion interactions, making their application in the regime of molecular dissociation questionable. Recently, the use of DFT functionals with an improved description of dispersion interactions has been employed in the study of the liquid and solid molecular phases in the neighborhood of molecular dissociation. It was found that the dissociation density changed when compared to calculations using PBE [77,80]. Since these functionals were not designed for materials at high density, and because dispersion interactions are clearly important in dense molecular hydrogen, there is a crucial need for accurate calculations that can be used to benchmark the different exchange-correlation functionals employed in first-principles simulations.

Since sufficient experimental data is not available to validate the quality of functionals in the high-pressure high-temperature regime of the phase diagram, we used fixed-node diffusion Monte Carlo (DMC) to benchmark the accuracy of several DFT functionals over a range of densities near the liquid-liquid phase transition at a temperature of T = 1000 K. Henceforth, we will refer to densities using the parameter r_s. First, we ran PIMD simulations with the PBE functional for $N = 54$ hydrogen atoms at three densities: $r_s = 1.30, 1.45, 1.60$. In this range of densities, the liquid goes from an insulating molecular state at $r_s = 1.60$ to a conducting atomic liquid at $r_s = 1.30$. The density $r_s = 1.45$ is intermediate and close to the LLPT for this functional. After equilibration, we sampled 100 ionic configurations from uncorrelated PIMD time slices for each density. For each configuration at each density, we calculated the DMC energy, and then computed $E_{DFT}(\mathbf{R})$ for the following functionals: LDA, PBE, vdW-DF [50], vdW-DF2 [51,52,88], and HSE [49].

All QMC calculations were performed with the QMCPACK [89–91] software package. We used a Slater-Jastrow trial wavefunction with twist-averaged boundary conditions [92], employing a $3 \times 3 \times 3$ grid of boundary conditions. For the Jastrow functions, we used real space b-splines with optimizable knots. We included spin-independent one-body proton-electron terms; a short-ranged term with the appropriate cusp condition, and a long-ranged term. We also included two long-ranged spin-dependent electron-electron functions with appropriate cusp conditions. For each configuration, linear optimization with VMC was performed for all Jastrow parameters at a single twist-angle, these parameters were subsequently used for all twists in the DMC calculations. For the DMC run, a timestep of $\tau = 0.05$ Ha^{-1} and 6000 walkers were used. The orbitals were obtained from DFT using the Quantum Espresso software package [46], using the PBE functional. We used a plane wave cutoff of 210 Ry. DFT calculations were performed with a Troullier-Martins norm conserving pseudo-potential [93] with a cutoff radius of $r_c = 0.5a_0$, DMC calculations were performed with the Coulomb potential. Based on the scale of the energy differences, we found a statistical error of

0.02 mHa/particle to be sufficient for present purposes. Since we were interested in measuring the spread of energy errors in this presentation, constant energy offsets were removed from our error assessments. This means that we did not have to include energetic finite size effects, although more detailed assessments will certainly call for this.

An example of the comparison between QMC and DFT is given in Figure 4. Shown is a histogram of the energy difference between the results of DMC and the PBE functional at the three densities: $\Delta E_{DFT} = E_{DFT} - E_{DMC}$. Given that $r_s = 1.30$ corresponds to the atomic liquid, and $r_s = 1.60$ to the molecular liquid, we immediately see that the errors incurred by using the PBE functional are not consistent across the LLPT. As expected, PBE offers a much better description of the atomic liquid compared to the molecular phase, where self-interaction errors are larger and dispersion interactions are important. This is a well-known failure of most semi-local density functionals, which tend to favor delocalized states.

Figure 4. Histograms of ΔE_{DFT} for the PBE functional for dense hydrogen at densities $r_s = 1.30, 1.45, 1.60$ at $T = 1,000K$. ΔE_{DFT} refers to the absolute energy difference per hydrogen atom between the DFT and QMC for a given configuration. There were 54 atoms per configuration.

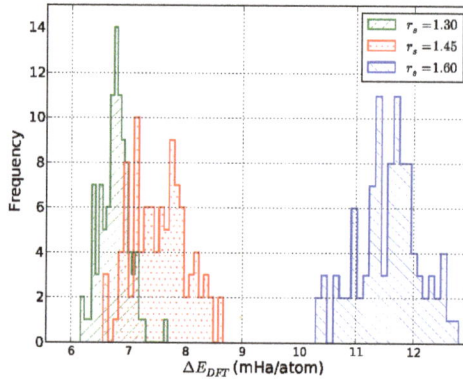

To better quantify and compare the quality of functionals, we have computed the mean absolute error (MAE) from data similar to that shown in Figure 4. This quantity is defined as $MAE_{func} = \langle |\Delta E_{DFT} - \langle \Delta E_{DFT} \rangle| \rangle$, where the average is taken over all configurations at a particular density. Notice that we subtract the average energy difference in the definition of the MAE, since the zero of energy of each functional is modified by the use of pseudopotentials. Fluctuations of the energy differences are more significant since the structure of the liquid is only sensitive to differences. The MAE gives us one measure of the quality, or predictive capability, of a given functional as defined by the reference method, in this case DMC. We have tabulated our results in Figure 5.

There are several interesting features in Figure 5 directly related to the expected performance of these functionals in the description of hydrogen near molecular dissociation in the liquid. First, the two semi-local functionals in the comparison, LDA and PBE, have considerably different errors in the molecular and atomic regimes. As described above, the atomic regime is more accurately described in comparison to the molecular phase, leading to a potentially strong underestimation of dissociation transition pressures in both solid and liquid phases. This is consistent with recently reported simulations [77]. On the other hand, both the hybrid HSE and the functionals with improved dispersion vdW-DF and vdW-DF2 offer a more consistent level of description between the two regimes. The mean absolute errors of the HSE and vdW-DF functionals are approximately half that of the PBE functional for all densities, which indicates that these functionals more accurately capture energy *differences* between various liquid configurations.

Figure 5. Mean absolute error of energy/atom *vs.* functional for dense liquid hydrogen at 1000 K. For each functional, we computed the mean absolute error for three different densities, denoted by the different colored bars.

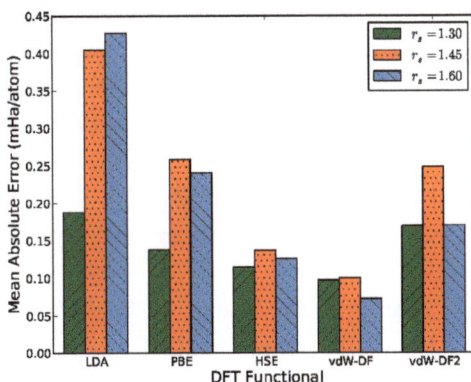

3.2.2. Liquid Water

Water plays a central role in many scientific fields [94]. It is a critical component to almost all chemical, biological, and geophysical processes. As a result, it is one of the most studied substances in science, both from an experimental and a theoretical point of view. Despite such broad importance, water's most basic property, its local structure at ambient conditions, characterized by the geometry of its underlying hydrogen-bond (H-bond) network, has remained a matter of debate for over a century [95–97]. Challenges arise because water is only ≈ 25 K (at room temperature) from the melting temperature of ice, where a variety of subtle and complex effects become important. While the structure is dominated by H bond between neighboring molecules, both van der Waals (vdW) interactions (which, in this context, refers to dispersion forces resulting from dynamical nonlocal electron correlations) and nuclear quantum effects (NQEs) influence the topology of the H-bond network. In fact, it is precisely these seemingly subtle effects (compared to H bonding) that are key to accurately describing ambient water, but have been (until recently) difficult or impossible to model.

Atomistic simulations have the potential to resolve these issues, particularly using first-principles methods. Providing an accurate theoretical description has been a central topic and open challenge in physical chemistry for many decades. Despite considerable focus over the last decade, to date DFT has proven insufficient for the accurate description of liquid water [4,98]. Nonetheless, much progress has occurred during the last several years. The main advances include the use of functionals that properly describe dispersion interactions in the liquid [50,52,99,100], the use of hybrid functionals [101], and the direct treatment of nuclear quantum effects [102]. The combination of all of these advances in first-principles simulations of liquid water could lead to an accurate description of its interesting properties, including its local structure. At the same time, the choice of exchange-correlation functional in DFT is still a source of complication, mainly due to the large number of possibilities and the inability to test their predictive capabilities without resorting to full first-principles calculations of a large set of observables. As in the case of hydrogen, an accurate first-principles description almost certainly requires the use of path integral methods in order to directly treat nuclear quantum effects, which makes the calculations quite computationally intensive. What is needed is a way to assess the quality of a given functional without having to resort to first-principles calculations of the liquid at the PIMD level, and if possible, a way to systematically improve them using high quality reference calculations from accurate many-body methods.

In this section, we present QMC calculations of configurations of molecules extracted from PIMD simulations of liquid water. QMC has been shown to be a reliable benchmark in the study of small water clusters [103–105], and should provide an accurate reference method to measure the quality of typical density functionals used in simulations of water. All DMC calculations were performed with the QMCPack software package [89–91]. A Troullier-Martins norm-conserving pseudo-potential [93] was used to represent both hydrogen and oxygen. In particular, we used the pseudo-potentials from the CASINO database [106,107], which were recently shown to produce accurate results in the study of small water clusters. A Slater-Jastrow trial wave-function was used. The orbitals in the Slater determinant were obtained from DFT calculations employing the PBE exchange-correlation functional. We do not expect a strong dependence of the resulting comparison on the functional used to generate the orbitals. The Jastrow term contains electron-ion, electron-electron and electron-electron-ion terms, the variational parameters were optimized at the VMC level using a variant of the linear method of Umrigar, *et al.* [108]. A time-step of 0.01 Ha^{-1} was found to be sufficiently small to produce accurate total energies and approximately 4800 walkers were used in the DMC calculations. Casula's T-moves [109] were used to reduce locality errors, while the Model Coulomb Potential [110] and Chiesa's [78] correction scheme were used to estimate finite-size corrections to the potential and kinetic energies respectively.

DFT calculations were performed with both Quantum Espresso (QE) [46] and VASP [111–113] simulation packages. In the case of QE calculations we employed norm-conserving Troullier-Martins pseudo-potentials, while in the case of VASP calculations we employed the Projector Augmented Wave method (PAW) [114,115]. A single pseudo-potential (constructed with PBE) was chosen in order to make a homogeneous comparison of all DFT functionals, since some of the functionals employed in this work do not yet allow for the production of pseudo-potentials. All simulations were performed at the Γ point of the supercell in order to be consistent with the corresponding DMC calculations; errors due to the lack of k-point integration were small enough to be safely discarded. We carefully tested the convergence with the plane-wave cutoff in all DFT calculations.

We present calculations for 3 different configuration sets. The first two sets, which we called *TIP5P-PI-0C-ICE* and *TIP5P-PI-0C-LIQ*, were generated with PIMD calculations on simulation cells using the semi-empirical TIP5P water model and 32 molecules [116]. As the name suggests, the PIMD calculations used to generate the configuration set were performed at T = 0 C, from stable solid and liquid phases. The third configuration set was obtained from PIMD calculations of 64 water molecules, at room temperature and density of 1 g/cm^3, with the vdW-DF2 functional, which has been recently shown to provide an accurate description of the structure of water when combined with a path integral representation [117]. The number of configurations in each set is 20, 47, 50, respectively. The three configuration sets sample different aspects of the potential energy surface of liquid water. While TIP5P is a rigid molecule model, the first-principles simulations with vdW-DF2 are fully flexible, which allows us to emphasize different ranges of the molecular interactions in the liquid. On the other hand, the simulations with TIP5P in both liquid and solid phases at T = 0 C sample the configurations that either strongly favor hydrogen bonding in the solid, with those where the hydrogen-bond network has been destabilized in the liquid.

Figure 6. Mean absolute error in the total energy between DMC and DFT with various exchange correlation functionals for a supercell containing water molecules. Results presented correspond to calculations using the PAW formulation with VASP. X-D, where X represents a given density functional, designates results using the empirical dispersion corrections of Grimme *et al.*, [118], in particular the DFT-D2 correction scheme as implemented in VASP. Statistical errors on the presented results are on the order of 0.003 mHa and 0.005 mHa for rigid and flexible molecule configurations respectively. They are not shown on the figure for clarity.

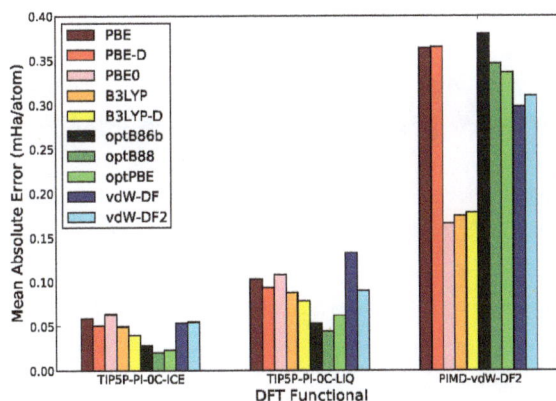

Figure 6 shows the mean absolute difference in the total energy between DMC and DFT calculations, results are separated by configuration sets in order to allow for a more clear comparison between them. Several functionals are considered including the semi-local functionals: PBE [87]; the hybrid functionals: PBE0 [49], B3LYP [119,120]; the non-local van der Waals functionals: optB88 [121], optPBE [121], optB86b [122], vdW-DF [50] and vdW-DF2 [52]; and finally functionals with the empirical van der Waals correction of Grimme, *et al.*, (DFT-D2) [118]. While there are many interesting results in this comparison, the most noticeable feature is the large difference in the scale of the MAE between rigid and flexible molecule configurations. This is not unexpected since the larger energy fluctuations in the system are found coupled to the intramolecular degrees of freedom of the molecule. In the case of flexible molecule configurations, hybrid functionals offer a much better agreement with DMC results, producing errors typically a factor of 2 smaller than non-hybrid functionals. This result shows the fact that hybrid functionals do a much better job at describing the intramolecular potential energy surface. This is consistent with the recent calculations of Alfe, *et al.* [104] and with the recent calculations of the absorption spectra of bulk water at ambient conditions of Zhang, *et al.* [101]. On the other hand, the functionals that include an appropriate description of dispersion interactions offer a clearly better comparison with QMC in the rigid-molecule configuration sets. In this case, the intermolecular interactions are the dominant energy contribution and the lack of appropriate dispersion leads to a larger error. In this case, we can also see a small but finite improvement with the inclusion of empirically corrected vdW functionals (PBE-D, B3LYP-D), but the gain is small and can not compete with non-local vdW functionals. Notice also that the performance of hybrids in the rigid-molecule sets is comparable to the performance of semi-local functionals, due to the fact that neither of these type of functionals can properly describe dispersion interactions. Finally, the configuration set with the smallest overall MAE is the one obtained from the calculations in the solid phase close to melting, showing the fact that most of these functionals can describe hydrogen bonded configurations fairly well.

4. Discussion

Direct first-principles simulations with QMC accuracy of condensed phases systems are nowadays possible but restricted so far to the simplest first few elements of the periodic table, namely hydrogen, helium and their mixtures. Even for those simple systems, challenges are present and the computational demand is large. Nonetheless, CEIMC predictions for the liquid-liquid phase transition in hydrogen remains today the target for less accurate but faster DFT-based FP methods. While much work remains to be done in developing QMC-based FP methods, the calculations presented here show one possible use of accurate many-body calculations: using QMC to benchmark the accuracy of DFT functionals. Not only does this allow us to make a judgment of the quality of a functional before its use in first-principles simulations, but it also shows us a path for the systematic improvement of the functionals by adjusting free parameters to minimize the errors. DFT users will often point to experimental data to validate the quality of a chosen functional. What we have shown is that we can use highly-accurate QMC methods to benchmark functionals around the liquid-liquid transition of hydrogen from first-principles. In addition, this set of reference energies for the bulk system can be used to optimize the free parameters in the DFT functional to minimize the errors, and in the limit of a large data set, reproduce the quality of the more accurate many-body method in first-principles calculations using DFT. This approach will be increasingly necessary as we continue to explore matter under extreme pressures, since experimental data is often insufficient or nonexistent at geophysical/planetary scales. It will also be necessary for other situations where DFT functionals have difficulties, such as near metal-insulator transitions.

Let us consider a more general point. We suggest that, in general, it is superior to use total energies to find an interatomic potential (force field). The traditional approach is to fit experimental data, for example, the melting temperature of ice, the density of water versus temperature, *etc.* Clearly this procedure was necessary in the past since experimental data was all that was available. However, using this approach requires very extensive calculations including free energy or equivalent computations and ultimately only gives a few constraints. We can invoke "The Allegory of the Cave" from Plato's *The Republic*. We should not look to fit the atomic potentials using the projections of the energy surface onto thermodynamic properties, but, instead to fit directly the energy surface. Thus we will obtain an interatomic potential suitable for all properties. The situation has changed since QMC methods have matured and much more computational power is available. We note that scanning potential energy surface is a task very well suited to massively parallel computers. Including total energy QMC benchmarks into the fitting procedure in addition to experimental data, can allow for much more systematic improvements. QMC thus can provide a unique role in giving total energies and is applicable to large enough systems to approximate condensed matter.

Water and hydrogen show an additional complication of using experimental data: namely because of the importance of quantum zero-point effects of the protons, fitting of the experimental data becomes particularly problematic. A common approach is to do a simulation of the classical system and assume that the effective classical system includes the effects of zero-point energy; clearly this then becomes quite approximate since the zero-point effects are not small. A complication is that the interatomic potential that results can become temperature and density dependent with all known pathologies related to the use of state dependent potentials [123]. One may need to do full PIMD simulations of the system in order to determine the best empirical potential, thus increasing the, already large, computational requirements considerably.

One aspect in determining good force fields is to find an appropriate basis set to parameterize the force field. Traditionally, these have contained few functions with very few parameters, e.g., the Lennard Jones potential with only two parameters: ϵ and σ. It is feasible today to calculate the energy and forces for millions of independent arrangements of ions. Using QMC techniques, each would come with an error estimate. Hence, we can envision fitting this data set to a force field with potentially tens of thousands of independent parameters. This will allow us to determine a completely general pair potential (say with a spline basis), a three-body potential, four-body potential, *etc.* However, the investigation into effective basis sets to describe these potentials becomes very important. We can imagine an integrated set of tools: QMC simulations of systems with thousands of electrons to produce data sets of energies and forces. These can be used either to tailor a DFT to a particular system, or to determine a force field. The DFT simulations and the effective force field simulations can then be used to model much larger systems. Thus simulations can thereby become much more predictive, and produce not just universal properties, but details important to applications and experiment.

Acknowledgments

Miguel Angel Morales was supported by the U.S. Department of Energy at the Lawrence Livermore National Laboratory under Contract DE-AC52-07NA27344, by LDRD Grant No. 13-LW-004 and by the Basic Energy Science (BES), DOE through the Predictive Theory and

88

Modeling for Materials and Chemical Science program, D. M. C. and R. C. were supported by the DOE grant DE-NA0001789 and C. P. by the Italian Institute of Technology (IIT) under the SEED project grant number 259 SIMBEDD Advanced Computational Methods for Biophysics, Drug Design and Energy Research. Computer resources have been provided by the US DOE INCITE program, Lawrence Livermore National Laboratory through the 7th Institutional Unclassified Computing Grand Challenge program and PRACE Project No. 2011050781.

Conflicts of Interest

The authors declare no conflict of interest.

References

1. Ballone, P. Modelling potential energy surfaces: From first-principle approaches to empirical force fields. *Entropy* **2014**, *16*, 322-349.
2. Car, R.; Parinello, M. Unified approach for molecular dynamics and density-functional theory. *Phys. Rev. Lett.* **1985**, *55*, 2471–2474.
3. Cohen, A.J.; Mori-Sánchez, P.; Yang, W. Challenges for density functional theory. *Chem. Rev.* **2012**, *112*, 289–320.
4. Schwegler, E.; Grossman, J.C.; Gygi, F.; Galli, G. Towards an assessment of the accuracy of density functional theory for first principles simulations of water. II. *J. Chem. Phys.* **2004**, *121*, 5400–5409.
5. Ceperley, D.M.; Alder, B.J. Ground state of the electron gas by a stochastic method. *Phys. Rev. Lett.* **1980**, *45*, 566–569.
6. Brown, E.W.; Clark, B.K.; DuBois, J.L.; Ceperley, D.M. Path-integral monte carlo simulation of the warm dense homogeneous electron gas. *Phys. Rev. Lett.* **2013**, *110*, 146405.
7. Ercolessi, F.; Adams, J.B. Interatomic potentials from first-principles calculations: The force-matching method. *Europhys. Lett.* **1994**, *26*, 583–588.
8. Pierleoni, C.; Ceperley, D.M. The Coupled Electron-Ion Monte Carlo Method. In *Computer Simulations in Condensed Matter Systems: From Materials to Chemical Biology*; Ferrario, M., Ciccotti, G., Binder, K., Eds.; Springer: Berlin/Heidelberg, Germany, 2006; Volume 703, pp. 641–683.
9. Attaccalite, C.; Sorella, S. Stable liquid hydrogen at high pressure by a novel abinitio molecular-dynamics calculation. *Phys. Rev. Lett.* **2008**, *100*, 114501.
10. McMahon, J.; Morales, M.A.; Pierleoni, C.; Ceperley, D.M. The properties of hydrogen and helium under extreme conditions. *Rev. Mod. Phys.* **2012**, *84*, 1607–1653.
11. McMillan, W.L. Ground state of liquid He^4. *Phys. Rev.* **1965**, *138*, A442–A451.
12. Hammond, B.L.; Lester, W.A.; Reynolds, P.J. *Monte Carlo Methods in Ab Initio Quantum Chemistry*; World Scientific Publishing Co. Pte. Ltd.: Singapore, 1994; p. 304.
13. Ceperley, D.; Chester, G.V.; Kalos, M. Monte Carlo simulation of a many-fermion system. *Phys. Rev. B* **1977**, *16*, 3081–3099.

14. Ceperley, D. Ground state of the fermion one-component plasma: A Monte Carlo study in two and three dimensions. *Phys. Rev. B* **1978**, *18*, 3126–3138.

15. Ceperley, D.; Alder, B. The calculation of the properties of metallic hydrogen using Monte Carlo. *Physica B+C* **1981**, *108*, 875–876.

16. Foulkes, W.M.C.; Mitas, L.; Needs, R.J.; Rajagopal, G. Quantum Monte Carlo simulations of solids. *Rev. Mod. Phys.* **2001**, *73*, 33–83.

17. Morales, M.A.; McMinis, J.; Clark, B.K.; Kim, J.; Scuseria, G.E. Multideterminant wave functions in quantum Monte Carlo. *J. Chem. Theory Comput.* **2012**, *8*, 2181–2188.

18. Casula, M.; Attaccalite, C.; Sorella, S. Correlated geminal wave function for molecules: An efficient resonating valence bond approach. *J. Chem. Phys.* **2004**, *121*, 7110–7126.

19. Neuscamman, E.; Umrigar, C.; Chan, G. Optimizing large parameter sets in variational quantum Monte Carlo. *Phys. Rev. B* **2012**, *85*, 1–6.

20. Clark, B.K.; Morales, M.A.; McMinis, J.B.; Kim, J.; Scuseria, G.E. Computing the energy of a water molecule using multideterminants: A simple, efficient algorithm. *J. Chem. Phys.* **2011**, *135*, 244105.

21. Grimm, R.; Storer, R. Monte-Carlo solution of Schrödinger's equation. *J. Comput. Phys.* **1971**, *7*, 134–156.

22. Ceperley, D.; Kalos, M. Quantum Many-Body Problems. In *Monte Carlo Methods in Statistical Physics*; Binder, K., Ed.; Springer: Berlin/Heidelberg, Germany, 1979; Volume 7, p. 145.

23. Reynolds, P.; Ceperley, D.; Alder, B.J.; Lester, W., Jr. Fixed-node quantum Monte Carlo for molecules. *J. Chem. Phys.* **1982**, *77*, 5593–5603.

24. Anderson, J.B. A random-walk simulation of the Schroĺdinger equation: H + 3. *J. Chem. Phys.* **1975**, *63*, 1499.

25. Caffarel, M.; Claverie, P. Development of a pure diffusion quantum Monte Carlo method using a full generalized FeynmanâĂŞKac formula. I. Formalism. *J. Chem. Phys.* **1988**, *88*, 1088.

26. Assaraf, R.; Caffarel, M.; Khelif, A. Diffusion Monte Carlo methods with a fixed number of walkers. *Phys. Rev. E* **2000**, *61*, 4566–4575.

27. Reynolds, P.J.; Barnett, R.N.; Hammond, B.L.; Lester, W.A. Molecular physics and chemistry applications of quantum Monte Carlo. *J. Stat. Phys.* **1986**, *43*, 1017–1026.

28. Pierleoni, C.; Ceperley, D.M. Computational methods in coupled electron-ion Monte Carlo simulations. *Chem. Phys. Chem.* **2005**, *6*, 1872–1878.

29. Baroni, S.; Moroni, S. Reptation quantum Monte Carlo: A method for unbiased ground-state averages and imaginary-time correlations. *Phys. Rev. Lett.* **1999**, *82*, 4745–4748.

30. Ahuja, K.; Clark, B.K.; de Sturler, E.; Ceperley, D.M.; Kim, J. Improved scaling for quantum Monte Carlo on insulators. *SIAM J. Sci. Comput.* **2011**, *33*, 1837–1859.

31. Ceperley, D.M. Path integrals in the theory of condensed helium. *Rev. Mod. Phys.* **1995**, *67*, 279–355.

32. Ceperley, D.M. Path Integral Monte Carlo Methods for Fermions. In *Monte Carlo and Molecular Dynamics of Condensed Matter Systems*; Binder, K., Ciccotti, G., Eds.; Editrice Compositori: Bologna, Italy, 1996.

33. Ceperley, D.M. Fermion nodes. *J. Stat. Phys.* **1991**, *63*, 1237–1267.

34. Ceriotti, M.; Manolopoulos, D.E.; Parrinello, M. Accelerating the convergence of path integral dynamics with a generalized Langevin equation. *J. Chem. Phys.* **2011**, *134*, 084104.

35. Ceperley, D.M.; Alder, B.J. Ground state of solid hydrogen at high pressures. *Phys. Rev. B* **1987**, *36*, 2092–2106.

36. Natoli, V.; Martin, R.M.; Ceperley, D.M. Crystal structure of atomic hydrogen. *Phys. Rev. Lett.* **1993**, *70*, 1952–1955.

37. Natoli, V.; Martin, R.M.; Ceperley, D.M. Crystal structure of molecular hydrogen at high pressure. *Phys. Rev. Lett.* **1995**, *74*, 1601–1604.

38. Pierleoni, C.; Ceperley, D.M.; Bernu, B.; Magro, W.R. Equation of state of the hydrogen plasma by path integral Monte Carlo simulation. *Phys. Rev. Lett.* **1994**, *73*, 2145–2149.

39. Magro, W.R.; Ceperley, D.M.; Pierleoni, C.; Bernu, B. Molecular dissociation in hot, dense hydrogen. *Phys. Rev. Lett.* **1996**, *76*, 1240–1243.

40. Militzer, B.; Ceperley, D.M.; Kress, J.D.; Johnson, J.D.; Collins, L.A.; Mazevet, S. Calculation of a deuterium double shock hugoniot from Ab Initio simulations. *Phys. Rev. Lett.* **2001**, *87*, 275502.

41. Ceperley, D.M.; Dewing, M. The penalty method for random walks with uncertain energies. *J. Chem. Phys.* **1999**, *110*, 9812.

42. Ceperley, D.M.; Dewing, M.; Pierleoni, C. The Coupled Electronic-Ionic Monte Carlo Simulation Method. In *Bridging Time Scales: Molecular Simulations for the Next Decade SE-17*; Nielaba, P., Mareschal, M., Ciccotti, G., Eds.; Springer: Berlin/Heidelberg, Germany, 2002; Volume 605, pp. 473–500.

43. Allen, M.P.; Tildesley, D.J. *Computer Simulation of Liquids*; Oxford Science Publication, Oxford University Press: Oxford, UK, 1989.

44. Tuckerman, M. *Statistical Mechanics and Molecular Simulations*; Oxford Graduate Texts, Oxford University Press: Oxford, UK, 2008.

45. Liberatore, E.; Morales, M.A.; Ceperley, D.M. Free energy methods in coupled electron ion Monte Carlo. *Mol. Phys.* **2011**, *109*, 3029–3036.

46. Giannozzi, P.; Baroni, S.; Bonini, N.; Calandra, M.; Car, R.; Cavazzoni, C.; Ceresoli, D.; Chiarotti, G.L.; Cococcioni, M.; Dabo, I.; *et al.* QUANTUM ESPRESSO: A modular and open-source software project for quantum simulations of materials. *J. Phys. Condens. Matter* **2009**, *21*, 395502.

47. Holzmann, M.; Ceperley, D.M.; Pierleoni, C.; Esler, K. Backflow correlations for the electron gas and metallic hydrogen. *Phys. Rev. E* **2003**, *68*, 046707.

48. Pierleoni, C.; Delaney, K.T.; Morales, M.A.; Ceperley, D.M.; Holzmann, M. Trial wave functions for high-pressure metallic hydrogen. *Comput. Phys. Commun.* **2008**, *179*, 89–97.

49. Heyd, J.; Scuseria, G.E.; Ernzerhof, M. Hybrid functionals based on a screened Coulomb potential. *J. Chem. Phys.* **2003**, *118*, 8207.

50. Dion, M.; Rydberg, H.; Schröder, E.; Langreth, D.C.; Lundqvist, B.I. Van der Waals density functional for general geometries. *Phys. Rev. Lett.* **2004**, *92*, 246401.

51. Román-Pérez, G.; Soler, J. Efficient implementation of a van der Waals density functional: Application to double-wall carbon nanotubes. *Phys. Rev. Lett.* **2009**, *103*, 096102.

52. Lee, K.; Murray, E.D.; Kong, L.; Lundqvist, B.I.; Langreth, D.C. Higher-accuracy van der Waals density functional. *Phys. Rev. B* **2010**, *82*, 081101.

53. Pickard, C.J.; Needs, R.J. Structure of phase III of solid hydrogen. *Nat. Phys.* **2007**, *3*, 473–476.

54. Silvera, I. The solid molecular hydrogens in the condensed phase: Fundamentals and static properties. *Rev. Mod. Phys.* **1980**, *52*, 393–452.

55. Mao, H.K.; Hemley, R. Ultrahigh-pressure transitions in solid hydrogen. *Rev. Mod. Phys.* **1994**, *66*, 671–692.

56. Deemyad, S.; Silvera, I. Melting line of hydrogen at high pressures. *Phys. Rev. Lett.* **2008**, *100*, 155701.

57. Bonev, S.A.; Schwegler, E.; Ogitsu, T.; Galli, G. A quantum fluid of metallic hydrogen suggested by first-principles calculations. *Nature* **2004**, *431*, 669–672.

58. Scandolo, S. Liquid-liquid phase transition in compressed hydrogen from first-principles simulations. *Proc. Natl. Acad. Sci. USA* **2003**, *100*, 3051–3053.

59. Morales, M.A.; Pierleoni, C.; Schwegler, E.; Ceperley, D.M. Evidence for a first-order liquid-liquid transition in high-pressure hydrogen from ab initio simulations. *Proc. Natl. Acad. Sci. USA* **2010**, *107*, 12799–12803.

60. Pierleoni, C.; Magro, W.R.; Ceperley, D.M.; Berne, B.J. Path Integral Monte Carlo Simulation of Hydrogen Plasma.pdf. In Proceedings of the International Conference on the Physics of Strongly Coupled Plasmas, Binz, Germany, 11-15 September 1995; Kraeft, W.D., Schlanges, M., Eds.; World Scientific Publishing Co. Pte. Ltd.: Singapore, 1996.

61. Militzer, B.; Ceperley, D.M. Path integral Monte Carlo calculation of the deuterium hugoniot. *Phys. Rev. Lett.* **2000**, *85*, 1890–1893.

62. Da Silva, L.B.; Celliers, P.; Collins, G.W.; Budil, K.S.; Holmes, N.C.; Barbee, T.W., Jr.; Hammel, B.A.; Kilkenny, J.D.; Wallace, R.J.; Ross, M.; *et al.* Absolute equation of state measurements on shocked liquid deuterium up to 200 GPa (2 Mbar). *Phys. Rev. Lett.* **1997**, *78*, 483–486.

63. Collins, G.W. Measurements of the equation of state of deuterium at the fluid insulator-metal transition. *Science* **1998**, *281*, 1178–1181.

64. Celliers, P.M.; Collins, G.W.; da Silva, L.; Gold, D.M.; Cauble, R.; Wallace, R.J.; Foord, M.E.; Hammel, B.A. Shock-induced transformation of liquid deuterium into a metallic fluid. *Phys. Rev. Lett.* **2000**, *84*, 5564–5567.

65. Knudson, M.D.; Hanson, D.L.; Bailey, J.E.; Hall, C.A.; Asay, J.R. Equation of state measurements in liquid deuterium to 70 GPa. *Phys. Rev. Lett.* **2001**, *87*, 225501.

66. Knudson, M.; Hanson, D.; Bailey, J.; Hall, C.; Asay, J. Use of a Wave Reverberation Technique to Infer the Density Compression of Shocked Liquid Deuterium to 75 GPa. *Phys. Rev. Lett.* **2003**, *90*, 035505.

67. Knudson, M.D.; Hanson, D.L.; Bailey, J.E.; Hall, C.A.; Asay, J.R.; Deeney, C. Principal Hugoniot, reverberating wave, and mechanical reshock measurements of liquid deuterium to 400 GPa using plate impact techniques. *Phys. Rev. B* **2004**, *69*, 144209.

68. Bailey, J.E.; Knudson, M.D.; Carlson, A.L.; Dunham, G.S.; Desjarlais, M.P.; Hanson, D.L.; Asay, J.R. Time-resolved optical spectroscopy measurements of shocked liquid deuterium. *Phys. Rev. B* **2008**, *78*, 144107.

69. Hicks, D.G.; Boehly, T.R.; Celliers, P.M.; Eggert, J.H.; Moon, S.J.; Meyerhofer, D.D.; Collins, G.W. Laser-driven single shock compression of fluid deuterium from 45 to 220 GPa. *Phys. Rev. B* **2009**, *79*, 014112.

70. Knudson, M.D.; Desjarlais, M.P. Shock compression of quartz to 1.6 TPa: Redefining a pressure standard. *Phys. Rev. Lett.* **2009**, *103*, 225501.

71. Boriskov, G.; Bykov, A.; IlâĂŹkaev, R.; Selemir, V.; Simakov, G.; Trunin, R.; Urlin, V.; Shuikin, A.; Nellis, W. Shock compression of liquid deuterium up to 109 GPa. *Phys. Rev. B* **2005**, *71*, 092104.

72. Grishechkin, S.K.; Gruzdev, S.K.; Gryaznov, V.K.; Zhernokletov, M.V.; IlâĂŹkaev, R.I.; Iosilevskii, I.L.; Kashintseva, G.N.; Kirshanov, S.I.; Manachkin, S.F.; Mintsev, V.B.; *et al.* Experimental measurements of the compressibility, temperature, and light absorption in dense shock-compressed gaseous deuterium. *J. Exp. Theor. Phys. Lett.* **2004**, *80*, 398–404.

73. Hu, S.X.; Militzer, B.; Goncharov, V.N.; Skupsky, S. FPEOS: A first-principles equation of state table of deuterium for inertial confinement fusion applications. *Phys. Rev. B* **2011**, *84*, 224109.

74. Militzer, B. Equation of state calculations of hydrogen-helium mixtures in solar and extrasolar giant planets. *Phys. Rev. B* **2013**, *87*, 014202.

75. Pierleoni, C.; Ceperley, D.M.; Holzmann, M. Coupled electron-ion monte carlo calculations of dense metallic hydrogen. *Phys. Rev. Lett.* **2004**, *93*, 146402.

76. Morales, M.A.; Pierleoni, C.; Ceperley, D.M. Equation of state of metallic hydrogen from coupled electron-ion Monte Carlo simulations. *Phys. Rev. E* **2010**, *81*, 1–9.

77. Morales, M.A.; McMahon, J.M.; Pierleoni, C.; Ceperley, D.M. Nuclear quantum effects and nonlocal exchange-correlation functionals applied to liquid hydrogen at high pressure. *Phys. Rev. Lett.* **2013**, *110*, 065702.

78. Chiesa, S.; Ceperley, D.; Martin, R.; Holzmann, M. Finite-size error in many-body simulations with long-range interactions. *Phys. Rev. Lett.* **2006**, *97*, 6–9.

79. Drummond, N.; Needs, R.J.; Sorouri, A.; Foulkes, W.M.C. Finite-size errors in continuum quantum Monte Carlo calculations. *Phys. Rev. B* **2008**, *78*, 125106.

80. Morales, M.A.; McMahon, J.M.; Pierleoni, C.; Ceperley, D.M. Towards a predictive first-principles description of solid molecular hydrogen with density functional theory. *Phys. Rev. B* **2013**, *87*, 184107.

81. Pickard, C.J.; Martinez-Canales, M.; Needs, R.J. Density functional theory study of phase IV of solid hydrogen. *Phys. Rev. B* **2012**, *85*, 214114.

82. Desjarlais, M.P. Density-functional calculations of the liquid deuterium Hugoniot, reshock, and reverberation timing. *Phys. Rev. B* **2003**, *68*, 064204.

83. Vorberger, J.; Tamblyn, I.; Militzer, B.; Bonev, S. Hydrogen-helium mixtures in the interiors of giant planets. *Phys. Rev. B* **2007**, *75*, 024206.

84. Morales, M.A.; Schwegler, E.; Ceperley, D.; Pierleoni, C.; Hamel, S.; Caspersen, K. Phase separation in hydrogen-helium mixtures at Mbar pressures. *Proc. Natl. Acad. Sci. USA* **2009**, *106*, 1324–1329.

85. McMahon, J.M.; Ceperley, D.M. Ground-state structures of atomic metallic hydrogen. *Phys. Rev. Lett.* **2011**, *106*, 165302.

86. Perdew, J.P.; Zunger, A. Self-interaction correction to density-functional approximations for many-electron systems. *Phys. Rev. B* **1981**, *23*, 5048–5079.

87. Perdew, J.; Burke, K.; Ernzerhof, M. Generalized gradient approximation made simple. *Phys. Rev. Lett.* **1996**, *77*, 3865–3868.

88. Thonhauser, T.; Cooper, V.R.; Li, S.; Puzder, A.; Hyldgaard, P.; Langreth, D.C. Van der Waals density functional: Self-consistent potential and the nature of the van der Waals bond. *Phys. Rev. B* **2007**, *76*, 125112.

89. Kim, J.; Esler, K.P.; McMinis, J.; Morales, M.A.; Clark, B.K.; Shulenburger, L.; Ceperley, D.M. Hybrid algorithms in quantum Monte Carlo. *J. Phys.* **2012**, *402*, 012008.

90. Esler, K.; Kim, J.; Ceperley, D.; Shulenburger, L. Accelerating quantum monte carlo simulations of real materials on GPU clusters. *Comput. Sci. Eng.* **2012**, *14*, 40–51.

91. Esler, K.; Kim, J.; McMinis, J. QMCPACK at http://qmcpack.cmscc.org/.

92. Lin, C.; Zong, F.; Ceperley, D. Twist-averaged boundary conditions in continuum quantum Monte Carlo algorithms. *Phys. Rev. E* **2001**, *64*, 016702.

93. Troullier, N.; Martins, J.L. Efficient pseudopotentials for plane-wave calculations. *Phys. Rev. B* **1991**, *43*, 1993–2006.

94. Franks, F. *Water: A Matrix of Life*, 2nd ed.; Royal Society of Chemistry Paperbacks, Royal Society of Chemistry: Cambridge, UK, 2000.

95. Ball, P. Water: Water—an enduring mystery. *Nature* **2008**, *452*, 291–292.

96. Clark, G.N.; Cappa, C.D.; Smith, J.D.; Saykally, R.J.; Head-Gordon, T. The structure of ambient water. *Mol. Phys.* **2010**, *108*, 1415–1433.

97. Nilsson, A.; Pettersson, L. Perspective on the structure of liquid water. *Chem. Phys.* **2011**, *389*, 1–34.

98. Grossman, J.C.; Schwegler, E.; Draeger, E.W.; Gygi, F.; Galli, G. Towards an assessment of the accuracy of density functional theory for first principles simulations of water. *J. Chem. Phys.* **2004**, *120*, 300–311.

99. Zhang, C.; Wu, J.; Galli, G.; Gygi, F. Structural and vibrational properties of liquid water from van der Waals density functionals. *J. Chem. Theory Comput.* **2011**, *7*, 3054–3061.

100. Møgelhøj, A.; Kelkkanen, A.K.; Wikfeldt, K.T.; Schiøtz, J.; Mortensen, J.J.R.; Pettersson, L.G.M.; Lundqvist, B.I.; Jacobsen, K.W.; Nilsson, A.; Nørskov, J.K. Ab initio van der Waals interactions in simulations of water alter structure from mainly tetrahedral to high-density-like. *J. Phys. Chem. B* **2011**, *115*, 14149–14160.

101. Zhang, C.; Donadio, D.; Gygi, F.; Galli, G. First principles simulations of the infrared spectrum of liquid water using hybrid density functionals. *J. Chem. Theory Comput.* **2011**, *7*, 1443–1449.

102. Morrone, J.; Car, R. Nuclear quantum effects in water. *Phys. Rev. Lett.* **2008**, *101*, 017801.

103. Santra, B.; Michaelides, A.; Fuchs, M.; Tkatchenko, A.; Filippi, C.; Scheffler, M. On the accuracy of density-functional theory exchange-correlation functionals for H bonds in small water clusters. II. The water hexamer and van der Waals interactions. *J. Chem. Phys.* **2008**, *129*, 194111.

104. Gillan, M.J.; Manby, F.R.; Towler, M.D.; Alfè, D. Assessing the accuracy of quantum Monte Carlo and density functional theory for energetics of small water clusters. *J. Chem. Phys.* **2012**, *136*, 244105.

105. Alfè, D.; Bartok, A.P.; Csanyi, G.; Gillan, M.J. Communication: Energy benchmarking with quantum Monte Carlo for water nano-droplets and bulk liquid water. *J. Chem. Phys.* **2013**, *138*, 221102.

106. Trail, J.R.; Needs, R.J. Smooth relativistic Hartree-Fock pseudopotentials for H to Ba and Lu to Hg. *J. Chem. Phys.* **2005**, *122*, 174109.

107. Trail, J.R.; Needs, R.J. Norm-conserving Hartree-Fock pseudopotentials and their asymptotic behavior. *J. Chem. Phys.* **2005**, *122*, 14112.

108. Umrigar, C.J.; Toulouse, J.; Filippi, C.; Sorella, S.; Hennig, R.G. Alleviation of the fermion-sign problem by optimization of many-body wave functions. *Phys. Rev. Lett.* **2007**, *98*, 110201.

109. Casula, M. Beyond the locality approximation in the standard diffusion Monte Carlo method. *Phys. Rev. B* **2006**, *74*, 161102.

110. Fraser, L.; Foulkes, W.; Rajagopal, G.; Needs, R.; Kenny, S.; Williamson, A. Finite-size effects and Coulomb interactions in quantum Monte Carlo calculations for homogeneous systems with periodic boundary conditions. *Phys. Rev. B* **1996**, *53*, 1814–1832.

111. Kresse, G.; Hafner, J. Ab initio molecular-dynamics simulation of the liquid-metal–amorphous-semiconductor transition in germanium. *Phys. Rev. B* **1994**, *49*, 14251–14269.

112. Kresse, G.; Hafner, J. Ab initio molecular dynamics for liquid metals. *Phys. Rev. B* **1993**, *47*, 558–561.

113. Kresse, G.; Furthmüller, J. Efficiency of ab-initio total energy calculations for metals and semiconductors using a plane-wave basis set. *Comput. Mater. Sci.* **1996**, *6*, 15–50.

114. Blöchl, P.E. Projector augmented-wave method. *Phys. Rev. B* **1994**, *50*, 17953–17979.

115. Kresse, G.; Joubert, D. From ultrasoft pseudopotentials to the projector augmented-wave method. *Phys. Rev. B* **1999**, *59*, 1758–1775.

116. Mahoney, M.W.; Jorgensen, W.L. A five-site model for liquid water and the reproduction of the density anomaly by rigid, nonpolarizable potential functions. *J. Chem. Phys.* **2000**, *112*, 8910.

117. McMahon, J.M.; Morales, M.A.; Kolb, B.; Thonhauser, T. Competing nuclear quantum effects and van der Waals interactions in water. *J. Phys. Chem. Letters* **2013**, submitted.

118. Grimme, S. Semiempirical GGA-type density functional constructed with a long-range dispersion correction. *J. Comput. Chem.* **2006**, *27*, 1787–1799.

119. Kim, K.; Jordan, K.D. Comparison of density functional and MP2 calculations on the water monomer and dimer. *J. Phys. Chem.* **1994**, *98*, 10089–10094.

120. Stephens, P.J.; Devlin, F.J.; Chabalowski, C.F.; Frisch, M.J. Ab Initio calculation of vibrational absorption and circular dichroism spectra using density functional force fields. *J. Phys. Chem.* **1994**, *98*, 11623–11627.

121. Klimeš, J.; Bowler, D.R.; Michaelides, A. Chemical accuracy for the van der Waals density functional. *J. Phys. Condens. Matter* **2010**, *22*, 022201.

122. Klimeš, J.; Bowler, D.R.; Michaelides, A. Van der Waals density functionals applied to solids. *Phys. Rev. B* **2011**, *83*, 195131.

123. D'Adamo, G.; Pelissetto, A.; Pierleoni, C. Predicting the thermodynamics by using state-dependent interactions. *J. Chem. Phys.* **2013**, *138*, 234107.

Reprinted from *Entropy*. Cite as: Bou-Rabee, N. Time Integrators for Molecular Dynamics. *Entropy* **2014**, *16*, 138–162.

Article

Time Integrators for Molecular Dynamics

Nawaf Bou-Rabee

Department of Mathematical Sciences, Rutgers University—Camden, 311 N 5th Street, Camden, NJ 08102, USA; E-Mail: nawaf.bourabee@rutgers.edu; Tel.: +1-856-225-6093; Fax: +1-856-225-6602

Received: 19 September 2013; in revised form: 20 November 2013 / Accepted: 4 December 2013 / Published: 27 December 2013

Abstract: This paper invites the reader to learn more about time integrators for Molecular Dynamics simulation through a simple MATLAB implementation.   An overview of methods is provided from an algorithmic viewpoint that emphasizes long-time stability and finite-time dynamic accuracy.   The given software simulates Langevin dynamics using an explicit, second-order (weakly) accurate integrator that exactly reproduces the Boltzmann-Gibbs density. This latter feature comes from adding a Metropolis acceptance-rejection step to the integrator. The paper discusses in detail the properties of the integrator. Since these properties do not rely on a specific form of a heat or pressure bath model, the given algorithm can be used to simulate other bath models including, e.g., the widely used v-rescale thermostat.

Keywords: explicit integrators; Metropolis algorithm; ergodicity; weak accuracy

Classification: MSC 82C80 (Primary); 82C31, 65C30, 65C05 (Secondary)

1. Introduction

Molecular Dynamics (MD) simulation refers to the time integration of Hamilton's equations often coupled to a heat or pressure bath [1–5]. From its early use in computing equilibrium dynamics of homogeneous molecular systems [6–13] and pico- to nano-scale protein dynamics [14–23], the method has evolved into a general purpose tool for simulating statistical properties of heterogeneous molecular systems [24]. Accessible time horizons have increased remarkably: the time line in Figure 1 attempts to capture this nearly billion-fold improvement in capability over the last forty or so years. To put this speedup in perspective, though, computing power has increased by about eight powers of ten over this time period as predicted by Moore's law.

To be clear, the selection of applications and methods shown in Figure 1 is not comprehensive and heavily biased towards the specific ideas and methods that inform this paper. The applications highlighted are simulations of liquid argon [6], water [11], protein dynamics without solvent [14,15] and biopolymer dynamics with solvent [25–31]. The methods include the following "upgrades" to MD simulation: Verlet integrator and neighbor lists [7], cell linked list [32], the SHAKE integrator for constraints [33], stochastic heat baths via Langevin dynamics [34,35], a library of empirical potentials [36], a deterministic heat bath via Nosé-Hoover dynamics [37,38], the fast multipole method [39], multiple time steps [40], splitting methods for Langevin dynamics [41–43], quasi-symplectic integrators [44,45], (fast) combined neighbor and cell lists [46], the v-rescale thermostat [47] and the stochastic Nosé-Hoover Langevin thermostat [48–50].

Near future applications of MD simulation include micro- to milli-scale simulations of biomolecular processes, like protein folding, ligand binding, membrane transport and biopolymer conformational changes [51–53]. In addition, atomistic MD simulations are used more sparingly in multiscale models [54–58] and rare event simulation, such as the finite temperature string method and milestoning [59–62]. Given this continuous development and generalization of MD, it is not a stretch to suppose that MD will play a transformative role in medicine, technology and education in the twenty-first century.

Figure 1. A time line of selected developments in MD simulation.

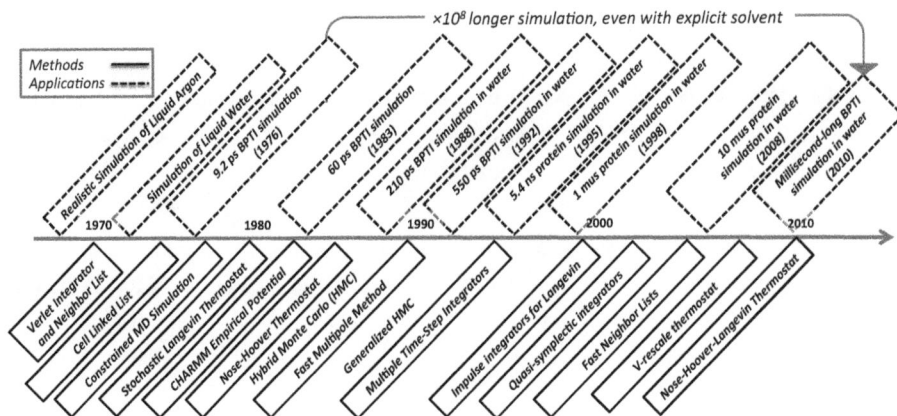

In its standard form, the method inputs a random initial condition, physical and numerical parameters and outputs a long discrete path of the molecular system. Statistical quantities, like velocity correlation or mean radius of gyration, are usually computed online, *i.e.*, as points along this trajectory are produced. MD simulation is built atop a cheap forward Euler-like integrator that requires only a single interactomic force field evaluation per step. Even though MD seems straightforward, software implementations of MD are typically optimized for performance [36,63,64], and as a side effect, make it cumbersome for non-experts to learn and modify.

Also, besides this issue, due to the interplay between stochastic Brownian and molecular forces, infinitely long trajectories of existing MD integrators do not have the right distribution. What happens is that the Brownian force can cause the integrator to enter regions where its approximation to the molecular force is inaccurate and possibly destabilizing. In the latter case, the approximation spends a disproportionate amount of time at higher energies, and thus, the invariant measure of the approximation, if it even exists, is not correct. This phenomenon is a well-known shortcoming of explicit integrators for nonlinear diffusions [65–69].

Recently, a probabilistic approach was proposed to solve this problem, which questions the notion that Monte Carlo methods and MD have different aims: the former strictly samples probability distributions, and the latter estimates dynamics. The basic idea is to combine a standard MD integrator with a Metropolis-Hastings algorithm targeted to the Boltzmann-Gibbs distribution [70–72]. Because the scheme is a Monte Carlo method, it exactly preserves the desired distribution [71,72]. This property implies numerical stability over long-time simulations. However, the price to be paid for this stability is a loss of accuracy whenever a move is rejected and some overhead in evaluating the Metropolis acceptance-rejection step. Still, a Metropolized integrator is dynamically accurate on finite-time intervals [72,73], and so, even though a Metropolized integrator involves a Monte Carlo step, its aim and philosophy are very different from Monte Carlo methods, whose only goal is to sample a target distribution with no concern for the dynamics [71,74–82]. In principle, this approach offers a simple alternative to costly implicit integrators, but are Metropolized integrators ready for daily use in MD simulation? The answer to this question is unclear, since this approach is new and has not been tested on enough examples.

Motivated by these issues, this paper builds a software system for MD simulation with a Metropolis step built in and applies it to a homogeneous molecular system. The algorithm and its properties are introduced in a step-by-step fashion. In particular, we show that the integrator is second-order weakly accurate on finite-time intervals and converges to the Boltzmann-Gibbs distribution in the long-time limit. The software version of the algorithm is written in the latest version of MATLAB with plenty of comments, variables that are descriptively named and operations that can be easily translated into mathematical expressions [83]. Since MATLAB is widely available, this design ensures that the software will be easy-to-use and cross-platform. The following MATLAB-specific file formats will be used.

(F1) MATLAB script and function files are written in the MATLAB language and can be run from the MATLAB command line without ever compiling them.

(F2) MATLAB executable (MEX) files are written in the "C" language and compiled using the MATLAB `mex` function. The resulting executable is comparable in efficiency to a "C" code and can be called directly from the MATLAB command line. We will use MEX-files for performance-critical routines [84].

(F3) MATLAB binary (MAT) files will be used to store simulation data.

The paper is organized as follows. We begin with an overview of integrators that have been proposed in MD simulation in Section 2. We explain how to Metropolize each of these schemes to make them long-time stable in Section 3, and as an application, we use a Metropolized scheme to generate a long trajectory of a Lennard-Jones fluid in Section 4. Generalizations of corrected MD integrators to other molecular models are discussed in Section 5. The paper closes by discussing some potential pitfalls in high dimension and tricks to get the integrator to scale well in Section 6.

2. Algorithmic Introduction to Time Integrators for MD Simulation

For pedagogical reasons, we will start with Langevin dynamics of a system of N molecules. Then, we show in Section 5 how to simulate more general models of molecular systems. Denote by $m_j > 0$ and q_j the mass and position of the j-th molecule, respectively. The governing Langevin equation is given by:

$$\begin{cases} \frac{dq_j}{dt}(t) = m_j^{-1}p_j(t) \, , \\ dp_j(t) = -\frac{\partial U}{\partial q_j}(q(t))dt - \gamma p_j(t)dt + \sqrt{2kT\gamma m_j}dw_j \, , \end{cases} \quad j = 1, \cdots, N \qquad (1)$$

where $q = (q_1, \cdots, q_N)$ and $p = (p_1, \cdots, p_N)$ denote the positions and momenta of the particles, kT is the temperature factor, and $\{w_j\}_{j=1}^N$ are N-independent Brownian motions. The last two terms in the second equation in (1) represent the effect of a heat bath with parameter γ. In Langevin dynamics, positions are differentiable, and due to the irregularity of the Brownian force, momenta are just continuous, but not differentiable. This difference in regularity explains why the first equation in (1) is written as an ordinary differential equation (ODE) and the second equation is written as a stochastic differential equation (SDE).

The bath-free dynamics is a Hamiltonian system with the following Hamiltonian energy function:

$$H(q,p) = \sum_{j=1}^N \frac{1}{2m_j}|p_j|^2 + U(q) \qquad (2)$$

Since the masses are constant, this Hamiltonian nicely separates into a kinetic and potential energy that are purely functions of p and q, respectively. The stationary probability density of the solution to Equation (1) is the Boltzmann-Gibbs density given by:

$$\nu(q,p) = Z^{-1}\exp\left(-\frac{1}{kT}H(q,p)\right) \, , \quad Z = \int \exp\left(-\frac{1}{kT}H(q,p)\right)dqdp \qquad (3)$$

Let h be a given time step size and $m = \text{diag}(m_1, \cdots, m_N)$. Let (Q_0, P_0) denote the position and momentum of the molecular system at time $t > 0$. The simplest approximation to Equation (1) is a forward Euler discretization or Euler-Maruyama scheme [85] that computes an updated position and momentum (Q_1, P_1) at $t + h$ using:

$$\begin{aligned} Q_1 &= Q_0 + hm^{-1}P_0 \\ P_1 &= P_0 - h\nabla U(Q_0) - h\gamma P_0 + \sqrt{h}\sqrt{2kT\gamma}m^{1/2}\xi \end{aligned} \qquad \text{(forward Euler)}$$

Here, $\boldsymbol{\xi} \in \mathbb{R}^n$ denotes a Gaussian random vector with mean zero and covariance $\mathbb{E}(\boldsymbol{\xi}_i \boldsymbol{\xi}_j) = \delta_{ij}$. The problem with this approximation is that the forward Euler method is known to diverge in finite-time when the derivatives of the potential are unbounded, which is the norm in MD simulation. The precise statement and proof of divergence in a general setting can be found in [86]. By far the most computationally intensive part of the time-stepping algorithm is the evaluation of the potential force. Thus, we will restrict our discussion to schemes that, like Euler, only require a single force field evaluation per step.

An improvement to the forward Euler method is the following two-step scheme:

$$\boldsymbol{Q}_2 = (1 + e^{-\gamma h})\boldsymbol{Q}_1 - e^{-\gamma h}\boldsymbol{Q}_0 + \frac{1 - e^{-\gamma h}}{\gamma} \boldsymbol{m}^{-1}\left(-h\nabla U(\boldsymbol{Q}_1) + \sqrt{h}\sqrt{2kT\gamma}\boldsymbol{m}^{1/2}\boldsymbol{\xi}\right) \quad \text{(BBK)}$$

In the limit, $\gamma \to 0$, this scheme reduces to the well-known Verlet integrator for MD simulation [7]. Just like Verlet, this integrator defines a map on pairs of molecular system configurations. Substituting the approximation, $e^{-\gamma h} \approx (1 - \gamma h/2)/(1 + \gamma h/2)$, into the above yields the Brünger-Brooks-Karplus (BBK) scheme, as appearing in [35]. Like the forward Euler method, this method is explicit and only requires one new force evaluation per step.

Second-order accurate schemes that generalize the Velocity Verlet integrator to Langevin dynamics were proposed in a sequence of papers [42–44,87,88]. Here, we mention two of these schemes that are both Strang splittings of Equation (1). The first was proposed by Ricci and Ciccotti [42] and consists of the following sub-steps:

$$\underbrace{\begin{pmatrix} \dot{\boldsymbol{q}}(t) = \boldsymbol{m}^{-1}\boldsymbol{p}(t) \\ d\boldsymbol{p}(t) = 0 \end{pmatrix}}_{\text{exactly evolve by } 1/2 \text{ a step}} \circ \underbrace{\begin{pmatrix} \dot{\boldsymbol{q}}(t) = 0 \\ d\boldsymbol{p}(t) = -\nabla U(\boldsymbol{q}(t))dt - \gamma \boldsymbol{p}(t)dt + \sqrt{2kT\gamma}\boldsymbol{m}^{1/2}d\boldsymbol{W} \end{pmatrix}}_{\text{exactly evolve by a step}} \circ \underbrace{\begin{pmatrix} \dot{\boldsymbol{q}}(t) = \boldsymbol{m}^{-1}\boldsymbol{p}(t) \\ d\boldsymbol{p}(t) = 0 \end{pmatrix}}_{\text{exactly evolve by } 1/2 \text{ a step}}$$

Each step in this decomposition can be exactly solved. Clearly, the half-steps are easy to solve, since momentum is constant over each of these half-steps. The SDE appearing in the inner step can also be exactly solved, since it is linear in momentum (see Chapter 5 in [89]). This splitting is quite natural, since it treats the heat bath forces in the same way as the potential forces.

A related, but different, splitting method was proposed by Bussi and Parinello in [43] and is given by:

$$\underbrace{\begin{pmatrix} \dot{\boldsymbol{q}}(t) = 0 \\ d\boldsymbol{p}(t) = -\gamma \boldsymbol{p}(t)dt + \sqrt{2kT\gamma}\boldsymbol{m}^{1/2}d\boldsymbol{W} \end{pmatrix}}_{\text{exactly evolve by } 1/2 \text{ a step}} \circ \underbrace{\begin{pmatrix} \dot{\boldsymbol{q}}(t) = \boldsymbol{m}^{-1}\boldsymbol{p}(t) \\ \dot{\boldsymbol{p}}(t) = -\nabla U(\boldsymbol{q}(t)) \end{pmatrix}}_{\substack{\text{approximately evolve} \\ \text{using a step of Verlet}}} \circ \underbrace{\begin{pmatrix} \dot{\boldsymbol{q}}(t) = 0 \\ d\boldsymbol{p}(t) = -\gamma \boldsymbol{p}(t)dt + \sqrt{2kT\gamma}\boldsymbol{m}^{1/2}d\boldsymbol{W} \end{pmatrix}}_{\text{exactly evolve by } 1/2 \text{ a step}}$$

Notice that this decomposition splits the Langevin dynamics into its Hamiltonian and heat bath parts, which makes it easy to analyze the structural properties of the scheme. A Velocity Verlet integrator is used to approximate the Hamiltonian dynamics. This approximation exactly preserves phase space volume and preserves energy to third-order accuracy per step. Moreover, the solution to the SDE appearing in the half-steps exactly preserves the Boltzmann-Gibbs density.

Since the Velocity Verlet integrator does not exactly preserve energy, the composition above does not exactly preserve the stationary distribution with density in Equation (3). In [90], it was shown that if the derivatives of the potential are all bounded, the Bussi and Parinello integrator possesses an invariant measure that is $\mathcal{O}(h^2)$ close to the Boltzmann-Gibbs distribution. In this same context, the leading order error term in the integrator's approximation to the invariant measure was explicitly determined [91]. Technically speaking, however, these results do not directly apply to MD simulation, since real MD simulation involves potentials whose derivatives are unbounded, e.g., Lennard-Jones forces. As a consequence of this irregularity in the force fields and discretization error, explicit schemes, like this one, may either not detect features of the potential energy properly, which leads to unnoticed, but large errors in dynamic quantities such as the mean first passage time, or may mishandle soft- or hard-core potentials, which leads to numerical instabilities; see the numerical examples in [92]. These numerical artifacts motivate adding a Metropolis accept/refusal sub-step to the integrator. In the next section, we show how to Metropolize all of the MD integrators presented in this section. In Section 5, we explain how to generalize the Metropolis-corrected Bussi and Parinello algorithm to a larger class of diffusion processes.

3. Metropolis-Corrected MD Integrators

Here, we show how to add a Metropolis acceptance-rejection step to a BBK-type scheme and the Bussi and Parinello splitting scheme and then precisely state the properties of these integrators. We start with a detailed description of each algorithm. Both algorithms require evaluating the acceptance probability given by the usual Metropolis ratio:

$$\alpha(\boldsymbol{q},\boldsymbol{p},\boldsymbol{Q},\boldsymbol{P}) = \min\left(1,\exp\left(-\frac{1}{kT}(H(\boldsymbol{Q},\boldsymbol{P})-H(\boldsymbol{q},\boldsymbol{p}))\right)\right) \tag{4}$$

The procedure to Metropolize the Ricci and Ciccotti scheme can be found in Section 2 of [70].

Algorithm 3.1 (First-order BBK-type integrator). Given the current state $(\boldsymbol{Q}_0,\boldsymbol{P}_0)$ at time t, the algorithm proposes a new state $(\boldsymbol{Q}_1^\star,\boldsymbol{P}_1^\star)$ at time $t+h$ for some time step $h>0$ via:

$$\begin{pmatrix}\boldsymbol{Q}_1^\star \\ \boldsymbol{P}_1^\star\end{pmatrix} = \begin{pmatrix}\boldsymbol{Q}_0+\boldsymbol{m}^{-1}\left(h\boldsymbol{P}_0-\frac{h^2}{2}\nabla U(\boldsymbol{Q}_0)\right) \\ \boldsymbol{P}_0-\frac{h}{2}\left(\nabla U(\boldsymbol{Q}_0)+\nabla U(\boldsymbol{Q}_1^\star)\right)\end{pmatrix} \tag{Step 1}$$

This "proposal move" $(\boldsymbol{Q}_1^\star,\boldsymbol{P}_1^\star)$ is then accepted or rejected:

$$\begin{pmatrix}\tilde{\boldsymbol{Q}}_1 \\ \tilde{\boldsymbol{P}}_1\end{pmatrix} = x\begin{pmatrix}\boldsymbol{Q}_1^\star \\ \boldsymbol{P}_1^\star\end{pmatrix} + (1-x)\begin{pmatrix}\boldsymbol{Q}_0 \\ -\boldsymbol{P}_0\end{pmatrix} \tag{Step 2}$$

where x is a Bernoulli random variable with parameter $\alpha(\boldsymbol{Q}_0,\boldsymbol{P}_0,\boldsymbol{Q}_1^\star,\boldsymbol{P}_1^\star)$ given by Equation (4). The actual update of the system is taken to be:

$$\begin{pmatrix}\boldsymbol{Q}_1 \\ \boldsymbol{P}_1\end{pmatrix} = \begin{pmatrix}\tilde{\boldsymbol{Q}}_1 \\ \exp(-\gamma h)\tilde{\boldsymbol{P}}_1+\sqrt{kT}\sqrt{1-\exp(-2\gamma h)}\boldsymbol{m}^{1/2}\boldsymbol{\xi}\end{pmatrix} \tag{Step 3}$$

Here, $\boldsymbol{\xi}\in\mathbb{R}^n$ denotes a Gaussian random vector with mean zero and covariance $\mathbb{E}(\boldsymbol{\xi}_i\boldsymbol{\xi}_j)=kT\delta_{ij}$.

The momenta of the molecules gets reversed if a move is rejected in Step 2 of Algorithm 3.1. This momentum flip is necessary for the algorithm to preserve the correct stationary distribution [70,71], but results in an $O(1)$ error in dynamics. High acceptance rates are therefore needed to ensure that the time lag between successive rejections is frequently long enough for the approximation to capture the desired dynamics. Since the acceptance rate in Equation (4) is related to how well the Verlet integrator in (Step 1) preserves energy after a single step, this rejection rate is $O(h^3)$. Thus, in practice, we find that the time step required to obtain a sufficiently high acceptance rate is often automatically fulfilled by a time step that sufficiently resolves the desired dynamics. Each step of this algorithm requires: evaluating the atomic force field once in the third equation of (Step 1), generating a Bernoulli random variable with parameter α in (Step 2) and generating an n-dimensional Gaussian vector in (Step 3). We stress that (Step 2) in Algorithm 3.1 is all that is needed to get MD integrators to exactly preserve the Boltzmann-Gibbs density in Equation (3).

Next, we show how to Metropolize the Bussi and Parinello splitting integrator.

Algorithm 3.2 (Second-order Bussi and Parinello integrator). Let $\xi, \eta \in \mathbb{R}^n$ be two independent Gaussian random vectors with mean zero and covariance $\mathbb{E}(\xi_i \xi_j) = \mathbb{E}(\eta_i \eta_j) = \delta_{ij}$. Given a time step size h and the current state (Q_0, P_0) at time t, the algorithm takes a half-step of the heat bath dynamics:

$$\begin{pmatrix} \tilde{Q}_0 \\ \tilde{P}_0 \end{pmatrix} = \begin{pmatrix} Q_0 \\ \exp(-\gamma h/2) P_0 + \sqrt{kT}\sqrt{1 - \exp(-\gamma h)} m^{1/2} \xi \end{pmatrix} \qquad \text{(Step 1)}$$

Followed by a full step of Verlet to compute a proposal move $(\tilde{Q}_1^\star, \tilde{P}_1^\star)$:

$$\begin{pmatrix} \tilde{Q}_1^\star \\ \tilde{P}_1^\star \end{pmatrix} = \begin{pmatrix} \tilde{Q}_0 + m^{-1}\left(h\tilde{P}_0 - \frac{h^2}{2}\nabla U(\tilde{Q}_0)\right) \\ P_0 - \frac{h}{2}\left(\nabla U(\tilde{Q}_0) + \nabla U(\tilde{Q}_1^\star)\right) \end{pmatrix} \qquad \text{(Step 2)}$$

This proposal move $(\tilde{Q}_1^\star, \tilde{P}_1^\star)$ is then accepted or rejected:

$$\begin{pmatrix} \tilde{Q}_1 \\ \tilde{P}_1 \end{pmatrix} = x \begin{pmatrix} \tilde{Q}_1^\star \\ \tilde{P}_1^\star \end{pmatrix} + (1-x) \begin{pmatrix} \tilde{Q}_0 \\ -\tilde{P}_0 \end{pmatrix} \qquad \text{(Step 3)}$$

where x is a Bernoulli random variable with parameter $\alpha(\tilde{Q}_0, \tilde{P}_0, \tilde{Q}_1^\star, \tilde{P}_1^\star)$ given by Equation (4). The actual update of the system at time $t + h$ is taken to be:

$$\begin{pmatrix} Q_1 \\ P_1 \end{pmatrix} = \begin{pmatrix} \tilde{Q}_1 \\ \exp(-\gamma h/2)\tilde{P}_1 + \sqrt{kT}\sqrt{1 - \exp(-\gamma h)} m^{1/2}\eta \end{pmatrix} \qquad \text{(Step 4)}$$

This algorithm requires generating two independent n-dimensional Gaussian vectors per step. Thus, it is more costly than Algorithm 3.1. However, the advantage of doing this is that the resulting Metropolis corrected algorithm is second-order weakly accurate, as the following Proposition states.

Proposition 3.3. *Let* (Q_n, P_n) *represent the numerical approximation produced by Algorithm 3.2 at time* nh *with the same initial condition as the true solution:* $(Q_0, P_0) = (q(0), p(0))$. *For every time interval* $T > 0$ *and for suitable observables* $f(q, p)$, *there exists a* $C(T) > 0$, *such that:*

$$|\mathbb{E}f(q(\lfloor t/h \rfloor h), p(\lfloor t/h \rfloor h)) - \mathbb{E}f(Q_{\lfloor t/h \rfloor}, P_{\lfloor t/h \rfloor})| \leq C(T)h^2 \tag{5}$$

for all $t < T$.

This accuracy concept is sufficient for computing means and correlation functions at finite-time and equilibrium correlations. Figure 2 verifies this Proposition by checking the weak accuracy of Algorithms 3.1 and 3.2 on a harmonic oscillator test problem.

Figure 2. Langevin dynamics of a harmonic oscillator.

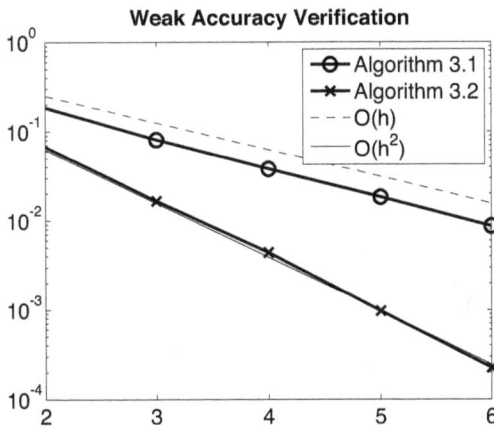

To be specific, Figure 2 plots the weak accuracy of the Metropolis-corrected MD integrators with respect to the true solution of the Langevin dynamics of a harmonic oscillator: $\dot{q}(t) = p(t)$, $dp(t) = -q(t) - p(t) + \sqrt{2}dw(t)$, with initial condition $q(0) = 1.0$, $p(0) = 0$. The time steps tested are $h = 2^{-n}$, where n is given on the x-axis. The quantity monitored for the error is the estimate of $\mathbb{E}(q(1)^2 + p(1)^2) = 1.699445410$ computed analytically. The dashed and solid curves are the graphs of $2^{-n}(= h)$ and $2^{-2n}(= h^2)$ *versus* n, respectively.

Proof. The desired single-step error estimate can be obtained from an application of the triangle inequality:

$$|\mathbb{E}f(q(h), p(h)) - \mathbb{E}f(Q_1, P_1)| \leq |\mathbb{E}f(q(h), p(h)) - \mathbb{E}f(\hat{Q}_1, \hat{P}_1)| + |\mathbb{E}f(\hat{Q}_1, \hat{P}_1) - \mathbb{E}f(Q_1, P_1)| \tag{6}$$

104

where $(\hat{Q}_1, \hat{P}_1)$ denotes one step of the uncorrected Bussi and Parinello scheme with $(\hat{Q}_0, \hat{P}_0) = (q(0), p(0))$. The first term in the upper bound in Equation (6) is $O(h^3)$, since the unadjusted scheme is a Strang splitting of Equation (1). To bound the second term in Equation (6), note that:

$$\mathbb{E}f(Q_1, P_1) - \mathbb{E}f(\hat{Q}_1, \hat{P}_1) = \mathbb{E}\left\{ \left(\bar{f}(\tilde{Q}_1^\star, \tilde{P}_1^\star) - \bar{f}(\tilde{Q}_0, -\tilde{P}_0) \right) \left(\alpha(\tilde{Q}_0, \tilde{P}_0, \tilde{Q}_1^\star, \tilde{P}_1^\star) - 1 \right) \right\}$$

where we have introduced the auxilary function:

$$\bar{f}(q, p) = \mathbb{E}f(q, \exp(-\gamma h/2)p + \sqrt{kT}\sqrt{1 - \exp(-\gamma h)}m^{1/2}\eta)$$

Since the rejection rate is $\mathcal{O}(h^3)$, it follows from the above expression that the second term in the upper bound of Equation (6) is also $O(h^3)$. Standard results in numerical analysis for SDEs then imply that the algorithm converges weakly on finite-time intervals with global order two; see, for instance, [93] (Chapter 2.2). □

For completeness sake, we also provide a statement that both algorithms are ergodic.

Proposition 3.4. *Let (Q_n, P_n) be the numerical approximation produced by Algorithms 3.1 or 3.2 at time nh. Then, for suitable observables $f(q, p)$:*

$$\lim_{T \to \infty} \frac{1}{T} \int_0^T f(Q_{\lfloor t/h \rfloor}, P_{\lfloor t/h \rfloor})dt \to \int_{\mathbb{R}^{2n}} f(q, p)\nu(q, p)dqdp \tag{7}$$

Here, $\nu(q, p)$ denotes the Boltzmann-Gibbs density defined in Equation (3).

A proof of this Proposition can be found in [72].

4. Application to Lennard-Jones Fluid

Listing 1 translates Algorithm 3.2 into the MATLAB language. Intrinsically defined MATLAB functions appear in boldface. The algorithm uses MATLAB's built in random number generators to carry out Step 1, Step 3 and Step 4. In particular, the Bernoulli random variable, x, in Step 3 is generated in Line 20, and the Gaussian vectors in Step 1 and Step 4 are generated on Line 9 and Line 29, respectively. In addition to updating the positions and momenta of the system, the program also stores the previous value of the potential energy and force, so that the force and potential energy is evaluated in Line 15 just once per simulation step. This evaluation calls a MEX function, which inputs the current position of the molecular system and outputs the force field and potential energy at that position. We use a MEX function, because the atomistic force field evaluation cannot be easily vectorized and is, by far, the most computationally demanding step in MD. The `PreProcessing` script file called in Line 2 defines the physical and numerical parameters, sets the initial condition and allocates space for storing simulation data. Sample averages are updated as new points on the trajectory are produced in the `UpdateSampleAverages` script file invoked in Line 35. Finally, the outputs produced by the algorithm are handled by the `PostProcessing` script file in Line 39.

Let us consider a concrete example: a Lennard-Jones fluid that consists of N identical atoms [1–3]. The configuration space of this system is a fixed cubic box with periodic boundary conditions. The distance between the i-th and j-th particle is defined according to the minimum image convention, which states that the distance between q_i and q_j in a cubic box of length ℓ is:

$$d_{MD}(q_i, q_j) \stackrel{\text{def}}{=} |(q_i - q_j) - \ell \lfloor (q_i - q_j)/\ell \rceil| \tag{8}$$

where $\lfloor \cdot \rceil$ is the nearest integer function. In terms of this distance, the total potential energy is a sum over all pairs:

$$U(q) = \sum_{i=1}^{n-1} \sum_{j=i+1}^{n} U_{LJ}(d_{MD}(q_i, q_j)) \tag{9}$$

where $U_{LJ}(r)$ is the following truncated Lennard-Jones potential function:

$$U_{LJ}(r) = \begin{cases} f(r) - f(r_c), & r < r_c \\ 0, & \text{otherwise} \end{cases} \tag{10}$$

Listing 1. Metropolized MD Integrator: `MDintegrator.m`

```
1
2 PreProcessing;
3
4 for i = 1:Ns
5
6     %--- Step 1 --- Heat Bath Step
7
8     tQ0=Q0;
9     tP0=f1*P0+f2*randn(3*Nm,1);
10
11    %--- Step 2 --- Velocity Verlet Proposal
12
13    Ppt5=tP0+0.5*h*F0;
14    tQ1star=tQ0+h*Ppt5;
15    [tF1star,tU1star]=ForceFieldmex(tQ1star,Nm,rcut2,ell);
16    tP1star=Ppt5+0.5*h*tF1star;
17
18    %--- Step 3 --- Accept or Refuse Step
19
20    x=(rand<exp(-(0.5*tP1star'*tP1star-0.5*tP0'*tP0+tU1star-U0)/kT));
21
22    tP1=x*P1star-(1-x)*P0;
23    tQ1=x*Q1star+(1-x)*Q0;
24    F1=x*tF1star+(1-x)*F0;  U1=x*tU1star+(1-x)*U0;
25
26    %--- Step 4 --- Heat Bath Step
```

```
27
28      Q1=tQ1;
29      P1=f1*tP1+f2*randn(3*Nm,1);
30
31      %--- iterate
32
33      Q0=Q1; P0=P1; F0=F1; U0=U1;
34
35      UpdateSampleAverages;
36
37  end
38
39  PostProcessing;
```

Listing 2. Metropolized MD Integrator: `PreProcessing.m`

```
1   %--- seed random # generator
2
3   rng(123);
4
5   %--- physical parameters
6
7   rho=0.6;                    % density
8   kT=0.5;                     % temperature factor
9   gama=0.1;                   % heat bath parameter
10  Nm=500;                     % # of molecules
11  T=2.0;                      % time span for velocity correlation
12  ell=(Nm/rho)^(1/3);         % length of cubic box
13
14  %--- simulation parameters
15
16  h=0.005;                    % time-step size
17  Ns=1e3;                     % # of steps
18  rcut = 2.0^(1/6);           % cutoff radius
19  rcut2 = rcut*rcut;
20
21  f1=exp(-0.5*gama*h); f2=sqrt((1.0-exp(-gama*h))*kT);
22
23  %--- initial condition
24
25  A=fcclattice(Nm,ell);
```

```
26 Q0=reshape(A, [3*Nm 1]);        % atoms on an fcc lattice
27 P0=zeros(3*Nm,1);               % atoms at rest
28
29 %--- initialize statistics
30
31 NA=ceil(T/h)+1;                 % preallocate space for
32 acf=zeros(NA,1);                % online correlation computation
33 varacf=zeros(NA,1);
34 pivot=zeros(NA,3*Nm);
35 nacf=zeros(NA,1);
36
37 AP=zeros(Ns,1);                 % vector of acceptance probabilities
38
39 [F0,U0]=ForceFieldmex(Q0,Nm,rcut2,ell);    % initial force & energy
```

Here, $f(r) = 4(1/r^{12} - 1/r^6)$ and r_c is the cutoff radius, which is bounded above by the size of the simulation box; and we have used dimensionless units to describe this system, where energy is rescaled by the depth of the Lennard-Jones potential energy and length by the point where the potential energy is zero. The error introduced by the truncation in Equation (10) is proportional to the density of the molecular system and can be made arbitrarily small by selecting the cutoff distance to be sufficiently large. A direct evaluation of the potential force, $\nabla U(\boldsymbol{q})$, scales like $O(N^2)$, and typically dominates the total computational cost. In practice, neighbor/cell lists, also called Verlet lists, are used in order to obtain a force evaluation that scales linearly with system size. Since the system we consider will have just a few hundred atoms, there is, however, little advantage to using these data structures, or using a fast force field evaluation, and thus, ForceFieldmex evaluates the force and energy using a sum over all particle pairs.

Table 1. Simulation parameters.

	Parameter	Description	Value
	ρ	density	$\{0.6, 0.7, 0.8, 0.9, 1.0, 1.1\}$
	kT	temperature factor	0.5
Physical Parameters	γ	heat bath parameter	0.01
	N_m	# of molecules	512
	T	time-span for autocorrelation	2
	h	time step	0.005
Numerical Parameters	N_s	# of simulation steps	10^5
	r_c	Lennard-Jones force cutoff radius	$2^{1/6}$

Listing 2 shows the PreProcessing script, which sets the parameters provided in Table 1 and constructs the initial condition, where the N atoms are assumed to be at rest and on the sites of a face-centered cubic lattice. The command, rng(123), on *Line 3* sets the seed of the random number generator functions, RAND and RANDN. The acceptance rates at every step and the velocity autocorrelation are updated in the UpdateSampleAverages script shown in Listing 3. The mean acceptance rate, which is outputted in the PostProcessing script shown in Listing 4, must be high enough to ensure that the dynamics is accurately represented. To compute the autocorrelation of an observable over a time interval of length T, the value of that observable along the entire trajectory is not needed. In fact, it suffices to use the values of this observable along a piece of trajectory over a moving time-window $[t_i, t_i + T]$, where $t_i = i \times h$. This storage space is allocated in PreProcessing and is updated in UpdateSampleAverages. More precisely, the molecular velocities are stored in the pivot array from $i - N_a$ to i, where i is the index of the current position and $N_a = \lceil T/h \rceil + 1$. Notice that velocity autocorrelations are not computed until after the index, i, exceeds 10^4. This *equilibration time* removes some of the statistical bias that may arise from using a non-random initial condition. Short-time trajectories of this molecular system are plotted in Figure 3 from an initial condition where atoms are placed on the sites of a face-centered cubic lattice and at rest. The trajectory is computed using the numerical and physical parameters indicated in Table 1, with the exception of the number of steps, which is set equal to $N_s = 1000$. Notice that at lower densities particle trajectories are more diffusive and less localized. Using the parameters provided in Table 1, we compute velocity autocorrelations for a range of density values in Figure 4. Since the heat bath parameter is set to a small value, these figures are in qualitative agreement with those obtained by simulating the molecular system with no heat bath as shown in Figure 5.2 of [3].

Listing 3. Metropolized MD Integrator: UpdateSampleAverages.m

```
1   %--- store acceptance probability
2
3   AP(i)=x;
4
5   %--- update correlation function
6
7   if (i>1e4)
8
9       pp=mod(i-1,NA)+1;
10      pivot(pp,:)=P0;
11
12      for j=1:min(i,NA)
13          nacf(j)=nacf(j)+1;
14          mui=acf(j);
15          vari=varacf(j);
16          n_samples=nacf(j);
17          xip1=pivot(mod(pp-j,NA)+1,:)*pivot(pp,:)'/(3.0*Nm);
18          acf(j)=mui+(xip1-mui)/n_samples;
19          varacf(j)=((n_samples-1)*vari+...
```

```
20        (xip1-mui)*(xip1-acf(j)))/n_samples;
21    end
22
23 end
```

Listing 4. Metropolized MD Integrator: `PostProcessing.m`

```
1 %--- output results
2
3 disp(['h=' num2str(h)   ',<AP>=' num2str(mean(AP))]);
4
5 figure(2); clf; hold on; tt=0:h:T;
6 errorbar(tt,acf,1.96*sqrt(varacf)./sqrt(nacf));
7
8 save('VelocityAutocorrelation.mat', 'tt', 'acf', 'varacf');
```

Figure 3. Atomic trajectories in a simulation box.

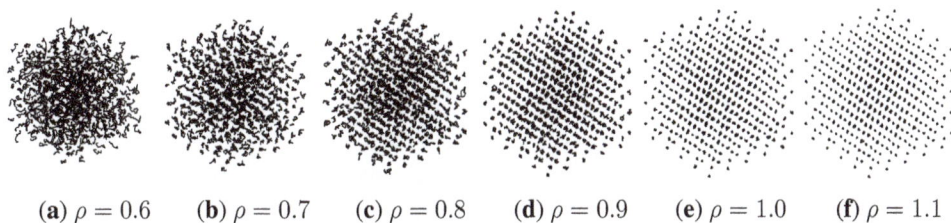

(**a**) $\rho = 0.6$ (**b**) $\rho = 0.7$ (**c**) $\rho = 0.8$ (**d**) $\rho = 0.9$ (**e**) $\rho = 1.0$ (**f**) $\rho = 1.1$

Figure 4. Soft-sphere velocity autocorrelation functions. A reproduction of Figure 5.2 of [3] using Langevin dynamics with heat bath parameter $\gamma = 0.01$. The remaining parameters are set equal to those provided in Table 1. The negative correlations at higher densities are consistent with what has been found in the literature [6,8].

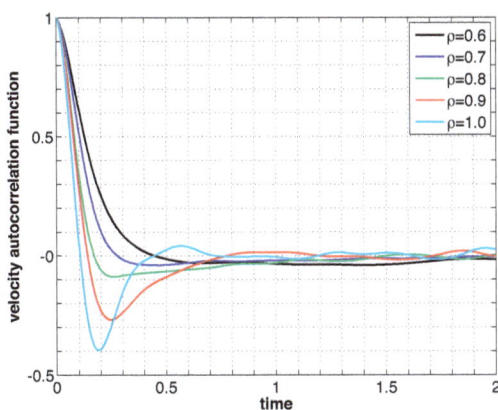

5. General Case

Here, we show how the preceding ideas extend to other molecular systems that obey stochastic differential equations. In the process, we generalize the Metropolized Bussi and Parinello integrator (Algorithm 3.2) to a big class of diffusion processes, including the v-rescale thermostat. We begin with the underlying Hamiltonian dynamics of a molecular system.

5.1. Bath-Free Dynamics

MD is based on Hamilton's equations for a Hamiltonian $H : \mathbb{R}^{2d} \to \mathbb{R}$:

$$\dot{z}(t) = J \nabla H(z(t)) , \quad z(0) \in \mathbb{R}^{2d} \tag{11}$$

where $z(t) = (q(t), p(t))$ is a vector of molecular positions $q(t) \in \mathbb{R}^d$ and momenta $p(t) \in \mathbb{R}^d$ and J is the $2d \times 2d$ skew-symmetric matrix defined as:

$$J = \begin{pmatrix} 0_{d \times d} & I_{d \times d} \\ -I_{d \times d} & 0_{d \times d} \end{pmatrix} \tag{12}$$

The Hamiltonian, $H(z)$, represents the total energy of the molecular system and is typically "separable", meaning that it can be written as:

$$H(z) = K(p) + U(q) , \quad z = (q, p) \tag{13}$$

where $K(p)$ and $U(q)$ are the kinetic and potential energy functions, respectively [94]. In MD, the kinetic energy function is a positive definite quadratic form, and the potential energy function involves "fudge factors" determined from experimental or quantum mechanical studies of pieces of the molecular system of interest [36]. The accuracy of the resulting energy function must be systematically verified by comparing MD simulation data to experimental data [95]. The flow that Equation (11) determines has the following structure:

(S1) volume-preserving (since the vector-field in Equation (11) is divergenceless); and
(S2) energy-preserving (since J is skew-symmetric and constant).

Explicit *symplectic integrators*, like the Verlet scheme, exploit these properties to obtain long-time stable schemes for Hamilton's equations [96,97].

5.2. Governing Stochastic Dynamics

In order to mimic experimental conditions, Equation (11) is often coupled to a bath that puts the system at constant temperature and/or pressure. The standard way to do this is to assume that the system with a bath is governed by a stochastic ordinary differential equation (SDE) of the type:

$$dY(t) = \underbrace{A(Y(t))dt}_{\text{deterministic drift}} + \underbrace{(\operatorname{div} D)(Y(t))dt + \sqrt{2kT} B(Y(t))dW(t)}_{\text{heat bath}} \tag{14}$$

Here, we have introduced the following notation.

$Y(t) \in \mathbb{R}^n$	state of the (extended) system
$A(x) \in \mathbb{R}^n$	deterministic drift vector field
$B(x) \in \mathbb{R}^{n \times n}$	noise-coefficient matrix
$D(x) \in \mathbb{R}^{n \times n}$	diffusion matrix
$W(t) \in \mathbb{R}^n$	n-dimensional Brownian motion
kT	temperature factor

The $n \times n$ *diffusion matrix*, $D(x)$, is defined in terms of the noise coefficient matrix, $B(x)$, as:

$$D(x) \stackrel{\text{def}}{=} kTB(x)B(x)^T , \quad \text{for all } x \in \mathbb{R}^n \tag{15}$$

where $B(x)^T$ denotes the transpose of the real matrix, $B(x)$. The diffusion matrix is symmetric and nonnegative definite. Depending on the particular bath that is used, the dimension, n, of $Y(t)$ in Equation (14) is related to the dimension, $2d$, of $z(t)$ in Equation (11) by the inequality: $n \geq 2d$. For example, in Nosé-Hoover Langevin dynamics, a single bath degree of freedom is added to Equation (11), so that $n = 2d + 1$, while in Langevin dynamics, the effect of the bath is modeled by added friction and Brownian forces that keep $n = 2d$. The Langevin Equation (1) can be put in the form of Equation (14) by letting $x = (q, p)$,

$$A(x) = \begin{pmatrix} m^{-1}p \\ -\nabla U(q) - \gamma p \end{pmatrix} , \quad B = \sqrt{\gamma} \begin{pmatrix} 0 & 0 \\ 0 & m^{1/2} \end{pmatrix} , \quad \text{and} \quad W = (w_1, \cdots, w_N) \tag{16}$$

where $m = \text{diag}(m_1, \cdots, m_N)$.

Equation (14) generates a stochastic process, $Y(t)$, that is a Markov diffusion process. We assume that this diffusion process admits a *stationary* distribution $\mu(dx)$, *i.e.*, a probability distribution preserved by the dynamics [98,99]. We denote by $\nu(x)$ the density of this distribution. Even though the diffusion matrix in Equation (15) is not necessarily positive definite, one can use the Hörmander's condition to prove that the process, $Y(t)$, is an ergodic process with a unique stationary distribution [100,101]. By the ergodic theorem, it then follows that:

$$\frac{1}{T} \int_0^T f(Y(t))dt \to \int_{\mathbb{R}^n} f(x)\nu(x)dx , \quad \text{as } T \to \infty, \quad \text{a.s.} \tag{17}$$

where $f(x)$ is a suitable test function.

The evolution of the probability density of the law of $Y(t)$ at time t, $\rho(t, x)$, satisfies the Fokker-Planck equation:

$$-\frac{\partial \rho}{\partial t} + L\rho = 0 \tag{18}$$

where $\rho(0, \cdot)$ is the density of the initial distribution, $Y(0) \sim \rho(0, \cdot)$, and L is defined as the following second-order partial differential operator:

$$(Lf)(x) \stackrel{\text{def}}{=} \text{div} \left(\text{div}(D(x)f(x)) - A(x)f(x) \right) \tag{19}$$

Since $\mu(dx) = \nu(x)dx$ is a stationary distribution of $Y(t)$, the probability density, $\nu(x)$, is a steady-state solution of Equation (18), *i.e.*, it satisfies:

$$(L\nu)(x) = 0 \tag{20}$$

Define the *probability current* as the vector field:

$$\boldsymbol{j}(\boldsymbol{x}) \stackrel{\text{def}}{=} \operatorname{div}(\boldsymbol{D}(\boldsymbol{x})\nu(\boldsymbol{x})) - \boldsymbol{A}(\boldsymbol{x})\nu(\boldsymbol{x}) \tag{21}$$

The stationarity condition in Equation (20) implies that $\boldsymbol{j}(\boldsymbol{x})$ is divergenceless. In the zero-current case, the diffusion process, $\boldsymbol{Y}(t)$, is *reversible*, and the stationary density $\nu(\boldsymbol{x})$ is called the equilibrium probability density of the diffusion [102].

In this case, the operator, L, is self-adjoint, in the sense that:

$$\langle Lf, g\rangle_\nu = \langle f, Lg\rangle_\nu \qquad \text{for all suitable test functions } f, g \tag{22}$$

where $\langle \cdot, \cdot\rangle_\nu$ denotes an L^2 inner product weighted by the density, $\nu(\boldsymbol{x})$. This property implies that the diffusion is ν-symmetric [103]:

$$\nu(\boldsymbol{x})p_t(\boldsymbol{x}, \boldsymbol{y}) = \nu(\boldsymbol{y})p_t(\boldsymbol{y}, \boldsymbol{x}) \qquad \text{for all } t > 0 \tag{23}$$

where $p_t(\boldsymbol{x}, \boldsymbol{y})$ denotes the transition probability density of $\boldsymbol{Y}(t)$. Indeed, Equation (22) is simply an infinitesimal version of Equation (23), which is referred to as the detailed balance condition. In the self-adjoint case, the drift is uniquely determined by the diffusion matrix and the stationary density $\nu(\boldsymbol{x})$:

$$\boldsymbol{j}(\boldsymbol{x}) = 0 \implies \boldsymbol{A}(\boldsymbol{x}) = \frac{1}{\nu(\boldsymbol{x})} \operatorname{div}(\boldsymbol{D}(\boldsymbol{x})\nu(\boldsymbol{x})) \tag{24}$$

Long-time stable explicit schemes adapted to this structure have been recently developed [92].

5.3. Splitting Approach to MD Simulation

We are now in a position to explain our general approach for deriving a long-time stable scheme for Equation (14). Crucial to our approach is that in MD simulation, we usually have a formula for a function proportional to the stationary density $\nu(\boldsymbol{x})$. Following [90], we can split Equation (14) into:

$$d\boldsymbol{Y} = -\boldsymbol{D}(\boldsymbol{Y})\nabla H_\nu(\boldsymbol{Y})dt + \operatorname{div} \boldsymbol{D}(\boldsymbol{Y})dt + \sqrt{2kT}B(\boldsymbol{Y})dW \tag{25}$$

$$\dot{\boldsymbol{Y}} = \boldsymbol{A}(\boldsymbol{Y}) + \boldsymbol{D}(\boldsymbol{Y})\nabla H_\nu(\boldsymbol{Y}) \tag{26}$$

where we have introduced $H_\nu(\boldsymbol{x}) = -(\log \nu)(\boldsymbol{x})$. An *exact splitting method* preserves $\mu(d\boldsymbol{x})$. It is formed by taking the exact solution (in law) of Equation (25) in composition with the exact flow of Equation (26). The process produced by Equation (25) is self-adjoint with respect to $\nu(\boldsymbol{x})$. Moreover, the stationarity of $\nu(\boldsymbol{x})$ implies that the flow of the ODE (26) preserves it. Since each step is preservative, their composition is, too.

In place of the exact splitting, a Metropolized explicit integrator can be used for Equation (25) [92], and a measure-preserving scheme can be designed to solve the ODE [72,104]. In [92], explicit schemes are introduced for Equation (25) that: (i) sample the exact equilibrium probability density of the SDE when this density exists (*i.e.*, whenever $\nu(\boldsymbol{x})$ is normalizable); (ii) generates a weakly accurate approximation to the solution of Equation (14) at constant kT;

(iii) acquire higher order accuracy in the small noise limit, $kT \to 0$; and (iv) avoid computing the divergence of the diffusion matrix $D(x)$. Compared to the methods in [72], the main novelty of these schemes stems from (iii) and (iv). The resulting explicit splitting method is accurate, since it is an additive splitting of Equation (14); and typically ergodic when the continuous process is ergodic [72].

This type of splitting of Equation (14) is quite natural and has been used before in MD [43,87], dissipative particle dynamics [105,106] and the simulation of inertial particles [107]. Other closely related schemes for Equation (14) include Brünger-Brooks-Karplus (BBK) [35], van Gunsteren and Berendsen (vGB) [108] and the Langevin-Impulse (LI) methods [41] and quasi-symplectic integrators [44]. However, for general MD force fields, none of these explicit integrators are long-time stable. Our framework to stabilize explicit MD integrators is the Metropolis-Hastings algorithm.

5.4. Metropolis-Hastings Algorithm

A Metropolis-Hastings method is a Monte Carlo method for producing samples from a probability distribution, given a formula for a function proportional to its density [74,75]. The algorithm consists of two sub-steps: firstly, a proposal move is generated according to a transition density, $g(x, y)$; and secondly, this proposal move is accepted or rejected with a probability:

$$\alpha(x, y) = 1 \wedge \frac{g(y, x)\nu(y)}{g(x, y)\nu(x)} \tag{27}$$

Standard results on Metropolis-Hastings methods can be used to classify this algorithm as ergodic [100,109,110].

6. Conclusions

This paper provided an algorithmic introduction to time integrators for MD simulation. A quick overview of existing algorithms was given. When the derivatives of the potential are bounded, it is well known that these integrators work just fine: they are convergent on finite-time intervals and possess an invariant measure that is nearby the Boltzmann-Gibbs density. However, in realistic MD simulation, the derivatives of the potential are unbounded. This lack of regularity can cause numerical instabilities or artifacts in explicit integrators. The paper demonstrated how a Metropolis acceptance-rejection step can be added to explicit MD integrators to mitigate some of these problems and, in principle, obtain long-time stable and finite-time accurate schemes. A MATLAB implementation of Metropolis-corrected MD integrators was provided and used to compute the velocity autocorrelation of a sea of Lennard-Jones particles at various densities between the solid and liquid phases. The paper did not provide an in-depth review of the theory of Metropolis integrators, which can be found elsewhere [72,73].

Calculating the force field at every step dominates the overall computational cost of MD simulation. These force fields involve: bonded interactions and non-bonded Lennard-Jones and electrostatic interactions. The calculation of bonded interactions is straightforward to vectorize and scales like $O(N)$. In addition, Lennard-Jones forces rapidly decay with interatomic distance. To a good approximation, every atom interacts only with neighbors within a sufficiently large ball. By using data structures, like neighbor lists and cell linked lists, these interactions can be calculated in $O(N)$ steps, and therefore, the Lennard-Jones interactions can be calculated in $O(N)$ steps [46]. On the other hand, the electrostatic energy between particles decays, like $1/r$, where r denotes an interatomic distance, which leads to long-range interactions between atoms. Unlike Lennard-Jones interaction, this interaction cannot be cutoff without introducing large errors. In this case, one can use sophisticated techniques, like the fast multipole method, to rigorously handle such interactions in $\mathcal{O}(N)$ steps [39,58].

However, the effect of these 'mathematical tricks' for fast calculation of the force field can become muted if the time step requirement for stability or accuracy becomes more severe in high dimension. This can happen in the Metropolis integrator, if the acceptance probability in Step 2 of Algorithm 3.1 or Step 3 of Algorithm 3.2 deteriorates in high dimension. The scaling of Metropolis algorithms has been quantified for the random walk Metropolis, hybrid Monte Carlo and Metropolis-adjusted Langevin algorithm (MALA) [111–115]. Since the acceptance probability is a function of an extensive quantity, the acceptance rate can artificially deteriorate with increasing system size, unless the time step is reduced. Because high acceptance rates are required to maintain dynamic accuracy, the dependence of the time step on system size limits the application of Metropolized schemes to large-scale systems. Fortunately, this scalability issue can often be resolved by using local, rather than global proposal moves, because the change in energy induced by a local move is typically an intensive quantity. For molecular dynamics calculations, this approach was pursued in [73]. Using dynamically consistent local moves (a so-called J-splitting [116]), it was shown that in certain situations, a scalable Metropolis integrator can be designed; however, the extent to which this strategy remedies the issue of high rejection rate in high dimension is not clear at this point and should be tested in applications.

Acknowledgments

The author wishes to acknowledge Eric Vanden-Eijnden for useful comments on an earlier version of this paper. The research that led to this paper was funded by the US National Science Foundation through Division of Mathematical Sciences (DMS) grant # DMS-1212058.

Conflicts of Interest

The authors declare no conflict of interest.

References

1. Allen, M.P.; Tildesley, D.J. *Computer Simulation of Liquids*; Clarendon Press: Oxford, UK, 1987.
2. Frenkel, D.; Smit, B. *Understanding Molecular Simulation: From Algorithms to Applications*, 2nd ed.; Academic Press: Waltham, MA, USA, 2002.

3. Rapaport, D.C. *The Art of Molecular Dynamics Simulation*; Cambridge University Press: Cambridge, UK, 2004.

4. Tuckerman, M. *Statistical Mechanics and Molecular Simulations*; Oxford University Press: Oxford, UK, 2008.

5. Schlick, T. *Molecular Modeling and Simulation: An Interdisciplinary Guide*; Springer: Berlin/Heidelberg, Germany, 2010; Volume 21.

6. Rahman, A. Correlations in the motion of atoms in liquid argon. *Phys. Rev.* **1964**, *136*, A405.

7. Verlet, L. Computer "experiments" on classical fluids. I. Thermodynamical properties of Lennard-Jones molecules. *Phys. Rev.* **1967**, *159*, 98–103.

8. Alder, B.J.; Wainwright, T.E. Velocity autocorrelations for hard spheres. *Phys. Rev. Lett.* **1967**, *18*, 988–990.

9. Alder, B.J.; Gass, D.M.; Wainwright, T.E. Studies in molecular dynamics. VIII. The transport coefficients for a hard-sphere fluid. *J. Chem. Phys.* **1970**, *53*, 3813–3826.

10. Harp, G.D.; Berne, B.J. Time-correlation functions, memory functions, and molecular dynamics. *Phys. Rev. A* **1970**, *2*, 975–996.

11. Rahman, A.; Stillinger, F.H. Molecular dynamics study of liquid water. *J. Chem. Phys.* **1971**, *55*, 3336–3359.

12. Stillinger, F.H.; Rahman, A. Improved simulation of liquid water by molecular dynamics. *J. Chem. Phys.* **1974**, *60*, 1545–1557.

13. Stillinger, F.H. Water revisited. *Science* **1980**, *209*, 451–457.

14. McCammon, A.J.; Gelin, B.R.; Karplus, M. Dynamics of folded proteins. *Nature* **1977**, *267*, 585–590.

15. Van Gunsteren, W.F.; Berendsen, H.J.C. Algorithms for macromolecular dynamics and constraint dynamics. *Mol. Phys.* **1977**, *34*, 1311–1327.

16. McCammon, J.A.; Karplus, M. Simulation of protein dynamics. *Annu. Rev. Phys. Chem.* **1980**, *31*, 29–45.

17. Van Gunsteren, W.F.; Karplus, M. Protein dynamics in solution and in a crystalline environment: A molecular dynamics study. *Biochemistry* **1982**, *21*, 2259–2274.

18. Karplus, M.; McCammon, A.J. Dynamics of proteins: Elements and function. *Annu. Rev. Biochem.* **1983**, *52*, 263–300.

19. Van Gunsteren, W.F.; Berendsen, H.J.C. Computer simulation of molecular dynamics: Methodology, applications, and perspectives in chemistry. *Angew. Chem. Int. Ed. Engl.* **1990**, *29*, 992–1023.

20. Karplus, M.; McCammon, A.J. Molecular dynamics simulations of biomolecules. *Nat. Struct. Mol. Biol.* **2002**, *9*, 646–652.

21. Case, D.A. Molecular dynamics and NMR spin relaxation in proteins. *Acc. Chem. Res.* **2002**, *35*, 325–331.

22. Adcock, A.S.; McCammon, A.J. Molecular dynamics: Survey of methods for simulating the activity of proteins. *Chem. Rev.* **2006**, *106*, 1589–1615.

23. Van Gunsteren, W.F.; Dolenc, J.; Mark, A.E. Molecular simulation as an aid to experimentalists. *Curr. Opin. Struct. Biol.* **2008**, *18*, 149–153.

24. Kapral, R.; Ciccotti, G. Molecular dynamics: An Account of its Evolution. In *Theory and Applications of Computational Chemistry: The First Forty Years*; Elsevier: New York, NY, USA, 2005; pp. 425–441.

25. Levitt, M. Molecular dynamics of native protein: I. Computer simulation of trajectories. *J. Mol. Biol.* **1983**, *168*, 595–617.

26. Levitt, M.; Sharon, R. Accurate simulation of protein dynamics in solution. *Proc. Natl. Acad. Sci. USA* **1988**, *85*, 7557–7561.

27. Daggett, V.; Levitt, M. A model of the molten globule state from molecular dynamics simulations. *Proc. Natl. Acad. Sci. USA* **1992**, *89*, 5142–5146.

28. Li, A.; Daggett, V. Investigation of the solution structure of chymotrypsin inhibitor 2 using molecular dynamics: Comparison to X-ray crystallographic and NMR data. *Protein Eng.* **1995**, *8*, 1117–1128.

29. Duan, Y.; Kollman, P.A. Pathways to a protein folding intermediate observed in a 1-microsecond simulation in aqueous solution. *Science* **1998**, *282*, 740–744.

30. Freddolino, P.L.; Liu, F.; Gruebele, M.; Schulten, K. Ten-microsecond molecular dynamics simulation of a fast-folding WW domain. *Biophys. J.* **2008**, *94*, 75–77.

31. Shaw, D.E.; Maragakis, P.; Lindorff-Larsen, K.; Piana, S.; Dror, R.O.; Eastwood, M.P.; Bank, J.A.; Jumper, J.M.; Salmon, J.K.; Shan, Y. Atomic-level characterization of the structural dynamics of proteins. *Science* **2010**, *330*, 341–346.

32. Quentrec, B.; Brot, C. New method for searching for neighbors in molecular dynamics computations. *J. Comput. Phys.* **1973**, *13*, 430–432.

33. Ryckaert, J.P.; Ciccotti, G.; Berendsen, H.J.C. Numerical integration of the Cartesian equations of motion of a system with constraints: Molecular dynamics of n-Alkanes. *J. Comput. Phys.* **1977**, *23*, 327–341.

34. Schneider, T.; Stoll, E. Molecular dynamics study of a three-dimensional one-component model for distortive phase transitions. *Phys. Rev. B* **1978**, *17*, 1302–1322.

35. Brünger, A.; Brooks, C.L.; Karplus, M. Stochastic boundary conditions for molecular dynamics simulations of ST2 water. *Chem. Phys. Lett.* **1984**, *105*, 495–500.

36. Brooks, B.R.; Bruccoleri, R.E.; Olafson, B.D.; States, D.J.; Swaminathan, S.; Karplus, M. CHARMM: A program for macromolecular energy, minimization, and dynamics calculations. *J. Comput. Chem.* **1983**, *4*, 187–217.

37. Nosé, S. A unified formulation for constant temperature molecular dynamics methods. *J. Chem. Phys.* **1984**, *81*, 511–519.

38. Hoover, W.G. Canonical dynamics: Equilibrium phase-space distributions. *Phys. Rev. A* **1985**, *31*, 1695–1697.

39. Greengard, L.; Rokhlin, V. A fast algorithm for particle simulations. *J. Comput. Phys.* **1987**, *73*, 325–348.

40. Tuckerman, M.E.; Berne, B.J.; Martyna, G. Reversible multiple time scale molecular dynamics. *J. Chem. Phys.* **1992**, *97*, 1990–2001.

41. Skeel, R.D.; Izaguirre, J. An impulse integrator for Langevin dynamics. *Mol. Phys.* **2002**, *100*, 3885–3891.

42. Ricci, A.; Ciccotti, G. Algorithms for Brownian dynamics. *Mol. Phys.* **2003**, *101*, 1927–1931.

43. Bussi, G.; Parrinello, M. Accurate sampling using Langevin dynamics. *Phys. Rev. E* **2007**, *75*, 056707.

44. Milstein, G.N.; Tretyakov, M.V. Quasi-symplectic methods for Langevin-type equations. *IMA J. Num. Anal.* **2003**, *23*, 593–626.

45. Milstein, G.N.; Tretyakov, M.V. Computing ergodic limits for Langevin equations. *Phys. D* **2007**, *229*, 81–95.

46. Yao, Z.; Wang, J.S.; Liu, G.R.; Cheng, M. Improved neighbor list algorithm in molecular simulations using cell decomposition and data sorting method. *Comput. Phys. Commun.* **2004**, *161*, 27–35.

47. Bussi, G.; Donadio, D.; Parrinello, M. Canonical sampling through velocity rescaling. *J. Chem. Phys.* **2007**, *126*, 014101.

48. Samoletov, A.A.; Chaplain, M.A.; Dettmann, C.P. Thermostats for "Slow" configurational modes. *J. Stat. Phys.* **2008**, *128*, 1321–1336.

49. Leimkuhler, B.; Noorizadeh, E.; Theil, F. A gentle stochastic thermostat for molecular dynamics. *J. Stat. Phys.* **2009**, *135*, 261–277.

50. Leimkuhler, B.; Reich, S. A Metropolis adjusted Nosé-Hoover thermostat. *Math. Model. Num. Anal.* **2009**, *43*, 743–755.

51. Scheraga, H.A.; Khalili, M.; Liwo, A. Protein-folding dynamics: Overview of molecular simulation techniques. *Annu. Rev. Phys. Chem.* **2007**, *58*, 57–83.

52. Dror, R.O.; Dirks, R.M.; Grossman, J.P.; Xu, H.; Shaw, D.E. Biomolecular simulation: A computational microscope for molecular biology. *Annu. Rev. Biophys.* **2012**, *41*, 429–452.

53. Lane, T.J.; Shukla, D.; Beauchamp, K.A.; Pande, V.S. To milliseconds and beyond: Challenges in the simulation of protein folding. *Curr. Opin. Struct. Biol.* **2013**, *23*, 58–65.

54. Nielsen, S.O.; Lopez, C.F.; Srinivas, G.; Klein, M.L. Coarse grain models and the computer simulation of soft materials. *J. Phys. Condens. Matter* **2004**, *16*, R481.

55. Tozzini, V. Coarse-grained models for proteins. *Curr. Opin. Struct. Biol.* **2005**, *15*, 144–150.

56. Clementi, C. Coarse-grained models of protein folding: Toy models or predictive tools? *Curr. Opin. Struct. Biol.* **2008**, *18*, 10–15.

57. Sherwood, P.; Brooks, B.R.; Sansom, M.S.P. Multiscale methods for macromolecular simulations. *Curr. Opin. Struct. Biol.* **2008**, *18*, 630–640.

58. Weinan, E. *Principles of Multiscale Modeling*; Cambridge University Press: Cambridge, UK, 2011.

59. Weinan, E.; Vanden-Eijnden, E. Metastability, Conformation Dynamics, and Transition Pathways in Complex Systems. In *Multiscale Modelling and Simulation*; Attinger, S., Koumoutsakos, P., Eds.; Lecture Notes in Computational Science and Engineering; Springer: Berlin/Heidelberg, Germany, 2004; pp. 35–68.

60. Vanden-Eijnden, E.; Venturoli, M. Markovian milestoning with Voronoi tessellations. *J. Chem. Phys.* **2009**, *130*, 194101.

61. Vanden-Eijnden, E.; Venturoli, M. Exact rate calculations by trajectory parallelization and twisting. *J. Chem. Phys.* **2009**, *131*, 044120.

62. Weinan, E.; Vanden-Eijnden, E. Transition-path theory and path-finding algorithms for the study of rare events. *Annu. Rev. Phys. Chem.* **2010**, *61*, 391–420.

63. Nelson, M.T.; Humphrey, W.; Gursoy, A.; Dalke, A.; Kalé, L.V.; Skeel, R.D.; Schulten, K. NAMD: A parallel, object-oriented molecular dynamics program. *Int. J. High Perform. Comput. Appl.* **1996**, *10*, 251–268.

64. Scott, W.R.P.; Hünenberger, P.H.; Tironi, I.G.; Mark, A.E.; Billeter, S.R.; Fennen, J.; Torda, A.E.; Huber, T.; Krüger, P.; van Gunsteren, W.F. The GROMOS biomolecular simulation program package. *J. Phys. Chem. A* **1999**, *103*, 3596–3607.

65. Talay, D. Stochastic Hamiltonian systems: Exponential convergence to the invariant measure, and discretization by the implicit Euler scheme. *Markov Process. Relat. Fields* **2002**, *8*, 1–36.

66. Higham, D.J.; Mao, X.; Stuart, A.M. Strong convergence of Euler-type methods for nonlinear stochastic differential equations. *IMA J. Num. Anal.* **2002**, *40*, 1041–1063.

67. Milstein, G.N.; Tretyakov, M.V. Numerical integration of stochastic differential equations with nonglobally Lipschitz coefficients. *IMA J. Num. Anal.* **2005**, *43*, 1139–1154.

68. Higham, D.J. Stochastic ordinary differential equations in applied and computational mathematics. *IMA J. Appl. Math.* **2011**, *76*, 449–474.

69. Hutzenthaler, M.; Jentzen, A.; Kloeden, P.E. Strong convergence of an explicit numerical method for SDEs with non-globally Lipschitz continuous coefficients. *Ann. Appl. Probab.* **2012**, *22*, 1611–1641.

70. Scemama, A.; Lelièvre, T.; Stoltz, G.; Cancés, E.; Caffarel, M. An efficient sampling algorithm for variational Monte Carlo. *J. Chem. Phys.* **2006**, *125*, 114105.

71. Akhmatskaya, E.; Bou-Rabee, N.; Reich, S. A comparison of generalized hybrid Monte Carlo methods with and without momentum flip. *J. Comput. Phys.* **2009**, *228*, 2256–2265.

72. Bou-Rabee, N.; Vanden-Eijnden, E. Pathwise accuracy and ergodicity of Metropolized integrators for SDEs. *Commun. Pure Appl. Math.* **2010**, *63*, 655–696.

73. Bou-Rabee, N.; Vanden-Eijnden, E. A patch that imparts unconditional stability to explicit integrators for Langevin-like equations. *J. Comput. Phys.* **2012**, *231*, 2565–2580.

74. Metropolis, N.; Rosenbluth, A.W.; Rosenbluth, M.N.; Teller, A.H.; Teller, E. Equations of state calculations by fast computing machines. *J. Chem. Phys.* **1953**, *21*, 1087–1092.

75. Hastings, W.K. Monte-Carlo methods using Markov chains and their applications. *Biometrika* **1970**, *57*, 97–109.

76. Rossky, P.J.; Doll, J.D.; Friedman, H.L. Brownian dynamics as smart Monte Carlo simulation. *J. Chem. Phys.* **1978**, *69*, 4628.

77. Duane, S.; Kennedy, A.D.; Pendleton, B.J.; Roweth, D. Hybrid Monte-Carlo. *Phys. Lett. B* **1987**, *195*, 216–222.

78. Horowitz, A.M. A generalized guided Monte-Carlo algorithm. *Phys. Lett. B* **1991**, *268*, 247–252.

79. Kennedy, A.D.; Pendleton, B. Cost of the generalized hybrid Monte Carlo algorithm for free field theory. *Nucl. Phys. B* **2001**, *607*, 456–510.

80. Liu, J.S. *Monte Carlo Strategies in Scientific Computing*, 2nd ed.; Springer: Berlin/Heidelberg, Germany, 2008.

81. Akhmatskaya, E.; Reich, S. GSHMC: An efficient method for molecular simulation. *J. Comput. Phys.* **2008**, *227*, 4937–4954.

82. Lelièvre, T.; Rousset, M.; Stoltz, G. *Free Energy Computations: A Mathematical Perspective*, 1st ed.; Imperial College Press: London, UK, 2010.

83. *MATLAB*, Version 8.0.0 (R2012b); The MathWorks Inc.: Natick, MA, USA, 2012.

84. Introducing MEX-Files. Available online: http://www.mathworks.com/help/matlab/matlab_external/introducing-mex-files.html (accessed on 19 September 2013).

85. Kloeden, P.E.; Platen, E. *Numerical Solution of Stochastic Differential Equations*; Springer: Berlin, Germany, 1992.

86. Hutzenthaler, M.; Jentzen, A.; Kloeden, P.E. Strong and weak divergence in finite time of Euler's method for stochastic differential equations with non-globally Lipschitz continuous coefficients. *Proc. R. Soc. A: Math. Phys. Eng. Sci.* **2011**, *467*, 1563–1576.

87. Vanden-Eijnden, E.; Ciccotti, G. Second-order integrators for Langevin equations with holonomic constraints. *Chem. Phys. Lett.* **2006**, *429*, 310–316.

88. Leimkuhler, B.; Matthews, C. Robust and efficient configurational molecular sampling via Langevin dynamics. *J. Chem. Phys.* **2013**, *138*, 174102.

89. Evans, L. *An Introduction to Stochastic Differential Equations*; American Mathematical Society: Providence, RI, USA, 2013.

90. Bou-Rabee, N.; Owhadi, H. Long-run accuracy of variational integrators in the stochastic context. *SIAM J. Numer. Anal.* **2010**, *48*, 278–297.

91. Leimkuhler, B.; Matthews, C.; Stoltz, G. The computation of averages from equilibrium and nonequilibrium Langevin molecular dynamics. **2013**, arXiv:1308.5814.

92. Bou-Rabee, N.; Donev, A.; Vanden-Eijnden, E. Metropolized integration schemes for self-adjoint diffusions. **2013**, arXiv:1309.5037.

93. Milstein, G.N.; Tretyakov, M.V. *Stochastic Numerics for Mathematical Physics*; Springer: Berlin, Germany, 2004.

94. Marsden, J.E; Ratiu, T.S. *Introduction to Mechanics and Symmetry: A Basic Exposition of Classical Mechanical Systems*; Springer: Berlin/Heidelberg, Germany, 1999.

95. Van Gunsteren, W.F.; Mark, A.E. Validation of molecular dynamics simulation. *J. Chem. Phys.* **1998**, *108*, 6109–6116.

96. Leimkuhler, B.; Reich, S. *Simulating Hamiltonian Dynamics*; Cambridge Monographs on Applied and Computational Mathematics; Cambridge University Press: Cambridge, UK, 2004.

97. Hairer, E.; Lubich, C.; Wanner, G. *Geometric Numerical Integration*; Springer: Berlin/Heidelberg, Germany, 2010.

98. Ikeda, N.; Watanabe, S. *Stochastic Differential Equations and Diffusion Processes*; North-Holland: Amsterdam, The Netherlands, 1989.

99. Klebaner, F.C. *Introduction to Stochastic Calculus with Applications*; Imperial College Press: London, UK, 2005.

100. Mengersen, K.L.; Tweedie, R.L. Rates of convergence of the Hastings and Metropolis algorithms. *Ann. Stat.* **1996**, *24*, 101–121.

101. Prato, G.D.; Zabczyk, J. *Ergodicity for Infinite Dimensional Systems*; Cambridge University Press: Cambridge, UK, 1996.

102. Haussman, U.G.; Pardoux, E. Time reversal for diffusions. *Ann. Probab.* **1986**, *14*, 1188–1205.

103. Kent, J. Time-reversible diffusions. *Adv. Appl. Prob.* **1978**, *10*, 819–835.

104. Ezra, G.S. Reversible measure-preserving integrators for non-Hamiltonian systems. *J. Chem. Phys.* **2006**, *125*, 034104.

105. Shardlow, T. Splitting for dissipative particle dynamics. *SIAM J. Sci. Comput.* **2003**, *24*, 1267–1282.

106. Serrano, M.; de Fabritiis, G.; Espanol, P.; Coveney, P.V. A stochastic Trotter integration scheme for dissipative particle dynamics. *Math. Comput. Simulat.* **2006**, *72*, 190–194.

107. Pavliotis, G.A.; Stuart, A.M.; Zygalakis, K.C. Calculating effective diffusivities in the limit of vanishing molecular diffusion. *J. Comput. Phys.* **2008**, *228*, 1030–1055.

108. Van Gunsteren, W.F.; Berendsen, H.J.C. Algorithms for Brownian dynamics. *Mol. Phys.* **1982**, *45*, 637–647.

109. Nummelin, E. *General Irreducible Markov Chains and Non-Negative Operators*; Cambridge University Press: New York, NY, USA, 1984.

110. Tierney, L. Markov chains for exploring posterior distributions. *Ann. Stat.* **1994**, *22*, 1701–1728.

111. Gelman, A.; Gilks, W.R.; Roberts, G.O. Weak convergence and optimal scaling of random walk Metropolis algorithms. *Ann. Appl. Probab.* **1997**, *7*, 110–120.

112. Roberts, G.O.; Rosenthal, J.S. Optimal scaling of discrete approximations to Langevin diffusions. *J. R. Stat. Soc. Ser. B* **1998**, *60*, 255–268.

113. Beskos, A.; Roberts, G.O.; Stuart, A.M. Optimal scalings for local Metropolis-Hastings chains on non-product targets in high dimensions. *Ann. Appl. Probab.* **2009**, *19*, 863–898.

114. Beskos, A.; Pillai, N.S.; Roberts, G.O.; Sanz-Serna, J.M.; Stuart, A.M. Optimal tuning of hybrid Monte-Carlo algorithm. **2010**, arXiv:1001.4460.

115. Mattingly, J.C.; Pillai, N.S.; Stuart, A.M. Diffusion limits of the random walk Metropolis algorithm in high dimensions. *Ann. Appl. Probab.* **2012**, *22*, 881–930.

116. Kang, F.; Dao-Liu, W. Dynamical Systems and Geometric Construction of Algorithms. In *Computational Mathematics in China*; Contemporary Mathmatics, Volume 163; Shi, Z.-C., Yang, C.C., Eds.; American Mathmatical Society: New York, NY, USA,1994; pp. 1–32.

Reprinted from *Entropy*. Cite as: Abrams, C.; Bussi, G. Enhanced Sampling in Molecular Dynamics Using Metadynamics, Replica-Exchange, and Temperature-Acceleration. *Entropy* **2014**, *16*, 163–199.

Article

Enhanced Sampling in Molecular Dynamics Using Metadynamics, Replica-Exchange, and Temperature-Acceleration

Cameron Abrams [1,]* **and Giovanni Bussi** [2]

[1] Department of Chemical and Biological Engineering, Drexel University, 3141 Chestnut Street, Philadelphia, PA 19104, USA

[2] Scuola Internazionale Superiore di Studi Avanzati (SISSA), via Bonomea 265, Trieste 34136, Italy; E-Mail:bussi@sissa.it

* Author to whom correspondence should be addressed; E-Mail: cfa22@drexel.edu; Tel.: +1-215-895-2231.

Received: 13 September 2013; in revised form: 7 November 2013 / Accepted: 11 November 2013/ Published: 27 December 2013

Abstract: We review a selection of methods for performing enhanced sampling in molecular dynamics simulations. We consider methods based on collective variable biasing and on tempering, and offer both historical and contemporary perspectives. In collective-variable biasing, we first discuss methods stemming from thermodynamic integration that use mean force biasing, including the adaptive biasing force algorithm and temperature acceleration. We then turn to methods that use bias potentials, including umbrella sampling and metadynamics. We next consider parallel tempering and replica-exchange methods. We conclude with a brief presentation of some combination methods.

Keywords: collective variables; free energy; blue-moon sampling; adaptive-biasing force algorithm; temperature-acceleration; umbrella sampling; metadynamics

1. Introduction

The purpose of molecular dynamics (MD) is to compute the positions and velocities of a set of interacting atoms at the present time instant given these quantities one time increment in the past.

Uniform sampling from the discrete trajectories one can generate using MD has long been seen as synonymous with sampling from a statistical-mechanical ensemble; this just expresses our collective wish that the ergodic hypothesis holds at finite times. Unfortunately, most MD trajectories are not ergodic and leave many relevant regions of configuration space unexplored. This stems from the separation of high-probability "metastable" regions by low-probability "transition" regions and the inherent difficulty of sampling a $3N$-dimensional space by embedding into it a one-dimensional dynamical trajectory.

This review concerns a selection of methods to use MD simulation to enhance the sampling of configuration space. A central concern with any enhanced sampling method is guaranteeing that the statistical weights of the samples generated are known and correct (or at least correctable) while simultaneously ensuring that as much of the relevant regions of configuration space are sampled. Because of the tight relationship between probability and free energy, many of these methods are known as "free-energy" methods. To be sure, there are a large number of excellent reviews of free-energy methods in the literature (e.g., [1–5]). The present review is in no way intended to be as comprehensive. As the title indicates, we will mostly focus on enhanced sampling methods of three flavors: tempering, metadynamics, and temperature-acceleration. Along the way, we will point out important related methods, but in the interest of brevity we will not spend much time explaining these. The methods we have chosen to focus on reflect our own preferences to some extent, but they also represent popular and growing classes of methods that find ever more use in biomolecular simulations and beyond.

We divide our review into three main sections. In the first, we discuss enhanced sampling approaches that rely on *collective variable biasing*. These include the historically important methods of thermodynamic integration and umbrella sampling, and we pay particular attention to the more recent approaches of the adaptive-biasing force algorithm, temperature-acceleration, and metadynamics. In the second section, we discuss approaches based on *tempering*, which is dominated by a discussion of the parallel tempering/replica exchange approaches. In the third section, we briefly present some relatively new methods derived from either collective-variable-based or tempering-based approaches, or their combinations.

2. Approaches Based on Collective-Variable Biasing

2.1. Background: Collective Variables and Free Energy

For our purposes, the term "collective variable" or CV refers to any multidimensional function $\boldsymbol{\theta}$ of $3N$-dimensional atomic configuration $\boldsymbol{x} \equiv (x_i | i = 1 \ldots 3N)$. The functions $\theta_1(\boldsymbol{x})$, $\theta_2(\boldsymbol{x}), \ldots, \theta_M(\boldsymbol{x})$ map configuration $\boldsymbol{x}$ onto an M-dimensional CV space $\boldsymbol{z} \equiv (z_j | j = 1 \ldots M)$, where usually $M \ll 3N$. At equilibrium, the probability of observing the system at CV-point $\boldsymbol{z}$ is the weight of all configurations $\boldsymbol{x}$ which map to $\boldsymbol{z}$:

$$P(\boldsymbol{z}) = \langle \delta[\boldsymbol{\theta}(\boldsymbol{x}) - \boldsymbol{z}] \rangle \tag{1}$$

The Dirac delta function picks out only those configurations for which the CV $\boldsymbol{\theta}(\boldsymbol{x})$ is $\boldsymbol{z}$, and $\langle \cdot \rangle$ denotes averaging its argument over the equilibrium probability distribution of $\boldsymbol{x}$. The probability can be expressed as a *free energy*:

$$F(\boldsymbol{z}) = -k_B T \ln \langle \delta[\boldsymbol{\theta}(\boldsymbol{x}) - \boldsymbol{z}] \rangle \qquad (2)$$

Here, k_B is Boltzmann's constant and T is temperature.

Local minima in F are metastable equilibrium states. F also measures the energetic cost of a maximally efficient (*i.e.*, reversible) transition from one region of CV space to another. If, for example, we choose a CV space such that two well-separated regions define two important allosteric states of a given protein, we could perform a free-energy calculation to estimate the change in free energy required to realize the conformational transition. Indeed, the promise of being able to observe with atomic detail the transition states along some pathway connecting two distinct states of a biomacromolecule is strong motivation for exploring these transitions with CVs.

Given the limitations of standard MD, how does one "discover" such states in a proposed CV space? A perfectly ergodic (infinitely long) MD trajectory would visit these minima much more frequently than it would the intervening spaces, allowing one to tally how often each point in CV space is visited; normalizing this histogram into a probability $P(\boldsymbol{z})$ would be the most straightforward way to compute F via Equation (2). In all too many actual cases, MD trajectories remain close to only one minimum (the one closest to the initial state of the simulation) and only very rarely, if ever, visit others. In the CV sense, we therefore speak of standard MD simulations failing to overcome *barriers* in free energy. "Enhanced sampling" in this context refers then to methods by which free-energy barriers in a chosen CV space are surmounted to allow as broad as possible an extent of CV space to be explored and statistically characterized with limited computational resources.

In this section, we focus on methods of enhanced sampling of CVs based on MD simulations that are directly biased on those CVs; that is, we focus on methods in which an investigator must identify the CVs of interest as an input to the calculation. We have chosen to limit discussion to two broad classes of biasing: those whose objective is direct computation of the gradient of the free energy $(\partial F / \partial \boldsymbol{z})$ at local points throughout CV space, and those in which non-Boltzmann sampling with bias potentials is used to force exploration of otherwise hard-to-visit regions of CV space. The canonical methods in these two classes are *thermodynamic integration* and *umbrella sampling*, respectively, and a discussion of these two methods sets the stage for discussion of three relatively modern variants: the Adaptive-Biasing Force Algorithm [6], Temperature-Accelerated MD [7] and Metadynamics [8].

2.2. Gradient Methods: Blue-Moon Sampling, Adaptive-Biasing Force Algorithm, and Temperature-Accelerated Molecular Dynamics

2.2.1. Overview: Thermodynamic Integration

Naively, one way to have an MD system visit a hard-to-reach point $\boldsymbol{z}$ in CV space is simply to create a realization of the configuration $\boldsymbol{x}$ at that point (*i.e.*, such that $\boldsymbol{\theta}(\boldsymbol{x}) = \boldsymbol{z}$). This is an inverse problem, since the number of degrees of freedom in $\boldsymbol{x}$ is usually much larger than in $\boldsymbol{z}$. One way to perform this inversion is by introducing external forces that guide the configuration to the desired point from some easy-to-create initial state; both targeted MD [9] and steered MD [10] are ways to do this. Of course, one would like MD to explore CV space in the vicinity of z, so after creating the configuration x, one would just let it run. Unfortunately, this would likely result in the system drifting away from z rather quickly, and there would be no way from such calculations to estimate the likelihood of observing an unbiased long MD simulation visit z. However, there is information in the fact that the system drifts away; if one knows *on average* which direction and how strongly the system would like to move if initialized at z, this would be a measure of negative gradient of the free energy, $-(\partial F/\partial z)$, or the "mean force". We have then a glimpse of a three-step method to compute F (*i.e.*, the statistics of CVs) over a meaningfully broad extent of CV space:

(1) visit a select number of local points in that space, and at each one,
(2) compute the mean force, then
(3) use numerical integration to reconstruct F from these local mean forces; formally expressed as

$$F(z) - F(z_0) = \int_{z_0}^{z} \left(\frac{\partial F}{\partial z} \right) dz \qquad (3)$$

Inspired by Kirkwood's original suggestion involving switching parameters [11], such an approach is generally referred to as "thermodynamic integration" or TI. TI allows us to reconstruct the statistical weights of any point in CV space by accumulating information on the gradients of free energy at selected points.

2.2.2. Blue-Moon Sampling

The discussion so far leaves open the correct way to compute the local free-energy gradients. A gradient is a local quantity, so a natural choice is to compute it from an MD simulation localized at a point in CV space by a constraint. Consider a long MD simulation with a holonomic constraint fixing the system at the point z. Uniform samples from this constrained trajectory $x(t)$ then represent *an* ensemble at fixed z over which the averaging needed to convert gradients in potential energy to gradients in free energy could be done. However, this constrained ensemble has the undesired property that the velocities $\dot{\theta}(x)$ are zero. This is a bit problematic because virtually none of the samples plucked from a long unconstrained MD simulation (as is implied by Equation (1)), would have $\dot{\theta} = 0$, and $\dot{\theta} = 0$ acts as a set of M unphysical constraints on the system velocities $\dot{x}$, since $\dot{\theta}_j = \sum_i (\partial \theta_j/\partial x_i) \dot{x}_i$. Probably the best-known example of a method to correct for this bias is the so-called "blue-moon" sampling method [12–15] or the constrained ensemble method [16,17]. The essence of the method is a decomposition of free energy gradients into components along the CV gradients and thermal components orthogonal to them:

$$\frac{\partial F}{\partial z_j} = \langle b_j(x) \cdot \nabla V(x) - k_B T \nabla \cdot b_j(x) \rangle_{\theta(x)=z} \qquad (4)$$

where $\langle \cdot \rangle_{\theta(x)=z}$ denotes averaging across samples drawn uniformly from the MD simulation constrained at $\theta(x) = z$, and the $b_j(x)$ is the vector field orthogonal to the gradients of every component k of θ for $k \neq j$:

$$b_j(x) \cdot \nabla \theta_k(x) = \delta_{jk} \tag{5}$$

where δ_{jk} is the Kroenecker delta. (For brevity, we have omitted the consideration of holonomic constraints other than that on the CV; the reader is referred to the paper by Ciccotti *et al.* for details [15].) The vector fields b_j for each θ_j can be constructed by orthogonalization. The first term in the angle brackets in Equation (4) implements the chain rule one needs to account for how energy V changes with z through all the ways z can change with x. The second term corrects for the thermal bias imposed by the constraint.

Although nowhere near exhaustive, below is a listing of common types of problems to which blue-moon sampling has been applied with some representative examples:

(1) sampling conformations of small flexible molecules and peptides [18–20];
(2) environmental effects on covalent bond formation/breaking (usually in combination with *ab initio* MD) [21–27];
(3) solvation and non-covalent binding of small molecules in solvent [28–32];
(4) protein dimerization [33,34].

2.2.3. The Adaptive Biasing Force Algorithm

The blue-moon approach requires multiple independent constrained MD simulations to cover the region of CV space in which one wants internal statistics. The care taken in choosing these quadrature points can often dictate the accuracy of the resulting free energy reconstruction. It is therefore sometimes advantageous to consider ways to avoid having to choose such points ahead of time, and adaptive methods attempt to address this problem. One example is the adaptive-biasing force (ABF) algorithm of Darve *et al.* [6,35] The essence of ABF is two-fold: (1) recognition that external bias forces of the form $\nabla_x \theta_j (\partial F / \partial z_j)$ for $j = 1, \ldots, M$ exactly oppose mean forces and should lead to more uniform sampling of CV space; and (2) that these bias forces can be converged upon adaptively during a single unconstrained MD simulation.

The first of those two ideas is motivated by the fact that "forces" that keep normal MD simulations effectively confined to free energy minima are mean forces on the collective variables projected onto the atomic coordinates, and balancing those forces against their exact opposite should allow for thermal motion to take the system out of those minima. The second idea is a bit more subtle; after all, in a running MD simulation with no CV constraints, the constrained ensemble expression for the mean force (Equation (4)) does not directly apply, because a constrained ensemble is not what is being sampled. However, Darve *et al.* showed how to relate these ensembles so that the samples generated in the MD simulation could be used to build mean forces [35]. Further, they showed using a clever choice of the fields of Equation (4) an equivalence between (*i*) the spatial gradients needed to computed forces, and (*ii*) time-derivatives of the CVs [6]:

$$\frac{\partial F}{\partial z_i} = -k_B T \left\langle \frac{d}{dt} \left(M_\theta \frac{d\theta_i}{dt} \right) \right\rangle_{\theta=z} \tag{6}$$

where M_θ is the transformed mass matrix given by

$$M_\theta^{-1} = J_\theta M^{-1} J_\theta \qquad (7)$$

where J_θ is the $M \times 3N$ matrix with elements $\partial\theta_i/\partial x_j$ ($i = 1\ldots M$, $j = 1\ldots 3N$), and M is the diagonal matrix of atomic masses. Equation (7) is the result of a particular choice for the fields $b_j(x)$. This reformulation of the instantaneous mean forces computed on-the-fly makes ABF exceptionally easy to implement in most modern MD packages. Darve *et al.* present a clear demonstration of the ABF algorithm in a pseudocode [6] that attests to this fact.

ABF has found rather wide application in CV-based free energy calculations in recent years. Below is a representative sample of some types of problems subjected to ABF calculations in the recent literature:

(1) Peptide backbone angle sampling [36,37];
(2) Nucleoside [38], protein [39] and fullerene [40,41] insertion into a lipid bilayer;
(3) Interactions of small molecules with polymers in water [42,43];
(4) Molecule/ion transport through protein complexes [44–47] and DNA superstructures [48];
(5) Calculation of octanol-water partition coefficients [49,50];
(6) Large-scale protein conformational changes [51];
(7) Protein-nanotube [52] and nanotube-nanotube [53] association.

2.2.4. Temperature-Accelerated Molecular Dynamics

Both blue-moon sampling and ABF are based on statistics in the constrained ensemble. However, estimation of mean forces need not only use this ensemble. One can instead relax the constraint and work with a "mollified" version of the free energy:

$$F_\kappa(z) = -k_B T \ln \langle \delta_\kappa \left[\theta(x) - z \right] \rangle \qquad (8)$$

where δ_κ refers to the Gaussian (or "mollified delta function"):

$$\delta_\kappa = \sqrt{\frac{\beta\kappa}{2\pi}} \exp\left[-\frac{1}{2}\beta\kappa \left| \theta(x) - z \right|^2 \right] \qquad (9)$$

where β is just shorthand for $1/k_B T$. Since $\lim_{\beta\kappa \to \infty} \delta_\kappa = \delta$, we know that $\lim_{\beta\kappa \to \infty} F_\kappa = F$. One way to view this Gaussian is that it "smoothes out" the true free energy to a tunable degree; the factor $1/\sqrt{\beta\kappa}$ is a length-scale in CV space below which details are smeared.

Because the Gaussian has continuous gradients, it can be used directly in an MD simulation. Suppose we have a CV space $\theta(x)$, and we extend our MD system to include variables z such that the combined set (x, z) obeys the following extended potential:

$$U(x, z) = V(x) + \sum_{j=1}^{M} \frac{1}{2}\kappa \left| \theta_j(x) - z_j \right|^2 \qquad (10)$$

where $V(x)$ is the interatomic potential, and κ is a constant. Clearly, if we fix z, then the resulting free energy is to within an additive constant the mollified free energy of Equation (8). (The additive constant is related to the prefactor of the mollified delta function and has nothing to do with the number of CVs.) Further, we can directly express the gradient of this mollified free energy with respect to z: [54]

$$\nabla_z F_\kappa = - \langle \kappa \left[\boldsymbol{\theta}(\boldsymbol{x}) - \boldsymbol{z} \right] \rangle \tag{11}$$

This suggests that, instead of using constrained ensemble MD to accumulate mean forces, we could work in the *restrained* ensemble and get very good approximations to the mean force. By "restrained", we refer to the fact that the term giving rise to the mollified delta function in the configurational integral is essentially a harmonic restraining potential with a "spring constant" κ. In this restrained-ensemble approach, no velocities are held fixed, and the larger we choose κ the more closely we can approximate the true free energy. Notice however that large values of κ could lead to numerical instabilities in integrating equations of motion, and a balance should be found. (In practice, we have found that for CVs with dimensions of length, values of κ less than about 1,000 kcal/mol/Å^2 can be stably handled, and values of around 100 kcal/mol/Å^2 are typically adequate.)

Temperature-accelerated MD (TAMD) [7] takes advantage of the restrained-ensemble approach to directly evolve the variables z in such a way to accelerate the sampling of CV space. First, consider how the atomic variables x evolve under the extended potential (assuming Langevin dynamics):

$$m_i \ddot{x}_i = -\frac{\partial V(\boldsymbol{x})}{\partial x_i} - \kappa \sum_{j=1}^{m} \left[\theta_j(\boldsymbol{x}) - z_j \right] \frac{\partial \theta_j(\boldsymbol{x})}{\partial x_i} - \gamma m_i \dot{x}_i + \eta_i(t; \beta) \tag{12}$$

Here, m_i is the mass of x_i, γ is the friction coefficient for the Langevin thermostat, and $\boldsymbol{\eta}$ is the thermostat white noise satisfying the fluctuation-dissipation theorem at physical temperature β^{-1}:

$$\langle \eta_i(t; \beta) \eta_j(t'; \beta) \rangle = \beta^{-1} \gamma m_i \delta_{ij} \delta(t - t') \tag{13}$$

Key to TAMD is that the z are treated as slow variables that evolve according to their own equations of motion, which here we take as diffusive (though other choices are possible [7]):

$$\bar{\gamma} \bar{m}_j \dot{z}_j = \kappa \left[\theta_j(\boldsymbol{x}) - z_j \right] + \xi_j(t; \bar{\beta}) \tag{14}$$

Here, $\bar{\gamma}$ is a fictitious friction, $\bar{m}_j$ is a mass, and the first term on the right-hand side represents the instantaneous force on variable z_j, and the second term represents thermal noise at the fictitious thermal energy $\bar{\beta}^{-1} \neq \beta^{-1}$.

The advantage of TAMD is that if (1) $\bar{\gamma}$ is chosen sufficiently large so as to guarantee that the slow variables indeed evolve slowly relative to the fundamental variables; *and* (2) κ is sufficiently large such that $\boldsymbol{\theta}(\boldsymbol{x}(t)) \approx \boldsymbol{z}(t)$ at any given time, then the force acting on z is approximately equal to minus the gradient of the free energy (Equation (11)) [7]. This is because the MD integration repeatedly samples $\kappa \left[\boldsymbol{\theta}(\boldsymbol{x}) - \boldsymbol{z} \right]$ for an essentially fixed (but actually very slowly moving) z, so z evolution effectively feels these samples as a mean force. In other words, the dynamics of $z(t)$ is effectively

$$\bar{\gamma} \bar{m}_j \dot{z}_j = -\frac{\partial F(\boldsymbol{z})}{\partial z_j} + \xi_j(t; \bar{\beta}) \tag{15}$$

This shows that the z-dynamics describes an equilibrium constant-temperature ensemble at *fictitious* temperature $\bar{\beta}^{-1}$ acted on by the "potential" $F(z)$, which is the free energy evaluated at the *physical* temperature β^{-1}. That is, under TAMD, z conforms to a probability distribution of the form $\exp\left[-\bar{\beta}F(z;\beta)\right]$, whereas under normal MD it would conform to $\exp\left[-\beta F(z;\beta)\right]$. The all-atom MD simulation (at β) simply serves to approximate the *local gradients* of $F(z)$. Sampling is enhanced by taking $\bar{\beta}^{-1} > \beta^{-1}$, which has the effect of attenuating the ruggedness of F. TAMD therefore can accelerate a trajectory $z(t)$ through CV space by increasing the likelihood of visiting points with relatively low physical Boltzmann factors. This borrows directly from the main idea of adiabatic free-energy dynamics [55] (AFED), in that one deliberately makes some variables hot (to overcome barriers) but slow (to keep them adiabatically separated from all other variables). In TAMD, however, the use of the mollified free energy means no cumbersome variable transformations are required. (The authors of AFED refer to TAMD as "driven"-AFED, or d-AFED [56].) It is also worth mentioning in this review that TAMD borrows heavily from an early version of metadynamics [57], which was formulated as a way to evolve the auxiliary variables z on a mollified free energy. However, unlike metadynamics (which we discuss below in Section 2.3.3), there is no history-dependent bias in TAMD.

Unlike TI, ABF, and the methods of umbrella sampling and metadynamics discussed in the next section, TAMD is not a method for direct calculation of the free energy. Rather, it is a way to overcome free energy barriers in a chosen CV space quickly without visiting irrelevant regions of CV space. (However, we discuss briefly a method in Section 4.2.2 in which TAMD gradients are used in a spirit similar to ABF to reconstruct a free energy.) That is, we consider TAMD a way to efficiently explore relevant regions CV space that are practically inaccessible to standard MD simulation. It is also worth pointing out that, unlike ABF, TAMD does not operate by opposing the natural gradients in free energy, but rather by using them to guide accelerated sampling. ABF can only use forces in locations in CV space the trajectory has visited, which means nothing opposes the trajectory going to regions of very high free energy. However, under TAMD, an acceleration of $\bar{\beta}^{-1} = 6$ kcal/mol on the CVs will greatly accelerate transitions over barriers of 6-12 kcal/mol, but will still not (in theory) accelerate excursions to regions requiring climbs of hundreds of kcal/mol. TAMD and ABF have in common the ability to handle rather high-dimensional CVs.

Although it was presented theoretically in 2006 [7], TAMD was not applied directly to large-scale MD until much later [58]. Since then, there has been growing interest in using TAMD in a variety of applications requiring enhanced sampling:

(1) TAMD-enhanced flexible fitting of all-atom protein and RNA models into low-resolution electron microscopy density maps [59,60];
(2) Large-scale (interdomain) protein conformational sampling [58,61,62];
(3) Loop conformational sampling in proteins [63];
(4) Mapping of diffusion pathways for small molecules in globular proteins [64,65];
(5) Vacancy diffusion [66];
(6) Conformational sampling and packing in dense polymer systems [67].

Finally, we mention briefly that TAMD can be used as a quick way to generate trajectories from which samples can be drawn for subsequent mean-force estimation for later reconstruction of a multidimensional free energy; this is the essence of the single-sweep method [68], which is an efficient means of computing multidimensional free energies. Rather than using straight numerical TI, single sweep posits the free energy as a basis function expansion and uses standard optimization methods to find the expansion coefficients that best reproduce the measured mean forces. Single-sweep has been used to map diffusion pathways of CO and H_2O in myoglobin [64,65].

2.3. Bias Potential Methods: Umbrella Sampling and Metadynamics

2.3.1. Overview: Non-Boltzmann Sampling

In the previous section, we considered methods that achieve enhanced sampling by using mean forces: in TI, these are integrated to reconstruct a free energy; in ABF, these are built on-the-fly to drive uniform CV sampling; and in TAMD, these are used on-the-fly to guide accelerated evolution of CVs. In this section, we consider methods that achieve enhanced sampling by means of controlled bias potentials. As a class, we refer to these as *non-Boltzmann sampling* methods.

Non-Boltzmann sampling is generally a way to derive statistics on a system whose energetics differ from the energetics used to perform the sampling. Imagine we have an MD system with bare interatomic potential $V(\boldsymbol{x})$, and we add a bias $\Delta V(\boldsymbol{x})$ to arrive at a biased total potential:

$$V_b(\boldsymbol{x}) = V(\boldsymbol{x}) + \Delta V(\boldsymbol{x}) \tag{16}$$

The statistics of the CVs on this biased potential are then given as

$$P_b(\boldsymbol{z}) = \frac{\int d\boldsymbol{x}\, e^{-\beta V_0(\boldsymbol{x})} e^{-\beta \Delta V(\boldsymbol{x})} \delta\left[\boldsymbol{\theta}(\boldsymbol{x}) - \boldsymbol{z}\right]}{\int d\boldsymbol{x}\, e^{-\beta V_0(\boldsymbol{x})} e^{-\beta \Delta V(\boldsymbol{x})}}$$

$$= \frac{\int d\boldsymbol{x}\, e^{-\beta V_0(\boldsymbol{x})} e^{-\beta \Delta V} \delta\left[\boldsymbol{\theta}(\boldsymbol{x}) - \boldsymbol{z}\right]}{\int d\boldsymbol{x}\, e^{-\beta V_0(\boldsymbol{x})}} \frac{\int d\boldsymbol{x}\, e^{-\beta V_0(\boldsymbol{x})}}{\int d\boldsymbol{x}\, e^{-\beta V_0(\boldsymbol{x})} e^{-\beta \Delta V(\boldsymbol{x})}}$$

$$= \frac{\left\langle e^{-\beta \Delta V(\boldsymbol{x})} \delta\left[\boldsymbol{\theta}(\boldsymbol{x}) - \boldsymbol{z}\right]\right\rangle}{\left\langle e^{-\beta \Delta V(\boldsymbol{x})}\right\rangle} \tag{17}$$

where $\langle \cdot \rangle$ denotes ensemble averaging on the unbiased potential $V(\boldsymbol{x})$. Further, if we take the bias potential ΔV to be explicitly a function only of the CVs $\boldsymbol{\theta}$, then it becomes invariant in the averaging of the numerator thanks to the delta function, and we have

$$P_b(\boldsymbol{x}) = \frac{e^{-\beta \Delta V(\boldsymbol{z})} \left\langle \delta\left[\boldsymbol{\theta}(\boldsymbol{x}) - \boldsymbol{z}\right]\right\rangle}{\left\langle e^{-\beta \Delta V[\boldsymbol{\theta}(\boldsymbol{x})]}\right\rangle} \tag{18}$$

Finally, since the unbiased statistics are $P(\boldsymbol{z}) = \left\langle \delta\left[\boldsymbol{\theta}(\boldsymbol{x}) - \boldsymbol{z}\right]\right\rangle$, we arrive at

$$P(\boldsymbol{z}) = P_b(\boldsymbol{z}) e^{\beta \Delta V(\boldsymbol{z})} \left\langle e^{-\beta \Delta V[\boldsymbol{\theta}(\boldsymbol{x})]}\right\rangle \tag{19}$$

Taking samples from an ergodic MD simulation on the biased potential V_b, Equation (19) provides the recipe for reconstructing the statistics the CVs *would* present were they generated using the *unbiased* potential V. However, the probability $P(z)$ is implicit in this equation, because

$$\left\langle e^{-\beta\Delta V}\right\rangle = \int dz\, P(z) e^{-\beta\Delta V[\theta(x)]} \tag{20}$$

This is not really a problem, since we can treat $\left\langle e^{-\beta\Delta V}\right\rangle$ as a constant we can get from normalizing $P_b(z)e^{\beta\Delta V(z)}$.

How does one choose ΔV so as to enhance the sampling of CV space? Evidently, from the standpoint of non-Boltzmann sampling, the closer the bias potential is to the negative free energy $-F(z)$, the more uniform the sampling of CV space will be. To wit: if $\Delta V[\theta(x)] = -F[\theta(x)]$, then $e^{\beta\Delta V(z)} = e^{-\beta F(z)} = P(z)$, and Equation (19) can be inverted for P_b to yield

$$P_b(z) = \frac{1}{\langle e^{\beta F(z)}\rangle} = \frac{1}{\int dz\, P(z)e^{\beta F(z)}} = \frac{1}{\int dz\, e^{-\beta F}e^{\beta F}} = \frac{1}{\int dz} \tag{21}$$

So we see that taking the bias potential to be the negative free energy makes all states z in CV space equiprobable. This is indeed the limit to which ABF strives by applying negative mean forces, for example [6].

We usually do not know the free energy ahead of time; if we did, we would already know the statistics of CV space and no enhanced sampling would be necessary. Moreover, perfectly uniform sampling of the entire CV space is usually far from necessary, since most CV spaces have many irrelevant regions that should be ignored. And in reference to the mean-force methods of the last section, uniform sampling is likely not necessary to achieve accurate mean force values; how good an estimate of ∇F is at some point z_0 should not depend on how well we sampled at some other point z_1. Yet achieving uniform sampling is an idealization since, if we do, this means we know the free energy. We now consider two other biasing methods that aim for this ideal, either in relatively small regions of CV space using fixed biases, or over broader extents using adaptive biases.

2.3.2. Umbrella Sampling

Umbrella sampling is the standard way of using non-Boltzmann sampling to overcome free energy barriers. In its debut [69], umbrella sampling used a function $w(x)$ that weights hard-to-sample configurations, equivalent to adding a bias potential of the form

$$\Delta V(x) = -k_B T \ln w(x) \tag{22}$$

w is found by trial-and-error such that configurations that are easy to sample on the unbiased potential are still easy to sample; that is, w acts like an "umbrella" covering both the easy- and hard-to-sample regions of configuration space. Nearly always, w is an explicit function of the CVs, $w(x) = W[\theta(x)]$.

Coming up with the umbrella potential that would enable exploration of CV space with a single umbrella sampling simulation that takes the system far from its initial point is not straightforward.

Akin to TI, it is therefore advantageous to combine results from several independent trajectories, each with its own umbrella potential that localizes it to a small volume of CV space that overlaps with nearby volumes. The most popular way to combine the statistics of such a set of independent umbrella sampling runs is the weighted-histogram analysis method (WHAM) [70].

To compute statistics of CV space using WHAM, one first chooses the points in CV space that define the little local neighborhoods, or "windows" to be sampled and chooses the bias potential used to localize the sampling. Not knowing how the free energy changes in CV space makes the first task somewhat challenging, since more densely packed windows are preferred in regions where the free energy changes rapidly; however, since the calculations are independent, more can be added later if needed. A convenient choice for the bias potential is a simple harmonic spring that tethers the trajectory to a reference point z_i in CV space:

$$\Delta V_i(x) = \frac{1}{2}\kappa \left| \theta(x) - z_i \right|^2 \tag{23}$$

which means the dynamics of the atomic variables x are identical to Equation (12) at fixed $z = z_i$. The points $\{z_i\}$ and the value of κ (which may be point-dependent) must be chosen such that $\theta[x(t)]$ from any one window's trajectory makes excursions into the window of each of its nearest neighbors in CV space.

Each window-restrained trajectory is directly histogrammed to yield apparent (*i.e.*, biased) statistics on θ; let us call the biased probability in the ith window $P_{b,i}(z)$. Equation (19) again gives the recipe to reconstruct the unbiased statistics $P_i(z)$ for z in the window of z_i:

$$P_i(z) = P_{b,i}(z) e^{\frac{1}{2}\beta\kappa|z-z_i|^2} \left\langle e^{-\beta\frac{1}{2}\kappa|\theta(x)-z_i|^2} \right\rangle \tag{24}$$

We could use Equation (24) directly assuming the biased MD trajectory is ergodic, but we know that regions far from the reference point will be explored very rarely and thus their free energy would be estimated with large uncertainty. This means that, although we can use sampling to compute $P_{b,i}$ *knowing* it effectively vanishes outside the neighborhood of z_i, we cannot use sampling to compute $\left\langle e^{-\beta\frac{1}{2}\kappa|\theta(x)-z_i|^2} \right\rangle$.

WHAM solves this problem by renormalizing the probabilities in each window into a single composite probability. Where there is overlap among windows, WHAM renormalizes such that the statistical variance of the probability is minimal. That is, it treats the factor $\left\langle e^{-\beta\frac{1}{2}\kappa|\theta(x)-z_i|^2} \right\rangle$ as an undetermined constant C_i for each window, and solves for specific values such that the composite unbiased probability $P(z)$ is continuous across all overlap regions with minimal statistical error. An alternative to WHAM, termed "umbrella integration", solves the problem of renormalization across windows by constructing the composite mean force [71,72].

The literature on umbrella sampling is vast (by simulation standards), so we present here a very condensed listing of some of its more recent application areas with representative citations:

(1) Small molecule conformational sampling [73–76];
(2) Protein-folding [77–79] and large-scale protein conformational sampling [80–83];
(3) Protein-protein/peptide-peptide interactions [84–92];

(4) DNA conformational changes [93] and DNA-DNA interactions [94–96];

(5) Binding and association free-energies [97–107];

(6) Adsorption on and permeation through lipid bilayers [108–117];

(7) Adsorption onto inorganic surfaces/interfaces [118,119];

(8) Water ionization [120,121];

(9) Phase transitions [122,123];

(10) Enzymatic mechanisms [124–132];

(11) Molecule/ion transport through protein complexes [133–140] and other macromolecules [141,142].

2.3.3. Metadynamics

As already mentioned, one of the difficulties of the umbrella sampling method is the choice and construction of the bias potential. As we already saw with the relationship among TI, ABF, and TAMD, an adaptive method for building a bias potential in a running MD simulation may be advantageous. Metadynamics [8,143] represents just such a method.

Metadynamics is rooted in the original idea of "local elevation" [144], in which a supplemental bias potential is progressively grown in the dihedral space of a molecule to prevent it from remaining in one region of configuration space. However, at variance with metadynamics, local elevation does not provide any means to reconstruct the unbiased free-energy landscape and as such it is mostly aimed at fast generation of plausible conformers.

In metadynamics, configurational variables x evolve in response to a biased total potential:

$$V(x) = V_0(x) + \Delta V(x, t) \tag{25}$$

where V_0 is the bare interatomic potential and $\Delta V(x, t)$ is a time-dependent bias potential. The key element of metadynamics is that the bias is built as a sum of Gaussian functions centered on the points in CV space already visited:

$$\Delta V[\theta(x), t] = w \sum_{\substack{t' = \tau_G, 2\tau_G, \dots \\ t' < t}} \exp\left(-\frac{|\theta[x(t)] - \theta[x(t')]|^2}{2\delta\theta^2}\right) \tag{26}$$

Here, w is the height of each Gaussian, τ_G is the size of the time interval between successive Gaussian depositions, and $\delta\theta$ is the Gaussian width. It has been first empirically [145] then analytically [146] demonstrated that in the limit in which the CVs evolve according to a Langevin dynamics, the bias indeed converges to the negative of the free energy, thus providing an optimal bias to enhance transition events. Multiple simulations can also be used to allow for a quicker filling of the free-energy landscape [147].

The difference between the metadynamics estimate of the free energy and the true free energy can be shown to be related to the diffusion coefficient of the collective variables and to the rate at which the bias is grown. A possible way to decrease this error as a simulation progresses is to decrease the growth rate of the bias. Well-tempered metadynamics [148] used an optimized schedule to decrease the deposition rate of bias by modulating the Gaussian height:

$$w = \omega_0 \tau_G e^{-\frac{\Delta V(\theta,t)}{k_B \Delta T}} \tag{27}$$

Here, ω_0 is the initial "deposition rate", measured Gaussian height per unit time, and ΔT is a parameter that controls the degree to which the biased trajectory makes excursions away from free-energy minima. It is possible to show that using well-tempered metadynamics the bias does not converge to the negative of the free-energy but to a fraction of it, thus resulting in sampling the CVs at an effectively higher temperature $T + \Delta T$, where normal metadynamics is recovered for $\Delta T \to \infty$. We notice that other deposition schedules can be used aimed, e.g., at maximizing the number of round-trips in the CV space [149]. Importantly, it is possible to recover equilibrium Boltzmann statistics of *unbiased* collective variables from samples drawn throughout a well-tempered metadynamics trajectory [150]; it does not seem clear that one can do this from an ABF trajectory. Finally, it is possible to tune the shape of the Gaussians on the fly using schemes based on the geometric compression of the phase space or on the variance of the CVs [151].

In the well-tempered ensemble, the parameter ΔT can be used to tune the size of the explored region, in a fashion similar to the fictitious temperature in TAMD. So both TAMD and well-tempered metadynamics can be used to explore *relevant* regions of CV space while surmounting *relevant* free energy barriers. However, there are important distictions between the two methods. First, the main source of error in TAMD rests with how well mean-forces are approximated, and adiabatic separation, realizable only when the auxiliary variables z never move, is the only way to guarantee they are perfectly accurate. In practical application, TAMD never achieves perfect adiabatic separation. In contrast, because the deposition rate of decreases as a well-tempered trajectory progresses, errors related to poor adiabatic separation are progressively damped. Second, as already mentioned, TAMD alone cannot report the free energy, but it also is therefore not practically limited by the dimensionality of CV space; multicomponent gradients are just as accurately calculated in TAMD as are single-component gradients. Metadynamics, as a histogram-filling method, must exhaustively sample a finite region around any point to know the free energy and its gradients are correct, which can sometimes limit its utility.

Metadynamics is a powerful method whose popularity continues to grow. In either its original formulation or in more recent variants, metadynamics has been employed successfully in several fields, some of which we point out below with some representative examples:

(1) Chemical reactions [57,152];
(2) Peptide backbone angle sampling [153–155];
(3) Protein folding [156–159];
(4) Protein aggregation [160];
(5) Molecular docking [161–163] ;
(6) Conformational rearrangement of proteins [164];
(7) Crystal structure prediction [165];
(8) Nucleation and crystal growth [166,167];

(9) and proton diffusion [168].

2.4. Some Comments on Collective Variables

2.4.1. The Physical Fidelity of CV-Spaces

Given a potential $V(x)$, any multidimensional CV $\theta(x)$ has a mathematically determined free energy $F(z)$, and in principle the free-energy methods we describe here (and others) can use and/or compute it. However, this does not guarantee that F is meaningful, and a poor choice for $\theta(x)$ can render the results of even the most sophisticated free-energy methods useless for understanding the nature of actual metastable states and the transitions among them. This puts two major requirements on any CV space:

(1) Metastable states and transition states must be unambiguously identified as *energetically* separate regions in CV space.
(2) The CV space must not contain hidden barriers.

The first of these may seem obvious: CVs are chosen to provide a low-dimensional description of some important process, say a conformational change or a chemical reaction or a binding event, and one can not describe a process without being able to discriminate states. However, it is not always easy to find CVs that do this. Even given representative configurations of two distinct metastable states, standard MD from these two different initial configurations may sample partially overlapping regions of CV space, making ambiguous the assignation of an arbitrary configuration to a state. It may be in this case that the two representative configurations actually belong to the same state, or that if there are two states, that no matter what CV space is overlaid, the barrier separating them is so small that, on MD timescales, they can be considered rapidly exchanging substates of some larger state.

However, a third possibility exists: the two MD simulations mentioned above may in fact represent very different states. The overlap might just be an artifact of neglecting to include one or more CVs that are truly necessary to distinguish those states. If there is a significant free energy barrier along this neglected variable, an MD simulation will not cross it, yet may still sample regions in CV space also sampled by an MD simulation launched from the other side of this hidden barrier. And it is even worse: if TI or umbrella sampling is used along a pathway in CV space that neglects an important variable, the free-energy barriers along that pathway might be totally meaningless.

Hidden barriers can be a significant problem in CV-based free-energy calculations. Generally speaking, one only learns of a hidden barrier after postulating its existence and testing it with a new calculation. Detecting them is not straightforward and often involves a good deal of CV space exploration. Methods such as TAMD and well-tempered metadynamics offer this capability, but much more work could be done in the automated detection of hidden barriers and the "right" CVs (e.g., [169–171]).

An obvious way of reducing the likelihood of hidden barriers is to use increase the dimensionality of CV space. TAMD is well-suited to this because it is a gradient method, but standard metadynamics, because it is a histogram-filling method, is not. A recent variant of metadynamics termed "reconnaissance metadynamics" [172] does have the capability of handling high-dimensional CV spaces. In reconnaissance metadynamics, bias potential kernels are deposited at the CV space points identified as centers of clusters detected and measured by an on-the-fly clusterization scheme. These kernels are hyperspherically symmetric but grow as cluster sizes grow and are able to push a system out of a CV space basin to discover other basins. As such, reconnaissance metadynamics is an automated way of identifying free-energy minima in high-dimensional CV spaces. It has been applied the identification of configurations of small clusters of molecules [173] and identification of protein-ligand binding poses [162].

2.4.2. Some Common and Emerging Types of CVs

There are very few "best practices" codified for choosing CVs for any given system. Most CVs are developed ad hoc based on the processes that investigators would like to study, for instance, center-of-mass distance between two molecules for studying binding/unbinding, or torsion angles for studying conformational changes, or number of contacts for studying order-disorder transitions. Cartesian coordinates of centers of mass of groups of atoms are also often used as CVs, as they are functions of these coordinates.

The potential energy $V(x)$ is also an example of a 1-D CV, and there have been several examples of using it in CV-based enhanced sampling methods, such as umbrella sampling [174], metadynamics [175] well-tempered metadynamics [176]. In a recent work based on steered MD, it has been shown that also relevant reductions of the potential energy (e.g., the electrostatic interaction free-energy) can be used as effective CVs [177]. The basic rationale for enhanced sampling of V is that states with higher potential energy often correspond to transition states, and one need make no assumptions about precise physical mechanisms. Key to its successful use as a CV, as it is for any CV, is a proper accounting for its entropy; *i.e.*, the classical density-of-states.

Coarse-graining of particle positions onto Eulerian fields was used early on in enhanced sampling [178]; here, the value of the field at any Cartesian point is a CV, and the entire field represents a very high-dimensional CV. This idea has been put to use recently in the "indirect umbrella sampling" method of Patel *et al.* [179] for computing free energies of solvation, and string method (Section 4.2.) calculations of lipid bilayer fusion [180]. In a similar vein, there have been recent attempts at variables designed to count the recurrency of groups of atoms positioned according to given templates, such as α-helices paired β-strands in proteins [181].

We finally mention the possibility of building collective variables based on set of frames which might be available from experimental data or generated by means of previous MD simulations. Some of these variables are based on the idea of computing the distances between the present configuration and a set of precomputed snapshots. These distances, here indicated with d_i, where i is the index of the snapshot, are then combined to obtain a coarse representation of the present configuration, which is then used as a CV. As an example, one might combine the distances as

$$s = \frac{\sum_i e^{-\lambda d_i} i}{\sum_i e^{-\lambda d_i}} \qquad (28)$$

If the parameter λ is properly chosen, this function returns a continuous interpolation between the indexes of the snapshots which are closer to the present conformation. If the snapshots are disposed along a putative path connecting two experimental structures, this CV can be used as a path CV to monitor and bias the progression along the path [182]. A nice feature of path CVs is that it is straightforward to also monitor the distance from the putative path. The standard way to do it is by looking at the distance from the closest reference snapshot, which can be approximately computed with the following continuous function:

$$z = -\lambda^{-1} \log \sum_i e^{-\lambda d_i} \qquad (29)$$

This approach, modified to use internal coordinates, was used recently by Zinovjev *et al.* to study the aqueous phase reaction of pyruvate to salycilate, and in the CO bond-breaking/proton transfer in PchB [183].

A generalization to multidimensional paths (*i.e.*, sheets) can be obtained by assigning a generic vector v_i to each of the precomputed snapshots and computing its average [184]:

$$s = \frac{\sum_i e^{-\lambda d_i} v_i}{\sum_i e^{-\lambda d_i}} \qquad (30)$$

3. Tempering Approaches

"Tempering" refers to a class of methods based on increasing the temperature of an MD system to overcome barriers. Tempering relies on the fact that according to the Arrhenius law the rate at which activated (barrier-crossing) events happen is strongly dependent on the temperature. Thus, an annealing procedure where the system is first heated and then cooled allows one to produce quickly samples which are largely uncorrelated. The root of all these ideas indeed lies in the simulated annealing procedure [185], a well-known method successfully used in many optimization problems.

3.1. Simulated Tempering

Simulated annealing is a form of Markov-chain Monte Carlo sampling where the temperature is artificially modified during the simulation. In particular, sampling is initially done at a temperature high enough that the simulation can easily overcome high free-energy barriers. Then, the temperature is decreased as the simulation proceeds, thus smoothly bringing the simulation to a local energy minimum. In simulated annealing, a critical parameter is the cooling speed. Indeed, the probability to reach the global minimum grows as this speed is decreased.

The search for the global minimum can be interpreted in the same way as sampling an energy landscape at zero temperature. One could thus imagine to use simulated annealing to generate conformations at, e.g., room temperature by slowly cooling conformations starting at high temperature. However, the resulting ensemble will strongly depend on the cooling speed, thus possibly providing a biased result. A better approach consists of the the so-called simulated tempering methods [186]. Here, a discrete list of temperatures T_i, with $i \in 1 \dots N$ are chosen *a priori*, typically spanning a range going from the physical temperature of interest to a temperature which is high enough to overcome all relevant free energy barriers. (Note that we do not have to stipulate a CV-space in which those barriers live.) Then, the index i, which indicates at which temperature the system should be simulated, is evolved with time. Two kind of moves are possible: (a) normal evolution of the system at fixed temperature, which can be done with a usual Markov Chain Monte Carlo or molecular dynamics and (b) change of the index i at fixed atomic coordinates. It is easy to show that the latter can be performed as a Monte Carlo step with acceptance equal to

$$\alpha = \min\left(1, \frac{Z_j}{Z_i} e^{-\frac{U(x)}{k_B T_j} + \frac{U(x)}{k_B T_i}}\right) \tag{31}$$

where i and j are the indexes corresponding to the present temperature and the new one. The weights Z_i should be choosen so as to sample equivalently all the value of i. It must be noticed that also within molecular dynamics simulations only the potential energy usually appears in the acceptance. This is due to the fact that the velocities are typically scaled by a factor $\sqrt{\frac{T_j}{T_i}}$ upon acceptance. This scaling leads to a cancellation of the contribution to the acceptance coming from the kinetic energy. Ultimately, this is related to the fact that the ensemble of velocities is analytically known *a priori*, such that it is possible to adapt the velocities to the new temperature instantaneously.

Estimating these weights Z_i is nontrivial and typically requires a preliminary step. Moreover, if this estimate is poor the system could spend no time at the physical temperature, thus spoiling the result. Iterative algorithms for adjusting these weights have been proposed (see e.g., [187]). We also observe that since the temperature sets the typical value of the potential energy, an effect much similar to that of simulated tempering with adaptive weights can be obtained by performing a metadynamics simulation using the potential energy as a CV (Section 2.4.3).

3.2. Parallel Tempering

A smart way to alleviate the issue of finding the correct weights is that of simulating several replicas at the same time [188,189]. Rather that changing the temperature of a single system, the defining move proposal in parallel tempering consists of a coordinate swap between two T-replicas with acceptance probability

$$\alpha = \min\left(1, e^{\left(\frac{1}{k_B T_j} - \frac{1}{k_B T_i}\right)[U(x_j) - U(x_i)]}\right) \tag{32}$$

This method is the root of a class of techniques collectively known as "replica exchange" methods, and the latter name is often used as a synonimous of parallel tempering. Notably, within this framework it is not necessary to precompute a set of weights. Indeed, the equal time spent by each replica at each temperature is enforced by the constraint that only pairwise swaps are allowed. Moreover, parallel tempering has an additional advantage: since the replicas are weakly coupled and only interact when exchanges are attempted, they can be simulated on different computers without the need of a very fast interconnection (provided, of course, that a single replica is small enough to run on a single node).

The calculation of the acceptance is very cheap as it is based on the potential energy which is often computed alongside force evaluation. Thus, one could in theory exploit also a large number of virtual, rejected exchanges so as to enhance statistical sampling [190,191]. Since efficiency of parallel tempering simulation can deteriorate if the stride between subsequent exchanges is too large [192,193], a typical recipe is to choose this stride as small as possible, with the only limitation of avoiding extra costs due to replica synchronization. One can push this idea further and implement asynchronous versions of parallel tempering, where overhead related to exchanges is minimized [193,194]. One should be however aware that, especially at high exchange rate, artifacts coming from e.g., the use of wrong thermostating schemes could spoil the results [195,196].

Parallel tempering is popular in simulations of protein conformational sampling [197,198], protein folding [189,199–203] and aggregation [204,205], due at least in part to the fact that one need not choose CVs to use it, and CVs for describing these processes are not always straightforward to determine.

3.3. Generalized Replica Exchange

The difference between the replicas is not restricted to be a change in temperature. Any control parameter can be changed, and even the expression of the Hamiltonian can be modified [206]. In the most general case every replica is simulated at a different temperature (and or pressure) and a different Hamiltonian, and the acceptance reads

$$\alpha = \min\left(1, \frac{e^{-\left(\frac{U_i(x_j)}{k_B T_i} + \frac{U_j(x_i)}{k_B T_j}\right)}}{e^{-\left(\frac{U_i(x_i)}{k_B T_i} + \frac{U_j(x_j)}{k_B T_j}\right)}}\right) \tag{33}$$

Several recipes for choosing the modified Hamiltonian have been proposed in the literature [207–219]. Among these, a notable idea is that of solute tempering [208,217] which is used for the simulation of solvated biomolecules. Here, only the Hamiltonian of the solute is modified. More precisely, one could notice that a scaling of the Hamiltonian by a factor λ is completely equivalent to a scaling of the temperature by a factor λ^{-1}. Hamiltonian scaling however can take advantage of the fact that the total energy of the system is an extensive property. Thus, one can limit the scaling to the portion of the system which is considered to be interesting and which has the relevant bottlenecks. With solute tempering, the solute energy is scaled whereas the solvent energy is left unchanged. This is equivalent to keeping the solute at a high effective temperature and the solvent at the physical temperature. Since in the simulation of solvated molecules most of the atoms belong to the solvent, this turns in a much smaller modification to the explored ensemble when compared with parallel tempering. In spite of this, the effect on the solute resemble much that of increasing the physical temperature.

A sometimes-overlooked subtlety in solute tempering is the choice for the treatment of solvent-solute interactions. Indeed, whereas solute-solute interactions are scaled with a factor $\lambda < 1$ and solvent-solvent interactions are not scaled, any intermediate choice (scaling factor between λ and 1) could intuitively make sense for solvent-solute coupling. In the original formulation, the authors used a factor $(1 + \lambda)/2$ for the solute-solvent interaction. This choice however was later shown to be suboptimal [217,220], and refined to be $\sqrt{\lambda}$. This latter choice appears to be more physically sound, since it allows one to just simulate the biased replicas with a modified force-field. Indeed, if one scales the charges of the solute by a factor $\sqrt{\lambda}$, electrostatic interactions are changed by a factor λ for solute-solute coupling and $\sqrt{\lambda}$ for solute-solvent coupling. The same is true for Lennard-Jones terms, albeit in this case it depends on the specific combination rules used. Notably, the same rules for scaling were used in a previous work [209]. As a final remark, we point out that solute tempering can be also used in a serial manner *a là* simulated tempering, in a simulated solute tempering scheme [221].

3.4. General Comments

In general, the advantage of these tempering methods over straighforward sampling can be rationalized as follows. A simulation is evolved so as to sample a modified ensemble by e.g., raising temperature or artificially modifying the Hamiltonian. The change in the ensemble could be drastic, so that trying to extract canonical averages by reweighting from such a simulation would be pointless. For this reason, a ladder of intermediate ensembles is built, interpolating between the physical one (*i.e.*, room temperature, physical Hamiltonian) and the modified one. Then, transitions between consecutive steps in this ladder (or, in parallel schemes, coordinate swaps) are performed using a Monte Carlo scheme. Assuming that the dynamics of the most modified ensemble is ergodic, independent samples will be generated every time a new simulation reaches the highest step of the ladder. Thus, efficiency of these methods is often based on the evaluation of the round trip time required for a replica to traverse the entire ladder.

Tempering methods are thus relying on the ergodicity of the most modified ensemble. This assumption is not always correct. A very simple example is parallel tempering used to accelerate the sampling over an entropic barrier. Since the height of an entropic barrier grows with the temperature, in this conditions the barrier in the most modified ensembles are unaffected [222]. Moreover, since a lot of time is spent in sampling states in non-physical situations (e.g., high temperature), the overall computational efficiency could even be lower than that of straightforward sampling. Real applications are often in an intermediate situation, and usefulness of parallel tempering should be evaluated case by case.

The number of intermediate steps in the ladder can be shown to grow with the square root of the specific heat of the system in the case of parallel tempering simulations. No general relationship can be drawn in the case of Hamiltonian replica exchange, but one can expect approximately that the number of replicas should be proportional to the square root of the number of degrees of freedom affected by the modification of the Hamiltonian. Thus, Hamiltonian replica exchange methods could be much more effective than simple parallel tempering as they allow the effort to be focused and the number of replicas to be minimized.

Parallel tempering has the advantage that all the replicas can be analyzed to obtain meaningful results, e.g., to predict the melting curve of a molecule. This procedure should be used with caution, especially with empirically parametrized potentials, which are often tuned to be realistic only at room temperature. On the other hand, Hamiltonian replica exchange often relies on unphysically modified ensembles which have no interest but for the fact that they increase ergodicity.

As a final note, we observe that data obtained at different temperature (or with modified Hamiltonians) could be combined to enhance statistics at the physical temperature [223]. However, the effectiveness of this data recycling is limited by the fact that high temperature replicas visit very rarely low energy conformations, thus decreasing the amount of additional information that can be extracted.

4. Combinations and Advanced Approaches

4.1. Combination of Tempering Methods and Biased Sampling

The algorithms presented in Section 3 and based on tempering are typically considered to be simpler to apply when compared with those discussed in Section 2 and based on biasing the sampling of selected collective variables. Indeed, by avoiding the problem of choosing collective variables which properly describe the reaction path, most of the burden of setting up a simulation is removed. However, this comes at a price: considering the computational cost, tempering methods are extremely expensive. This cost is related to the fact that they are able to accelerate all degrees of freedom to the same extent, without an *a priori* knowledge of the sampling bottlenecks. In this sense, Hamiltonian replica exchange methods are in an intermediate situation, since they are typically less expensive than parallel tempering but allow to embed part of the knowledge of the system in the simulation set up.

Because of the conceptual difference between tempering methods and CV-based methods, these approaches can be easily and efficiently combined. As an example, the combination of metadynamics and parallel tempering can be used to take advantage of the known bottlenecks with biased collective variables at the same time accelerating the overall sampling with parallel tempering [156]. In that work, the free energy landscape for the folding of a small hairpin was computed by biasing a small number of selected CVs (gyration radius and the number of hydrogen bonds). These CVs alone are not enough to describe folding, as can be easily shown by performing a metadynamics simulation using these CVs. However, the combination with parallel tempering allowed acceleration of all the degrees of freedom blindly and reversible folding of the hairpin. This combined approach also improves the results when compared with parallel tempering alone, since it accelerates exploration of phase-space. Moreover, since parallel tempering samples the unbiased canonical distribution, it is very difficult to use it to compute free-energy differences which are larger than a few $k_B T$. The metadynamics bias can be used to disfavor, e.g., the folded state so as to better estimate the free-energy difference between the folded and unfolded states.

It is also possible to combine metadynamics with the solute tempering method so as to decrease the number of required replicas and the computational cost [224]. As an alternative to solute tempering, metadynamics in the well-tempered ensemble can be effectively used to enhance the acceptance in parallel tempering simulations and to decrease the number of necessary replicas [176]. This combination of parallel tempering with well-tempered ensemble can be pushed further and combined with metadynamics on a few selected degrees of freedom [225]. As a final note, bias exchange metadynamics [226] combines metadynamics and replica echange in a completely different spirit: every replica is run using a different CV, thus allowing many CVs to be tried at the same time.

142

This technique has been succesfully applied to several problems. For a recent review, we refer the reader to [227].

4.2. Some Methods Based on TAMD

4.2.1. String Method in Collective Variables

The string method is generally an approach to find pathways of minimal energy connecting two points in phase space [228]. When working in CVs, the string method is used to find minimal free-energy paths (MFEP's) [229]. String method calculations involve multiple replicas, each representing a point z_s in CV space at position s along a discretized string connecting two points of interest (reactant and product states, say). The forces on each replica's z_s are computed and their z_s's updated, as in TAMD, with the addition of forces that act to keep the z's equidistant along the string (so-called reparameterization forces):

$$\bar{\gamma}\dot{z}_j(s,t) = \sum_k \left[\tilde{M}_{jk}(\mathbf{x}(s,t))\kappa[\theta_k(\mathbf{x}(s,t)) - z_k(s,t)] \right] + \eta_z(t) + \lambda(s,t)\frac{\partial z_j}{\partial s} \qquad (34)$$

Here, $\tilde{M}_{jk}$ is the metric tensor mapping distances on the manifold of atomic coordinates to the manifold of CV space, η is thermal noise and $\lambda(s,t)\frac{\partial z_j}{\partial s}$ represents the reparameterization force tangent to the string that is sufficient to maintain equidistant images along the string. String method has been used to study activation of the insulin-receptor kinase [63], docking of insulin to its receptor [230], and myosin [231]. In these examples, the update of the string coordinates is done at a lower frequency than the atomic variables in each image.

In contrast, in the on-the-fly variant of string method in CVs, the friction on the z_s's is set high enough to make the effective averaging of the forces approach the true mean forces, and the z updates occur in lockstep with the x updates of the MD system [232]. Just as in TAMD, the atomic variables obey an equation of motion like Equation (12) tethering them to the z_s. Stober and Abrams recently demonstrated an implementation of on-the-fly string method to study the thermodynamics of the normal-to-amyloidogenic transition of $\beta 2$-microglobulin [233]. Unique in this approach was the construction of a single composite MD system containing 27 individual $\beta 2$ molecules restrained to points on $3 \times 3 \times 3$ grid inside a single large solvent box. Zinovjev *et al.* used a combination of the on-the-fly string method and of path-collective variables (see Equations (28) and (29)) in a quantum-mechanics/molecular-mechanics approach to study a methyltransferase reaction [234].

4.2.2. On-the-Fly Free Energy Parameterization

Because TAMD provides mean-force estimates as it is exploring CV space, it stands to reason that those mean forces could be used to compute a free energy. In contrast, in the single-sweep method [68], the TAMD forces are only used in the CV space exploration phase, not the free-energy calculation itself. Recently, Abrams and Vanden-Eijnden proposed a method for using TAMD directly to *parameterize* a free energy; that is, to determine the best set of some parameters $\boldsymbol{\lambda}$ on which a free energy of known functional form depends [235]:

$$F(\boldsymbol{z}) = F(\boldsymbol{z}; \boldsymbol{\lambda}^*) \qquad (35)$$

The approach, termed "on-the-fly free energy parameterization", uses forces from a running TAMD simulation to progressively optimize $\boldsymbol{\lambda}$ using a time-averaged gradient error:

$$E(\boldsymbol{\lambda}) = \frac{1}{2t} \int_0^t |\nabla_z F\left[\boldsymbol{z}(s), \boldsymbol{\lambda}(t)\right] + \kappa\left[\theta(\boldsymbol{x}(s)) - \boldsymbol{z}(s)\right]|^2 \, ds \qquad (36)$$

If constructed so that F is linear in $\boldsymbol{\lambda} = (\lambda_1, \lambda_2, \ldots, \lambda_M)$, minimization of E can be expressed as a simple linear algebra problem

$$\sum_j A_{ij}\lambda_j = b_i, \quad i = 1, \ldots, M \qquad (37)$$

and the running TAMD simulation provides progressively better estimates of A and b until the $\boldsymbol{\lambda}$ converge. In the cited work, it was shown that this method is an efficient way to derive potentials of mean force between particles in coarse-grained molecular simulations as basis-function expansions. It is currently being investigated as a means to parameterize free energies associated with conformational changes of proteins.

Chen, Cuendet, and Tuckermann developed a very similar approach that in addition to parameterizing a free energy using d-AFED-computed gradients uses a metadynamics-like bias on the potential [236]. These authors demonstrated efficient reconstruction of the four-dimensional free-energy of vacuum alanine dipeptide with this approach.

5. Conclusions

In this review, we have summarized some of the current and emerging enhanced sampling methods that sit atop MD simulation. These have been broadly classified as methods that use collective variable biasing and methods that use tempering. CV biasing is a much more prevalent approach than tempering, due partially to the fact that it is perceived to be cheaper, since tempering simulations are really only useful for enhanced sampling of configuration space when run in parallel. CV-biasing also reflects the desire to rein in the complexity of all-atom simulations by projecting configurations into a much lower dimensional space. (Parallel tempering can be thought of as increasing the dimensionality of the system by a factor equal to the number of simulated replicas.) But the drawback of all CV-biasing approaches is the risk that the chosen CV space does not provide the most faithful representation of the true spectrum of metastable subensembles and the barriers that separate them. Guaranteeing that sampling of CV space is not stymied by hidden barriers must be of paramount concern in the continued evolution of such methods. For this reason, methods that specifically allow broad exploration of CV space, like TAMD (which can handle large numbers of CVs) and well-tempered metadynamics will continue to be valuable. So too will parallel tempering because its broad sampling of configuration space can be used to inform the choice of better CVs. Accelerating development of combined CV-tempering methods bodes well for enhanced sampling generally.

Although some of these methods involve time-varying forces (ABF, TAMD, and metadynamics), all methods we've discussed have the underlying rationale of the equilibrium ensemble. TI uses the constrained ensemble, ABF and metadynamics ideally converge to an ensemble in which a bias erases free-energy variations, and TAMD samples an attenuated/mollified equilibrium ensemble. There is an entirely separate class of methods that inherently rely on *non-equilibrium* thermodynamics. We have not discussed at all the several free-energy methods based on non-equilibrium MD simulations; we refer interested readers to the article by Christoph Dellago and Gerhard Hummer in this issue.

Finally, we have also not really touched on any of the practical issues of implementing and using these methods in conjunction with modern MD packages (e.g., NAMD [237], LAMMPS [238], Gromacs [239], Amber [240], and CHARMM [241], to name a few). At least two packages (NAMD and CHARMM) have native support for collective variable biasing, and NAMD in particular offers both native ABF and a TcL-based interface which has been used to implement TAMD [58]. The native collective variable module for NAMD has been recently ported to LAMMPS [242]. Gromacs offers native support for parallel tempering. Generally speaking, however, modifying MD codes to handle CV-biasing and multiple replicas is not straightforward, since one would like access to the data structures that store coordinates and forces. A major help in this regard is the PLUMED package [243,244], which patches a variety of MD codes to enable users to use many of the techniques discussed here.

Acknowledgments

CFA would like to acknowledge support of NSF (DMR-1207389) and NIH (1R01GM100472). GB would like to acknowledge the European Research Council (Starting Grant S-RNA-S, no. 306662) for financial support. Both authors would like to acknowledge NSF support of a recent Pan-American Advanced Studies Institute Workshop "Molecular-based Multiscale Modeling and Simulation" (OISE-1124480; PI: W. J. Pfaednter, U. Washington) held in Montevideo, Uruguay, 12–15 September 2012, where the authors met and began discussions that influenced the content of this review.

Conflicts of Interest

The authors declare no conflict of interest.

References

1. Kollman, P. Free-energy calculations—Applications to chemical and biochemical phenomena. *Chem. Rev.* **1993**, *93*, 2395–2417.
2. Trzesniak, D.; Kunz, A.P.E.; van Gunsteren, W.F. A comparison of methods to compute the potential of mean force. *Chem. Phys. Chem* **2007**, *8*, 162–169.

3. Vanden-Eijnden, E. Some recent techniques for free energy calculations. *J. Comput. Chem.* **2009**, *30*, 1737–1747.

4. Dellago, C.; Bolhuis, P.G. Transition path sampling and other advanced simulation techniques for rare events. In *Advanced Computer Simulation Approaches for Soft Matter Sciences III*; Springer: Berlin/Heidelberg, Germany, 2009; pp. 167–233.

5. Christ, C.D.; Mark, A.E.; van Gunsteren, W.F. Basic ingredients of free energy calculations: A review. *J. Comput. Chem.* **2010**, *31*, 1569–1582.

6. Darve, E.; Rodriguez-Gomez, D.; Pohorille, A. Adaptive biasing force method for scalar and vector free energy calculations. *J. Chem. Phys.* **2008**, *128*, doi:10.1063/1.2829861.

7. Maragliano, L.; Vanden-Eijnden, E. A temperature-accelerated method for sampling free energy and determining reaction pathways in rare events simulations. *Chem. Phys. Lett.* **2006**, *426*, 168–175.

8. Laio, A.; Parrinello, M. Escaping free-energy minima. *Proc. Natl. Acad. Sci. USA* **2002**, *99*, 12562–12566.

9. Schlitter, J.; Engels, M.; Kruger, P.; Jacoby, E.; Wollmer, A. Targeted molecular-dynamics simulation of conformational change—Application to the T[–]T transition in insulin. *Mol. Sim.* **1993**, *10*, 291–308.

10. Grubmüller, H.; Heymann, B.; Tavan, P. Ligand binding: Molecular mechanics calculation of the streptavidin biotin rupture force. *Science* **1996**, *271*, 997–999.

11. Kirkwood, J.G. Statistical mechanics of fluid mixtures. *J. Chem. Phys.* **1935**, *3*, 300–313.

12. Carter, E.; Ciccotti, G.; Hynes, J.T.; Kapral, R. Constrained reaction coordinate dynamics for the simulation of rare events. *Chem. Phys. Lett.* **1989**, *156*, 472–477.

13. Sprik, M.; Ciccotti, G. Free energy from constrained molecular dynamics. *J. Chem. Phys.* **1998**, *109*, 7737–7744.

14. Ciccotti, G.; Ferrario, M. Blue moon approach to rare events. *Mol. Sim.* **2004**, *30*, 787–793.

15. Ciccotti, G.; Kapral, R.; Vanden-Eijnden, E. Blue moon sampling, vectorial reaction coordinates, and unbiased constrained dynamics. *Chem. Phys. Chem.* **2005**, *6*, 1809–1814.

16. Den Otter, W.; Briels, W. The calculation of free-energy differences by constrained molecular-dynamics simulations. *J. Chem. Phys.* **1998**, *109*, 4139–4146.

17. Schlitter, J.; Klähn, M. A new concise expression for the free energy of a reaction coordinate. *J. Chem. Phys.* **2003**, *118*, 2057–2060.

18. Depaepe, J.; Ryckaert, J.; Paci, E.; Ciccotti, G. Sampling of molecular-conformations by molecular-dynamics techniques. *Mol. Phys.* **1993**, *79*, 515–522.

19. Zhao, X.; Rignall, T.R.; McCabe, C.; Adney, W.S.; Himmel, M.E. Molecular simulation evidence for processive motion of Trichoderma reesei Cel7A during cellulose depolymerization. *Chem. Phys. Lett.* **2008**, *460*, 284–288.

20. Kim, H.; Goddard, W.A., III; Jang, S.S.; Dichtel, W.R.; Heath, J.R.; Stoddart, J.F. Free energy barrier for molecular motions in bistable [2]rotaxane molecular electronic devices. *J. Phys. Chem. A* **2009**, *113*, 2136–2143.

21. Hytha, M.; Stich, I.; Gale, J.; Terakura, K.; Payne, M. Thermodynamics of catalytic formation of dimethyl ether from methanol in acidic zeolites. *Chem. Eur. J.* **2001**, *7*, 2521–2527.

22. Fois, E.; Gamba, A.; Spano, E. Competition between water and hydrogen peroxide at Ti center in Titanium zeolites. An ab initio study. *J. Phys. Chem. B* **2004**, *108*, 9557–9560.

23. Ivanov, I.; Klein, M. Dynamical flexibility and proton transfer in the arginase active site probed by ab initio molecular dynamics. *J. Am. Chem. Soc.* **2005**, *127*, 4010–4020.

24. Stubbs, J.; Marx, D. Aspects of glycosidic bond formation in aqueous solution: Chemical bonding and the role of water. *Chem.-A Eur. J.* **2005**, *11*, 2651–2659.

25. Trinh, T.T.; Jansen, A.P.J.; van Santen, R.A.; Meijer, E.J. The role of water in silicate oligomerization reaction. *Phys. Chem. Chem. Phys.* **2009**, *11*, 5092–5099.

26. Liu, P.; Cai, W.; Chipot, C.; Shao, X. Thermodynamic insights into the dynamic switching of a cyclodextrin in a bistable molecular shuttle. *J. Phys. Chem. Lett.* **2010**, *1*, 1776–1780.

27. Bucko, T.; Hafner, J. Entropy effects in hydrocarbon conversion reactions: Free-energy integrations and transition-path sampling. *J. Phys.-Cond. Mat.* **2010**, *22*, doi:10.1088/0953-8984/22/38/384201.

28. Paci, E.; Marchi, M. Membrane crossing by a polar molecule—a molecular-dynamics simulation. *Mol. Sim.* **1994**, *14*, 1–10.

29. Sa, R.; Zhu, W.; Shen, J.; Gong, Z.; Cheng, J.; Chen, K.; Jiang, H. How does ammonium dynamically interact with benzene in aqueous media? A first principle study using the Car-Parrinello molecular dynamics method. *J. Phys. Chem. B* **2006**, *110*, 5094–5098.

30. Mugnai, M.; Cardini, G.; Schettino, V.; Nielsen, C.J. Ab initio molecular dynamics study of aqueous formaldehyde and methanediol. *Mol. Phys.* **2007**, *105*, 2203–2210.

31. Chunsrivirot, S.; Trout, B.L. Free energy of binding of a small molecule to an amorphous polymer in a solvent. *Langmuir* **2011**, *27*, 6910–6919.

32. Sato, M.; Yamataka, H.; Komeiji, Y.; Mochizuki, Y. FMO-MD simulations on the hydration of formaldehyde in water solution with constraint dynamics. *Chem. Eur. J.* **2012**, *18*, 9714–9721.

33. Sergi, A.; Ciccotti, G.; Falconi, M.; Desideri, A.; Ferrario, M. Effective binding force calculation in a dimeric protein by molecular dynamics simulation. *J. Chem. Phys.* **2002**, *116*, 6329–6338.

34. Maragliano, L.; Ferrario, M.; Ciccotti, G. Effective binding force calculation in dimeric proteins. *Mol. Sim.* **2004**, *30*, 807–816.

35. Darve, E.; Pohorille, A. Calculating free energies using average force. *J. Chem. Phys.* **2001**, *115*, 9169–9183.

36. Fogolari, F.; Corazza, A.; Varini, N.; Rotter, M.; Gumral, D.; Codutti, L.; Rennella, E.; Viglino, P.; Bellotti, V.; Esposito, G. Molecular dynamics simulation of beta(2)-microglobulin in denaturing and stabilizing conditions. *Proteins* **2011**, *79*, 986–1001.

37. Faller, C.E.; Reilly, K.A.; Hills, R.D., Jr.; Guvench, O. Peptide backbone sampling convergence with the adaptive biasing force algorithm. *J. Phys. Chem. B* **2013**, *117*, 518–526.

38. Wei, C.; Pohorile, A. Permeation of nucleosides through lipid bilayers. *J. Phys. Chem. B* **2011**, *115*, 3681–3688.

39. Vivcharuk, V.; Kaznessis, Y.N. Thermodynamic analysis of protegrin-1 insertion and permeation through a lipid bilayer. *J. Phys. Chem. B* **2011**, *115*, 14704–14712.

40. Kraszewski, S.; Tarek, M.; Ramseyer, C. Uptake and translocation mechanisms of cationic amino derivatives functionalized on pristine C-60 by lipid membranes: A molecular dynamics simulation study. *ACS Nano* **2011**, *5*, 8571–8578.

41. Kraszewski, S.; Bianco, A.; Tarek, M.; Ramseyer, C. Insertion of short amino-functionalized single-walled carbon nanotubes into phospholipid bilayer occurs by passive diffusion. *PLoS One* **2012**, *7*, e40703.

42. Liu, X.; Lu, X.; Meijer, E.J.; Wang, R.; Zhou, H. Acid dissociation mechanisms of Si(OH)(4) and Al(H2O)(6)(3+) in aqueous solution. *Geochim. Cosmochim. Acta* **2010**, *74*, 510–516.

43. Caballero, J.; Poblete, H.; Navarro, C.; Alzate-Morales, J.H. Association of nicotinic acid with a poly(amidoamine) dendrimer studied by molecular dynamics simulations. *J. Mol. Graph. Model.* **2013**, *39*, 71–78.

44. Wilson, M.A.; Wei, C.; Bjelkmar, P.; Wallace, B.A.; Pohorille, A. Molecular dynamics simulation of the antiamoebin ion channel: Linking structure and conductance. *Biophys. J.* **2011**, *100*, 2394–2402.

45. Cheng, M.H.; Coalson, R.D. Molecular dynamics investigation of Cl- and water transport through a eukaryotic CLC transporter. *Biophys. J.* **2012**, *102*, 1363–1371.

46. Wang, S.; Orabi, E.A.; Baday, S.; Berneche, S.; Lamoureux, G. Ammonium transporters achieve charge transfer by fragmenting their substrate. *J. Am. Chem. Soc.* **2012**, *134*, 10419–10427.

47. Tillman, T.; Cheng, M.H.; Chen, Q.; Tang, P.; Xu, Y. Reversal of ion-charge selectivity renders the pentameric ligand-gated ion channel GLIC insensitive to anaesthetics. *Biochem. J.* **2013**, *449*, 61–68.

48. Akhshi, P.; Acton, G.; Wu, G. Molecular dynamics simulations to provide new insights into the asymmetrical ammonium ion movement inside of the [d(G(3)T(4)G(4))](2) G-quadruplex DNA structure. *J. Phys. Chem. B* **2012**, *116*, 9363–9370.

49. Kamath, G.; Bhatnagar, N.; Baker, G.A.; Baker, S.N.; Potoff, J.J. Computational prediction of ionic liquid 1-octanol/water partition coefficients. *Phys. Chem. Chem. Phys.* **2012**, *14*, 4339–4342.

50. Bhatnagar, N.; Kamath, G.; Chelst, I.; Potoff, J.J. Direct calculation of 1-octanol-water partition coefficients from adaptive biasing force molecular dynamics simulations. *J. Chem. Phys.* **2012**, *137*, doi:10.1063/1.4730040.

51. Wereszczynski, J.; McCammon, J.A. Nucleotide-dependent mechanism of Get3 as elucidated from free energy calculations. *Proc. Natl. Acad. Sci. USA* **2012**, *109*, 7759–7764.

52. Jana, A.K.; Sengupta, N. Adsorption mechanism and collapse propensities of the full-length, monomeric a beta(1-42) on the surface of a single-walled carbon nanotube: A molecular dynamics simulation study. *Biophys. J.* **2012**, *102*, 1889–1896.

53. Uddin, N.M.; Capaldi, F.; Farouk, B. Molecular dynamics simulations of carbon nanotube interactions in water/surfactant systems. *J. Eng. Mater.-T. ASME* **2010**, *132*, doi:10.1115/1.4000231.

54. Kaestner, J. Umbrella integration in two or more reaction coordinates. *J. Chem. Phys.* **2009**, *131*, doi:10.1063/1.3175798.

55. Rosso, L.; Tuckerman, M. An adiabatic molecular dynamics method for the calculation of free energy profiles. *Mol. Sim.* **2002**, *28*, 91–112.

56. Abrams, J.B.; Tuckerman, M.E. Efficient and direct generation of multidimensional free energy surfaces via adiabatic dynamics without coordinate transformations. *J. Phys. Chem. B* **2008**, *112*, 15742–15757.

57. Iannuzzi, M.; Laio, A.; Parrinello, M. Efficient exploration of reactive potential energy surfaces using Car-Parrinello molecular dynamics. *Phys. Rev. Lett.* **2003**, *90*, 238302.

58. Abrams, C.; Vanden-Eijnden, E. Large-scale conformational sampling of proteins using temperature-accelerated molecular dynamics. *Proc. Natl. Acad. Sci. USA* **2010**, *107*, 4961–4966.

59. Vashisth, H.; Skiniotis, G.; Brooks, C.L., III. Using enhanced sampling and structural restraints to refine atomic structures into low-resolution electron microscopy maps. *Structure* **2012**, *20*, 1453–1462.

60. Vashisth, H.; Skiniotis, G.; Brooks, C.L., III. Enhanced sampling and overfitting analyses in structural refinement of nucleic acids into electron microscopy maps. *J. Phys. Chem. B* **2013**, *117*, 3738–3746.

61. Vashisth, H.; Brooks, C.L., III. Conformational sampling of maltose-transporter components in cartesian collective variables is governed by the low-frequency normal modes. *J. Phys. Chem. Lett.* **2012**, *3*, 3379–3384.

62. Hu, Y.; Hong, W.; Shi, Y.; Liu, H. Temperature-accelerated sampling and amplified collective motion with adiabatic reweighting to obtain canonical distributions and ensemble averages. *J. Chem. Theory Comput.* **2012**, *8*, 3777–3792.

63. Vashisth, H.; Abrams, C.F. DFG-flip in the insulin receptor kinease is facilitated by a helical intermediate state of the activation loop. *Biophys. J.* **2012**, *102*, 1979–1987.

64. Maragliano, L.; Cottone, G.; Ciccotti, G.; Vanden-Eijnden, E. Mapping the network of pathways of CO diffusion in myoglobin. *J. Am. Chem. Soc.* **2010**, *132*, 1010–1017.

65. Lapelosa, M.; Abrams, C.F. A computational study of water and CO migration sites and channels inside myoglobin. *J. Chem. Theory Comput.* **2013**, *9*, 1265–1271.

66. Geslin, P.A.; Ciccotti, G.; Meloni, S. An observable for vacancy characterization and diffusion in crystals. *J. Chem. Phys.* **2013**, *138*, doi:10.1063/1.4796322.

67. Lucid, J.; Meloni, S.; MacKernan, D.; Spohr, E.; Ciccotti, G. Probing the structures of hydrated nafion in different morphologies using temperature-accelerated molecular dynamics simulations. *J. Phys. Chem. C* **2013**, *117*, 774–782.

68. Maragliano, L.; Vanden-Eijnden, E. Single-sweep methods for free energy calculations. *J. Chem. Phys.* **2008**, *128*, 184110.

69. Torrie, G.M.; Valleau, J.P. Nonphysical sampling distributions in Monte Carlo free-energy estimation: Umbrella sampling. *J. Comput. Phys.* **1977**, *23*, 187–199.

70. Kumar, S.; Rosenberg, J.M.; Bouzida, D.; Swendsen, R.H.; Kollman, P.A. The weighted histogram analysis method for free-energy calculations on biomolecules. I. The method. *J. Comput. Chem.* **1992**, *13*, 1011–1021.

71. Kaestner, J.; Thiel, W. Bridging the gap between thermodynamic integration and umbrella sampling provides a novel analysis method: "Umbrella integration". *J. Chem. Phys.* **2005**, *123*, 144104.

72. Kaestner, J. Umbrella sampling. *Wires. Comput. Mol. Sci.* **2011**, *1*, 932–942.

73. Schaefer, M.; Bartels, C.; Karplus, M. Solution conformations and thermodynamics of structured peptides: Molecular dynamics simulation with an implicit solvation model. *J. Mol. Biol.* **1998**, *284*, 835–848.

74. Banavali, N.; MacKerell, A. Free energy and structural pathways of base flipping in a DNA GCGC containing sequence. *J. Mol. Biol.* **2002**, *319*, 141–160.

75. Cruz, V.; Ramos, J.; Martinez-Salazar, J. Water-mediated conformations of the alanine dipeptide as revealed by distributed umbrella sampling simulations, quantum mechanics based calculations, and experimental data. *J. Phys. Chem. B* **2011**, *115*, 4880–4886.

76. Islam, S.M.; Richards, M.R.; Taha, H.A.; Byrns, S.C.; Lowary, T.L.; Roy, P.N. Conformational analysis of oligoarabinofuranosides: Overcoming torsional barriers with umbrella sampling. *J. Chem. Theory Comput.* **2011**, *7*, 2989–3000.

77. Young, W.; Brooks, C. A microscopic view of helix propagation: N and C-terminal helix growth in alanine helices. *J. Mol. Biol.* **1996**, *259*, 560–572.

78. Sheinerman, F.; Brooks, C. Calculations on folding of segment B1 of streptococcal protein G. *J. Mol. Biol.* **1998**, *278*, 439–456.

79. Bursulaya, B.; Brooks, C. Folding free energy surface of a three-stranded beta-sheet protein. *J. Am. Chem. Soc.* **1999**, *121*, 9947–9951.

80. Rick, S.; Erickson, J.; Burt, S. Reaction path and free energy calculations of the transition between alternate conformations of HIV-1 protease. *Proteins* **1998**, *32*, 7–16.

81. Allen, T.; Andersen, O.; Roux, B. Structure of gramicidin A in a lipid bilayer environment determined using molecular dynamics simulations and solid-state NMR data. *J. Am. Chem. Soc.* **2003**, *125*, 9868–9877.

82. Shams, H.; Golji, J.; Mofrad, M.R.K. A molecular trajectory of alpha-actinin activation. *Biophys. J.* **2012**, *103*, 2050–2059.

83. Yildirim, I.; Park, H.; Disney, M.D.; Schatz, G.C. A dynamic structural model of expanded RNA CAG repeats: A refined X-ray structure and computational investigations using molecular dynamics and umbrella sampling simulations. *J. Am. Chem. Soc.* **2013**, *135*, 3528–3538.

84. Masunov, A.; Lazaridis, T. Potentials of mean force between ionizable amino acid side chains in water. *J. Am. Chem. Soc.* **2003**, *125*, 1722–1730.

150

85. Tarus, B.; Straub, J.; Thirumalai, D. Probing the initial stage of aggregation of the a beta(10-35)-protein: Assessing the propensity for peptide dimerization. *J. Mol. Biol.* **2005**, *345*, 1141–1156.

86. Makowski, M.; Liwo, A.; Sobolewski, E.; Scheraga, H.A. Simple physics-based analytical formulas for the potentials of mean force of the interaction of amino-acid side chains in water. V. Like-charged side chains. *J. Phys. Chem. B* **2011**, *115*, 6119–6129.

87. Casalini, T.; Salvalaglio, M.; Perale, G.; Masi, M.; Cavallotti, C. Diffusion and aggregation of sodium fluorescein in aqueous solutions. *J. Phys. Chem. B* **2011**, *115*, 12896–12904.

88. Wanasundara, S.N.; Krishnamurthy, V.; Chung, S.H. Free energy calculations of gramicidin dimer dissociation. *J. Phys. Chem. B* **2011**, *115*, 13765–13770.

89. Zhang, B.W.; Brunetti, L.; Brooks, C.L., III. Probing pH-dependent dissociation of HdeA dimers. *J. Am. Chem. Soc.* **2011**, *133*, 19393–19398.

90. Periole, X.; Knepp, A.M.; Sakmar, T.P.; Marrink, S.J.; Huber, T. Structural determinants of the supramolecular organization of G protein-coupled receptors in bilayers. *J. Am. Chem. Soc.* **2012**, *134*, 10959–10965.

91. Vijayaraj, R.; van Damme, S.; Bultinck, P.; Subramanian, V. Molecular dynamics and umbrella sampling study of stabilizing factors in cyclic peptide-based nanotubes. *J. Phys. Chem. B* **2012**, *116*, 9922–9933.

92. Mahdavi, S.; Kuyucak, S. Why the drosophila shaker K+ channel is not a good model for ligand binding to voltage-gated Kv1 channels. *Biochemistry* **2013**, *52*, 1631–1640.

93. Banavali, N.; Roux, B. Free energy landscape of A-DNA to B-DNA conversion in aqueous solution. *J. Am. Chem. Soc.* **2005**, *127*, 6866–6876.

94. Giudice, E.; Varnai, P.; Lavery, R. Base pair opening within B-DNA: Free energy pathways for GC and AT pairs from umbrella sampling simulations. *Nucl. Acids Res.* **2003**, *31*, 1434–1443.

95. Matek, C.; Ouldridge, T.E.; Levy, A.; Doye, J.P.K.; Louis, A.A. DNA cruciform arms nucleate through a correlated but asynchronous cooperative mechanism. *J. Phys. Chem. B* **2012**, *116*, 11616–11625.

96. Bagai, S.; Sun, C.; Tang, T. Potential of mean force of polyethylenimine-mediated DNA attraction. *J. Phys. Chem. B* **2013**, *117*, 49–56.

97. Czaplewski, C.; Rodziewicz-Motowidlo, S.; Liwo, A.; Ripoll, D.; Wawak, R.; Scheraga, H. Molecular simulation study of cooperativity in hydrophobic association. *Protein Sci.* **2000**, *9*, 1235–1245.

98. Lee, M.; Olson, M. Calculation of absolute protein-ligand binding affinity using path and endpoint approaches. *Biophys. J.* **2006**, *90*, 864–877.

99. Peri, S.; Karim, M.N.; Khare, R. Potential of mean force for separation of the repeating units in cellulose and hemicellulose. *Carbohyd. Res.* **2011**, *346*, 867–871.

100. St-Pierre, J.F.; Karttunen, M.; Mousseau, N.; Rog, T.; Bunker, A. Use of umbrella sampling to calculate the entrance/exit pathway for Z-Pro-Prolinal inhibitor in prolyl oligopeptidase. *J. Chem. Theory Comput.* **2011**, *7*, 1583–1594.

101. Rashid, M.H.; Kuyucak, S. Affinity and selectivity of ShK toxin for the Kv1 potassium channels from free energy simulations. *J. Phys. Chem. B* **2012**, *116*, 4812–4822.

102. Chen, R.; Chung, S.H. Conserved functional surface of antimammalian scorpion beta-toxins. *J. Phys. Chem. B* **2012**, *116*, 4796–4800.

103. Wilhelm, M.; Mukherjee, A.; Bouvier, B.; Zakrzewska, K.; Hynes, J.T.; Lavery, R. Multistep drug intercalation: Molecular dynamics and free energy studies of the binding of daunomycin to DNA. *J. Am. Chem. Soc.* **2012**, *134*, 8588–8596.

104. Louet, M.; Martinez, J.; Floquet, N. GDP release preferentially occurs on the phosphate side in heterotrimeric G-proteins. *PLoS Comput. Biol.* **2012**, *8*, doi:10.1371/journal.pcbi.1002595.

105. Zhang, H.; Tan, T.; Feng, W.; van der Spoel, D. Molecular recognition in different environments: Beta-cyclodextrin dimer formation in organic solvents. *J. Phys. Chem. B* **2012**, *116*, 12684–12693.

106. Kessler, J.; Jakubek, M.; Dolensky, B.; Bour, P. Binding energies of five molecular pincers calculated by explicit and implicit solvent models. *J. Comput. Chem.* **2012**, *33*, 2310–2317.

107. Mascarenhas, N.M.; Kaestner, J. How maltose influences structural changes to bind to maltose-binding protein: Results from umbrella sampling simulation. *Proteins* **2013**, *81*, 185–198.

108. MacCallum, J.; Tieleman, D. Computer simulation of the distribution of hexane in a lipid bilayer: Spatially resolved free energy, entropy, and enthalpy profiles. *J. Am. Chem. Soc.* **2006**, *128*, 125–130.

109. Tieleman, D.P.; Marrink, S.J. Lipids out of equilibrium: Energetics of desorption and pore mediated flip-flop. *J. Am. Chem. Soc.* **2006**, *128*, 12462–12467.

110. Kyrychenko, A.; Sevriukov, I.Y.; Syzova, Z.A.; Ladokhin, A.S.; Doroshenko, A.O. Partitioning of 2,6-Bis(1H-Benzimidazol-2-yl)pyridine fluorophore into a phospholipid bilayer: Complementary use of fluorescence quenching studies and molecular dynamics simulations. *Biophys. Chem.* **2011**, *154*, 8–17.

111. Lemkul, J.A.; Bevan, D.R. Characterization of interactions between PilA from pseudomonas aeruginosa strain K and a model membrane. *J. Phys. Chem. B* **2011**, *115*, 8004–8008.

112. Paloncyova, M.; Berka, K.; Otyepka, M. Convergence of free energy profile of coumarin in lipid bilayer. *J. Chem. Theory Comput.* **2012**, *8*, 1200–1211.

113. Samanta, S.; Hezaveh, S.; Milano, G.; Roccatano, D. Diffusion of 1,2-Dimethoxyethane and 1,2-Dimethoxypropane through phosphatidycholine bilayers: A molecular dynamics study. *J. Phys. Chem. B* **2012**, *116*, 5141–5151.

114. Grafmueller, A.; Lipowsky, R.; Knecht, V. Effect of tension and curvature on the chemical potential of lipids in lipid aggregates. *Phys. Chem. Chem. Phys.* **2013**, *15*, 876–881.

115. Cerezo, J.; Zuniga, J.; Bastida, A.; Requena, A.; Ceron-Carrasco, J.P. Conformational changes of beta-carotene and zeaxanthin immersed in a model membrane through atomistic molecular dynamics simulations. *Phys. Chem. Chem. Phys.* **2013**, *15*, 6527–6538.

116. Tian, J.; Sethi, A.; Swanson, B.I.; Goldstein, B.; Gnanakaran, S. Taste of sugar at the membrane: Thermodynamics and kinetics of the interaction of a disaccharide with lipid bilayers. *Biophys. J.* **2013**, *104*, 622–632.

117. Karlsson, B.C.G.; Olsson, G.D.; Friedman, R.; Rosengren, A.M.; Henschel, H.; Nicholls, I.A. How Warfarin's structural diversity influences its phospholipid bilayer membrane permeation. *J. Phys. Chem. B* **2013**, *117*, 2384–2395.

118. Euston, S.R.; Bellstedt, U.; Schillbach, K.; Hughes, P.S. The adsorption and competitive adsorption of bile salts and whey protein at the oil-water interface. *Soft Matter* **2011**, *7*, 8942–8951.

119. Doudou, S.; Vaughan, D.J.; Livens, F.R.; Burton, N.A. Atomistic simulations of calcium Uranyl(VI) carbonate adsorption on calcite and stepped-calcite surfaces. *Environ. Sci. Tech.* **2012**, *46*, 7587–7594.

120. Pomes, R.; Roux, B. Free energy profiles for H+ conduction along hydrogen-bonded chains of water molecules. *Biophys. J.* **1998**, *75*, 33–40.

121. Jagoda-Cwiklik, B.; Cwiklik, L.; Jungwirth, P. Behavior of the eigen form of hydronium at the air/water interface. *J. Phys. Chem. A* **2011**, *115*, 5881–5886.

122. Calvo, F.; Mottet, C. Order-disorder transition in Co-Pt nanoparticles: Coexistence, transition states, and finite-size effects. *Phys. Rev. B* **2011**, *84*.

123. Sharma, S.; Debenedetti, P.G. Free energy barriers to evaporation of water in hydrophobic confinement. *J. Phys. Chem. B* **2012**, *116*, 13282–13289.

124. Ridder, L.; Rietjens, I.; Vervoort, J.; Mulholland, A. Quantum mechanical/molecular mechanical free energy Simulations of the glutathione S-transferase (M1-1) reaction with phenanthrene 9,10-oxide. *J. Am. Chem. Soc.* **2002**, *124*, 9926–9936.

125. Kaestner, J.; Senn, H.; Thiel, S.; Otte, N.; Thiel, W. QM/MM free-energy perturbation compared to thermodynamic integration and umbrella sampling: Application to an enzymatic reaction. *J. Chem. Theory Comput.* **2006**, *2*, 452–461.

126. Wang, S.; Hu, P.; Zhang, Y. Ab initio quantum mechanical/molecular mechanical molecular dynamics simulation of enzyme catalysis: The case of histone lysine methyltransferase SET7/9. *J. Phys. Chem. B* **2007**, *111*, 3758–3764.

127. Ke, Z.; Guo, H.; Xie, D.; Wang, S.; Zhang, Y. Ab initio QM/MM free-energy studies of arginine deiminase catalysis: The protonation state of the Cys nucleophile. *J. Phys. Chem. B* **2011**, *115*, 3725–3733.

128. Yan, S.; Li, T.; Yao, L. Mutational effects on the catalytic mechanism of cellobiohydrolase I from Trichoderma reesei. *J. Phys. Chem. B* **2011**, *115*, 4982–4989.

129. Mujika, J.I.; Lopez, X.; Mulholland, A.J. Mechanism of C-terminal intein cleavage in protein splicing from QM/MM molecular dynamics simulations. *Org. Biomol. Chem.* **2012**, *10*, 1207–1218.

130. Lonsdale, R.; Hoyle, S.; Grey, D.T.; Ridder, L.; Mulholland, A.J. Determinants of reactivity and selectivity in soluble epoxide hydrolase from quantum mechanics/molecular mechanics modeling. *Biochemistry* **2012**, *51*, 1774–1786.

131. Rooklin, D.W.; Lu, M.; Zhang, Y. Revelation of a catalytic calcium-binding site elucidates unusual metal dependence of a human apyrase. *J. Am. Chem. Soc.* **2012**, *134*, 15595–15603.

132. Lior-Hoffmann, L.; Wang, L.; Wang, S.; Geacintov, N.E.; Broyde, S.; Zhang, Y. Preferred WMSA catalytic mechanism of the nucleotidyl transfer reactionin human DNA polymerase kappa elucidates error-free bypass of a bulky DNA lesion. *Nucl. Acids Res.* **2012**, *40*, 9193–9205.

133. Crouzy, S.; Berneche, S.; Roux, B. Extracellular blockade of K+ channels by TEA: Results from molecular dynamics simulations of the KcsA channel. *J. Gen. Physiol.* **2001**, *118*, 207–217.

134. Allen, T.; Bastug, T.; Kuyucak, S.; Chung, S. Gramicidin a channel as a test ground for molecular dynamics force fields. *Biophys. J.* **2003**, *84*, 2159–2168.

135. Hub, J.S.; de Groot, B.L. Mechanism of selectivity in aquaporins and aquaglyceroporins. *Proc. Natl. Acad. Sci. USA* **2008**, *105*, 1198–1203.

136. Xin, L.; Su, H.; Nielsen, C.H.; Tang, C.; Torres, J.; Mu, Y. Water permeation dynamics of AqpZ: A tale of two states. *BBA-Biomembranes* **2011**, *1808*, 1581–1586.

137. Furini, S.; Domene, C. Selectivity and permeation of alkali metal ions in K+-channels. *J. Mol. Biol.* **2011**, *409*, 867–878.

138. Kim, I.; Allen, T.W. On the selective ion binding hypothesis for potassium channels. *Proc. Natl. Acad. Sci. USA* **2011**, *108*, 17963–17968.

139. Domene, C.; Furini, S. Molecular dynamics simulations of the TrkH membrane protein. *Biochemistry* **2012**, *51*, 1559–1565.

140. Zhu, F.; Hummer, G. Theory and simulation of ion conduction in the pentameric GLIC channel. *J. Chem. Theor. Comput.* **2012**, *8*, 3759–3768.

141. Zhongjin, H.; Jian, Z. Steered molecular dynamics simulations of ions traversing through carbon nanotubes. *Acta Chim. Sin.* **2011**, *69*, 2901–2907.

142. Nalaparaju, A.; Jiang, J. Ion exchange in metal-organic framework for water purification: Insight from molecular simulation. *J. Phys. Chem. C* **2012**, *116*, 6925–6931.

143. Barducci, A.; Bonomi, M.; Parrinello, M. Metadynamics. *WIREs Comput. Mol. Sci.* **2011**, *1*, 826–843.

144. Huber, T.; Torda, A.; van Gunsteren, W.F. Local elevation: A method for improving the searching properties of molecular dynamics simulation. *J. Comput.-Aided Mol. Des.* **1994**, *8*, 695–708.

145. Laio, A.; Rodriguez-Fortea, A.; Gervasio, F.L.; Ceccarelli, M.; Parrinello, M. Assessing the accuracy of metadynamics. *J. Phys. Chem. B* **2005**, *109*, 6714–6721.

146. Bussi, G.; Laio, A.; Parrinello, M. Equilibrium free energies from nonequilibrium metadynamics. *Phys. Rev. Lett.* **2006**, *96*, 090601.

147. Raiteri, P.; Laio, A.; Gervasio, F.L.; Micheletti, C.; Parrinello, M. Efficient reconstruction of complex free energy landscapes by multiple walkers metadynamics. *J. Phys. Chem. B* **2006**, *110*, 3533–3539.

148. Barducci, A.; Bussi, G.; Parrinello, M. Well-tempered metadynamics: A smoothly converging and tunable free-energy method. *Phys. Rev. Lett.* **2008**, *100*, 020603.

149. Singh, S.; Chiu, C.C.; de Pablo, J.J. Flux tempered metadynamics. *J. Stat. Phys.* **2011**, *145*, 932–945.

150. Bonomi, M.; Barducci, A.; Parrinello, M. Reconstructing the equilibrium boltzmann distribution from well-tempered metadynamics. *J. Comput. Chem.* **2009**, *30*, 1615–1621.

151. Branduardi, D.; Bussi, G.; Parrinello, M. Metadynamics with adaptive Gaussians. *J. Chem. Theory Comput.* **2012**, *8*, 2247–2254.

152. McGrath, M.J.; Kuo, I.F.W.; Hayashi, S.; Takada, S. ATP hydrolysis mechanism in kinesin studied by combined quantum-mechanical/molecular-mechanical metadynamics simulations. *J. Am. Chem. Soc.* **2013**, *135*, 8908–8919.

153. Mantz, Y.A.; Branduardi, D.; Bussi, G.; Parrinello, M. Ensemble of transition state structures for the Cis- trans isomerization of N-Methylacetamide. *J. Phys. Chem. B* **2009**, *113*, 12521–12529.

154. Leone, V.; Lattanzi, G.; Molteni, C.; Carloni, P. Mechanism of action of cyclophilin a explored by metadynamics simulations. *PLoS Comput. Biol.* **2009**, *5*, e1000309.

155. Melis, C.; Bussi, G.; Lummis, S.C.; Molteni, C. Trans-cis switching mechanisms in proline analogues and their relevance for the gating of the 5-HT3 receptor. *J. Phys. Chem. B* **2009**, *113*, 12148–12153.

156. Bussi, G.; Gervasio, F.L.; Laio, A.; Parrinello, M. Free-energy landscape for β hairpin folding from combined parallel tempering and metadynamics. *J. Am. Chem. Soc.* **2006**, *128*, 13435–13441.

157. Gangupomu, V.; Abrams, C. All-atom models of the membrane-spanning domain of HIV-1 gp41 from metadynamics. *Biophys. J.* **2010**, *99*, 3438–3444.

158. Berteotti, A.; Barducci, A.; Parrinello, M. Effect of urea on the β-Hairpin conformational ensemble and protein denaturation mechanism. *J. Am. Chem. Soc.* **2011**, *133*, 17200–17206.

159. Granata, D.; Camilloni, C.; Vendruscolo, M.; Laio, A. Characterization of the free-energy landscapes of proteins by NMR-guided metadynamics. *Proc. Natl. Acad. Sci. USA* **2013**, *110*, 6817–6822.

160. Baftizadeh, F.; Biarnes, X.; Pietrucci, F.; Affinito, F.; Laio, A. Multidimensional view of amyloid fibril nucleation in atomistic detail. *J. Am. Chem. Soc.* **2012**, *134*, 3886–3894.

161. Gervasio, F.L.; Laio, A.; Parrinello, M. Flexible docking in solution using metadynamics. *J. Am. Chem. Soc.* **2005**, *127*, 2600–2607.

162. Soederhjelm, P.; Tribello, G.A.; Parrinello, M. Locating binding poses in protein-ligand systems using reconnaissance metadynamics. *Proc. Natl. Acad. Sci. USA* **2012**, *109*, 5170–5175.

163. Limongelli, V.; Bonomi, M.; Parrinello, M. Funnel metadynamics as accurate binding free-energy method. *Proc. Natl. Acad. Sci. USA* **2013**, *110*, 6358–6363.

164. Sutto, L.; Gervasio, F.L. Effects of oncogenic mutations on the conformational free-energy landscape of EGFR kinase. *Proc. Natl. Acad. Sci. USA* **2013**, doi:10.1073/pnas.1221953110.

165. Martonak, R.; Donadio, D.; Oganov, A.R.; Parrinello, M. Crystal structure transformations in SiO_2 from classical and ab initio metadynamics. *Nat. Mater.* **2006**, *5*, 623–626.

166. Trudu, F.; Donadio, D.; Parrinello, M. Freezing of a Lennard-Jones fluid: From nucleation to spinodal regime. *Phys. Rev. Lett.* **2006**, *97*, 105701.

167. Stack, A.G.; Raiteri, P.; Gale, J.D. Accurate rates of the complex mechanisms for growth and dissolution of minerals using a combination of rare-event theories. *J. Am. Chem. Soc.* **2011**, *134*, 11–14.

168. Zhang, C.; Knyazev, D.G.; Vereshaga, Y.A.; Ippoliti, E.; Nguyen, T.H.; Carloni, P.; Pohl, P. Water at hydrophobic interfaces delays proton surface-to-bulk transfer and provides a pathway for lateral proton diffusion. *Proc. Natl. Acad. Sci. USA* **2012**, *109*, 9744–9749.

169. Das, P.; Moll, M.; Stamati, H.; Kavraki, L.E.; Clementi, C. Low-dimensional, free-energy landscapes of protein-folding reactions by nonlinear dimensionality reduction. *Proc. Natl. Acad. Sci. USA* **2006**, *103*, 9885–9890.

170. Perilla, J.R.; Woolf, T.B. Towards the prediction of order parameters from molecular dynamics simulations in proteins. *J. Chem. Phys.* **2012**, *136*, 164101.

171. Ceriotti, M.; Tribello, G.A.; Parrinello, M. Simplifying the representation of complex free-energy landscapes using sketch-map. *Proc. Natl. Acad. Sci. USA* **2011**, *108*, 13023–13028.

172. Tribello, G.A.; Ceriotti, M.; Parrinello, M. A self-learning algorithm for biased molecular dynamics. *Proc. Natl. Acad. Sci. USA* **2010**, *107*, 17509–17514.

173. Tribello, G.A.; Cuny, J.; Eshet, H.; Parrinello, M. Exploring the free energy surfaces of clusters using reconnaissance metadynamics. *J. Chem. Phys.* **2011**, *135*, doi:10.1063/1.3628676.

174. Bartels, C.; Karplus, M. Probability distributions for complex systems: Adaptive umbrella sampling of the potential energy. *J. Phys. Chem. B* **1998**, *102*, 865–880.

175. Micheletti, C.; Laio, A.; Parrinello, M. Reconstructing the density of states by history-dependent metadynamics. *Phys. Rev. Lett.* **2004**, *92*, 170601.

176. Bonomi, M.; Parrinello, M. Enhanced sampling in the well-tempered ensemble. *Phys. Rev. Lett.* **2010**, *104*, 190601.

177. Do, T.N.; Carloni, P.; Varani, G.; Bussi, G. RNA/peptide binding driven by electrostatic insight from bidirectional pulling simulations. *J. Chem. Theory Comput.* **2013**, *9*, 1720–1730.

178. Roitberg, A.; Elber, R. Modeling side-chains in peptides and proteins—application of the locally enhanced sampling and the simulated annealing methods to find minimum energy conformations. *J. Chem. Phys.* **1991**, *95*, 9277–9287.

179. Patel, A.J.; Varilly, P.; Chandler, D.; Garde, S. Quantifying density fluctuations in volumes of all shapes and sizes using indirect umbrella sampling. *J. Stat. Phys.* **2011**, *145*, 265–275.

180. Mueller, M.; Smirnova, Y.G.; Marelli, G.; Fuhrmans, M.; Shi, A.C. Transition path from two apposed membranes to a stalk obtained by a combination of particle simulations and string method. *Phys. Rev. Lett.* **2012**, *108*, doi:10.1103/PhysRevLett.108.228103.

181. Pietrucci, F.; Laio, A. A collective variable for the efficient exploration of protein beta-sheet structures: Application to sh3 and gb1. *J. Chem. Theory Comput.* **2009**, *5*, 2197–2201.

182. Branduardi, D.; Gervasio, F.L.; Parrinello, M. From A to B in free energy space. *J. Chem. Phys.* **2007**, *126*, 054103.

183. Zinovjev, K.; Martí, S.; Tuñón, I. A collective coordinate to obtain free energy profiles for complex reactions in condensed phases. *J. Chem. Theory Comput.* **2012**, *8*, 1795–1801.

184. Spiwok, V.; Králová, B. Metadynamics in the conformational space nonlinearly dimensionally reduced by Isomap. *J. Chem. Phys.* **2011**, *135*, 224504–224504.

185. Kirkpatrick, S.; Gelatt, C.D., Jr.; Vecchi, M.P. Optimization by simulated annealing. *Science* **1983**, *220*, 671–680.

186. Marinari, E.; Parisi, G. Simulated tempering: A new Monte Carlo scheme. *Europhys. Lett.* **1992**, *19*, doi:10.1209/0295-5075/19/6/002.

187. Park, S.; Pande, V.S. Choosing weights for simulated tempering. *Phys. Rev. E* **2007**, *76*, 016703.

188. Hansmann, U.H. Parallel tempering algorithm for conformational studies of biological molecules. *Chem. Phys. Lett.* **1997**, *281*, 140–150.

189. Sugita, Y.; Okamoto, Y. Replica-exchange molecular dynamics method for protein folding. *Chem. Phys. Lett.* **1999**, *314*, 141–151.

190. Frenkel, D. Speed-up of Monte Carlo simulations by sampling of rejected states. *Proc. Natl. Acad. Sci. USA* **2004**, *101*, 17571–17575.

191. Coluzza, I.; Frenkel, D. Virtual-move parallel tempering. *Chem. Phys. Chem.* **2005**, *6*, 1779–1783.

192. Sindhikara, D.; Meng, Y.; Roitberg, A.E. Exchange frequency in replica exchange molecular dynamics. *J. Chem. Phys.* **2008**, *128*, 024103.

193. Bussi, G. A simple asynchronous replica-exchange implementation. *Nuovo Cimento della Societa Italiana di Fisica C* **2009**, *32*, 61–65.

194. Gallicchio, E.; Levy, R.M.; Parashar, M. Asynchronous replica exchange for molecular simulations. *J. Comput. Chem.* **2008**, *29*, 788–794.

195. Rosta, E.; Buchete, N.V.; Hummer, G. Thermostat artifacts in replica exchange molecular dynamics simulations. *J. Chem. Theory Comput.* **2009**, *5*, 1393–1399.

196. Sindhikara, D.J.; Emerson, D.J.; Roitberg, A.E. Exchange often and properly in replica exchange molecular dynamics. *J. Chem. Theory Comput.* **2010**, *6*, 2804–2808.

197. Vreede, J.; Crielaard, W.; Hellingwerf, K.; Bolhuis, P. Predicting the signaling state of photoactive yellow protein. *Biophys. J.* **2005**, *88*, 3525–3535.

198. Zhang, T.; Mu, Y. Initial binding of ions to the interhelical loops of divalent ion transporter CorA: Replica exchange molecular dynamics simulation study. *PLoS One* **2012**, *7*, e43872.

199. Zhou, R. Trp-cage: Folding free energy landscape in explicit water. *Proc. Natl. Acad. Sci. USA* **2003**, *100*, 13280–13285.

200. Garcia, A.; Onuchic, J. Folding a protein in a computer: An atomic description of the folding/unfolding of protein A. *Proc. Natl. Acad. Sci. USA* **2003**, *100*, 13898–13903.

201. Im, W.; Feig, M.; Brooks, C. An implicit membrane generalized born theory for the study of structure, stability, and interactions of membrane proteins. *Biophys. J.* **2003**, *85*, 2900–2918.

202. Mei, Y.; Wei, C.; Yip, Y.M.; Ho, C.Y.; Zhang, J.Z.H.; Zhang, D. Folding and thermodynamic studies of Trp-cage based on polarized force field. *Theor. Chem. Acc.* **2012**, *131*, doi:10.1007/s00214-012-1168-0.

203. Berhanu, W.M.; Jiang, P.; Hansmann, U.H.E. Folding and association of a homotetrameric protein complex in an all-atom Go model. *Phys. Rev. E* **2013**, *87*, 014701.

204. Kokubo, H.; Okamoto, Y. Self-assembly of transmembrane helices of bacteriorhodopsin by a replica-exchange Monte Carlo simulation. *Chem. Phys. Lett.* **2004**, *392*, 168–175.

205. Oshaben, K.M.; Salari, R.; McCaslin, D.R.; Chong, L.T.; Horne, W.S. The native GCN4 leucine-zipper domain does not uniquely specify a dimeric oligomerization state. *Biochemistry* **2012**, *51*, 9581–9591.

206. Sugita, Y.; Okamoto, Y. Replica-exchange multicanonical algorithm and multicanonical replica-exchange method for simulating systems with rough energy landscape. *Chem. Phys. Lett.* **2000**, *329*, 261–270.

207. Fukunishi, H.; Watanabe, O.; Takada, S. On the Hamiltonian replica exchange method for efficient sampling of biomolecular systems: Application to protein structure prediction. *J. Chem. Phys.* **2002**, *116*, 9058–9067.

208. Liu, P.; Kim, B.; Friesner, R.A.; Berne, B. Replica exchange with solute tempering: A method for sampling biological systems in explicit water. *Proc. Natl. Acad. Sci. USA* **2005**, *102*, 13749–13754.

209. Affentranger, R.; Tavernelli, I.; Di Iorio, E.E. A novel Hamiltonian replica exchange MD protocol to enhance protein conformational space sampling. *J. Chem. Theory Comput.* **2006**, *2*, 217–228.

210. Fajer, M.; Hamelberg, D.; McCammon, J.A. Replica-exchange accelerated molecular dynamics (REXAMD) applied to thermodynamic integration. *J. Chem. Theory Comput.* **2008**, *4*, 1565–1569.

211. Xu, C.; Wang, J.; Liu, H. A hamiltonian replica exchange approach and its application to the study of side-chain type and neighbor effects on peptide backbone conformations. *J. Chem. Theory Comput.* **2008**, *4*, 1348–1359.

212. Zacharias, M. Combining elastic network analysis and molecular dynamics simulations by hamiltonian replica exchange. *J. Chem. Theory Comput.* **2008**, *4*, 477–487.

213. Vreede, J.; Wolf, M.G.; de Leeuw, S.W.; Bolhuis, P.G. Reordering hydrogen bonds using Hamiltonian replica exchange enhances sampling of conformational changes in biomolecular systems. *J. Phys. Chem. B* **2009**, *113*, 6484–6494.

214. Itoh, S.G.; Okumura, H.; Okamoto, Y. Replica-exchange method in van der Waals radius space: Overcoming steric restrictions for biomolecules. *J. Chem. Phys.* **2010**, *132*, 134105.

215. Meng, Y.; Roitberg, A.E. Constant pH replica exchange molecular dynamics in biomolecules using a discrete protonation model. *J. Chem. Theory Comput.* **2010**, *6*, 1401–1412.

158

216. Terakawa, T.; Kameda, T.; Takada, S. On easy implementation of a variant of the replica exchange with solute tempering in GROMACS. *J. Comput. Chem.* **2011**, *32*, 1228–1234.

217. Wang, L.; Friesner, R.A.; Berne, B. Replica exchange with solute scaling: A more efficient version of replica exchange with solute tempering (REST2). *J. Phys. Chem. B* **2011**, *115*, 9431–9438.

218. Zhang, C.; Ma, J. Folding helical proteins in explicit solvent using dihedral-biased tempering. *Proc. Natl. Acad. Sci. USA* **2012**, *109*, 8139–8144.

219. Bussi, G. Hamiltonian replica-exchange in GROMACS: A flexible implementation. *Mol. Phys.* **2013**, doi:10.1080/00268976.2013.824126.

220. Huang, X.; Hagen, M.; Kim, B.; Friesner, R.A.; Zhou, R.; Berne, B. Replica exchange with solute tempering: Efficiency in large scale systems. *J. Phys. Chem. B* **2007**, *111*, 5405–5410.

221. Denschlag, R.; Lingenheil, M.; Tavan, P.; Mathias, G. Simulated solute tempering. *J. Chem. Theory Comput.* **2009**, *5*, 2847–2857.

222. Zuckerman, D.M.; Lyman, E. A second look at canonical sampling of biomolecules using replica exchange simulation. *J. Chem. Theory Comput.* **2006**, *2*, 1200–1202.

223. Chodera, J.D.; Swope, W.C.; Pitera, J.W.; Seok, C.; Dill, K.A. Use of the weighted histogram analysis method for the analysis of simulated and parallel tempering simulations. *J. Chem. Theory Comput.* **2007**, *3*, 26–41.

224. Camilloni, C.; Provasi, D.; Tiana, G.; Broglia, R.A. Exploring the protein G helix free-energy surface by solute tempering metadynamics. *Proteins* **2008**, *71*, 1647–1654.

225. Deighan, M.; Bonomi, M.; Pfaendtner, J. Efficient simulation of explicitly solvated proteins in the well-tempered ensemble. *J. Chem. Theory Comput.* **2012**, *8*, 2189–2192.

226. Piana, S.; Laio, A. A bias-exchange approach to protein folding. *J. Phys. Chem. B* **2007**, *111*, 4553–4559.

227. Baftizadeh, F.; Cossio, P.; Pietrucci, F.; Laio, A. Protein folding and ligand-enzyme binding from bias-exchange metadynamics simulations. *Curr. Phys. Chem.* **2012**, *2*, 79–91.

228. Weinan, E.; Ren, W.; Vanden-Eijnden, E. String method for the study of rare events. *Phys. Rev. B* **2002**, *66*, 052301.

229. Maragliano, L.; Fischer, A.; Vanden-Eijnden, E.; Ciccotti, G. String method in collective variables: Minimum free energy paths and isocommittor surfaces. *J. Chem. Phys.* **2006**, *125*, 024106.

230. Vashisth, H.; Abrams, C.F. All-atom structural models of insulin binding to the insulin receptor in the presence of a tandem hormone-binding element. *Proteins* **2013**, *81*, 1017–1030.

231. Ovchinnikov, V.; Karplus, M.; Vanden-Eijnden, E. Free energy of conformational transition paths in biomolecules: The string method and its application to myosin VI. *J. Chem. Phys.* **2011**, *134*, 085103.

232. Maragliano, L.; Vanden-Eijnden, E. On-the-fly string method for minimum free energy paths calculation. *Chem. Phys. Lett.* **2007**, *446*, 182–190.

233. Stober, S.T.; Abrams, C.F. Energetics and mechanism of the normal-to-amyloidogenic isomerization of b2-microglobulin: On-the-fly string method calculations. *J. Phys. Chem. B* **2012**, *116*, 9371–9375.

234. Zinovjev, K.; Ruiz-Pernia, J.; Tuñón, I. Toward an automatic determination of enzymatic reaction mechanisms and their activation free energies. *J. Chem. Theory Comput.* **2013**, *9*, 3740–3749.

235. Abrams, C.F.; Vanden-Eijnden, E. On-the-fly free energy parameterization via temperature accelerated molecular dynamics. *Chem. Phys. Lett.* **2012**, *547*, 114–119.

236. Chen, M.; Cuendet, M.A.; Tuckerman, M.E. Heating and flooding: A unified approach for rapid generation of free energy surfaces. *J. Chem. Phys.* **2012**, *137*, 024102.

237. Phillips, J.C.; Braun, R.; Wang, W.; Gumbart, J.; Tajkhorshid, E.; Villa, E.; Chipot, C.; Skeel, R.D.; Kalé, L.; Schulten, K. Scalable molecular dynamics with NAMD. *J. Comput. Chem.* **2005**, *26*, 1781–1802.

238. Plimpton, S. Fast parallel algorithms for short-range molecular-dynamics. *J. Comput. Phys.* **1995**, *117*, 1–19.

239. Hess, B.; Kutzner, C.; van der Spoel, D.; Lindahl, E. GROMACS 4: Algorithms for highly efficient, load-balanced, and scalable molecular simulation. *J. Chem. Theory Comput.* **2008**, *4*, 435–447.

240. Case, D.; Cheatham, T.; Darden, T.; Gohlke, H.; Luo, R.; Merz, K.; Onufriev, A.; Simmerling, C.; Wang, B.; Woods, R. The Amber biomolecular simulation programs. *J. Comput. Chem.* **2005**, *26*, 1668–1688.

241. Brooks, B.; Bruccoleri, R.; Olafson, B.; States, D.; Swaminathan, S.; Karplus, M. CHARMM—A program for macromolecular energy, minimization, and dynamics calculations. *J. Comput. Chem.* **1983**, *4*, 187–217.

242. Fiorin, G.; Klein, M.L.; Hénin, J. Using collective variables to drive molecular dynamics simulations. *Mol. Phys.* **2013**, doi:10.1080/00268976.2013.813594.

243. Bonomi, M.; Branduardi, D.; Bussi, G.; Camilloni, C.; Provasi, D.; Raiteri, P.; Donadio, D.; Marinelli, F.; Pietrucci, F.; Broglia, R.A.; Parrinello, M. PLUMED: A portable plugin for free-energy calculations with molecular dynamics. *Comput. Phys. Comm.* **2009**, *180*, 1961–1972.

244. Tribello, G.; Bonomi, M.; Branduardi, D.; Camilloni, C.; Bussi, G. PLUMED 2: New feathers for an old bird. *Comput. Phys. Commun.* **2013**, doi:10.1016/j.cpc.2013.09.018.

160

Reprinted from *Entropy*. Cite as: Dellago, C.; Hummer, G. Computing Equilibrium Free Energies Using Non-Equilibrium Molecular Dynamics. *Entropy* **2014**, *16*, 41–61.

Article

Computing Equilibrium Free Energies Using Non-Equilibrium Molecular Dynamics

Christoph Dellago [1,*] **and Gerhard Hummer** [2]

[1] Faculty of Physics, University of Vienna, Boltzmanngasse 5, 1090 Vienna, Austria
[2] Department of Theoretical Biophysics, Max Planck Institute of Biophysics, Max-von-Laue-Str. 3, 60438 Frankfurt am Main, Germany; E-Mail: gerhard.hummer@biophys.mpg.de

* Author to whom correspondence should be addressed; E-Mail: Christoph.Dellago@univie.ac.at; Tel.: +43-1-4277-51260.

Received: 10 October 2013; in revised form: 12 November 2013 / Accepted: 19 November 2013 / Published: 27 December 2013

Abstract: As shown by Jarzynski, free energy differences between equilibrium states can be expressed in terms of the statistics of work carried out on a system during non-equilibrium transformations. This exact result, as well as the related Crooks fluctuation theorem, provide the basis for the computation of free energy differences from fast switching molecular dynamics simulations, in which an external parameter is changed at a finite rate, driving the system away from equilibrium. In this article, we first briefly review the Jarzynski identity and the Crooks fluctuation theorem and then survey various algorithms building on these relations. We pay particular attention to the statistical efficiency of these methods and discuss practical issues arising in their implementation and the analysis of the results.

Keywords: fast switching simulations; non-equilibrium work theorem; fluctuation theorem; non-equilibrium molecular dynamics

1. Introduction

The calculation of free energies from atomistic simulations is of great importance in many applications, ranging from the prediction of the phase behavior of a certain substance to the calculation of ligand affinities in drug design. Since the computation of free energies (or, more precisely, of free energy differences) involves the determination of entropic contributions and, hence, the estimation of phase space volumes [1], free energy calculations are computationally very demanding in most cases. Therefore, a significant effort has been devoted to the development of more efficient free energy calculation algorithms. This endeavor has received new momentum with Jarzynski's discovery of a very general relation between equilibrium free energies and non-equilibrium work [2,3], which has inspired several molecular dynamics-based algorithms for free energy computations. In this article, we will give an overview of these methods.

According to the maximum work theorem, a consequence of the second law of thermodynamics, the amount of work W performed on a system during a non-equilibrium transformation is larger than the free energy difference ΔF between the equilibrium states corresponding to the transition end points:

$$\langle W \rangle \geq \Delta F \tag{1}$$

Equivalently, the amount of work that can be extracted from a system is bounded from above by the free energy difference. In the above equation, the equal sign holds only if the transformation is carried out reversibly, maintaining equilibrium at all times. The angular brackets on the left-hand side of the maximum work theorem indicate an average over many realizations of the non-equilibrium process. If one considers a macroscopic system, for instance, a piston compressing a gas enclosed in a cylinder, the average is not necessary, because every realization of the process yields, for all practical purposes, the same amount of work W, if the transformation is carried out following the same protocol. This is essentially a consequence of the central limit theorem for thermal fluctuations. In the case of a microscopic system, however, fluctuations become important, and different realizations of the transformation typically produce different work values, leading to a statistical distribution of W. For instance, stretching a biomolecule with atomic force microscopes or optical tweezers will cost a different amount work for each repetition of the experiment. In some cases, the work expended on the system might even be smaller than the free energy difference, seemingly violating the maximum work theorem and, hence, the second law of thermodynamics.

As shown by Jarzynski in 1997 [2,3], the work fluctuations resulting for microscopic systems can be accounted for in an exact way, transforming the maximum work theorem into an equality:

$$\langle e^{-\beta W} \rangle = e^{-\beta \Delta F} \tag{2}$$

Here, $\beta = 1/k_B T$ is the reciprocal temperature of the equilibrium state from which the transformation is started, and k_B is the Boltzmann constant. Remarkably, this result, now commonly referred to as Jarzynski equation or Jarzynski non-equilibrium work theorem, relates the statistics of irreversible work carried out on the system, while it is driven away from equilibrium, to an equilibrium free energy difference. A closely connected result is the Crooks fluctuation theorem [4–6], which relates the equilibrium free energy difference to the work distributions of the forward and reversed process.

In general, processes during which work is performed on or by the system drive the system away from equilibrium, such that the phase space distribution obtained at the end of the process may differ strongly from the equilibrium distribution to which the system relaxes after the external perturbation has been stopped. For instance, a piston pushed quickly into a gas-filled cylinder generates non-equilibrium states with strong flows markedly different from the static equilibrium state to which the gas eventually relaxes after the piston has reached its final state. At first sight, it is therefore surprising that equilibrium properties, such as free energy differences, can be extracted from non-equilibrium trajectories. As discussed in the following sections of this paper, a closer analysis reveals that averaging over the work exponential is equivalent to removing the bias introduced during the driving process. It is this unbiasing that ultimately permits the extraction of equilibrium properties (as we will discuss in Section 5, in principle, one can determine the entire equilibrium distribution and not only the free energy) from non-equilibrium trajectories. Thus, the non-equilibrium work theorem can be viewed as a prescription of how to compensate for the effects of manipulations that drive the system into non-equilibrium rather than a tool that illuminates the nature of non-equilibrium processes. Nevertheless, it is remarkable that the bias has a very simple exponential form and can be expressed in terms of the work only.

The Jarzynski non-equilibrium work theorem, as well as the Crooks fluctuation theorem provide the framework for the interpretation of single-molecule pulling experiments [7–9], in which non-equilibrium effects can never be fully avoided. These exact results can also be exploited to devise computer simulation algorithms for the calculation of free energies. In this article, we review several computational approaches based on the collection of work statistics in a fast-switching non-equilibrium setting, paying particular attention to the accuracy and efficient implementation of these methods compared to conventional free energy computation methods (see [10–12]). In the remainder of this article, we will first state the Jarzynski and Crooks theorems more explicitly and discuss the conditions under which they apply. After that, we will survey several fast switching algorithms in which free energies are determined from sets of molecular dynamics trajectories obtained while changing a control parameter, thereby exerting work on the system. We conclude with a brief summary and outlook to future possibilities and applications.

2. Jarzynski Identity and Crooks Fluctuation Theorem

To set the notation, consider a classical system with energy $H(x, \lambda)$ depending on the microscopic state x of the system, as well as on a parameter λ. The microscopic state x is specified by the positions of all particles in the system and, if necessary, also by all momenta. The parameter λ is a control parameter that can be changed externally, for instance, the volume of the cylinder containing the particles or an external field. According to the basic laws of statistical mechanics, the free energy difference between the two equilibrium states A and B corresponding to the values λ_A and λ_B, respectively, of the order parameter is given by:

$$\Delta F = F_B - F_A = -k_{\mathrm{B}} T \ln \frac{Z_B}{Z_A} \tag{3}$$

where $Z_A = \int dx \, \exp\{-\beta H(x, \lambda_A)\}$ and $Z_B = \int dx \, \exp\{-\beta H(x, \lambda_B)\}$ are the canonical partition functions of the two equilibrium states (up to a combinatorial prefactor irrelevant for our considerations). The free energy difference ΔF is the work required to change the external parameter from λ_A to λ_B in a *reversible* process. Such a reversible transformation could be realized, for instance, by changing the parameter λ infinitely slowly, while keeping the system in contact with a heat bath. In this case, the free energy difference is equal to the work of the system.

Instead of changing the control parameter λ very slowly, one could change it at a finite rate over a time interval τ, following a certain protocol $\lambda(t)$, where $\lambda(0) = \lambda_A$ and $\lambda(\tau) = \lambda_B$. In general, such a fast switching of the control parameter drives the system away from equilibrium in an *irreversible* way, such that the work required to do the change exceeds the free energy difference, as posited by the maximum work theorem of Equation (1). To be more specific, the work performed on the system along a particular trajectory $x(t)$ is the energy change caused by changes of the control parameter accumulated along the trajectory:

$$W[x(t), \lambda(t)] = \int_0^\tau \frac{\partial H(x(t), \lambda)}{\partial \lambda}\bigg|_{\lambda=\lambda(t)} \dot{\lambda}(t)\, dt \tag{4}$$

where $\dot{\lambda}(t)$ is the time derivative of $\lambda(t)$. Note that this work depends both on the protocol $\lambda(t)$ as well as on the particular trajectory $x(t)$ followed by the system. The average appearing on the left-hand side of Equation (1) is over many repetitions of the switching process starting from initial conditions distributed according to the equilibrium distribution $\rho(x) \propto \exp(-\beta H(x, \lambda_A))$ for control parameter λ_A. In a computer simulation, one could realize such a process by sampling initial conditions from a canonical distribution and then integrating the underlying equations of motion, while at the same time changing the control parameter λ according to the protocol $\lambda(t)$.

Jarzynski has shown [2,3] that averaging over the exponential of the work $\exp(-\beta W(\tau))$ rather than the work, turns the maximum work theorem into an equality, $\langle \exp\{-\beta W[x(t), \lambda(t)]\}\rangle = \exp\{-\beta \Delta F\}$. It is important to realize that the average over the work exponential involves two averages, one over the distribution of initial conditions and another one over the set of trajectories that originate from a particle initial condition. For deterministic dynamics, the initial condition determines the entire trajectory, $x(t)$, but for stochastic dynamics, the system evolves in different ways, even if one repeatedly starts from the same initial condition. Hence, for stochastic dynamics, the average appearing in the Jarzynski equation also requires an average over noise histories.

The Jarzynski equation is an exact result that holds under very general conditions. The requirements are that initially, the system must be in equilibrium and that for a fixed control parameter, the dynamics conserves the equilibrium distribution corresponding to that value of the control parameter. The latter condition is satisfied by most types of dynamics usually used in computer simulations, including Newtonian, thermostated, Langevin and Monte Carlo dynamics. It is worth pointing out that it is not necessary that the system be in an equilibrium state at the end of the transformation process or relax towards equilibrium after the control parameter switching is completed. Furthermore, it is interesting that the Jarzynski equation holds, even if the switching is carried out according to different (though prescribed) protocols provided that $\lambda(0) = \lambda_A$ and $\lambda(\tau) = \lambda_B$, *i.e.*, all protocols start at λ_A at time 0 and finish at λ_B at time τ. After Jarzynski's seminal work [2], in which the Jarzynski equality was derived for systems evolving deterministically with and without coupling to a heat bath, several other proofs were provided, for instance, based on a master equation [3], for Markovian dynamics satisfying detailed balance [5,13], for dynamical systems conserving the canonical distribution [14] or from the Feynman–Kac theorem [7].

In the limiting cases of infinitely fast switching and infinitely slow switching, the Jarzynski equality reduced to two well-known results. For instantaneous switching, $\tau \to 0$, the initial and final point of a trajectory are identical, as the system has no time to evolve. In this case, the work carried out on the system at a particular microscopic state x equals the difference in energy evaluated for the two values of the control parameter:

$$W(x) = H(x, \lambda_B) - H(x, \lambda_A) \tag{5}$$

The Jarzynski equation then becomes:

$$e^{-\beta \Delta F} = \left\langle e^{-\beta[H(x,\lambda_B) - H(x,\lambda_A)]} \right\rangle_{\lambda_A} \tag{6}$$

where the subscript next to the angular bracket indicates that the average has to be carried out with respect to the equilibrium distribution at λ_A. The above equation is the central result of free energy perturbation theory [15] and is often used to compute free energy differences. In the opposite limit of infinitely slow switching, $\tau \to \infty$, the system has time to equilibrate for every intermediate value of the control parameter, such that the Jarzynski equation together with Equation (4) implies:

$$\Delta F = \int_{\lambda_A}^{\lambda_B} \left\langle \frac{\partial H(x, \lambda)}{\partial \lambda} \right\rangle_\lambda d\lambda \tag{7}$$

This expression provides the basis for the thermodynamic integration method [16], in which equilibrium simulations are carried out for different, but fixed values of the control parameter λ to compute the average energy derivatives $\langle \partial H / \partial \lambda \rangle_\lambda$. The free energy difference is then obtained by numerical integration, for instance, by using the Simpson rule or more sophisticated integration schemes. The maximum work theorem of Equation (1) also immediately follows from the Jarzynski equation by virtue of Jensen's inequality, $\langle \exp(-x) \rangle \geq \exp(-\langle x \rangle)$.

As mentioned in the introduction, the Jarzynski equation can be viewed as a way to remove the bias introduced by the switching process into the phase space distribution obtained at the end of the process. Following similar considerations as those used to derive the Jarzynski equality, one can prove that for any phase space function $A(x)$ the following equation holds [4,7,17]:

$$\langle A(x) \rangle_{eq,\lambda_B} = \langle A(x(\tau)) e^{-\beta[W(\tau) - \Delta F]} \rangle_{non\text{-}eq} \tag{8}$$

Here, the angular brackets on the left-hand side indicate an equilibrium average for the control parameter fixed at λ_B, and the average on the right-hand side is an average over non-equilibrium pathways generated with protocol $\lambda(t)$ just as in the Jarzynski equations. To make this difference even more explicit, we have added the subscripts eq and non-eq to the equilibrium and non-equilibrium average, respectively. In the above equation, $x(\tau)$ refers to the endpoints of the non-equilibrium trajectories. The Jarzynski equation is simply obtained by setting $A(x) = 1$. Equation (8) implies that equilibrium averages can be computed by reweighting the non-equilibrium distribution obtained as a result of the switching procedure by $\exp(-\beta W + \beta \Delta F)$. In particular, the equilibrium distribution for λ_B is obtained by setting $A(x) = \delta(x - x(\tau))$, where $\delta(x)$ is the Dirac delta function:

$$\rho_{eq}(x, \lambda_B) = \langle \delta(x - x(\tau)) e^{-\beta[W(\tau) - \Delta F]} \rangle_{non\text{-}eq} \tag{9}$$

Hence, in principle, all equilibrium properties for λ_B (and with appropriate modifications, also for all intermediate values $\lambda(t)$ of the control parameter) can be extracted from a set of non-equilibrium trajectories obtained from simulation or experiment.

If the dynamics of the system not only conserves the equilibrium distribution for a fixed control parameter, but is also microscopically reversible, *i.e.*, if it satisfies detailed balance, the work distribution for the forward process is simply related to that of the process carried out with the time reversed protocol. More specifically, the distribution $P(W)$ of work W observed in repeated realizations of the switching process is given by:

$$P(W) = \langle \delta(W - W[x(t), \lambda(t)]) \rangle_A \qquad (10)$$

where the average is over initial conditions of the equilibrium ensemble A and over pathways starting from these initial conditions under the action of the protocol $\lambda(t)$. Now, consider the time inverted protocol $\lambda_R(t) = \lambda(\tau - t)$. The distribution $P_R(W)$, observed for the reverse process, in which the control parameter is changed from λ_B back to λ_A, can be written as:

$$P_R(W) = \langle \delta(W - W[x(t), \lambda_R(t)]) \rangle_B \qquad (11)$$

where, now, the average is over initial conditions from the equilibrium ensemble B with trajectories evolving, while the control parameter follows the inverted protocol $\lambda_R(t)$. Crooks has shown that for dynamics that is microscopically reversible, the work distributions $P(W)$ and $P_R(W)$ for the forward and reverse process, respectively, are related by [5,6]:

$$P(W) = P_R(-W)e^{\beta(W - \Delta F)} \qquad (12)$$

This exact result, known as the Crooks fluctuation theorem, also serves as a basis for various free energy calculation methods, as explained in detail in subsequent sections.

3. Implementing Fast Switching Simulations

Jarzynski's non-equilibrium work theorem and the Crooks fluctuation theorem suggest interesting algorithms for the calculation of free energy differences. The power of these algorithms derives from the fact that all quantities appearing in these relations can be easily determined. The simplest of these algorithms consists in the following steps. First, one needs to prepare initial conditions distributed according to the Boltzmann–Gibbs distribution. This can be achieved using a variety of methods, for instance, canonical Monte Carlo simulation, possibly combined with enhanced sampling methods, such as parallel replica sampling, or with thermostated molecular dynamics. To improve the efficiency of the free energy calculation, it is important to make sure that these initial conditions are sufficiently decorrelated.

From these initial conditions, one then starts trajectories of the desired length that are integrated, while, at the same time, changing the control parameter according to the protocol $\lambda(t)$. Both the choice of the parameter λ used to drive the transformation, as well as the shape of the protocol influence the efficiency of the calculation, as described in detail below. One can compute the dynamics of the system based on stochastic equations of motion, such as the Langevin equation, or deterministic equations of motion, such as Newton's equations with or without thermostat. Along the computed trajectories, one then has to compute the work W carried out on the system by changing the control parameter. This is most easily done by dividing the basic molecular dynamics steps into two sub-steps. In the first sub-step, the state $x(t + \Delta t)$ of the system at time $t + \Delta t$ is computed by carrying out an integration step with the control parameter fixed at value $\lambda(t)$. In the second sub-step, one then changes the control parameter from $\lambda(t)$ to $\lambda(t + \Delta t)$, while keeping the state $x(t + \Delta t)$ of the system unchanged. Only in this second sup-step is work carried out on the system. In this two-step procedure, the work carried out on the system along a particular trajectory up to time $t + \Delta t$ is given by:

$$W(t + \Delta t) = W(t) + H(x_{t+\Delta t}, \lambda_{t+\Delta t}) - H(x_{t+\Delta t}, \lambda_t) \tag{13}$$

where $x_t \equiv x(t)$ and $\lambda_t \equiv \lambda(t)$ are the state of the system and the value of the control parameter at time t, respectively. From the work values collected in this way for the forward process, and possibly also for the backward process, one can then determine the free energy difference by applying the types of analyses discussed in the next section.

An important choice one has to make in the context of fast switching free energy computations is how to allocate computing time. In particular, one has to decide whether to generate many short trajectories with a large switching rate or fewer and longer trajectories along which the system is driven more gently. Without enhanced sampling schemes, as those discussed in subsequent sections, one generally expects the slow switching regime to give more accurate free energy estimates for a given amount of computing time [18]. As a rule of thumb, one should carry out the switching slowly enough, such that the standard deviation of the work values does not exceed $k_\mathrm{B}T$. In this slow switching regime, the statistical error obtained with a given amount of computing time grows slowly with the switching rate. It is nevertheless more advantageous to compute several trajectories at a moderate switching rate than one single long trajectory, because then, an error estimate for the free energy can be obtained in a straightforward manner. Furthermore, multiple trajectories can be run in parallel to exploit the capabilities of parallel processing machines. Another important choice to make in fast switching simulations concerns the direction in which the transformation is carried out. Interestingly, it can be shown that the direction in which more work is dissipated is computationally beneficial [19]. This formal result is consistent with experience in free energy calculations using perturbation theory. In the calculation of chemical potentials, for instance, test particle insertion typically produces a larger variation in the energy change compared to particle removal and leads to more accurate estimates of the chemical potential [1].

As discussed above, the statistical error of a free energy computed via fast switching strongly depends on the rate at which the system is driven out of equilibrium. However, while the switching rate is certainly the most important parameter, also the particular shape of the protocol $\lambda(t)$ for a given total switching time τ plays an important role in determining the accuracy of the free energy estimate. Since the Jarzynski equality and the Crooks fluctuation theorem hold for arbitrary protocols, one can exploit this freedom to design protocols that optimize the free energy computation. Recently, Schmiedl and Seifert have addressed a related question, asking how the protocol should be designed to minimize the average work expended during the non-equilibrium transformation for a given total of τ [20]. Their analysis, carried out for a particle dragged through a fluid and for a particle in a harmonic trap with changing strength, indicates that, surprisingly, the optimum protocol has discontinuous jumps, both at the beginning and at the end of the process. This result is in contrast to an earlier linear-response analysis [21], which implied that the optimum protocol is smooth and free of jumps. In the cases studied by Schmiedl and Seifert, the optimum protocol with jumps led to a reduction of the dissipated work by up to 12% compared to the case with a continuous protocol changing linearly in time. A subsequent numerical study of a non-linear system carried out by Then and Engel [22] showed that the optimum protocol can have one, two or even more jumps. Steps occur also in the optimum protocol for underdamped Langevin dynamics, for which also delta-like singularities appear at the start and the end of the switching process, effectively kicking the system discontinuously [23].

While, in general, protocols in which the dissipated work is small are expected to yield a more accurate free energy estimate, there is no simple relation between the average work and the statistical error in the free energy. Hence, a protocol optimized with respect to the work does not necessarily minimize the statistical error. However, numerical protocol optimizations conducted for various models indicate that control parameter steps at the start and the end of the protocol (but never in between) are beneficial also for free energy computations [24]. These steps are most pronounced in the fast switching regime and disappear for slow switching. For small switching rates, the minimum work protocol and the minimum error protocol are identical, but for large switching rates, that may differ. In some cases the minimum error protocol even yields an average work that is larger than that of a linear protocol without steps. While appropriate steps in the protocol can lead to a considerable reduction of the computational cost of fast switching free energy calculations, such large savings typically occur only in switching regimes where the straightforward application of the Jarzynski equality is impractical. Whether work biased sampling schemes (discussed in Section 6) may serve to leverage the potential power of discontinuous protocols is currently an open question.

4. Analysis of Non-Equilibrium Free Energy Calculations

The simplest, but also most error-prone, method to obtain free energies from one-sided non-equilibrium simulations is a direct evaluation of the exponential estimator:

$$\Delta F = -k_{\mathrm{B}}T \ln \left\langle e^{-\beta W} \right\rangle \approx -k_{\mathrm{B}}T \ln \sum_{i=1}^{n} e^{-\beta W_i}/n \tag{14}$$

where W_i are the work values obtained in n independent non-equilibrium runs. If the work distribution is broad, with a variance $\mathrm{var}(W) \gg (k_{\mathrm{B}}T)^2$, then the estimate will tend to be dominated by only a few trajectories [19]. All others have negligible weight, resulting not only in sampling inefficiency, but also a systematic bias of the free energy estimate (*i.e.*, the average of ΔF, obtained in repeated sampling with a fixed number n of trajectories, deviates from the exact value [25]). The resulting systematic errors can be estimated and at least partly corrected [17,26–28]. Alternatively, the width of the work distribution can be reduced by breaking the transformation up

168

into segments [18,29,30]. However, the computational cost of re-equilibration at intermediate stages can be significant. The bias can also be eliminated by using cumulant estimators [2,18], in particular, the second-order approximation:

$$\Delta F \approx \langle W \rangle - \beta \, \mathrm{var}(W)/2 \tag{15}$$

However, while eliminating the bias of the exponential estimator, the cumulant approximation is only approximate and, thus, has a systematic error if the work distribution deviates from a Gaussian. Other approaches using the tail statistics of work values have also been proposed [31,32]. In closing the discussion of the direct estimator, we note that the width of the work distribution is closely related to the amount of energy dissipated in the non-equilibrium transformation:

$$\langle W \rangle - \Delta F \approx \beta \, \mathrm{var}(W)/2 \tag{16}$$

Large variance, and, thus, large dissipation, arises from hysteresis effects and can be minimized by optimising the transformation protocol with respect to its time dependence [20] and the choice of control parameter.

More accurate and asymptotically unbiased free energy estimates can be obtained from two-sided simulations by using the Crooks relation. By exploiting the analogy between equilibrium perturbation theory and non-equilibrium simulations, one can adapt Bennett's acceptance ratio as the estimator [33,34]. It requires solving an implicit relation:

$$\sum_{i=1}^{n_f} \frac{1}{1 + \frac{n_f}{n_b} e^{\beta(W_i - \Delta F)}} = \sum_{i=1}^{n_b} \frac{1}{1 + \frac{n_b}{n_f} e^{\beta(\underline{W}_i + \Delta F)}} \tag{17}$$

where W_i and $\underline{W}_i$ are the work values obtained on the n_f and n_b forward and reverse transformations, respectively. This equation can be solved numerically, e.g., by using the Newton–Raphson method. Note that the work values, $\underline{W}_i$, on the reversed path have the opposite sign.

The analogy to the equilibrium method also allows us to adapt two-sided cumulant estimators [35] to non-equilibrium work distributions [18] or to use Bennett's overlapping histogram method [33]. While less efficient as a free energy estimator than the acceptance ratio method, the histogram method provides us with a test of consistency between forward and reverse transformations. According to Equation (12), a plot of the logarithm of $P(W)/P_R(-W)$ should be a straight line as a function of W with slope β. Deviations point to sampling issues or other problems. Another approach [36] for the calculation of free energies from non-equilibrium switching simulations relies on the ideas of waste-recycling Monte Carlo [37].

5. Calculating Potentials of Mean Force

Potentials of mean force (PMF) $G(q)$ along a chosen coordinate $q = q(x)$ are defined as:

$$G(q) = -k_B T \ln \int dx e^{-\beta H(x)} \delta[q - q(x)] \tag{18}$$

up to an arbitrary constant. The coordinate q depends on the phase space coordinate x and, thus, fluctuates along a trajectory. To apply the Jarzynski equality, one would need to make q a control parameter equivalent to λ. However, in molecular simulations, one may not be able to (or want to) control q explicitly, e.g., by applying a holonomic constraint. Instead, it may be easier to restrain q, for instance, by imposing harmonic biasing potentials, as in umbrella sampling. Even if such bias potentials are explicit functions of time, e.g., by moving the center of the harmonic bias, one can obtain equilibrium PMFs from an extension of the Jarzynski equality [7]. The central relation is Equation (9), which allows us to obtain an estimate of the equilibrium phase space density by reweighting trajectory data. If the time-dependent biasing potential is of the form $V = V[q(x), t]$, then the equilibrium PMF in the absence of the bias V, up to a time-dependent constant, can be recovered by weighting trajectory points $q[x(t)]$ with the Boltzmann factor of the work minus the energy stored in the pulling spring:

$$G(q) = -k_{\mathrm{B}} T \ln \left\langle \delta[q - q[x(t)]] e^{-\beta[W(t) - V[q[x(t)], t]]} \right\rangle \tag{19}$$

In principle, this relation applies at every time, t. In practice, q values at time t will be concentrated in a narrow region, whose location depends on the bias, V, and its history. Therefore, to obtain a complete PMF over a range of q values, one should combine results at different times t. In the original derivation, the histogram-reweighting procedure of Ferrenberg and Swendsen [38] was adapted for non-equilibrium PMF calculations [7,17]:

$$G(q) = -k_{\mathrm{B}} T \ln \frac{\sum_t \frac{\langle \delta[q - q(t)] \exp[-\beta W(t)] \rangle}{\langle \exp[-\beta q(t)] \rangle}}{\sum_t \frac{\exp[-\beta V(q, t)]}{\langle \exp[-\beta W(t)] \rangle}} \tag{20}$$

where the sums extend over different time points t. This is not the only possible way to combine histograms obtained at different times, and other procedures have been suggested [39–41].

In many practical applications, the biasing potentials V are harmonic. In such "steered molecular dynamics" simulations and similar approaches [42–45], one can obtain estimates of the PMF using approximate formalisms that involve the system's free energy difference $\Delta F(t)$ and its time dependence. In the limit of very stiff pulling springs $V(q, t) = k[q - z(t)]^2/2$, constraining q to a prescribed path $z(t)$ with large k, one can use the "stiff-spring approximation" of Park et al. [46]. In this limit, q is almost a control parameter, which results in an approximate relation between the system free energy difference $\Delta F(t)$ and the PMF $G(q)$:

$$G[q = q(t)] \approx \Delta F(t) - \frac{1}{2kv^2} \left(\frac{\Delta \ddot{F}(t)}{\beta} - \Delta \dot{F}(t)^2 \right) \tag{21}$$

where we assumed, for simplicity, that the spring moves at a constant velocity v, i.e., $z(t) = vt$, and $\Delta \dot{F} = dF(t)/dt$. More accurate approaches using the same information, $\Delta F(t)$ and its first two time derivatives, have been derived on the basis of the Weierstrass transform [17,47]:

$$G\left(q = vt - \frac{\Delta \dot{F}(t)}{kv} \right) \approx \Delta F(t) - \frac{\Delta \dot{F}(t)^2}{2kv^2}$$

$$+ \frac{1}{2\beta} \ln \left(1 - \frac{\Delta \ddot{F}(z)}{kv^2} \right) \tag{22}$$

Note that the PMF is calculated at a shifted position and that the argument of the logarithm is positive by definition, being proportional to a variance [47]. In practical applications of Equation (21) or (22), $\Delta F(t)$ can be obtained from either unidirectional simulations using the Jarzynski equality or from bidirectional sampling using, e.g., the method of Minh and Adib [48], building on the Crooks fluctuation theorem. Minh and Adib [48] have also developed histogram-based PMF reconstructions that combine information from simulations starting at different transition endpoints, *i.e.*, with initial biases $V(q, 0)$ and $V(q, \tau)$ and evolving as $V(q, t)$ and $V(q, \tau - t)$.

6. Importance Sampling of Fast-Switching Trajectories

Fast switching simulations carried out at large switching rates typically generate work distributions that lead to large statistical uncertainties in the free energy estimate. As discussed earlier, the reason is that trajectories with typical work values contribute little to the exponential average of the Jarzynski equation, while trajectories with work values dominating the average are very rare. As a consequence, the convergence of the computed free energy is impractically slow for overly fast switching. A solution to this problem consists in favoring the generation of trajectories with important work values. In this section, we discuss how path sampling techniques can be used for this purpose.

To introduce computational methods for realizing this idea, we rewrite the exponential work average as an explicit sum over pathways:

$$e^{-\beta \Delta F} = \int \mathcal{D}x(t) P[x(t), \lambda(t)] e^{-\beta W[x(t), \lambda(t)]} \tag{23}$$

where the notation $\int \mathcal{D}x(t)$ implies an integral over all pathways $x(t)$ and $P[x(t), \lambda(t)]$ is the probability to observe the trajectory $x(t)$ for given protocol $\lambda(t)$. Note that the path probability $P[x(t), \lambda(t)]$ also includes the probability of the initial condition x_0. As suggested by Ytreberg and Zuckerman [49] and by Athènes [50], one way to enhance the sampling of important trajectories consists in introducing an explicit bias function $\pi[x(t)]$ (assumed to be integrable and positive everywhere) in the average:

$$e^{-\beta \Delta F} = \frac{\int \mathcal{D}x(t) P[x(t)] \pi[x(t)] e^{-\beta W[x(t)]} / \pi[x(t)]}{\int \mathcal{D}x(t) P[x(t)] \pi[x(t)] / \pi[x(t)]} \tag{24}$$

where we have dropped the explicit dependence on the protocol $\lambda(t)$ in the arguments of $P[x(t)]$ and $W[x(t)]$ to simplify the notation. The right-hand side of this equation, obtained by simply dividing and multiplying by the (so far unspecified) bias function $\pi[x(t)]$ can be viewed as the ratio of two averages taken in a biased ensemble, leading to:

$$e^{-\beta \Delta F} = \frac{\langle e^{-\beta W[x(t)]} / \pi[x(t)] \rangle_\pi}{\langle 1 / \pi[x(t)] \rangle_\pi} \tag{25}$$

Here, the angular brackets $\langle \cdots \rangle_\pi$ denote an average over pathways distributed according to the biased ensemble $P_\pi[x(t)] \propto P[x(t)] \pi[x(t)]$. Since, in general, the bias function $\pi[x(t)]$ depends on the entire pathway $x(t)$, the biased ensemble cannot be sampled by preparing initial conditions according to a certain distribution and running fast switching trajectories from them. Instead, one can use trajectory sampling algorithms (such as the shooting algorithm) adapted from transition path sampling, a methodology originally developed for the simulation of rare events occurring in complex systems [51–53]. In this approach, the bias function appears in the acceptance probability of the path sampling scheme, steering the simulation towards the desired regions of trajectory space.

Since the bias function should enhance the sampling of important, but rare, work values, a bias function depending on the path $x(t)$ only through the work $W[x(t)]$ suffices, $\pi[x(t)] = \pi[W[x(t)]]$. The accuracy of a free energy calculation carried out with biased path sampling now crucially depends on the particular choice of this bias function. It is evident that to obtain an accurate estimate of ΔF, the bias function should be selected, such that the statistical error is small both in the numerator and in the denominator of the fraction on the right-hand side of Equation (25). This implies that the work distribution in the biased ensemble should have a large overlap with the work distribution $P(W)$ in the unbiased ensemble, as well as with the integrand $P(W) \exp(-\beta W)$ appearing in the Jarzynski equality. It has been shown [49,50] that large efficiency increases can be obtained using the bias function $\pi(W) = \exp(-\beta W/2)$, which produces a work distribution in between the two distributions $P(W)$ and $P(W) \exp(-\beta W)$ [54]. A more systematic investigation [55] of the statistical error in the free energy estimate obtained by biased path sampling yields the optimum bias $\pi(W) = |\exp(-\beta(W - \Delta F)) - 1|$. This result implies that the expected statistical error in the free energy is smallest if typical and dominant work values are sampled with high frequency. Interestingly, sampling work values $W \approx \Delta F$ near the free energy difference is not important. Unfortunately, the practical usefulness of this optimum bias function is limited, because its application requires prior knowledge of the free energy difference, i.e., the very quantity one wants to compute. However, iterative schemes, in which the bias function is adapted as the simulation goes on, might make productive use of the functional form of the optimum bias. A recently suggested approach [36] based on the waste-recycling estimator [37] effectively introduces a bias that covers both peaks of the optimum bias, $\pi(W)$.

Another way of realizing work biased path sampling of fast-switching trajectories for the computation of free energies was suggested by Sun [56,57]. In this approach, which can be viewed as a thermodynamic integration procedure in path space, a parameter α is introduced into the exponential average:

$$e^{-\beta \Delta \tilde{F}(\alpha)} = \int \mathcal{D}x(t) P[x(t)] e^{-\beta \alpha W[x(\tau)]} \tag{26}$$

The right-hand side defines, in effect, the generating function of the work distribution at the end of the transformation. The free energy difference $\Delta \tilde{F}(\alpha)$ defined by the above equation depends on this parameter α. While for $\alpha = 0$ one obtains $\Delta \tilde{F}(0) = 0$ due to the normalization of the path distribution, for $\alpha = 1$ one recovers the original free energy difference $\Delta \tilde{F}(\alpha) = \Delta F$. One can thus compute ΔF by taking the derivative of $\Delta \tilde{F}(\alpha)$ with respect to α and then integrate over α from zero to one [18,56,57]:

$$\Delta F = \int_0^1 d\alpha \frac{d\Delta \tilde{F}(\alpha)}{d\alpha} \tag{27}$$

The advantage of writing the free energy difference in this way is that the derivative of $\Delta \tilde{F}(\alpha)$ with respect to α yields a simple average over the work:

$$\frac{d\Delta \tilde{F}(\alpha)}{d\alpha} = \langle W \rangle_\alpha \tag{28}$$

where the notation $\langle \cdots \rangle_\alpha$ indicates a path average over the work weighted path ensemble:

$$P_\alpha[x(t)] \propto P[x(t)]e^{-\beta \alpha W[x(t)]} \tag{29}$$

The work average $\langle W \rangle_\alpha$ is not affected by the type of statistical errors that make the computation of the exponential work average difficult, and it can be evaluated efficiently in a path sampling simulation. By repeating such a calculation for different values of α and integrating the work average numerically, one finally obtains the desired free energy difference. Furthermore, in this method, the statistical errors are kept low by making sure that pathways with both dominant and typical work values are sampled with sufficient frequency. This can be seen explicitly by noting that in the work biased ensemble corresponding to a particular value of the bias parameter, α, the work, W, is distributed according to $P_\alpha(W) \propto P(W)\exp(-\beta \alpha W)$. Thus, by gradually changing α from zero to one, one switches the work distribution from $P(W)$ to $P(W)\exp(-\beta W)$, sweeping over all important work values in the course of the thermodynamic integration procedure.

One can show that in the limit of infinitely short trajectories, Sun's method reduces to conventional thermodynamic integration. This result raises the question of which trajectory length leads to the most efficient free energy calculations and, in particular, if work biased path sampling algorithms perform better then conventional methods, such as thermodynamic integration or umbrella sampling. Extensive calculations carried out for various models indicate [58,59] that work biased fast switching path algorithms are generally less efficient than standard methods, such as thermodynamic integration, thermodynamic perturbation or umbrella sampling. There are however cases, such as an ideal gas compressed by a piston moving in a cylinder, where fast switching is advantageous [59]. In this particular case, the work distribution does not converge to a limiting form for increasing switching speed, and the typical work values keep growing. As a consequence, the optimum switching rate is finite in this case, even if an optimum work bias is applied [59].

7. Fast Switching with Large Time Steps

Molecular dynamics simulations are usually carried out with time steps that are a compromise between accuracy (often assessed in terms of energy conservation) and computing speed. Small time steps yield accurate trajectories with good energy conservation, but require a larger computational effort, because the cost of a trajectory of a given length is proportional to the number of steps and, hence, inversely proportional to the size of the time step. Larger time steps reduce the computing time, but corrupt the accuracy, resulting in poor energy conservation. In general, using such low-accuracy trajectories for free energy computations introduces a systematic error into the free energy estimate. It is, however, possible to devise exact expressions akin to the Jarzynski equation to compute free energy differences from crude trajectories calculated with large time steps [13,60].

Using this approach, which is based on a generalization of the Jarzynski equation for phase space mappings [61], can help to considerably increase the efficiency of fast switching simulations, due to the reduced computational cost of the large time step trajectories.

As mentioned earlier, in the limit of instantaneous switching, the Jarzynski equation reduces to the perturbation identity of Equation (6). Free energy computation methods relying on this equation perform well if there is a large overlap between the ensembles A and B, corresponding to the control parameters λ_A and λ_B, respectively. If, however, these ensembles strongly differ, the free energy calculation converges poorly, because important contributions to the average are rarely sampled. To remedy this situation, Jarzynski has devised the targeted free energy perturbation method [61] based on a generalization of the Jarzynski equality. The basic ideas underlying this approach is to improve the efficiency of the perturbative calculation by applying a mapping that transforms the equilibrium ensemble A into an ensemble A' that overlaps more strongly with ensemble B. The mapping $\phi(x)$ considered in this approach is required to be invertible and differentiable, but is arbitrary otherwise. By starting from the definition of the free energy difference (Equation (3)) and carrying out a variable transformation from x to $x' = \phi(x)$, one can then show that:

$$e^{-\beta \Delta F} = \left\langle e^{-\beta W_\phi(x)} \right\rangle \tag{30}$$

where the "work" function is defined as:

$$W_\phi(x) = H(\phi(x), \lambda_B) - H(x, \lambda_A) - k_{\mathrm{B}} T \ln \left| \frac{\partial \phi}{\partial x} \right| \tag{31}$$

The last term in the work function results from the Jacobian of the transformation and vanishes for phase space volume preserving maps. If the mapping is chosen to be the propagator of Newtonian dynamics, Equation (30) reduces to the Jarzynski equation for isolated systems evolving at constant energy. By using the inverse map, ϕ^{-1}, with the corresponding work definition, one can also use this mapping approach together with the Crooks fluctuation theorem.

Equation (30) suggests the following algorithm for free energy computation. One first samples phase space points x from the equilibrium ensemble A. Then, to each of these points, one applies the mapping and computes W_ϕ. Finally, the average of $\exp(-\beta W_\phi(x))$ carried out over all points x yields the free energy difference. Now, the efficiency of this method crucially depends on the ability to devise appropriate mapping $\phi(x)$. The closer the ensemble resulting from the transformation resembles B, the higher is the efficiency. No general methods exists to derive $\phi(x)$, but a well-chosen mapping can substantially reduce the cost of a free energy computation.

One possible strategy to exploit Equation (30) consists in choosing a sequence of molecular dynamics steps as phase space mapping. Each of these steps, designed to approximate the time evolution of the system over a small interval Δt maps a phase point x_i into the next phase point x_{i+1} along the molecular dynamics trajectory. Hence, a sequence of n molecular dynamics steps may also be considered as a phase space mapping that takes the initial point x_0 into the final point x_n. The expression for the work W_ϕ is particularly simple for integrators, such as the Verlet algorithm, that conserve phase space volume. Then, the Jacobian of the mapping is unity, and Equation (30) turns into:

$$e^{-\beta \Delta F} = \left\langle e^{-\beta [H(x_n, \lambda_B) - H(x_0, \lambda_A)]} \right\rangle \tag{32}$$

Interestingly, this relation holds exactly independently of the size of the time step Δt used in the integration algorithm. Hence, fast switching simulations can be carried out with large time steps, producing only approximate trajectories. Nevertheless, the free energies obtained in this way are in principle exact. Since trajectories computed with a large time step require a smaller number of integration steps, such fast switching simulation holds the promise to improve the efficiency of the free energy calculation. Whether this is indeed the case, depends on how the work distribution changes due to the large time step. Calculations carried out for several model systems indicate that while the molecular dynamics trajectories generated with large time steps are approximate, they still reproduce the essential physics of the process, such that the work distributions are not affected adversely. As a consequence, for optimum efficiency, time steps of fast switching free energy computations can be increased up to the stability limit of the simulation. Note that this large time step approach can be used also using integrators that do not conserve phase space volume [60], but this unnecessarily complicate the simulations, because one has to keep track of the Jacobian while computing the molecular dynamics trajectories.

The large time step formalism can also be used for the calculation of potentials of mean force [62]. In such a simulation, the work based reweighing of Equation (30) is applied at each stage of the time evolution with a work function that accumulates along the trajectory. Fast switching simulations were carried out for the force induced unfolding of a decalanine molecule [62]. The free energy profile obtained for a time step of 3.2 fs, *i.e.*, close to the stability limit, agrees well with that calculated using a conservative time step of 0.5 fs. An efficiency analysis reveals that the optimum time step for the unfolding simulations lies in the range 1–3 fs. It is interesting to note that the fast-switching trajectories may show unphysical features, such as a redistribution from potential to kinetic energy, due to the conserved shadow Hamiltonian belonging to the integrator used in the simulation [62]. Nevertheless, the obtained free energy profile is exact up to statistical errors.

8. Applications

Arguably the most important practical application of non-equilibrium work theorems has been to experiments. Almost immediately after the connection between non-equilibrium single-molecule pulling experiments and Jarzynski's identity was rigorously established [7], experimental studies of the folding and unfolding of nucleic acids using optical tweezers followed [8,63]. It is often difficult, if not impossible, to conduct pulling experiments sufficiently slowly to maintain near-equilibrium conditions. Nonetheless, the use of non-equilibrium free energy reconstruction has made it possible to extract thermodynamic information.

Applications to pulling have been mirrored on the simulation side. Simulated pulling methods mimicking experiments have been developed, initially to probe mechanical perturbations on biomolecules [42–44]. Non-equilibrium pulling methods have been applied not only to protein unfolding, but also to many other complex molecular processes, including ligand dissociation [64–66] and channel translocation [67,68]. To analyze such "steered molecular dynamics" simulations and extract PMFs, the stiff-spring approximation is widely used [46], though Equation (22) offers a more accurate method using the same information [47] that produce results comparable to full histogram reweighting. In molecular simulations, non-equilibrium methods tend to be less efficient than optimized equilibrium methods as a tool to calculate free energies [18,58]. However, as discussed above, the optimization of non-equilibrium sampling methods is an area of active research, in particular, using importance sampling methods involving path reweighting [49,50,55–59] and nonlinear maps [69,70]. Moreover, non-equilibrium methods can provide valuable insight into the mechanism underlying a process. By forcing the system through a transition and monitoring the resulting bottlenecks [71], one may be able to devise improved control variables that result in a smoother transition and improved sampling efficiency, both in non-equilibrium and equilibrium simulations.

9. Conclusions and Outlook

The Jarzynski non-equilibrium work theorem and the Crooks fluctuation theorem are fundamental exact relations that link the irreversible work carried out on a system during a non-equilibrium transformation to the system's equilibrium statistics. To date, the most significant application of these relations lies in the interpretation of single-molecule pulling experiments, in which forces exerted by atomic force microscopes or optical tweezers are used to probe the properties of individual molecules. Due to technological limitations, such experiments are necessarily carried out at a finite pulling rate, leading to non-equilibrium effects that cannot be neglected. The theorems of Jarzynski and Crooks provide a practical tool for the interpretation of such single-molecule pulling experiments and permit one to extract equilibrium information, such as potentials of mean force, from data obtained under inherently non-equilibrium conditions [7–9,72].

From a computational point of view, the Jarzynski and Crooks theorems have provided a new and powerful framework for the calculation of free energies using computer simulations. Apart from putting earlier slow-growth free energy simulations on a firm theoretical footing, these results have spawned the development of several new free energy algorithms based on non-equilibrium, fast-switching trajectories.

Depending on the rate at which the system is driven away from equilibrium, fast switching free energy computations can be plagued by large statistical errors. For strong driving, *i.e.*, for large switching rates, work distributions are broad, with typical work values by far exceeding the free energy difference. As a consequence, the exponential work average of the Jarzynski equation is dominated by a few rare contributions, leading to large statistical uncertainties and a bias in the free energy estimate. Such errors can outweigh the computational advantage of running inexpensive short trajectories rather than one single long trajectory [18,29,58]. In fact, it has been shown that in the slow switching regime, one obtains more accurate results from few slow simulations than from many faster ones [18]. Numerical simulations carried out for various model systems [58,59] indicate that conventional free energy computation methods, such as thermodynamic integration or free energy perturbation theory, are more efficient than fast switching simulations, even if work biasing techniques are employed. Fast switching methods may, however, be advantageous for systems in which the states of interest are connected by several distinct pathways. In such a case, conventional methods may fail to sample all important transition routes while multiple fast switching

176

trajectories have the chance to probe all important pathways. Such a situation was indeed observed for transitions between low-energy configurations of Lennard-Jones clusters [41], which could be sampled successfully only with non-equilibrium path sampling, but not with other approaches. Compared to standard methods, fast switching algorithms appeared on the scene only recently, such that substantial improvements and new developments are to be expected [13,21,57,73–78]. It is worth noting that fast switching ideas have not only been applied to the calculation of free energies, but have also been combined with existing sampling methods to enhance the efficiency of the simulation. For instance, non-equilibrium switches have been used to improve the acceptance probability of replica exchange simulations [79,80] and to generate trial configurations for Monte Carlo simulations [81,82]. Conversely, waste-recycling Monte Carlo [37] can be adapted for the calculation of free energies from non-equilibrium switching simulations [36].

One aspect of fast switching simulations that has not been fully exploited is the freedom in choosing the transformation protocol. While the optimization of the time dependence of the driving parameter has been the subject of previous numerical and analytical studies [23,24], the extension of such optimizations to multiple control parameters is unexplored to date. The control parameter at the start and the end of the transformation are given, but in between, additional parameters can be subjected to a change as well, without affecting the validity of the relations that provide the basis for fast switching simulation. As an early example, an external pressure has been heuristically adjusted to maintain reasonable box sizes and prevent phase separation in a transformation between liquid and ideal gas states [54]. Defining parameter spaces of higher dimension and determining optimum parameter pathways in these spaces may offer efficient ways to control the work distribution and, hence, reduce the computational cost of fast switching simulations.

Acknowledgments

We acknowledge the financial support of the Austrian Science Fund (FWF, Fonds zur Förderung der Wissenschaftlichen Forschung) within the SFB ViCoM (Spezialforschungsbereich Vienna Computational Materials Laboratory), Grant F41, as well as Project P24681-N20 (C.D.) and the Max Planck Society (G.H.).

Conflicts of Interest

The authors declare no conflict of interest.

References

1. Frenkel, D. Free-Energy Computation and First-Order Phase Transitions. In *Molecular Dynamics Simulations of Statistical Mechanical Systems*, Proceedings of the Enrico Fermi Summer School, Varenna, 1985; Ciccotti, G., Hoover, W.G., Eds.; North-Holland Elsevier Science Publisher: Amsterdam, The Netherlands, 1987; pp. 151–188.
2. Jarzynski, C. Nonequilibrium equality for free energy differences. *Phys. Rev. Lett.* **1997**, *78*, 2690–2693.

3. Jarzynski, C. Equilibrium free energy differences from nonequilibrium measurements: A master-equation approach. *Phys. Rev. E* **1997**, *56*, 5018–5035.

4. Crooks, G.E. Path-ensemble averages in systems driven far from equilibrium. *Phys. Rev. E* **2000**, *61*, 2361–2366.

5. Crooks, G.E. Nonequilibrium measurements of free energy differences for microscopically reversible Markovian systems. *J. Stat. Phys.* **1998**, *90*, 1481–1487.

6. Crooks, G.E. Entropy production fluctuation theorem and the nonequilibrium work relation for free energy differences. *Phys. Rev. E* **1999**, *60*, 2721–2726.

7. Hummer, G.; Szabo, A. Free energy reconstruction from nonequilibrium single-molecule pulling experiments. *Proc. Natl. Acad. Sci. USA* **2001**, *98*, 3658–3661.

8. Liphardt, J.; Dumont, S.; Smith, S.B.; Tinoco, I.; Bustamante, C. Equilibrium information from nonequilibrium measurements in an experimental test of Jarzynski's equality. *Science* **2002**, *296*, 1832–1835.

9. Noy, A. Direct determination of the equilibrium unbinding potential profile for a short DNA duplex from force spectroscopy data. *Appl. Phys. Lett.* **2004**, *85*, 4792–4794.

10. Chipot, C.; Pohorille, A., Eds. *Free Energy Calculations*; Spinger Series in Chemical Physics 86; Springer: Berlin/Heidelberg, Germany, 2007.

11. Lelièvre, T.; Rousset, M.; Stoltz, G. *Free Energy Computations*; Imperial College Press: London, UK, 2010.

12. Frenkel, D.; Smit, B. *Understanding Molecular Simulation: From Algorithms to Applications*; Academic Press: San Diego, CA, USA, 2001.

13. Lechner, W.; Oberhofer, H.; Dellago, C.; Geissler, P.L. Equilibrium free energies from fast-switching trajectories with large time steps. *J. Chem. Phys.* **2006**, *124*, 044113.

14. Schöll-Paschinger, E.; Dellago, C. A proof of Jarzynski's nonequilibrium work theorem for dynamical systems that conserve the canonical distribution. *J. Chem. Phys.* **2006**, *125*, 054105.

15. Zwanzig, R.W. High-temperature equation of state by a perturbation method. I. Nonpolar gases. *J. Chem. Phys.* **1954**, *22*, 1420–1426.

16. Kirkwood, J. Statistical mechanics of fluid mixtures. *J. Chem. Phys.* **1935**, *3*, 300.

17. Hummer, G.; Szabo, A. Free energy surfaces from single-molecule force spectroscopy. *Acc. Chem. Res.* **2005**, *38*, 504–513.

18. Hummer, G. Fast-growth thermodynamic integration: Error and efficiency analysis. *J. Chem. Phys.* **2001**, *114*, 7330–7337.

19. Jarzynski, C. Rare events and the convergence of exponentially averaged work values. *Phys. Rev. E* **2006**, *73*, 046105.

20. Schmiedl, T.; Seifert, U. Optimal finite-time processes in stochastic thermodynamics. *Phys. Rev. Lett.* **2007**, *98*, 108301.

21. De Koning, M. Optimizing the driving function for nonequilibrium free-energy calculations in the linear regime: A variational approach. *J. Chem. Phys.* **2005**, *122*, 104106.

22. Then, H.; Engel, A. Computing the optimal protocol for finite-time processes in stochastic thermodynamics. *Phys. Rev. E* **2008**, *77*, 041105.

23. Gomez-Marin, A.; Schmiedl, T.; Seifert, U. Optimal protocols for minimal work processes in underdamped stochastic thermodynamics. *J. Chem. Phys.* **2008** , *129*, 024114.

24. Geiger, P.; Dellago, C. Optimum protocol for fast switching free energy calculations. *Phys. Rev. E* **2010**, *81*, 021127.

25. Wood, R.H.; Mühlbauer, W.C.F.; Thompson, P.T. Systematic errors in free energy perturbation calculations due to a finite sample of configuration space. Sample-size hysteresis. *J. Phys. Chem.* **1991**, *95*, 6670–6675.

26. Gore, J.; Ritort, F.; Bustamante, C. Bias and error in estimates of equilibrium free-energy differences from nonequilibrium measurements. *Proc. Natl. Acad. Sci. USA* **2003**, *100*, 12564–12569.

27. Zuckerman, D.M.; Woolf, T.B. Theory of a systematic computational error in free energy differences. *Phys. Rev. Lett.* **2002**, *89*, 180602.

28. Wu, D.; Kofke, D.A. Asymmetric bias in free-energy perturbation measurements using two hamiltonian-based models. *Phys. Rev. E* **2004**, *70*, 066702.

29. Rodriguez-Gomez, D.; Darve, E.; Pohorille, A. Assessing the efficiency of free energy calculation methods. *J. Chem. Phys.* **2004**, *120*, 3563–3578.

30. Ozer, G.; Quirk, S.; Hernandez, R. Thermodynamics of decaalanine stretching in water obtained by adaptive steered molecular dynamics simulations. *J. Chem. Theory Comput.* **2012**, *8*, 4837–4844.

31. Zuckerman, D.M.; Woolf, T.B. Overcoming finite-sampling errors in fast-switching free-energy estimates. Extrapolative analysis of a molecular system. *Chem. Phys. Lett.* **2002**, *351*, 445–453.

32. Ytreberg, F.M.; Zuckerman, D.M. Efficient use of nonequilibrium measurement to estimate free energy differences for molecular systems. *J. Comp. Chem.* **2004**, *25*, 1749–1759.

33. Bennett, C.H. Efficient estimation of free energy differences from Monte Carlo data. *J. Comput. Phys.* **1976**, *22*, 245–268.

34. Shirts, M.R.; Bair, E.; Hooker, G.; Pande, V.S. Equilibrium free energies from nonequilibrium measurements using maximum-likelihood methods. *Phys. Rev. Lett.* **2003**, *91*, 140601.

35. Hummer, G.; Szabo, A. Calculation of free energy differences from computer simulations of initial and final states. *J. Chem. Phys.* **1996**, *105*, 2004–2010.

36. Adjanor, G.; Athènes, M.; J. Rodgers, J. Waste-recycling Monte Carlo with optimal estimates: Application to free energy calculations in alloys. *J. Chem. Phys.* **2011**, *135*, 044127.

37. Frenkel, D. Speed-up of Monte Carlo simulations by sampling of rejected states. *Proc. Natl. Acad. Sci. USA* **2004**, *101*, 17571–17575.

38. Ferrenberg, A.M.; Swendsen, R.H. Optimized Monte Carlo data analysis. *Phys. Rev. Lett.* **1989**, *63*, 1195–1198.

39. Oberhofer, H.; Dellago, C. Efficient extraction of free energy profiles from non-equilibrium experiments. *J. Comput. Chem.* **2009**, *30*, 1726–1736.

40. Imparato, A.; Peliti, L. Evaluation of free energy landscapes from manipulation experiments. *J. Stat. Mech.* **2006**, *2006*, P03005.

41. Athènes, M.; Marinica, M.-C. Free energy reconstruction from steered dynamics without post-processing. *J. Comput. Phys.* **2010**, *229*, 7129–7146.

42. Grubmüller, H.; Heymann, B.; Tavan, P. Ligand binding molecular mechanics calculation of the streptavidin biotin rupture force. *Science* **1996**, *271*, 997–999.

43. Izrailev, S.; Stepaniants, S.; Balsera, M.; Oono, Y.; Schulten, K. Molecular dynamics study of unbinding of the avidin-biotin complex. *Biophys. J.* **1997**, *72*, 1568–1581.

44. Paci, E.; Karplus, M. Forced unfolding of fibronectin Type 3 modules: An analysis by biased molecular dynamics simulations. *J. Mol. Biol.* **1999**, *288*, 441–459.

45. Park, S.; Schulten, K. Calculating potentials of mean force from steered molecular dynamics simulations. *J. Chem. Phys.* **2004**, *120*, 5946–5961.

46. Park, S.; Khalili-Araghi, F.; Tajkhorshid, E.; Schulten, K. Free energy calculation from steered molecular dynamics simulations using Jarzynski's equality. *J. Chem. Phys.* **2003**, *119*, 3559–3566.

47. Hummer, G.; Szabo, A. Free energy profiles from single-molecule pulling experiments. *Proc. Natl. Acad. Sci. USA* **2010**, *107*, 21441–21446.

48. Minh, D.D.L.; Adib, A.B. Optimized free energies from bidirectional single-molecule force spectroscopy. *Phys. Rev. Lett.* **2008**, *100*, 180602.

49. Ytreberg, F.M.; Zuckerman, D.M. Single-ensemble nonequilibrium path-sampling estimates of free energy differences. *J. Chem. Phys.* **2004**, *120*, 10876–10879.

50. Athènes, M. A path-sampling scheme for computing thermodynamic properties of a many-body system in a generalized ensemble. *Eur. Phys. J. B* **2004**, *38*, 651.

51. Dellago, C.; Bolhuis, P.G.; Csajka, F.S.; Chandler, D. Transition path sampling and the calculation of rate constants. *J. Chem. Phys.* **1998**, *108*, 1964.

52. Dellago, C.; Bolhuis, P.G.; Geissler, P.L. Transition path sampling. *Adv. Chem. Phys.* **2002**, *123*, 1–84.

53. Dellago, C.; Bolhuis, P.G.; Geissler, P.L. Transition Path Sampling Methods. In *Computer Simulations in Condensed Matter: From Materials to Chemical Biology*; Ciccotti, G., Binder, K., Eds.; Springer: Berlin/Heidelberg, Germany, 2006.

54. Adjanor, G.; Athènes, M. Gibbs free-energy estimates from direct path-sampling computations. *J. Chem. Phys.* **2005**, *123*, 234104.

55. Oberhofer, H.; Dellago, C. Optimum bias for fast-switching free energy calculations. *Comput. Phys. Commun.* **2008**, *179*, 41–45.

56. Sun, S.X. Equilibrium free energies from path sampling of nonequilibrium trajectories. *J. Chem. Phys.* **2003**, *118*, 5769–5775.

57. Atilgan, E.; Sun, S.X. Equilibrium free energy estimates based on nonequilibrium work relations and extended dynamics. *J. Chem. Phys.* **2004**, *121*, 10392–10400.

58. Oberhofer, H.; Dellago, C.; Geissler, P.L. Biased sampling of nonequilibrium trajectories: Can fast switching simulations outperform conventional free energy calculation methods? *J. Phys. Chem. B* **2005**, *109*, 6902–6915.

59. Lechner, W.; Dellago, C. On the efficiency of path sampling methods for the calculation of free energies from non-equilibrium simulations. *J. Stat. Mech.* **2007**, *2007*, P04001.

60. Oberhofer, H.; Dellago, C. Large timestep fast-switching simulations with non-volume preserving integrators for free energy calculations. *Isr. J. Chem.* **2007**, *47*, 215.

61. Jarzynski, C. Targeted free energy perturbation. *Phys. Rev. E* **2002**, *65*, 046122.

62. Oberhofer, H.; Dellago, C.; Boresch, S. Single molecule pulling with large time steps. *Phys. Rev. E* **2007**, *75*, 061106.

63. Collin, D.; Ritort. F.; Jarzynski, C.; Smith, S.B.; Tinoco, I.; Bustamante, C. Verification of the Crooks fluctuation theorem and recovery of RNA folding free energies. *Nature* **2005**, *437*, 231–234.

64. Vashisth, H.; Abrams, C. F. Ligand escape pathways and (un)binding free energy calculations for the hexameric insulin-phenol complex. *Biophys. J.* **2008**, *95*, 4193–4204.

65. Cuendet, M.A.; Michielin, O. Protein-protein interaction investigated by steered molecular dynamics the Tcr-Pmhc complex. *Biophys. J.* **2008**, *95*, 3575–3590.

66. Zhang, D.Q. ; Gullingsrud, J.; McCammon, J.A. Potentials of mean force for acetylcholine unbinding from the alpha7 nicotinic acetylcholine receptor ligand-binding domain. *J. Am. Chem. Soc.* **2006**, *128*, 3019–3026.

67. Jensen, M.O.; Park, S.; Tajkhorshid, E.; Schulten, K. Energetics of glycerol conduction through aquaglyceroporin Glpf. *Proc. Natl. Acad. Sci. USA* **2002**, *99*, 6731–6736.

68. Amaro, R.; Luthey-Schulten, Z. Molecular dynamics simulations of substrate channeling through an alpha-beta barrel protein. *Chem. Phys.* **2004**, *307*, 147–155.

69. Vaikuntanathan, S.; Jarzynski, C. Escorted free energy simulations: Improving convergence by reducing dissipation. *Phys. Rev. Lett* **2008**, *100*, 190601.

70. Vaikuntanathan, S.; Jarzynski, C. Escorted free energy simulations. *J. Chem. Phys.* **2011**, *134*, 054107.

71. Chelli, R. Local sampling in steered monte carlo simulations decreases dissipation and enhances free energy estimates via nonequilibrium work theorems. *J. Chem. Theory Comput.* **2012**, *8*, 4040.

72. Trepagnier, E.H.; Jarzynski, C.; Ritort, F.; Crooks, G.E.; Bustamante, C.J.; Liphardt, J. Experimental test of Hatano and Sasa's nonequilibrium steady-state equality. *Proc. Natl. Acad. Sci. USA* **2004**, *101*, 15038–15041.

73. Shirts, M.R.; Pande, V.S. Comparison of efficiency and bias of free energies computed by exponential averaging, the Bennett acceptance ratio and thermodynamic integration. *J. Chem. Phys.* **2005**, *122*, 144107.

74. Ytreberg, F.M.; Zuckerman, D.M. Peptide conformational equilibria computed via a single-stage shifting protocol. *J. Phys. Chem. B* **2005**, *109*, 9096–9103.

75. Chernyak, V.; Chertkov, M.; Jarzynski, C. Dynamical generalization of nonequilibrium work relation. *Phys. Rev. E* **2005**, *71*, 025102.

76. Rodinger, T.; Pomès, R. Enhancing the accuracy the efficiency and the scope of free energy simulations. *Curr. Opin. Struct. Biol.* **2005**, *15*, 164–170.

77. Lua, R.C.; Grosberg, A.Y. Practical applicability of the Jarzynski relation in statistical mechanics: A pedagogical example. *J. Phys. Chem. B* **2005**, *109*, 6805–6811.

78. Adib, A.B. Entropy and density of states from isoenergetic nonequilibrium processes. *Phys. Rev. E* **2005**, *71*, 056128.

79. Ballard, A.J.; Jarzynski, C. Replica exchange with nonequilibrium switches. *Proc. Natl. Acad. Sci. USA* **2009**, *106*, 12224–12229.

80. Ballard, A.J.; Jarzynski, C. Replica exchange with nonequilibrium switches: Enhancing equilibrium sampling by increasing replica overlap. *J. Chem. Phys.* **2012**, *136*, 194101.

81. Athènes, M. Computation of a chemical potential using a residence weight algorithm. *Phys. Rev. E* **2002**, *66*, 046705.

82. Nilmeier, J.P.; Crooks, G.E.; Minh, D.L.; Chodera, J.D. Nonequilibrium candidate Monte Carlo is an efficient tool for equilibrium simulation. *Proc. Natl. Acad. Sci. USA* **2011**, *108*, E1009.

Reprinted from *Entropy*. Cite as: Ciccotti, G.; Ferrario, M. Dynamical Non-Equilibrium Molecular Dynamics. *Entropy* **2014**, *16*, 233–257.

Review

Dynamical Non-Equilibrium Molecular Dynamics

Giovanni Ciccotti [1] **and Mauro Ferrario** [2,*]

[1] Dipartimento di Fisica, Università di Roma La Sapienza, P.le A. Moro 2, 00185 Roma, Italy;
   E-Mail: giovanni.ciccotti@roma1.infn.it
[2] Dipartimento di Scienze Fisiche, Informatiche e Matematiche, Università di Modena e Reggio
   Emilia, Via G. Campi 213/A, 41125 Modena, Italy

* Author to whom correspondence should be addressed; E-Mail: mauro.ferrario@unimore.it;
   Tel.: +39-592055291; Fax: +39-592055235.

Received: 10 November 2013; in revised form: 26 November 2013 / Accepted: 16 December 2013 / Published: 27 December 2013

Abstract: In this review, we discuss the Dynamical approach to Non-Equilibrium Molecular Dynamics (D-NEMD), which extends stationary NEMD to time-dependent situations, be they responses or relaxations. Based on the original Onsager regression hypothesis, implemented in the nineteen-seventies by Ciccotti, Jacucci and MacDonald, the approach permits one to separate the problem of dynamical evolution from the problem of sampling the initial condition. D-NEMD provides the theoretical framework to compute time-dependent macroscopic dynamical behaviors by averaging on a large sample of non-equilibrium trajectories starting from an ensemble of initial conditions generated from a suitable (equilibrium or non-equilibrium) distribution at time zero. We also discuss how to generate a large class of initial distributions. The same approach applies also to the calculation of the rate constants of activated processes. The range of problems treatable by this method is illustrated by discussing applications to a few key hydrodynamic processes (the "classical" flow under shear, the formation of convective cells and the relaxation of an interface between two immiscible liquids).

Keywords: non-equilibrium; molecular dynamics; dynamical relaxation; hydrodynamics

1. Introduction

The most widespread use of Molecular Dynamics (MD) [1,2], in the same spirit of Monte Carlo (MC) [3,4], is to compute the thermodynamic or statistical behavior of molecular systems at equilibrium. This means that, starting from the assumption of the validity of the ergodic hypothesis, dynamical (MD) or fictitious-time (MC) trajectories are used to sample the equilibrium distribution in phase space (MD) or in configurational space (MC). "Time" averages over the generated trajectories will thereafter provide the statistical properties of the system.

At variance with Monte Carlo, the dynamical approach of Molecular Dynamics can be directly extended to sample distributions corresponding to stationary non-equilibrium conditions, where there exists a stationary distribution but, at variance with equilibrium, its expression is not explicitly known. However, the statistical problem of sampling a time-dependent ensemble cannot be solved by generating states along a single dynamical non-equilibrium trajectory, as long as time cannot be taken as homogeneous and averages over time make no sense.

Generally, to compute macroscopic dynamical behaviors, as, e.g., in hydrodynamics, the assumption of time-scale separation is made and rigorous ensemble averages are substituted with short-time averages equivalent to local smoothing. This may not be the case, sometimes. Moreover, the statistical error implied by this procedure cannot be made as small as desirable and possible. These difficulties can be faced and solved.

In the nineteen-thirties, Lars Onsager [5] observed that an induced (non-equilibrium) relaxation towards equilibrium could be obtained by studying the regression of the corresponding spontaneous fluctuations at equilibrium. Later, in the nineteen-fifties, Kubo [6] provided a mathematical formulation of Onsager's ideas by showing how the (linear) response of a system, initially at equilibrium, to a time-dependent (external) physical perturbation could be obtained by convoluting it with an appropriate equilibrium time-correlation function [7–9]. Kubo also derived the formal expression for the complete (linear and nonlinear) response.

In the case of Kubo's procedure one does not need to make reference to an initial *equilibrium* state, but can, rather, refer to an arbitrary initial distribution at time $t_0 = 0$ of the system. This result has an important consequence for Molecular Dynamics simulations, since it allows one to separate the problem of dynamical evolution from the problem of sampling the initial condition.

Starting from the mid-nineteen-seventies, the direct numerical simulation of the response was used in conjunction with a sample of initial conditions extracted from an equilibrium trajectory [10,11]. In this context, the problem of achieving a reasonable signal-to-noise ratio, even for weak perturbations, was solved for short times by introducing the so-called *subtraction technique* [12], which permitted one to verify, with surprising results [13], the range of the validity of linearity.

Some time later on, it was realized that the same approach could be used to calculate dynamical properties for rare events (e.g., transmission coefficients) by averaging the dynamical response over time-dependent trajectories started from initial conditions sampled from a constrained/conditional equilibrium ensemble [14–18].

Quite recently, finally, the idea of creating a large sample of non-equilibrium trajectories starting from a given initial distribution has been extended to cover whatever distribution that can be sampled starting from an equilibrium or a non-equilibrium, but stationary, dynamics. In particular stationary non-equilibrium ensembles can be generated by suitably restraining standard MD simulations.

In particular, we will illustrate the approach by reporting the results of a study of the time evolution of classical fields, including the onset of convective cells and the relaxation of hydrodynamic interfaces in simple liquids. In this context, we will also briefly address a conceptual difficulty of the approach, due to the possible existence of more than one macroscopic state associated with specific perturbations. In particular cases the problem can be circumvented.

The structure of the paper is as follows. In Section 2 we derive the general framework and specify the possible forms for the initial ensemble. In Section 3 we present a few successful applications of the method. Finally, in Section 4 we try to assess the situation and sketch an outlook.

2. Dynamical Approach to Non-Equilibrium: Theoretical Background

2.1. General Formulation

We start considering, in a very general way, a (classical) dynamical system with n degrees of freedom, whose time evolution is described by a set of first order differential equations in a phase space of dimension $2n$. We will refer to the phase space variables in a collective way with the vector formalism $\vec{\Gamma} = \{q_1, p_1, q_2, p_2, \ldots, q_n, p_n\}$, where the q's and the p's reduce to the usual coordinate-momentum pairs for Hamiltonian dynamics. The equations of motion can be written in the compact form

$$\dot{\Gamma}_j = \dot{\Gamma}_j(\vec{\Gamma}; t) = \dot{\Gamma}_j(q_1, p_1, q_2, p_2, \ldots, q_n, p_n; t), \quad j = 1, 2, \ldots, 2n \tag{1}$$

The above equations could be the usual Hamiltonian equations of motion for an isolated system of N particles [19], contain a number of holonomic constraints [20] or represent the more general case of an "extended" system, possibly non-Hamiltonian [21,22], including couplings of the system to a thermal and/or pressure bath by means of a few extra degrees of freedom, so that, in general, $n > 3N$ (see, also, [23]). We will only assume that the dynamics described by Equation (1) are ergodic, *i.e.*, if we wait long enough, all regions of the phase space available to the system, in accord with the imposed conditions, will be explored by the dynamic evolution. With this in mind, the statistical mechanics description of the system requires the introduction of the invariant measure $d\mu(\vec{\Gamma}, d^{2n}\Gamma)$ in phase space [23]. We start by introducing the generator of time translations in terms of the Liouville operator, $\hat{\mathcal{L}}$

$$i\hat{\mathcal{L}}(\vec{\Gamma}; t) = \sum_{j=1}^{2n} \dot{\Gamma}_j(\vec{\Gamma}; t) \cdot \frac{\partial}{\partial \Gamma_j} = \sum_{j=1}^{n} \dot{q}_j(\vec{\Gamma}; t) \cdot \frac{\partial}{\partial q_j} + \sum_{j=1}^{n} \dot{p}_j(\vec{\Gamma}; t) \cdot \frac{\partial}{\partial p_j} \tag{2}$$

so that the equations of motion can be rephrased in the operator form and formally solved. As the Liouville operator depends explicitly on time, integrating Equation (2) from some initial time

t_0 to time t, one obtains an implicit integral equation that can be solved by iteration for each $j = 1, 2, \ldots, 2n$

$$\Gamma_j(t) = \Gamma_j(t_0) + \int_{t_0}^{t} ds \left(\imath \hat{\mathcal{L}}(s) \right) \Gamma_j(t_0) + \int_{t_0}^{t} ds_1 \int_{t_0}^{s_1} ds_2 \left(\imath \hat{\mathcal{L}}(s_1) \right) \left(\imath \hat{\mathcal{L}}(s_2) \right) \Gamma_j(t_0) + \cdots . \quad (3)$$

The results can be expressed in closed "operatorial" form

$$\dot{\Gamma}_j(t) = \imath \hat{\mathcal{L}}(t) \Gamma_j(t) \quad \longrightarrow \quad \Gamma_j(t) = \hat{S}(t, t_0) \Gamma_j(t_0), \quad j = 1, 2, \ldots, 2n \quad (4)$$

with the introduction of the evolution operator

$$\hat{S}(t, t_0) = \hat{T} \exp \left[\int_{t_0}^{t} \imath \hat{\mathcal{L}}(s) \, ds \right] \quad (5)$$

where $\hat{T}$ is the time-ordering operator.

Time evolution in phase space can be alternatively expressed in term of the Jacobian $J\left(\vec{\Gamma}(t), \vec{\Gamma}(t_0) \right)$ of the time transformation from $\vec{\Gamma}(t_0)$ to $\vec{\Gamma}(t)$. The phase space element $d^{2n}\Gamma(t_0)$ at time t_0 transforms into the volume element $d^{2n}\Gamma(t) = J\left(\vec{\Gamma}(t), \vec{\Gamma}(t_0) \right) d^{2n}\Gamma(t_0)$ at time t, where $J = \det \mathbf{J}$ obeys the differential equation [23]

$$\frac{dJ\left(\vec{\Gamma}(t), \vec{\Gamma}(t_0) \right)}{dt} = \hat{\kappa}(\vec{\Gamma}(t); t) \, J\left(\vec{\Gamma}(t), \vec{\Gamma}(t_0) \right) \quad (6)$$

and the phase space compressibility $\hat{\kappa}$ is defined by

$$\hat{\kappa}(\vec{\Gamma}; t) = \sum_{j=1}^{2n} \frac{\partial}{\partial \Gamma_j} \cdot \dot{\Gamma}_j(\vec{\Gamma}; t) \quad (7)$$

For a Hamiltonian system the compressibility $\hat{\kappa}$ vanishes, $J\left(\vec{\Gamma}(t), \vec{\Gamma}(t_0) \right) = 1$ and the dynamics preserves volume in phase space (Liouville Theorem). More generally, when $\hat{\kappa}$ does not vanish, $d^{2n}\Gamma$ is no longer a dynamical invariant and one needs to introduce a metric factor to define the invariant measure of the phase space under the dynamical evolution. Starting from the general expression for the Jacobian determinant, one gets

$$J\left(\vec{\Gamma}(t), \vec{\Gamma}(t_0) \right) = \exp \left[\int_{t_0}^{t} \hat{\kappa}(\vec{\Gamma}(s); s) ds \right] = e^{w(\vec{\Gamma}(t);t) - w(\vec{\Gamma}(t_0);t_0)} = \frac{Z(\vec{\Gamma}(t_0); t_0)}{Z(\vec{\Gamma}(t); t)} \quad (8)$$

where w is the indefinite time integral of $\hat{\kappa}$ and $Z(\vec{\Gamma}(t); t) = \exp \left[-w(\vec{\Gamma}(t); t) \right]$. The dynamically invariant volume element in phase space can be defined as

$$\begin{aligned} d\mu \left(\vec{\Gamma}(t), d^{2n}\Gamma \right) &= Z(\vec{\Gamma}(t); t) d^{2n}\Gamma(t) = Z(\vec{\Gamma}(t); t) J\left(\vec{\Gamma}(t); \vec{\Gamma}(t_0) \right) d^{2n}\Gamma(t_0) \\ &= Z(\vec{\Gamma}(t_0); t_0) d^{2n}\Gamma(t_0) = d\mu \left(\vec{\Gamma}(t_0), d^{2n}\Gamma(t_0) \right) \end{aligned} \quad (9)$$

Consider, now, an ensemble of systems whose dynamical evolution is defined by Equation (1). The statistical mechanics is described by the time-dependent probability distribution function in phase space $f(\vec{\Gamma}; t)$ which must obey the global conservation law for probabilities

$$\int d\mu(\vec{\Gamma}, d^{2n}\Gamma) f(\vec{\Gamma}; t) = 1$$

The corresponding local, differential, conservation law can be derived by transforming the integral back to the phase space element $d^{2n}\Gamma$, by using

$$f(\vec{\Gamma}; t) d\mu\left(\vec{\Gamma}, d^{2n}\Gamma\right) = f(\vec{\Gamma}; t) Z(\vec{\Gamma}; t) d^{2n}\Gamma = \rho(\vec{\Gamma}; t) d^{2n}\Gamma \tag{10}$$

and introducing the phase space density $\rho(\vec{\Gamma}; t) = Z(\vec{\Gamma}; t) f(\vec{\Gamma}; t)$. The continuity equation to be satisfied is

$$\frac{\partial \rho(\vec{\Gamma}; t)}{\partial t} + \sum_{j=1}^{2n} \frac{\partial}{\partial \Gamma_j} \cdot \left(\dot{\Gamma}_j(\vec{\Gamma}) \rho(\vec{\Gamma}; t)\right) = 0 \tag{11}$$

which when expressed in terms of the Liouville operator $\hat{\mathcal{L}}$ and the phase space compressibility $\hat{\kappa}$ becomes the "generalized" Liouville equation

$$\frac{\partial \rho(\vec{\Gamma}; t)}{\partial t} + \left[\imath\hat{\mathcal{L}}(\vec{\Gamma}; t) + \hat{\kappa}(\vec{\Gamma}; t)\right] \rho(\vec{\Gamma}; t) = 0 \tag{12}$$

and reduces to the more "familiar" equation for the probability density $f(\vec{\Gamma}; t)$

$$\frac{\partial f(\vec{\Gamma}; t)}{\partial t} + \imath\hat{\mathcal{L}} f(\vec{\Gamma}; t) = 0 \tag{13}$$

However, we must point out that this last equation may lead to confusion if one does not keep in mind that, while the Liouville operator $\hat{\mathcal{L}}$ defines the dynamical evolution of the time-dependent probability density in phase space f, the not-vanishing compressibility $\hat{\kappa}$, hidden in the phase space invariant volume, defines the time evolution of the phase space volume $d^{2n}\Gamma$.

The solution of Equation (12) can be retrieved along the same lines followed for Equation (2) and the results can be formally written in closed "operatorial" form

$$\rho(\vec{\Gamma}; t) = \hat{S}^{\dagger}(t, t_0) \rho(\vec{\Gamma}; t_0), \quad \hat{S}^{\dagger}(t, t_0) = \hat{T} \exp\left[\int_{t_0}^{t} -\left(\imath\hat{\mathcal{L}}(s) + \hat{\kappa}(s)\right) ds\right] \tag{14}$$

where we have introduced the adjoint $\hat{S}^{\dagger}(t, t_0)$ of the previously defined time evolution operator $\hat{S}(t, t_0)$ acting on the phase space variables $\vec{\Gamma}$ and the phase density $\rho_0 = \rho(\vec{\Gamma}; t_0)$ at the initial time t_0.

The average over the (non-)equilibrium ensemble of a physical observable $O(t) = \langle\hat{O}(\vec{\Gamma})\rangle_t$ or, more generally, of a macroscopic field $O(\vec{x}, t) = \langle\hat{O}(\vec{x}, \vec{\Gamma})\rangle_t = \left\langle\sum_j \hat{O}(\vec{\Gamma})\delta(\vec{x} - \vec{R}_j)\right\rangle_t$ (the sum is over the particles) can be defined as

$$O(t) = \int \hat{O}(\vec{\Gamma}) f(\vec{\Gamma}; t) d\mu(\vec{\Gamma}, d^{2n}\Gamma) = \int \hat{O}(\vec{\Gamma}) \rho(\vec{\Gamma}; t) d^{2n}\Gamma \tag{15}$$

$$O(\vec{x}, t) = \int \hat{O}(\vec{x}, \vec{\Gamma}) f(\vec{\Gamma}; t) d\mu(\vec{\Gamma}, d^{2n}\Gamma) = \int \hat{O}(\vec{x}, \vec{\Gamma}) \rho(\vec{\Gamma}; t) d^{2n}\Gamma \tag{16}$$

We can make the time evolution explicit by means of the adjoint time evolution operator $\rho(\vec{\Gamma}; t) = \hat{S}^\dagger(t, t_0)\rho(\vec{\Gamma}; t_0)$ and then, by taking advantage of the fact that $\hat{S}^\dagger$ is the adjoint of the dynamics, we can transfer the effect of time evolution to the physical observables

$$
\begin{aligned}
O(t) &= \int \hat{O}(\vec{\Gamma})\,\hat{S}^\dagger(t, t_0)\rho(\vec{\Gamma}; t_0)d^{2n}\Gamma = \int \left(\hat{S}(t, t_0)\hat{O}(\vec{\Gamma}) \right) \rho(\vec{\Gamma}; t_0)d^{2n}\Gamma \\
&= \int \hat{O}(\vec{\Gamma}; t)\,\rho(\vec{\Gamma}; t_0)d^{2n}\Gamma \quad\Rightarrow\quad O(t) = \langle\, \hat{O}(\vec{\Gamma}; t)\, \rangle_{\rho_0}
\end{aligned}
\tag{17}
$$

$$
\begin{aligned}
O(\vec{x}, t) &= \int \hat{O}(\vec{x}, \vec{\Gamma})\,\hat{S}^\dagger(t, t_0)\rho(\vec{\Gamma}; t_0)d^{2n}\Gamma = \int \left(\hat{S}(t, t_0)\hat{O}(\vec{x}, \vec{\Gamma}) \right) \rho(\vec{\Gamma}; t_0)d^{2n}\Gamma \\
&= \int \hat{O}(\vec{x}, \vec{\Gamma}; t)\,\rho(\vec{\Gamma}; t_0)d^{2n}\Gamma \quad\Rightarrow\quad O(\vec{x}, t) = \langle\, \hat{O}(\vec{x}, \vec{\Gamma}; t)\, \rangle_{\rho_0}
\end{aligned}
\tag{18}
$$

where $\hat{O}(\vec{\Gamma}; t) = \hat{S}(t, t_0)\,\hat{O}(\vec{\Gamma})$, *i.e.*, the time evolution along the dynamical trajectory of the system starting from the initial condition $\vec{\Gamma}(t_0)$ at time t_0. We have introduced the shorthand notation, $\langle \cdots \rangle_{\rho_0}$, for the averages over the ensemble described by the space density ρ_0 at the initial time t_0.

Despite the apparent complexity of the time evolution operator $\hat{S}(t, t_0)$ in Equation (5), its action is a task that can be simply accomplished by MD, *i.e.*, by the numerical integration of the evolution defined by Equation (1). Note that all this is possible thanks to the fact that the Liouville equation can be integrated by the method of characteristics.

In the following, we will deal with fluid systems where the relevant macroscopic fields are [24] the density field $\varrho(\vec{x}, t)$, the velocity field $\vec{v}(\vec{x}, t)$ and the temperature field $T(\vec{x}, t)$:

$$
\begin{aligned}
\varrho(\vec{x}, t) &= \int \sum_{j=1}^{N} m_j \delta\left(\vec{x} - \vec{R}_j \right) \hat{S}^\dagger(t, t_0)\rho(\vec{\Gamma}; t_0)\,d^{2n}\Gamma \\
&\Rightarrow \left\langle \sum_{j=1}^{N} m_j \delta\left(\vec{x} - \vec{R}_j(t) \right) \right\rangle_{\rho_0}
\end{aligned}
\tag{19}
$$

$$
\begin{aligned}
\vec{v}(\vec{x}, t) &= \frac{1}{\varrho(\vec{x}, t)} \int \sum_{j=1}^{N} \vec{P}_j \delta\left(\vec{x} - \vec{R}_j \right) \hat{S}^\dagger(t, t_0)\rho(\vec{\Gamma}; t_0)\,d^{2n}\Gamma \\
&\Rightarrow \frac{1}{\varrho(\vec{x}, t)} \left\langle \sum_{j=1}^{N} \vec{P}_j(t) \delta\left(\vec{x} - \vec{R}_j(t) \right) \right\rangle_{\rho_0}
\end{aligned}
\tag{20}
$$

$$
\begin{aligned}
\left(\frac{f}{N}\right) k_B T(\vec{x}, t) &= \frac{1}{\varrho(\vec{x}, t)} \int \sum_{j=1}^{N} \left[\vec{P}_j - m_j \vec{v}(\vec{x}, t) \right]^2 \delta\left(\vec{x} - \vec{R}_j \right) \hat{S}^\dagger(t, t_0)\rho(\vec{\Gamma}; t_0)\,d^{2n}\Gamma \\
&\Rightarrow \frac{1}{\varrho(\vec{x}, t)} \left\langle \sum_{j=1}^{N} \left[\vec{P}_j(t) - m_j \vec{v}(\vec{x}, t) \right]^2 \delta\left(\vec{x} - \vec{R}_j(t) \right) \right\rangle_{\rho_0}
\end{aligned}
\tag{21}
$$

where N is the number of particles and the factor f, usually equal to $3N$, counts the number of degrees of freedom in the presence of constraints.

2.2. Ensembles at t_0

Equations (17) and (18) express what we like to call the Onsager–Kubo relations and state that we can obtain the time evolution of a macroscopic observable or of a macroscopic field as the average of the time evolved corresponding microscopic expression over the initial-time-ensemble described by the phase space density $\rho_0 = \rho(\vec{\Gamma}; t_0)$.

If the ensemble at the initial time t_0 can be simulated by a dynamical system in stationary conditions, then such a probability density function can be sampled by MD, generating a set of (possibly independent) phase space points distributed according to ρ_0. From each of these points, one can then start an independent dynamical trajectory along which the observables $\hat{O}(\vec{\Gamma}; t)$ and $\hat{O}(\vec{x}, \vec{\Gamma}; t)$ can be computed. Finally, by averaging over all the trajectories, the values of the involved observables at time t, one can obtain the macroscopic time-dependent behavior of the system as visualized in Figure 1.

In order to use MD to sample the appropriate initial ensemble at time t_0, one needs to define, for any specific problem, the dynamical evolution, Equation (1), and the auxiliary conditions to which the systems is subjected. Sometimes, but not always, this will be possible within the Hamiltonian formulation of the dynamics.

Figure 1. Phase space representation of the ensemble of dynamical side-trajectories providing the non-equilibrium statistical averages: in blue, the Molecular Dynamics (MD) trajectory sampling the ensemble at time t_0; in black, the individual non-equilibrium trajectories sampling the Non-Equilibrium Molecular Dynamics (D-NEMD) ensemble, over which one can average the time behavior of the observable $\hat{O}$, as a function of the time t.

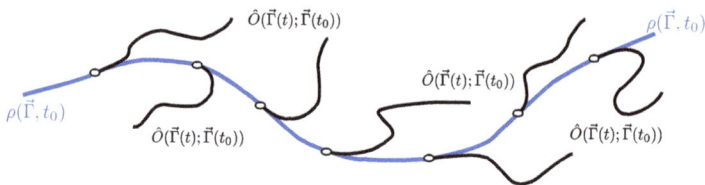

3. D-NEMD Selected Applications

We will now list a number of cases, which will later be illustrated with the corresponding application. Transport properties, like viscosity, thermal conductivity, *etc.*, have been computed and their linearity range investigated by non-equilibrium MD since the 1970s [10–12,25–34]. These results were obtained by measuring on a computer the mechanical response when switching on the external (at the beginning Hamiltonian and later on, more generally, also non-Hamiltonian) perturbation applied to a model system initially at equilibrium. In other words, we identify in the present case the ensemble at time t_0 with the statistical mechanics equilibrium ensemble, while the dynamical trajectories are carried out under the influence of an external (time-dependent) force field.

More generally, we can generate (and sample) initial ensembles by less trivial procedures, e.g., in the case of the formation of convective cells, gravity is considered as the external perturbation to be applied on a system initially in a steady state under the effect of a thermal gradient. The ensemble at time t_0 no longer corresponds to the equilibrium one, but it is set up by introducing a stationary boundary perturbation which, in the specific case, is just an *ad hoc* boundary condition, which models a thermal wall stochastically. Moreover, a confining wall, present in the form of an external field acting at the boundary on each particle, confines the system in the simulation box. This boundary condition is perfectly compatible with the presence of a gravity field.

Another possible case we will consider is the relaxation to equilibrium of an interface between two immiscible liquids, starting from an imposed, non-equilibrium, condition in which the curvature of the interface is maintained by a macroscopic restraint fixing the shape of the initial interface. The ensemble at time t_0 is described by a conditional probability density in which an *ad hoc* restraint is imposed on a field-like observable. The sample is generated by using an advanced MD sampling technique, where the dynamical trajectory evolves under the effect of a suitable restraining potential, from which we can extract an unbiased sample of the conditional probability density function. Time-dependent averages are then taken over dynamical trajectories generated according to the un-restrained dynamics of the systems. The different situations described are summarized in Figure 2.

Figure 2. We distinguish three different classes for the sampling of the initial distribution: equilibrium, direct stationary non-equilibrium simulations and advanced conditional sampling. They are shown to be associated with the corresponding sampling techniques and test-case applications.

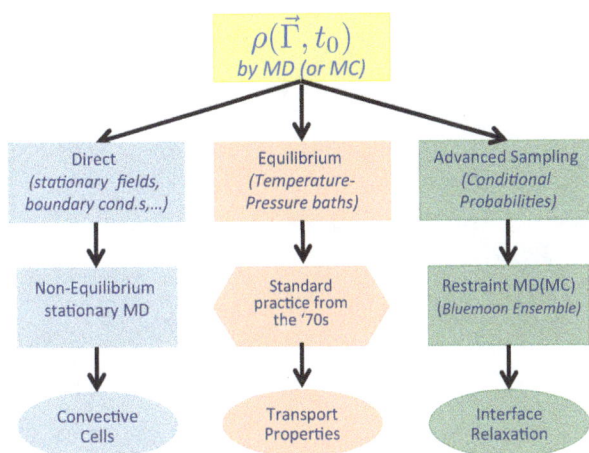

3.1. Transport and Linear Response

Linear Response Theory is a nice result of the nineteen-fifties in the theory of irreversible processes [6], where well-defined microscopic expressions for all transport coefficients have been derived in terms of a properly chosen perturbation [7,8,35,36]. In the Dynamical approach to

Non-Equilibrium Molecular Dynamics (D-NEMD) framework it has been possible to investigate the linear and, more generally, the non-linear response by making reference to the canonical ensemble for sampling the initial conditions at time t_0.

3.1.1. Hamiltonian Perturbations

For a system of particles in three dimensions described by the usual set of Cartesian coordinates and momenta, $\{\vec{R}_j, \vec{P}_j, \; j = 1, 2, \dots \}$, the perturbation can be put in Hamiltonian form by choosing a physical property $\hat{A}(\vec{x}|\vec{\Gamma}) = \sum_j A_j(\vec{\Gamma})\delta(\vec{x} - \vec{R}_j)$ that describes the coupling of the system to the applied external local field $\psi(\vec{x}, t) = \varphi(\vec{x})\chi(t)$, whose time-dependent intensity $\chi(t)$ can be constant or periodic or even arbitrary, generating corresponding flux conditions. Especially important are the cases in which the perturbation is either a step function $\theta(t - t_0)$ ($\theta(t > t_0) = 1$, $\theta(t < t_0) = 0$) or a Dirac delta impulse $\delta(t - t_0)$, at $t = t_0$, after which the system is left free to relax. In the linear regime, the general response can be computed as the superposition of these impulsive responses. One then derives the equations of motion using the standard Hamiltonian route, where we start by separating in the Hamiltonian $\mathcal{H}(\vec{\Gamma}, t) = \mathcal{H}_0(\vec{\Gamma}) + \mathcal{H}_p(\vec{\Gamma}, t)$ the time-dependent perturbation term

$$\mathcal{H}_p(t) = -\int d\vec{x}\,\hat{A}(\vec{x}|\vec{\Gamma})\psi(\vec{x}, t) = -\left(\sum_j A_j \varphi(\vec{R}_j)\right)\chi(t) = -h_p\chi(t) \tag{22}$$

where the Hamiltonian $\mathcal{H}_0$ is the equilibrium Hamiltonian to which one can possibly add the coupling to a thermostat or a barostat, something that can be done in a variety of ways that we do not need to specify here. Indicating generically the possible presence of such couplings to different baths with ellipses, the equations of motion for particle j can be written

$$
\begin{aligned}
\dot{\vec{R}}_j &= +\frac{\partial \mathcal{H}_0}{\partial \vec{P}_j} + \frac{\partial \mathcal{H}_p}{\partial \vec{P}_j} = \left(\frac{\vec{P}_j}{m_j} + \cdots\right) - \frac{\partial h_p}{\partial \vec{P}_j}\chi(t) \\
\dot{\vec{P}}_j &= -\frac{\partial \mathcal{H}_0}{\partial \vec{R}_j} - \frac{\partial \mathcal{H}_p}{\partial \vec{R}_j} = \left(\vec{F}_j + \cdots\right) + \frac{\partial h_p}{\partial \vec{R}_j}\chi(t)
\end{aligned} \tag{23}
$$

The structure of the equations of motion can be broken into the two terms of the Liouville operator defined in Equation (2), $i\hat{\mathcal{L}}(\vec{\Gamma}; t) = i\hat{\mathcal{L}}_0(\vec{\Gamma}) + i\hat{\mathcal{L}}_p(\vec{\Gamma}; t)$, with the partial Liouville operator $i\hat{\mathcal{L}}_0$ defining the dynamical evolution in phase space for the sampling of the ensemble at time t_0. Accordingly, the corresponding evolution operator for the stationary dynamics will be called $\hat{S}_0(t)$. The dynamics of the time-dependent trajectories will be generated by the t_0-(time dependent) evolution operator $\hat{S}(t, t_0)$, obeying the (usual) Dyson equation

$$\hat{S}(t, t_0) = \hat{S}_0(t) + \int_{t_0}^t \hat{S}_0(t - s)i\hat{\mathcal{L}}_p(s)\hat{S}(s, t_0)\,ds \tag{24}$$

which (if of interest) can be taken as the basis to develop the perturbative approach, whose first term leads to the Linear Response Theory approach. However, in many cases of interest, for example for constrained systems with a Hamiltonian or non-Hamiltonian structure, it becomes very difficult, if not impossible, to carry out the standard manipulations leading to the correlation function

191

expressions for the linear response [17,18]. Nevertheless, a linear (or non-linear) response can always be computationally investigated using the procedure defined by Equations (17) and (18), as outlined in Figure 1.

3.1.2. Non-Hamiltonian Perturbations

A more general scheme has also been used for bulk perturbations, where the new equations of motion, which cannot be derived from a time-dependent Hamiltonian in a way that remains consistent with applied (periodic) boundary conditions, are obtained from Equation (23) by substituting the terms derived from the Hamiltonian perturbation h_p, with two sets of "*ad hoc*" phase space functions $\{\vec{C}_j(\vec{\Gamma}), \vec{D}_j(\vec{\Gamma}), j = 1, 2, \ldots\}$:

$$\dot{\vec{R}}_j = \left(\frac{\vec{P}_j}{m_j} + \cdots\right) + \vec{C}_j(\vec{\Gamma}) \cdot \chi(t)$$

$$\dot{\vec{P}}_j = \left(\vec{F}_j + \cdots\right) + \vec{D}_j(\vec{\Gamma}) \cdot \chi(t) \tag{25}$$

A specific, notable, example is the one known under the name of "SLLOD tensor" dynamics [37], where $\vec{C}_j \chi(t) = -(\vec{R}_j \cdot \underline{\kappa}) \chi(t)$ and $\vec{D}_j \chi(t) = (\vec{P}_j \cdot \underline{\kappa}) \chi(t)$ are coupled with specific, synchronized, Lees–Edwards periodic boundary conditions [38] (see Figure 3), which are needed to establish the tensor $\underline{\kappa}$ expressing the desired velocity gradient in the non-equilibrium simulation of viscous flows by molecular dynamics [39–42].

Figure 3. The Lees–Edwards periodic boundary conditions (Panel **A**) used to establish a stationary Couette flow (Panel **B**). In the case of a step function perturbation, periodic images above and below the reference MD cell are translated by an amount $\pm v\delta t$ at each time step, starting from time t_0. Periodic boundary conditions can be effectively imposed using the equivalent non-orthogonal reference cell, highlighted in red (the actual inclination increases uniformly with time).

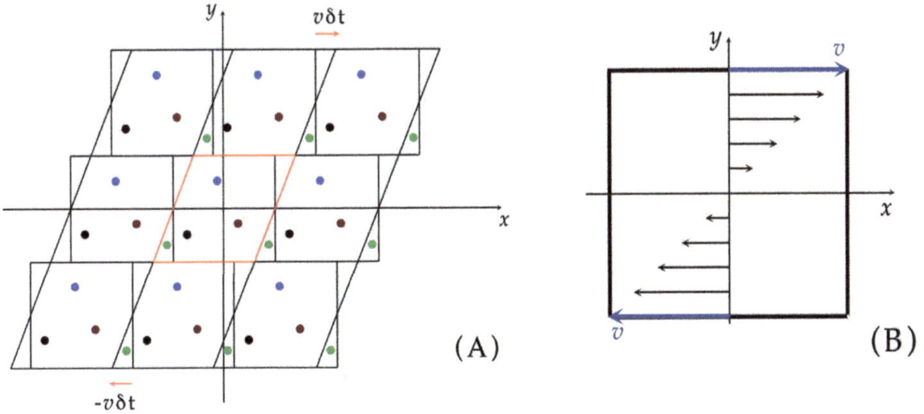

In the typical setup for a planar Couette flow, one establishes a gradient of the x-component of the velocity along the y-axis of the simulation and measures the response using as an observable the xy component of the pressure tensor σ_{xy}, which can be written, for a system where the potential U is given by a sum of pair interactions, as

$$\sigma_{x,y} = \frac{1}{V} \left[\sum_j \frac{\vec{P}_j^{(x)} \vec{P}_j^{(y)}}{m_j} + \sum_{i<j} \left(\vec{R}_{ij}^{(x)} \right) \frac{\partial U}{\partial \vec{R}_{ij}^{(y)}} \right] \qquad (26)$$

where $\vec{R}_{ij} = (\vec{R}_i - \vec{R}_j)$. In the D-NEMD approach, if the external field term is switched on with a step function perturbation in time at $t = t_0 = 0$, one can measure the viscous time-dependent response $\eta(t) = -\langle \sigma_{xy}(t) \rangle_{\rho 0}/\gamma$, where γ is the applied shear rate and the asymptotic value η at long times of $\eta(t)$ gives the viscosity of the fluid.

For the purpose of illustrating the method in the original applications, when the ensemble at the initial time t_0 is an equilibrium ensemble, we will restrict ourselves to the simple case of shear (Couette) flow. We would like to mention, however, that also elongational flows [41,43–46] and, later on, mixed shear-elongational flows [47–49] have been simulated both in atomic and molecular fluids. In these cases, it becomes technically much more difficult to maintain for an indefinite length of time the periodic boundary conditions and, for that, we refer the interested reader to [50,51].

Figure 4. Panel (**A**). Comparison of shear viscosity values as a function of the shear rate for the planar Couette flow: (a) D-NEMD asymptotic values from Reference [52]; (b) and (c) average values from stationary non-equilibrium calculations from Reference [54] and Reference [55] respectively. The solid line is the Lorentzian best-fit to the data and the dashed line is the Ree-Eyring-Eu prediction [56]; Panel (**B**). The running-time integral (solid line) of the D-NEMD viscous dynamical response to a $\delta(t - t_0)$ perturbation with $\gamma = 10^{-4}$, averaged over 4000 trajectories versus the running-time integral (dashed line) of the stress autocorrelation function shows the agreement of D-NEMD results with the Green-Kubo linear reponse theory [52]. The error bars, extrapolated using the mean square fluctuations over the 4000 trajectories, increase with time restricting the time range over which the response can be computed. (nb: the same kind of time dependent behavior for $\eta(t)$ is observed directly when using a step function perturbation).

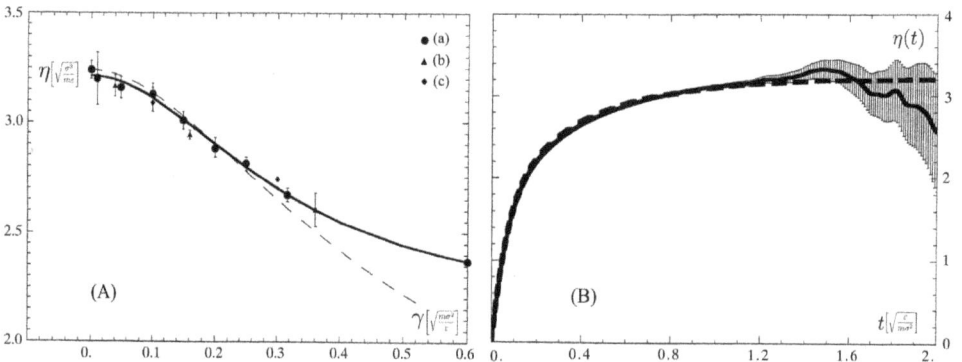

In Panel (A) of Figure 4, we show the results of a calculation [52] with a step function perturbation on a Lennard–Jones (LJ) fluid at the triple point, $\varrho = 0.8442$ and $k_B T_p = 0.725$ in reduced LJ units, *i.e.*, ε for energy, σ for distances and the particle mass m for masses. The temperature of a system of 2,048 particles was controlled using a Nosé–Hoover thermostat [22], both on the long equilibrium trajectory, which samples the (independent) initial conditions from a canonical ensemble at temperature T_p and on the non-equilibrium trajectories to handle the heat produced, especially at high shear rates. The behavior of the time-dependent viscous response for the case of a $t_0 = 0$ impulsive perturbation with a $\delta(t - t_0)$ term was used to investigate the range of validity of the Linear Response Theory for very small shear rates by comparison with the running time integral of the corresponding stress autocorrelation at equilibrium [53].

3.2. Non-Equilibrium (Steady State) Initial Conditions at Time t_0

The D-NEMD approach can be used also to follow the transient evolution of a system, which, starting from an out-of-equilibrium state under the effect of a stationary thermodynamic field, reaches a final (different) non-equilibrium state in response to an additional external perturbation. Below, we illustrate the approach with a case worked out in [57]. This is the case of the build up of a convective roll in a two-dimensional (2D) model fluid kept in an out-of-equilibrium condition by the presence of a thermal gradient when an external gravity field is (instantaneously) switched on.

The 2D system is composed of $N = 5,401$ identical particles in a square box of size L in the xz plane with periodic boundary conditions along the x direction and a pair of confining walls along z obtained by means of an external field $\psi(z)$, acting at the top and the bottom of the simulation box to avoid the drifting away of the particles, which interact with each other via a purely repulsive (2D) Weeks–Chandler–Andersen (WCA) [58] pair potential obtained by truncating the Lennard–Jones potential at its minimum $r_m = 2^{1/6}\sigma$ and shifting its value by ε in such a way that both the force and potential are continuous and equal to zero for $r \geqslant r_m$.

$$V_{WCA} = 4\varepsilon \left[\frac{1}{4} + \left(\frac{\sigma}{r}\right)^{12} - \left(\frac{\sigma}{r}\right)^6 \right], \quad r \leqslant r_m; \quad V_{WCA} = 0, \quad r \geqslant r_m \tag{27}$$

The size of the MD box is $L = 84.9$ in reduced LJ units, which leads to a density $\varrho_f = 0.75$, on average. The confining potential V_{wall} is constructed as the result of a (2D) LJ fluid with continuous constant density ϱ_w filling the two half planes above and below the periodically replicated MD boxes and has a 10-4 power dependence with parameters defined in [57]:

$$V_{wall}(z) = U(z) - U(z_m), \quad z \leqslant z_m; \quad U(z) = \left(\rho_w \sigma^2\right) 4\varepsilon \left[V_{10} \left(\frac{\sigma}{z}\right)^{10} - V_4 \left(\frac{\sigma}{z}\right)^4 \right]$$

$$V_{wall}(z) = 0, \quad z \geqslant z_m;$$

where z is the distance from the box edge, z_m is the value at which $U(z)$ has its minimum and $V_n = 2\pi n! / (2^{n+1} (n/2)! n)$, obtaining, in analogy with WCA, a purely repulsive wall. The thermal gradient along the x-direction is obtained by means of two stochastic reservoirs, which are implemented in the two stripe regions at the x-extremities of the MD box (see Panel (A) of Figure 5).

The velocities of each particle located in these two stripes are sampled from a 2D Maxwellian distribution $f(\vec{v}) = e^{-m\vec{v}^2/(2k_BT_i)}/(2\pi mk_BT_i)$ at the temperature T_i of the stripe, with $i = 1,2$ labeling the two reservoirs. Periodic boundary conditions along the x-direction mean that a particle can actually travel from the first (cold) to the second (hot) reservoir. To avoid that a non-thermalized particle in the inner region interacts with both reservoirs, a stripe thickness $\Delta x_T = 1.68 > r_m$ was chosen. For the system in stationary conditions, each reservoir contains an average of around 100 particles. The reservoir temperatures were chosen to be $T_1 = 1.5$ and $T_2 = 9.9$, corresponding to a thermal gradient $\nabla T = 0.1$. Using these conditions, in the absence of gravity, a long stationary trajectory was generated and, from it, a set of 1,000 initial conditions at time t_0 was sampled. Time-dependent trajectories have been generated and then suitable properties averaged at times $0 \leqslant s \leqslant t$, switching on a gravity field with acceleration $g = 0.1$ in LJ reduced units (while huge compared with Earth gravity, this is a very small value when compared to the accelerations coming from the interatomic interactions). The behavior of the system was analyzed by coarse graining the MD box into a 15×15 mesh of square cells of sides $\ell = 5.66$ that was used to compute local macroscopic fields. Coarse graining is applied by approximating $\delta(\vec{x}-\vec{R}_i) = \delta(x-X_i)\delta(z-Z_i)$ with the value $1/\ell^2$ for particles inside the cell labeled by (j,k) and centered on the mesh point (x_j, z_k) and zero otherwise. The velocity field is calculated as an average over the D-NEMD trajectories

$$\vec{v}(x_j, z_k; t) = \frac{\langle \vec{p}(x_j, z_k; t) \rangle_{\rho_0}}{\varrho_m(x_j, z_k; t)} \tag{28}$$

where ϱ_m, the mass density, is given in terms of the average of the number of particles $n_{jk}(t)$ inside the cell (j,k) at time t

$$\varrho_m(x_j, z_k; t) = m\langle n_{jk}(t) \rangle_{\rho_0} \tag{29}$$

Figure 5. The simulation setup (Panel **A**) for sampling the initial distribution showing the regions where the confining field $\psi(z)$ and the two temperature reservoirs act on the particles. In Panel (**B**) the average evolution of the circulation of the velocity field is shown after averaging over 200 independent initial conditions (in the inset, we show the path along which the circulation was calculated).

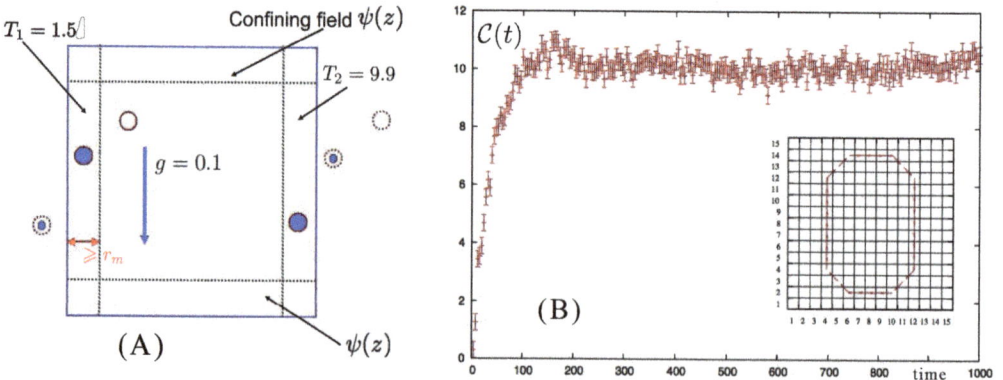

The collective behavior of the velocity field can be monitored by calculating its circulation $\mathcal{C}(t) = \oint_P \vec{v}(x(s), z(s); t)ds$ on a closed path P, as a function of time. Its evolution is shown in Panel (B) of Figure 5 as a function of the time t after the ignition of the gravity field along a path located in the bulk region, but far enough from the center of the box. The circulation starts from zero at $t_0 = 0$; its value grows with time and, after a small overshooting, reaches its plateau, stationary, value at $t \approx 200$. This is the time at which also both temperature and density fields become stationary. The transient is characterized by correlated oscillations of temperature and density with a very similar period $\tau = 18$ in all cells, but with opposite phases for the cells at the bottom of the box with respect to the ones at the top. The velocity field, shown in Figure 6, is initially null, in Panel (A), acquires first at $t \approx 4.5$, in Panel (B), an almost uniform downward component in the direction of the force as a consequence of the ignition of gravity, then it is almost null again at $t \approx 9.0$; it shows a maximum reaction to compression at $t \approx 13.5$, in Panel (C), and almost vanishes again at $t \approx 18$. The cycle restarts and at $t \approx 22.5$, in Panel (D), the field is again predominantly in the downward direction, although one can start to see the building up of a convective flow, which is shown in its stationary condition at $t \approx 205$, in Panel (E).

We have seen how D-NEMD can be used to illustrate the build up of a convective roll when a gravity field is instantaneously switched on in a system where a stationary (non-equilibrium) thermal gradient was already present. This is not the only case in which a convective roll can be observed. Indeed, keeping the same geometry for the system, *i.e.*, with the gravity field orthogonal to the thermal gradient (Panel A of Figure 5), one could alternatively start from initial conditions in which the system is at equilibrium in the presence of the gravity field and, then, follow the dynamics when the thermal gradient is instantaneously switched on or even start from a homogeneous fluid at equilibrium and instantaneously switch on both the gravity field and the thermal gradient [57]. In all these cases, although following different paths, the system eventually reaches the same (macroscopic) final steady state with the formation of a clockwise rotating convective roll, centered at the center of the box.

Figure 6. The build up of the convective flow is shown by visualizing the local velocity field averaged over 1,000 independent initial configurations as a function of time: **(A)** $t = 0$; **(B)** $t \approx 4.5$; **(C)** $t \approx 13.5$; **(D)** $t \approx 22.5$; **(E)** $t \approx 205$.

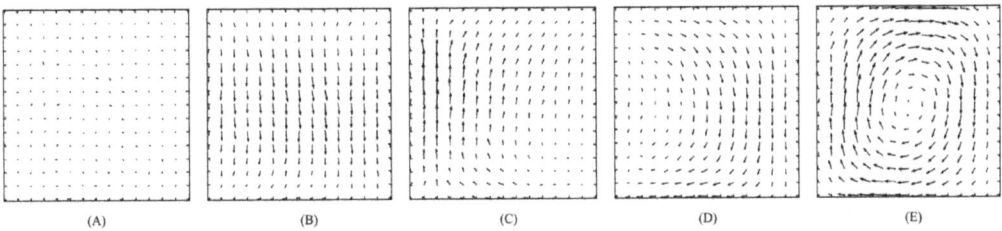

(A) (B) (C) (D) (E)

Complications arise if the system is setup with a different geometry, e.g., in the case of Rayleigh–Bénard convection when the direction of the thermal gradient is parallel to the direction of the gravity field. The system has a higher symmetry and rolls rotating both in the counterclockwise and clockwise directions are possible. This implies that the D-NEMD averages cannot be carried out directly, as in the case we have described so far. In fact, now, in the ensemble of individual trajectories, one samples with equal probability the initial conditions leading to clockwise or to counterclockwise rolls. Performing ensemble averages without paying attention to the direction of rotation would give a wrong result. One needs either to enforce a mechanism that breaks this symmetry or to weight trajectories differently, according to the direction of rotation of the convective roll. The latter was the choice applied in [57], where it was simply impossible to fix *a priori*, by tweaking the initial conditions, the final rotation direction of each trajectory. Instead, the direction of the convective roll rotation was analyzed in the steady-state part of each trajectory by time averaging the velocity field over the last 10,000 steps. The ensemble averages were consistently computed, afterwards, by taking the specular image of the fields when the roll rotation was opposite to the one (arbitrarily) chosen as the reference.

3.3. Sampling from Conditional Distributions at Time t_0

Probably the most interesting application of the D-NEMD procedure is when the non-equilibrium dynamical trajectories start from states corresponding to a very unlikely fluctuation and we want to follow dynamically the way in which the system relaxes back to equilibrium. An efficient sampling of the points in phase space representing the initial condition cannot be achieved just by waiting long enough for the desired event to occur during a standard MD trajectory, but more advanced methods are required to enhance the sampling. Whenever the conditions can be described using an "order parameter", *i.e.*, an appropriate phase space function or field, one can define a macroscopic constraint that applied to the system will allow one to explore the interesting, but unlikely, region of phase space. A viable method in many cases is the well-known Blue Moon approach [15,59], where the conditional probability density is constructed by augmenting the dynamical system with a set of holonomic constraints that force the system to explore states on the specific hypersurface of interest in phase space. The points sampled on a dynamical trajectory subject to constraints, however, cannot be directly used to start the time-dependent non-equilibrium trajectories, because of the unphysical additional conditions enforced on velocities to keep the dynamics on the constrained hypersurface. In the Blue Moon approach, such caveats are overcome, the correct procedure is outlined in [15], requiring, first, an apt resampling of the velocities and, then, an appropriate reweighting when computing the time-dependent averages. The Blue Moon approach was initially devised and successfully applied to the calculation of rate constants for activated processes. In particular, the rate constants are defined in terms of the product of two terms: a transition state theory term and a "transmission coefficient" (*i.e.*, the plateau value, in the intermediate time scale, reached following the transient behavior of the time-dependent values of the reaction coordinate(s) [14,60]). The calculation of the transmission coefficient is a task that can be accomplished exactly along the same lines of the proposed D-NEMD approach.

As a more advanced illustration of D-NEMD when sampling from a conditional probability density, we describe the case of the hydrodynamic relaxation to equilibrium of the interface between two immiscible liquids [61]. The relaxation can be described by following the time evolution of the difference of the density fields of the two species A and B, $\Delta\varrho(\vec{x};t) = \varrho^A(\vec{x};t) - \varrho^B(\vec{x};t)$, and the associated velocity field $\vec{v}(\vec{x};t)$. The distribution at time t_0 corresponds to the stationary conditions of the system subject to a macroscopic restraint that forces a non-equilibrium geometry for the interface. This requires the implementation of a method, like the Blue Moon one, that allows one to sample the conditional probability density associated with the constraint. However, using the Blue Moon approach for vector or field-like constraints can become considerably cumbersome and rather inconvenient in practice, especially for molecular systems where constraints are already used in the force field to impose molecular geometries. A much more practical alternative is to use restrained MD, where one substitutes the constraint with an equivalent restraining potential in terms of an additional coupling parameter, asymptotically reproducing unbiased constrained conditions. Let us summarize the restraint MD approach for the case in which the constraint is imposed on a field-like observable, as for the density difference $\Delta\varrho(\vec{x},t)$. Constraining the shape of the interface S corresponds to specifying the set of spatial points $\{\vec{x}_S\}$ where $\Delta\varrho(\vec{x}_S, t) = 0$.

According to Irving and Kirkwood [24], we associate with this macroscopic field a microscopic observable

$$\Delta\hat{\varrho}(\vec{x}, \vec{\Gamma}) = \sum_{j=1}^{N_A} m_j^A \delta(\vec{x} - \vec{R}_j^A) - \sum_{j=1}^{N_B} m_j^B \delta(\vec{x} - \vec{R}_j^B) \tag{30}$$

on which we need to impose the condition $\Delta\varrho(\vec{x}, t_0) = \Delta\tilde{\varrho}(\vec{x}) = 0$ on all points $\{\vec{x}\}$ in the domain corresponding to the desired geometrical surface at the time t_0. However, in the numerical approach, one cannot deal directly with a continuous (vector) variable $\vec{x}$, therefore, the volume available to the system needs to be discretized over a mesh. With the choice of subdividing the volume in elementary cubic cells, one introduces the space discretization $\{\vec{x}_\alpha, \alpha = 1, 2, \ldots\}$, where the reference point $\vec{x}_\alpha$ coincides with the center of the $\alpha - th$ cell and the microscopic observable field at this point $\vec{x}_\alpha$ is defined as the average $\hat{F}$ of $\Delta\hat{\varrho}$ over the volume Ω_α of the $\alpha - th$ cell:

$$\hat{F}(\vec{x}_\alpha, \vec{\Gamma}) = \frac{1}{\Omega_\alpha} \int_{\Omega_\alpha} d^3x \left[\sum_{j=1}^{N_A} m_j^A \delta(\vec{x} - \vec{R}_j^A) - \sum_{j=1}^{N_B} m_j^B \delta(\vec{x} - \vec{R}_j^B) \right], \quad \alpha = 1, 2, \ldots \tag{31}$$

on which we now need to impose the condition $\hat{F}(\vec{x}_\alpha, \vec{\Gamma}) = \tilde{F}(\vec{x}_\alpha) = 0$ at each of the m points $\{\vec{x}_\alpha, \alpha = 1, 2, \ldots, m\}$, which correspond to the subset of cells that make up the discretized representation of the chosen interface between the two immiscible liquids.

Consider, now, a system described by the Hamiltonian

$$\mathcal{H}_k(\vec{\Gamma}) = \mathcal{H}(\vec{\Gamma}) + \frac{k}{2} \sum_{\alpha=1}^{m} \left[\hat{F}(\vec{x}_\alpha, \vec{\Gamma}) - \tilde{F}(\vec{x}_\alpha) \right]^2 \tag{32}$$

where $\mathcal{H}(\vec{\Gamma})$ is the Hamiltonian of the unconstrained system and k is a tunable parameter that defines the strength of the (harmonic) restraining potential, *i.e.*, the last term on the right-hand side of Equation (32). This Hamiltonian can be used to drive either an MC simulation or an MD simulation at a fixed temperature T generating trajectories, stochastic or dynamic, which sample the phase space of the system according to the canonical probability density $\rho_0^{(k)}(\vec{\Gamma})$ at time t_0

$$\rho_0^{(k)}(\vec{\Gamma}) = \frac{e^{-\beta\mathcal{H}_k(\vec{\Gamma})}}{\mathcal{Z}_0^{(k)}} = \frac{e^{-\beta\mathcal{H}(\vec{\Gamma})}\prod_{\alpha=1}^m e^{-\frac{\beta k}{2}[\hat{F}(\vec{x}_\alpha,\vec{\Gamma})-\tilde{F}(\vec{x}_\alpha)]^2}}{\mathcal{Z}}\frac{1}{P^{(k)}(\tilde{F})} \equiv \frac{P^{(k)}(\vec{\Gamma},\tilde{F})}{P^{(k)}(\tilde{F})} = P^{(k)}(\vec{\Gamma}|\tilde{F}) \quad (33)$$

where $\mathcal{Z}_0^{(k)} = \int d^{2n}\Gamma\, e^{-\beta\mathcal{H}_k(\vec{\Gamma})}$ and $\mathcal{Z} = \int d^{2n}\Gamma\, e^{-\beta\mathcal{H}(\vec{\Gamma})}$ are the canonical partition functions and

$$P^{(k)}(\tilde{F}(\vec{x}_1), \tilde{F}(\vec{x}_2), \ldots \tilde{F}(\vec{x}_m)) = \frac{1}{\mathcal{Z}}\int d^{2n}\Gamma\left\{e^{-\beta\mathcal{H}(\vec{\Gamma})}\prod_{\alpha=1}^m e^{-\frac{\beta k}{2}[\hat{F}(\vec{x}_\alpha,\vec{\Gamma})-\tilde{F}(\vec{x}_\alpha)]^2}\right\} \quad (34)$$

We see, then, more explicitly, that, thanks to the restraint potential, at a given value k of the tunable coupling parameter, we are sampling the conditional probability density of $\vec{\Gamma}$ given $\tilde{F}$, whose limit for $k \to \infty$ is just the ensemble associated with that given fluctuation.

The idea of using a biasing potential to sample unlikely points in configuration space was pioneered by Torrie and Valleau for MC simulation [62], then presented in this form in [63]. However, while in their case of *umbrella/window sampling*, the bias is tuned in such a way to sample, in a statistically significant manner, a wider portion of the configuration space, in the restraint MD approach, one considers high enough values of the tunable parameter k with the aim of sampling the conditional probability associated with the portion of phase space representing a rare region of interest. Indeed, using that $\lim_{a\to\infty} \exp\left\{-\frac{a}{2}(y-\tilde{y})^2\right\} \longrightarrow \frac{\sqrt{2\pi}}{a}\delta(y-\tilde{y})$, one recovers, in the limit $\beta k \to \infty$, the joint probability density

$$\lim_{\beta k\to\infty} \rho_0^{(k)}(\vec{\Gamma}) = \rho_0(\vec{\Gamma}|\{\tilde{F}\}) = \frac{e^{-\beta\mathcal{H}(\vec{\Gamma})}}{\mathcal{Z}}\prod_{\alpha=1}^m \delta\left(\hat{F}(\vec{x}_\alpha,\vec{\Gamma})-\tilde{F}(\vec{x}_\alpha)\right)/\mathcal{P}(\{\tilde{F}\}) \quad (35)$$

where, in the normalizing factor, the probability density of the "condition" $\{\tilde{F}\}$ is given by

$$\mathcal{P}(\{\tilde{F}\}) = \mathcal{P}\left(F(\vec{x}_1) = \tilde{F}(\vec{x}_1), F(\vec{x}_2) = \tilde{F}(\vec{x}_2), \ldots, F(\vec{x}_m) = \tilde{F}(\vec{x}_m)\right) \quad (36)$$

$$= \frac{1}{\mathcal{Z}}\int d^{2n}\Gamma\left\{e^{-\beta\mathcal{H}(\vec{\Gamma})}\prod_{\alpha=1}^m \delta\left(\hat{F}(\vec{x}_\alpha,\vec{\Gamma})-\tilde{F}(\vec{x}_\alpha)\right)\right\} \quad (37)$$

The choice of a restraining potential, which depends only on the coordinates of the particles, as in this case, does not influence the probability density in the momentum space, which remains the Maxwellian (equilibrium) distribution and, at variance with the Blue Moon approach, independent points along the stationary restrained MD trajectory can be directly taken as initial configurations representative of the probability density at time t_0. Moreover, if needed, the restrained MD approach can be further generalized to enforce a more general macroscopic constraint affecting also the momenta of the particles, for example, coupling it with the *ad hoc* boundary conditions and the localized velocity sampling described in Section 3.2 to impose a non-uniform macroscopic velocity field or a temperature gradient in the system.

The definition of the microscopic field in Equation (31) has still one important drawback, which can prompt major issues in particular conditions. In fact, because of the presence of δ-functions in the definition, as a particle crosses the border between one cell and the neighboring one, the integrals that define the macroscopic field at the two corresponding points in space change by, plus or minus, respectively, a finite value introducing discontinuities in the restraining potential in Equation (32). This results in the (highly undesirable) appearance of impulsive terms in the forces on the atoms. Consider the $\delta(\vec{x} - \vec{R}_j)$ function contribution to the integral in Equation (31) for a specific particle j. This is given by the products of three terms, corresponding to the orthogonal directions in space, where each of them is the difference of the values of the cumulative distribution at the edges of the $\alpha - th$ cell. The cumulative distribution for the delta function $\delta(\xi)$ is the step function $\theta(\xi)$, $\{\theta(\xi < 0) = 0, \theta(\xi > 0) = 1\}$, so that each of the above three terms can only be either zero or one. In order to smooth the restraining potential, we need to replace the step function with a smoother function, like a sigmoid, resulting in a continuous variation, between zero and one, of each term in the product. This is equivalent to giving a finite extension to the particle size, resulting in the possibility that a particle contributes, fractionally, to the density field of more than one cell at the same time. In this way, the restraining potential changes smoothly with the motion of the particles in time, without discontinuities when particles cross the cell borders. One possible choice for such function is given by the error function, which corresponds to replacing the delta function by the equivalent Gaussian, $e^{-\frac{\xi^2}{2a}}/\sqrt{2\pi a} \longrightarrow \delta(\xi)$, in the limit $a \longrightarrow 0$, where the parameter a gives the order of magnitude for the (1D) size of the particle. Within such an approximation, Equation (31) becomes

$$\hat{F}(\vec{x}_\alpha, \vec{\Gamma}) = \frac{1}{\Omega_\alpha} \left[\sum_{j=1}^{N_A} m_j^A \Theta(a, \vec{x}_\alpha, \vec{R}_j^A) - \sum_{j=1}^{N_B} m_j^B \Theta(a, \vec{x}_\alpha, \vec{R}_j^B) \right], \quad \alpha = 1, 2, \ldots \quad (38)$$

where the function $\Theta(a, \vec{x}_\alpha, \vec{R}_j)$ is the product of three terms corresponding to the integrals along the three spatial components ($x^1 \equiv x, x^2 \equiv y$ and $x^3 \equiv z$) each involving the evaluation of two values of the error function relative to the border of the cubic cell of length ℓ:

$$\Theta(a, \vec{x}_\alpha, \vec{R}_j) = \prod_{\nu=1}^{3} \left[\text{erf} \left(\frac{x_\alpha^\nu + \ell/2 - R_j^\nu}{\sqrt{a}} \right) - \text{erf} \left(\frac{x_\alpha^\nu - \ell/2 - R_j^\nu}{\sqrt{a}} \right) \right] \quad (39)$$

The two fluids, A and B, are modeled using identical Lennard–Jones particles with mass m and (unique) parameters $\sigma = \sigma_{AA} = \sigma_{BB} = \sigma_{AB}$ and $\varepsilon = \varepsilon_{AA} = \varepsilon_{BB} = \varepsilon_{AB}$ for the LJ potential. The immiscibility is obtained by removing the attractive term for the pair interactions between a particle of type A and a particle of type B, keeping only the purely repulsive part, *i.e.*, taking $u(r = |\vec{R}^A - \vec{R}^B|) = 4\varepsilon [\sigma/r]^{12}$. The simulation was performed at a fixed temperature $k_B T = 1.5\varepsilon$ on a system totaling 171,500 particles, of which 88,889 for Fluid A and 82,611 for Fluid B, in a rectangular parallelepiped box with the same width and height $w = h \approx 44$ and double length $d \approx 88$, corresponding to an average particle density $n = 1.024$, where all figures are in reduced LJ units. The density and temperature are in the fluid region of the phase space of a pure LJ fluid. In order to follow the behavior of the density and velocity fields, the space was discretized using 5,488 cubic cells arranged on a $14 \times 14 \times 28$ grid. In this way, local field values are obtained averaging out on roughly 30 particles in each cell. For the ensemble at time t_0, the initial configuration for the interface between Fluid A and Fluid B is defined by selecting the m cells (centered on the mesh points, $\vec{x}_\alpha$, $\alpha = 1, 2, \ldots, m$), which are cut across by the ideal cylindrical surface, $\tilde{S}$,

$$\tilde{S} = \left\{ \vec{x} : z = \mathcal{A} \sin \left(\frac{\pi x}{w} \right) + \frac{h - \mathcal{A}}{2}, \quad 0 \leqslant x \leqslant w, \quad 0 \leqslant y \leqslant h \right\} \tag{40}$$

where $\mathcal{A} = 50$ is the amplitude that determines the curvature of the surface, which is approximatively placed halfway along the z-direction in the simulation box. The restraint potential is completely defined by the choice of the coupling parameter $k = 0.004$ in LJ units and the imposed values $\tilde{F}(\vec{x}_\alpha) = 0$, $\alpha = 1, 2, \ldots, m$. The initial configuration was first prepared with the equilibration of a pure Type A fluid at the target temperature and density and then identifying, within the simulation box, as particles of Type A all the particles that are on the red side of the surface $\tilde{S}$, as particles of Type B all the particles that are on the blue side (see the left panel in Figure 7), taking care of having exactly half of the particles of Type A and half of Type B in the m cells that make up the discretized interface at time t_0. Periodic boundary conditions are applied in all directions, so that a second flat interface (the condition that minimizes the surface tension) is created at the same time at the sides of the box along the z-direction. Then, the system was equilibrated running restrained MD with a time step $\delta t = 4.56 \cdot 10^{-4}$ in LJ units. Such a rather small value for δt was used to ensure a proper numerical integration of the "stiff" restraining forces. A typical snapshot of the isosurface $\Delta \varrho(\vec{x}) = 0$ is shown in the left panel of Figure 7.

A long, 10^6 MD restrained trajectory was then carried out, with that same time step, taking out, at regular intervals of 25,000 steps, the configuration of the system in phase space. A set of 40 independent initial conditions was collected in this way, and from each of them was started, now with a regular time step $\delta t = 4.56 \cdot 10^{-3}$, a 25,000-steps unrestrained MD trajectory at constant energy, *i.e.*, using the equations of motion derived only from the Hamiltonian $\mathcal{H}(\vec{\Gamma})$, given in Equation (32). The D-NEMD averaging procedure was used with those 40 trajectories to compute the time-dependent behavior of the macroscopic density and velocity fields, discretized on the previously mentioned cubic mesh.

In the right panel of Figure 7, the dynamical behavior of the interface is shown by plotting the isosurfaces $\Delta \varrho(\vec{x}) = 0$ for four successive times.

One can see that the interface curvature diminishes progressively towards the flat, equilibrium condition, while maintaining (approximatively) both the initial uniformity along the direction of the y-axis and the initial mirror symmetry with respect to the middle yz-plane. Small deviations are present, as expected, considering that this macroscopic field results from averaging over a relatively small sample of 40 independent trajectories. They are compatible with the expected amplitudes of the equilibrium fluctuations of the interface. Relaxation of the initially curved surface reaches the flat equilibrium condition fully in a time lapse of approximatively 20,000 steps, *i.e.*, something just short of 100 LJ time units. One can use this information to estimate the order of the relaxation time and, from it, a maximum value $v_{max} \approx 0.5$ in LJ units for the average velocity field at the mid-point along x of the interface, *i.e.*, in the region corresponding to the maximum displacement at time t_0 of the interface. If one takes the LJ parameters of argon (for which $\sigma = 3.405 \cdot 10^{-10}$ m and the unit of time corresponds to $\tau = 2.156 \cdot 10^{-12}$ s), this value translates to an experimentally convincing velocity of ≈ 80 m $\cdot$ s^{-1}.

Figure 7. (**Left** panel) A sampled initial condition for the S isosurface $\Delta\varrho(\vec{x}) = 0$ separating the two liquids, A and B (the second planar interface at the long edge of the simulation box is not shown). (**Right** panel) The evolution in time of the initially curved interface (purple) towards the relaxed planar condition (green). The snapshots are the D-NEMD results averaged over a sample of 40 initial conditions.

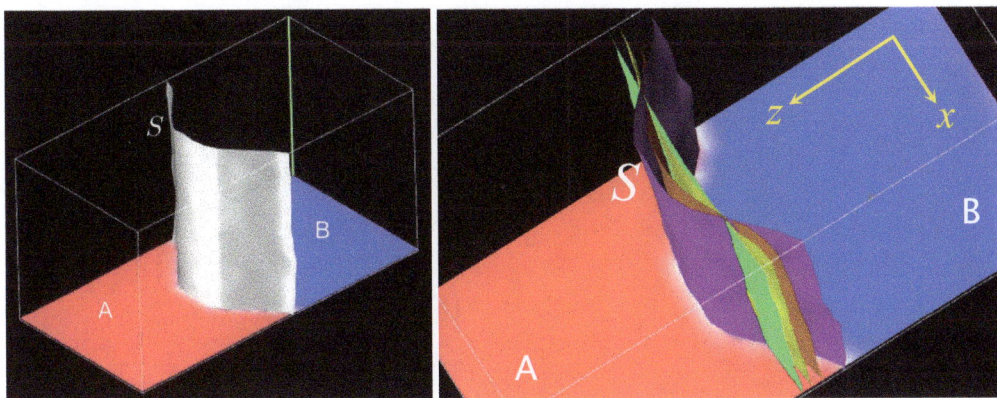

The second interface on the sides of the MD box remains flat, with even smaller deviations, all along the 25,000 time steps. There seem to be no significant effects on it as a result of the relaxation process, which takes place in the middle of the box. The reason for this will become evident after looking at the time-dependent behavior of the velocity field. We have shown, in fact, how the D-NEMD approach provides very detailed information on the hydrodynamic behavior of the system and unravels the underlying physical mechanisms.

Starting from Equation (20), the discretized velocity field is calculated accordingly to:

$$\vec{v}(\vec{x}, t) = \frac{\left\langle \sum_{j=1}^{N_A} \vec{P}_j^A(t)\Theta(a, \vec{x}_\alpha, \vec{R}_j^A(t)) + \sum_{j=1}^{N_B} \vec{P}_j^B(t)\Theta(a, \vec{x}_\alpha, \vec{R}_j^B(t)) \right\rangle_{\rho_0}}{\left\langle \sum_{j=1}^{N_A} m\Theta(a, \vec{x}_\alpha, \vec{R}_j^A(t)) + \sum_{j=1}^{N_B} m\Theta(a, \vec{x}_\alpha, \vec{R}_j^B(t)) \right\rangle_{\rho_0}} \tag{41}$$

where we made explicit use of the fact that the particles have identical masses m, regardless of whether they are of Type A or B, and the D-NEMD average is taken over the 40 trajectories with initial conditions sampled from the ρ_0 probability density along the restrained MD trajectory. The results of the calculation are shown in the left panels of Figure 8. The direction of the projections of the velocity field in each cell on the xz-plane (after averaging along the translational symmetry axis y) is represented by a small arrow whose length is proportional to its modulus.

Figure 8. The behavior of the velocity field in the two liquid regions as a function of time: comparison of the results obtained using the D-NEMD approach averaging over 40 initial conditions (Panels **A–C**) and the local time averaging procedure (Panels **D–F**). The results emphasize how the latter approach returns a much flatter picture of the velocity field, exposing features of the relaxation mechanism that are in marked contrast with the underlying symmetries of the process. (reproduced from [61] with permission from the Physical Chemistry Chemical Physics Owner Societies.)

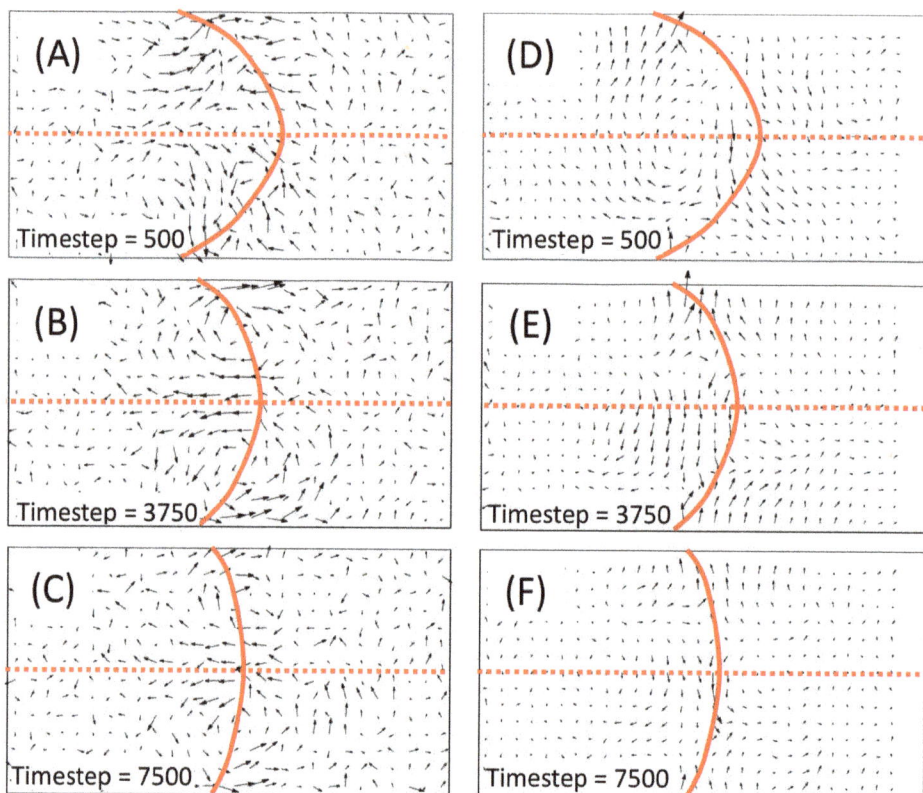

Snapshots at three successive times are shown. In the snapshot taken 500 time steps after t_0, one can notice the build up of some coherence in the velocity field, which becomes structured in a region that is, along the z-direction, about twice the size of the width of the curved interface, the projection of which is represented by the contour line at the value $\Delta\varrho = 0$. For the sake of clarity, the symmetry yz plane is as well highlighted by a dashed straight line. One can distinguish a quasi-symmetric two-tail profile, where the push towards the edges of the interface appears to be more pronounced than the one, in the opposite direction, in the region near the center of the interface (Panel A). This behavior marks the initial build up of a more stable two-roll velocity profile, of roughly the same width across the interface region in the A and B fluids, which becomes very evident after 3,750 steps

(Panel B) and is still neatly visible and qualitatively unchanged, even after 7,500 steps from t_0, when the curvature of the interface is significantly reduced (Panel C).

One can notice that the size and the intensity of the field has decreased, but the profile remains highly symmetrical along the mirror symmetry plane. All along, one can notice also that the field in the extreme sides of the simulation box remains essentially unperturbed, which explains why the second flat interface at the boundary is not affected in any significant way and remains stable during the whole relaxation process of the curved interface. It is very interesting to compare these insights on the hydrodynamic processes underlying the interface relaxation, as given by the time-dependent behavior of the dynamical response calculated using the D-NEMD procedure, with the standard approach used on single trajectory simulations of hydrodynamic processes. Starting from the local equilibrium hypothesis of hydrodynamic theory, based on the assumption of a time scale separation between the fast microscopic motion of the particles and the slower hydrodynamic processes, the macroscopic fields are computed as local time averages on the short time scale τ of the atomistic processes [64].

By applying this approach to the time evolution from a single initial condition, one obtains the results shown in the right panels of Figure 8. At first glance, the velocity field appears to be much smoother than the D-NEMD result, which is relatively noisy due to the limited size (40) of the sample of initial conditions used. However, the picture that is returned is quite different, with the physical mechanism exposed by the D-NEMD procedure effectively washed out by the local time averaging, which also presents a velocity field that does seem to violate the mirror plane symmetry initially imposed on the system at time t_0, contrary to the more convincing evidence, given from the D-NEMD results, of a relaxation mechanism satisfying, on average, such symmetry at all times along the dynamical trajectory.

4. Conclusions and Perspectives

In this paper, we have presented a dynamical approach to non-equilibrium MD, which makes it possible to compute, numerically, but, otherwise, rigorously, time-dependent non-equilibrium responses, *i.e.*, to observe directly transient responses in non-stationary regimes. We have shown that using a proper simulation setup, it is possible to go beyond the usual situation of initial equilibrium conditions to treat interesting cases in which the initial condition is either a stationary non-equilibrium or a constrained equilibrium condition dictated by means of a macroscopic constraint, which can be expressed, in a general way, either as a scalar or field-like observable, outlining also the connections of the D-NEMD approach to the Blue Moon method [15] to compute the transmission coefficient contribution to the rate constants of activated processes.

We illustrated a few applications of the method starting from the early, historical, approach to the calculation of transport properties in the lines of Linear Response Theory and beyond, to a couple of recent atomistic simulations of hydrodynamic processes: the establishing of a convective cell, when gravity is switched on in the presence of a stationary thermal gradient, and the relaxation of an initially curved interface between two immiscible liquids. We have shown that the method generates rigorous time-dependent non-equilibrium averages, providing valuable insights on the mechanisms of hydrodynamic processes that can be missed using a method like the local time average, which cannot have a rigorous justification, presents a statistical error that cannot be reduced at will and, finally, as we have seen above, can bias the statistical response. A word of caution is needed, though. The time-dependent ensemble averages are meaningful only if the thermodynamical response is unique [65]. Whenever this is not the case, the meaning of the statistical averages becomes questionable. To our knowledge, in these cases a systematic answer does not exist for the non-equilibrium thermodynamic response and problems have to be treated on a one-by-one basis.

In summary, with the outlined exceptions, D-NEMD is a method ready for challenging applications, by which it is possible to study complex time-dependent phenomena using only the fundamental laws of Statistical Mechanics, *i.e.*, without using empirical approaches as, for example, in the case of continuum hydrodynamic theories. Work is in progress in this direction.

Acknowledgments

We acknowledge the Science Foundation Ireland SFI Grant No. 08-IN.1-I1869 and the Istituto Italiano di Tecnologia under the SEED Project grant No. 259 SIMBEDD-Advanced Computational Methods for Biophysics, Drug Design and Energy Research for financial support.

Conflicts of Interest

The authors declare no conflict of interest.

References

1. Alder, B.J.; Wainwright, T.E. Phase transition for a hard sphere system. *J. Chem. Phys.* **1957**, *27*, 1208–1209.
2. Rahman, A. Correlations in the motion of atoms in liquid argon. *Phys. Rev.* **1964**, *136*, A405–A411.
3. Metropolis, N.; Rosenbluth, A.W.; Rosenbluth, M.N.; Teller, A.H.; Teller, E. Equation of state calculations by fast computing machines. *J. Chem. Phys.* **1953**, *21*, 1087–1092.
4. Wood, W.W.; Parker, F.R. Monte Carlo equation of state of molecules interacting with the Lennard-Jones potential. I. A supercritical isotherm at about twice the critical temperature. *J. Chem. Phys.* **1957**, *27*, 720–733.
5. Onsager, L. Reciprocal relations in irreversible processes. I. *Phys. Rev.* **1931**, *37*, 405–426.
6. Kubo, R. Statistical-mechanical theory of irreversible processes. I. General theory and simple applications to magnetic and conduction problems. *J. Phys. Soc. Jpn.* **1957**, *12*, 570–586.
7. Kadanoff, L.P.; Martin, P.C. Hydrodynamic equations and correlation functions. *Ann. Phys.* **1963**, *24*, 419–469.
8. Zwanzig, R. Time-correlation functions and transport coefficients in statistical mechanics. *Annu. Rev. Phys. Chem.* **1965**, *16*, 67–102.
9. Kubo, R. The Fluctuation-dissipation theorem. *Rep. Prog. Phys.* **1966**, *29*, 255–284.

10. Ciccotti, G.; Jacucci, G. Direct computation of dynamical response by molecular dynamics: The mobility of a charged Lennard-Jones particle. *Phys. Rev. Lett.* **1975**, *35*, 789–792.

11. Ciccotti, G.; Jacucci, G.; McDonald, I.R. Transport properties of molten alkali halides. *Phys. Rev. A* **1976**, *13*, 426–436.

12. Ciccotti, G.; Jacucci, G.; McDonald, I.R. "Thought-Experiments" by molecular dynamics. *J. Stat. Phys.* **1979**, *21*, 1–22.

13. It was found that the linear regime normally remains valid in what, on a macroscopic scale, is an enormous range, going up to a perturbation almost of the order of the intermolecular characteristics. Therefore, for example, for charged systems, the perturbing electrical field could go up to the order of 10^6 V·cm^{-1} or thermal gradients up to the order of 10^8 K·cm^{-1} and, finally, for viscous phenomena, the shear rate applicable to simple fluids up to the order of 10^{12} s^{-1}, as compared with an intercollisional frequency of the order of 10^{13} s^{-1}.

14. Chandler, D. Statistical mechanics of isomerization dynamics in liquids and the transition state approximation. *J. Chem. Phys.* **1978**, *68*, 2959–2970.

15. Carter, E.A.; Ciccotti, G.; Hynes, J.T.; Kapral, R. Constrained reaction coordinate dynamics for the simulation of rare events. *Chem. Phys. Lett.* **1989**, *156*, 472–477.

16. Ciccotti, G.; Ferrario, M.; Hynes, J.T.; Kapral, R. Dynamics of ion pair interconversion in a polar solvent. *J. Chem. Phys.* **1990**, *93*, 7137–7147.

17. Ciccotti, G.; Kapral, R.; Sergi, A. Non-equilibrium molecular dynamics. In *Handbook of Materials Modeling. Volume I: Methods and Model*; Yip, S., Ed.; Springer: Dordrecht, The Netherlands, 2005; pp. 1–17.

18. Hartmann, C.; Schütte, C.; Ciccotti, G. On the linear response of mechanical systems with constraints. *J. Chem. Phys.* **2010**, *132*, 111103.

19. Goldstein, H.; Poole, C.; Safko, J. *Classical Mechanics*, 2nd ed.; Addison-Wesley: Reading, MA, USA, 1980.

20. Ryckaert, J.P.; Ciccotti, G.; Berendsen, H.J.C. Numerical integration of the Cartesian equation of motion of a system with constraints: Molecular dynamics of N-alkanes. *J. Comput. Phys.* **1977**, *23*, 327–341.

21. Andersen, H.C. Molecular dynamics simulations at constant pressure and/or temperature. *J. Chem. Phys.* **1980**, *72*, 2384–2393.

22. Nosé, S. Constant temperature molecular-dynamics methods. *Prog. Theor. Phys. Suppl.* **1991**, *103*, 1–46.

23. Tuckerman, M.E.; Liu, Y.; Ciccotti, G.; Martyna, G.J. Non-Hamiltonian molecular dynamics: Generalizing Hamiltonian phase space principles to non-Hamiltonian systems. *J. Chem. Phys.* **2001**, *115*, 1678–1702.

24. Irving, J.H.; Kirkwood, J.G. The statistical mechanical theory of transport processes. IV. The equations of hydrodynamics. *J. Chem. Phys.* **1950**, *18*, 817–829.

25. Ciccotti, G.; Jacucci, G.; McDonald, I.R. Thermal response to a weak external field. *J. Phys. C* **1978**, *11*, L509–L513.

26. Singer, K.; Singer, J.V.; Fincham, D. Determination of the shear viscosity of atomic liquids by nonequilibrium molecular dynamics molecular-dynamics study. *Mol. Phys.* **1980**, *40*, 515–519.

27. Allen, M.P.; Kivelson, D. Non equilibrium molecular dynamics simulation and generalized hydrodynamics of transverse modes in molecular fluids. *Mol. Phys.* **1981**, *44*, 945–965.

28. Heyes, D.M. Self-diffusion and shear viscosity of simple fluids. A molecular-dynamics study. *J. Chem. Soc., Faraday Trans. 2* **1983**, *79*, 1741–1758.

29. Massobrio, C.; Ciccotti, G. Lennard-Jones triple-point conductivity via weak external fields. *Phys. Rev. A* **1984**, *30*, 3191–3197.

30. Hoover, W.G.; Ciccotti, G.; Paolini, G.V.; Massobrio, C. Lennard-Jones triple-point conductivity via weak external fields: Additional calculations. *Phys. Rev. A* **1985**, *32*, 3765–3767.

31. Paolini, G.V.; Ciccotti, G.; Massobrio, C. Non-Linear thermal response of a LJ fluid near its triple point. *Phys. Rev. A* **1986**, *34*, 1355–1362.

32. Paolini, G.V.; Ciccotti, G. Cross thermotransport in Liquid mixtures by nonequilibrium molecular dynamics. *Phys. Rev. A* **1987**, *35*, 5156–5166.

33. Pierleoni, C.; Ciccotti, G.; Bernu, B. Thermal conductivity of the classical one-component plasma by non-equilibrium molecular-dynamics. *Europhys. Lett.* **1987**, *4*, 1115–1120.

34. Morriss, G.P.; Evans, D.J. Application of transient correlation functions to shear flow far from equilibrium. *Phys. Rev. A* **1987**, *35*, 792–797.

35. Jackson, J.L.; Mazur, P. On the statistical mechanics derivation of the correlation formula for the viscosity. *Physica* **1964**, *30*, 2295–2304.

36. Luttinger, J.M. Theory of thermal transport coefficients. *Phys. Rev.* **1964**, *135*, A1505–A1514.

37. Ladd, A.J. Equations of motion for non-equilibrium molecular dynamics simulations of viscous flow in molecular fluids. *Mol. Phys.* **1984**, *53*, 459–463.

38. Lees, A.W.; Edwards, F. The computer study of transport process under extreme conditions. *J. Phys. C* **1972**, *5*, 1921–1929.

39. Evans, D.J.; Morris, G.P. Shear thickening and turbulence in simple fluids. *Phys. Rev. Lett.* **1986**, *56*, 2172–2175.

40. Ciccotti, G.; Pierleoni, C.; Ryckaert, J.P. Theoretical foundation and rheological application of non-equilibrium molecular dynamics. In *Microscopic Simulation of Complex Hydrodynamics Phenomena*; Mareschal, M., Holian, B., Eds.; Plenum Press: New York, NY, USA, 1992; pp. 25–45.

41. Pierleoni, C.; Ryckaert, J.P. Non-Newtonian viscosity of atomic fluids in shear and shear-free flows. *Phys. Rev. A* **1991**, *44*, 5314–5317.

42. Palla, P.L.; Pierleoni, C.; Ciccotti, G. Bulk viscosity of the Lennard-Jones system at the triple point by dynamical nonequilibrium molecular dynamics. *Phys. Rev. E* **2008**, *78*, 021204.

43. Hounkonnou, M.N.; Pierleoni, C.; Ryckaert, J.P. Liquid chlorine in shear and elongational flows: A nonequilibrium molecular dynamics study. *J. Chem. Phys.* **1992**, *97*, 9335–9344.

44. Todd, B.D. Application of transient-time correlation functions to nonequilibrium molecular-dynamics simulations of elongational flow. *Phys. Rev. E* **1997**, *56*, 6723–6728.

45. Todd, B.D. Nonlinear response theory for time-periodic elongational flows. *Phys. Rev. E* **1998**, *58*, 4587–4593.

46. Ryckaert, J.P.; Pierleoni, C. Polymer solutions in flow: A non-equilibrium molecular dynamics approach. In *Flexible Polymer Chains in Elongational Flow*; Nguyen, T., Kausch, H.H., Eds.; Springer: Berlin/Heidelberg, Germany, 1999; pp. 5–40.

47. Todd, B.D.; Daivis, P.J. Homogeneous non-equilibrium molecular dynamics simulations of viscous flow: Techniques and applications. *Mol. Simul.* **2007**, *33*, 189–229.

48. Hunt, T.A.; Todd, B.D. Diffusion of linear polymer melts in shear and extensional flows. *J. Chem. Phys.* **2009**, *131*, 054904.

49. Hartkamp, R.; Bernardi, S.; Todd, B.D. Transient-time correlation function applied to mixed shear and elongational flows. *J. Chem. Phys.* **2012**, *136*, 064105.

50. Kraynik, A.; Reinelt, D. Extensional motions of spatially periodic lattices. *Int. J. Multiph. Flow* **1992**, *18*, 1045–1059.

51. Hunt, T.A.; Todd, B.D. On the Arnold cat map and periodic boundary conditions for planar elongational flow. *Mol. Phys.* **2003**, *101*, 3445–3454.

52. Ferrario, M.; Ciccotti, G.; Holian, B.L.; Ryckaert, J.P. Shear rate dependence of the viscosity of the Lennard-Jones liquid at the triple point. *Phys. Rev. A* **1991**, *44*, 6936–6939.

53. Ryckaert, J.P.; Bellemans, A.; Ciccotti, G.; Paolini, G.V. Evaluation of transport coefficients of simple fluids by molecular dynamics: Comparison of Green-Kubo and nonequilibrium approaches for shear viscosity. *Phys. Rev. A* **1988**, *39*, 259–267.

54. Evans, D.J.; Morriss, G.P.; Hood, L.M. On the number dependence of viscosity in three dimensional fluids. *Mol. Phys.* **1989**, *68*, 637–646.

55. Heyes, D.M. Shear thinning and thickening of the Lennard-Jones liquid. A molecular dynamics study. *J. Chem. Soc. Faraday Trans. 2* **1986**, *82*, 1365–1383.

56. Eu, B.C. Shear-rate dependence of viscosity for simple fluids. *Phys. Lett. A* **1983**, *96*, 29–32.

57. Mugnai, M.L.; Caprara, S.; Ciccotti, G.; Pierleoni, C.; Mareschal, M. Transient hydrodynamical behavior by dynamical nonequilibrium molecular dynamics: The formation of convective cells. *J. Chem. Phys.* **2009**, *131*, 064106.

58. Weeks, J.D.; Chandler, D.; Andersen, H.C. Role of repulsive forces in determining the equilibrium structure of simple liquids. *J. Chem. Phys.* **1971**, *54*, 5237–5247.

59. Ciccotti, G.; Ferrario, M.; Hynes, J.T.; Kapral, R. Constrained molecular dynamics and the mean potential for an ion pair in a polar solvent. *Chem. Phys.* **1989**, *129*, 241–251.

60. Vanden-Eijnden, E.; Tal, F.A. Transition state theory: Variational formulation, dynamical corrections, and error estimates. *J. Chem. Phys.* **2005**, *123*, 184103.

61. Orlandini, S.; Meloni, S.; Ciccotti, G. Hydrodynamics from statistical mechanics: Combined dynamical-NEMD and conditional sampling to relax an interface between two immiscible liquids. *Phys. Chem. Chem. Phys.* **2011**, *13*, 13177–13181.

62. Torrie, G.; Valleau, J. Nonphysical sampling distributions in Monte Carlo free-energy estimation: Umbrella sampling. *J. Comput. Phys.* **1977**, *23*, 187–199.

63. Maragliano, L.; Fischer, A.; Vanden-Eijnden, E.; Ciccotti, G. String method in collective variables: Minimum free energy paths and isocommittor surfaces. *J. Chem. Phys.* **2006**, *125*, 024106.

64. Puhl, A.; Malek Mansour, M.; Mareschal, M. Quantitative comparison of molecular dynamics with hydrodynamics in Rayleigh-Benard convection. *Phys. Rev. A* **1989**, *40*, 1999–2012.

65. Callen, H.B. *Thermodynamics: An Introduction to the Physical Theories of Equilibrium Thermostatics and Irreversible Thermodynamics*; Wiley: New York, NY, USA, 1960.

Reprinted from *Entropy*. Cite as: Warren, P.B.; Allen, R.J. Malliavin Weight Sampling: A Practical Guide. *Entropy* **2014**, *16*, 221–232.

Article

Malliavin Weight Sampling: A Practical Guide

Patrick B. Warren [1,*] **and Rosalind J. Allen** [2,*]

[1] Unilever R&D Port Sunlight, Quarry Road East, Bebington, Wirral, CH63 3JW, UK
[2] Scottish Universities Physics Alliance (SUPA), School of Physics and Astronomy,
   the University of Edinburgh, the Kings Buildings, Mayfield Road, Edinburgh, EH9 3JZ, UK

* Authors to whom correspondence should be addressed;
  E-Mails: patrick.warren@unilever.com (P.B.W.); rallen2@ph.ed.ac.uk (R.J.A.);
  Tel.: +44-151-641-3352 (P.B.W.); +44-131-651-7197 (R.J.A.).

Received: 25 September 2013; in revised form: 9 October 2013 / Accepted: 18 October 2013 / Published: 27 December 2013

Abstract: Malliavin weight sampling (MWS) is a stochastic calculus technique for computing the derivatives of averaged system properties with respect to parameters in stochastic simulations, without perturbing the system's dynamics. It applies to systems in or out of equilibrium, in steady state or time-dependent situations, and has applications in the calculation of response coefficients, parameter sensitivities and Jacobian matrices for gradient-based parameter optimisation algorithms. The implementation of MWS has been described in the specific contexts of kinetic Monte Carlo and Brownian dynamics simulation algorithms. Here, we present a general theoretical framework for deriving the appropriate MWS update rule for any stochastic simulation algorithm. We also provide pedagogical information on its practical implementation.

Keywords: stochastic calculus; Brownian dynamics

1. Introduction

Malliavin weight sampling (MWS) is a method for computing derivatives of averaged system properties with respect to parameters in stochastic simulations [1,2]. The method has been used in quantitative financial modelling to obtain the "Greeks" (price sensitivities) [3], and as the Girsanov transform, in kinetic Monte Carlo simulations for systems biology [4]. Similar ideas have been used to study fluctuation-dissipation relations in supercooled liquids [5]. However, MWS appears

210

to be relatively unknown in the fields of soft matter, chemical and biological physics, perhaps because the theory is relatively impenetrable for non-specialists, being couched in the language of abstract mathematics (e.g., martingales, Girsanov transform, Malliavin calculus, *etc.*); an exception in financial modelling is [6].

MWS works by introducing an auxiliary stochastic quantity, the Malliavin weight, for each parameter of interest. The Malliavin weights are updated alongside the system's usual (unperturbed) dynamics, according to a set of rules. The derivative of any system function, A, with respect to a parameter of interest is then given by the average of the product of A with the relevant Malliavin weight; or in other words, by a weighted average of A, in which the weight function is given by the Malliavin weight. Importantly, MWS works for non-equilibrium situations, such as time-dependent processes or driven steady states. It thus complements existing methods based on equilibrium statistical mechanics, which are widely used in soft matter and chemical physics.

MWS has so far been discussed only in the context of specific simulation algorithms. In this paper, we present a pedagogical and generic approach to the construction of Malliavin weights, which can be applied to any stochastic simulation scheme. We further describe its practical implementation in some detail using as our example one dimensional Brownian motion in a force field.

2. The Construction of Malliavin Weights

The rules for the propagation of Malliavin weights have been derived for the kinetic Monte-Carlo algorithm [4,7], for the Metropolis Monte-Carlo scheme [5] and for both underdamped and overdamped Brownian dynamics [8]. Here we present a generic theoretical framework, which encompasses these algorithms and also allows extension to other stochastic simulation schemes.

We suppose that our system evolves in some state space, and a point in this state space is denoted as S. Here, we assume that the state space is continuous, but our approach can easily be translated to discrete or mixed discrete-continuous state spaces. Since the system is stochastic, its state at time t is described by a probability distribution, $P(S)$. In each simulation step, the state of the system changes according to a propagator, $W(S \to S')$, which gives the probability that the system moves from point S to point S' during an application of the update algorithm. The propagator has the property that

$$P'(S') = \int_S dS \, W(S \to S') \, P(S) \qquad (1)$$

where $P'(S)$ is the probability distribution after the update step has been applied and the integral is over the whole state space. We shall write this in a shorthand notation as

$$P' = \int WP. \qquad (2)$$

Integrating Equation (1) over S', we see that the propagator must obey $\int_{S'} W(S \to S') = 1$. It is important to note, however, that we do *not* assume the detailed balance condition $P_{eq}(S) \, W(S \to S') = P_{eq}(S') \, W(S' \to S)$ (for some equilibrium $P_{eq}(S)$). Thus, our results apply to systems whose dynamical rules do not obey detailed balance (such as chemical models of gene regulatory networks [9]), as well as to systems out of steady state. We observe that the (finite) product

$$\mathbb{W}(S_1, \ldots, S_n) = W(S_1 \to S_2) \times \cdots \times W(S_{n-1} \to S_n) \tag{3}$$

is proportional to the probability of occurrence of a trajectory of states, $\{S_1, \ldots, S_n\}$, and can be interpreted as a *trajectory weight*.

Let us now consider the average of some quantity $A(S)$ over the state space, in shorthand

$$\langle A \rangle = \int A P. \tag{4}$$

The quantity A might well be a complicated function of the state of the system: for example the extent of crystalline order in a particle-based simulation, or a combination of the concentrations of various chemical species in a simulation of a biochemical network. We suppose that we are interested in the sensitivity of $\langle A \rangle$ to variations in some parameter of the simulation, which we denote as λ. This might be one of the force field parameters (or the temperature) in a particle-based simulation or a rate constant in a kinetic Monte Carlo simulation. We are interested in computing $\partial \langle A \rangle / \partial \lambda$. This quantity can be written as

$$\frac{\partial \langle A \rangle}{\partial \lambda} = \int A P Q_\lambda, \tag{5}$$

where

$$Q_\lambda = \frac{\partial \ln P}{\partial \lambda} \tag{6}$$

(using the fact that $\partial \ln P / \partial \lambda = (1/P)\partial P / \partial \lambda$).

Let us now suppose that we track in our simulation not only the physical state of the system, but also an auxiliary stochastic variable, which we term q_λ. At each simulation step q_λ is updated according to a rule that depends on the system state; this does not perturb the system's dynamics, but merely acts as a "readout". By tracking q_λ, we *extend* the state space, so that S becomes $\{S, q_\lambda\}$. We can then define the average $\langle q_\lambda \rangle_S$, which is an average of the value of q_λ in the extended state space, with the constraint that the original (physical) state space point is fixed at S (see further below).

Our aim is to define a set of rules for updating q_λ, such that $\langle q_\lambda \rangle_S = Q_\lambda$, *i.e.*, such that the average of the auxiliary variable, for a particular state space point, measures the *derivative* of the probability distribution with respect to the parameter of interest, λ. If this is the case then, from Equation (5)

$$\frac{\partial \langle A \rangle}{\partial \lambda} = \langle A q_\lambda \rangle. \tag{7}$$

The auxiliary variable q_λ is the Malliavin weight corresponding to the parameter λ.

How do we go about finding the correct updating rule? If the Malliavin weight exists, we should be able to derive its updating rule from the system's underlying stochastic equations of motion. We obtain an important clue from differentiating Equation (1) with respect to λ. Extending the shorthand notation, one finds

$$P' Q'_\lambda = \int W P \left(Q_\lambda + \frac{\partial \ln W}{\partial \lambda} \right). \tag{8}$$

This strongly suggests that the rule for updating the Malliavin weight should be

$$q'_\lambda = q_\lambda + \frac{\partial \ln W}{\partial \lambda} . \tag{9}$$

In fact, this is correct. The proof is not difficult and, for the case of Brownian dynamics, can be found in the supplementary material for [8]. It involves averaging Equation (9) in the extended state space $\{S, q_\lambda\}$.

From a practical point of view, for each time step, we implement the following procedure:

- propagate the system from its current state S to a new state S' using the algorithm that implements the stochastic equations of motion (Brownian, kinetic Monte-Carlo, *etc.*);
- with knowledge of S and S', and the propagator $W(S \to S')$, calculate the change in the Malliavin weight $\Delta q_\lambda = \partial \ln W(S \to S')/\partial \lambda$;
- update the Malliavin weight according to $q_\lambda \to q'_\lambda = q_\lambda + \Delta q_\lambda$.

At the start of the simulation, the Malliavin weight is usually initialised to $q_\lambda = 0$.

Let us first suppose that our system is not in steady state. However rather the quantity $\langle A \rangle$ in which we are interested is changing in time and likewise $\partial \langle A(t) \rangle / \partial \lambda$ is a time-dependent quantity. To compute $\partial \langle A(t) \rangle / \partial \lambda$ we run N independent simulations, in each one tracking as a function of time $A(t)$, $q_\lambda(t)$ and the product $A(t) q_\lambda(t)$. The quantities $\langle A(t) \rangle$ and $\partial \langle A(t) \rangle / \partial \lambda$ are then given by

$$\langle A(t) \rangle \approx \frac{1}{N} \sum_{i=1}^{N} A_i(t), \quad \frac{\partial \langle A(t) \rangle}{\partial \lambda} \approx \frac{1}{N} \sum_{i=1}^{N} A_i(t) q_{\lambda,i}(t) , \tag{10}$$

where $A_i(t)$ is the value of $A(t)$ recorded in the ith simulation run (and likewise for $q_{\lambda,i}(t)$). Error estimates can be obtained from the variance among the replicate simulations.

If, instead, our system is in steady state, the procedure needs to be modified slightly. This is because the variance in the values of $q_\lambda(t)$ across replicate simulations increases linearly in time (this point is discussed further below). For long times, computation of $\partial \langle A \rangle / \partial \lambda$ using Equation (10) therefore incurs a large statistical error. Fortunately, this problem can easily be solved by computing the correlation function

$$C(t, t') = \langle A(t) [q_\lambda(t) - q_\lambda(t')] \rangle . \tag{11}$$

In steady state, $C(t, t') = C(t - t')$, with the property that $C(\Delta t) \to \partial A / \partial \lambda$ as $\Delta t \to \infty$. In a single simulation run, we simply measure $q_\lambda(t)$ and $A(t)$ at time intervals separated by Δt (which is typically multiple simulation steps). At each measurement, we compute $A(t) [q_\lambda(t) - q_\lambda(t - \Delta t)]$. We then average this latter quantity over the whole simulation run to obtain an estimate of $\partial \langle A \rangle / \partial \lambda$. For this estimate to be accurate, we require that Δt is long enough that $C(\Delta t)$ has reached its plateau value; this typically means that Δt should be longer than the typical relaxation time of the system's dynamics. The correlation function approach is discussed in more detail in [7,8].

Returning to a more theoretical perspective, it is interesting to note that the rule for updating the Malliavin weight, Equation (9), depends deterministically on S and S'. This implies that the value of the Malliavin weight at time t is completely determined by the trajectory of system states during the time interval $0 \rightarrow t$. In fact, it is easy to show that

$$q_\lambda = \frac{\partial \ln \mathbb{W}}{\partial \lambda} \tag{12}$$

where $\mathbb{W}$ is the trajectory weight defined in Equation (3). Similar expressions are given in [5,7]. Thus, the Malliavin weight q_λ is not fixed by the state point S but by the entire trajectory of states that have led to state point S. Since many different trajectories can lead to S, many values of q_λ are possible for the same state point S. The average $\langle q_\lambda(t)\rangle_S$ is actually the expectation value of the Malliavin weight, averaged over all trajectories that reach state point S at time t. This can be used to obtain an alternative proof that $\langle q_\lambda \rangle_S = \partial \ln P / \partial \lambda$. Suppose we sample N trajectories, of which N_S end up at state point S (or a suitably defined vicinity thereof, in a continuous state space). We have $P(S) = \langle N_S \rangle / N$. Then, the Malliavin property implies $\partial P / \partial \lambda = \langle N_S q_\lambda \rangle / N$, and hence $\partial \ln P / \partial \lambda = \langle N_S q_\lambda \rangle / \langle N_S \rangle = \langle q_\lambda \rangle_S$.

3. Multiple Variables, Second Derivatives and the Algebra of Malliavin Weights

Up to now, we have assumed that the quantity A does not depend explicitly on the parameter λ. There may be cases, however, when A does have an explicit λ-dependence. In these cases, Equation (7) should be replaced by

$$\frac{\partial \langle A \rangle}{\partial \lambda} = \left\langle \frac{\partial A}{\partial \lambda} \right\rangle + \langle A q_\lambda \rangle . \tag{13}$$

If we set A to be a constant in this, we immediately obtain the general result that $\langle q_\lambda \rangle = 0$. Equation (13) reveals a kind of 'algebra' for Malliavin weights: we see that the operations of taking an expectation value and taking a derivative can be commuted, provided the Malliavin weight is introduced as the commutator.

We can also extend our analysis further to allow us to compute higher derivatives with respect to the parameters. These may be useful, for example, for increasing the efficiency of gradient-based parameter optimisation algorithms. Taking the derivative of Equation (13) with respect to a second parameter μ gives

$$\frac{\partial^2 \langle A \rangle}{\partial \lambda \partial \mu} = \frac{\partial}{\partial \mu} \left\langle \frac{\partial A}{\partial \lambda} \right\rangle + \frac{\partial \langle A q_\lambda \rangle}{\partial \mu}$$

$$= \left\langle \frac{\partial^2 A}{\partial \lambda \partial \mu} \right\rangle + \left\langle \frac{\partial A}{\partial \lambda} q_\mu \right\rangle + \left\langle A \frac{\partial q_\lambda}{\partial \mu} \right\rangle + \left\langle \frac{\partial A}{\partial \mu} q_\lambda \right\rangle + \langle A q_\lambda q_\mu \rangle$$

$$= \langle A(q_{\lambda\mu} + q_\lambda q_\mu) \rangle + \left\langle \frac{\partial A}{\partial \lambda} q_\mu \right\rangle + \left\langle \frac{\partial A}{\partial \mu} q_\lambda \right\rangle + \left\langle \frac{\partial^2 A}{\partial \lambda \partial \mu} \right\rangle , \tag{14}$$

where in the second line we iterate the commutation relation, and in the third line we collect like terms and introduce

$$q_{\lambda\mu} = \frac{\partial q_\lambda}{\partial \mu} . \tag{15}$$

In the case where A is independent of the parameters, this result simplifies to

$$\frac{\partial^2 \langle A \rangle}{\partial \lambda \partial \mu} = \langle A \left(q_{\lambda \mu} + q_\lambda q_\mu \right) \rangle . \tag{16}$$

The quantity $q_{\lambda \mu}$ here is a new, second order, Malliavin weight which from Equations (12) and (15) satisfies

$$q_{\lambda \mu} = \frac{\partial^2 \ln \mathbb{W}}{\partial \lambda \partial \mu} . \tag{17}$$

To compute second derivatives with respect to the parameters, we should therefore track these second order Malliavin weights in our simulation, updating them alongside the existing Malliavin weights by the rule

$$q'_{\lambda \mu} = q_{\lambda \mu} + \frac{\partial^2 \ln W(S \rightarrow S')}{\partial \lambda \partial \mu} . \tag{18}$$

Setting A as a constant in Equation (16), we also obtain the interesting result that $\langle q_{\lambda \mu} \rangle = - \langle q_\lambda q_\mu \rangle$.

Steady state problems can be approached by extending the correlation function method to second order weights. Define, *cf.* Equation (11),

$$C(t, t') = \langle A(t) \left\{ [q_{\lambda \mu}(t) + q_\lambda(t) q_\mu(t)] - [q_{\lambda \mu}(t') + q_\lambda(t') q_\mu(t')] \right\} \rangle . \tag{19}$$

As in the first order case, in steady state we expect $C(t, t') = C(t - t')$ with the property that $C(\Delta t) \rightarrow \partial^2 \langle A \rangle / \partial \lambda \partial \mu$ as $\Delta t \rightarrow \infty$.

4. One-Dimensional Brownian Motion in a Force Field

We now demonstrate this machinery by way of a practical but very simple example, namely one-dimensional (overdamped) Brownian motion in a force field. In this case, the state space is specified simply by the particle position x which evolves according to the Langevin equation

$$\frac{dx}{dt} = f(x) + \eta \tag{20}$$

where $f(x)$ is the force field and η is Gaussian white noise of amplitude $2T$, where T is temperature. Without loss of generality, we have chosen units so that there is no prefactor multiplying the force field. We discretise the Langevin equation to the following updating rule:

$$x' = x + f(x) \, \delta t + \xi , \tag{21}$$

where δt is the time step and ξ is a Gaussian random variate with zero mean and variance $2T \, \delta t$. Corresponding to this updating rule is an explicit expression for the propagator,

$$W(x \rightarrow x') = \frac{1}{\sqrt{4 \pi T \, \delta t}} \exp \left(-\frac{(x' - x - f(x) \, \delta t)^2}{4 T \, \delta t} \right) . \tag{22}$$

This follows from the statistical distribution of ξ. Let us suppose that the parameter of interest λ enters into the force field (the temperature T could also be chosen as a parameter). Making this assumption

$$\frac{\partial \ln W(x \rightarrow x')}{\partial \lambda} = \frac{(x' - x - f \, \delta t)}{2T} \frac{\partial f}{\partial \lambda} . \tag{23}$$

We can simplify this result by noting that from Equation (21), $x' - x - f\,\delta t = \xi$. Making use of this, the final updating rule for the Malliavin weight is

$$q'_\lambda = q_\lambda + \frac{\xi}{2T}\frac{\partial f}{\partial \lambda} \tag{24}$$

where ξ is the *exact same* value that was used for updating the position in Equation (21). Because the value of ξ is the same for the updates of position and of q_λ, the change in q_λ is completely determined by the end points, x and x'. The derivative $\partial f/\partial \lambda$ should be evaluated at x since that is the position at which the force is computed in Equation (21). Since ξ in Equation (21) is a random variate uncorrelated with x, averaging Equation (24) shows that $\langle q'_\lambda \rangle = \langle q_\lambda \rangle$. As the initial condition is $q_\lambda = 0$, this means that $\langle q_\lambda \rangle = 0$, as predicted in the previous section. Equation (24) is essentially the same as that derived in [8].

If we differentiate Equation (23) with respect to a second parameter μ we get

$$\frac{\partial^2 \ln W(x \to x')}{\partial \lambda \partial \mu} = \frac{(x' - x - f\,\delta t)}{2T}\frac{\partial^2 f}{\partial \lambda \partial \mu} - \frac{\delta t}{2T}\frac{\partial f}{\partial \lambda}\frac{\partial f}{\partial \mu}. \tag{25}$$

Hence, the updating rule for the second order Malliavin weight can be written as

$$q'_{\lambda\mu} = q_{\lambda\mu} + \frac{\xi}{2T}\frac{\partial^2 f}{\partial \lambda \partial \mu} - \frac{\delta t}{2T}\frac{\partial f}{\partial \lambda}\frac{\partial f}{\partial \mu}, \tag{26}$$

where again ξ is the exact same value as that used for updating the position in Equation (21). If we average Equation (26) over replicate simulation runs, we find $\langle q'_{\lambda\mu} \rangle = \langle q_{\lambda\mu} \rangle - (\delta t/2T)(\partial f/\partial \lambda)(\partial f/\partial \mu)$. Hence the mean value $\langle q_{\lambda\mu} \rangle$ drifts in time, unlike $\langle q_\lambda \rangle$ or $\langle q_\mu \rangle$. However, one can show that the mean value of the sum $\langle (q_{\lambda\mu} + q_\lambda q_\mu) \rangle$ is constant in time and equal to zero as long as initially $q_\lambda = q_\mu = 0$.

Now let us consider the simplest case of a particle in a linear force field $f = -\kappa x + h$ (also discussed in [8]). This corresponds to a harmonic trap with the potential $U = \frac{1}{2}\kappa x^2 - hx$. We let the particle start from x_0 at $t = 0$ and track its time-dependent relaxation to the steady state. We shall set $T = 1$ for simplicity. The Langevin equation can be solved exactly for this case, and the mean position evolves according to

$$\langle x(t) \rangle = x_0 e^{-\kappa t} + \frac{h}{\kappa}(1 - e^{-\kappa t}). \tag{27}$$

We suppose that we are interested in derivatives with respect to both h and κ, for a "baseline" parameter set in which κ is finite but $h = 0$. Taking derivatives of Equation (27) and setting $h = 0$, we find

$$\frac{\partial \langle x(t) \rangle}{\partial h} = \frac{1 - e^{-\kappa t}}{\kappa}, \quad \frac{\partial \langle x \rangle(t)}{\partial \kappa} = -x_0 t e^{-\kappa t}, \quad \frac{\partial^2 \langle x(t) \rangle}{\partial h \partial \kappa} = \frac{t e^{-\kappa t}}{\kappa} - \frac{1 - e^{-\kappa t}}{\kappa^2}. \tag{28}$$

We now show how to compute these derivatives using Malliavin weight sampling. Applying the definitions in Equations (24) and (26), the Malliavin weight increments are

$$q'_h = q_h + \frac{\xi}{2}, \quad q'_\kappa = q_\kappa - \frac{x\xi}{2}, \quad q'_{h\kappa} = q_{h\kappa} + \frac{x\,\delta t}{2}, \tag{29}$$

and the position update itself is

$$x' = x - \kappa x \, \delta t + \xi \,. \tag{30}$$

We track these Malliavin weights in our simulation and use them to calculate derivatives according to

$$\frac{\partial \langle x(t) \rangle}{\partial h} = \langle x(t) q_h(t) \rangle \,, \quad \frac{\partial \langle x(t) \rangle}{\partial \kappa} = \langle x(t) q_\kappa(t) \rangle \,, \quad \frac{\partial^2 \langle x(t) \rangle}{\partial h \partial \kappa} = \langle x(t) (q_{h\kappa}(t) + q_h(t) q_\kappa(t)) \rangle \,. \tag{31}$$

Equations (29)–(31) have been coded up as a MATLAB script, described in Section 5. A typical result generated by running this script is shown in Figure 1. Equations (29) and (30) are iterated with $\delta t = 0.01$ up to $t = 5$, for a trap strength $\kappa = 2$ and initial position $x_0 = 1$. The weighted averages in Equation (31) are evaluated as a function of time for $N = 10^5$ samples as in Equation (10). These results are shown as the solid lines in Figure 1. The dashed lines are theoretical predictions for the time dependent derivatives from Equation (28). As can be seen, the agreement between the time-dependent derivatives and the Malliavin weight averages is very good.

Figure 1. Time-dependent derivatives, $\partial \langle x \rangle / \partial h$ (top curve, blue), $\partial \langle x \rangle / \partial \kappa$ (middle curve, green), and $\partial^2 \langle x \rangle / \partial h \partial \kappa$ (bottom curve, red). Solid lines (slightly noisy) are the Malliavin weight averages as indicated in the Figure, generated by running the MATLAB script described in Section 5. Dashed lines are theoretical predictions from Equation (28).

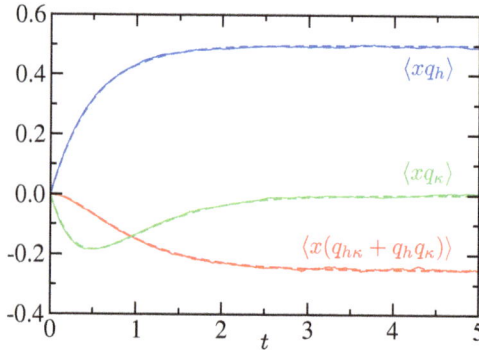

As discussed briefly above, in this procedure the sampling error in the computation of $\partial \langle A(t) \rangle / \partial \lambda$ is expected to grow with time. Figure 2 shows the mean square Malliavin weight as a function of time for the same problem. For the first order weights q_h and q_κ the growth rate is typically linear in time. Indeed, from Equation (29), one can prove that in the limit $\delta t \to 0$ (see Section 5)

$$\frac{d \langle q_h^2 \rangle}{dt} = \frac{1}{2} \,, \quad \frac{d \langle q_\kappa^2 \rangle}{dt} = \frac{\langle x^2 \rangle}{2} \,. \tag{32}$$

Thus q_h behaves exactly as a random walk, as should be obvious from the updating rule. The other weight q_κ also ultimately behaves as a random walk since $\langle x^2 \rangle = 1/\kappa$ in steady state (from equipartition). Figure 2 also shows that the second order weight $q_{h\kappa}$ grows superdiffusively; one can show that eventually $\langle (q_{h\kappa} + q_h q_\kappa)^2 \rangle \sim t^2$, although the transient behaviour is complicated. Full expressions are given in Section 5. This suggests that computation of second order derivatives is likely to suffer more severely from statistical sampling problems than the computation of first order derivatives.

Figure 2. Growth of mean square Malliavin weights with time. The solid lines are from simulations and the dashed lines are from Equation (35) in the Appendix. Parameters are as for Figure 1.

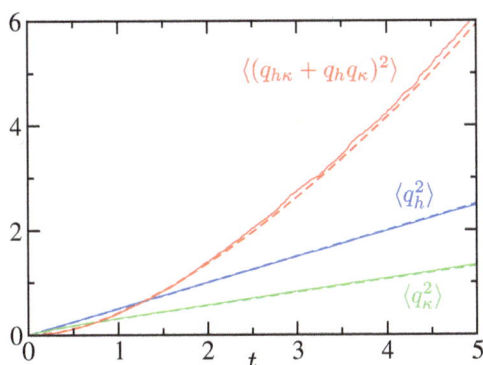

5. Conclusions

In this paper, we have provided an outline of the generic use of Malliavin weights for sampling derivatives in stochastic simulations, with an emphasis on practical aspects. The usefulness of MWS for a particular simulation scheme hinges on the simplicity or otherwise of constructing the propagator $W(S \rightarrow S')$ which fixes the updating rule for the Malliavin weights according to Equation (9). The propagator is determined by the algorithm used to implement the stochastic equations of motion; MWS may be easier to implement for some algorithms than for others. We note however that there is often some freedom of choice about the algorithm, such as the choice of a stochastic thermostat in molecular dynamics, or the order in which update steps are implemented. In these cases, a suitable choice may simplify the construction of the propagator and facilitate the use of Malliavin weights.

Acknowledgments

Rosalind J. Allen is supported by a Royal Society University Research Fellowship.

Conflicts of Interest

The authors declare no conflict of interest.

References

1. Bell, D.R. *The Malliavin Calculus*; Dover: Mineola, NY, USA, 2006.
2. Nualart, D. *The Malliavin Calculus and Related Topics*; Springer: New York, NY, USA, 2006.

3. Fournié, E.; Lasry, J.M.; Lebuchoux, J.; Lions, P.L.; Touzi, N. Applications of Malliavin calculus to Monte Carlo methods in finance. *Financ. Stoch.* **1999**, *3*, 391–412.

4. Plyasunov, A.; Arkin, A.P. Efficient stochastic sensitivity analysis of discrete event systems. *J. Comput. Phys.* **2007**, *221*, 724–738.

5. Berthier, L. Efficient measurement of linear susceptibilities in molecular simulations: Application to aging supercooled liquids. *Phys. Rev. Lett.* **2007**, *98*, 220601.

6. Chen, N.; Glasserman, P. Malliavin Greeks without Malliavin calculus. *Stoch. Proc. Appl.* **2007**, *117*, 1689–1723.

7. Warren, P.B.; Allen, R.J. Steady-state parameter sensitivity in stochastic modeling via trajectory reweighting. *J. Chem. Phys.* **2012**, *136*, 104106.

8. Warren, P.B.; Allen, R.J. Malliavin weight sampling for sensitivity coefficients in Brownian dynamics simulations. *Phys. Rev. Lett.* **2012**, *109*, 250601.

9. Warren, P.B.; ten Wolde, P.R. Chemical models of genetic toggle switches. *J. Phys. Chem. B* **2005**, *109*, 6812–6823.

Appendix

MATLAB Script

The MATLAB script in Listing 1 was used to generate the results shown in Figure 1. It implements Equations (29)–(31) above, making extensive use of the compact MATLAB syntax for array operations, for instance invoking '.*' for element-by-element multiplication of arrays.

Listing 1. MATLAB script used to generate Figure 1.

```
1  clear all
2  randn('seed', 12345);
3  kappa = 2; x0 = 1; tend = 5; dt = 0.01; nsamp = 10^5;
4  npt = round(tend/dt) + 1;
5  t = (0:npt-1)' * dt;
6  x = zeros(npt, 1); xi = zeros(npt, 1);
7  qh = zeros(npt, 1); qk = zeros(npt, 1); qhk = zeros(npt, 1);
8  x_av = zeros(npt, 1); xqh_av = zeros(npt, 1);
9  xqk_av = zeros(npt, 1); xqhk_av = zeros(npt, 1);
10 for samp = 1:nsamp
11   x(1) = x0; qh(1) = 0; qk(1) = 0; qhk(1) = 0;
12   xi = randn(npt, 1) * sqrt(2*dt);
13   for i = 1:npt-1
14     x(i+1) = x(i) - kappa*x(i)*dt + xi(i);
15     qh(i+1) = qh(i) + 0.5*xi(i);
16     qk(i+1) = qk(i) - 0.5*x(i)*xi(i);
17     qhk(i+1) = qhk(i) + 0.5*x(i)*dt;
18   end
19   x_av = x_av + x;
20   xqh_av = xqh_av + x.*qh;
21   xqk_av = xqk_av + x.*qk;
22   xqhk_av = xqhk_av + x.*(qhk + qh.*qk);
23 end
24 x_av = x_av / nsamp; xqh_av = xqh_av / nsamp;
25 xqk_av = xqk_av / nsamp; xqhk_av = xqhk_av / nsamp;
26 hold on
27 plot(t, x_av, 'k'); plot(t, xqh_av, 'b');
28 plot(t, xqk_av, 'g'); plot(t, xqhk_av, 'r');
29 plot(t, x0*exp(-kappa*t), 'k--')
30 plot(t, (1-exp(-kappa*t))/kappa, 'b--')
31 plot(t, -x0*t.*exp(-kappa*t), 'g--')
32 plot(t, t.*exp(-kappa*t)/kappa-(1-exp(-kappa*t))/(kappa^2), 'r--')
33 result = [t x_av xqh_av xqk_av xqhk_av];
34 save('result.dat', '-ascii', 'result');
```

Here is a brief explanation of the script. *Lines 1–3* initialise the problem and the parameter values. *Lines 4* and *5* calculate the number of points in a trajectory and initialise a vector containing the time coordinate of each point. *Lines 6–9* set aside storage for the actual trajectory, Malliavin weights and cumulative statistics. *Lines 10–23* implement a pair of nested loops, which are the kernel of the simulation. Within the outer (trajectory sampling) loop, *Line 11* initialises the particle position and Malliavin weights, *Line 12* precomputes a vector of random displacements (Gaussian random variates), and *Lines 13–18* generate the actual trajectory. Within the inner (trajectory generating loop), *Lines 14–17* are a direct implementation of Equations (29) and (30). After each individual trajectory has been generated, the cumulative sampling step implied by Equation (31) is done in *Lines 19–22*; after all the trajectories have been generated, these quantities are normalised in *Lines 24* and *25*. Finally, *Lines 26–32* generate a plot similar to Figure 1 (albeit with the addition of $\langle x \rangle$), and *Lines 33* and *34* show how the data can be exported in tabular format for replotting using an external package.

Listing 1 is complete and self-contained. It will run in either MATLAB or Octave. One minor comment is perhaps in order. The choice was made to precompute a vector of Gaussian random variates, which are used as random displacements to generate the trajectory and update the Malliavin weights. One could equally well generate random displacements on-the-fly, in the inner loop. For this one-dimensional problem storage is not an issue and it seems more elegant and efficient to exploit the vectorisation capabilities of MATLAB. For a more realistic three-dimensional problem, with many particles (and a different programming language), it is obviously preferable to use an on-the-fly approach.

Selected Analytic Results

Here, we present analytic results for the growth in time of the mean square Malliavin weights. We can express the rate of growth of the mean of a generic function $f(x, q_h, q_\kappa, q_{h\kappa})$ as

$$\frac{d\langle f \rangle}{dt} = \lim_{\delta t \to 0} \frac{\langle f(x', q_h', q_\kappa', q_{h\kappa}') \rangle - f(x, q_h, q_\kappa, q_{h\kappa}) \rangle}{\delta t}, \tag{33}$$

where on the right-hand side (RHS) the values of x', q_h', q_κ' and $q_{h\kappa}'$ are substituted from the updating rules in Equations (29) and (30). In calculating the RHS average, we note that the distribution of ξ is a Gaussian independent of the position and Malliavin weights and thus one can substitute $\langle \xi \rangle = 0$, $\langle \xi^2 \rangle = 2\,\delta t$, $\langle \xi^3 \rangle = 0$, $\langle \xi^4 \rangle = 12\,\delta t^2$, *etc.* Proceeding in this way, with judicious choices for f, one can obtain the following set of coupled ordinary differential equations (ODEs):

$$\frac{d\langle q_h^2 \rangle}{dt} = \frac{1}{2}, \quad \frac{d\langle q_\kappa^2 \rangle}{dt} = \frac{\langle x^2 \rangle}{2}, \quad \frac{d\langle x^2 \rangle}{dt} + 2\kappa\langle x^2 \rangle = 2, \quad \frac{d\langle xq_h \rangle}{dt} + \kappa\langle xq_h \rangle = 1,$$

$$\frac{d\langle x^2 q_h^2 \rangle}{dt} + 2\kappa\langle x^2 q_h^2 \rangle = 2\langle q_h^2 \rangle + 4\langle xq_h \rangle + \frac{\langle x^2 \rangle}{2}, \quad \frac{d\langle xq_h q_\kappa \rangle}{dt} + \kappa\langle xq_h q_\kappa \rangle = -\langle xq_h \rangle - \frac{\langle x^2 \rangle}{2},$$

$$\frac{d\langle (q_{h\kappa} + q_h q_\kappa)^2 \rangle}{dt} = \frac{\langle q_\kappa^2 \rangle}{2} - \langle xq_h q_\kappa \rangle + \frac{\langle x^2 q_h^2 \rangle}{2} \quad \left(= \frac{\langle (q_\kappa - xq_h)^2 \rangle}{2} \right). \tag{34}$$

Some of these have already been encountered in the main text. The last one is for the desired mean square second order weight. The ODEs can be solved with the initial conditions that at $t = 0$ all averages involving Malliavin weights vanish but $\langle x^2 \rangle = x_0^2$. The results include *inter alia*

$$\langle q_h^2 \rangle = \frac{t}{2}, \quad \langle q_\kappa^2 \rangle = \frac{t}{2\kappa} + \frac{(\kappa x_0^2 - 1)(1 - e^{-2\kappa t})}{4\kappa^2},$$

$$\langle (q_{h\kappa} + q_h q_\kappa)^2 \rangle = \frac{2\kappa^2 t^2 + (19 + \kappa x_0^2)\kappa t + 2\kappa x_0^2 - 34}{8\kappa^3} + \frac{2\kappa t + 10 - \kappa x_0^2}{2\kappa^3} e^{-\kappa t}$$

$$+ \frac{(1 - \kappa x_0^2)\kappa t + 2\kappa x_0^2 - 6}{8\kappa^3} e^{-2\kappa t}. \tag{35}$$

These are shown as the dashed lines in Figure 2. The leading behaviour of the last as $t \to \infty$ is

$$\langle (q_{h\kappa} + q_h q_\kappa)^2 \rangle = \frac{t^2}{4\kappa} + \text{subdominant terms}, \tag{36}$$

however the approach to the pure asymptotic limit is slow.

222

Reprinted from *Entropy*. Cite as: Hartmann, C.; Banisch, R.; Sarich, M.; Badowski, T.; Schütte, C. Characterization of Rare Events in Molecular Dynamics. *Entropy* **2014**, *16*, 350–376.

Article

Characterization of Rare Events in Molecular Dynamics

Carsten Hartmann [1,*]**, Ralf Banisch** [1]**, Marco Sarich** [1]**, Tomasz Badowski** [1]
and Christof Schütte [1,2]

[1] Institut für Mathematik, Freie Universität Berlin, Arnimallee 6, 14195 Berlin, Germany;
  E-Mails: ralf.banisch@fu-berlin.de (R.B.); sarich@mi.fu-berlin.de (M.S.);
  tomasz.badowski@gmail.com (T.B.); schuette@mi.fu-berlin.de (C.S.)
[2] Konrad-Zuse Zentrum, Takustraße 7, 14195 Berlin, Germany

* Author to whom correspondence should be addressed; E-Mail: chartman@fu-berlin.de;
  Tel.: +49-(0)30-838-75286.

Received: 13 September 2013; in revised form: 8 October 2013 / Accepted: 22 November 2013 / Published: 30 December 2013

Abstract: A good deal of molecular dynamics simulations aims at predicting and quantifying rare events, such as the folding of a protein or a phase transition. Simulating rare events is often prohibitive, especially if the equations of motion are high-dimensional, as is the case in molecular dynamics. Various algorithms have been proposed for efficiently computing mean first passage times, transition rates or reaction pathways. This article surveys and discusses recent developments in the field of rare event simulation and outlines a new approach that combines ideas from optimal control and statistical mechanics. The optimal control approach described in detail resembles the use of Jarzynski's equality for free energy calculations, but with an optimized protocol that speeds up the sampling, while (theoretically) giving variance-free estimators of the rare events statistics. We illustrate the new approach with two numerical examples and discuss its relation to existing methods.

Keywords: rare events; molecular dynamics; optimal pathways; stochastic control; dynamic programming; change of measure; cumulant generating function

1. Introduction

Rare but important transition events between long-lived states are a key feature of many systems arising in physics, chemistry, biology, *etc*. Molecular dynamics (MD) simulations allow for analysis and understanding of the dynamical behavior of molecular systems. However, realistic simulations for interesting (large) molecular systems in solution on timescales beyond microseconds are still infeasible even on the most powerful general purpose computers. This significantly limits the MD-based analysis of many biological equilibrium processes, because they often are associated with rare events. These rare events require prohibitively long simulations because the average waiting time between the events is orders of magnitude longer than the timescale of the transition characterizing the event itself. Therefore, the straightforward approach to such a problem via direct numerical simulation of the system until a reasonable number of events has been observed is impractically excessive for most interesting systems. As a consequence, rare event simulation and estimation are among the most challenging topics in molecular dynamics.

In this article, we consider typical rare events in molecular dynamics for which conformation changes or protein folding may serve as examples. They can be described in the following abstract way: The molecular system under consideration has the ability to go from a reactant state given by a set A in its state space (e.g., an initial conformation) to a product state described by another set B (e.g., the target conformation). Dynamical transitions from A to B are rare. The general situation we will address is as follows:

- The system is (meta)stable, with the sets A and B being two of its metastable sets in the sense that if the system is put there, it will remain there for a long time; transitions between A and B are rare events.
- The sets A and B are separated by an unknown and, in general, rough or diffusive energy landscape (that will be denoted by V).

In addition, we will assume that the system under consideration is in equilibrium with respect to the stationary Gibbs-Boltzmann density

$$\mu(x) = \frac{1}{Z} \exp(-\beta V(x)). \tag{1}$$

We are interested in characterizing the transitions leading from A into B, that is, we are interested in the statistical properties of the ensemble of *reactive trajectories* that go *directly* from A to B (*i.e.*, start in A without returning to A before going to B). In other words, we are interested in all trajectories comprising the actual transition. We would like to:

- know which parts of state space such reactive trajectories visit most likely, *i.e.*, where in state space do we find transition pathways or transition channels through which most of the probability current generated by reactive trajectories flows and
- characterize the rare event statistically, *i.e.*, compute the transition rate, the free energy barrier, the mean first passage time or even more elaborated statistical quantities.

The molecular dynamics literature on rare event simulations is rich. Since the 1930s, transition state theory (TST) [1,2] and extensions thereof based on the reactive flux formalism have provided the main theoretical framework for the description of transition events. TST can, however, at best deliver rates and does not allow one to characterize transition channels. It is based on partitioning the state space into two sets with a dividing surface in between, leaving set A on one side and the target set B on the other, and the theory only tells how this surface is crossed during the reaction. Often, it is difficult to choose a suitable dividing surface, and a bad choice will lead to a very poor estimate of the rate. The TST estimate is then extremely difficult to correct, especially if the rare event is of the diffusive type, where many different reaction channels co-exist. Therefore, many techniques have been proposed that try to go beyond TST.

These different strategies approach the problem by sampling the ensemble of reactive trajectories or by directly searching for the transition channels of the system. Most notable among these techniques are (1) Transition Path Sampling (TPS) [3]; (2) the so-called String Methods [4], or optimal path approaches [5–7] and variants thereof; (3) techniques that follow the progress of the transition through interfaces, like Forward-Flux Simulation (FFS) [8], Transition Interface Sampling (TIS) [9] or the Milestoning techniques [10,11]; and (4) methods that drive the molecular system by external forces with the aim of making the required transition more frequent while still allowing one to compute the exact rare event statistics for the unforced system, e.g., based on Jarzynski's and Crook's identity [12,13]. All of these methods consider the problem in continuous state space, *i.e.*, through reactive trajectories or transition channels in the original state space of the molecular system. They all face substantial problems, e.g., if the ensemble of reactive trajectories and/or transition channels of the system under consideration are too complicated (multi-modal, irregular, essentially high dimensional) or they suffer from too large variance of the underlying statistical estimators. We should moreover stress that each of these methods has its specific scope of application; some methods are mainly useful for computing transition rates, whereas others can be used to compute transition pathways or free energy differences.

Our aim is (A) to review some of these methods based on a joint theoretical basis and (B) to outline a new approach to the estimation of rare event statistics based on a combination of ideas from optimal control and statistical mechanics. In principle, this approach allows for a *variance-free* estimation of rare event statistics in combination with *much reduced simulation time*. The rest of the article is organized as follows: We start with a precise characterization of reactive trajectories, transition channels and related quantities in the framework of Transition Path Theory (TPT) in Section 2. Then, in Sections 3 and 4, we discuss the methods from classes (1)–(3) and characterize their potential problems in more detail. In Section 5, we consider methods of type (4) as an introduction to the presentation of the new optimal control approach that is outlined in detail in Sections 6 and 7, including some numerical experiments.

Alternative, inherently discrete methods, like Markov State Modeling, that discretize the state space appropriately and try to compute transition channels and rates *a posteriori* based on the resulting discrete model of the dynamics will not be discussed herein and are considered in the article [14] in a way related to the presentation at hand. We should further mention that not all rare event problems in molecular dynamics are related to sampling the underlying Gibbs–Boltzmann statistics, e.g., nucleation events under shear [15] or genuine nonequilibrium systems without a stationary probability distribution [16].

2. Reactive Trajectories, Transition Rates and Transition Channels

Since our results are rather general, it is useful to set the stage somewhat abstractly. To this end, we borrow some notation from [17] and consider a system whose state space is $\mathbb{R}^n$ and denote by X_t the current state of the system at time t. For example, X_t may be the set of instantaneous positions and momenta of the atoms of a molecular system. We assume that the system is ergodic with respect to a probability (equilibrium) distribution μ and that we can generate an infinitely long equilibrium trajectory $\{X_t\}_{t\in\mathbb{R}}$ where, for technical reasons, we let the trajectory start at time $t = -\infty$. The trajectory will go infinitely many times from A to B and each time the reaction happens. This reaction involves reactive trajectories that can be defined as follows: Given the trajectory $\{X(t)\}_{t\in\mathbb{R}}$, we say that its reactive pieces are the segments during which X_t is neither in A or B, came out of A last and will go to B next. To formalize things, let

$$t^+_{AB}(t) = \text{smallest } s \geq t \text{ such that } X(s) \in A \cup B,$$
$$t^-_{AB}(t) = \text{largest } s \leq t \text{ such that } X(s) \in A \cup B.$$

Then, the trajectory $\{X(t)\}_{t\geq0}$ is reactive for all $t \in R$ where $R \subset [0, \infty)$ is defined by the requirements

$$X_t \notin A \cup B, \quad X_{t^+_{AB}(t)} \in B \quad \text{and} \quad X_{t^-_{AB}(t)} \in A$$

and the ensemble of reactive trajectories is given by the set

$$\mathcal{R} = \{X_t : t \in R\}$$

where each specific continuous piece of trajectory going directly from A to B in the ensemble belongs to a specific interval $[t_1, t_2] \subset R$.

Given the ensemble of reactive trajectories, we want to characterize it statistically by answering the following questions:

(Q1) What is the probability of observing a trajectory at $x \notin (A \cup B)$ at time t, conditional on $t \in R$?

(Q2) What is the probability current of reactive trajectories? This probability current is the vector field $j_{AB}(x)$ with the property that given any separating surface S between A and B (i.e., the boundary of a region that contains A but not B), the surface integral of j_{AB} over S gives the probability flux of reactive trajectories between A and B across S.

(Q3) What is the transition rate of the reaction, i.e., what is the mean frequency k_{AB} of transitions from A to B?

(Q4) Where are the main transition channels used by most of the reactive trajectories?

Question (Q1) can be answered easily, at least theoretically: The probability density to observe any trajectory (reactive or not) at point x is $\mu(x)$. Let $q(x)$ be the so-called committor function, that is the probability that the trajectory starting from x reaches first B rather than A. If the dynamics are reversible, then the probability that a trajectory we observe at state x is reactive is $q(x)(1 - q(x))$, where the first factor appears since the trajectory must go to B rather than A next, and the second factor appears since it needs to come from A rather than B last. Now, the Markov property of the dynamics implies that the probability density to observe a *reactive* trajectory at point x is

$$\mu_{AB}(x) \propto q(x)(1 - q(x))\,\mu(x)\,,$$

which is the probability of observing any trajectory in x times the probability that it will be reactive (the proportionality symbol $\propto$ is used to indicate identity up to normalization).

2.1. Transition Path Theory (TPT)

In order to give answers to the other questions, we will exploit the framework of *transition path theory* (TPT), which has been developed in [17–20] in the context of diffusions and has been generalized to discrete state spaces in [21,22]. In order to review the key results of TPT, let us consider diffusive molecular dynamics in an energy landscape $V: \mathbb{R}^n \to \mathbb{R}$:

$$dX_t = -\nabla V(X_t)dt + \sqrt{2\epsilon}\,dB_t\,, \quad X_0 = x\,. \tag{2}$$

Here, B_t denotes standard n-dimensional Brownian motion, and $\epsilon > 0$ is the temperature of the system. Under mild conditions on the energy landscape function V, we have ergodicity with respect to the stationary distribution $\mu(x) = Z^{-1} \exp(-\beta V(x))$ with $\beta = 1/\epsilon$. The dynamics are reversible with respect to this distribution, *i.e.*, the detailed balance condition holds. We assume throughout that the temperature is small relative to the largest energy barriers, *i.e.*, $\epsilon \ll \Delta V_{\max}$. As a consequence, the relaxation of the dynamics towards equilibrium is dominated by the rare transitions over the largest energy barriers.

For these kind of dynamics, Questions (Q2) and (Q3) have surprisingly simple answers: The reactive probability current is given by

$$j_{AB}(x) = \epsilon\mu(x)\,\nabla q(x)$$

where ∇q denotes the gradient of the committor function q. Based on this, the transition rate can be computed by the total reactive current across an arbitrary separating surface S:

$$k_{AB} = \int_S n_S(x)j_{AB}(x)d\sigma_S(x)$$

where n_S denotes the unit normal vector on S pointing towards B and σ_S the associated surface element. The rate can also be expressed by

$$k_{AB} = \epsilon \int_{(A \cup B)^c} (\nabla q(x))^2 \mu(x)dx$$

where $(A \cup B)^c$ denotes the entire state space excluding A and B. Given the reactive current, we can even answer Question (Q4): The transition channels of the reaction $A \to B$ are the regions of $(A \cup B)^c$ in which the streamlines of the reactive current, *i.e.*, the solutions of

$$\frac{d}{dt}x_{AB}(t) = j_{AB}\Big(x_{AB}(t)\Big), \quad x_{AB}(0) \in A$$

are exceptionally dense.

Figure 1. (**Top left panel**) Three-well energy landscape V as described in the text. (**Top right panel**) Typical reactive trajectory in the three-well landscape. (**Middle left panel**) Committor functions q_{AB} for diffusion molecular dynamics with relatively high temperature $\epsilon = 0.6$ for the sets A (main well, right-hand side) and B (main well, left-hand side). (**Middle right panel**) Committor q_{AB} for the low temperature case $\epsilon = 0.15$. (**Bottom left panel**) Transition channels for $\epsilon = 0.6$. (**Bottom right panel**) Transition channels for $\epsilon = 0.15$. For details of the computations underlying the pictures, see [22].

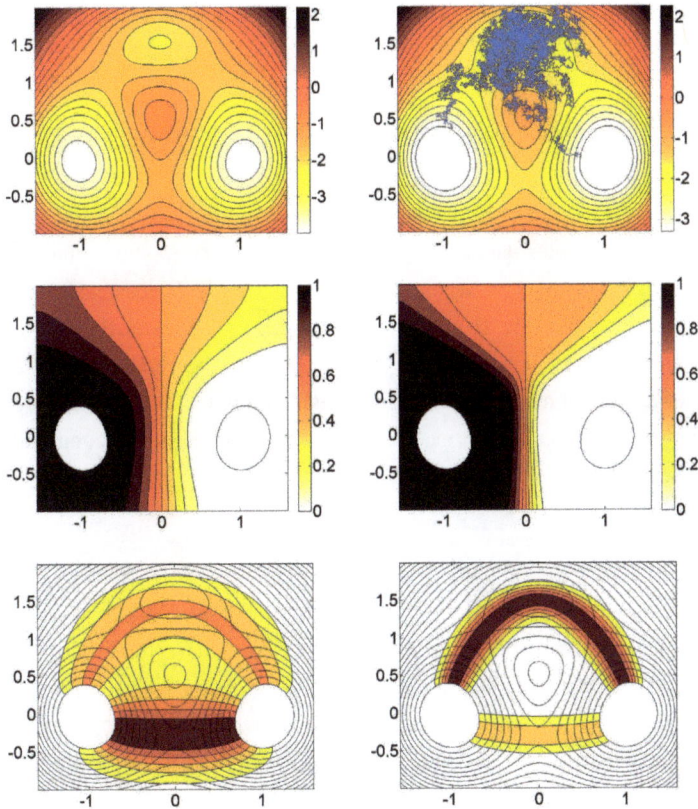

Figure 1 illustrates these quantities for the case of a 2D three well potential with two main wells (the bottoms of which we take as A and B in the following) and a less significant third well. The three main saddle points separating the wells are such that the two saddle points between the main wells and the third well are lower in energy than the saddle point between the main wells, such that in the zero temperature limit, we expect that almost all reactive trajectories take the route through the third well across the two lower saddle points. We observe that the committor functions for low and higher temperatures exhibit smooth isocommittor lines separating the sets A and B, as expected. The transition channels computed from the associated reactive current also show what one should expect: For a lower temperature, the channel through the third well and across the two lower saddle points is dominant, while for a higher temperature, the direct transition from A to B across the higher saddle point is preferred.

These considerations can be generalized to a wide range of different kinds of dynamics in continuous state spaces, including, e.g., full Langevin dynamics, see [17–20].

This example illustrates that TPT in principle allows us to quantify all aspects of the transition behavior underlying a rare event. We can compute transition rates exactly and even characterize the transition mechanisms if we can compute the committor function. Deeper insight using the Feynman–Kac formula yields that the committor function can be computed as the solution of a linear boundary value problem, which for diffusive molecular dynamics reads

$$Lq_{AB} = 0 \quad \text{in } (A \cup B)^c, \quad q_{AB} = 0 \text{ in } A, \quad q_{AB} = 1 \text{ in } B$$

where the generator L has the following form

$$L = \epsilon \Delta - \nabla V(x) \cdot \nabla \tag{3}$$

where $\Delta = \sum_i \partial^2 / \partial x_i^2$ denotes the Laplace operator. This equation allows the computation of q_{AB} in relatively low-dimensional spaces, where the discretization of L is possible based on finite element methods or comparable techniques. In realistic biomolecular state spaces, this is infeasible because of the curse of dimensionality. Therefore, TPT gives a complete theoretical background for rare event simulation, but its application in high dimensional situations is still problematic. As a remedy, a discrete version of TPT has been developed [21,22], which can be used in combination with Markov State Modeling; see [23].

2.2. Transition Path Sampling (TPS)

TPS has been developed in order to sample from the probability distribution of reactive trajectories in so-called "path space", which means nothing else than the space of all discrete or continuous paths starting in A and ending up in B equipped with the probability distribution generated by the dynamics through the ensemble of associated reactive trajectories. Let P_T denote the path measure on the space of discrete or continuous trajectories $\{X_t\}_{0 \leq t \leq T}$ of length T. The *path measure of reactive trajectories* then is

$$P_T^{AB}(\{X_t\}_{0 \leq t \leq T}) = \frac{1}{Z_{AB}} \mathbf{1}_A(X_0) \, P_T(\{X_t\}_{0 \leq t \leq T}) \, \mathbf{1}_B(X_T) \tag{4}$$

where $\mathbf{1}_A$ denotes the indicator function of set A (that is, $\mathbf{1}_A(x) = 0$ if $x \notin A$ and $= 1$ otherwise).

TPS is a Metropolis Monte-Carlo (MC) method for sampling $P_T^{AB}(\{X_t\}_{0 \leq t \leq T})$ that uses explicit information regarding the path measure P_T, such as Equation (5), with MC moves that are based on a perturbation of a precomputed reactive trajectory [3,24]. It delivers an ensemble of reactive trajectories of length T that (under the assumption of convergence of the MC scheme) is representative for P_T^{AB} and thus allows one to compute respective expectation values, like the probability to observe a reactive trajectory or the reactive current. However, its potential drawbacks are obvious: (1) A typical reactive trajectory is very long and rather uninformative (*cf.* Figure 1), *i.e.*, the computational effort of generating an entire ensemble of long reactive trajectories can be prohibitive; (2) convergence of the MC scheme in the infinite dimensional path space can be very poor; and (3) the limitation to a pre-defined trajectory length T can lead to biased statistics of the TPS ensemble. Advanced TPS schemes try to remedy these drawbacks by combining the original TPS idea with interface methods [9]. Even though TPS can be used no matter whether the underlying dynamics is deterministic or stochastic, the algorithm is usually used in connection with deterministic Hamiltonian dynamics [3].

3. Finding Transition Channels

Whenever a transition channel exists, one can try to approximate the center curve of the transition channel instead of sampling the ensemble of reactive trajectories. If the center curve (also: *principal curve*) is a rather smooth object, then such a method would not suffer from the extensive length of reactive trajectories. Several such methods have been introduced; they differ with respect to the definition of the transition channel and the corresponding center or principal curve.

3.1. Action-Based Methods

Rather than sampling the probability distribution of reactive pathways, such as Equation (4), one can try to obtain a representative or *dominant* pathway, e.g., by computing the pathway that has maximum probability under P_T. For the case of diffusive molecular dynamics, the path measure P_T has a probability density relative to a (fictitious) uniform measure on the space of all continuous paths in $\mathbb{R}^n$ of length T that are generated by Brownian motion; the relative density reads

$$\ell(\varphi) = \exp\left(-\frac{1}{2\epsilon} I_\epsilon(\varphi)\right)$$

where I_ϵ is the Onsager–Machlup action

$$I_\epsilon(\varphi) = \int_0^T \left\{\frac{1}{2}|\dot{\varphi}(s)|^2 + \frac{1}{2}|\nabla V(\varphi(s))|^2 - \epsilon \Delta V(\varphi(s))\right\} dt. \tag{5}$$

More precisely, $\ell(\varphi)$ is the limiting ratio between the probability that the solution of Equation (2) remains in a small tubular neighborhood of a smooth path $\varphi(\cdot)$ and the probability that $\sqrt{2\epsilon} B_t$ remains in a small neighborhood of the initial value $x = \varphi(0)$, as the size of the neighborhoods go to zero [25].

The fact that the Euler discretization of the path density ℓ, with I_ϵ interpreted in the sense of Itô integrals, corresponds to the probability density of the Euler-discretized reaction path with respect to Lebesgue measure has led to the idea that by minimizing the Onsager–Machlup action over all continuous paths $\varphi \colon [0, T] \to \mathbb{R}^n$ going from A to B, one can find the dominant reactive path

$\varphi^* = \text{argmin}_\varphi I^\epsilon(\varphi)$ in the sense of a maximum likelihood estimator. The hope is that this path, often also called the *optimal path* or *most probable path*, on the one hand, contains information on the transition mechanism and, on the other hand, is much smoother and easier to interpret than a typical reactive trajectory. Note, however, that the actual *probability* that the solution of Equation (2) remains in a small neighborhood of a given path $\varphi(\cdot)$ is exponentially small in the size of the neighborhood.

In [7], a comparison between the Onsager–Machlup action and its zero temperature limit has been given using gradient descent methods, raising issues regarding the correct interpretation of the minimizers of I_ϵ (that need not exist) as *most probable paths*. In [5], the *dominant reaction pathway method* has been outlined, which uses a simplified version of the Onsager–Machlup functional that leads to a computationally simpler optimization problem and is applicable to large-scale problems, e.g., protein folding [6]. However, even if the globally dominant pathways can be computed, such that the optimization does not get stuck in local minima, and even if we ignore the issues regarding the correct interpretation of minimizers, the resulting pathways in general do *not* allow one to gain statistical information on the transition (like rates, currents, mean first passage times).

Another action-based method that has been introduced in [26] is the *MaxFlux* method, which seeks the path that carries the highest reactive flux among all reactive trajectories of a certain length. The idea is to compute the path of least resistance by minimizing the functional

$$L(\varphi) = \int_0^T \exp\left(\epsilon^{-1} V(\varphi(s))\right) ds .$$

Several algorithmic approaches for the minimization of the resistance functional L have been proposed, e.g., a path-based method [27], discretization of the corresponding Euler–Lagrange equation based on a mean-field approximation of it [28] or a Hamilton–Jacobi-based approach using the method of characteristics [29]. Minimizing L for different values of T then yields a collection of paths, each of which carries a certain percentage of the total reactive flux. The method is useful if the temperature is small, so that the reactive flux concentrates around a sufficiently small number of reactive pathways.

3.2. String Method and Variants

There are several other methods that entirely avoid the computation of reactive trajectories, but try to reconstruct the less complex transition channels or pathways instead, analyzing the energy landscape of the system. One group of such techniques, like the Zero Temperature String method [4], the Geometric Minimum Action method [30] or the Nudged Elastic Band method [31], concentrate on the computation of the *minimal energy path* (MEP), *i.e.*, the path of lowest potential energy between (a point in) A and (a point in) B. Under diffusive molecular dynamics and for vanishing temperature, the MEP is the path that transitions take with probability one [32]. It turns out that the MEP in this case is the minimizer of the Onsager–Machlup action (5) in the limit $\epsilon \to 0$. For non-zero temperature and a rugged energy landscape, the MEP will in general be not very informative and must be replaced by a finite-temperature transition channel. This is done by the finite-temperature string (FTS) method [33] based on the following considerations: Firstly, the isocommittor surfaces Γ_α,

$\alpha \in [0,1]$, of the committor q are taken as natural interfaces that separate A from B. Secondly, each Γ_α is weighted with the stationary distribution μ to find reactive trajectories crossing it at a certain point $x \in \Gamma_\alpha$,

$$\rho_\alpha(x) = \frac{1}{Z_\alpha} q(x)(1 - q(x)) \, \mu(x), \qquad Z_\alpha = \int_{\Gamma_\alpha} q(x)(1 - q(x)) \, \mu(x) d\sigma_\alpha(x)$$

The idea of the FTS method is that the ensemble of reactive trajectories can be characterized by this distribution on the isocommittor surfaces. Third, one assumes that for each α, the probability density ρ_α is peaked in just one point $\varphi(\alpha)$ and that the curve $\varphi = \varphi(\alpha)$, $\alpha \in [0,1]$ defined by the sequence of these points forms the center of the (single) transition channel. More precisely, one defines $\varphi(\alpha) = \langle x \rangle_{\Gamma_\alpha}$ where the average is taken according to ρ_α along the respective isocommittor surface Γ_α. Fourth, it is assumed that the covariance $C_\alpha = \langle (x - \varphi(\alpha)) \otimes (x - \varphi(\alpha)) \rangle_{\Gamma_\alpha}$, which defines the width of the transition channel, is small, which implies that the isocommittor surfaces can be locally approximated by hyperplanes P_α. The computation of the FTS string φ then is done by approximating it via $\varphi(\alpha) = \langle x \rangle_{P_\alpha}$, where the average is computed by running constrained dynamics on P_α while iteratively refining the hyperplanes P_α; see [34] for details. Later extensions [35] remove the restrictions resulting from the hyperplanes by using Voronoi tessellations instead.

The FTS method allows one to compute single transition channels in rugged energy landscapes as long as these are not too extended and rugged. Compared to methods that sample the ensemble of reactive trajectories, it has the significant advantage that the string, that is, the principal curve inside the transition channel, is rather smooth and short, as compared to the typical reactive trajectories. The FTS further allows one to compute the free energy profile $F = F(\alpha)$ along the string,

$$F(\alpha) = -\beta^{-1} \log \int_{P_\alpha} \mu(x) d\sigma_\alpha(x)$$

that characterizes the transition rates associated with the transition channel (at least in the limits of the approximations invoked by the FTS).

4. Computing Transition Rates

The computation of transition rates can be performed without computing the dominant transition channels or similar objects. There is a list of rather general techniques, with Forward Flux Sampling (FFS) [8], Transition Interface Sampling (TIS) [9] and Milestoning [10] as examples, that approximate transition rates by exploring how the transition progresses from one to the next interface that separates A from B.

4.1. Forward Flux Sampling (FFS)

The first step of FFS is the choice of a finite sequence of interfaces I_k, $k = 1, \ldots, N$, in state space between A and $B = I_N$. The transition rate k_{AB} comes as the product of two factors: (1) the probability current J_A of *all* trajectories leaving A and hitting I_1; and (2) the probability

$$\mathbb{P}(B|I_1) = \prod_{j=1}^{N-1} \mathbb{P}(I_{k+1}|I_k)$$

that a trajectory that leaves I_1 makes it to B before it returns to A; here, $\mathbb{P}(I_{k+1}|I_k)$ denotes the probability that a trajectory starting in I_k makes it to I_{k+1} before it returns to A. FFS first performs a brute-force simulation starting in A, which yields an ensemble of points at the first interface I_1, yielding an estimate for the flux J_A (the number of trajectories hitting I_1 per unit of time). Second, a point from this ensemble on I_1 is selected at random and used to start a trajectory, which is followed until it either hits the next interface I_2 or returns to A; this gives $\mathbb{P}(I_2|I_1)$. This procedure then is iterated from interface to interface. Finally, the rate $k_{AB} = J_A \cdot \mathbb{P}(B|I_1)$ is computed. Variants of this algorithm are described in [36,37], for example.

FFS has been demonstrated to be quite general in approximating the flux of reactive trajectories through a given set of interfaces; it can be applied to equilibrium, as well as nonequilibrium systems, and its implementation is easy (see [16,38]). The interfaces used in FFS are, in principle, arbitrary. However, the efficiency of the sampling of the reactive hitting probabilities $\mathbb{P}(I_{k+1}|I_k)$ crucially depends on the choice of the interfaces. In practice, the efficiency of FFS will drop dramatically if one does not use appropriate surfaces, and totally misleading rates may result from this. Ideally, one would like to choose these surfaces, so that the computational gain offered by FFS in optimized, but in practice, this is not a trivial task; see [39]. The same is true for TIS that couples TPS with progressing from interface to interface.

4.2. Milestoning

Milestoning [10] is similar to FFS in so far as it also uses a set of interfaces I_k, $k = 1, \ldots, N$ that separate A and $B = I_N$. In contrast to FFS and TIS, the fundamental quantities in Milestoning are the hitting time distributions $K_i^\pm(\tau)$, $i = 1, \ldots, N - 1$, where $K_i^\pm(\tau)$ is the probability that a trajectory starting at $t = 0$ at interface I_i hits $I_{i\pm1}$ before time τ. Trajectories that make it to milestone I_i must come from milestones $I_{i\pm1}$ and *vice versa*. In the original algorithm, these distributions are approximated as follows [10]: For each milestone I_i, one first samples the distribution μ constrained to I_i. Based on the resulting sample, we start a trajectory from each point, which is terminated when it reaches one of its two neighboring milestones $I_{i\pm1}$. The hitting times are recorded and collected into two distributions $K_i^\pm(\tau)$.

These local kinetics are then compiled into the global kinetics of the process: For each i, one defines $P_i(t)$ as the probability that the process is found between I_{i-1} and I_{i+1} at time t and that the last milestone hit was I_i. Milestoning is based on a (non-Markovian) construction of $P_i(t)$ from the $K_i^\pm(\tau)$. Its efficiency comes from two sources: (1) It does not require the computation of long reactive trajectories but only short ones between milestones (which therefore should be 'close enough'); (2) It is easily parallelizable. Its disadvantage is the dependence on the milestones that have to be chosen in advance: It can be shown that Milestoning with perfect sampling allows one to compute exact transition rates or mean first passage times if the interfaces are given by the isocommittor surfaces (which in general are not known in advance) [40]; if the interfaces are chosen inappropriately, the results can be rather misleading.

5. Nonequilibrium Forcing and Jarzynski's Identity

The computation of reliable rare event statistics suffers from the enormous lengths of reactive trajectories. One obvious way to overcome this obstacle is to force the system to exhibit the transition of interest on shorter timescales. Therefore, can we *drive* the molecular system to make the required transition more frequently but still compute the exact rare event statistics for the unforced system?

As was shown by Jarzynski and others, nonequilibrium forcing can in fact be used to obtain equilibrium rare event statistics. The advantage seems to be that the external force can speed up the sampling of the rare events by biasing the equilibrium distribution towards a distribution under which the rare event is no longer rare. We will shortly review Jarzynski's identity before discussing the matter in more detail.

5.1. Jarzynski's Identity

Jarzynski's and Crook's formulae [12,13] relate the equilibrium Helmholtz free energy to the nonequilibrium work exerted under external forcing: Given a system with energy landscape $V(x)$, the total Helmholtz free energy can be defined as

$$F = -\beta^{-1} \log Z \quad \text{with} \quad Z = \int \exp(-\beta V(x)) dx \,.$$

Jarzynski's equality [12] then relates the free energy difference $\Delta F = -\beta^{-1} \log(Z_1/Z_0)$ between two equilibrium states of a system given by an unperturbed energy V_0 and its perturbation V_1 with the work W applied to the system under the perturbation: Suppose we set $V_\xi = (1 - \xi)V_0 + \xi V_1$ with $\xi \in [0, 1]$, and assume we set a protocol that describes how the system evolves from $\xi = 0$ to $\xi = 1$. If, initially, the system is distributed according to $\exp(-\beta V_0)$, then, by the second law of thermodynamics, it follows that $\mathbf{E}(W) \geq \Delta F$ where W is the total work applied to the system and $\mathbf{E}$ denotes the average overall possible realizations of the transition from $\xi = 0$ to $\xi = 1$; equality is attained if the transition is infinitely slow (*i.e.*, adiabatic). Jarzynski's identity now asserts that the free energy is always equal to the exponential average of the nonequilibrium work,

$$\Delta F = -\beta^{-1} \log \mathbf{E}\left[\exp(-\beta W) \right]$$

arbitrarily far away from the adiabatic regime. Many generalizations exist: In [13], a generalized version of this fluctuation theorem, the so-called Crook's formula, for stochastic, microscopically reversible dynamics, is derived. In [41,42], it is shown how one can compute conditional free energy profiles along a reaction coordinate for the *unperturbed* system, rather than total free energy differences between a perturbed and unperturbed system.

*Algorithmic application prohibitive.*Despite the fact that Jarzynski's and Crook's formulae are used in molecular dynamics applications [43], their algorithmic usability is limited by the fact that the likelihood ratio between equilibrium and nonequilibrium trajectories is highly degenerate, and the overwhelming majority of nonequilibrium forcings generate trajectories that have almost zero weight with respect to the equilibrium distribution that is relevant for the rare event. This leads to the fact that most rare event sampling algorithms based on Jarzynski's identity have *prohibitively large variance*. Recent developments have reduced this problem by sampling just the reversible work processes based on Crook's formula, but could not fully remove the problem of large variance [44]; see also [45]. Because of this, we will approach the problem of variance reduction subsequently.

5.2. Cumulant Generating Functions

In order to demonstrate how to improve approaches based on the idea of driving molecular systems to make rare events frequent, we first have to introduce some concepts and notation from statistical mechanics: Let W be a random variable that depends on the sample paths of $(X_t)_{t \geq 0}$, *i.e.*, on molecular dynamics trajectories of the system under investigation. Further, let P be the underlying probability measure on the space of continuous trajectories as introduced in Section 2.2 (but without the restriction to a given length T). We define the *cumulant generating function* (CGF) of W by

$$\gamma(\sigma) = -\sigma^{-1} \log \mathbf{E}[\exp(-\sigma W)] \tag{6}$$

where σ is a non-zero scalar parameter and $\mathbf{E}[f] = \int f \, dP$ denotes the expectation value with respect to P. Note that the CGF is basically the free energy at inverse temperature β as in Jarzynski's formula, but here, it is considered as a function of the independent parameter σ. (Definition (6) differs from the standard CGF only by the prefactor σ^{-1} in front.) Taylor expanding the CGF about $\sigma = 0$, we observe that $\gamma(\sigma) \approx \mathbf{E}[W] - \frac{\sigma}{2}\mathbf{E}[(W - \mathbf{E}[W])^2]$; hence, for sufficiently small σ, the variance is decoupled from the mean. Moreover, it follows by Jensen's inequality that

$$\gamma(\sigma) \leq \mathbf{E}[W]$$

where equality is achieved if and only if W is almost surely constant, in accordance with the second law of thermodynamics. (This is the case, e.g., when W is the work associated with an adiabatic transition between thermodynamic equilibrium states.)

Optimal reweighting.

The CGF admits a variational characterization in terms of relative entropies. To this end, let Q be another probability measure, so that P is absolutely continuous with respect to Q, *i.e.*, the likelihood ratio dP/dQ exists and is Q-integrable. Then, using Jensen's inequality again,

$$-\sigma^{-1} \log \int e^{-\sigma W} \, dP = -\sigma^{-1} \log \int e^{-\sigma W + \log\left(\frac{dP}{dQ}\right)} \, dQ$$
$$\leq \int \left\{ W + \sigma^{-1} \log \left(\frac{dQ}{dP} \right) \right\} dQ,$$

which, noting that the logarithmic term is the relative entropy (or Kullback–Leibler divergence) between Q and P, can be recast as

$$\gamma(\sigma) \leq \int W \, dQ + H(Q \| P) \tag{7}$$

where

$$H(Q \| P) = \sigma^{-1} \int \log \left(\frac{dQ}{dP} \right) dQ, \tag{8}$$

and we declare that $H(Q\|P) = \infty$ if Q does not have a density with respect to P. Again, it follows from the strict convexity of the exponential function that equality is achieved if and only if the new random variable

$$Z = W + \sigma^{-1} \log \left(\frac{dQ}{dP} \right)$$

is Q-almost surely constant. This gives us the following variational characterization of the cumulant generating function that is due to [46]: *Variational formula for the cumulant generating function.*

Let W be bounded from above, with $\mathbf{E}[\exp(-\sigma W)] < \infty$. Then

$$\gamma(\sigma) = \inf_{Q \ll P} \left\{ \int W \, dQ + H(Q\|P) \right\} \tag{9}$$

where the infimum runs over all probability measures Q that have a density with respect to P. Moreover, the minimizer Q^* exists and is given by

$$dQ^* = e^{\gamma(\sigma) - \sigma W} \, dP \, .$$

6. Optimal Driving from Control Theory

When X_t denotes stochastic dynamics, such as Equation (2), the above variational formula admits a nice interpretation in terms of an optimal control problem with a quadratic cost. To reveal it, we first need some technical assumptions.

(A1) We define $Q = [0, T) \times O$ where $T \in [0, \infty]$ and $O \subset \mathbb{R}^n$ is a bounded open set with smooth boundary ∂O. Further, let $\tau < \infty$ be the stopping time

$$\tau = \inf\{t > t_0 : (t, X_t) \notin Q\} \, ,$$

i.e., τ is the stopping time that either $t = T$ or X_t leaves the set O, whichever comes first.

(A2) The random variable W is of the form

$$W = \frac{1}{\epsilon} \int_0^\tau f(X_t) \, dt + \frac{1}{\epsilon} g(X_\tau)$$

for some continuous and nonnegative functions $f, g : \mathbb{R}^n \to \mathbb{R}$, which are bounded from above and at most polynomially growing in x (compare Jarzynski's formula).

(A3) The potential $V : \mathbb{R}^n \to \mathbb{R}$ in Equation (2) is smooth, bounded below and satisfies the usual local Lipschitz and growth conditions.

We consider the conditioned version of the moment generating function (which is just the exponential of the cumulant generating function):

$$\psi_\sigma(x, t) = \mathbf{E}[\exp(-\sigma W)|X_t = x] \, . \tag{10}$$

By the Feynman–Kac theorem, ψ_σ solves the linear boundary value problem

$$\left(A - \frac{\sigma}{\epsilon} f \right) \psi_\sigma = 0$$

$$\psi_\sigma|_{E^+} = \exp\left(-\frac{\sigma}{\epsilon} g \right) \tag{11}$$

where E^+ is the terminal set of the augmented process (t, X_t), precisely $E^+ = ([0, T) \times \partial O) \cup (\{T\} \times O)$, and

$$\mathcal{A} = \frac{\partial}{\partial t} + L$$

is the backward evolution operator associated with X_t, with the shorthand

$$L = \epsilon \Delta - \nabla V \cdot \nabla$$

introduced in Equation (3). Assumptions (A1)–(A3) guarantee that Equation (11) has a unique smooth solution ψ_σ for all $\sigma > 0$. Moreover, the stopping time τ is almost surely finite, which implies that

$$0 < c \leq \psi_\sigma \leq 1$$

for some constant $c \in (0, 1)$.

Log transformation of the cumulant generating function.

In order to arrive at the optimal control version of the variational formula (9), we introduce the logarithmic transformation of ψ_σ as

$$v_\sigma(x, t) = -\frac{\epsilon}{\sigma} \log \psi_\sigma(x, t),$$

which is analogous to the CGF γ, except for the leading factor ϵ and the dependence on the initial condition x. As we will show below, v_σ is related to an optimal control problem. To see this, remember that ψ_σ is bounded away from zero and note that

$$-\frac{\epsilon}{\sigma} \psi_\sigma^{-1} \mathcal{A} \psi_\sigma = \mathcal{A} v_\sigma - \sigma |\nabla v_\sigma|^2,$$

which implies that Equation (11) is equivalent to

$$\mathcal{A} v_\sigma - \sigma |\nabla v_\sigma|^2 + f = 0$$
$$v_\sigma|_{E^+} = g.$$

Equivalently,

$$\min_{\alpha \in \mathbb{R}^n} \{ \mathcal{A} v_\sigma + \alpha \cdot \nabla v_\sigma + \frac{1}{4\sigma} |\alpha|^2 + f \} = 0 \tag{12}$$
$$v_\sigma|_{E^+} = g$$

where we have used that

$$-\sigma |y|^2 = \min_{\alpha \in \mathbb{R}^n} \left\{ \alpha \cdot y + \frac{1}{4\sigma} |\alpha|^2 \right\}.$$

(For the general framework of change-of-measure techniques and Girsanov transformations and their relation to logarithmic transformations, we refer to ([47] (Section VI.3)).)

Optimal control problem. Equation (12) is a Hamilton–Jacobi–Bellman (HJB) equation and is recognized as the dynamic programming equation of the following optimal control problem: minimize

$$J(u) = \mathbf{E} \left[\int_0^\tau \left\{ f(X_t) + \frac{1}{4\sigma} |u_t|^2 \right\} dt + g(X_\tau) \right] \tag{13}$$

over a suitable space of admissible control functions $u \colon [0, \infty) \to \mathbb{R}^n$ and subject to the dynamics

$$dX_t = (u_t - \nabla V(X_t)) \, dt + \sqrt{2\epsilon} dW_t . \tag{14}$$

Form of optimal control. In more detail, one can show (e.g., see ([47] (Section IV.2))) that assumptions (A1)–(A3) above imply that Equation (12) has a classical solution (*i.e.*, twice differentiable in x, differentiable in t and continuous at the boundaries). Moreover, v_σ satisfies

$$v_\sigma(x, t) = \mathbf{E} \left[\int_t^\tau \left\{ f(X_s) + \frac{1}{4\sigma} |u_s^*|^2 \right\} ds + g(X_\tau) \Big| X_t = x \right] \tag{15}$$

where u^* is the unique minimizer of $J(u)$ that is given by the Markovian feedback law

$$u_t^* = \alpha^*(X_t, t) ,$$

with

$$\alpha^* = \operatorname*{argmin}_{\alpha \in \mathbb{R}^n} \left\{ \alpha \cdot \nabla v_\sigma + \frac{1}{4\sigma} |\alpha|^2 \right\} .$$

The function v_σ is called the *value function* or *optimal-cost-to-go* for the optimal control problems (13) and (14). Specifically, $v_\sigma(x, t)$ measures the minimum cost needed to drive the system to the terminal state when started at x at time t. We briefly mention the two most relevant special cases of (13) and (14).

6.1. Case I: The Exit Problem

We want to consider the limit $T \to \infty$. To this end, call $\tau_O = \inf\{t > 0 \colon X_t \notin O\}$ the first exit time of the set $O \subset \mathbb{R}^n$. The stopping time $\tau = \min\{T, \tau_O\}$ then converges to τ_O, *i.e.*,

$$\min\{T, \tau_O\} \to \tau_O .$$

As a consequence (using monotone convergence), v_σ converges to the value function of an optimal control problem with cost functional

$$J_\infty(u) = \mathbf{E} \left[\int_0^{\tau_O} \left\{ f(X_t) + \frac{1}{4\sigma} |u_t|^2 \right\} dt + g(X_{\tau_O}) \right] . \tag{16}$$

It can be shown that the value function

$$v_\sigma(x, t) = \mathbf{E} \left[\int_t^{\tau_O} \left\{ f(X_s) + \frac{1}{4\sigma} |u_s^*|^2 \right\} ds + g(X_\tau) \Big| X_t = x \right]$$

with $u^* = \operatorname{argmin} J_\infty(u)$ is *independent* of the initial time t; hence, we can drop the dependence on t and redefine $v_\sigma(x) := v_\sigma(x, t)$. The value function now solves the boundary value HJB equation

$$\min_{\alpha \in \mathbb{R}^n} \left\{ L v_\sigma + \alpha \cdot \nabla v_\sigma + \frac{1}{4\sigma} |\alpha|^2 + f \right\} = 0 \tag{17}$$

$$v_\sigma|_{\partial O} = g .$$

6.2. Case II: Finite Time Horizon Optimal Control

If we keep $T < \infty$ fixed while letting O grow, such that $\mathrm{diam}(O) \to \infty$, where $\mathrm{diam}(O) = \sup\{r > 0 \colon \mathcal{B}_r(x) \subset O,\, x \in O\}$ is understood as the maximum radius $r > 0$ that an open ball $\mathcal{B}_r(\cdot)$ contained in O can have, it follows that

$$\min\{T, \tau_O\} \to T.$$

In this case, v_σ converges to the value function with a finite time horizon and cost functional

$$J_T(u) = \mathbf{E}\left[\int_0^T \left\{ f(X_t) + \frac{1}{4\sigma}|u_t|^2 \right\} dt + g(X_T) \right]. \tag{18}$$

Now, v_σ is again a function on $\mathbb{R}^n \times [0, T]$ and given by

$$v_\sigma(x, t) = \mathbf{E}\left[\int_t^T \left\{ f(X_s) + \frac{1}{4\sigma}|u_s^*|^2 \right\} ds + g(X_\tau) \,\bigg|\, X_t = x \right],$$

with u^* being the minimizer of $J_T(u)$. The value function solves the backward evolution HJB equation

$$\min_{\alpha \in \mathbb{R}^n}\{\mathcal{A}v_\sigma + \alpha \cdot \nabla v_\sigma + \frac{1}{4\sigma}|\alpha|^2 + f\} = 0$$
$$v_\sigma(x, T) = g(x), \tag{19}$$

with a terminal condition at time $t = T$.

6.3. Optimal Control Potential and Optimally Controlled Dynamics

The optimal control u^* that minimizes the functional in Equation (13) is again of gradient form and given by

$$u_t^* = -2\sigma \nabla v_\sigma(X_t, t)$$

as can be readily checked by minimizing the corresponding expression in Equation (12) over α. Given v_σ, the *optimally controlled dynamics* reads

$$dX_t = -\nabla U(X_t, t)dt + \sqrt{2\epsilon}dW_t, \tag{20}$$

with the *optimal control potential*

$$U(x, t) = V(x) + 2\sigma v_\sigma(x, t). \tag{21}$$

In the case when $T \to \infty$ (Case I, above), the biasing potential is independent of t.

Remarks. Some remarks are in order.

(a) Monte-Carlo estimators of the conditional CGF

$$\gamma(\sigma; x) = -\sigma^{-1}\log \mathbf{E}[\exp(-\sigma W)|X_0 = x]$$

that are based on the optimally controlled dynamics have zero variance. This is so because the optimal control minimizes the variational expression in Equation (9), but at the minimum, the random variable inside the expectation must be almost surely constant (as a consequence of Jensen's inequality and the strict convexity of the exponential function). Hence, we have a *zero-variance estimator* of the conditional CGF.

(b) The reader may now wonder as to whether it is possible to extract single moments from the CGF (e.g., mean first passage times). In general, this question is not straightforward to answer. One of the difficulties is that extracting moments from the CGF requires one to take derivatives at $\sigma = 0$, but small values of σ imply strong penalization, which renders the control inactive and, thus, makes the approach inefficient. Another difficulty is that reweighting the controlled trajectories back to the original (equilibrium) path measure can increase the variance of a rare event estimator, as compared to the corresponding estimator based on the uncontrolled dynamics. As yet, the efficient calculation of moments from the CGF by either extrapolation methods or reweighing is an open question and currently a field of active research (see, e.g., [48,49]).

(c) Jarzynski's identity relates equilibrium free energies to averages that are taken over an ensemble of trajectories generated by controlled dynamics, and the reader may wonder whether the above zero-variance property can be used in connection with free energy computations à la Jarzynski (*cf.* [45]). Indeed, we can interpret the CGF as the free energy of the nonequilibrium work

$$W_\xi = \int_0^T f(X_t, \xi_t)\, dt$$

where f is the nonequilibrium force exerted on the system under driving it with some prescribed protocol $\xi\colon [0, T] \to \mathbb{R}$; in this case, the dynamics X_t depend on ξ_t, as well, and writing down the HJB equation according to Equation (19) is straightforward. However, even if we can solve Equation (19), we do not get zero-variance estimators for the free energy

$$F(\xi_T) - F(\xi_0) = -\beta^{-1} \log \mathbf{E}[\exp(-\beta W_\xi)]\,.$$

The reason for this is simple: Jarzynski's formula requires that the initial conditions are chosen from an equilibrium distribution, say, π_0 the equilibrium distribution corresponding to the initial value ξ_0 of the protocol, but optimal controls are defined point-wise for each state (t, X_t) and

$$-\beta^{-1} \log \int_{\mathbb{R}^n} \mathbf{E}[\exp(-\beta W_\xi)|X_0 = x]\, d\pi_0(x)$$

$$\neq -\beta^{-1} \int_{\mathbb{R}^n} \log \mathbf{E}[\exp(-\beta W_\xi)|X_0 = x]\, d\pi_0(x)\,.$$

In other words:

$$F(\xi_T) - F(\xi_0) \neq \int_{\mathbb{R}^n} v_\beta(x, 0)\, d\pi_0(x)\,.$$

(d) A similar argument as the one underlying the derivation of the HJB equation from the linear boundary value problem yields that Jarzynski's formula can be interpreted as a two-player zero-sum differential game (*cf.* [50]).

7. Characterize Rare Events by Optimally Controlled MD

Now, we illustrate how to use the results of the last section in practice. We will mainly consider the case discussed in Section 6.1 regarding the statistical characterization of hitting a certain set.

7.1. First Passage Times

Roughly speaking, the CGF encodes information about the moments of any random variable W that is a functional of the trajectories $(X_t)_{t \geq 0}$. For example, for $f = \epsilon$ and $T \to \infty$, we obtain the CGF of the first exit time from O, i.e.,

$$-\sigma^{-1} \log \mathbf{E}_x[\exp(-\sigma \tau_O)] = \min_u \mathbf{E}_x^u \left[\tau_O + \frac{1}{4\sigma} \int_0^{\tau_O} |u_t|^2 \, dt \right]$$

where we have introduced the shorthand $\mathbf{E}_x[\cdot] = \mathbf{E}[\cdot | X_0 = x]$ to denote the conditional expectation when starting at $X_0 = x$ and the superscript "u" to indicate that the expectation is understood with respect to the controlled dynamics

$$dX_t = (u_t - \nabla V(X_t)) \, dt + \sqrt{2\epsilon} dW_t$$

where $\mathbf{E} = \mathbf{E}^0$ denotes expectation with respect to the unperturbed dynamics.

7.2. Committor Probabilities Revisited

It is not only possible to use the moment generating function to collect statistics about rare events in terms of the cumulant generating function, but also to express the committor function directly in terms of an optimal control problem (see Section 2.1 for the definition of the committor q_{AB} between to sets A and B). To this end, let $\sigma = 1$, and suppose we divide ∂O into two sets $B \subset \partial O$ and $A = \partial O \setminus B$ (i.e., τ_O is the stopping time that is defined by hitting either A or B). Setting

$$f = 0 \quad \text{and} \quad g(x) = -\epsilon \log \mathbf{1}_B(x)$$

reduces the moment generating function (10) to

$$\psi_1(x) = \mathbf{E}_x[\mathbf{1}_B(X_{\tau_O})]$$

or, in more familiar terms,

$$\psi_1(x) = \mathbf{P}[X_{\tau_O} \in B \wedge X_{\tau_O} \notin A | X_0 = x] = q_{AB}(x).$$

According to Equation (16) the corresponding optimal control problem has the cost functional

$$J(u) = \mathbf{E} \left[\frac{1}{4} \int_0^{\tau_O} |u_s|^2 \, ds - \epsilon \log \mathbf{1}_B(X_{\tau_O}) \right],$$

which amounts to a control problem with zero terminal cost when ending up in B and an infinite terminal cost for hitting A. Therefore, the HJB equation for $v = v_1$ has a singular boundary value at A; it reads

$$\min_{\alpha \in \mathbb{R}^n} \{ Lv + \alpha \cdot \nabla v + \frac{1}{4} |\alpha|^2 \} = 0$$

$$v|_A = \infty, \quad v|_B = 0.$$

Setting $v(x) = -\epsilon \log q_{AB}(x)$ yields the equality

$$- \log q_{AB}(x) = \min_{u} \mathbf{E}_x^u \left[\frac{1}{4\epsilon} \int_0^{\tau_O} |u_s|^2 ds - \log \mathbf{1}_B(X_{\tau_O}) \right].$$

In this case, the optimally controlled dynamics (20) is of the form

$$dX_t = -\nabla U_{AB}(X_t)dt + \sqrt{2\epsilon}dW_t,$$

with optimal control potential

$$U_{AB}(x) = V(x) - 2\epsilon \log q_{AB}(x).$$

Remarks. Some remarks on the committor equation follow:

(a) The logarithmic singularity of the value function at "reactant state" A has the effect that the control will try to avoid running back into A, for there is an infinite penalty on hitting A. In other words, by controlling the system, we condition it on hitting the "product state" B at time $t = \tau_O$. Conditioning a diffusion (or general Markov) process on an exit state has a strong connection with Doob's h-transform, which can be considered a change-of-measure transformation of the underlying path measure that forces the diffusion to hit the exit state with probability one [51].

(b) The optimally controlled dynamics has a stationary distribution with a density proportional to

$$\exp(-\beta U_{AB}(x)) = q_{AB}^2(x) \exp(-\beta V(x))$$

where we used $\beta = 1/\epsilon$.

7.3. Algorithmic Realization

For the exit problem ("Case I", above), one can find an efficient algorithm for computing the conditional CGF $\gamma(\sigma; x)$ or, equivalently, the value function $v_\sigma(x)$ in [52]. The idea of the algorithm is to exploit that, according to Equations (20) and (21), the optimal control is of gradient form. The latter implies that the value function can be represented as a minimization of the cost functional over time-homogeneous candidate functions C for the optimal bias potential, in other words,

$$v_\sigma(x) = \min_{C} \mathbf{E}_x \left[\int_0^{\tau_O} \left\{ f(X_t) + \frac{1}{4\sigma} |\nabla C_t|^2 \right\} dt + g(X_{\tau_O}) \right] \tag{22}$$

where the expectation $\mathbf{E}$ is understood with respect to the path measure generated by

$$dX_t = -\left(\nabla C(X_t) + \nabla V(X_t) \right) dt + \sqrt{2\epsilon}dW_t.$$

Once the optimal C has been computed, both value function and CGF can be recovered by setting

$$v_\sigma(x) = -\frac{C(x)}{2\sigma} \quad \text{and} \quad \gamma(\sigma; x) = -\frac{C(x)}{2\epsilon\sigma}.$$

242

The algorithm that finds the optimal C works by iteratively minimizing the cost functional for potentials C from a finite-dimensional ansatz space, *i.e.*,

$$C(x) = \sum_{j=1}^{M} a_j \varphi_j(x) \,,$$

with appropriately chosen ansatz functions φ_j. The iterative minimization is then carried out on the M-dimensional coefficient space of the $a_1, \ldots, a_M$. With this algorithm, we are able to compute the optimal control potential for the exit problem in the two interesting cases: first passage times and committor probabilities (as outlined in Sections 7.1 and 7.2).

Remarks. Let us briefly comment on some aspects of the gradient descent algorithm.

(a) The minimization algorithm for the value function belongs to the class of expectation-maximization algorithms (although, here, we carry out a minimization rather than a maximization), in that each minimization step is followed by a function evaluation that involves computing an expectation. In connection with rare events sampling and molecular dynamics problems, a close relative is the *adaptive biasing force* (ABF) method for computing free energy profiles, the latter being intimately linked with cumulant generating functions or value functions (*cf.* Section 5). In ABF methods (or its variants, such as *metadynamics* or *Wang–Landau dynamics*), the gradient of the free energy is estimated on the fly, running a molecular dynamics simulation, and then added as a biasing force to accelerate the sampling in the direction of the relevant coordinates [53,54]. The biasing force eventually converges to the derivative of the free energy, which is the optimal bias for passing over the relevant energy barriers that are responsible for the rare events [55].

(b) The number of basis functions needed depends mainly on the roughness of the value function, but is independent of the system dimension. For systems with clear time scale separation, it has been moreover shown [56] that the optimal control is independent of the fast variables; hence, we expect that the algorithm can be efficient, even for large-scale systems, provided that some information about the relevant collective variables and a reasonable initial guess are available. Yet, the question remains: How many basis functions are needed to approximate the optimal control up to a given accuracy? Controlling the error in the value function and the resulting optimal control is particularly important, as a wrong (e.g., suboptimal) bias potential may lead to Monte-Carlo estimators that may have a larger variance than the vanilla rare event estimator, as has been pointed out in [57,58]. The first results in this direction have been obtained in [59], in which error bounds for the CGF for suboptimal controls have been derived, and [60], which discusses the approximation error of the Galerkin approximation of the corresponding HJB equations; see also [61] for a related discussion regarding so-called log-efficient estimators for rare events.

7.4. Numerical Examples

In our first example, we consider diffusive molecular dynamics as in Equation (2) with $\epsilon = 0.1$ and V being the five-well potential shown in Figure 2. We first compute the CGF of the first passage time as discussed in Section 7.1, using the gradient descent algorithm described in Section 7.3 with 10 Gaussian ansatz functions that are centered around the critical points of the potential energy function. The resulting optimal control potential (21) after roughly 20 iterations of the gradient descent is displayed in Figure 2 for different values of σ. As the set O, we take the whole state space, except a small neighborhood of its global minimum of V, so that its complement O^c is identical to the vicinity of the global minimum and the exit time τ_O is the first passage time to O^c. Figure 2 shows that the optimal control potential alters the original potential V significantly in the sense that for $\sigma > 0$, the set O^c is the bottom of the only well of the potential, so that all trajectories starting somewhere else will quickly enter O^c.

Figure 2. Five-well potential (**left**) and associated optimal control potential for the first passage time to the target set O^c given by a small interval around the main minimum x_1 (**right**) for different values of σ (**right**). $\epsilon = 0.1$; the gradient descent solution fully agrees with the reference finite element solution (that is not shown) in the "eye-norm".

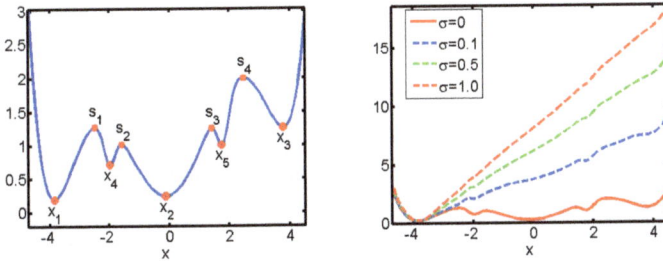

This case is instructive: For the unperturbed original dynamics, the mean first passage time $\mathbf{E}_x(\tau_O)$ takes values of around 10^4 for $x > -2$. For the optimally controlled dynamics, the mean first passage times into O^c are less than five for $\sigma = 0.1, 0.5, 1.0$, so that the estimation of $\mathbf{E}_x(\tau_O)$ resulting from the optimal control approach requires trajectories that are a factor of at least 10^3 shorter than the ones we would have to use by direct numerical simulation of the unperturbed dynamics.

Figure 3 shows the optimal control potentials for computation of the committor q_{AB}, as described in Section 7.2. We observe that the optimal control potential exhibits a singularity at the boundary of the basin of attraction of the set A. That is, it prevents the optimally controlled dynamics from entering the basin of attraction of A and, thus, avoids the waste of computational effort by unproductive returns to A.

In our second example, we consider two-dimensional diffusive molecular dynamics as in Equation (2) with the energy landscape V being the three-well potential shown in Figure 1. In Figure 4, the optimal control potential for computing the committors q_{AB} between the two main wells for two different temperatures $\epsilon = 0.15$ and $\epsilon = 0.6$ are displayed. The numerical solution is based on a Galerkin approximation of the log-transformed HJB equation, using precomputed committor functions as the basis set; see [60] for details.

Figure 3. Optimally-corrected potential for the case of J being the committor q_{AB} for B being the ± 0.1-interval around the main minimum x_1 of the potential. (**Left panel**) $A =]x_3 - 0.1, x_3 + 0.1[$ the ± 0.1 interval around the highest minimum x_3. (**Right panel**) $A =]x_2 - 0.1, x_2 + 0.1[$ the ± 0.1 interval around the second lowest minimum x_2.

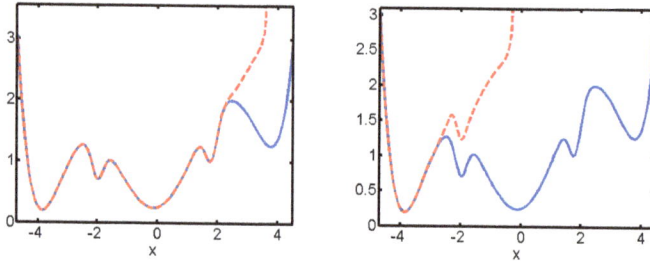

As in our former experiment, we observe that the optimal control potential prevents the dynamics from returning to A; in addition, it flattens the third well significantly, such that the optimally controlled dynamics in any case quickly goes into B. For $\epsilon = 0.15$, a TPS sampling of reactive trajectories between the two main wells, precisely from A to B with A and B, as indicated in Figure 4, results in an average length of 367 for reactive trajectories based on the original dynamics. For the optimally controlled dynamics, we found an average length of 1.3.

Figure 4. Optimally-corrected potential for the three-well potential shown in Figure 1 for the committor q_{AB} for the medium temperature $\epsilon = 0.6$ case (**left**), the low temperature $\epsilon = 0.15$ case (**right**) and for the sets A (ellipse in main well, right-hand side) and B (ellipse in main well, left-hand side). Note that the committor basis is not smooth at the boundaries of the initial and target sets (see Figure 1 for comparison), which explains the roughness of the control potential in the neighborhood of the sets A and B.

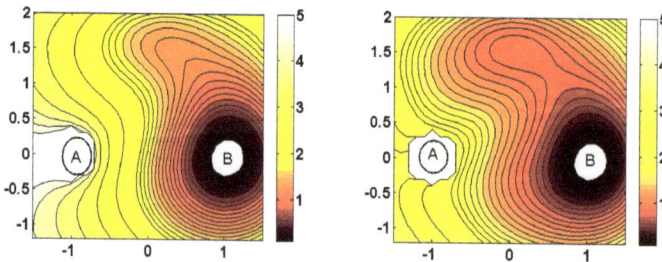

8. Conclusions

We have surveyed various techniques for the characterization and computation of rare events occurring in molecular dynamics. Roughly, the approaches fall into two categories: (a) methods that approach the problem by characterizing the ensemble of reactive trajectories between metastable states or (b) path-based methods that target dominant transition channels or pathways by minimization of suitable action functionals. Methods of the first type, e.g., Transition Path Theory, Transition Path Sampling, Milestoning or variants thereof, are predominantly Monte-Carlo-type methods for generating one very long or many short trajectories, from which the rare event statistics can then be estimated. Methods that belong to the second category, e.g., MaxFlux, Nudged-Elastic Band or the String Method, are basically optimization methods (sometimes combined with a Monte-Carlo scheme); here, the objectives are few (single or multiple) smooth pathways that describe, e.g., a transition event. It is clear that this classification is not completely unambiguous, in that action-based methods for computing most probable pathways can be also used to sample an ensemble of reactive trajectories. Another possible classification (with its own drawbacks) is along the lines of the biased-unbiased dichotomy that distinguishes between methods that characterize rare events based on the original dynamics and methods that bias the underlying equilibrium distribution towards a new probability distribution under which the rare events are no longer rare. Typical representatives of the second class range from biasing force methods, such as ABF or metadynamics, up to genuine nonequilibrium approaches based on Jarzynski's identity for computing free energy profiles. The problem often is that rare event estimators based on an ensemble of nonequilibrium trajectories suffer from large variances, unless the bias is cleverly chosen.

We have described a strategy to find such a cleverly chosen perturbation, based on ideas from optimal control. The idea rests on the fact that the cumulant generating function of a certain observable, e.g., the first exit time from a metastable set, can be expressed as the solution to an optimal control problem, which yields a zero variance estimator for the cumulant generating function. The control acting on the system has essentially two effects: (1) Under the controlled dynamics, the rare events are no longer rare, as a consequence of which the simulations become much shorter; (2) The variance of the statistical estimators is small (or even zero if the optimal control is known exactly). We should stress that, depending on the type of observable, the approach only appears to be a nonequilibrium method, for the optimal control is an exact gradient of a biasing potential; hence, the optimally perturbed system satisfies a detailed balance, which is one criterion for thermodynamic equilibrium. Future research should address the question as to whether the approach is competitive for realistic molecular systems, how to efficiently and robustly extract information about specific moments rather than cumulant generating functions and how to extend it to the more general observables or the calculation of free energy profiles.

Acknowledgments

The authors are grateful to Eric Vanden-Eijnden, Giovanni Ciccotti, Frank Pinski and Christoph Dellago for valuable discussions and comments. Ralf Banisch and Tomasz Badowski hold scholarships from the Berlin Mathematical School (BMS). This work was supported by the DFG Research Center "Mathematics for key technologies" (MATHEON) in Berlin.

Conflicts of Interest

The authors declare no conflict of interest.

References

1. Eyring, H. The activated complex in chemical reactions. *J. Chem. Phys.* **1935**, *3*, 107–115.
2. Wigner, E. Calculation of the rate of elementary association reactions. *J. Chem. Phys.* **1937**, *5*, 720–725.
3. Bolhuis, P.G.; Chandler, D.; Dellago, C.; Geissler, P. Transition path sampling: Throwing ropes over rough mountain passes, in the dark. *Annu. Rev. Phys. Chem.* **2002**, *59*, 291–318.
4. Weinan, E.; Ren, W.; Vanden-Eijnden, E. String method for the study of rare events. *Phys. Rev. B* **2002**, *66*, 052301.
5. Beccara, S.; Skrbic, T.; Covino, R.; Faccioli, P. Dominant folding pathways of a WW domain. *Proc. Natl. Acad. Sci. USA* **2012**, *109*, 2330–2335.
6. Faccioli, P.; Lonardi, A.; Orland, H. Dominant reaction pathways in protein folding: A direct validation against molecular dynamics simulations. *J. Chem. Phys.* **2010**, *133*, 045104.
7. Pinski, F.; Stuart, A. Transition paths in molecules: Gradient descent in path space. *J. Chem. Phys.* **2010**, *132*, 184104.
8. Allen, R.; Warren, P.; ten Wolde, P.R. Sampling rare switching events in biochemical networks. *Phys. Rev. Lett.* **2005**, *94*, 018104.
9. Moroni, D.; van Erp, T.; Bolhuis, P. Investigating rare events by transition interface sampling. *Physica A* **2004**, *340*, 395–401.
10. Faradjian, A.K.; Elber, R. Computing time scales from reaction coordinates by milestoning. *J. Chem. Phys.* **2004**, *120*, 10880–10889.
11. Olender, R.; Elber, R. Calculation of classical trajectories with a very large time step: Formalism and numerical examples. *J. Chem. Phys.* **1996**, *105*, 9299–9315.
12. Jarzynski, C. Nonequilibrium equality for free energy differences. *Phys. Rev. Lett.* **1997**, *78*, 2690–2693.
13. Crooks, G. Entropy production fluctuation theorem and the nonequilibrium work relation for free energy differences. *Phys. Rev. E* **1999**, *60*, 2721–2726.
14. Sarich, M.; Bansich, R.; Hartmann, C.; Schütte, C. Markov state models for rare events in molecular dynamics. *Entropy* **2014**, *16*, 258–286.
15. Allen, R.J.; Valeriani, C.; Tanase-Nicola, S.; ten Wolde, P.R.; Frenkel, D. Homogeneous nucleation under shear in a two-dimensional Ising model: Cluster growth, coalescence, and breakup. *J. Chem. Phys.* **2008**, *129*, 134704.
16. Berryman, J.T.; Schilling, T. Sampling rare events in nonequilibrium and nonstationary systems. *J. Chem. Phys.* **2010**, *133*, 244101.

17. Weinan, E.; Vanden-Eijnden, E. Towards a theory of transition paths. *J. Stat. Phys.* **2006**, *123*, 503–523.

18. Weinan, E.; Vanden-Eijnden, E. Metastability, Conformation Dynamics, and Transition Pathways in Complex Systems. In *Multiscale Modeling and Simulation*; Springer: Berlin, Germany, 2004; pp. 35–68.

19. Metzner, P.; Schütte, C.; Vanden-Eijnden, E. Illustration of transition path theory on a collection of simple examples. *J. Chem. Phys.* **2006**, *125*, 084110.

20. Weinan, E.; Vanden-Eijnden, E. Transition-path theory and path-finding algorithms for the study of rare events. *Annu. Rev. Phys. Chem.* **2010**, *61*, 391–420.

21. Metzner, P.; Schütte, C.; Vanden-Eijnden, E. Transition path theory for Markov jump processes. *Multiscale Model. Sim.* **2009**, *7*, 1192–1219.

22. Metzner, P. Transition Path Theory for Markov Processes: Application to Molecular Dynamics. Ph.D. Thesis, Freie Universität Berlin, Berlin, Germany, 2007.

23. Noé, F.; Schütte, C.; Vanden-Eijnden, E.; Reich, L.; Weikl, T. Constructing the full ensemble of folding pathways from short off-equilibrium trajectories. *Proc. Natl. Acad. Sci. USA* **2009**, *106*, 19011–19016.

24. Chandler, D. *Finding Transition Pathways: Throwing Ropes over Rough Montain Passes, in the Dark*; World Scientific: Singapore, Singapore, 1998.

25. Dürr, D.; Bach, A. The Onsager-Machlup function as Lagrangian for the most probable path of a diffusion process. *Commun. Math. Phys.* **1978**, *60*, 153–170.

26. Berkowitz, M.; Morgan, J.D.; McCammon, J.A.; Northrup, S.H. Diffusion-controlled reactions: A variational formula for the optimum reaction coordinate. *J. Chem. Phys.* **1983**, *79*, 5563–5565.

27. Huo, S.; Straub, J.E. The MaxFlux algorithm for calculating variationally optimized reaction paths for conformational transitions in many body systems at finite temperature. *J. Chem. Phys.* **1997**, *107*, 5000–5006.

28. Zhao, R.; Shen, J.; Skeel, R.D. Maximum flux transition paths of conformational change. *J. Chem. Theory Comput.* **2010**, *6*, 2411–2423.

29. Cameron, M. Estimation of reactive fluxes in gradient stochastic systems using an analogy with electric circuits. *J. Comput. Phys.* **2013**, *247*, 137–152.

30. Vanden-Eijnden, E.; Heymann, M. The geometric minimum action method for computing minimum energy paths. *J. Chem. Phys.* **2008**, *128*, 061103.

31. Jonsson, H.; Mills, G.; Jacobsen, K. Nudged Elastic Band Method for Finding Minimum Energy Paths of Transitions. In *Classical and Quantum Dynamics in Condensed Phase Simulations*; Berne, B.J., Ciccotti, G., Coker, D.F., Eds.; World Scientific: Singapore, Singapore, 1998.

32. Freidlin, M.; Wentzell, A.D. *Random Perturbations of Dynamical Systems*; Springer, New York, NY, USA, 1998.

33. Weinan, E.; Ren, W.; Vanden-Eijnden, E. Finite temperature string method for the study of rare events. *J. Phys. Chem. B* **2005**, *109*, 6688–6693.

34. Ren, W.; Vanden-Eijnden, E.; Maragakis, P; Weinan, E. Transition pathways in complex systems: Application of the finite-temperature string method to the alanine dipeptide. *J. Phys. Chem.* **2005**, *123*, 134109.

35. Vanden-Eijnden, E.; Venturoli, M. Revisiting the finite temperature string method for the calculation of reaction tubes and free energies. *J. Phys. Chem.* **2009**, *130*, 194103.

36. Allen, R.; Frenkel, D.; ten Wolde, P.R. Simulating rare events in equilibrium or nonequilibrium stochastic systems. *J. Chem. Phys.* **2006**, *124*, 024102.

37. Allen, R.; Frenkel, D.; ten Wolde, P.R. Forward flux sampling-type schemes for simulating rare events: Efficiency analysis. *J. Chem. Phys.* **2006**, *124*, 194111.

38. Becker, N.B.; Allen, R.J.; ten Wolde, P.R. Non-stationary forward flux sampling. *J. Chem. Phys.* **2012**, *136*, 174118.

39. Kratzer, K.; Arnold, A.; Allen, R.J. Automatic, optimized interface placement in forward flux sampling simulations. *J. Chem. Phys.* **2013**, *138*, 164112.

40. Vanden-Eijnden, E.; Venturoli, M.; Ciccotti, G.; Elber, R. On the assumptions underlying milestoning. *J. Chem. Phys.* **2008**, *129*, 174102.

41. Latorre, J.; Hartmann, C.; Schütte, C. Free energy computation by controlled Langevin processes. *Procedia Comput. Sci.* **2010**, *1*, 1591–1600.

42. Lelièvre, T.; Stoltz, G.; Rousset, M. *Free Energy Computations: A Mathematical Perspective*; Imperial College Press: London, UK, 2010.

43. Isralewitz, B.; Gao, M.; Schulten, K. Steered molecular dynamics and mechanical functions of proteins. *Curr. Opin. Struct. Biol.* **2001**, *11*, 224–230.

44. Vaikuntanathan, S.; Jarzynski, C. Escorted free energy simulations: Improving convergence by reducing dissipation. *Phys. Rev. Lett.* **2008**, *100*, 190601.

45. Oberhofer, H.; Dellago, C. Optimum bias for fast-switching free energy calculations. *Comput. Phys. Commun.* **2008**, *179*, 41–45.

46. Dai Pra, P.; Meneghini, L.; Runggaldier, W. Connections between stochastic control and dynamic games. *Math. Control Signals Syst.* **1996**, *9*, 303–326.

47. Fleming, W.; Soner, H. *Controlled Markov Processes and Viscosity Solutions*; Springer: New York, NY, USA, 2006.

48. Awad, H.; Glynn, P.; Rubinstein, R. Zero-variance importance sampling estimators for markov process expectations. *Math. Oper. Res.* **2013**, *38*, 358–388.

49. Badowski, T. Importance Sampling Using Discrete Girsanov Estimators. Ph.D. Thesis, Freie Universität Berlin, Berlin, Germany, **2013**, in preparation.

50. Fleming, W. Risk sensitive stochastic control and differential games. *Commun. Inf. Syst.* **2006**, *6*, 161–178.

51. Day, M. Conditional exits for small noise diffusions with characteristic boundary. *Ann. Probab.* **1992**, *20*, 1385–1419.

52. Hartmann, C.; Schütte, C. Efficient rare event simulation by optimal nonequilibrium forcing. *J. Stat. Mech. Theor. Exp.* **2012**, *2012*, P11004.

53. Darve, E.; Rodriguez-Gomez, D.; Pohorille, A. Adaptive biasing force method for scalar and vector free energy calculations. *J. Chem. Phys.* **2008**, *128*, 144120.

54. Lelièvre, T.; Rousset, M.; Stoltz, G. Computation of free energy profiles with parallel adaptive dynamicss. *J. Chem. Phys.* **2007**, *126*, 134111.

55. Lelièvre, T.; Rousset, M.; Stoltz, G. Long-time convergence of an adaptive biasing force methods. *Nonlinearity* **2008**, *21*, 1155–1181.

56. Zhang, W.; Latorre, J.; Pavliotis, G.; Hartmann, C. Optimal control of multiscale diffusions using reduced-order models. *J. Comput. Dynam.* **2013**, submitted.

57. Dupuis, P.; Wang, H. Importance sampling, large deviations, and differential games. *Stochast. Stochast. Rep.* **2004**, *76*, 481–508.

58. Dupuis, P.; Wang, H. Subsolutions of an isaacs equation and efficient schemes for importance sampling. *Math. Oper. Res.* **2007**, *32*, 723–757.

59. Zhang, W.; Hartmann, C.; Weber, M.; Schütte, C. Importance sampling in path space for diffusion processes. *Multiscale Model. Sim.* **2013**, submitted.

60. Banisch, R.; Hartmann, C. Meshless Discretizations of LQ-type stochastic control problems. *SIAM J. Control Optim.* **2013**, submitted.

61. Vanden-Eijnden, E.; Weare, J. Rare event simulation of small noise diffusions. *Commun. Pure Appl. Math.* **2012**, *65*, 1770–1803.

Reprinted from *Entropy*. Cite as: Sarich, M.; Banisch, R.; Hartmann, C.; Schütte, C. Markov State Models for Rare Events in Molecular Dynamics. *Entropy* **2014**, *16*, 258–286.

Article

Markov State Models for Rare Events in Molecular Dynamics

Marco Sarich [1,*]**, Ralf Banisch** [1]**, Carsten Hartmann** [1] **and Christof Schütte** [1,2]

[1] Department of Mathematics and Computer Science, Freie Universität Berlin, Arnimallee 6,
   Berlin 14195, Germany; E-Mails: ralf.banisch@fu-berlin.de (R.B.);
   chartman@mi.fu-berlin.de (C.H.); christof.schuette@fu-berlin.de (C.S.)
[2] Zuse Institute Berlin, Takustr. 7, Berlin 14195, Germany

* Author to whom correspondence should be addressed; E-Mail: sarich@math.fu-berlin.de;
   Tel.: +49-30-838-75-322.

Received: 18 September 2013; in revised form: 3 December 2013 / Accepted: 9 December 2013 / Published: 30 December 2013

Abstract: Rare, but important, transition events between long-lived states are a key feature of many molecular systems. In many cases, the computation of rare event statistics by direct molecular dynamics (MD) simulations is infeasible, even on the most powerful computers, because of the immensely long simulation timescales needed. Recently, a technique for spatial discretization of the molecular state space designed to help overcome such problems, so-called Markov State Models (MSMs), has attracted a lot of attention. We review the theoretical background and algorithmic realization of MSMs and illustrate their use by some numerical examples. Furthermore, we introduce a novel approach to using MSMs for the efficient solution of optimal control problems that appear in applications where one desires to optimize molecular properties by means of external controls.

Keywords: rare events; Markov State Models; long timescales; optimal control

1. Introduction

Stochastic processes are widely used to model physical, chemical or biological systems. The goal is to approximately compute interesting properties of the system by analyzing the stochastic model. As soon as randomness is involved, there are mainly two options for performing this analysis: (1) Direct sampling and (2) the construction of a discrete coarse-grained model of the system. In a direct sampling approach, one tries to generate a statistically significant amount of events that characterize the property of the system one in which is interested. For this purpose, computer simulations of the model are a powerful tool. For example, an event could refer to the transition between two well-defined macroscopic states of the system. In chemical applications, such transitions can often be interpreted as reactions or, in the context of a molecular system, as conformational changes. Interesting properties are, e.g., average waiting times for such reactions or conformational changes and along which pathways the transitions typically occur. The problem with a direct sampling approach is that many interesting events are so-called rare events. Therefore, the computational effort for generating sufficient statistics for reliable estimates is very high, and particularly if the state space is continuous and high dimensional, estimation by direct numerical simulation is infeasible.

Available techniques for rare event simulations in continuous state space are discussed in [1]. In this article, we will discuss approach (2) to the estimation of rare event statistics via *discretization* of the state space of the system under consideration. That is, instead of dealing with the computation of rare events for the original, continuous process, we will approximate them by a so-called Markov State Model (MSM) with *discrete* finite state space. The reason is that for such a discrete model, one can numerically compute many interesting properties without simulation, mostly by solving linear systems of equations as in discrete transition path theory (TPT) [2]. We will see that this approach, called Markov State Modeling, *avoids* the combinatorial explosion of the number of discretization elements with the increasing size of the molecular system in contrast to other methods for spatial discretization.

The actual construction of an MSM requires one to sample certain transition probabilities of the underlying dynamics between sets. The idea is: (1) to choose the sets such that the sampling effort is much lower than the direct estimation of the rare events under consideration; and (2) to compute all interesting quantities for the MSM from its transition matrix, *cf.* [2,3]. There are many examples for the successful application of this strategy. In [4], for example, it was used to compute dominant folding pathways for the PinWW domain in explicit solvent. However, we have to make sure that the Markov State Model approximates the original dynamics well enough. For example, the MSM should correctly reproduce the timescales of the processes of interest. These approximation issues have been discussed since more than a decade now [5,6]; in this article, we will review the present state of research on this topic. In the algorithmic realization of Markov State Modeling for realistic molecular systems, the transition probabilities and the respective statistical uncertainties are estimated from short molecular dynamics (MD) trajectories only, *cf.* [7]. This makes Markov State Modeling applicable to many different molecular systems and processes, *cf.* [8–13].

In the first part of this article, we will discuss the approximation quality of two different types of Markov State Models that are defined with respect to a full partition of state space or with respect to so-called core sets. We will also discuss the algorithmic realization of MSMs and provide references to the manifold of realistic applications to molecular systems in equilibrium that are available in the literature today.

The second part will show how to use MSMs for optimizing particular molecular properties. In this type of application, one wants to steer the molecular system at hand by external controls in a way such that a pre-selected molecular property is optimized (minimized or maximized). That is, one wants to compute a specific external control from a family of admissible controls that optimizes the property of interest under certain side conditions. The property to be optimized can be quite diverse: For example, it can be (1) the population of a certain conformation that one wants to maximize under a side condition that limits the total work done by the external control or (2) the mean first passage time to a certain conformation that one wants to minimize (in order to speed up a rare event), but under the condition that one can still safely estimate the mean first passage time of the uncontrolled system. The theoretical background of case (1) has been considered in [14], for example, and of case (2) in [1,15]. There, one finds the mathematical problem that has to be solved in order to compute the optimal control. Here, we will demonstrate that one can use MSMs for the efficient solution of such a mathematical problem (for both cases). We will see that the spatial discretization underlying an MSM turns the high-dimensional continuous optimal control problem into a rather low-dimensional discrete optimal control problem of the same form that can be solved efficiently. Based on these insights, MSM discretization yields an efficient algorithm for solving the optimal control problem, whose performance we will outline in some numerical examples, including an application to alanine dipeptide.

2. MSM Construction

Let $(X_t)_{t \geq 0}$ be a time-continuous Markov process on a continuous state space, E, e.g., $E \subset R^d$. That is, X_t is the state of the molecular system at time t resulting from any usually used form of molecular dynamics simulation, be it based on Newtonian dynamics with thermostats or resulting from Langevin dynamics or other diffusion molecular dynamics models. The idea of Markov State Modeling is to derive a Markov chain, $(\hat{X}_k)_{k \in \mathbb{N}}$, on a finite and preferably small state space $\hat{E} = \{1, ..., n\}$ that models characteristic dynamics of the continuous process, (X_t). For example, in molecular dynamics applications, such characteristic dynamics could refer to protein folding processes [16,17], conformational rearrangements between native protein substates [18,19], or ligand binding processes [20]. Since the approximating Markov chain, $(X_k)_{k \in \mathbb{N}}$, lives on a finite state space, the construction of an MSM boils down to the computation of its transition matrix, P:

$$P_{ij} = \mathbb{P}[\hat{X}_{k+1} = j | \hat{X}_k = i] \tag{1}$$

The main benefit is that for a finite Markov chain, one can compute many interesting dynamical properties directly from its transition matrix, e.g., timescales and the metastability in the system [5,21,22], a hierarchy of important transition pathways [2] or mean first passage times between selected states. With respect to an MSM, these computations should be used afterwards to answer related questions for the original continuous process. To do this, we must be able to link the states of the Markov chain back to the spatial information of the original process, and the approximation of the process (X_t) by the MSM must be valid in some sense.

Having this in mind, the first natural idea is to let the states of an MSM correspond to sets $A_1, ..., A_n \subset E$ in continuous state space that form a full partition, *i.e.,*:

$$A_i \cap A_j = \emptyset \text{ for } i \neq j, \qquad \bigcup_{i=1}^{n} A_i = E \qquad (2)$$

Typical choices for such sets are box discretizations or Voronoi tessellations [23]. For such a full partition, it is trivial to also define a corresponding discretized process by the original switching dynamics between the sets. For a given lag time, $\tau > 0$, we can define the index process:

$$\tilde{X}_k = i \Leftrightarrow X_{k\tau} \in A_i \qquad (3)$$

It is well known that this process is not Markovian, mainly due to the so-called recrossing problem. This refers to the fact that the original process typically crosses the boundary between two sets, A_i and A_j, several times when transitions take place, as illustrated in Figure 1. This results in cumulative transitions between indices i and j for the index process, that is, a not memoryless transition behavior.

Figure 1. Cumulative transitions between two sets along boundaries are typical.

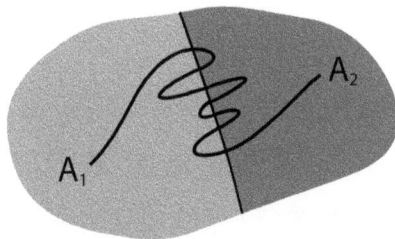

The non-Markovianity of the index process is often seen as a problem in Markov State Modeling, because many arguments assume that $\tilde{X}_k$ is a Markov process. In this article, we will *not* make this assumption. We interpret the process $(\tilde{X}_k)$ as a tool to construct the following transition matrix, P^τ:

$$P_{ij}^\tau = \mathbb{P}[\tilde{X}_{k+1} = j | \tilde{X}_k = i] = \mathbb{P}[X_{(k+1)\tau} \in A_j | X_{k\tau} \in A_i] \qquad (4)$$

and, hence, the MSM as the Markov chain, $(\hat{X}_k)_{k \in \mathbb{N}}$, associated with this transition matrix. From above, it is clear that, in general, we have $\hat{X}_k \neq \tilde{X}_k$, and in [24] it was analyzed how these two processes relate in terms of density propagation. In the following, we will show under which assumptions and in which sense the MSM $(\hat{X}_k)$ will be a good approximation of the original dynamics given by (X_t). For convenience, we will usually write $P^\tau \equiv P$ and leave the τ-dependence implicit.

3. Analytical Results

In order to compare the MSM to the continuous process, we introduce one of the key objects for our analysis, the *transfer operator* of a Markov process. We assume that the Markov process (X_t)

has a unique, positive invariant probability measure, μ, and that it is time-reversible. Then, for any time-step, $t \geq 0$, we define the transfer operator, T_t, via the property:

$$\int_A T_t v(y)\mu(dy) = \int_E v(x)p(t, x, A)\mu(dx) \qquad \text{for all measurable } A \qquad (5)$$

as an operator $T_t : L^2(\mu) \to L^2(\mu)$. Here, $p(t, x, A) = \mathbb{P}[X_t \in A | X_0 = x]$ defines the transition probability measure and $L^2(\mu)$ denotes the Hilbert space of functions v with:

$$\int_E v(y)^2 \mu(dy) \leq \infty \qquad (6)$$

and the scalar product:

$$\langle v, w \rangle = \int_E v(y)w(x)\mu(dy) \qquad (7)$$

Note that T_t is nothing else other than the propagator of densities under the dynamics, but the densities are understood as densities with respect to the measure, μ. That is, if the Markov process is initially distributed according to:

$$\mathbb{P}[X_0 \in A] = \int_A v_0(x)\mu(dx) \qquad (8)$$

its probability distribution at time t is given by:

$$\mathbb{P}[X_t \in B] = \int_B v_t(x)\mu(dx), \qquad v_t = T_t v_0 \qquad (9)$$

The benefit of working with μ-weighted densities is that the transfer operator, T_t, becomes essentially self-adjoint on $L^2(\mu)$ for all cases of molecular dynamics satisfying some form of detailed balance condition. Hence, it has real eigenvalues and orthogonal eigenvectors with respect to Equation (7) (or, at least, the dominant spectral elements are real-valued). Moreover, the construction of an MSM can be seen as a projection of the transfer operator [25]. Assume Q is an orthogonal projection in $L^2(\mu)$ onto an n-dimensional subspace, $D \subset L^2(\mu)$, with $\mathbb{1} \in D$, and $\chi_1, ..., \chi_n$ is a basis of D. Then, the so-called projected transfer operator, $QT_\tau Q : D \to D$, has the matrix representation:

$$P_Q = PM^{-1} \qquad (10)$$

with the non-negative, invertible mass matrix, $M \in \mathbb{R}^{n,n}$, with entries:

$$M_{ij} = \frac{\langle \chi_i, \chi_j \rangle}{\langle \chi_i, \mathbb{1} \rangle} \qquad (11)$$

The matrix, $P \in \mathbb{R}^{n,n}$, is also non-negative and has entries:

$$P_{ij} = \frac{\langle \chi_i, T_\tau \chi_j \rangle}{\langle \chi_i, \mathbb{1} \rangle} \qquad (12)$$

Full Partition MSM. If we choose $\chi_i = \mathbb{1}_{A_i}$ to be the characteristic function of set A_i for $i = 1, ..., n$, one can easily check that we get $M = I$ to be the identity matrix and:

$$P_{ij} = \mathbb{P}_\mu[X_\tau \in A_j | X_0 \in A_i] \tag{13}$$

as in Equation (4). The subscript, μ, shall indicate that $X_0 \sim \mu$. Therefore, the transition probabilities are evaluated along equilibrium paths.

The previously constructed transition matrix of the MSM based on a full partition can be interpreted as a projection onto a space of densities that are constant on the partitioning sets. This interpretation of an MSM is useful, since it allows one to analyze its approximation quality. For example, in [25,26], it is proven that we can reproduce an eigenvalue, λ, of a self-adjoint transfer operator, T_t, by the MSM by choosing the subspace appropriately. That is, if u is a corresponding normalized eigenvector, Q the orthogonal projection to a subspace, D, with $\mathbb{1} \in D$, then there exists an eigenvalue, $\hat{\lambda}$, of the projected transfer operator, QT_tQ, with:

$$|\lambda - \hat{\lambda}| \leq \lambda_1 \delta(1 - \delta^2)^{-\frac{1}{2}}$$

where $\lambda_1 < 1$ is the largest non-trivial eigenvalue of T_t and $\delta = \|u - Qu\|$.

In particular, for $\delta \leq \frac{3}{4}$, one can simplify the equation to:

$$|\lambda - \hat{\lambda}| \leq 2\lambda_1\delta \tag{14}$$

An eigenvalue, λ_i, of the transfer operator directly relates to an implied timescale, $\mathcal{T}_i$, of the system via:

$$\mathcal{T}_i = -\frac{\tau}{\log(\lambda_i)} \tag{15}$$

Therefore, the transition matrix Equation (4) that we construct from transitions between the sets, $A_1, ..., A_n$, will generate a Markov chain that will reproduce the original timescales well if the partitioning sets are chosen such that the corresponding eigenvectors are almost constant on these sets. In this case, $\delta = \|u - Qu\|$; that is, the approximation error of the eigenvector by a piecewise constant function on the sets will be small.

The projection error, δ, depends on our choice of the discretizing sets. As an example, let us consider a diffusion in the potential that is illustrated in Figure 2, that is, the reversible Markov process given by the stochastic differential equation:

$$dX_t = -\nabla V(X_t)dt + \sqrt{2\varepsilon}dB_t \tag{16}$$

where V is the potential, B_t denotes a Brownian motion and $\varepsilon > 0$.

The figure also shows a choice of three sets that form a full partition of state space. The computation of the transition matrix Equation (4) for $\sigma = 0.7$ and a lag time $\tau = 1$ yields:

$$P_Q = P = \begin{pmatrix} 0.9877 & 0.0123 & 0.0000 \\ 0.0420 & 0.9160 & 0.0419 \\ 0.0000 & 0.0123 & 0.9877 \end{pmatrix}$$

that has three eigenvalues $\lambda_0 = 1, \lambda_1 = 0.9877, \lambda_2 = 0.9037$. Table 1 shows the two resulting implied timescales Equation (15) in comparison to the timescales of the original system.

Figure 2. A potential with three wells and a choice of three sets, A_1, A_2, A_3.

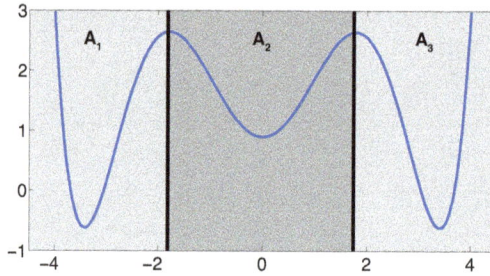

Table 1. Comparison of implied timescales

	T_1	T_2
Original	103.7608	11.9566
Full partition 3 sets	80.6548	9.8784

As one can see, the timescales are strongly underestimated. This is a typical phenomenon. From a statistical point of view, the recrossing problem will lead to cumulatively appearing transition counts when one computes the transition probabilities, $\mathbb{P}_\mu[X_\tau \in A_j | X_0 \in A_i]$, from a trajectory (X_t), as discussed above. Therefore, on average, transitions between sets seem to become too likely, and hence, the processes in the coarse-grained system get accelerated. We have seen in Equation (14) that this cannot happen if the associated eigenvectors can be approximated well by the subspace that corresponds to the MSM. Figure 3 shows the first non-trivial eigenvector, u_1, belonging to the timescale $T_1 = 103.7608$ and its best-approximation by a step function.

Figure 3. The first non-trivial eigenvector, u_1 (solid blue), and its projection, Qu_1 (dashed red), onto step functions that are constant on A_1, A_2, A_3.

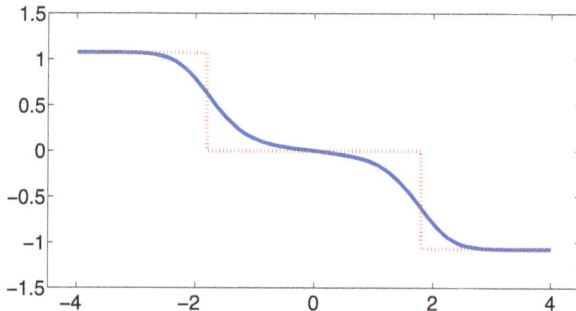

The eigenvector is indeed almost constant in the vicinity of the wells, but within the transition region between the wells, the eigenvector is varying and the approximation by a step function is not accurate. Therefore, we have two explanations of why the main error is introduced in the region close to shared boundaries of neighboring sets: (1) because of recrossing issues; and (2) because of the main projection error of the associated eigenvector. Of course, one solution would be an adaptive refinement of the discretization, that is, one could choose a larger number of smaller sets, such that the eigenvector is better approximated by a step function on these sets. In the following section, we will present an alternative solution for overcoming the recrossing problem and reducing the projection error without refining the discretization.

4. The Core Set Approach

From Equation (10), we know how to compute a matrix representation for a projected transfer operator for an arbitrary subspace, $D \subset L^2(\mu)$. For a given basis, $\chi_1, ..., \chi_n$, we have to compute Equations (11) and (12), so:

$$M_{ij} = \frac{\langle \chi_i, \chi_j \rangle}{\langle \chi_i, \mathbb{1} \rangle}, \qquad P_{ij} = \frac{\langle \chi_i, T_\tau \chi_j \rangle}{\langle \chi_i, \mathbb{1} \rangle} \qquad (17)$$

In general, the evaluation of these scalar products for arbitrary basis functions is a non-trivial task. On the other hand, we have seen that for characteristic functions $\chi_i = \mathbb{1}_{A_i}$ on a full partition, we do not have to compute the scalar products numerically, since the matrix entries have a stochastic interpretation in terms of transition probabilities between set Equation (13). This means they can be directly estimated from a trajectory of the process, which is a strong computational advantage, particularly in high-dimensional state spaces.

Now, the question is if there is another basis other than characteristic functions that: (a) is more adapted to the eigenvectors of the transfer operator; and (b) still leads to a probabilistic interpretation of the matrix entries Equation (17), such that scalar products never have to be computed. The basic idea is to stick to a set-oriented definition of the basis, but to relax the full partition constraint. We will define our basis with respect to so-called *core sets*, $C_1, ..., C_n \subset E$, that are still disjoint, so $C_i \cap C_j = \emptyset$, but they do not have to form a full partition. Figure 4 suggests that this could lead to a reduction of the recrossing phenomenon, since the sets do not share boundaries anymore.

Figure 4. Core sets do not have to share boundaries anymore. This can reduce the recrossing effect.

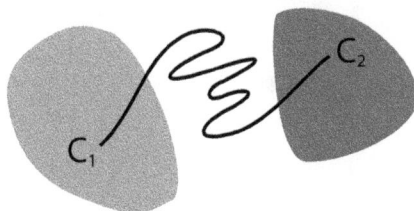

Now, we use the core sets to define our basis functions, $\chi_1, ..., \chi_n$. Assume T_τ is, again, a self-adjoint transfer operator and consider n core sets $C_1, ..., C_n$. For every i, take the committor function, χ_i, of the process with respect to core set C_i; that is, $\chi_i(x)$ denotes the probability to hit the core set, C_i, next, rather than the other core sets, when starting the process in x. If we now study the projection, Q, onto the space spanned by these committor functions, the two following properties hold [25,27].

(P1) The matrices, M and P, in Equation (10) can be written as:

$$M_{ij} = \mathbb{P}_\mu[\tilde{X}_k^+ = j | \tilde{X}_k^- = i], \qquad P_{ij} = \mathbb{P}_\mu[\tilde{X}_{k+1}^+ = j | \tilde{X}_k^- = i] \qquad (18)$$

where $(\tilde{X}_k^+)$ and $(\tilde{X}_k^-)$ are forward and backward milestoning processes [25,28]; that is, $\tilde{X}_k^- = i$ if the process came at time $t = k\tau$, last from core set C_i and $\tilde{X}_k^+ = j$ if the process went next to core set C_j after time $t = k\tau$.

(P2) Let u_i be an eigenvector of T_τ that is almost constant on the core sets. Let the region $C = E \setminus \bigcup_i C_i$ that is not assigned to a core set be left quickly enough, so $\mathbb{E}_x[\tau(C^c)] \ll \mathcal{T}_i$ for all $x \in C$, where $\mathcal{T}_i$ is the timescale associated with u_i and $\mathbb{E}_x[\tau(C^c)]$ is the expected hitting time of $C^c = \bigcup_i C_i$ when starting in $x \in C$. Then, $\|u_i - Qu_i\|$ is small; so, the committor approximation to the eigenvector is accurate.

The message behind (P1) is that it is possible to relax the full partition constraint and use a core set discretization that does not cover the whole state space. We can still define a basis for a projection of the transfer operator that leads to a matrix representation that can be interpreted in terms of transition probabilities.

Important Remark: The construction of the projection onto the committors is only necessary for theoretical purposes. In practice, neither the committor functions nor scalar products between the committors have to be computed numerically, since the matrix entries of M and P can be estimated from trajectories again.

Property (P2) yields that the relaxation of the full partition constraint should also lead to an improvement of the MSM if the region, C, between the core sets is typically left on a faster timescale than the processes of interest taking place. Let us get back to the example from above. We will see that we can achieve a strong improvement of the approximation by simply excluding a small part of state space from our discretization. In Figure 5, we have turned our initial full partition into a core set discretization by removing parts of the transition region between the wells.

The matrix $P_Q = PM^{-1}$ that represents the projection, $QT_\tau Q$, of the transfer operator onto the committor space associated with the core sets is given by:

$$P_Q = \begin{pmatrix} 0.9897 & 0.0103 & 0.0000 \\ 0.0352 & 0.9298 & 0.0351 \\ 0.0000 & 0.0103 & 0.9897 \end{pmatrix}$$

Comparing to the MSM for the full partition one can see that transitions between indices i and j, $i \neq j$ are less likely. Table 2 shows this leads to a far more accurate reproduction of the timescales in the system.

From the discussion above, this has to be expected, because the eigenvectors are almost constant in the vicinity of the wells, and we removed a part of state space from the discretization that is typically left quickly compared to the timescales, T_1 and T_2. Therefore, the committor functions should deliver a good approximation of the first two eigenvectors. Figure 6 underlines this theoretical result.

Figure 5. Excluding a small region of state space from the sets, $A_1, A_2, A3$, as in Figure 2, to form core sets C_1, C_2, C_3 that do not share boundaries anymore.

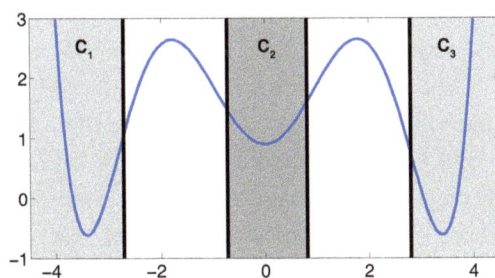

Figure 6. (**Upper panel**) The first non-trivial eigenvector, u_1 (solid blue), and its projection, $Q_f u_1$ (finely dashed red), onto step functions (full partition) and its projection, $Q_c u_1$ (dashed green), onto committors (core sets). (**Lower panel**) The same plot for the second non-trivial eigenvector, u_2.

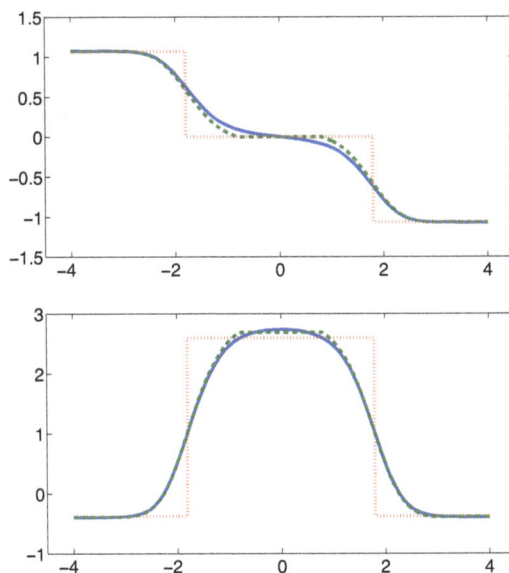

Table 2. More accurate approximation if implied timescales.

	T_1	T_2
Original	103.7608	11.9566
3 core sets	100.8066	11.9145
Full partition 3 sets	80.6548	9.8784

5. Practical Considerations and MD Applications

In the previous sections, we have interpreted the construction of an MSM as a projection of the dynamics onto some finite dimensional ansatz space. We have discussed two types of spaces that both have been defined on the basis of a set discretization. First, we chose a full partition of state space and the associated space of step functions, and second, we analyzed a discretization by core sets and the associated space spanned by committor functions. These two methods have the advantage that the resulting projections lead to transition matrices for the MSM with entries that are given in terms of transition probabilities between the sets. That is, one can compute estimates for the transition matrices from simulation data. This is an important property for practical applications, because it means that we never need to compute committor functions or scalar products between committors or step functions. We rather generate trajectories $x_0, x_1, ... x_N$ of the process (X_t), let us say, for a time step $h > 0$, so $x_i = X_{hi}$. For example, we can then define for a full partition, $A_1, ..., A_m$, and a lag time $\tau = nh$ the discrete trajectory $s_k = i \Leftrightarrow x_k \in A_i$ and compute the matrix, $\hat{P}$:

$$\hat{P}_{ij} = \frac{C_{ij}}{\sum\limits_{j} C_{ij}}, \qquad C_{ij} = \sum_{k=0}^{N-n} \mathbb{1}_{\{s_k=i\}} \mathbb{1}_{\{s_{k+n}=j\}} \qquad (19)$$

It is well known [7] that $\hat{P}$ is a maximum likelihood estimator for the full partition MSM transition matrix Equation (4). Similarly one can also compute estimates for a core set MSM by using the definition of milestoning processes [27,28]. That is, if we have core sets $C_1, ..., C_m$, a lag time $\tau = nh$ as before, and we define discrete milestoning trajectories by:

$$s_k^- = i \Leftrightarrow x_k \in A_i \text{ or came last from } A_i \text{ before time } k$$
$$s_k^+ = i \Leftrightarrow x_k \in A_i \text{ or went next to } A_i \text{ after time } k$$

we can compute an estimator $\hat{P}_Q = \hat{P}\hat{M}^{-1}$ of the core set MSM matrix Equation (10) by counting transitions:

$$\hat{P}_{ij} = \frac{C_{ij}}{\sum\limits_{j} C_{ij}}, \qquad C_{ij} = \sum_{k=0}^{N-n} \mathbb{1}_{\{s_k^-=i\}} \mathbb{1}_{\{s_{k+n}^+=j\}} \qquad (20)$$

$$\hat{M}_{ij} = \frac{N_{ij}}{\sum\limits_{j} N_{ij}}, \qquad N_{ij} = \sum_{k=0}^{N} \mathbb{1}_{\{s_k^-=i\}} \mathbb{1}_{\{s_k^+=j\}} \qquad (21)$$

Since, in practice, we will only have a finite amount of data available, we will have statistical errors when constructing an MSM. This is an additional error to the projection error related to the discretization that we have discussed above. On the other hand, one should note that these errors are *not independent* of each other. For example, it is clear that if we take a full partition of state space, and we let the partition become arbitrarily fine by letting the number of sets go to infinity, the discretization error will vanish. At the same time, for a fixed amount of statistics, the statistical error will become arbitrarily large, because we will need to compute more and more estimators for transition events between the increasing number of sets. For more information on statistical errors, we refer to the literature [7,29].

Besides the choice of discretization and the available statistics, the estimates above also depend on a lag time, τ. This dependence can be used to validate an MSM by a Chapman–Kolmogorov test [7]. This is based on the fact that the MSM matrices approximately form a semi-group for all large enough lag times $\tau > \tau^*$; although, for small lag times, this is typically not true, due to memory effects. These facts also motivate one to look at something, like an infinitesimal generator, that approximately generates these MSM transition matrices for large enough lag times. In [27], two types of generator constructions have been compared for a core set setting. The first generator, K, is simply constructed from the transition rates between the core sets in the milestoning sense, that is:

$$K_{ij} = \lim_{T \to \infty} \frac{N_{ij}^T}{R_i^T}, i \neq j \qquad K_{ii} = -\sum_{j \neq i} K_{ij} \tag{22}$$

where N_{ij}^T is the amount of time in $[0, T]$ the process has spent on its way from core set C_i to C_j and R_i^T is the total time in $[0, T]$ the process came last from C_i. On the other hand, one can see [27,30] that $K^* = K M^{-1}$ with the mass matrix, M, from above Equation (18), can be interpreted as a projection of the original generator of the process and, also, as a derivative of the core set MSM from above, *i.e.,*:

$$K^* = \lim_{\tau \to 0} \frac{P M^{-1} - I}{\tau} \tag{23}$$

where P depends on τ Equation (17).

Let us now analyze how the choice of core sets, particularly the size of the core sets, influences the resulting approximation. Therefore, we consider an MD example that was discussed in [27], namely one molecule of alanine dipeptide monitored via its ϕ and ψ backbone dihedral angles. Two core sets are defined as balls with radius r around the two points with angular coordinates $x_\alpha = (-80, -60)$ and $x_\beta = (-80, 170)$. The stationary distribution of the process and the two centers of the core sets, x_α, x_β, in the angular space are shown in Figure 7.

For computing a reference timescale, several MSMs based on three different full partitions using 10, 15 and 250 sets have been constructed for increasing lag times. In [27], it is shown that in each setting, the estimate for the longest implied timescale of the process converged to ≈ 19 ps for large enough τ. Now, the implied timescales for the two different generators, K Equation (22) and K^* Equation (23), are computed. In Figure 8, the resulting timescales are plotted against the reference timescale ≈ 19 ps for varying size of the core sets.

Figure 7. The stationary distribution of alanine dipeptide and the two centers of the core sets, x_α, x_β, in the angular space as white dots.

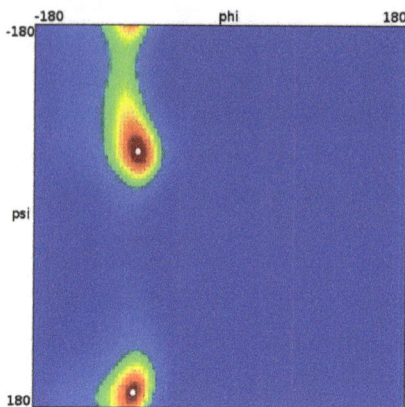

Figure 8. Estimate of the implied timescales from K Equation (22), the projected generator K^* Equation (23) and the reference computed from several full partition Markov State Models (MSMs).

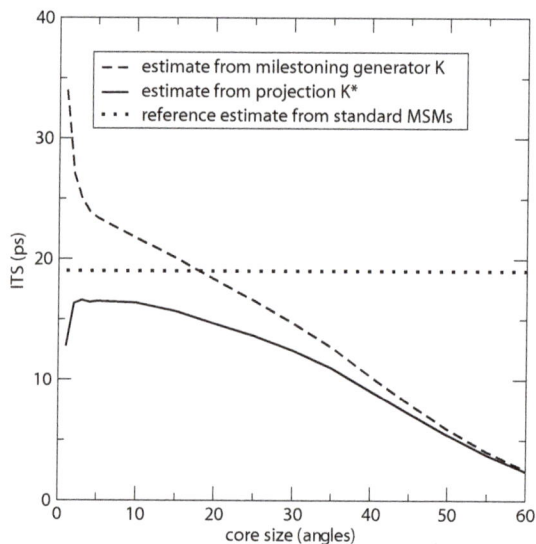

One can see that the estimate by the milestoning generator, K, is rather sensitive to the size of core sets. It overestimates the timescales for small core sizes and underestimates it for larger core sizes. On the other hand, the projected generator, K^*, can never overestimate the timescale, due to its interpretation as projection. It is also rather robust against the choice of size of the core sets until the core sets become too large, e.g., $r > 15$. Then, the discretization becomes close to a full partition discretization using only two sets. In this case, the timescales have to be underestimated heavily, because of recrossing phenomena. On the other hand, the underestimation for very small core sets has to be explained by a lack of statistics. When the core sets are chosen as arbitrarily small, it is clearly more difficult for the process to hit the sets, and therefore, transition events become rare. Note that for the straightforward milestoning generator, K, the processes seem to become very slow, but for the projected generator $K^* = KM^{-1}$, this effect is theoretically corrected by the mass matrix, M. Nevertheless, in both cases, the generation of enough statistics will be problematic for too small core sets.

6. Further Applications in MD

Markov State Modeling has been show to apply successfully to many different molecular systems, like peptides, including time-resolved spectroscopic experiments [10–12], proteins and protein folding [4,9,13], DNA [31] and ligand-receptor interaction [32]. In most of the respective publications, full partition MSMs are used, and the underlying discretization is based on cluster finding methods (see [7] for a review), while the sampling issues are tackled by means of ideas from enhanced sampling [33] and based on ensembles of rather short trajectories instead of one long one, *cf.* [4]. Core set-based approaches have been used just recently [10,27]; related algorithms are less well developed. However, recent work has shown that and how every full partition MSM can be easily transformed into a core set-based MSM with significantly *improved* approximation quality [34], making core set MSMs the most promising next generation MSM tools.

Very Rare Transitions between Discretization Sets. When constructing a full partition or a core set MSM, we have to estimate transition probabilities between sets in state space, and it can happen that we cannot avoid that some of these transitions are *very* rare. That is, the transition probabilities for a lag time, τ, between some sets can be non-zero, but small even, if compared to the remaining transition probabilities that are small already. This is why it is important to note that neglecting these very rare transitions during the construction of an MSM does *not* harm its approximation quality. For example, we can define for a transition matrix, P, another transition matrix, $\tilde{P}$, by:

$$\tilde{P}_{ij} = \begin{cases} P_{ij}, & i \neq j, (i,j) \notin R \\ 0, & i \neq j, (i,j) \in R \end{cases} \tag{24}$$

where R denotes the set of pairs of indices for which the transition are very rare and for which we set the transition probability to zero. If the Markov chain is reversible and $(i,j) \in R \Leftrightarrow (j,i) \in R$, one can show that for all ordered eigenvalues, $\lambda_k(P)$ and $\lambda_k(\tilde{P})$, it holds that:

$$|\lambda_k(P) - \lambda_k(\tilde{P})| \leq \max_i \sum_{j \neq i, (i,j) \in R} P_{ij} \tag{25}$$

That is, if we cannot estimate a very small transition probability, P_{ij}, for a very rare transition event between two sets, A_i and A_j, and even totally neglect this probability by setting it to zero, the timescales of the MSM remain almost unaffected. Thus, if we compute the set of the "first order" transition probability of a system correctly enough and ignore all "higher order" ones, then our accuracy will not be spoiled. This nicely illustrates the main advantage of MSM modeling compared to classical long-term simulation: since only neighboring core sets have to be connected by accurately estimated rates, the long residence time of long-term trajectories between and in core sets can be avoided, thus cutting down total simulation time.

Computation from Trajectories. Clearly, constructing and analyzing a core set MSM will only have a computational advantage compared to the direct sampling of a rare event if the transition events between neighboring core sets occur on a much shorter timescale than the rare event itself. One should note that from the purely theoretical point of view, it would be optimal to choose only very few core sets in the most metastable regions of state space, because this would minimize the projection error $\delta = \|u - Qu\|$ for each dominant eigenvector u, as discussed in Section 3. On the other hand, when estimating the MSM from trajectories, only a finite amount of statistics will be available, so there will also be a statistical error. In order to keep the total error small, additional core sets in less metastable parts of state space typically have to be introduced. In the end, this makes the estimation of a core set MSM possible without having to sample rare events. Note that the projection error is still under control, as long as there is a transition region between the core sets that is typically left very quickly (see Property (P2) in Section 4).

In practice, the statistics of the transition events between core sets will preferably be estimated from many short trajectories using milestoning techniques [27,28] and parallel computing. However, any algorithm for the construction of a core set MSM has to find a balance between sampling issues (not too many too long trajectories needed) and discretization issues (not too many core sets). Construction of such an algorithm still is ongoing research.

This article cannot give a detailed review on the algorithmic realization of MSMs for realistic molecular systems and on the findings resulting from such applications, since this is discussed to some extent elsewhere; see [7] for a recent review of the algorithmic aspects and [32,35] for ligand-receptor interaction.

7. MSM for Optimal Control Problems

In this section, we will borrow ideas from the previous section and explain how MSMs can be used to discretize optimal control problems that are linear-quadratic in the control variables and which appear in, e.g., the sampling of rare events. Specifically, we consider the case that $(X_t)_{t \geq 0}$ is the solution of:

$$dX_t = (\sqrt{2}u_t - \nabla V(X_t))dt + \sqrt{2\varepsilon}dB_t \tag{26}$$

with potential V, Brownian motion B_t and temperature $\varepsilon > 0$, as in Equation (16), and an unknown control variable, $u \colon [0, \infty) \to \mathbb{R}^d$, that is chosen so as to minimize the cost function:

$$J(u; x) = \mathbb{E}\left[\int_0^\tau \left(f(X_s) + \frac{1}{2}|u_t|^2\right) ds \,\Big|\, X_0 = x\right] \tag{27}$$

(The factors of $1/2$ and $\sqrt{2}$ in front of the control terms are for notational convenience.) Here, $f \geq 0$ is a bounded continuous function called *running cost* and $\tau < \infty$ (almost surely) is a random stopping time that is determined by X_t hitting a given target set, $A \subset E$, *i.e.*, $\tau = \inf\{t > 0 \colon X_t \in A\}$, in other words, we are interested in controlling $X_t = X_t^u$ until it reaches A. As an example, consider the case $f = 1$ with the potential considered in Figure 5 and the target region, A, around the left well. This situation is illustrated in Figure 9 and amounts to the situation that one seeks to minimize the time to reach A by tilting the potential towards A; tilting the potential too much is prevented by the quadratic penalization term in the cost functional that grows when too much force is applied.

Figure 9. The potential from Figure 5 (blue) and a tilted potential to minimize the time required to hit the target set, A (green).

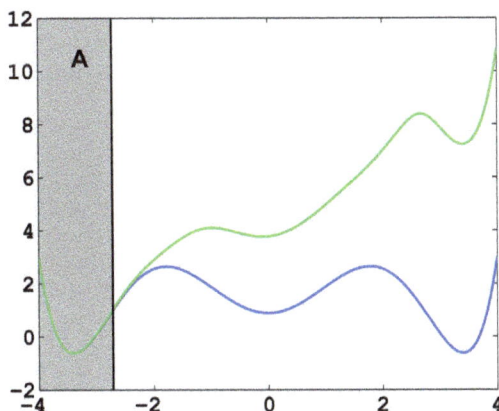

Other choices of f in Equation (26) result in alternative applications. One obvious application would be to set $\tau = T$ to a fixed time and f to the characteristic function of the complement of a conformation set C, $f = \mathbb{1}_{E \backslash C}$. In this case, minimization of J wrt. the control u_t would mean maximization of the probability to find the system in the conformation, C, until time T under a penalty on the external work done to the system. See [14] for more details on such applications.

There are other types of cost functions, J, one might consider, e.g., control until a deterministic finite time $\tau = T$ is reached or, even, $\tau \to \infty$, and the construction would follow analogously. For compactness, we consider here only cost functions as in Equation (27).

Optimal Control and Equilibrium Expectation Values. It turns out that when minimizing J, it is sufficient to consider control strategies that are Markovian and depend only on X_t, *i.e.*, we consider feedback laws of the form $u_t = \alpha(X_t)$ for some smooth function, $\alpha \colon E \to \mathbb{R}^d$. Moreover, only controls with finite energy are considered, for otherwise, $J(u; x) = \infty$. For control problems of the form Equations (26) and (27), the optimal feedback function can be shown to be $\alpha^*(x) = -\sqrt{2}\nabla W$, where W is the value function or optimal-cost-to-go [1,15]:

$$W(x) = \min_u J(u; x) \tag{28}$$

with the minimum running over all admissible Markovian feedback strategies. It can be shown that W satisfies the following dynamic programming equation of the Hamilton–Jacobi–Bellman type (see [36]):

$$LW(x) - |\nabla W(x)|^2 + f = 0$$
$$W|_A = 0 \qquad (29)$$

with the second-order differential operator:

$$L = \varepsilon\Delta - \nabla V \cdot \nabla$$

that is the infinitesimal generator of the process, X_t, for $u = 0$. If the value function, W, is known, it can be plugged into the equation of motion, which then turns out to be of the form:

$$dX_t^* = -\nabla U(X_t^*)dt + \sqrt{2\varepsilon}dB_t \qquad (30)$$

with the new potential:

$$U(x) = V(x) + 2W(x)$$

The difficulty is that Equation (29) is a nonlinear partial differential equation and for realistic high-dimensional systems, it is not at all obvious how to discretize it, employing any kind of state space partitioning. It has been demonstrated in [14,15] that Equation (29) can be transformed into a linear equation by a logarithmic transformation. Setting: $W(x) = -\varepsilon \log \phi(x)$ it readily follows, using chain rule and Equation (29), that ϕ solves the linear equation:

$$(L - \varepsilon^{-1}f)\phi = 0$$
$$\phi|_A = 1 \qquad (31)$$

The last equation is linear and can be solved by using MSMs, as we will show below. Moreover, by the Feynman–Kac theorem [37], the solution to Equation (31) can be expressed as:

$$\phi(x) = \mathbb{E}\left[\exp\left(-\frac{1}{\varepsilon}\int_0^\tau f(X_t)dt\right)\Big|X_0 = x\right] \qquad (32)$$

where X_t solves the control-free equation:

$$dX_t = -\nabla V(X_t)dt + \sqrt{2\varepsilon}dB_t$$

That is, the optimal control for Equation (26) can be computed by solving Equation (31), which can be done *in principle* via Monte Carlo approximation of the expected value in Equation (32) if critical slowing down by rare events can be avoided.

Remark. The optimization problem Equation (28) admits an interpretation in terms of entropy minimization: let $Q = Q_x^u$ and $P = Q_x^0$ denote the path probability measures of controlled and uncontrolled trajectories starting at x at time $t = 0$, and set:

$$Z = \int_0^\tau f(X_s)\, ds$$

Then, it follows that we can write:

$$W(x) = \min_{Q \ll P} J(u; x), \qquad J(u; x) = \int \left\{ Z + \varepsilon \log\left(\frac{dQ}{dP}\right) \right\} dQ \qquad (33)$$

where the notation "$Q \ll P$" means that Q has a density (That is, the density function, dQ/dP, exists and is almost everywhere positive and normalized) with respect to P. It turns out that for every such Q, there is exactly one control strategy, u, such that $Q = Q_x^u$ is generated by Equation (26); in this sense, the notation in Equation (33) is meaningful. The second term:

$$H(Q\|P) = \varepsilon \int \log \left(\frac{dQ}{dP} \right) dQ$$

is the relative entropy or Kullback–Leibler divergence between Q and P. For details on this matter that are based on Girsanov transformations for stochastic differential equations, we refer to [38] or the article [1] in this special issue.

8. MSM Discretization of Optimal Control Problems

The basic idea is now to choose a subspace, $D \subset L^2(\mu)$, with basis $\chi_1, \ldots, \chi_n$ as in Markov State Modeling and then discretize the dynamic programming Equation (29) of our optimal control problem by projecting the equivalent log transformed Equation (31) onto that subspace. As we will see, the resulting discrete matrix equation can be transformed back into an optimal control problem for a discrete Markov jump process (MJP).

We will do this construction for the full partition case $\chi_i = \mathbb{1}_{A_i}$ and the core set case $\chi_i = q_i$ discussed earlier. We will see that in both cases, we arrive at a structure-preserving discretization of the original optimal control problem, where the states of the corresponding MJP will be related to the partition subsets, A_i. The first case will give us back a well-known lattice discretization for continuous control problems, the Markov chain approximation [39]. This is illustrated in the following diagram:

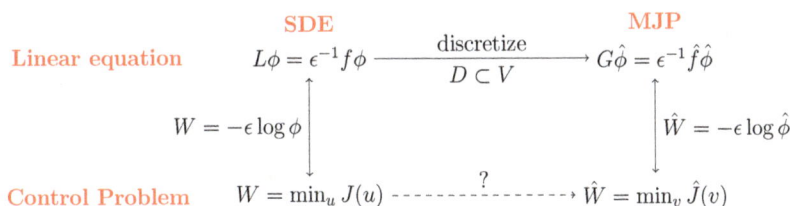

Subspace Projection. The key step for the discretization is that we pick a suitable subspace, $D \subset L^2(\mu)$, that is adapted to the boundary value problem Equation (31). Specifically, we require that the subspace contains the constant function, $\mathbb{1} \in D$, and that it gives a good representation of the most dominant metastable sets. To this end, we choose basis functions $\chi_1, \ldots, \chi_{n+1}$ with the following properties:

(S1) The χ_i form a partition of unity, that is $\sum_{i=1}^{n+1} \chi_i = \mathbb{1}$.

(S2) The χ_i are adapted to the boundary conditions in Equation (31), that is $\chi_{n+1}|_A = 1$ and $\chi_i|_A = 0$ for $i \in \{1, \ldots, n\}$.

Now, let Q be the orthogonal projection onto D, and define the matrices:

$$F_{ij} = \frac{\langle \chi_i, f\chi_j \rangle}{\langle \chi_i, 1 \rangle}, \qquad K_{ij} = \frac{\langle \chi_i, L\chi_j \rangle}{\langle \chi_i, 1 \rangle}$$

Now, if ϕ solves the linear boundary value problem Equation (31), then the coefficients, $\hat{\phi}_1, \ldots, \hat{\phi}_{n+1}$, of its finite-dimensional representation $Q\phi = \sum_j \hat{\phi}_j \chi_j$ on the subspace, D, satisfy the constrained linear system:

$$\sum_{j=1}^{n+1} \left(K_{ij} - \varepsilon^{-1} F_{ij} \right) \hat{\phi}_j = 0, \quad i \in \{1, \ldots, n\}$$

$$\hat{\phi}_{n+1} = 1$$

(34)

that is the discrete analogue of Equation (31). The discrete solution $\hat{\phi} = Q\phi$ is optimal in the sense of being the best approximation of ϕ in the energy norm, *i.e.*,:

$$\|\phi - \hat{\phi}\|_A = \inf_{\psi \in D} \|\phi - \psi\|_A$$

(35)

where:

$$\|\phi\|_A^2 = \langle \phi, (\varepsilon^{-1} f - L)\phi \rangle$$

is the energy norm on $L^2(\mu)$, and the infimum runs over all functions, $\psi \in L^2(\mu)$, that are of the form $\psi(x) = \sum_j \psi_j \chi_j(x)$ with coefficients $\psi_j \in \mathbb{R}$. This is a standard result about projections of PDEs; see [40] for details. (By the same argument as in the previous sections, $A = \varepsilon^{-1} f - L$ is symmetric and positive definite as an operator on the weighted Hilbert space, $L^2(\mu)$. Moreover, $\|\phi\|_A^2 = \varepsilon^{-1}\langle \phi, f\phi \rangle + \varepsilon \langle \nabla \phi, \nabla \phi \rangle$.) In analogy with Equation (14), we can use the above result to get the error estimate:

$$\|\phi - \hat{\phi}\|_\mu^2 \leq \left(1 + \frac{1}{\delta^2} \|QAQ^\perp\|^2 \right) \inf_{\psi \in D} \|\phi - \psi\|_\mu^2$$

(36)

where $A = \varepsilon^{-1} f - L$ is a shorthand for the operator appearing in Equation (31) and the constant $\delta > 0$ is defined, such that $\|v\|_A^2 \geq \delta \|v\|_\mu^2$ holds for all $v \in L^2(\mu)$; see [41]. The bottom line of Equation (35) shows that discretizing Equation (31) via Equation (34) minimizes the projection error measured in the energy norm. Since all functions are μ-weighted, the approximation will be good in regions visited with high probability and less good in regions with lower probability. The error estimate Equation (36) is along the lines of the MSM approximation result: if we switch to the norm on $L^2(\mu)$, the function $\hat{\phi} = Q\phi$ is still almost the best approximation of ϕ, provided that A leaves the subspace, D, almost invariant. As was pointed out earlier, this is exactly the case when the χ_i are close to the eigenfunctions of A (e.g., when the system is metastable).

The best approximation error $\|Q^\perp \phi\|_\mu = \inf_{\psi \in D} \|\phi - \psi\|_\mu$, which appears in Equation (36), will vanish if the χ_i form an arbitrarily fine full partition of E. If we follow the core set idea from Section 4 and choose the χ_i to be committor functions, we have good control over $\|Q^\perp \phi\|_\mu$. Due to [41]:

$$\|Q^\perp \phi\|_\mu \leq \|P^\perp \phi\|_\mu + \mu(C)^{1/2} \left[\kappa \|f\|_\infty + 2\|P^\perp \phi\|_\infty \right]$$

(37)

where $C = E \setminus \cup_i C_i$ is the transition region, $\kappa = \sup_{x \in C} \mathbb{E}_x \tau_{E \setminus C}$ is the maximum expected time of hitting the metastable set from outside (which is short) and P is the orthogonal projection onto the subspace $V = \{v \in L^2(\mu), v = const \text{ on every } C_i\} \subset L^2(\mu)$. Note that $P^\perp \phi = 0$ on C. The errors, $\|P^\perp \phi\|_\mu$ and $\|P^\perp \phi\|_\infty$, measure how constant the solution, ϕ, is on the core sets. Hence, Equation (37) together with Equation (36) gives us complete control over the approximation error of our projection method, even if we consider just a few core sets. In Section 9, we will investigate the full and core set partition cases further.

Properties of the Projected Problem. We introduce now the diagonal matrix, Λ, with entries $\Lambda_{ii} = \sum_j F_{ij}$ (zero otherwise) and the full matrix $G = K - \varepsilon^{-1}(F - \Lambda)$, and rearrange Equation (34) as follows:

$$\sum_{j=1}^{n+1} \left(G_{ij} - \varepsilon^{-1}\Lambda_{ij}\right) \hat\phi_j = 0, \quad i \in \{1, \ldots, n\}$$

(38)

$$\hat\phi_{n+1} = 1$$

This equation can be given a stochastic interpretation. To this end, let us introduce the vector, $\pi \in \mathbb{R}^{n+1}$, with nonnegative entries $\pi_i = \langle \chi_i, 1 \rangle$ and notice that $\sum_i \pi_i = 1$ follows immediately from the fact that the basis functions, χ_i, form a partition of unity, i.e., $\sum_i \chi_i = 1$. This implies that π is a probability distribution on the discrete state space $\hat E = \{1, \ldots, n+1\}$. We summarize properties of the matrices, K, F and G; see also [41]:

(M1) K is a generator matrix of an MJP $(\hat X_t)_{t\geq 0}$ (i.e., K is a real-valued square matrix with row sum zero and positive off-diagonal entries) with stationary distribution, π, that satisfies detailed balance:

$$\pi_i K_{ij} = \pi_j K_{ji}, \quad i, j \in \hat E$$

(M2) $F \geq 0$ (entry-wise) with $\pi_i F_{ij} = \pi_j F_{ji}$ for all $i, j \in \hat E$.

(M3) G has a row sum of zero and satisfies $\pi^T G = 0$ and $\pi_i G_{ij} = \pi_j G_{ji}$ for all $i, j \in \hat E$; furthermore, there exists a constant $0 < C < \infty$, such that $G_{ij} \geq 0$ for all $i \neq j$ if $\|f\|_\infty \leq C$. In this case, Equation (38) admits a unique and strictly positive solution $\hat\phi > 0$.

It follows that if the running costs, f, are such that (M3) holds, then G is a generator matrix of an MJP that we shall denote by $(\hat X_t)_{t\geq 0}$, and Equation (38) has a unique and positive solution. In this case, the logarithmic transformation $\hat W = -\varepsilon \log \hat\phi$ is well defined. It was shown in [42] that $\hat W$ can be interpreted as the value function of a Markov decision problem with cost functional (cf. also [36]):

$$\hat J(v; i) = \mathbb{E}\left[\int_0^\tau \left(\hat f(\hat X_s) + k(\hat X_s, v_s)\right) ds \Big| \hat X_0 = i\right]$$

(39)

that is minimized over the set of Markovian control strategies, $v \colon \hat E \to (0, \infty)$, subject to the constraint that the controlled process $\hat X_t = \hat X_t^v$ is generated by G^v, where:

$$G_{ij}^v = \begin{cases} v(i)^{-1} G_{ij} v(j), & i \neq j \\ -\sum_{j\neq i} G_{ij}^v, & i = j \end{cases}$$

(40)

with stopping time $\tau = \inf\{t > 0 \colon \hat X_t = n+1\}$ and running costs:

$$\hat f(i) = \Lambda_{ii}, \quad k(i, v) = \varepsilon \sum_{j\neq i} G_{ij} \left\{ \frac{v(j)}{v(i)} \left[\log \frac{v(j)}{v(i)} - 1\right] + 1\right\}$$

(41)

Properties of the Projected Problem, Continued. From [42], we know that the optimal cost:

$$\hat{W}(i) = \min_v \hat{J}(v; i)$$

is given by $\hat{W} = -\epsilon \log \hat{\phi}$, where $\hat{\phi}$ solves Equation (38), with the optimal feedback strategy given by $v^*(i) = \hat{\phi}_i$ (see [36]). We list additional properties:

(i) The v-controlled system has the unique invariant distribution:

$$\pi^v = (\pi_1^v, \ldots, \pi_{n+1}^v), \qquad \pi_i^v = \frac{v(i)^2 \pi_i}{Z_v}$$

with Z_v an appropriate normalization constant; in terms of the value function, $\pi^* = \pi^{v^*}$ reads:

$$\pi^* = (\pi_1^*, \ldots, \pi_{n+1}^*), \qquad \pi_i^* = \frac{1}{Z_*} e^{-2\varepsilon^{-1}\hat{W}(i)} \pi_i$$

(ii) G^v is reversible and stationary with respect to π^v, i.e., $\pi_i^v G_{ij}^v = \pi_j^v G_{ji}^v$ for all $i, j \in \hat{E}$.

(iii) $\hat{J}$ admits the same interpretation as Equation (33) in terms of the relative entropy:

$$\hat{W}(i) = \min_{Q \ll P} \hat{J}(v; i), \qquad \hat{J}(v; i) = \int \left\{ \hat{Z} + \varepsilon \log \left(\frac{dQ}{dP} \right) \right\} dQ$$

where P denotes expectation with respect to the uncontrolled MJP, $\hat{X}_t$, starting at $\hat{X}_0 = i$, Q denotes the path measure of the corresponding controlled process with generator G^v and:

$$\hat{Z} = \int_0^T \hat{f}(\hat{X}_s) \, ds$$

A few remarks seem in order: Item (i) of the above list is in accordance with the continuous setting, in which the optimally controlled dynamics is governed by the new potential $U = V + 2W$ and has the stationary distribution, $\mu^* \propto \exp(-2\epsilon^{-1}W)\mu$, with μ being the stationary distribution of the uncontrolled process. Hence, the effect of the control on the invariant distribution is the same in both cases. Further, note that optimal strategies change the jump rates according to:

$$G_{ij}^{v^*} = G_{ij} e^{-\varepsilon^{-1}\left(\hat{W}(j) - \hat{W}(i)\right)} \qquad (42)$$

that is, $\hat{W}$ acts as an effective potential as in the continuous case, and the change in the jump rates can be interpreted in terms of Kramer's law for this effective potential.

This completes our derivation of the discretized optimal control problem, and we now compare it with the continuous problem we started with for the case of a full partition of E and a core set partition of E.

9. Markov Chain Approximations and Beyond

Full Partitions. Let E be fully partitioned into disjoint sets, $A_1, \ldots, A_{n+1}$, with centers $x_1, \ldots, x_{n+1}$ and such that $A_{n+1} := A$, and define $\chi_i := \chi_{A_i}$. These χ_i satisfy Assumptions (S1) and (S2) discussed in Section 8. Since they are not overlapping, F is diagonal, and:

$$\hat{f}(i) = \frac{1}{\pi_i} \int_{A_i} f(x)\mu(x)dx = \mathbb{E}_\mu[f(X_t)|X_t \in A_i] \tag{43}$$

is just obtained by averaging $f(x)$ over the cell, A_i. Equation (43) is also a sampling formula for $\hat{f}(i)$. It follows directly that $G = K$, and in particular, (M3) holds for any f. One can show that K has components:

$$K_{ij} \approx \frac{1}{\Delta_{ij}} e^{-\beta(V(\bar{x}_{ij}) - V(x_i))}, \quad \Delta_{ij}^{-1} = \beta^{-1} \frac{m(S_{ij})}{m(h_{ij})m(A_i)} \tag{44}$$

if i and j are neighbors ($K_{ij} = 0$ otherwise). Here, m is the Lebesgue measure, and h_{ij}, S_{ij} and $\bar{x}_{ij}$ are defined as in Figure 10. K is the generator of an MJP on the cells, A_i, and coincides with the so-called *finite volume approximation* of L discussed in [43]. It is reversible with stationary distribution:

$$\pi_i = \int_{A_i} d\mu \approx m(A_i)e^{-\beta V(x_i)}$$

Figure 10. The mesh for the full partition.

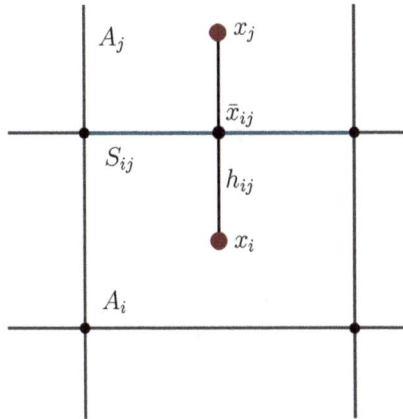

One can show that the approximation error vanishes for $n \to \infty$. K and π can be computed from the potential, V, and the geometry of the mesh. By inspecting Equations (12) and (13), we see that K is connected to the transition matrix, P^τ, of a full partition MSM with lag time τ by

$$\lim_{\tau \to 0} \frac{1}{\tau} \left(P_{ij}^\tau - M_{ij} \right) = \lim_{\tau \to 0} \frac{1}{\pi_i} \langle \chi_i, \frac{1}{\tau}(T_\tau - 1)\chi_j \rangle = \frac{1}{\pi_i} \langle \chi_i, L\chi_j \rangle = K_{ij},$$

thus K is the generator of the semigroup of transition matrices, P^τ. Therefore we could obtain K by sampling in the same way we obtained P^τ through Equation (19) in Section 5. This is difficult,

272

however, due to recrossing problems for small τ; see e.g., [44]. Finally, let us note in passing that we can drastically simplify k^v if the cells, A_i, are boxes of length h. Denote the elementary lattice vectors by e_n. Then,

$$k^v(i) = \frac{1}{2}|u^v(i)|^2 + \mathcal{O}(h), \quad u_n^v(i) := \frac{1}{\sqrt{2}}\frac{\epsilon}{2h}\left(\log v(i + e_n) - \log v(i - e_n)\right),$$

which establishes the connection to the continuous case. However, more is true: The whole discrete control problem reduces to first order in h to the well-known Markov chain approximation (MCA) [39], which allows us to use convergence theory for MCAs to conclude that, for $n \to \infty$, the optimal control and value function of the discrete control problem converge to their continuous counterparts. More details can be found in [41].

Core Set Partition. Now, we choose core sets $C_1, \ldots, C_{n+1}$ with $C_{n+1} = A$, and we let $\chi_i = q_i$ be the committor function of the process with respect to C_i, as in Section 4. These χ_i satisfy Assumptions (S1) and (S2) discussed in Section 8. The projection onto the committor basis also allows for a stochastic interpretation. Recall the definition of the forward and backward milestoning process, $\tilde{X}_t^\pm$, from Equation (18). The discrete costs can be written as:

$$\hat{f}(i) = \frac{1}{\pi_i}\langle q_i, f \sum_j q_j\rangle = \int \nu_i(x)f(x)dx = \mathbb{E}_\mu\left[f(X_t)\Big|\tilde{X}_t^- = i\right] \tag{45}$$

where $\nu_i(x) = \frac{q_i(x)\mu(x)}{\pi_i} = \mathbb{P}(X_t = x|\tilde{X}_t^- = i)$ is the probability density of finding the system in state x given that it came last from i. Hence, $\hat{f}(i)$ is the average costs conditioned on the information $\tilde{X}_t^- = i$, i.e., X_t came last from A_i, which is the natural extension to the full partition case, where $\hat{f}(i)$ was the average costs conditioned on the information that $X_t \in A_i$.

The matrix $K = \pi_i^{-1}\langle q_i, Lq_j\rangle$ is reversible with stationary distribution

$$\pi_i = \langle q_i, 1\rangle = \mathbb{P}_\mu(\tilde{X}_t^- = i)$$

and is related to core MSMs again:

$$K = \lim_{\tau \to 0}\frac{1}{\tau}(P^\tau - M)$$

where P^τ and M are now the matrices for core MSMs, as in Equation (18). Formally, K is the generator of the P^τ, but these do not form a semigroup, since $M \neq 1$. Therefore, we cannot interpret K directly as, e.g., the generator of $\tilde{X}_t^-$. Nevertheless, the entries of K are the transition rates between the core sets, as defined in transition path theory [45]. We can sample P^τ and M using Equations (20) and (21); because we used an incomplete partition, the recrossing problem is removed, and there is no difficulty in sampling P^τ for all lag times, τ, and therefore, K directly. It is worth noting that F can also be sampled:

$$F_{ij} = \mathbb{E}_\mu\left[f(X_t)\chi_{\{\tilde{X}_t^+=j\}}\Big|\tilde{X}_t^- = i\right]$$

Therefore, as in the construction of core MSMs, we do not need to compute committor functions explicitly. Note, however, that $G \neq L$, there is a reweighting, due to the overlap of the q_i's, which causes F to be non-diagonal. This reweighting is the surprising bit of this discretization. From properties (M1)–(M3) from Section 8, we see, however, that G and K are both reversible with stationary distribution, π. Finally, note that if the cost function, $f(x)$, does not satisfy $\|f\|_\infty \le C$ from (M3), G will not even be a generator matrix. In this case, (34) still has a solution, $\hat{\phi}$, which is the best approximation to ϕ, but this solution may not be unique; it may not satisfy $\hat{\phi} > 0$, and we have no interpretation as a discrete control problem.

10. Numerical Results

10.1. 1D Potential Revisited

Firstly, we study diffusion in the triple well potential, which is presented in Figure 2. This potential has three minima at approximately $x_{0/1} = \pm 3.4$ and $x_2 = 0$. We choose the three core sets $C_i = [x_i - \delta, x_i + \delta]$ around the minima with $\delta = 0.2$. Take τ to be the first hitting time of C_0. We are interested in the moment generating function $\phi(x) = \mathbb{E}\left[e^{-\epsilon^{-1}\sigma\tau}\right]$ of passages into C_0 and the cumulant generating function $W = \epsilon \log \phi$. This is of the form Equation (32) for $A = C_i$ and $f = \sigma$, a constant function.

In Figure 11a, the potential, V, and effective potential, U, are shown for $\beta = 2$ and $\sigma = 0.08$ (solid lines), cf. Equation (30). One can observe that the optimal control effectively lifts the second and third well up, which means that the optimal control will drive the system into C_0 very quickly. The reference computations here have been carried out using a full partition FEM (finite element method) discretization of Equation (31) with a lattice spacing of $h = 0.01$. Now, we study the MJP approximation constructed via the committor functions shown in Figure 11b. These span a three-dimensional subspace, but due to the boundary conditions, the subspace, D, of the method is actually two-dimensional. The dashed line in Figure 11a gives the approximation to U calculated by solving Equation (38). We can observe extremely good approximation quality, even in the transition region. In Figure 11c, the approximation to the optimal control, $\alpha^*(x)$ (solid line), and its approximation $\hat{\alpha}^* = -\sqrt{2}\nabla\hat{W}$ (dashed line) are shown. The core sets are shown in blue. We can observe jumps in $\hat{\alpha}^*$ at the left boundaries of the core sets. This is to be expected and comes from the fact that the committor functions are not smooth at the boundaries of the core sets, but only continuous. Therefore, the approximation to U is continuous, but the approximation to α^* is not.

Next, we construct a core MSM to sample the matrices, K and F. One hundred trajectories of length $T = 20,000$ were used to build the MSM. In Figure 11d, W and its estimate using the core MSM are shown for $\epsilon = 0.5$ and different values of σ. Each of the 100 trajectories has seen about four transitions. For comparison, a direct sampling estimate of W using the same data is shown (green). The direct sampling estimate suffers from a large bias and variance and is practically useless. In contrast, the MSM estimator for W performs well for all considered values of σ, and always, its variance is significantly small. The constant, C, which ensures $\hat{\phi} > 0$ when $\sigma \le C$, is approximately 0.2 in this case. This seems restrictive, but still allows one to capture all interesting information about ϕ and W.

Figure 11. Three well potential example for $\epsilon = 0.5$ and $\sigma = 0.08$. **(a)** Potential $V(x)$ (blue), effective potential $U = V + 2W$ (green) and approximation of U with committors (dashed red). **(b)** The three committors, $q_1(x)$, $q_2(x)$ and $q_3(x)$. **(c)** The optimal control $\alpha^*(x)$ (solid line) and its approximation (dashed line). Core sets are shown in blue; **(d)** Optimal cost W for $\beta = 2$ as a function of σ. Blue: Exact solution. Red: Core MSM estimate. Green: Direct sampling estimate.

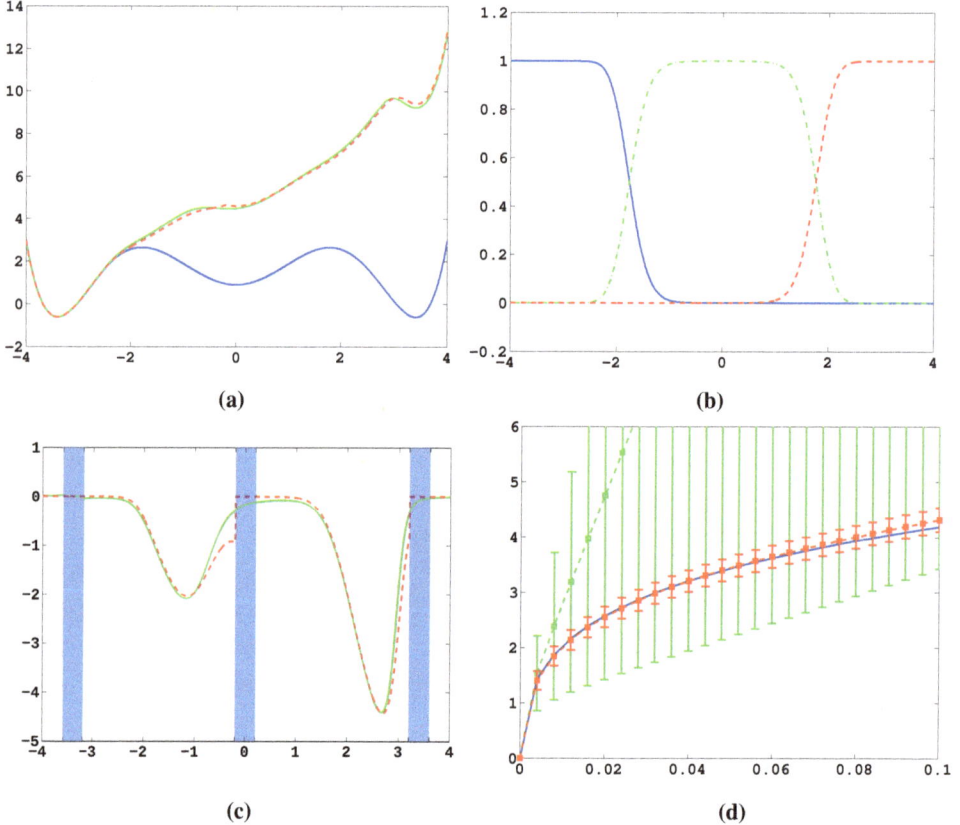

(a)

(b)

(c)

(d)

10.2. Alanine Dipeptide

Lastly, we study α-β-transitions in alanine dipeptide, a well-studied test system for molecular dynamics applications. We use a $1\mu s$ long trajectory simulated with the CHARMM (Chemistry at HARvard Molecular Mechanics) 27 force field. The conformational dynamics is monitored as usual via the backbone dihedral angles, ϕ and ψ. The data was first presented in [27]. We construct a full partition MSM with 250 clusters using k-means clustering. We are interested in the MFPT (mean first passage time) $\hat{t}(i) = \mathbb{E}_i[\tau_\alpha]$, where τ_α is the first hitting time of the α conformation, which we define as a circle with radius $r = 45$ around $(\phi_\alpha, \psi_\alpha) = (-80, -60)$. The MFPT vector, $\hat{t}$, solves the boundary value problem

$$K\hat{t} = -1 \text{ outside of } \alpha, \quad \hat{t} = 0 \text{ in } \alpha$$

but since K is not available directly via sampling, we have to consider the equation

$$\frac{1}{\tau} \left(P^\tau - 1 \right) \hat{t} = -1 \text{ outside of } \alpha, \quad \hat{t} = 0 \text{ in } \alpha$$

instead. The result will depend on the choice of lag time τ. In Figure 12a, the results are shown for $\tau = 5$; we can identify the β-structure as the red cloud of clusters where $\hat{t}(i)$ is approximately constant. In Figure 12b, $\hat{t}_{\beta\alpha} = \mathbb{E}(\hat{t}(i)|i \in \beta)$ is shown as a function of τ. We observe a linear behavior for large τ, which is due to the linear error introduced in the replacement of K with $\frac{1}{\tau}(P^\tau - 1)$, and a nonlinear drop for small τ, which is due to non-Markovianity. Our best guess is, therefore, a linear interpolation to $\tau = 0$, which is indicated by the solid line. The result is $\hat{t}_{\beta\alpha}^{(0)} = 35.5ps$. As a comparison, the reference value $\hat{t}_{\beta\alpha}^{ref} = 36.1ps$ from [27] is shown as a dashed line. It was computed in [27] as an inverse rate, using the slowest ITS (implied time scale) and information about the equilibrium weights of the α and β structure. We see very good agreement. The result is, of course, dependent, though, on the assignment of clusters to the α and β structure. Some tests show that $\hat{t}_{\beta\alpha}^{(0)}$ as computed with the interpolation method is fairly insensitive to this choice.

Figure 12. Dipeptide example. **(a)** MFPT from β to α in ϕ-ψ space for $\tau = 5$. The red cloud to the right is the β-structure. **(b)** MFPT as a function of τ (dashed line) and linear interpolation to $\tau = 0$ (solid line). Green dashed line: reference computed via the slowest ITS .

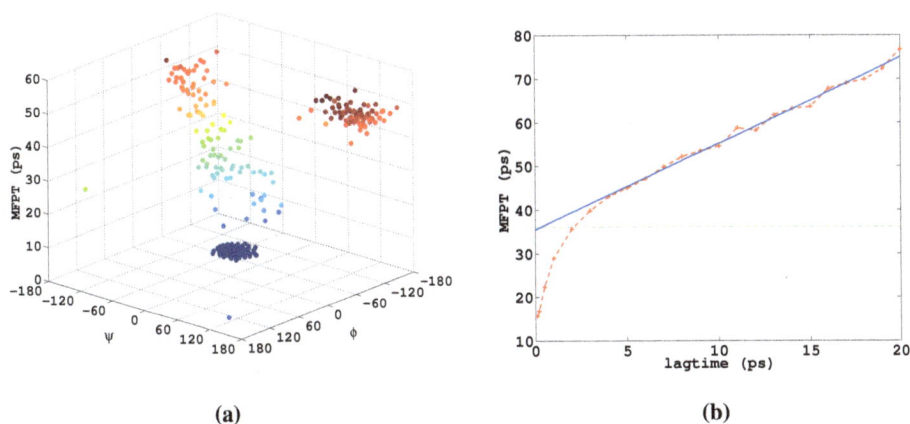

(a)   (b)

In [14], it is demonstrated how to use the method presented herein for maximizing the population of the α-conformation of alanine dipeptide based on the MSM used here.

11. Conclusions

In this article, we have discussed an approach to overcome direct sampling issues of rare events in molecular dynamics based on spatial discretization of the molecular state space. The strategy is to define a discretization by subsets of state space, such that the sampling effort with respect to transitions between the sets is much lower than the direct estimation of the rare events under consideration. That is, without having to simulate rare events, we construct a so-called Markov State Model, a Markov chain approximation to the original dynamics. Since the state space of the MSM is finite, we can then calculate the properties of interest by simply solving linear systems of equations. Of course, it is crucial that these properties of the MSM can be related to the rare event properties of the original process that we have not been able to sample directly.

This is why we have analyzed the approximation quality of MSMs in the first part of the article. We have used the interpretation of MSMs as projections of the transfer operator to: (1) derive conditions that guarantee an accurate reproduction of the dynamics; and (2) show how to construct models based on a core set discretization by leaving the state space partly undiscretized.

In the second part of the article, we have used the concept of MSM discretization to solve MD optimal control problems in which one computes the optimal external force that drives the molecular system to show an optimized behavior (maximal possible population in a conformation; minimal mean first passage time to a certain conformation) under certain constraints. We have demonstrated that the spatial discretization underlying an MSM turns the high-dimensional continuous optimal control problem into a rather low-dimensional discrete optimal control problem of the same form that can be solved efficiently. This result allows two different types of applications: (1) if one can construct an MSM for a molecular system in equilibrium, then one can use it to compute optimal controls that extremize a given costs criterion; (2) if an MSM can be computed based on transition probabilities between neighboring core sets alone, then the rare event statistics for transitions between strongly separated metastable states of the system can be computed from an associated optimal control problem that can be solved after discretization using the pre-computed MSM.

Acknowledgments

The authors have been supported by the DFG Research Center MATHEON.

Conflicts of Interest

The authors declare no conflict of interest.

References

1. Hartmann, C.; Banisch, R.; Sarich, M.; Badowski, T.; Schütte, C. Characterization of rare events in molecular dynamics. *Entropy* **2014**, *16*, 350–376.
2. Metzner, P.; Schütte, C.; Vanden-Eijnden, E. Transition path theory for markov jump processes. *Multiscale Model. Simul.* **2009**, *7*, 1192–1219.
3. Pan, A.C.; Roux, B. Building Markov state models along pathways to determine free energies and rates of transitions. *J. Chem. Phys.* **2008**, *129*, 064107.
4. Noé, F.; Schütte, C.; vanden-Eijnden, E.; Reich, L.; Weikl, T. Constructing the full ensemble of folding pathways from short off-equilibrium trajectories. *Proc. Natl. Acad. Sci. USA* **2009**, *106*, 19011–19016.

5. Schütte, C. Conformational Dynamics: Modelling, Theory, Algorithm, and Applications to Biomolecules. Habilitation Thesis, Fachbereich Mathematik und Informatik, Freie Universität Berlin, Berlin, Germany, 1998.

6. Schütte, C.; Fischer, A.; Huisinga, W.; Deuflhard, P. A direct approach to conformational dynamics based on hybrid monte Carlo. *J. Comput. Phys.* **1999**, *151*, 146–168.

7. Prinz, J.H.; Wu, H.; Sarich, M.; Keller, B.; Fischbach, M.; Held, M.; Chodera, J.D.; Schütte, C.; Noé, F. Markov models of molecular kinetics: Generation and validation. *J. Chem. Phys.* **2011**, *134*, 174105.

8. Noé, F.; Horenko, I.; Schütte, C.; Smith, J.C. Hierarchical analysis of conformational dynamics in biomolecules: Transition networks of metastable states. *J. Chem. Phys.* **2007**, *126*, 155102.

9. Chodera, J.D.; Dill, K.A.; Singhal, N.; Pande, V.S.; Swope, W.C.; Pitera, J.W. Automatic discovery of metastable states for the construction of Markov models of macromolecular conformational dynamics. *J. Chem. Phys.* **2007**, *126*, 155101.

10. Buchete, N.V.; Hummer, G. Coarse master equations for peptide folding dynamics. *J. Phys. Chem. B* **2008**, *112*, 6057–6069.

11. Prinz, J.H.; Keller, B.; Noé, F. Probing molecular kinetics with Markov models: Metastable states, transition pathways and spectroscopic observables. *Phys. Chem. Chem. Phys.* **2011**, *13*, 16912–16927.

12. Keller, B.; Prinz, J.H.; Noé, F. Markov models and dynamical fingerprints: Unraveling the complexity of molecular kinetics. *Chem. Phys.* **2011**, in press.

13. Bowman, G.; Volez, V.; Pande, V.S. Taming the complexity of protein folding. *Curr. Opin. Struct. Biol.* **2011**, *21*, 4–11.

14. Schütte, C.; Winkelmann, S.; Hartmann, C. Optimal control of molecular dynamics using Markov state models. *Math. Program. Ser. B* **2012**, *134*, 259–282.

15. Hartmann, C.; Schütte, C. Efficient rare event simulation by optimal nonequilibrium forcing. *J. Stat. Mech. Theor. Exp.* **2012**, doi:10.1088/1742-5468/2012/11/P11004.

16. Jäger, M.; Zhang, Y.; Bieschke, J.; Nguyen, H.; Dendle, M.; Bowman, M.E.; Noel, J.P.; Gruebele, M.; Kelly, J.W. Structure-function-folding relationship in a WW domain. *Proc. Natl. Acad. Sci. USA* **2006**, *103*, 10648–10653.

17. Kobitski, A.Y.; Nierth, A.; Helm, M.; Jäschke, A.; Nienhaus, G.U. Mg^{2+} dependent folding of a Diels-Alderase ribozyme probed by single-molecule FRET analysis. *Nucleic Acids Res.* **2007**, *35*, 2047–2059.

18. Fischer, S.; Windshuegel, B.; Horak, D.; Holmes, K.C.; Smith, J.C. Structural mechanism of the recovery stroke in the Myosin molecular motor. *Proc. Natl. Acad. Sci. USA* **2005**, *102*, 6873–6878.

19. Noé, F.; Krachtus, D.; Smith, J.C.; Fischer, S. Transition networks for the comprehensive characterization of complex conformational change in proteins. *J. Chem. Theory Comput.* **2006**, *2*, 840–857.

20. Ostermann, A.; Waschipky, R.; Parak, F.G.; Nienhaus, U.G. Ligand binding and conformational motions in myoglobin. *Nature* **2000**, *404*, 205–208.

21. Huisinga, W. Metastability of Markovian Systems a Transfer Operator Based Approach in Application to Molecular Dynamics. Ph.D Thesis, Fachbereich Mathematik und Informatik, Freie Universität Berlin, Berlin, Germany, 2001.

22. Bovier, A.; Eckhoff, M.; Gayrard, V.; Klein, M. Metastability in reversible diffusion processes. I. Sharp asymptotics for capacities and exit times. *J. Eur. Math. Soc.* **2004**, *6*, 399–424.

23. Voronoi, M.G. Nouvelles applications des parametres continus a la theorie des formes quadratiques. *J. Reine Angew. Math.* **1908**, *134*, 198–287.

24. Sarich, M.; Noé, F.; Schütte, C. On the approximation quality of markov state models. *Multiscale Model. Simul.* **2010**, *8*, 1154–1177.

25. Sarich, M. Projected Transfer Operators. Ph.D. Thesis, Freie Universität Berlin, Berlin, Germany, 2011.

26. Sarich, M.; Schütte, C. Approximating selected non-dominant timescales by Markov state models. *Commun. Math. Sci.* **2012**, *10*, 1001–1013.

27. Schütte, C.; Noé, F.; Lu, J.; Sarich, M.; Vanden-Eijnden, E. Markov state models based on milestoning. *J. Chem. Phys* **2011**, *134*, 204105.

28. Faradjian, A.K.; Elber, R. Computing time scales from reaction coordinates by milestoning. *J. Chem. Phys.* **2004**, *120*, 10880–10889.

29. Roeblitz, S. Statistical Error Estimation and Grid-free Hierarchical Refinement in Conformation Dynamics. Ph.D. Thesis, Freie Universität Berlin, Berlin, Germany, 2008.

30. Djurdjevac, N.; Sarich, M.; Schütte, C. On Markov State Models for Metastable Processes. In Proceedings of the ICM 2010 as Invited Lecture, Hyderabad, India, 19–27 August 2010. Available online: http://www.biocomputing-berlin.de/biocomputing/en/?cmd=publication (accessed on 29 November 2010).

31. Horenko, I.; Dittmer, E.; Lankas, F.; Maddocks, J.; Metzner, P.; Schütte, C. Macroscopic dynamics of complex metastable systems: Theory, algorithms, and application to B-DNA. *J. Appl. Dyn. Syst.* **2008**, *7*, 532–560.

32. Weber, M.; Bujotzek, A.; Haag, R. Quantifying the rebinding effect in multivalent chemical ligand-receptor systems. *J. Chem. Phys.* **2012**, *137*, 054111.

33. Bowmana, G.R.; Huangb, X.; Pande, V.S. Using generalized ensemble simulations and Markov state models to identify conformational states. *Methods* **2009**, *49*, 197–201.

34. Schütte, C.; Sarich, M. *Metastability and Markov State Models in Molecular Dynamics. Modeling, Analysis, Algorithmic Approaches (Courant Lecture Notes No. 24)*; AMS: Providence, RI, USA, 2013.

35. Weber, M.; Fackeldey, K. Computing the minimal rebinding effect included in a given kinetics. *Multiscale Model. Simul.* **2013**. Available onlin: http://www.zib.de/en/numerik/publications.html (accessed on 1 December 2013).

36. Fleming, W.; Soner, H. *Controlled Markov Processes and Viscosity Solutions*; 2nd ed.; Springer: New York, NY, USA, 2005.

37. Oksendal, B. *Stochastic Differential Equations*; Springer: Heidelberg, Germany, 2003.

38. Pra, P.; Meneghini, L.; Runggaldier, W. Connections between stochastic control and dynamic games. *Math. Control Signals Syst.* **1996**, *9*, 303–326.

39. Kushner, H.; Dupuis, P. *Numerical Methods for Stochastic Control Problems in Continuous Time*; Springer Verlag: New York, NY, USA, 1992.

40. Braess, D. *Finite Elements: Theory, Fast Solvers and Applications in Solid Mechanics*; Cambridge University Press: Camebridge, UK, **2007**.

41. Banisch, R.; Hartmann, C. A meshfree discretization for optimal control problems. *SIAM J. Control Optim.* **2013**, submitted.

42. Sheu, S. Stochastic control and exit probabilities of jump processes. *SIAM J. Control Optim.* **1985**, *23*, 306–328.

43. Latorre, J.; Metzner, P.; Hartmann, C.; Schütte, C. A Structure-preserving numerical discretization of reversible diffusions. *Commun. Math. Sci.* **2011**, *9*, 1051–1072.

44. Chodera, J.D.; Elms, P.J.; Swope, W.C.; Prinz, J.H.; Marqusee, S.; Bustamante, C.; Noé, F.; Pande, V.S. A robust approach to estimating rates from time-correlation Functions. **2011**, arXiv:1108.2304.

45. Vanden-Eijnden, E. Transition path theory. In *Computer Simulations in Condensed Matter Systems: From Materials to Chemical Biology Volume 1*; Ferrario, M., Ciccotti, G., Binder, K., Eds.; Springer: Berlin/Heidelberg, Gemary, 2006; Volume 703, pp. 353–493.

Reprinted from *Entropy*. Cite as: Delle Site, L. What is a Multiscale Problem in Molecular Dynamics?. *Entropy* **2014**, *16*, 23–40.

Concept Paper

What is a Multiscale Problem in Molecular Dynamics?

Luigi Delle Site

Institute for Mathematics, Freie Universität Berlin, Arnimallee 6, Berlin D-14195, Germany;
E-Mail: luigi.dellesite@fu-berlin.de; Tel.: +49-(0)-30-838-75775; Fax: +49-(0)-30-838-75412

Received: 25 June 2013; in revised form: 7 August 2013 / Accepted: 11 September 2013 / Published: 27 December 2013

Abstract: In this work, we make an attempt to answer the question of what a multiscale problem is in Molecular Dynamics (MD), or, more in general, in Molecular Simulation (MS). By introducing the criterion of separability of scales, we identify three major (reference) categories of multiscale problems and discuss their corresponding computational strategies by making explicit examples of applications.

Keywords: multiscale modeling; quantum; classical atomistic; coarse graining; adaptive resolution

1. Introduction

One of the major challenges of Molecular Dynamics (MD) over the last decade has been the development and application of techniques that allow the bridging of length and time scales in a physically consistent way. The relevance of such an effort is obvious: the understanding of the microscopic origin of large-scale properties leads to a deeper knowledge of physical phenomena and, when required, to the design of physical systems with specific properties on demand. The computational and conceptual progress in bridging scales, in condensed matter, material science, chemical physics and related fields, has been rather massive, and nowadays, the expression "*multiscale modeling*" has become almost routine. However, what is exactly meant by multiscale modeling is not clear yet. Obviously, one must go beyond the approach of combining, in a brute force fashion, different simulation techniques designed for different scales; computers will always give an answer; however, physical consistency must not be violated beyond the level of a controlled error/approximation. In this work, we discuss a possible classification of multiscale problems and relate them to the corresponding computational techniques and to the idea of physical consistency.

The paper is organized as follows. Based on the concept of scale separability, that is, how much scales can be separated in a system, we will identify three major categories: problems with "separated scales", those with "separable scales" and those with "highly-interconnected scales". Next, we will treat for each case some specific examples taken from applications and discuss the corresponding computational strategy. The final part will be dedicated to one (specific) emerging scale-coupling technique, that is, the adaptive resolution MD approach. This latter allows, in principle, one to describe in a unified simulation framework, the molecular (chemical) origin of large-scale properties and, thus, to interpret the multiscale idea in its full meaning.

2. Separability of Scales

Rather than providing a universal definition of multiscale, we instead introduce an objective criterion and define the idea of multiscale accordingly. The criterion in question is the "separability of scales", that is, how much in space and time, given some properties or phenomena of interest, scales can be separated in a system. Obviously, this separation is never sharp, and scales are never exactly disjoint; for this reason, this classification must be intended only as a general reference scheme. According to the concept above, we have identified three major categories: "separated scales", "separable scales" and "highly-interconnected scales"; below, we comment on each specific category.

- *Separated Scales*: This corresponds to the typical situation where simulation data from a scale (let us say), A, are used as input for modeling a molecular system at scale B. Next, simulations at scale B can proceed, and the connection to scale A is no longer required. The corresponding computational strategy goes under the name of "sequential strategy". Just to give an immediate idea, this is the typical case of molecular-based coarse graining: one needs an atomistic simulation to derive an effective coarse-grained model. Next, the coarse-grained simulation can run without any reference to atomistic details, provided that the properties of interest can be described by the coarse-grained model only.
- *Separable scales*: As for the situation above, scale A is used to model scale B. Next, scale B evolves, but differently from the case above, in this case one needs to go back from B to A, refine the model and start again. A loop-like strategy involves the two scales, *i.e.*, $A \rightarrow B \rightarrow A \rightarrow B$........ The corresponding computational strategy is usually called a "backmapping scheme".
- *Highly-interconnected scales*: Scales cannot be separated at all, and actually, it is exactly the interplay between different scales that characterizes the essential physics and chemistry of the system. The corresponding computational strategy requires that the different scales are treated on the same footing in a simultaneous coupling; for this reason, the technique is named "concurrent coupling".

Above, we have reported, for simplicity, only the case of a two scales problem; however, the extension to multiple scales is obvious. Moreover, in real applications, multiscale problems occupy the spectrum of categories given above in a sort of continuous way and, thus, require various combinations of the computational strategies reported above. It must be clarified that the separability of scales is not the only criterion possible for such a classification. An example of a complementary criterion is the one used by Berendsen: scales' hierarchy. This stems from deciding *a priori* which scale is more relevant in a problem rather than looking primarily at how scales are connected [1,2]. An overview of how the idea of multiscale is interpreted in the field of condensed matter, material science, chemical physics and related disciplines, together with recent method developments and applications, can be found in [3,4]. The fact that two relevant journals in the field dedicate an entire issue to the subject is an implicit confirmation of the relevance of the idea within the community of chemists and physicists (and even beyond, e.g., to mathematicians, engineers, biologists). In the next section, we will provide few examples of problems where the abstract classification defined above finds practical application in the field of MD.

3. Separated Scales and Sequential Strategy

Let us start by discussing an example where scales can be separated in a quite good approximation, and thus, the sequential coupling strategy is appropriate. The system we will discuss is that of macromolecular samples on solid inorganic surfaces. There are two main aspects in this problem:

- adhesion of the macromolecule at the surface, dominated by the specific local chemistry of the polymer moieties and by the specific reactivity of the surface;
- bulk properties of the macromolecular liquid, dominated by molecular packing and by slow equilibration processes.

The first aspect requires the detailed description of specific chemistry and its corresponding electronic properties; thus, a quantum mechanical description is mandatory. The second aspect is dominated by the entropic character of the chain entanglements in bulk and, thus, requires classical statistical methods, which properly sample the vast configurational space of a liquid. When we put the solid surface in contact with the macromolecular liquid, the properties of interest are those at the interface. These would emerge as a result of a non-trivial combination of adhesion and molecular packing. From a methodological point of view, this implies that one must combine, in a proper consistent way, quantum mechanics and classical statistical mechanics. As an example, this idea has been put in practice for a polycarbonate melt on a surface of nickel (111) [5,6]. Figure 1 gives a pictorial explanation of the idea. First, quantum mechanical calculations are performed for each isolated polymer subgroup. However, calculations are done by taking into account all possible allowed geometries at the surface consistent with the topological constraints of the large polymer; then, an effective moiety-surface potential is derived. In parallel, a coarse-grained (bead and spring) model for the polymer, which reproduces the bulk properties of the liquid, is derived from short full atomistic simulations. Finally, a coarse-grained simulation of a large system with the quantum-based surface-polymer interaction is performed [7]. In this specific case, we found that only phenolic chain ends experience a strong attraction; internal beads or other suitable chemical modifications of the chain ends, experience the surface as a hard wall. The results of this study allow one, then, to establish whether the interface properties are energy dominated (polymers are grafted onto the surface, and this leads to a sort of brush-like interface structure) or entropy dominated (polymers are topologically confined by a purely repulsive surface, and this leads to a parallel layering of the liquid) [8–10]. The same general idea has been later on extended to the case of the adsorption of large biomolecules out of solution on metal surfaces [11–15].

Figure 1. Pictorial representation of the computational strategy adopted for studying a melt of polycarbonate on a Ni(111) surface. On the left side, part (**c**), the explicit chemistry of the submolecular unit studied at the quantum level on the nickel surface. Part (**a**) illustrates the corresponding bead and spring coarse-grained model with the underlying atomistic structure. Part (**b**) represents one polymer out of the melt at the surface, interacting with the effective, quantum-based-derived potential (only the phenolic chain ends sticks onto the surface). On the right side, the cartoon summarizes the result of the simulation at a large scale; as a function of the chemical specificity of the chain ends, one can go from energy-dominated interface properties to entropy-dominated interface properties.

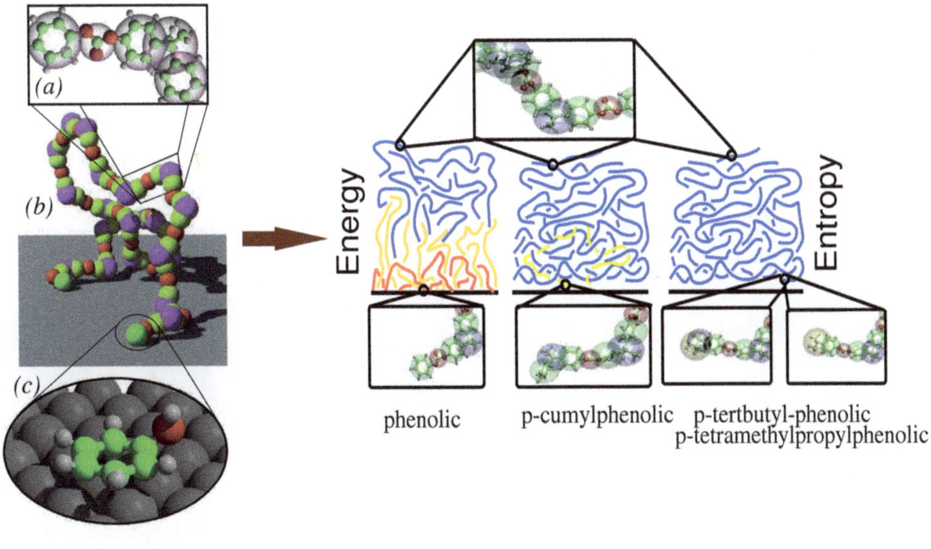

The case illustrated above is a sequential strategy in the "bottom-up" fashion, that is, from a finer to a coarser scale; however, the sequential strategy can be applied in the other direction, in a "top-down" fashion, that is, from a coarse scale down to a refined finer scale. For example, this was done by Zhu and Hummer (see, also, Figure 2 for a pictorial representation) when studying the gating transition in biological ion channels: the transition from opened to closed configuration (and *vice versa*) is done at a coarse-grained level and later refined at the atomistic level to study the specific role of hydration, which requires the atomistic resolution of water molecules [16,17].

In the next section, we will make a step forward and consider those cases where hopping between scales is required.

284

Figure 2. Pictorial representation of the top-down strategy adopted by Zhu and Hummer for studying the gating transition in biological ion channels. The first step consists of describing the large-scale conformational changes using a computationally affordable and physically consistent coarse-grained model. Next, atomistic details are reinserted, and finally, water is treated with atomistic resolution; this allows for investigating the role of channel hydration in the gating transition. This figure is adapted from Figure 2 of [17] with the permission of the corresponding author.

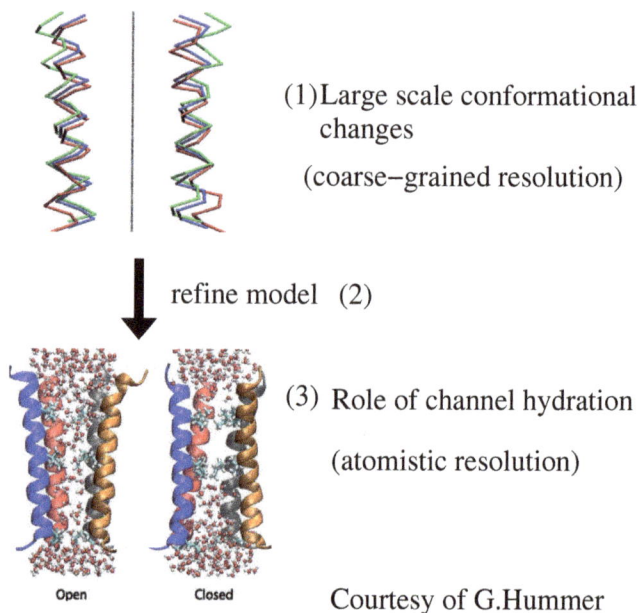

(1) Large scale conformational changes

(coarse–grained resolution)

refine model (2)

(3) Role of channel hydration

(atomistic resolution)

Courtesy of G.Hummer

4. Separable Scales and Backmapping Strategy

The sequential strategy of the chapter above cannot be applied when, despite a clear separation of scales in space and time, the evolution of the system requires a refinement of the coarser model as the simulation proceeds. This case of "separable", but not fully separated, scales is illustrated in this section via the example of photoswitchable liquid crystals. The physical ingredients of the problem are the following: liquid crystals containing azobenzene groups (see Figure 3) upon illumination can isomerize by changing conformation from the *trans* to the *cis* state.

This is the basic mechanism for light-induced mesoscopic transition: upon illumination, one can have a transition from the nematic to the isotropic phase, as shown by Ikeda and Tsutsumi [18]. From the computational point of view, the scales involved are the electronic/quantum, the classical atomistic and the coarse-grained. From the quantum mechanical point of view, one must describe the photoinduced electronic excitation and the possible consequent isomerization. The classical atomistic level is then required, because as the molecule is excited, a certain intermediate conformation is taken; however, the conformation resulting from the de-excitation (*i.e.*, staying in the *trans* or isomerizing in the *cis* state) strongly depends on the immediate atomistic environment; this, in turn, leads to a local rearrangement of the surrounding molecules. Finally, the coarse-grained scale is required to describe the large conformational response in the bulk involving the slow relaxation process of the photo-mechanical response. From the point of view of the computational strategy, one needs to link quantum, classical atomistic and coarse-grained in a consistent way [19–22]. Figure 4 illustrates the strategy. First, a coarse-grained model is derived from an all atom simulation of a relatively small system. Next, a large coarse-grained sample is simulated for long enough to have bulk equilibration. From the equilibrated system, a subsystem is cut out, and the atomistic degrees of freedom are mapped back. Next, one runs a simulation that ensures atomistic equilibration, and then, from the atomistic sample, a subsystem is cut out. Next, in this subsystem, the excitation of one molecule is allowed by treating the problem at quantum mechanical level. After the system decays from the excited state and the electronic degrees of freedom are equilibrated, the resulting configuration of the subsystem is reinserted into the the the classical atomistic sample, equilibrated at the atomistic level, then, reinserted into the coarse-grained sample and equilibrated at the coarse-grained level; at this point, the loop is repeated. The meaning of separable becomes clear: time scales are separated at least by one order of magnitude (order of femtoseconds for the quantum, at least picoseconds, for the atomistic, and at least nanoseconds, for the coarse grained). Length scales are obviously separated as well, thus space and time scales are separable. However, the process at each scale is intimately linked to the response at the other scale.

Figure 3. The azobenzene group in the *trans* (**top**) and *cis* (**bottom**) configuration. Upon illumination, the isomerization can take place; the molecule goes through an intermediate configuration corresponding to the excited state and, then, decays, either back into the *trans* or isomerizes into the *cis* state; this depends on the immediate atomistic neighborhood.

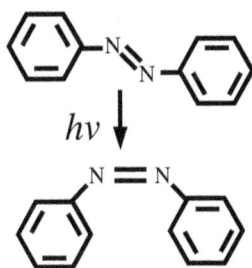

At this point, it must be clarified that the idea of the example discussed above can capture, at this stage, only the response of the system to the excitation/de-excitation of one single molecule per time and cannot directly address the question of how many molecules can concurrently undergo the *cis-trans* transition. In order to model this more realistic scenario, one would require treating larger quantum systems in order to understand how excited molecules influence each other. At the current state-of-the-art, this would imply a prohibitive computational effort, even for the simple case of three molecules treated at the quantum level. Tests (quantum calculations) are being performed for the case of two molecules in order to understand, at a basic level, the mutual influence of excited molecules. The idea, in perspective, is that of including the information obtained by the quantum studies of one molecule and two molecules into a classical model of an azobenzene molecule that can switch mechanically from *trans* to *cis* [23], under the hypothesis that, further, many-molecule effects, at the quantum level, may be negligible. Next, one would use the multiscale simulation, including the quantum subsystem, as a reference for a test of basic consistency of the classical model. If the test is satisfactory, then the question of how many molecules can concurrently undergo the *cis-trans* transition could be treated at the (classical) atomistic-coarse-grained level, keeping in mind that beyond two-molecule correlations, the quantum effects in the switching process are neglected. Obviously, when larger computational resources become available, one could systematically improve the classical model, adding information from larger quantum calculations. Anyway, in the current paper, the specific strategy reported above has to be intended as a typical example of a problem where going back and forward from one scale to another is the main characteristic of the modeling idea; however, from the practical point of view, it shows also that, because of the current computational limitations, the whole complexity of the problem can only be addressed in an approximate way. Nevertheless the relevant message here is that, because of the clear separation of time and length scales, the basic strategy of going back and forward is still the optimal one, even in the case that computational resources were available for studying macroscopic systems. Finally, in the next section, we will describe the case where scales cannot be separated and, thus, a simultaneous coupling of the corresponding computational techniques and models is required.

Figure 4. The backmapping loop. Following the black-arrowed line (**I**), derive a suitable coarse-grained model; (**II**) then, equilibrate a large coarse-grained sample; (**III**) next, cut out a subsystem and map back the atomistic degrees of freedom; (**IV**) the atomistic sample is then equilibrated; and finally, (**V**) a subsystem is cut out from it, and for this latter, the quantum process is allowed. Once the quantum process has occurred, the procedure continues by reinserting the subsystem into the atomistic sample, and then, the loop is repeated (red-arrowed line).

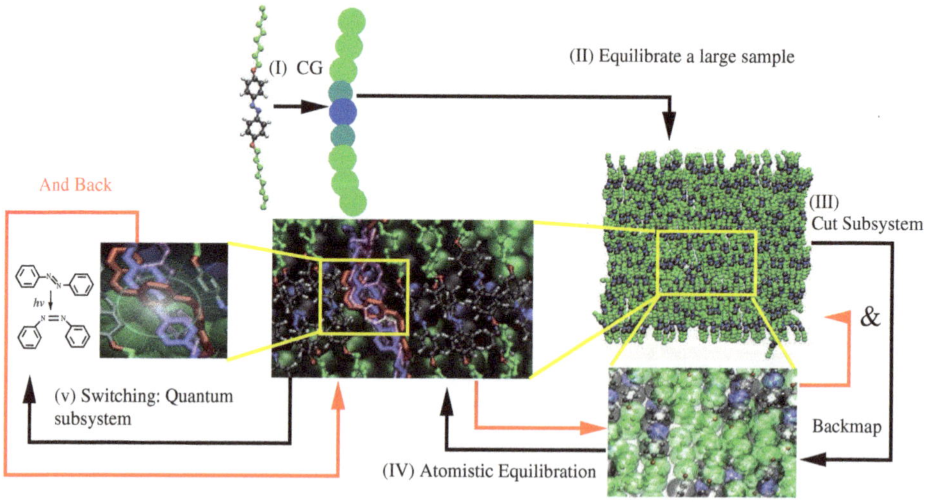

5. Highly-Interconnected Scales and Concurrent Coupling

As reported above, the backmapping strategy is rather efficient if there is a clear separation in length and time scales; however, when scales are intimately interconnected, in most of the cases, the relevant properties of the system are the expression of this interconnection. From the computational point of view, it is then mandatory to adopt strategies where, within a unified approach, all the scales are treated at the same time, *i.e.*, a simultaneous or concurrent coupling of scales. A typical example is that reported in Figure 5 of solvation of molecules in water. High resolution—at least classical atomistic—is needed in the first hydration shells of the molecule, where the explicit formation of the hydrogen bonding network is required, as this uniquely characterizes the solvation of the specific molecule. Far away in the bulk, water plays the role of a thermodynamic bath, and thus, it may be described by coarser models, which reproduce the essential thermodynamic features (e.g., temperature and particle density fluctuations). However, the scale at which the first hydration shells evolve cannot be separated by the thermodynamic scale of the bulk, since between the two regions, there is a simultaneous exchange of information in terms of energy and (eventually) particles.

For this reason, both scales must be treated in a simultaneous fashion, taking care that the overall thermodynamics is well preserved. Popular computational approaches of this kind are, for example, the Quantum Mechanical/Molecular Mechanical (QM/MM) set-ups, where a small quantum region is coupled to a large atomistic or coarse-grained classical region, allowing free exchange of energy (though not particles, see the discussion later on) (see, e.g., [24]). The set-up mentioned above allows then for studying systems where the local process can be linked to the global behavior; for example, *in situ* simulation of chemical reactions and molecular excitations in solution, where the chemical reaction or the molecular excitation occurs at the quantum level in a small region of the simulation space, while the bulk solvent can be treated at the classical atomistic level [25]. Another example, relevant for mechanical engineering, is that of the crack propagation in solids. The breaking of atomic bonds in the region of the crack must be treated at the quantum level, because this is an electronic property; then, the surrounding region can be described at the classical atomistic level, so that one can see the induced crystal distortion. Finally, the bulk material is described at the continuum (finite elements) level in order to detect its macroscopic mechanical changes. However, all the scales exchange information simultaneously as the crack propagates, and thus, they are coupled in a simultaneous fashion [26]. Although the dividing line between separable scales and interconnected scales is not sharp, one can see a clear difference between the examples reported in this section and that of the azobenzene systems of the previous section. In fact, let us suppose, for example, that we were interested only in the influence of the immediate molecular neighborhood onto the excitation and de-excitation of an azobenzene molecule. In this case, a QM/MM approach would be highly appropriate, because the local liquid structure and its local fluctuations would slowly follow (and at the same time, influence) the evolution of the electronic and conformational properties of the excited molecule. In fact, in the multiscale study of the azobenzene system, the QM/MM approach was used for quantum calculations; however, the macroscopic response of the bulk cannot be predicted only by QM/MM calculations, because it would require a size for the MM system and a time scale that are, at this stage, computationally prohibitive. For this reason, the coarse-grained approach for obtaining relaxed macroscopic configurations is in this case mandatory. The three categories of problems so far discussed provide an overview of what in the literature can be classified as multiscale. However, there is an underlying general message in all the examples made: multiscale essentially means the interplay between local and global aspects (in space and time). Thus, the detailed understanding of the molecular origin of macroscopic properties requires a step forward, beyond the strategies shown so far (or their possible combinations). This will be discussed in the next chapter, where we introduce the idea of adaptive resolution simulation.

Figure 5. Pictorial representation of a molecule solvated in water. The first hydration shells must be treated at least at the atomistic level, so that the formation of the hydrogen bonding network can be explicitly described. Far away in the bulk, a coarse model (e.g., spherical) can be employed to assure the proper thermodynamic bath. However, the two scales are coupled simultaneously. Note that in standard computational set-ups, such as the one of Quantum Mechanical/Molecular Mechanical (QM/MM) discussed in the text, the high resolution region is fixed and particles cannot be exchanged; thus, there is only an exchange of energy.

Bulk=Thermodynamic Bath

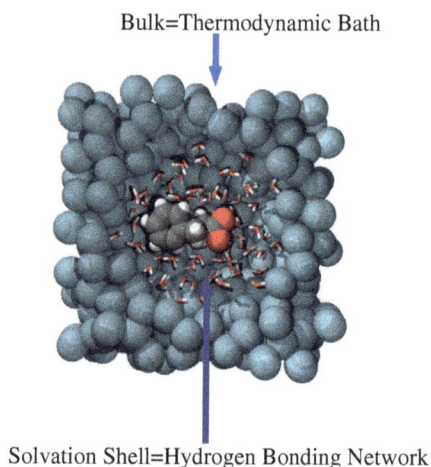

Solvation Shell=Hydrogen Bonding Network

6. Molecular Origin of Macroscopic Properties: Zooming in at the Molecular Scale

The molecular origin of macroscopic properties can be understood by zooming in (and out) in the region where the relevant microscopic physics and chemistry is taking place.

Figure 6 explains this idea for two examples previously discussed, namely, the adsorption of a large molecule on a solid surface and the solvation of a molecule in water. In the first case, while the molecule is far away from the surface, the only relevant physics is related to the proper sampling of the conformational space of the backbone; thus, a simplified bead and spring model would be sufficient for this purpose. Instead, as the molecule approaches the surface, one needs to zoom in (put the system under a magnifying glass) at the contact region and have an explicit atomistic description, so that the chemical recognition between the molecule and the surface can be properly described and be understood together with the consequent conformational rearrangement of the rest of the molecule (at a coarser scale). The same kind of idea applies to the solvation of molecules in water; the specific solvation structures of the liquid can be understood by zooming in at the hydration region around the solute: when a water molecule enters under the viewing region of the magnification glass, it must be described at the atomistic level; when it leaves, it then takes a coarse-grained description. This process requires that the high resolution region be open and allow for a free exchange of molecules. This, from the methodological point of view, implies that one must go beyond the idea of concurrent coupling to that of "adaptive resolution simulation".

Figure 6. On the left side, (**a**), zooming in on the contact region between the molecule and the surface. The magnifying glass is intended as a computational tool to introduce explicit chemistry and atomistic structure, so that the process of chemical recognition between the molecule and the solid surface can be properly described together with the simultaneous evolution of the large-scale conformational changes of the polymer. On the right side, (**b**), the same idea, but for the solvation process. All the molecules under the magnifying glass must have atomistic resolution, so that the hydrogen bonding network of the hydration shell can be properly described. Molecules that leave the solvation region loose atomistic degrees of freedom.

(a)

(b)

6.1. Beyond Concurrent Coupling: Adaptive Resolution Molecular Dynamics

From the methodological point of view, the essential requirements of an adaptive resolution molecular dynamics scheme are the following:

- (**i**) The algorithm should change molecular resolution in a subregion of the space, leaving the rest of the system at lower resolution.
- (**ii**) It should allow for free exchange of molecules from the high resolution to the low resolution region and *vice versa*.

- **(iii)** Finally, the process of (i) and (ii) should occur under conditions of thermodynamic equilibrium: *i.e.*, the same particle density, same temperature and same pressure all over the simulation box ($\rho_{atom} = \rho_{cg}$, $p_{atom} = p_{cg}$, $T_{atom} = T_{cg}$).

Of course, the thermodynamic state point must be the same as if the whole system was described at high resolution. Several adaptive resolution methods, which sometimes satisfy and sometime do not satisfy the requirements above, have been proposed in the last few years [27–31]. While most of the methods can switch only between two resolutions, the AdResS method has extended the idea of adaptivity from the quantum description of atoms [32–34] to the continuum description of a liquid [35,36]. The original idea was to have an on-the-fly interchange between atomistic and coarse-grained description of a liquid through a two-stage procedure. First, develop an effective, coarse-grained pair potential, U^{cm}, from the reference all atom simulation. Next, the atomistic and coarse-grained resolutions are coupled through an interpolation formula on the forces:

$$\mathbf{F}_{\alpha\beta} = w(X_\alpha)w(X_\beta)\mathbf{F}_{\alpha\beta}^{atom} + [1 - w(X_\alpha)w(X_\beta)]\mathbf{F}_{\alpha\beta}^{cm} \tag{1}$$

Here, α and β are the labels of two molecules, $\mathbf{F}_{\alpha\beta}^{atom}$ is the force derived from the atomistic potential and $\mathbf{F}_{\alpha\beta}^{cm}$ is the force derived from the coarse-grained potential. X_α and X_β are the coordinates of the center of mass of, respectively, the molecule α and the molecule β. The multiplicative function, $w(x)$, is zero in the coarse-grained region, one in the atomistic region and smooth and monotonic in an intermediate region, Δ. Figure 7 shows the idea for a test molecule (tetrahedral molecule), left, coarse grained, in Δ, a hybrid resolution according to $w(x)$ and, on the right, atomistic.

Figure 7. Schematic representation of the adaptive idea for tetrahedral molecules.

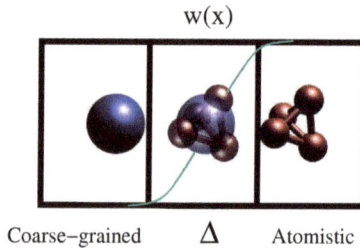

With this set-up, two atomistic molecules interact as atomistic, coarse-grained molecules interacts with all the others as coarse-grained pairs (coarse-grained molecules do not posses any atomistic degrees of freedom), while for the other cases, molecules interact according to their coupled value of $w(X_\alpha)w(X_\beta)$ with hybrid resolutions. This means that a molecule that goes from the atomistic to the coarse-grained region slowly looses its atomistic degrees of freedom (rotations and vibrations) and becomes an effective sphere, going through a continuous stage of hybrid resolutions in Δ. The same process, but in the opposite direction (acquiring degrees of freedom) occurs to a coarse-grained molecule moving towards the atomistic region. At this point, the "technical" meaning of "loosing" or "acquiring" degrees of freedom in the boundary (hybrid) region that divides the atomistic and the coarse-grained region must be clarified. All molecules of the system, independently of their position in space, retain the full atomistic structure. For a molecule, α, in the hybrid region, the force acting on each single atom (derived from the atomistic potential) is weighted by an amount, $w(X_\alpha)$, multiplied by the weight, $w(X_\beta)$, of the paired molecule, β, and, by construction, the remaining total force acting on the molecule (derived from the coarse-grained potential) is assigned to the center of mass by the weighting term, $[1 - w(X_\alpha)w(X_\beta)]$. Since $w(x)$ goes to zero in the coarse-grained region, the closer the molecule is to the coarse-grained region, the weaker is the force (derived from the atomistic potential) acting on each atom (and the larger the force on the center of mass). When the molecule enters into the coarse-grained region, the only force acting on the molecule is that on the center of mass, independently of the resolution of the paired molecule. The forces acting on each atomistic degrees of freedom coming from the atomistic potential are no more considered, although the underlying atomistic structure is artificially kept for technical reasons. In practice, the interactions of the atomistic degrees of freedom for this molecule are no more explicitly considered. This implies that their contributions to the dynamic evolution and to the energy of the system disappear. Moreover, the internal kinetic energy of the molecule (*i.e.*, kinetic energy associated with rotations and internal vibrations) is also not considered in the calculation of the properties of the system (thus, effectively, the molecule behaves as a sphere). The computational gain consists of a drastically reduced number of interactions that needs to be calculated. *Vice versa*, when a molecule enters into the transition region, the force acting on each single atom is slowly reactivated according to the weight, $w(X_\beta)$ (and to the weight of the paired molecule, $w(X_\beta)$). When the molecule reaches the atomistic region, we have $w(X_\alpha) = 1$, and thus, it interacts in the standard full atomistic manner with molecules of the atomistic region, while for the interaction with molecules in other regions, the weight on the atomistic force depends only on the position of the paired molecule (above is commented the extreme case of the paired molecule being in the coarse-grained region). A detailed explanation of all the technical details of the implementation of this idea can be found in [37]. Obviously, in the diving boundary region between the atomistic and coarse-grained region, one must properly deactivate/reactivate the atomistic degrees of freedom. This is done by introducing an external source of heat (thermostat), which acts locally on each degree of freedom in order to assure that despite the deactivated/reactivated degrees of freedom, the temperature in this region is the same as the target one (that is, the temperature of the atomistic and of the coarse-grained systems) [38–40]. Additionally an external force derived on the basis of the first principles of thermodynamics (*i.e.*, a force that balances the difference of the chemical potential between the atomistic region and the rest of the system) [41] is added to assure thermodynamic equilibrium. The natural question for a such molecular dynamics scheme is whether the force of Equation (1) is conservative, and the answer is negative. In fact, despite a different claim [29,42,43], it has been shown both analytically [44] and numerically [45] that within this scheme, there is no possibility of deriving Equation (1) from a potential. According to the statement above, a natural conclusion is that without a conservative force one cannot be in the microcanonical (or canonical) ensemble, and thus, even time-independent properties cannot be accurately calculated. However, it has been shown that the atomistic region (the most delicate case) is characterized by a Grand Canonical distribution [46], and thus, the method reproduces the same statistics of the equivalent region in a full atomistic system (the key property of any valid adaptive resolution method!). Moreover, an exact Hamiltonian can be written for the atomistic (and coarse-grained) region, where exact means that each term of the Hamiltonian is physically well defined without introducing any artificial (unphysical) quantity; this, then, allows for a full Grand Canonical-like formalization of the adaptive set-up [47,48]. In general, all the relevant properties (radial distribution functions, density distribution across the simulation box, molecular diffusion, particle density fluctuations and solvation structures) determined via full atomistic simulation for several liquids and solvated systems were properly reproduced (see, e.g., [27,38,41,49–53]). However, a key issue remains unresolved, that is, whether or not there exists a well-founded Hamiltonian route to adaptive resolution simulation. If a Hamiltonian approach exists, it must fulfill the following necessary requirements:

- It should lead to the correct limits. That is, it should be an atomistic Hamiltonian in the atomistic region and a coarse-grained Hamiltonian in the coarse grained region
- It should automatically lead, in the atomistic region, to a spatial probability distribution, which is the same as that in the corresponding subregion in a full atomistic simulation.

Recently, Potestio *et al.* proposed a global Hamiltonian approach of AdResS [54]. This is based on the interpolation in space of the atomistic and coarse-grained potential, that is, in this case, the potential is interpolated instead of the forces, as in the standard AdResS. The idea provides an elegant thermodynamic procedure of how to equilibrate the system, but, by construction, cannot satisfy the requirement of correct limit according to [44]. In fact, an additional force is generated by the gradient of the weighting function, $w(x)$. This force induces an unphysical flux of particles from one region to another. In order to balance this flux, an additional field must be added to the original Hamiltonian. In [54], an elegant thermodynamic procedure is used to determine this field. However, in [44], it has been shown that such a field is a solution of a first order partial differential equation, and in order to describe a proper adaptive system, it requires that the field satisfies two boundary conditions; that is, it must be zero in the atomistic and in the coarse-grained region. Since the equation is of the first order, only one boundary condition can be used to fix the solution. Let us suppose we fix the boundary condition in the atomistic region; then, inevitably, in the coarse-grained region, the original potential is changed by an artificial, unphysical additional term. This implies that if we, ideally, shrink the atomistic region to zero, we do not recover the original coarse-grained potential. The additional term is constant in space, but strongly depends on the size of the system; although this may turn out to be a non-relevant aspect from the technical point of view, however, from the conceptual point of view, this makes the Hamiltonian, and, thus, the corresponding adaptive process, artificial. Moreover, at a more practical level, according to [48], as it stands now, the method can assure that in the atomistic region (when compared to a reference full atomistic simulation), only the first order of the probability distribution of the system, that is, the molecular number density, can be obtained with high accuracy. Higher orders, as atom-atom (or even molecule-molecule) radial distribution functions or three-body correlations, do not come automatically. Nevertheless, the approach of [54] may represent the closest (somehow "first order") procedure for a truly adaptive Hamiltonian procedure with the further, major, advantage of the possibility of performing adaptive Monte Carlo simulations [55]. In conclusion, a truly, satisfactory, adaptive Hamiltonian has not been found yet, and my personal opinion is that this is not really needed; however, the cultural barrier in the MD community is such that, unfortunately, a method without a Hamiltonian is seen as a problem, rather than an alternative to the standard routes. The consequence is that often, the various additions to the original idea of AdResS, despite being based on clear thermodynamic arguments, are considered more at the level of practical patches, to cover, at the best, unavoidable conceptual holes, due to the lack of a Hamiltonian, rather than a natural methodological evolution *per se*.

On the other hand, one should keep in mind the following: the constraint that the atomistic region of the adaptive simulation should reproduce the probability distribution of the full atomistic system (that is, the key criterion to evaluate the quality of any adaptive method) can be easily implemented within the standard force-based AdResS, but (so far) not in a Hamiltonian one [48]. In [47,48], first principles, analytic/numerical conditions on the probability distribution of the system have been defined in such a way that the accuracy in the high resolution region can be controlled and systematically improved. Moreover, in [48], it has been also shown that the method is also a very powerful tool for the calculation of the chemical potential of complex liquids, at a computational cost that is orders of magnitude below that of the Insertion Particle Method, routinely used for calculating such a quantity.

7. Conclusions

We have addressed the question of how to provide a concrete meaning to the expression *multiscale modeling* in Molecular Dynamics (simulation). The use of the criterion of separability of scale allows for classifying multiscale problems in three main categories. Next, the most popular computational strategies adopted for each category have been discussed. In this perspective, we have underlined the role of the concurrent coupling of scales via the idea of adaptive molecular resolution. In principle, this approach can be considered a truly multiscale technique, which acts as a sort of computational microscope to determine the molecular origin of the large-scale behavior of matter. In fact, it can zoom in to the very microscopic detail and zoom out to the large-scale behavior while the simulation is running. Future developments should be focused on including electronic resolution and, thus, address the question of interfacing a quantum and a classical system in a physically-consistent way.

Acknowledgments

This work was supported by the Deutsche Forschungsgemeinschaft (DFG) within the Heisenberg program (grant code DE 1140/5-1).

Conflicts of Interest

The authors declare no conflict of interest.

294

References

1. Berendsen, H.J.C. *Simulating the Physical World, Hierarchical Modeling from Quantum Mechanics to Fluids*; Cambridge University Press: Cambridge, UK, 2007.
2. Berendsen, H.J.C. Concluding remarks. *Faraday Discuss.* 144, 467–481.
3. Themed issue: Multiscale modeling of soft matter. *Faraday Discuss.* 2010, *144*, 1–494.
4. Themed issue: Multiscale modeling. *Phys. Chem. Chem. Phys.* 2010, *13*, 10381–10583.
5. Delle Site, L.; Abrams, C.F.; Alavi, A.; Kremer, K. Polymers near metal surfaces: Selective adsorption and global conformations. *Phys. Rev. Lett.* 2002, *89*, 156103.
6. Delle Site, L.; Kremer, K. Multiscale modeling of polymers on a surface: From ab initio Density Functional calculations of molecular adsorption to large scale properties. *Int. J. Quant. Chem.* 2005, *101*, 733–739.
7. Abrams, C.F.; Delle Site, L.; Kremer, K. Dual-resolution coarse-grained/atomistic simulation of the bisphenol-a-polycarbonate/nickel interface. *Phys. Rev. E* 2003, *67*, 021807.
8. Delle Site, L.; Leon, S.; Kremer, K. BPA-PC on a Ni(111) surface: The interplay between adsorption energy and conformational entropy for different chain-end modifications. *J. Am. Chem. Soc.* 2004, *126*, 2944-2955 .
9. Delle Site, L.; Leon, S.; Kremer, K. Specific interaction of polymers with surface defects: Structure formation of polycarbonate on nickel. *J. Phys. Condens. Matter* 2005, *17*, L53–L60.
10. Andrienko, D.; Leon, S.; Delle Site L.; Kremer, K. Adhesion of polycarbonate blends on a nickel surface. *Macromolecules* 2005, *38*, 5810–5816.
11. Schravendijk, P.; van der Vegt, N.F.A.; Delle Site, L.; Kremer, K. Dual scale modeling of benzene adsorption onto Ni(111) and Au(111) surfaces in explicit water. *Chem. Phys. Chem.* 2005, *6*, 1866–1871.
12. Ghiringhelli, L.M.; Schravendijk, P.; Delle Site, L. Adsorption of alanine on a Ni(111) surface: A multiscale modeling oriented density functional study. *Phys. Rev. B* 2006, *74*, 035437 .
13. Schravendijk, P.; Ghiringhelli, L.M.; Delle Site, L.; van der Vegt, N.F.A. Interaction of hydrated amino acids with metal surfaces: A multiscale modeling description. *J. Phys. Chem. C* 2007, *111*, 2631–2642.
14. Ghiringhelli, L.M.; Delle Site, L. Phenylalanine near inorganic surfaces: Conformational statistics *vs.* specific chemistry. *J. Am. Chem. Soc.* 2008, *130*, 2634–2638.
15. Ghiringhelli, L.M.; Hess, B.; van der Vegt, N.F.A; Delle Site, L. Competing adsorption between hydrated peptides and water onto metal surfaces: From electronic to conformational properties. *J. Am. Chem. Soc.* 2008, *130*, 13460-13464.
16. Zhu, F.; Hummer, G. Gating transitions of pentameric ligand-gated ion channels. *Biophys. J.* 2009, *97*, 2456–2463.
17. Zhu, F.; Hummer, G. Pore opening and closing of a pentameric ligand-gated ion channel. *Proc. Natl. Acad. Sci. USA* 2010, *107*, 19814–19819.
18. Ikeda, T.; Tsutsumi, O. Optical switching and image storage by means of azobenzene liquid-crystal films. *Science* 1995, *268*, 1873–1875.

19. Böckmann, M.; Peter, C.; Delle Site, L.; Doltsinis, N.; Kremer, K.; Marx, D. Atomistic force field for azobenzene compounds adapted for QM/MM Simulations with applications to liquids and liquid crystals. *J. Chem. Theory Comput.* **2007**, *3*, 1789–1802.

20. Peter, C.; Delle Site, L.; Kremer, K. Classical simulations from the atomistic to the mesoscale and back: Coarse graining an azobenzene liquid crystal. *Soft Matter* **2008**, *4*, 859–869.

21. Böckmann, M.; Marx, D.; Peter, C.; Delle Site, L.; Kremer, K.; Doltsinis, N. Multiscale modelling of mesoscopic phenomena triggered by quantum events: Light-driven azo-materials and beyond. *Phys. Chem. Chem. Phys.* **2011**, *13*, 7604–7621.

22. Mukherjee, B.; Delle Site, L.; Kremer, K.; Peter, C. Derivation of coarse grained models for multiscale simulation of liquid crystalline phase transitions. *J. Phys. Chem. B* **2012**, *116*, 8474–8484.

23. Böckmann, M.; Doltsinis, N.; Marx, D. Azobenzene photoswitches in bulk materials. *Phys. Rev. E* **2008**, *78*, 036101.

24. Warshel, A.; Lewitt, M. Theoretical studies of enzymic reactions: Dielectric, electrostatic and steric stabilization of the carbonium ion in the reaction of lysozyme. *J. Mol. Biol.* **1976**, *103*, 227–249.

25. Röhrig,U.F.; Frank, I.; Hutter, J.; Laio, A.; VandeVondele, J.; Rothlisberger, U. QM/MM Car-Parrinello molecular dynamics study of the solvent effects on the ground state and on the first excited singlet state of acetone in water. *Chem. Phys. Chem.* **2003**, *4*, 1177–1182.

26. Lu, G.; Tadmor, E.B.; Kaxiras, E. From electrons to finite elements: A concurrent multiscale approach for metals. *Phys. Rev. B* **2006**, *73*, 024108.

27. Praprotnik, M.; Delle Site, L.; Kremer, K. Adaptive resolution molecular dynamics simulation: Changing the degrees of freedom on the fly. *J. Chem. Phys.* **2005**, *123*, 224106.

28. Praprotnik, M.; Delle Site, L.; Kremer, K. Multiscale simulation of soft matter: From scale bridging to adaptive resolution. *Annu. Rev. Phys. Chem.* **2008**, *59*, 545–571.

29. Ensing, B.; Nielsen, S.O.; Moore, P.B.; Klein, M.L.; Parrinello, M. Energy conservation in adaptive hybrid atomistic/coarse-grain molecular dynamics. *J. Chem. Theory Comput.* **2007**, *3*, 1100–1105.

30. Heyden, A.; Truhlar, D.G. Conservative algorithm for an adaptive change of resolution in mixed atomistic / coarse-grained multiscale simulations. *J. Chem. Theory Comput.* **2009**, *4*, 217–221.

31. S.Izvekov, I.; Voth, G.A. Mixed resolution modeling of interactions in condensed-phase systems. *J. Chem. Theory Comput.* **2009**, *5*, 3232–3244

32. Poma, A.B.; Delle Site, L. Classical to path-integral adaptive resolution in molecular simulation: Towards a smooth quantum-classical coupling. *Phys. Rev. Lett.* **2010**, *104*, 250201.

33. Poma, A.B.; Delle Site, L. Adaptive resolution simulation of liquid para-hydrogen: Testing the robustness of the quantum-classical adaptive coupling. *Phys. Chem. Chem. Phys.* **2011**, *13*, 10510–10519.

34. Potestio, R.; Delle Site, L. Quantum locality and equilibrium properties in low-temperature parahydrogen: A multiscale simulation study. *J. Chem. Phys.* **2012**, *136*, 054101.

35. Delgado-Buscalioni, R.; Kremer, K.; Praprotnik, M. Concurrent triple-scale simulation of molecular liquids. *J. Chem. Phys.* **2008**, *128*, 114110.
36. Delgado-Buscalioni, R.; Kremer, K.; Praprotnik, M. Coupling atomistic and continuum hydrodynamics through a mesoscopic model: Application to liquid water. *J. Chem. Phys.* **2009**, *131*, 244107.
37. Junghans, C; Poblete, S. A reference implementation of the adaptive resolution scheme in ESPResSo. *Comput. Phys. Commun.* **2010**, *181*, 1449–1454.
38. Praprotnik, M.; Delle Site, L.; Kremer, K. Adaptive resolution scheme (AdResS) for efficient hybrid atomistic/mesoscale molecular dynamics simulations of dense liquids. *Phys. Rev. E* **2006**, *73*, 066701.
39. Praprotnik, M.; Kremer, K.; Delle Site, L. Adaptive molecular resolution via a continuous change of the phase space dimensionality. *Phys. Rev. E* **2007**, *75*, 017701.
40. Praprotnik, M.; Kremer, K.; Delle Site, L. Fractional dimensions of phase space variables: A tool for varying the degrees of freedom of a system in a multiscale treatment. *J. Phys. A* **2007**, *40*, F281–F288.
41. Poblete, S.; Praprotnik, M.; Kremer, K.; Delle Site, L.; Coupling different levels of resolution in molecular simulations. *J. Chem. Phys.* **2010**, *132*, 114101.
42. Nielsen, S.O.; Bulo, R.E.; Moore, P.B.; Ensing, B. Recent progress in adaptive multiscale molecular dynamics simulations of soft matter. *Phys. Chem. Chem. Phys.* **2010**, *12*, 12401–12414.
43. Nielsen, S.O.; Moore, P.B.; Ensing, B. Adaptive multiscale molecular dynamics of macromolecular fluids. *Phys. Rev. Lett.* **2010**, *105*, 237802.
44. Delle Site, L. Some fundamental problems for an energy-conserving adaptive-resolution molecular dynamics scheme. *Phys. Rev. E* **2007**, *76*, 047701.
45. Praprotnik, M.; Poblete, S.; Delle Site, L.; Kremer, K. Comment on: Adaptive multiscale molecular dynamics of macromolecular fluids. *Phys. Rev. Lett.* **2011**, *107*, 099801.
46. Fritsch, S.; Poblete, S.; Junghans, C.; Ciccotti, G.; Delle Site, L.; Kremer, K. Adaptive resolution molecular dynamics simulation through coupling to an internal particle reservoir. *Phys. Rev. Lett.* **2012**, *108*, 170602.
47. Wang, H.; Schütte, C.; Delle Site, L. Adaptive resolution simulation (AdResS): A smooth thermodynamic and structural transition from atomistic to coarse grained resolution and vice versa in a grand canonical fashion. *J. Chem. Theory Comput.* **2012**, *8*, 2878–2887.
48. Wang, H.; Hartmann, C.; Schütte, C.; Delle Site, L. Grand-canonical-like molecular-dynamics simulations by using an adaptive-resolution technique. *Phys. Rev. X* **2013**, *3*, 011018.
49. Praprotnik, M.; Delle Site, L.; Kremer, K. A macromolecule in a solvent: Adaptive resolution molecular dynamics simulation. *J. Chem. Phys.* **2007**, *126*, 134902.
50. Praprotnik, M.; Matysiak, S.; Delle Site, L.; Kremer, K.; Clementi, C. Adaptive resolution simulation of liquid water. *J. Phys. Condens. Matter* **2007**, *19*, 292201.
51. Matysiak, S.; Clementi, C.; Praprotnik, M.; Kremer, K.; Delle Site, L. Modeling diffusive dynamics in adaptive resolution simulation of liquid water. *J. Chem. Phys.* **2008**, *128*, 024503.

52. Lambeth, B.P.; Junghans, C.; Kremer, K.; Clementi, C.; Delle Site, L. On the locality of hydrogen bond networks at hydrophobic interfaces. *J. Chem. Phys.* **2010**, *133*, 221101.

53. Mukherij, D.; van der Vegt, N.F.A.; Kremer, K.; Delle Site, L. Kirkwood-Buff analysis of liquid mixtures in an open boundary simulation. *J. Chem. Theory Comput.* **2012**, *8*, 375–379.

54. Potestio, R.; Fritsch, S.; Espanol, P.; Delgado-Buscalioni, R.; Kremer, K.; Everaers, R.; Donadio, D. Hamiltonian adaptive resolution simulation for molecular liquids. *Phys. Rev. Lett.* **2013**, *110*, 108301.

55. Potestio, R.; Espanol, P.; Delgado-Buscalioni, R.; Everaers, R.; Kremer, K.; Donadio, D. Monte Carlo adaptive resolution simulation of multicomponent molecular liquids. *Phys. Rev. Lett.* **2013**, *111*, 060601.

Reprinted from *Entropy*. Cite as: Hsieh, C.-Y.; Kapral, R. Correlation Functions in Open Quantum-Classical Systems. *Entropy* **2014**, *16*, 200–220.

Article

Correlation Functions in Open Quantum-Classical Systems

Chang-Yu Hsieh and Raymond Kapral *

Chemical Physics Theory Group, Department of Chemistry, University of Toronto, Toronto, ON M5S 3H6, Canada; E-Mail: kim.hsieh@utoronto.ca

* Author to whom correspondence should be addressed; E-Mail: rkapral@chem.utoronto.ca; Tel.: +1-416-978-6106 ; Fax: +1-416-978-5325.

Received: 25 September 2013; in revised form: 21 October 2013 / Accepted: 22 October 2013 / Published: 27 December 2013

Abstract: Quantum time correlation functions are often the principal objects of interest in experimental investigations of the dynamics of quantum systems. For instance, transport properties, such as diffusion and reaction rate coefficients, can be obtained by integrating these functions. The evaluation of such correlation functions entails sampling from quantum equilibrium density operators and quantum time evolution of operators. For condensed phase and complex systems, where quantum dynamics is difficult to carry out, approximations must often be made to compute these functions. We present a general scheme for the computation of correlation functions, which preserves the full quantum equilibrium structure of the system and approximates the time evolution with quantum-classical Liouville dynamics. Several aspects of the scheme are discussed, including a practical and general approach to sample the quantum equilibrium density, the properties of the quantum-classical Liouville equation in the context of correlation function computations, simulation schemes for the approximate dynamics and their interpretation and connections to other approximate quantum dynamical methods.

Keywords: quantum correlation functions; quantum-classical systems; nonadiabatic dynamics

1. Introduction

The dynamical properties of condensed-phase or complex systems are often investigated experimentally by applying external fields to weakly perturb a system and observe its relaxation back to the thermal equilibrium state. In such experiments, measurable quantities can be related to equilibrium time correlation functions via linear response theory [1,2]:

$$C_{AB}(t) = \frac{1}{Z_Q}\text{Tr}\left[e^{-\beta\hat{H}}\hat{A}(0)\hat{B}(t)\right] = \frac{1}{Z_Q}\text{Tr}\left[e^{-\beta\hat{H}}\hat{A}e^{\frac{i}{\hbar}\hat{H}t}\hat{B}e^{-\frac{i}{\hbar}\hat{H}t}\right] \tag{1}$$

where $\hat{A}$ and $\hat{B}$ are operators corresponding to some specific dynamical variables under investigation, $\hat{H}$ is the unperturbed Hamiltonian and Z_Q is the quantum canonical partition function associated with $\hat{H}$. Many experiments employing spectroscopic methods directly probe such time correlation functions.

Exact numerical evaluation of Equation (1) for real condensed phase quantum systems is prohibitive, since the computational cost scales exponentially with respect to the number of degrees of freedom (DOF). Various approaches have been developed to address this challenging problem. A common approach shared by many methods is to partition the entire system into a subsystem (whose dynamical properties are of interest) and an environment (or bath) in which the subsystem resides. Other recently developed schemes for computing quantum correlation functions do not rely on such a partition and instead utilize approximations to treat the quantum evolution of the entire system in conjunction with quantum equilibrium sampling [3–5]. In this paper, we focus on schemes based on the system-bath partition, and using this partition, the Hamiltonian reads: $\hat{H} = \hat{H}_b + \hat{h}_s + \hat{V}_c(\hat{R})$; where $\hat{H}_b = \frac{\hat{P}^2}{2M} + \hat{V}_b(\hat{R})$ and $\hat{h}_s$ represent the pure bath and subsystem Hamiltonians, respectively. The last term in $\hat{H}$ is a coupling potential that depends on the spatial coordinates of the bath wave functions. We shall always take the bath part of the Hamiltonian in the coordinate representation; however, we can represent $\hat{h}_s = \frac{\hat{p}^2}{2m} + \hat{V}_s(\hat{r})$ in some quantum basis: $\hat{h}_s = \sum_{ij} |i\rangle\langle i|\,\hat{h}_s\,|j\rangle\langle j|$.

Several methods based on various master equations [6–10] and path integral influence functional methods [11,12] provide approximate schemes, often in the weak coupling limit, to systematically project out the environmental DOF and yield a subsystem dynamics that incorporates dissipation and decoherence, due to coupling to the environment. However, for many applications, such as proton and electron transfer in condensed phases, it is desirable to explicitly simulate, even approximately, the bath dynamics, since specific local bath DOF may be crucial for a description of the dynamics of the quantum subsystem. For this purpose, several semiclassical [13–15] and mixed quantum-classical [16,17] (MQC) methods, which either treat the entire dynamics semiclassically or simulate the dynamics of the bath and subsystem with different levels of rigor (e.g., classical *versus* quantum mechanical), have been formulated. Many semiclassical and mixed quantum-classical approaches, adopting powerful classical simulation techniques, evaluate Equation (1) by combined Monte Carlo-molecular dynamics (MC-MD) techniques.

In this paper, we formulate MC-MD schemes to evaluate Equation (1) within the framework of the quantum-classical Liouville equation (QCLE) [18]. The QCLE employs a partial Wigner representation of the environmental (bath) DOF and may be derived from full quantum dynamics by truncating the quantum evolution operator to the first order in a small parameter related to the ratio of the characteristic masses of quantum and bath DOF [18]. In particular, we suppose that the quantum subsystem has a finite-dimension Hilbert space. Under this assumption, Equation (1) is cast in the following form [19,20]:

$$C_{AB}(t) = \frac{1}{Z_Q} \sum_{n_1,n_2} \int dX \left[\left(e^{-\beta\hat{H}} A \right)_W^{n_1 n_2} (X) B_W^{n_2 n_1}(X,t) \right] \tag{2}$$

where the n_j indices label the basis states (in some chosen quantum basis), $X = (R, P)$ represents the Wigner-transformed phase space point for the bath, N_B is the number of bath DOF and the subscript, W, on an operator indicates a partial Wigner transform on the bath DOF; e.g., an operator is partially Wigner transformed as $\hat{B}_W(X) = \int dZ \left\langle R - \frac{Z}{2} \right| \hat{B} \left| R + \frac{Z}{2} \right\rangle e^{\frac{i}{\hbar} P \cdot Z}$.

Two main tasks are involved in evaluating Equation (2) with an MC-MD algorithm. First, one needs to sample initial conditions (for an ensemble of trajectories) from the partially Wigner-transformed quantum density, $\left(\hat{\rho}_{eq} \hat{A} \right)_W (X)$ with $\rho_{eq} = e^{-\beta\hat{H}}/Z_Q$. There exist numerical algorithms to accomplish such a task [21,22]. Second, one needs to propagate the initial points in the phase space. These time-evolved trajectories may then be used to construct the matrix elements, $B_W^{nm}(X,t)$, needed to compute the correlation function. Various simulation methods, whose structure depends on the basis chosen to represent the quantum degrees of freedom in the QCLE, have been devised to simulate the mixed quantum-classical dynamics [23–31]. Simulation methods that utilize an adiabatic basis can be cast into the form of surface-hopping dynamics, but in a way that includes coherent evolution segments that account for creation and destruction of coherence in a proper manner. More recently, as in some semiclassical approaches [32], the mapping basis [33] was used to describe the quantum degrees of freedom in the QCLE in a continuous classical-like manner, leading to a trajectory description in the full system phase space [30,31,34–36].

The goals and outline of the paper are as follows: We first consider how the two ingredients, quantum equilibrium sampling and evolution of quantum operators, which are needed to compute quantum correlation functions, may be carried out. In Section 2, we describe a path-integral scheme to perform MC sampling from the partially Wigner transformed quantum density. In the Appendix, we also discuss a simplified, but approximate sampling scheme that is useful in the high-temperature limit. Another aim of this paper is to demonstrate how a recently-developed simulation method for the QCLE, the forward-backward trajectory solution (FBTS), can be used to efficiently obtain quantum correlation functions. To place these results in proper context, in Section 3, we sketch the important features and properties of the QCLE and discuss both the adiabatic Trotter-based surface-hopping (TBSH) algorithm and the FBTS, which is formulated in the mapping basis. In this section, we also present the explicit form of the N-level generalization of the TBSH algorithm. Comparisons of the trajectories that underlie these algorithms allow us to investigate how completely different ensembles of trajectories can be used to simulate the same observable correlation function. The implementation and utility of the simulation algorithms are illustrated on the dynamics in a two-level system coupled to a quartic oscillator embedded in a bath of independent harmonic oscillators, described in Section 4. Finally, in Section 5, we comment on the advantages, challenges and potential problems in adopting an approximate mixed quantum-classical dynamics for the computation of quantum time-correlation functions.

2. Sampling from the Partially Wigner-Transformed Density

In general, analytical expressions for the Wigner transform of the density matrix cannot be determined easily. In this section, we present a path-integral-based scheme to perform MC sampling from the Wigner-transformed density, $\left(\hat{\rho}_{eq}\hat{A}\right)_W^{n_1 n_2}(X)$, in Equation (2).

First, we recall the definition of partial Wigner transform:

$$\left(\hat{\rho}_{eq}\hat{A}\right)_W^{n_1 n_2}(X) = \frac{1}{Z_Q} \int dZ \left\langle n_1, R - \frac{Z}{2}\middle| e^{-\beta\hat{H}}\hat{A}\middle| n_2, R + \frac{Z}{2}\right\rangle e^{\frac{i}{\hbar}P\cdot Z} \tag{3}$$

where R represents the vector of bath coordinates, n denotes a basis state for the subsystem and $\hat{H} = \frac{\hat{P}^2}{2M} + \hat{V}_b(\hat{R}) + \hat{h}(\hat{R})$ with $\hat{h}(\hat{R}) \equiv \hat{h}_s + \hat{V}_c(\hat{R})$. One way to compute the integral on the right side of Equation (3) is to first factorize $e^{-\beta\hat{H}} = \prod e^{-\beta_L \hat{H}}$ into $L - 1$ pieces with $\beta_L = \beta/(L-1)$. Following the standard procedures for path integral calculations, we then insert resolutions of the identity, $\mathcal{I} = \int dR_i \sum_{m_i} |m_i, R_i\rangle \langle m_i, R_i|$, between every pair of factorized operators and apply the approximation, $e^{-\beta_L \hat{H}} \approx e^{-\beta_L \frac{\hat{P}^2}{2M}} e^{-\beta_L \left(\hat{V}_b(\hat{R}) + \hat{h}(\hat{R})\right)}$. The integrand on the right side of Equation (3) can then be written as follows:

$$\left\langle n_1, R - \frac{Z}{2}\middle| e^{-\beta\hat{H}}\hat{A}\middle| n_2, R + \frac{Z}{2}\right\rangle$$

$$= \int \prod_{i=1}^{L-1} dR_i \sum_{\{m_i\}} \left\langle n_1, R - \frac{Z}{2}\middle| e^{-\beta_L \hat{H}}\middle| m_1, R_1\right\rangle \left\langle m_1, R_1\middle| e^{-\beta_L \hat{H}}\middle| m_2, R_2\right\rangle \cdots$$

$$\times \left\langle m_{L-1}, R_{L-1}\middle| \hat{A}\middle| n_2, R + \frac{Z}{2}\right\rangle,$$

$$= \left(\frac{M}{2\pi\beta_L\hbar^2}\right)^{\frac{N_B(L-1)}{2}} \int \prod_{i=1}^{L-1} dR_i \sum_{\{m_i\}} \left\{ \prod_{i=1}^{L-2} \mathcal{M}_{i,i+1}(R_i)e^{-\beta_L V_b(R_i)}e^{-\frac{M}{2\beta_L\hbar^2}(R_i - R_{i+1})^2} \right\}$$

$$\times \left\langle n_1, R - \frac{Z}{2}\middle| \left(e^{-\beta_L \hat{H}}\middle| m_1, R_1\right\rangle \left\langle m_{L-1}, R_{L-1}\middle| \hat{A}\right)\middle| n_2, R + \frac{Z}{2}\right\rangle \tag{4}$$

where:

$$\mathcal{M}_{i,j} = \langle m_i| e^{-\beta_L \hat{h}(R_i)} |m_j\rangle = \begin{cases} e^{-\beta_L h^{ij}(R_i)}, & i = j \\ -\beta_L h^{ij}(R_i)e^{-\beta_L h^{ii}(R)}, & i \neq j \end{cases} \tag{5}$$

which is correct to order $\mathcal{O}(\beta_L^2)$. Substituting Equation (4) into Equation (3), the new integrand of the Wigner transform becomes $\hat{A} = \left(e^{-\beta_L \hat{H}} |m_1, R_1\rangle \langle m_{L-1}, R_{L-1}| \hat{A}\right)$, as shown in the last line of Equation (4). An analytical approximation for the Wigner transform of $\hat{A}$ can be obtained easily in most cases when $\hat{A}$ is a pure observable subsystem or if it depends on just one of the conjugate variables: R or P. Since $\beta_L \ll 1$, it is possible to replace the term, $e^{-\beta_L \hat{H}}$, inside $\hat{A}$ with its high-temperature approximation (discussed in the Appendix). Letting $\hat{A}_W(X)$ be the partial Wigner transform of $\hat{A}$, Equation (3) reads:

$$\left(\hat{\rho}_{eq}\hat{A}\right)_W^{n_1 n_2}(X) = \frac{\mathcal{G}^{N_B(L-1)/2}}{Z_Q} \int \prod_{i=1}^{L-1} dR_i \sum_{\{m_i\}} \left\{ \prod_{i=1}^{L-2} \mathcal{M}_{i,i+1}(R_i)e^{-\beta_L V_b(R_i)}e^{-\pi\mathcal{G}(R_i - R_{i+1})^2} \right\} \mathcal{A}_W^{n_1 n_2}(X) \tag{6}$$

where $\mathcal{G} = \left(\frac{M}{2\pi\beta_L\hbar^2}\right)$. Substituting Equation (6) into Equation (2), the time correlation function becomes:

$$C_{AB}(t) = \frac{\mathcal{G}^{N_B(L-1)/2}}{Z_Q} \sum_{n_1,n_2} \sum_{\{m_i\}} \int \prod_{i=1}^{L-1} dR_i \left\{ \prod_{i=1}^{L-2} \mathcal{M}_{i,i+1}(R_i) e^{-\beta_L V_b(R_i)} e^{-\pi\mathcal{G}(R_i-R_{i+1})^2} \right\}$$
$$\times \int dX \left(\hat{A}\right)_W^{n_1 n_2} (X) B_W^{n_2 n_1}(X,t) \tag{7}$$

Following [37], we remark that the initial phase space coordinate $X = (R,P)$ and auxiliary variables, $\{R_i\}$, can be sampled from probability densities constructed from $\hat{A}_W(X)$ and $|\mathcal{M}_{i,i+1}(R_i)|e^{-\beta_L V_b(R_i)}e^{-\pi\mathcal{G}(R_i-R_{i+1})^2}$, respectively.

3. Quantum-Classical Liouville Equation

In this section, we discuss how one can simulate the time-evolved matrix elements, $B_W^{n_2 n_1}(X,t)$, in Equation (2) using the QCLE:

$$\frac{\partial\hat{B}_W(X,t)}{\partial t} = \frac{i}{\hbar}[\hat{H}_W,\hat{B}_W] - \frac{1}{2}(\{\hat{H}_W,\hat{B}_W\} - \{\hat{B}_W,\hat{H}_W\})$$
$$= i\hat{\mathcal{L}}\hat{B}_W(X,t) = \frac{i}{\hbar}\left(\vec{\mathcal{H}}_\Lambda \hat{B}_W - \hat{B}_W \overleftarrow{\mathcal{H}}_\Lambda\right) \tag{8}$$

where $\Lambda = \overleftarrow{\nabla}_P\vec{\nabla}_R - \overleftarrow{\nabla}_R\vec{\nabla}_P$. The arrow on top of a differential operator indicates the direction in which it acts. In the first line, the square bracket and the curly brackets denote the quantum commutator and classical Poisson brackets, respectively. The two kinds of Lie bracket act together as the generator of the mixed quantum-classical dynamics. Due to the fact that $\hat{H}_W(X)$ and $\hat{B}_W(X,t)$ are quantum operators with respect to the subsystem DOF, two differently ordered Poisson brackets are needed to properly account for the mixed dynamics. However, in general, the dynamics described by the QCLE does not have a Lie algebraic structure, a feature that is common to mixed quantum-classical approaches [38]. In the second line, we introduce the abstract, quantum-classical Liouville (QCL) superoperator, $\hat{\mathcal{L}}$. Finally, the third equality is another equivalent representation of QCLE in terms of the forward and backward mixed quantum-classical Hamiltonians:

$$\vec{\mathcal{H}}_\Lambda = \hat{H}_W\left(1 + \frac{\hbar\Lambda}{2i}\right), \quad \overleftarrow{\mathcal{H}}_\Lambda = \left(1 + \frac{\hbar\Lambda}{2i}\right)\hat{H}_W \tag{9}$$

The QCLE has many desirable features, such as the conservation of energy, momentum and phase space volumes. Furthermore, the QCLE is equivalent to full quantum dynamics for arbitrary quantum subsystems, which are bilinearly coupled to a harmonic bath. For instance, commonly used spin boson models are of this type. In this circumstance, the combination of quantum and classical brackets in the QCLE does have a Lie algebraic structure. For the more general bath and coupling potentials, the QCLE provides an approximate description of the quantum dynamics. In this case, comparisons of simulations of QCL dynamics with exact quantum results have indicated that it is quantitatively accurate for a wide range of systems [36,39–48]

The QCLE equation can be simulated using ensembles of trajectories, which, in combination with the quantum initial condition sampling discussed above, provides a way to compute quantum correlation functions. As we shall see, the nature of the trajectories that enter in the simulations depends on the algorithm and should not be ascribed physical significance. It is only the observable, in this case, the correlation function, that has physical meaning and is independent of the manner in which it is simulated, provided the simulation algorithm is capable of exactly solving the QCLE, which is not always the case. One of the goals of this paper is to illustrate how a recently-developed FBTS [31] can be used to easily compute quantum correlation functions. For this purpose, it is interesting to contrast the solution using this scheme, and the trajectory description that underlies it, with the previously-developed and frequently-used TBSH algorithm [26]. Taking the adiabatic representation of the QCL superoperator is the key step in implementing the TBSH algorithm. The last representation of QCLE in Equation (8) resembles the quantum Liouville equation and forms the starting point of the FBTS.

3.1. Adiabatic Trotter-Based Surface Hopping

In order to discuss the nature of the trajectory description involved in the TBSH algorithm, we briefly describe how it is implemented and, in particular, present the explicit generalization to an N-level quantum subsystem, which was only outlined in [26]. We first consider the adiabatic representation of the QCLE, since the TBSH algorithm is cast in this basis. The adiabatic basis is defined by $\hat{h}_W(R)\,|\alpha; R\rangle = E_\alpha(R)\,|\alpha; R\rangle$, where $\hat{h}_W(R) = \hat{H}_W(R) - P^2/2M$ is taken to be the adiabatic Hamiltonian for a static configuration of R in this section. In the adiabatic basis, the QCLE reads:

$$\frac{\partial B_W^{\alpha\alpha'}}{\partial t} = i\mathcal{L}_{\alpha\alpha',\beta\beta'} B_W^{\beta\beta'}(X,t) \tag{10}$$

where the matrix elements of the QCL superoperator are given by:

$$i\mathcal{L}_{\alpha\alpha',\beta\beta'} = (i\omega_{\alpha\alpha'} + iL_{\alpha\alpha'})\,\delta_{\alpha\beta}\delta_{\alpha'\beta'} - \mathcal{J}_{\alpha\alpha',\beta\beta'} = i\mathcal{L}_{\alpha\alpha'}^0\delta_{\alpha\beta}\delta_{\alpha'\beta'} - \mathcal{J}_{\alpha\alpha',\beta\beta'} \tag{11}$$

with $\omega_{\alpha\alpha'} = (E_\alpha - E_{\alpha'})/\hbar$. (The Einstein summation convention will be used throughout the following sections, although sometimes, sums will be explicitly written if there is the possibility of confusion.) The Liouville operator, $i\mathcal{L}$, may be separated into two contributions: The classical propagator is defined as:

$$iL_{\alpha\alpha'} = \frac{P}{M} \cdot \frac{\partial}{\partial P} + \frac{1}{2}\,(F_\alpha + F_{\alpha'}) \cdot \frac{\partial}{\partial R} \tag{12}$$

where $F^\alpha = \langle\alpha; R|\frac{\partial \hat{h}_W(R)}{\partial R}|\alpha; R\rangle$ is the Hellmann-Feynman force. The superoperator, $\mathcal{J}_{\alpha\alpha',\beta\beta'}$, is responsible for nonadiabatic transitions and associated momentum changes in the bath. For an

N-level system, there exist $N(N-1)/2$ unique transitions. In the following, we define $\mathcal{J}$ as a sum of $\mathcal{J}_{\lambda\lambda'}$, which introduces transitions only between the specific pair of λ and λ' adiabatic states:

$$
\begin{aligned}
\mathcal{J}_{\alpha\alpha',\beta\beta'} &= \sum_{\lambda>\lambda'} (\mathcal{J}_{\lambda\lambda'})_{\alpha\alpha',\beta\beta'} \\
&= \sum_{\lambda>\lambda'} \Bigg\{ -d_{\lambda\lambda'} \cdot \frac{P}{M} \left((\delta_{\lambda\alpha}\delta_{\lambda'\beta} - \delta_{\lambda'\alpha}\delta_{\lambda\beta})\delta_{\alpha'\beta'} + (\delta_{\lambda\alpha'}\delta_{\lambda'\beta'} - \delta_{\lambda'\alpha'}\delta_{\lambda\beta'})\delta_{\alpha\beta} \right) \\
&\quad -\frac{1}{2}\hbar\omega_{\lambda\lambda'}d_{\lambda\lambda'} \cdot \frac{\partial}{\partial P} \left((\delta_{\lambda\alpha}\delta_{\lambda'\beta} + \delta_{\lambda'\alpha}\delta_{\lambda\beta})\delta_{\alpha'\beta'} + (\delta_{\lambda\alpha'}\delta_{\lambda'\beta'} + \delta_{\lambda'\alpha'}\delta_{\lambda\beta'})\delta_{\alpha\beta} \right) \Bigg\} \\
&= -\frac{P}{M}\cdot d_{\alpha\beta}\left(1 + \frac{1}{2}S_{\alpha\beta}\cdot\frac{\partial}{\partial P}\right)\delta_{\alpha'\beta'} + \frac{P}{M}\cdot d_{\beta'\alpha'}\left(1 - \frac{1}{2}S_{\beta'\alpha'}\cdot\frac{\partial}{\partial P}\right)\delta_{\alpha\beta} \quad (13)
\end{aligned}
$$

where $d_{\alpha\beta} = \langle\alpha; R|\partial/\partial R|\beta; R\rangle$ and $S_{\alpha\beta} = \hbar\omega_{\alpha\beta}d_{\alpha\beta}\left(\frac{P}{M}\cdot d_{\alpha\beta}\right)^{-1}$. The second equality gives the adiabatic representation of $\mathcal{J}_{\lambda\lambda'}$. We remark that it is difficult to exactly simulate the term, $\mathcal{J}$, involving bath momentum derivatives within the context of a trajectory-based algorithm. Using the identity that $\frac{1}{2}S_{\alpha\beta}\cdot\frac{\partial}{\partial P} = \hbar\omega_{\alpha\beta}M\cdot\partial/\partial(\hat{d}_{\alpha\beta}\cdot P)^2$, where M is a diagonal matrix of the masses of the bath particles and $\hat{d}_{\alpha\beta}$ is the unit vector along $d_{\alpha\beta}$, allows us to employ the momentum-jump approximation:

$$
\left(1 + \frac{c}{2}S_{\alpha\beta}\cdot\frac{\partial}{\partial P}\right)f(P) \approx e^{\frac{c}{2}S_{\alpha\beta}\cdot\frac{\partial}{\partial P}}f(P) = e^{c\hbar\omega_{\alpha\beta}M\cdot\partial/\partial(\hat{d}_{\alpha\beta}\cdot P)^2}f(P) = f(P + \Delta P_c) \quad (14)
$$

where $c = 1,2$ corresponding to single and double hops, respectively, and $\Delta P_c = \hat{d}_{\alpha\beta}\text{sgn}\left(\hat{d}\cdot P\right)\sqrt{(\hat{d}_{\alpha\beta}\cdot P)^2 + c\hbar\omega_{\alpha\beta}M} - \hat{d}\left(\hat{d}\cdot P\right)$. We have a translation operator with respect to the variable, $(\hat{d}_{\alpha\beta}\cdot P)^2$, in the above equation. Decomposing $P = P_\perp + P_\parallel = P_\perp + \hat{d}_{\alpha\beta}\text{sgn}\left(\hat{d}_{\alpha\beta}\cdot P\right)\sqrt{\left(\hat{d}_{\alpha\beta}\cdot P\right)^2}$, it becomes obvious that the translation operator updates $P_\parallel$ components by ΔP_c, as presented in Equation (14). This momentum update conserves the energy of surface-hopping trajectories. Apart from technical issues associated with sampling when the algorithm is implemented, this is the only approximation made to QCL evolution. In fact, it is this approximation that gives this algorithm a surface-hopping structure that has some features in common with Tully's surface-hopping method; however, coherence and decoherence are automatically incorporated in the evolution. The QCLE does not have such sudden momentum changes, and its evolution is described by continuous momentum changes in the course of the evolution. Comparisons of results using this algorithm with exact quantum solutions indicate that the momentum-jump is rarely the source of problems.

Equation (10) admits a formal solution:

$$
\hat{B}_W^{\alpha\alpha'}(X,t) = \left(e^{i\mathcal{L}t}\right)_{\alpha\alpha',\beta\beta'} B_W^{\beta\beta'}(X) \quad (15)
$$

Thus, our following discussion focuses on evaluating:

$$
\left(e^{i\mathcal{L}t}\right)_{\alpha\alpha',\alpha_K\alpha'_K} = \sum_{(\alpha_1\alpha'_1)\dots(\alpha_K\alpha'_K)}\prod_{j=1}^{K}\left(e^{i\hat{\mathcal{L}}\Delta t_j}\right)_{\alpha_{j-1}\alpha'_{j-1},\alpha_j\alpha'_j} \quad (16)
$$

In the above equation, we simply factorize the propagator into K pieces with $\Delta t_j = t_j - t_{j-1} = \Delta t$. In each small time slice, we perform the symmetric Trotter decomposition:

$$\left(e^{i\hat{\mathcal{L}}\Delta t_j}\right)_{\alpha\alpha',\beta\beta'} \approx \mathcal{W}_{\beta\beta'}\left(t_{j-1}, t_{j-1} + \frac{\Delta t}{2}\right) e^{iL_{\beta\beta'}\Delta t/2} \mathcal{Q}_{\beta\beta',\alpha\alpha'} \mathcal{W}_{\alpha\alpha'}\left(t_{j-1} + \frac{\Delta t}{2}, t_j\right) e^{iL_{\alpha\alpha'}\Delta t/2}$$
(17)

where: $\mathcal{W}_{\alpha\alpha'}(t_1, t_2) = e^{i\omega_{\alpha\alpha'}(t_2 - t_1)}$, and:

$$\mathcal{Q}_{\alpha\alpha',\beta\beta'} = \left(e^{\mathcal{J}\Delta t}\right)_{\alpha\alpha',\beta\beta'} = \left(e^{\sum_{\lambda > \lambda'} \mathcal{J}_{\lambda\lambda'}\Delta t}\right)_{\alpha\alpha',\beta\beta'} \approx \left(\prod_{\lambda > \lambda'} e^{\mathcal{J}_{\lambda\lambda'}\Delta t}\right)_{\alpha\alpha',\beta\beta'}$$
(18)

We observe that it is possible to express $e^{\mathcal{J}_{\lambda\lambda'}\Delta t}$ in the following block-diagonal matrix form:

$$e^{\mathcal{J}_{\lambda\lambda'}\Delta t} = \mathcal{M}^{\lambda\lambda'} \oplus \mathcal{K}^{\lambda\lambda'}_{\xi_1} \cdots \oplus \mathcal{K}^{\lambda\lambda'}_{\xi_{N-2}} \oplus \mathcal{N}^{\lambda\lambda'}$$
(19)

where ξ_i is one of the $N-2$ adiabatic states other than λ and λ'. In the above equation, $\mathcal{M}$ is a four by four matrix, defined with respect to the basis, $\{(\lambda, \lambda), (\lambda, \lambda'), (\lambda', \lambda), (\lambda', \lambda')\}$:

$$\mathcal{M}^{\lambda\lambda'} = \begin{pmatrix} \cos^2(a) & -\cos(a)\sin(a)\hat{j}_{\lambda\lambda'} & -\cos(a)\sin(a)\hat{j}_{\lambda\lambda'} & \sin^2(a)\hat{j}_{\lambda\to\lambda'} \\ \cos(a)\sin(a)\hat{j}_{\lambda\lambda'} & \cos^2(a) & -\sin^2(a) & -\sin(a)\cos(a)\hat{j}_{\lambda\lambda'} \\ \cos(a)\sin(a)\hat{j}_{\lambda\lambda'} & -\sin^2(a) & \cos^2(a) & -\sin(a)\cos(a)\hat{j}_{\lambda\lambda'} \\ \sin^2(a)\hat{j}_{\lambda\to\lambda'} & \cos(a)\sin(a)\hat{j}_{\lambda\lambda'} & \cos(a)\sin(a)\hat{j}_{\lambda\lambda'} & \cos^2(a) \end{pmatrix}$$
(20)

with $a = (P/M) \cdot d_{\lambda\lambda'}\Delta t$, and $\hat{j}_{\lambda\lambda'}$ and $\hat{j}_{\lambda\to\lambda'}$ are the momentum-jump operators, $e^{\frac{1}{2}S_{\lambda\lambda'}\frac{\partial}{\partial P}}$ and $e^{S_{\lambda\lambda'}\frac{\partial}{\partial P}}$, defined in Equation (14) with $c = 1, 2$, respectively. In Equation (19), there exists another set of four by four matrices, $\mathcal{K}^{\lambda\lambda'}_{\xi_i}$, with $i = 1, \ldots, N-2$. Each of these matrices is defined with respect to a basis of the form, $\{(\lambda, \xi_i), (\lambda', \xi_i)\} \oplus \{(\xi_i, \lambda), (\xi_i, \lambda')\}$:

$$\mathcal{K}^{\lambda\lambda'}_{\xi} = \begin{pmatrix} \cos(a) & -\sin(a)\hat{j}_{\lambda\lambda'} \\ \sin(a)\hat{j}_{\lambda\lambda'} & \cos(a) \end{pmatrix} \oplus \begin{pmatrix} \cos(a) & -\sin(a)\hat{j}_{\lambda\lambda'} \\ \sin(a)\hat{j}_{\lambda\lambda'} & \cos(a) \end{pmatrix}$$
(21)

Finally, there is a null matrix, $\mathcal{N}^{\lambda\lambda'}$, of a size of $(N-2)^2$, and the associated null space is spanned by basis vectors, (ξ_1, ξ_2), where $\xi_i \neq \lambda^{(')}$. We remark that one has to permute the basis vectors in order to construct these block-diagonal matrices [26].

At this point, we have specified all the necessary details in order to simulate the QCL dynamics in the adiabatic basis:

$$B_W^{\alpha\alpha'}(X, t) = \sum_{\substack{(\alpha_1\alpha'_1),\ldots, \\ (\alpha_K\alpha'_K)}} \left[\prod_{j=1}^{K} \mathcal{W}_{\alpha_{j-1}\alpha'_{j-1}} e^{iL_{\alpha_{j-1}\alpha'_{j-1}}\Delta t} \mathcal{Q}_{\alpha_{j-1}\alpha'_{j-1},\alpha_j\alpha'_j} \mathcal{W}_{\alpha_j\alpha'_j} e^{iL_{\alpha_j\alpha'_j}\Delta t}\right] B_W^{\alpha_K\alpha'_K}(X)$$
(22)

where $\alpha_0^{(')} = \alpha^{(')}$. The explicit summation over all quantum indices, $(\alpha_1\alpha'_1)\ldots(\alpha_K\alpha'_K)$, can also be evaluated stochastically. For instance, given a pair of indices, $(\alpha_j\alpha_{j-1})$, one can determine the next pair at the time slice, $j+1$, by drawing an MC sample from the transition probability:

$$P(\alpha_{j+1}, \alpha'_{j+1}|\alpha_j, \alpha'_j) = \frac{|\mathcal{Q}_{\alpha_j\alpha'_j, \alpha_{j+1}\alpha'_{j+1}}|}{\sum_{\beta_{j+1},\beta'_{j+1}} |\mathcal{Q}_{\alpha_j\alpha'_j, \beta_{j+1}\beta'_{j+1}}|}$$
(23)

If the sampled new pair of indices differs from the starting pair, then the sampled $\mathcal{Q}$ matrix element must contain the proper momentum-jump operators to update the energy of the trajectory after the jump. In any actual implementation of this algorithm, it is desirable to restrict to nonadiabatic transitions between one pair of states in every time slice. Under this assumption, one can then approximate:

$$
\mathcal{Q}_{\alpha\alpha',\beta\beta'} \approx \begin{cases} \delta_{\alpha\beta}\delta_{\alpha'\beta'} & \text{if no hop happens,} \\ \left(e^{\mathcal{J}_{\mu\gamma}}\right)_{\alpha\alpha',\beta\beta'} & \text{if } (\alpha,\alpha') \to (\beta,\beta') \text{ involves transition between } (\mu,\gamma) \text{ states,} \\ 0 & \text{if } (\alpha,\alpha') \to (\beta,\beta') \text{ involves transitions between two or more pairs of states.} \end{cases} \tag{24}
$$

In this algorithm, we see that the trajectories in the ensemble that are used to simulate the time evolution are non-Newtonian in character, consisting of Newtonian segments where the system evolves on adiabatic surfaces, or the mean of two adiabatic surfaces, interspersed with quantum transitions and momentum changes.

3.2. Forward-Backward Trajectory Solution

This scheme is motivated by another way of writing the formally exact solution [38] of the QCLE using the last line of Equation (8):

$$
\hat{B}_W(X,t) = \mathcal{S}\left(e^{i\vec{\mathcal{H}}_\Lambda t/\hbar}\hat{B}_W(X)e^{-i\overleftarrow{\mathcal{H}}_\Lambda t/\hbar}\right) \tag{25}
$$

The $\mathcal{S}$ operator [31,38] specifies the order in which the forward and backward evolution operators act on $\hat{B}_W(X)$. The ordering of evolution operators is critical because of the lack of an underlying Lie algebraic structure [38] of the QCLE.

One approach to solve Equation (25) is to apply the mapping transformation in which N discrete quantum states of the subsystem are represented by the continuous position and momenta of N fictitious harmonic oscillators. The properties of the original subsystem are then obtained via an ensemble average involving trajectories in the phase space of the fictitious oscillators. More precisely, in the mapping representation, a subsystem state, $|\lambda\rangle$, is replaced by $|m_\lambda\rangle = |0_1, \cdots, 1_\lambda, \cdots 0_N\rangle$, a product state specifying the occupation numbers (limited to zero or one) of N fictitious harmonic oscillators [33,49]. Creation and annihilation operators, $\hat{a}_\lambda^\dagger$ and $\hat{a}_\lambda$, satisfy the commutation relation $[\hat{a}_\lambda, \hat{a}_{\lambda'}^\dagger] = \delta_{\lambda,\lambda'}$ for harmonic oscillators. The actions of these operators on the single-excitation mapping states are $\hat{a}_\lambda^\dagger |0\rangle = |m_\lambda\rangle$ and $\hat{a}_\lambda |m_\lambda\rangle = |0\rangle$, where $|0\rangle = |0_1 \ldots 0_N\rangle$ is the ground state of the mapping basis.

Next, we define the mapping version of operators, $\hat{B}_m(X) = B_W^{\lambda\lambda'}(X)\hat{a}_\lambda^\dagger\hat{a}_{\lambda'}$, such that matrix elements of $\hat{B}_W$ in the subsystem basis are equal to the matrix elements of the corresponding mapping operator: $B_W^{\lambda\lambda'}(X) = \langle\lambda|\hat{B}_W(X)|\lambda'\rangle = \langle m_\lambda|\hat{B}_m(X)|m_{\lambda'}\rangle$. In particular, the mapping Hamiltonian is:

$$
\hat{H}_m = H_b(X) + h^{\lambda\lambda'}(R)\hat{a}_\lambda^\dagger\hat{a}_{\lambda'} \equiv H_b(X) + \hat{h}_m \tag{26}
$$

where we applied the mapping transformation only on the part of the Hamiltonian that involves the subsystem DOF in Equation (26). The mapping Hamiltonian, $\hat{h}_m$, is always a quadratic Hamiltonian with respect to the quantum DOF. The pure bath term, $\hat{H}_b(X)$, acts as an identity operator in the subsystem basis and is mapped onto the identity operator of the mapping space directly. The mapped formal solution of QCLE now reads:

$$\hat{B}_m(X,t) = \mathcal{S}\left(e^{i\overrightarrow{\mathcal{H}}^m_\Lambda t/\hbar} \hat{B}_m(X) e^{-i\overleftarrow{\mathcal{H}}^m_\Lambda t/\hbar}\right) \tag{27}$$

where $\overrightarrow{\mathcal{H}}^m_\Lambda$ is given by $\overrightarrow{\mathcal{H}}^m_\Lambda = \hat{H}_m(1 + \hbar\Lambda/2i)$, with an analogous definition for $\overleftarrow{\mathcal{H}}^m_\Lambda$.

We now introduce the coherent states, $|z\rangle$, in the mapping space, $\hat{a}_\lambda |z\rangle = z_\lambda |z\rangle$ and $\langle z| \hat{a}^\dagger_\lambda = z^*_\lambda \langle z|$, where $|z\rangle = |z_1,\ldots,z_N\rangle$, and the eigenvalue is $z_\lambda = (q_\lambda + ip_\lambda)/\sqrt{\hbar}$. The variables $q = (q_1,\ldots,q_N)$ and $p = (p_1,\ldots,p_N)$ are mean coordinates and momenta of the harmonic oscillators encoded in the coherent state, $|z\rangle$, respectively. The coherent states form an overcomplete basis with the inner product between any two such states, $\langle z| z'\rangle = e^{-(|z-z'|^2)-i(z\cdot z'^*-z^*\cdot z')}$. Finally, we remark that the coherent states provide the resolution of identity:

$$\mathcal{I} = \int \frac{d^2z}{\pi^N} |z\rangle\langle z| \tag{28}$$

where $d^2z = d(\Re(z))d(\Im(z)) = dqdp/(2\hbar)^N$.

Similar to the path integral approach for solving the quantum dynamics, we decompose the forward and backward evolution operators in Equation (27) into a concatenation of M short-time evolutions with $\Delta t_i = \tau$ and $M\tau = t$. In each short-time interval, Δt_i, we introduce two sets of coherent states, $|z_i\rangle$ and $|z'_i\rangle$, via Equation (28) to expand the forward and backward time evolution operators, respectively. The time evolution (generated by a quadratic Hamiltonian) of coherent states can be represented by trajectory evolution in the phase space of (q,p). After some algebra, the matrix elements of Equation (27) can be approximated by:

$$B^{\lambda\lambda'}_W(X,t) = \sum_{\mu\mu'} \int dxdx' \phi(x)\phi(x')\frac{1}{\hbar}(q_\lambda + ip_\lambda)(q'_{\lambda'} - ip'_{\lambda'})B^{\mu\mu'}_W(X_t)$$

$$\times \frac{1}{\hbar}(q_\mu(t) - ip_\mu(t))(q'_{\mu'}(t) + ip'_{\mu'}(t)) \tag{29}$$

where $x = (q,p)$ gives the real and imaginary parts of z, $dx = dqdp$ and $\phi(x) = (\hbar)^{-N} e^{-\sum_\nu (q^2_\nu + p^2_\nu)/\hbar}$ is the normalized Gaussian distribution function. In deriving Equation (29), we have invoked an orthogonality approximation on the inner product between subsequent coherent state variables, $\langle z_i| e^{\frac{i}{\hbar}\hat{h}t} |z_{i+1}\rangle = \langle z_i(t)|z_{i+1}\rangle \approx \pi^N \delta(z_{i+1} - z_i(t_i))$, with i being the time step index. This approximation is necessary to construct a continuous trajectory of $z(t)$. In the extended phase space of $(X(t), z(t), z'(t))$, the trajectories follow Hamiltonian dynamics:

$$\frac{d\chi_\mu}{dt} = \frac{\partial H_e(\chi,\pi)}{\partial \pi_\mu}, \qquad \frac{d\pi_\mu}{dt} = -\frac{\partial H_e(\chi,\pi)}{\partial \chi_\mu} \tag{30}$$

where $H_e(\chi,\pi) = P^2/2M + V_0(R) + \frac{1}{2\hbar}h^{\lambda\lambda'}(R)(q_\lambda q_{\lambda'} + p_\lambda p_{\lambda'} + q'_\lambda q'_{\lambda'} + p'_\lambda p'_{\lambda'})$ with $V_0(R) = V_b(R) - Tr\hat{h}(R)$, $\chi = (R,q,q')$ and $\pi = (P,p,p')$. We remark that the FBTS trajectories manifestly conserve energy. Furthermore, simulating the dynamics with a standard velocity Verlet type of symplectic integrator has a stationary solution proportional to $H_{pseudo} = H_e(\chi, \pi) + \Delta t^2 \delta H$, as discussed in [35].

The main approximation introduced in the derivation of the FBTS, Equation (29), is the orthogonality approximation. The simplest improvement to the algorithm is to refrain from applying this approximation at every time step. In [36], we outlined a practical approach to evaluate the set of selected integrals of z_i and z_i' (which could be evaluated analytically if the orthogonality approximation were applied). We termed this extension of FBTS as the jump FBTS (JFBTS). Since the computational cost grows quickly with respect to the number of jumps inserted, one needs to make a trade-off between numerical efficiency and accuracy.

In the simplest approach, one selects every (M/K) time step from a total of M steps to fully evaluate the coherent state integrals:

$$
B_W^{\lambda\lambda'}(X, t) = \sum_{\mu\mu'} \sum_{\substack{s_0 s_0' \cdots \\ s_{K-1} s_{K-1}'}} \int \prod_{v=0}^{K} dx dx' \phi(x_v) \phi(x_v')
$$

$$
\times \frac{1}{\hbar}(q_{0\lambda} + i p_{0\lambda})(q_{0\lambda'}' - i p_{0\lambda'}') B_W^{\mu\mu'}(X_t)
$$

$$
\times \frac{1}{\hbar} \left\{ \prod_{v=1}^{K} \left(q_{(v-1)s_{v-1}}(\tau_v) - i p_{(v-1)s_{v-1}}(\tau_v) \right) \left(q_{vs_v} + i p_{vs_v} \right) \right\}
$$

$$
\times \frac{1}{\hbar} \left\{ \prod_{v=1}^{K} \left(q_{(v-1)s_{v-1}}'(\tau_v) + i p_{(v-1)s_{v-1}}'(\tau_v) \right) \left(q_{vs_v}' - i p_{vs_v}' \right) \right\}
$$

$$
\times \frac{1}{\hbar}(q_{K\mu}(\tau_{K+1}) - i p_{K\mu}(\tau_{K+1}))(q_{K\mu'}'(\tau_{K+1}) + i p_{K\mu'}'(\tau_{K+1})) \tag{31}
$$

where the subscripts, v and s, refer to the v-th time step and the s-th component of the q and p vectors, respectively, and $\tau_v = t_{i_v} - t_{i_{v-1}}$ with $t_{i_0} = 0$ and $t_{i_{K+1}} = t$. According to this prescription, the continuous FB trajectories experience K discontinuous jumps in the (x, x') phase space. Between subsequent jumps, the evolution of the FB trajectory is governed by Equation (30). Simulations show that with a sufficient number of jumps, numerically exact solutions of the QCLE can be obtained [36].

3.3. Comparisons between Algorithms

The differences between the two QCLE simulation algorithms can be traced to the quantum basis that is used and the way that feedback between quantum and classical systems is treated. In the case of the TBSH algorithm, the trajectories are propagated through a Hellmann-Feynman force, or the mean of two Hellmann-Feynman forces [Equation (12)], with intermittent surface hops that switch the adiabatic surfaces on which the trajectories propagate. In the case of FBTS, one not only propagates the bath dynamical variables as trajectories, but also the quantum dynamical variables, which are associated with fictitious harmonic oscillators. In this extended phase space, we have exact Hamiltonian dynamics. In particular, the force acting on the bath particles simultaneously involves all N adiabatic surfaces, which is similar to, but different from, the Ehrenfest mean-field approach. The very different characteristics of the trajectories in two algorithms manifest the artificial character of the trajectory dynamics. Thus, one should not attach physical significance to single trajectories in the computation. All physical properties of the system can only be extracted from a proper ensemble average of a large set of trajectories, as implied in Equation (2). Nevertheless, insight into the trajectory dynamics of each algorithm will help to judge the simulation efficiency for various classes of models.

For certain problems, such as proton transfer reactions, where the time scales of the bath and subsystem are well-separated, even during nonadiabatic transitions, the TBSH algorithm can yield quantitatively accurate results with a few hops. There are also dynamical problems in which distinct bath motions can be explicitly correlated with the subsystem's quantum states. For instance, in the simple Tully I model [35,50], trajectories populated on the excited state will cross the avoided crossing point, while the ground state trajectories will eventually be reflected and retrace their paths in the opposite direction. This kind of behavior is, however, completely missed when one propagates trajectories in a single effective mean field. Again, the inherent multi-configuration nature of surface-hopping-like algorithms is a more appropriate choice for this case. However, a recent study [51] has indicated that the "jump" version of mean-field-like algorithms can improve the simulation results in cases of this type.

Alternatively, there are also many examples where one would expect FBTS to be the preferred simulation method. In general, the TBSH algorithm has convergence issues, as the MC weights associated with nonadiabatic hops grows rapidly. Even for the simple spin boson model, one can identify parameter regimes where this numerical instability is clearly observed. In these cases, the FBTS and JFBTS are certainly the alternatives that one should adopt for efficient simulations.

4. An Example: Quartic Oscillator in a Harmonic Bath

As a specific example to illustrate the formalism outlined above, we consider a two-level system coupled to a quartic bistable oscillator with a single pair of phase space coordinates $X_0 = (R_0, P_0)$. The quartic oscillator is, in turn, coupled to an Ohmic heat bath of N_b independent harmonic oscillators with phase space coordinates $X_i = (R_i, P_i)$ and $i = 1 \ldots N_b$. The partially Wigner transformed Hamiltonian, expressed in the diabatic basis, $\{|R\rangle, |L\rangle\}$, reads:

$$\hat{H}_W = \begin{pmatrix} \hbar\gamma_0 R_0 & -\hbar\Omega \\ -\hbar\Omega & -\hbar\gamma_0 R_0 \end{pmatrix} + \left(\frac{P_0^2}{2M_0} + V_n(R_0) + \sum_{j=1}^{N_b} \frac{P_j^2}{2M_j} + \frac{M_j\omega_j^2}{2}\left(R_j - \frac{\gamma_b c_j}{M_j\omega_j^2}R_0\right)^2 \right) \mathbf{I}$$

(32)

where $V_n(R_0) = -M_0\omega_0^2 R_0^2/2 + AR_0^4/4$ and $\mathbf{I}$ is an identity matrix. We take $N_b = 40$ harmonic oscillators for the discretization of the Ohmic heat bath. Following the discretization scheme introduced in [52], we set $\omega_j = \omega_c \ln(1 - j\omega_c/\delta\omega)$ and $c_j = (\xi\hbar\delta\omega M_j)^{1/2}\omega_j$ with $\delta\omega = (1 - \exp(\omega_{max}/\omega_c))/N_b$. The parameters, ω_c and ω_{max}, are the characteristic and cut-off frequencies for the Ohmic bath, respectively. The Kondo parameter is ξ.

The adiabatic states for the subsystem are:

$$|+; R_0\rangle = \frac{1}{\mathcal{N}(R_0)} [(1 - G)|R\rangle - (1 + G)|L\rangle]$$

$$|-; R_0\rangle = \frac{1}{\mathcal{N}(R_0)} [(1 + G)|R\rangle + (1 - G)|L\rangle] \tag{33}$$

where $\mathcal{N}(R_0) = \sqrt{2(1 + G^2(R_0))}$ and $G(R_0) = (\gamma_0 R_0)^{-1} \left[-\Omega + \sqrt{\Omega^2 + \gamma_0^2 R_0^2}\right]$. The adiabatic energies are given by $E_\pm = V_n(R_0) \pm \hbar\sqrt{\Omega^2 + \gamma_0^2 R_0^2} = V_n(R_0) \pm \epsilon_\pm(R_0)$.

We shall study the autocorrelation functions, C_{LL}, with $\hat{A} = \hat{B} = |L\rangle\langle L|$. The entire system is assumed to be in thermal equilibrium initially. Using the high-temperature approximation presented in the Appendix, the correlation function of interest can be given in a compact form:

$$C_{LL}(t) = \int dX_0 dX_b \, \mathcal{W}(R_0) \mathcal{G}\left(P_0; \frac{M_0}{\beta}\right) \prod_{j=1}^{N_b} \mathcal{G}\left(P_j; \frac{M_j}{\beta}\right) \mathcal{G}\left(R_j - \frac{\gamma_b c_j}{M_j \omega_j^2} R_0; \frac{1}{\beta M_j \omega_j^2}\right)$$

$$\times \sum_{n=L,R} \sum_{\alpha,\alpha'} F_{\alpha\alpha'}(X_0)\langle n|\alpha; R_0\rangle\langle \alpha'; R_0|L\rangle B_W^{Ln}(X_t) \tag{34}$$

where $\mathcal{G}(x; \sigma^2) = (2\pi\sigma^2)^{-1/2} e^{-x^2/2\sigma^2}$, and:

$$\mathcal{W}(R_0) = \frac{e^{-\beta\left(\frac{A}{4}R_0^4 - \frac{1}{2}M_0\omega_0^2 R_0^2\right)} \left(e^{-\beta\epsilon_+(R_0)} + e^{\beta\epsilon_-(R_0)}\right)}{\int dR_0 e^{-\beta\left(\frac{A}{4}R_0^4 - \frac{1}{2}M_0\omega_0^2 R_0^2\right)} \left(e^{-\beta\epsilon_+(R_0)} + e^{\beta\epsilon_-(R_0)}\right)} \tag{35}$$

An MC evaluation of the integrals can be done by sampling P_0, R_b, P_b from the Gaussian distributions and sampling R_0 from $\mathcal{W}(R_0)$, respectively. The time-evolved matrix element, $B_W^{nm}(X_t)$, will be computed using both the TBSH and the FBTS algorithms. Finally, we note that the path-integral-based sampling scheme introduced in Section 2 should be adopted to sample phase-space points from $(\hat{\rho}_{eq})_W(X)$ for more generalized situations, including cases of low-temperature, arbitrary subsystem-bath divisions of a composite system, strong subsystem-bath couplings and an arbitrary potential energy profile.

In this study, we report numerical results in the energy unit, $\hbar\omega_c$, and distance unit, $\sqrt{\hbar/M_j\omega_c}$, for each environmental DOF. We consider two sets of parameters. In the first case, we use the following parameter values, $a = 1.0, \omega_0 = 1.2, \gamma_0 = 0.05 \, \gamma_b = 1.0, \Omega = 0.3, \xi = 0.1, \omega_{max} = 3$ and $\beta = 0.2$, in the dimensionless units. Figure 1a presents the potential surface profiles [53], $W_\alpha(R_0)$. The two diabatic surfaces, $W_{L,R}(R_0)$, remain close to each other, and the two adiabatic surfaces, $W_\pm(R_0)$, share essentially the same characteristics. In this case, a mean-field-based algorithm, like FBTS, should be accurate and efficient. This problem can also be handled easily in the adiabatic basis, since the surface-hopping trajectories will be initialized in both the adiabatic ground and excited states, because the system is in a thermal equilibrium state at $t = 0$. Furthermore, the coupling parameter, γ_b, was purposely chosen to be small in order to minimize the number of nonadiabatic transitions (or hops) encountered in the TBSH algorithm. In panel (b), $C_{LL}(t)$ is computed using both algorithms. The agreement between these results is good.

Figure 1. (a) Potential surface profiles, $W_\alpha(R_0)$, for the ground adiabatic state (black, dotted), excited adiabatic state (black, dotted) and for the diabatic states, L (green) and R (red). (b) $C_{LL}(t)$ correlation function. These results are associated with the first set of parameters.

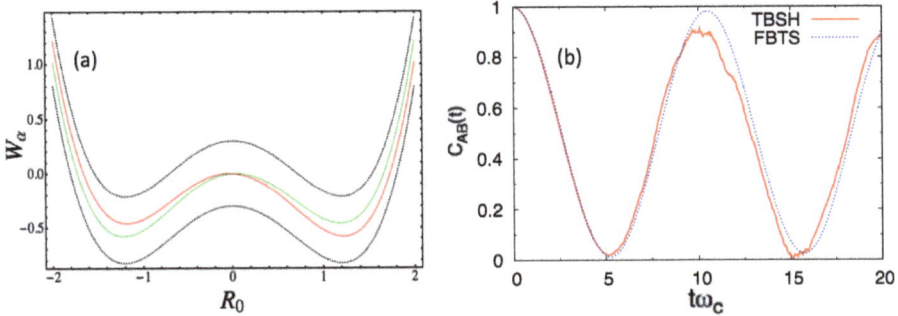

Figure 2. (a) Potential surface profiles, $W_\alpha(R_0)$, for the ground adiabatic state (black, dotted), excited adiabatic state (black, dotted) and for the diabatic states, L (green) and R (red). The blue curve is a plot of the un-normalized distribution function, $\mathcal{W}(R_0)$, Equation (35). (b) $C_{LL}(t)$ correlation functions. (**Inset**) Short-time $C_{LL}(t)$ computed by the FBTS (blue) and TBSH (red) algorithms. These results are associated with the second set of parameters.

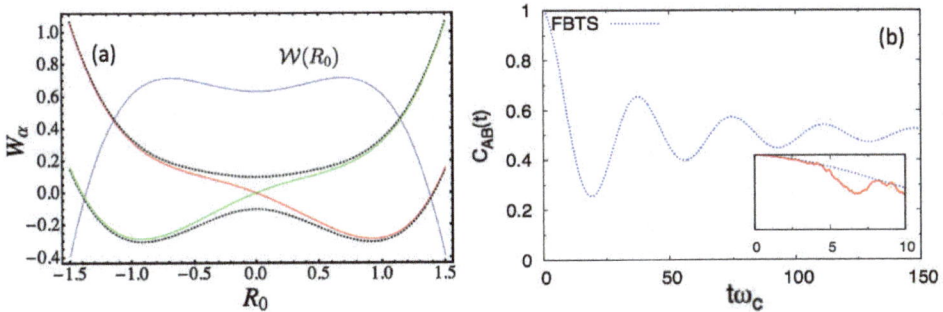

Next, we consider the following parameter set, $a = 0.8$, $\omega_0 = 0.6$, $\gamma_0 = 0.3$ $\gamma_b = 1.0$, $\Omega = 0.1$, $\xi = 0.1$, $\omega_{max} = 3$, and $\beta = 0.2$ in the dimensionless units. Figure 2a shows the potential surface profiles, $W_\alpha(R_0)$, obtained from this set of parameters. In this case, the adiabatic, $W_\pm(R_0)$, and diabatic surfaces, $W_{L,R}(R_0)$, only differ markedly near the region of the barrier top, where an avoided crossing point indicates significant mixing of the two diabatic states. Nonadiabatic effects should be most prominent near this barrier top. A stronger coupling, γ_0, is also chosen in this case. Figure 2b presents the autocorrelation functions. In the main figure of panel (b), the blue curves ($C_{LL}(t)$ computed by the FBTS) start with the full correlation at one, then gradually reduce to $1/2$, which implies that the subsystem is in an equal admixture of the two diabatic states in the asymptotic limit. The TBSH simulation results are only valid for very short times (as shown in the inset of the Figure 2b), due to instability arising from the accumulation of weights, even with filtering [54]. The thermal equilibrium distribution, $\mathcal{W}(R_0)$, has a bimodal distribution profile, as illustrated in Figure 2a; however, for the (inverse) temperature, $\beta = 0.2$, the double-peaked structure is very broad. The $\mathcal{W}(R_0)$ distribution profile (blue curve in Figure 2a) suggests that the thermal equilibrium state has a non-trivial contribution from the excited surface. Sampling from $\mathcal{W}(R_0)$ yields many R_0 values near the barrier top, where several hops immediately take place for this strong-coupling case, and the instability sets in early in the simulation. Lowering β will produce a more pronounced double-peak structure for $\mathcal{W}(R_0)$, but the quartic oscillator's momentum, P_0, will fluctuate with a larger variance in the presence of the heat bath in this case. Since nonadiabatic transitions depend non-trivially on $a = P_0 \cdot d_{12}(R_0)\Delta t$ in the TBSH algorithm, large momentum fluctuations will eventually affect the long-time result. This case shows some of the practical limitations of the TBSH algorithm for the computation of this correlation function.

5. Conclusions

The scheme for computing the quantum correlation function in Equation (2) combines a numerically exact quantum initial sampling method with dynamics described by the QCLE; thus, the approximations in the simulation method reside in the dynamics. It is easier to compute the equilibrium properties of a quantum system, for instance, by using the imaginary-time Feynman path integral method, than to obtain dynamical properties by using similar real-time Feynman path integrals without adopting further approximations. Since we approximate the quantum dynamics of the entire system, quantum subsystem plus bath, by QCL dynamics, it is appropriate to comment on some of its features.

It is known that the quantum-classical bracket, defined in terms of the commutator and Poisson brackets in Equation (8), does not possess a Lie algebraic structure, since it fails to satisfy the Jacobi identity [2,38]. This lack of a proper algebraic structure is shared by all known MQC methods and simply reflects the inconsistency in mixing classical and quantum mechanical dynamics. One consequence of this inconsistency is that the partial Wigner transform, $\hat{\rho}_{We}(R, P)$, of the full canonical equilibrium density function, $\hat{\rho}_{eq} = e^{-\beta \hat{H}}/Z_Q$, is not stationary under the QCLE; however, $\hat{\rho}_{We}(R, P)$ can be written as an expansion in the mass ratio (or $\hbar$), and it has been shown that the full quantum equilibrium density is conserved under the QCL dynamics up to $\mathcal{O}(\hbar)$. Therefore, the detailed balance relation is also satisfied to this order. The violation of a detailed balance is a common problem that affects all major MQC methods, including the two most popular approaches, Ehrenfest mean-field [55] and Tully's fewest switching surface hopping [56] (FSSH), to various degrees. Of course, as noted earlier, for the class of models where an arbitrary quantum system is bilinearly coupled to a harmonic bath, the dynamics is exact, and a detailed balance is exactly satisfied.

The dynamics described by the QCLE can be related to that prescribed by other methods. In [57], it was shown that one could derive both Ehrenfest mean-field dynamics and a version of surface-hopping dynamics starting from the QCLE. In the former case, one simply drops all the "correlations" (including entanglement) between the subsystem and bath densities in the QCLE [58]. In the later case, one projects out all the off-diagonal matrix elements of the density in the QCLE to obtain a generalized master equation for the subsystem alone. Then, one considers decoherence to suppress the coherences in order to recover a simple "surface hopping" dynamics [59] similar to that prescribed in the FSSH algorithm. Furthermore, it had been proven [60] that the QCLE and the partially linearized path integral (PLPI) method [61–64] share the same starting mathematical foundation. In particular, the most recent PLPI algorithm, called PLDM (Partially Linearized Density Matrix) method [64], is very similar to the FBTS presented in this paper [31]. One can also draw comparisons between methods based on the QCLE and semiclassical initial value representations. For instance, numerical schemes based on the Poisson bracket mapping equation (PBME) [30], an approximate equation derived from the mapping-transformed QCLE, and the linearized semiclassical initial value representations [65] share the same set of equations of motion for the trajectories.

Mixed quantum-classical methods are often the only feasible approach to explore the dynamics of large complex systems, such as condensed phase or biochemical systems, where only a few light-mass DOF need be treated quantum mechanically. In many rate processes of interest, such as electron transfer or proton transfer, the local polar solvent motions are responsible for important features of the reaction mechanism. As a result, it is essential that the dynamics of these environmental degrees of freedom be treated in detail. Open quantum system methods that trace out all bath details cannot capture important aspects of such dynamics.

Some recent work [48,66] has suggested interesting ways to combine the QCLE and the generalized master equation [67–69] approach. Simulation tests on spin boson models [48] and a two-level system coupled to an anharmonic bath [68] indicate that accurate, long-time dynamical properties of such systems can be efficiently calculated with an improved memory kernel (which takes the short-time QCLE computation of some bath correlation functions as the input) for the general master equation. This type of hybrid approach may eventually prove to be useful for studies of more complex systems.

Finally, we provide comments that may help in choosing between the two algorithms for simulations. The TBSH algorithm, without filtering, provides a very accurate QCL dynamics before the onset of the sign problem associated with its heavy reliance on Monte Carlo sampling. While filtering can be used to extend simulations to much longer times, the problems related to Monte Carlo sampling limits its usefulness in performing long-time simulations, as vividly illustrated in Section 4. However, the TBSH is found to be the preferred simulation method (in comparison to the FBTS) when one investigates bath dynamical properties of systems in the vicinity of conical intersections and avoided crossings. For instance, the TBSH results accurately capture the intricate geometric phases [46] and the bimodal structure in the momentum distribution [35] in the Tully 1 model (a single avoided crossing model), while the FBTS fails to reproduce these delicate features, even though it provides fairly accurate population dynamics, as reported in [36]. Since the FBTS trajectory dynamics is based on a mean-field description, one finds that the results are usually very accurate (even in the long-time limit) when the energy gap between diabatic energy surfaces is small in comparison to the typical subsystem-bath coupling strength. Another advantage of the FBTS is the availability of the JFBTS [36] algorithm, which implements systematic correction of FBTS

314

results towards the exact QCL dynamics and provides a simple method to gauge the sufficiency of the FBTS results.

Acknowledgments

Research was supported in part by a grant from the Natural Sciences and Engineering Research Council of Canada.

Conflicts of Interest

The authors declare no conflict of interest.

Appendix

High Temperature Limit

Many realistic chemical and biological processes take place at room temperature, in which case, it is often justified to apply a classical approximation to the bath. In this Appendix, we make two assumptions: As in most condensed phase models, we consider a pure subsystem observable, $\hat{A}$, such that $\left(\hat{\rho}_{eq}\hat{A}\right)_W = (\hat{\rho}_{eq})_W \hat{A}$. We also assume that the environment is further partitioned into an immediate part that can couple nonlinearly to the quantum subsystem and shield the subsystem from the larger set of environmental DOF, often modeled as a heat bath of independent harmonic oscillators. Furthermore, we write $X = \{X_b, X_n\}$, where n refers to the few DOF that couple directly to the quantum subsystem and b refers to the remainder of the large number of coordinates that only couple to the n-labeled coordinates. Similarly, we re-label different parts of the Hamiltonian as follows: $\hat{H} = \hat{H}_s + \hat{H}_n + \hat{V}_{sn} + \hat{H}_b + \hat{V}_{bn}$ with $\hat{H}_i = \hat{K}_i + \hat{V}_i$ and $i = s, b, n$. The quantities, $\hat{K}_i$ and $\hat{V}_i$, are the total kinetic energy and isolated potential, respectively, of the i-th system. Potential energy terms with a subscript of two letters imply a coupling potential between two components of the composite system. In addition, we introduce $\hat{h}_W(R_n) = \hat{H}_s + \hat{V}_{sn}(R_n) + V_n(R_n)$, $\hat{H}_{bn}(R_n) = \hat{H}_b + \hat{V}_{bn}(R_n)$ and $\hat{H}_{sn} = \hat{h}_W(R_n) + \hat{K}_n$. In the following, we express the distance in units of $\lambda_j = \sqrt{\hbar/M_j\omega_c}$ and energy in units of $\hbar\omega_c$, where ω_c is the cut-off frequency of the heat bath.

Under these assumptions, one needs to evaluate the partial Wigner transform of $e^{-\beta\hat{H}}$ alone. In the high temperature limit, we factorize the un-normalized equilibrium density matrix operator, $\hat{\rho} = e^{-\beta\hat{H}_{sn}}e^{-\beta\hat{H}_{bn}}$. The partial Wigner transform of this approximate density operator reads:

$$\hat{\rho}_W(X) = \int dZ e^{-iP_n\cdot Z}\left\langle R_n + \frac{Z}{2}\left|e^{-\beta\hat{H}_{sn}}\right|R_n - \frac{Z}{2}\right\rangle \rho_b(X_b; R_n) \tag{36}$$

where $\rho_b(X_b; R_n) = \int dZ_b e^{-iP_b\cdot Z_b}\left\langle R_b + \frac{Z_b}{2}\left|e^{-\beta\hat{H}_{bn}(R_n)}\right|R_b - \frac{Z_b}{2}\right\rangle$ is the Wigner transform of the un-normalized equilibrium density matrix for the heat bath.

We next apply a symmetric Trotter decomposition to the matrix element of Equation (36):

$$\left\langle R_n + \frac{Z}{2} \middle| e^{-\beta \hat{H}_{sn}} \middle| R_n - \frac{Z}{2} \right\rangle \approx \left\langle R_n + \frac{Z}{2} \middle| e^{-\frac{\beta}{2}\Delta \hat{H}_W (R_n + \frac{Z}{2})} e^{-\beta \hat{H}_{ho}} e^{-\frac{\beta}{2}\Delta \hat{H}_W (R_n - \frac{Z}{2})} \middle| R_n - \frac{Z}{2} \right\rangle$$

$$= \left(\frac{\omega}{2\pi \sinh(\omega\beta)} \right)^{N_n/2} \exp\left(-\coth\left(\frac{\omega\beta}{2}\right) \frac{\omega Z^2}{4} \right) \exp\left(-\tanh\left(\frac{\omega\beta}{2}\right) \omega R_n^2 \right)$$

$$\times e^{-\frac{\beta}{2}\Delta \hat{H}_W (R_n + \frac{Z}{2})} e^{-\frac{\beta}{2}\Delta \hat{H}_W (R - \frac{Z}{2})} \tag{37}$$

In this equation, the symmetric Trotter decomposition separates the subsystem potential in $\hat{h}_W(R_n)$ into harmonic, $\hat{V}_{ho} = \frac{1}{2}\omega^2 R_n^2$, and anharmonic, $\Delta\hat{H}(R_n) = \hat{h}_W(R_n) - V_{ho}(R_n)$, contributions; furthermore, we define $\hat{H}_{ho} = \hat{K}_n + \hat{V}_{ho}$.

The anharmonic term in Equation (37) can be approximated as follows:

$$e^{-\beta\Delta\hat{h}_W (R_n + \frac{Z}{2})} e^{-\beta\Delta\hat{h}_W (R_n - \frac{Z}{2})}$$

$$= \sum_{\alpha,\alpha'} e^{-\beta\tilde{E}_\alpha(R_n)} \left[\delta_{\alpha\alpha'} + \frac{Z}{2} O_{\alpha\alpha'}(R_n) d_{\alpha\alpha'}(R_n) \right] |\alpha; R_n\rangle \langle\alpha'; R_n| \tag{38}$$

where $|n\rangle$ is the subsystem basis and $|\alpha; R_n\rangle$ is the real-valued adiabatic state with adiabatic energy $E_\alpha(R_n)$ with respect to the Hamiltonian, $\hat{h}_W(R_n)$. The adjusted energy is $\tilde{E}_\alpha(R_n) = E_\alpha(R_n) - V_{ho}(R_n)$. The O function in Equation (38) reads:

$$O_{\alpha\alpha'}(R_n) = \left[1 - e^{-\frac{\beta}{2}(\tilde{E}_{\alpha'}(R_n) - \tilde{E}_\alpha(R_n))} \right]^2 \tag{39}$$

and $d_{\alpha\alpha'} = \langle\alpha; R_n| \nabla_{R_n} |\alpha'; R_n\rangle$. Details of a similar derivation for Equations (37) and (38) may be found in [70].

Substituting Equation (37) into Equation (36) and integrating out the Z variable, Equation (36) simplifies to:

$$\hat{\rho}_W(X) = \left(\frac{1}{2\pi\hbar \cosh(\frac{\omega\beta}{2})} \right)^{N_n} \rho_b(X_b; R_n) e^{-\frac{P_n^2}{\omega} \tanh(\frac{\omega\beta}{2})} \sum_\lambda e^{-\beta E_\lambda(R_n)}$$

$$\sum_{\alpha,\alpha'} |\alpha; R_n\rangle \langle\alpha'; R_n| F_{\alpha\alpha'}(X_n) \tag{40}$$

where:

$$F_{\alpha\alpha'}(X_n) = \frac{e^{-\beta E_\alpha(R_n)}}{\sum_\lambda e^{-\beta E_\lambda(R_n)}} \left[\delta_{\alpha\alpha'} - i\frac{P_n}{\omega} \tanh(\omega\beta/2) O_{\alpha\alpha'}(R_n) d_{\alpha\alpha'}(R_n) \right] \tag{41}$$

Now, the canonical partition function is determined by:

$$Z_Q = \sum_\alpha \int dX_n dX_b \rho_W^{\alpha\alpha}(X)$$

$$= \left(\frac{1}{\cosh(\frac{\omega\beta}{2})} \right)^{N_n} \left(\prod_{j=1}^{N_b} \frac{\pi}{\sinh(\omega_j\beta/2)} \right) \sqrt{\frac{\pi\omega}{\tanh(\omega\beta/2)}} \int dR_n \sum_\lambda e^{-\beta E_\lambda(R_n)}$$

$$= \left(\frac{1}{\cosh(\frac{\omega\beta}{2})} \right)^{N_n} Z_b Z_{sn} \tag{42}$$

316

where Z_b is defined by the expression in the second bracket on the second line and Z_{sn} is defined by the expression behind the second bracket on the second line. Z_b and Z_{sn} are the bath and subsystem (with its immediate environment) canonical partition functions, respectively. In summary, the time correlation function takes the following simple form:

$$C_{AB}(t) = \frac{1}{Z_Q} \sum_{n_1,n_2,n_3} \int dX \, \langle n_1| \, \hat{\rho}_W(X) \, |n_3\rangle \, \langle n_3| \, \hat{A} \, |n_2\rangle \, \langle n_2| \, \hat{B}_W(X,t) \, |n_1\rangle \qquad (43)$$

where $\hat{\rho}_W(X)$ and Z_Q are given by Equations (40) and (42), respectively.

References

1. Zwanzig, R. Time-correlation functions and transport coefficients in statistical mechanics. *Annu. Rev. Phys. Chem.* **1965**, *16*, 67–102.
2. Kapral, R.; Ciccotti, G. A Statistical Mechanical Theory of Quantum Dynamics in Classical Environments. In *Bridging Time Scales: Molecular Simulations for the Next Decade*; Nielaba, P., Mareschal, M., Ciccotti, G., Eds.; Springer: Berlin, Germany, 2002; pp. 445–472.
3. Bonella, S.; Monteferrante, M.; Pierleoni, C.; Ciccotti, G. Path integral based calculations of symmetrized time correlation functions. I. *J. Chem. Phys.* **2010**, *133*, 164104.
4. Bonella, S.; Monteferrante, M.; Pierleoni, C.; Ciccotti, G. Path integral based calculations of symmetrized time correlation functions. II. *J. Chem. Phys.* **2010**, *133*, 164105.
5. Monteferrante, M.; Bonella, S.; Ciccotti, G. Linearized symmetrized quantum time correlation functions calculation via phase pre-averaging. *Mol. Phys.* **2011**, *109*, 3015–3027.
6. Redfield, A. The Theory of Relaxation Processes. In *Advances in Magnetic Resonance*; Waugh, J., Ed.; Academic Press: New York, NY, USA, 1965; Volume 1.
7. Lindblad, G. On the generators of quantum dynamical semigroups. *Commun. Math. Phys.* **1976**, *48*, 119–130.
8. Weiss, U. *Quantum Dissipative Systems*; World Scientific: Singapore, Singapore, 1999.
9. Blum, K. *Density Matrix Theory and Applications*; Plenum: New York, NY, USA, 1981.
10. Breuer, H.P.; Petruccione, F. *The Theory of Open Quantum Systems*; Oxford University Press: Oxford, UK, 2007.
11. Feynman, R.P.; Vernon, J.F.L. The theory of a general quantum mechanical system interacting with a linear dissipative system. *Ann. Phys.* **1963**, *24*, 118–173.
12. Feynman, R.P.; Hibbs, A.R. *Quantum Mechanics and Path Integrals*; McGraw-Hill: New York, NY, USA, 1965.
13. Herman, M.F. Dynamics by semiclassical methods. *Annu. Rev. Phys. Chem.* **1994**, *45*, 83–111.
14. Thoss, M.; Wang, H.B. Semiclassical description of molecular dynamics based on initial-value representation methods. *Annu. Rev. Phys. Chem.* **2004**, *55*, 299–332.
15. Kay, K.G. Semiclassical initial value treatments of atoms and molecules. *Annu. Rev. Phys. Chem.* **2005**, *56*, 255–280.

16. Tully, J.C. Nonadiabatic Dynamics. In *Modern Methods for Multidimensional Dynamics Computations in Chemistry*; Thompson, D.L., Ed.; World Scientific: New York, NY, USA, 1998; p. 34.
17. Kapral, R. Progress in the theory of mixed quantum-classical dynamics. *Ann. Rev. Phys. Chem.* **2006**, *57*, 129–157.
18. Kapral, R.; Ciccotti, G. Mixed quantum-classical dynamics. *J. Chem. Phys.* **1999**, *110*, 8919–8929.
19. Sergi, A.; Kapral, R. Quantum-classical limit of quantum correlation functions. *J. Chem. Phys.* **2004**, *121*, 7565–7576.
20. Nassimi, A.; Kapral, R. Mapping approach for quantum-classical time correlation functions 1. *Can. J. Chem.* **2009**, *87*, 880–890.
21. Filinov, V.; Bonella, S.; Lozovik, Y.; Filinov, A.; Zacharov, I. Quantum Molecular Dynamics Using Wigner Representation. In *Classical Dynamics in Condensed Phase Simulations*; Berne, B.J., Ciccotti, G., Coker, D.C., Eds.; World ScientiïñАс: Singapore, Singapore, 1998.
22. Basire, M.; Borgis, D.; Vuilleumier, R. Computing Wigner distributions and time correlation functions using the quantum thermal bath method: Application to proton transfer spectroscopy. *Phys. Chem. Chem. Phys.* **2013**, *15*, 12591–12601.
23. Martens, C.C.; Fang, J.Y. Semiclassical-limit molecular dynamics on multiple electronic surfaces. *J. Chem. Phys.* **1996**, *106*, 4918–4930.
24. Donoso, A.; Martens, C.C. Simulation of coherent nonadiabatic dynamics using classical trajectories. *J. Phys. Chem. A* **1998**, *102*, 4291–4300.
25. MacKernan, D.; Kapral, R.; Ciccotti, G. Sequential short-time propagation of quantum-classical dynamics. *J. Phys. Condens. Matter* **2002**, *14*, 9069–9076.
26. MacKernan, D.; Ciccotti, G.; Kapral, R. Trotter-based simulation of quantum-classical dynamics. *J. Phys. Chem. B* **2008**, *112*, 424–432.
27. Horenko, I.; Salzmann, C.; Schmidt, B.; Schutte, C. Quantum-classical Liouville approach to molecular dynamics: Surface hopping Gaussian phase-space packets. *J. Chem. Phys.* **2002**, *117*, 11075–11088.
28. Wan, C.; Schofield, J. Mixed quantum-classical molecular dynamics: Aspects of the multithreads algorithm. *J. Chem. Phys.* **2000**, *113*, 7047–7054.
29. Wan, C.; Schofield, J. Solutions of mixed quantum-classical dynamics in multiple dimensions using classical trajectories. *J. Chem. Phys.* **2002**, *116*, 494–506.
30. Kim, H.; Nassimi, A.; Kapral, R. Quantum-classical Liouville dynamics in the mapping basis. *J. Chem. Phys.* **2008**, *129*, 084102.
31. Hsieh, C.Y.; Kapral, R. Nonadiabatic dynamics in open quantum-classical systems: Forward-backward trajectory solution. *J. Chem. Phys.* **2012**, *137*, 22A507.
32. Stock, G.; Thoss, M. Classical description of nonadiabatic quantum dynamics. *Adv. Chem. Phys.* **2005**, *131*, 243–375.
33. Schwinger, J. On Angular Momentum. In *Quantum Theory of Angular Momentum*; Biedenharn, L.C., Dam, H.V., Eds.; Academic Press: New York, NY, USA, 1965; p. 229.

34. Nassimi, A.; Bonella, S.; Kapral, R. Analysis of the quantum-classical Liouville equation in the mapping basis. *J. Chem. Phys.* **2010**, *133*, 134115.

35. Kelly, A.; van Zon, R.; Schofield, J.; Kapral, R. Mapping quantum-classical Liouville equation: Projectors and trajectories. *J. Chem. Phys.* **2012**, *136*, 084101.

36. Hsieh, C.Y.; Kapral, R. Analysis of the forward-backward trajectory solution for the mixed quantum-classical Liouville equation. *J. Chem. Phys.* **2013**, *138*, 134110.

37. Ananth, N.; Miller, T.F. Exact quantum statistics for electronically nonadiabatic systems using continuous path variables. *J. Chem. Phys.* **2010**, *133*, 234103.

38. Nielsen, S.; Kapral, R.; Ciccotti, G. Statistical mechanics of quantum-classical systems. *J. Chem. Phys.* **2001**, *115*, 5805–5815.

39. Kim, H.; Hanna, G.; Kapral, R. Analysis of kinetic isotope effects for nonadiabatic reactions. *J. Chem. Phys.* **2006**, *125*, 084509.

40. Hanna, G.; Kapral, R. Quantum-classical Liouville dynamics of nonadiabatic proton transfer. *J. Chem. Phys.* **2005**, *122*, 244505.

41. Kim, H.; Kapral, R. Solvation and proton transfer in polar molecule nanoclusters. *J. Chem. Phys.* **2005**, *125*, 234309.

42. Kim, H.; Kapral, R. Proton and deuteron transfer reactions in molecular nanoclusters. *ChemPhysChem* **2008**, *9*, 470–474.

43. Hanna, G.; Geva, E. Vibrational energy relaxation of a hydrogen-bonded complex dissolved in a polar liquid via the mixed quantum-classical lionville method. *J. Phys. Chem. B* **2008**, *112*, 4048–4058.

44. Horenko, I.; Schmidt, B.; Schutte, C. A theoretical model for molecules interacting with intense laser pulses: The floquet-based quantum-classical Liouville equation. *J. Chem. Phys.* **2001**, *115*, 5733–5743.

45. Morales, C.M.; Thompson, W.H. Mixed quantum-classical molecular dynamics analysis of the molecular-level mechanisms of vibrational frequency shifts. *J. Phys. Chem. A* **2007**, *111*, 5422–5433.

46. Kelly, A.; Kapral, R. Quantum-classical description of environmental effects on electronic dynamics at conical intersections. *J. Chem. Phys.* **2010**, *133*, 084502.

47. Kim, H.W.; Kelly, A.; Park, J.W.; Rhee, Y.M. All-atom semiclassical dynamics study of quantum coherence in photosynthetic fenna–matthews–olson complex. *J. Am. Chem. Soc.* **2012**, *134*, 11640–11651.

48. Kelly, A.; Markland, T.E. Efficient and accurate surface hopping for long time nonadiabatic quantum dynamics. *J. Chem. Phys.* **2013**, *139*, 014104.

49. Stock, G.; Thoss, M. Semiclassical description of nonadiabatic quantum dynamics. *Phys. Rev. Lett.* **1997**, *78*, 578–581.

50. Ananth, N.; Venkataraman, C.; Miller, W.H. Semiclassical description of electronically nonadiabatic dynamics via the initial value representation. *J. Chem. Phys.* **2007**, *127*, 084114.

51. Huo, P.; Coker, D.F. Consistent schemes for non-adiabatic dynamics derived from partial linearized density matrix propagation. *J. Chem. Phys.* **2012**, *137*, 22A535.

52. Makri, N.; Thompson, K. Semiclassical influence functionals for quantum systems in anharmonic environments. *Chem. Phys. Lett.* **1998**, *291*, 101–109.

53. Sergi, A.; Kapral, R. Quantum-classical dynamics of nonadiabatic chemical reactions. *J. Chem. Phys.* **2003**, *118*, 8566–8575.

54. Bonella, S.; Coker, D.F.; Kernan, D.M.; Kapral, R.; Ciccotti, G. Trajectory Based Simulations of Quantum-Classical Systems. In *Energy Transfer Dynamics in Biomaterial Systems*; Burghardt, I., May, V., Micha, D.A., Bittner, E.R., Eds.; Springer: Berlin/Heidelberg, Germany, 2009; Volume 93.

55. Parandekar, P.V.; Tully, J.C. Mixed quantum-classical equilibrium. *J. Chem. Phys.* **2005**, *122*, 094102.

56. Schmidt, J.R.; Parandekar, P.V.; Tully, J.C. Mixed quantum-classical equilibrium: Surface hopping. *J. Chem. Phys.* **2008**, *129*, 044104.

57. Grunwald, R.; Kelly, A.; Kapral, R. Quantum Dynamics in Almost Classical Environments. In *Energy Transfer Dynamics in Biomaterial Systems*; Springer: Berlin/Heidelberg, Germany, 2009; pp. 383–413.

58. Gerasimenko, V.I. Dynamical equations of quantum-classical systems. *Theor. Math. Phys.* **1982**, *50*, 49–55.

59. Grunwald, R.; Kapral, R. Decoherence and quantum-classical master equation dynamics. *J. Chem. Phys.* **2007**, *126*, 114109.

60. Bonella, S.; Ciccotti, G.; Kapral, R. Linearization approximation and Liouville Quantum-classical dynamics. *Chem. Phys. Lett.* **2010**, *484*, 399–404.

61. Bonella, S.; Coker, D.F. Semiclassical implementation of the mapping Hamiltonian approach for nonadiabatic dynamics using focused initial distribution sampling. *J. Chem. Phys.* **2003**, *118*, 4370–4385.

62. Bonella, S.; Coker, D.F. LAND-map, a linearized approach to nonadiabatic dynamics using the mapping formalism. *J. Chem. Phys.* **2005**, *122*, 194102.

63. Dunkel, E.; Bonella, S.; Coker, D.F. Iterative linearized approach to nonadiabatic dynamics. *J. Chem. Phys.* **2008**, *129*, 114106.

64. Huo, P.; Coker, D.F. Partial linearized density matrix dynamics for dissipative, non-adiabatic quantum evolution. *J. Chem. Phys.* **2011**, *135*, 201101.

65. Liu, J.; Miller, W.H. Real time correlation function in a single phase space integral beyond the linearized semiclassical initial value representation. *J. Chem. Phys.* **2007**, *126*, 234110.

66. Shi, Q.; Geva, E. A derivation of the mixed quantum-classical Liouville equation from the influence functional formalism. *J. Chem. Phys.* **2004**, *121*, 3393–3404.

67. Shi, Q.; Geva, E. A new approach to calculating the memory kernel of the generalized quantum master equation for an arbitrary systemâĂŞbath coupling. *J. Chem. Phys.* **2003**, *119*, 12063–12076.

68. Shi, Q.; Geva, E. A semiclassical generalized quantum master equation for an arbitrary system-bath coupling. *J. Chem. Phys.* **2004**, *120*, 10647–10658.

320

69. Zhang, M.L.; Ka, B.J.; Geva, E. Nonequilibrium quantum dynamics in the condensed phase via the generalized quantum master equation. *J. Chem. Phys.* **2006**, *125*, 044106.
70. Kim, H.; Kapral, R. Nonadiabatic quantum-classical reaction rates with quantum equilibrium structure. *J. Chem. Phys.* **2005**, *123*, 194108.

Reprinted from *Entropy*. Cite as: Bonella, S.; Ciccotti, G. Approximating Time-Dependent Quantum Statistical Properties. *Entropy* **2014**, *16*, 86–109.

Article

Approximating Time-Dependent Quantum Statistical Properties

Sara Bonella * and Giovanni Ciccotti

Department of Physics and CNISM Unit 1, University of Rome "La Sapienza", Ple A. Moro 5, 00185 Rome, Italy; E-Mail: giovanni.ciccotti@roma1.infn.it

* Author to whom correspondence should be addressed; E-Mail: sara.bonella@roma1.infn.it;
  Tel.: +39-06-49914208; Fax: +39-06-4957697.

Received: 11 November 2013; in revised form: 10 December 2013 / Accepted: 19 December 2013 / Published: 27 December 2013

Abstract: Computing quantum dynamics in condensed matter systems is an open challenge due to the exponential scaling of exact algorithms with the number of degrees of freedom. Current methods try to reduce the cost of the calculation using classical dynamics as the key ingredient of approximations of the quantum time evolution. Two main approaches exist, quantum classical and semi-classical, but they suffer from various difficulties, in particular when trying to go beyond the classical approximation. It may then be useful to reconsider the problem focusing on statistical time-dependent averages rather than directly on the dynamics. In this paper, we discuss a recently developed scheme for calculating symmetrized correlation functions. In this scheme, the full (complex time) evolution is broken into segments alternating thermal and real-time propagation, and the latter is reduced to classical dynamics via a linearization approximation. Increasing the number of segments systematically improves the result with respect to full classical dynamics, but at a cost which is still prohibitive. If only one segment is considered, a cumulant expansion can be used to obtain a computationally efficient algorithm, which has proven accurate for condensed phase systems in moderately quantum regimes. This scheme is summarized in the second part of the paper. We conclude by outlining how the cumulant expansion formally provides a way to improve convergence also for more than one segment. Future work will focus on testing the numerical performance of this extension and, more importantly, on investigating the limit for the number of segments that goes to infinity of the approximate expression for the symmetrized correlation function to assess formally its convergence to the exact result.

Keywords: semiclassical statistical properties; time correlation functions; mixed quantum classical dynamics

---

1. Introduction

Exact simulation methods to compute either the evolution of the wave function or dynamical statistical averages for quantum systems in the condensed phase are currently restricted to small sizes and short times. The exponential scaling of available algorithms with the number of degrees of freedom, in fact, limits calculations to ten–twenty particles (and this for Hamiltonians of relatively simple form) and to time scales of at most a few picoseconds. This situation is in striking contrast with analogous classical calculations, which, when empirical potentials are adopted, are nowadays routinely used to study high dimensional, complex systems for times reaching, on dedicated machines, microseconds. (*Ab initio* classical molecular dynamics is considerably more expensive, but, depending on the number of electrons that have to be included, even in this case moderately sized systems of up to a hundred particles can be integrated for hundreds of picoseconds.) Several approximate schemes have thus been proposed attempting to import, with appropriate modifications, methods from classical molecular dynamics to quantum dynamics. Two approaches, in particular, can be identified in which classical trajectories play a crucial role: semi-classical and mixed quantum classical.

In semi-classical schemes, originally developed for approximating wave function propagation, all degrees of freedom are treated on equal footings. To begin with, the quantum time propagator is expressed, in the path integral formalism [1], as a sum over all possible paths connecting the initial and final states, each path being weighted by a complex exponential, whose argument is the classical action along it. The approximate propagator is then obtained by expanding the action to second order around its stationary points, which are classical trajectories, and performing the remaining quadratic path integral analytically [2,3]. Different forms exist for the semi-classical propagator depending on the specific representation adopted for the path integral (most notably standard coordinates, usually followed by the so-called initial value transformation [4,5], hybrid coordinate momenta [6] or coherent states [7–9]), but the evolved wave function has always the same structure, which we illustrate with the most commonly used Herman Kluk expression [10–12]:

$$|\Psi(t)\rangle_{sc} = \int dpdq|q(t),p(t)\rangle W(t)e^{\frac{i}{\hbar}S_{cl}(t)}\langle p,q|\Psi(0)\rangle \qquad (1)$$

In the expression above, $|q,p\rangle$ indicates a coherent state (in the coordinate representation, $\langle r|q,p\rangle \propto e^{-\gamma(q-r)^2+ip(q-r)/\hbar}$), $S_{cl}(t)$ is the action computed along a classical trajectory propagated to time t from initial conditions (q,p), $(q(t),p(t))$ is the endpoint of the trajectory and $W(t)$, a known function, is the result of the integration over the quadratic fluctuations around the stationary paths. All the functions of time in the integrand are calculable using classical evolution algorithms and, once the ket has been saturated (for example, via a scalar product), the integral over the initial conditions can be estimated via Monte Carlo sampling of a probability density based on the absolute value of the wave function at $t = 0$. Although calculations based on this scheme have been used, the semi-classical wave function is still remarkably expensive, due mainly to the characteristics of $W(t)$. This function is in fact related to a linear combination of the monodromy matrices of the system (*i.e.*, the matrices of the derivatives of the endpoints of the trajectory with respect to the initial conditions), and, for chaotic dynamics, it can assume values varying over several orders of magnitude, hindering the convergence of the Monte Carlo sampling. Furthermore, the actual evaluation of the wave function requires one to project $|\Psi(t)\rangle_{sc}$ on a basis. While this is, in principle, straightforward (for example, one could choose the continuous coordinate representation and then discretize it on a grid), in practice it reinstates the exponential scaling of the numerical effort with the number of degrees of freedom. This last problem can be avoided focusing on expected values (observables):

$$_{sc}\langle\Psi(t)|\hat{A}|\Psi(t)\rangle_{sc} = \int d\tilde{p}d\tilde{q}dpdq\langle\Psi(0)|\tilde{q},\tilde{p}\rangle\tilde{W}^*(t)e^{-\frac{i}{\hbar}\tilde{S}_{cl}(t)}\langle\tilde{q}(t),\tilde{p}(t)|\hat{A}|q(t),p(t)\rangle W(t)e^{\frac{i}{\hbar}S_{cl}(t)}\langle p,q|\Psi(0)\rangle \quad (2)$$

($\hat{A}$ is a Hermitian operator) at the price of doubling the dimension of the Monte Carlo sampling. The expression above, however, requires averaging the product of the two unstable W functions (for common operators, the matrix element in the integrand is known analytically, thus posing no problem), and although some schemes exist to mitigate the problem [13,14], this approach is of limited practical use. Shifting the focus from the wave function to the observables proves considerably more effective moving to the Heisenberg representation and taking advantage of the presence of two propagators in their exact quantum expression to develop alternative approximations. This strategy is most commonly adopted when calculating time-dependent statistical properties, more specifically, time correlation functions of operators $\hat{A}$ and $\hat{B}$, usually defined as:

$$\begin{aligned} C_{A,B}(t;\beta) &= \frac{1}{Z}\text{Tr}\left\{e^{-\beta\hat{H}}\hat{A}e^{\frac{i}{\hbar}\hat{H}t}\hat{B}e^{-\frac{i}{\hbar}\hat{H}t}\right\} \\ &= \frac{1}{Z}\int drdr_N d\tilde{r}d\tilde{r}_N\langle r|e^{-\beta\hat{H}}\hat{A}|\tilde{r}\rangle\langle\tilde{r}|e^{\frac{i}{\hbar}\hat{H}t}|\tilde{r}_N\rangle\langle\tilde{r}_N|\hat{B}|r_N\rangle\langle r_N|e^{-\frac{i}{\hbar}\hat{H}t}|r\rangle \end{aligned} \quad (3)$$

where $\hat{H}$ is the Hamiltonian, Z the partition function, $\beta = 1/k_B T$ with k_B Boltzmann's constant and T temperature. (Throughout the paper, we consider distinguishable particles.) In the second line, the trace was expressed, in the coordinate representation, as the product of four matrix elements; from left to right, that of the product of the quantum Boltzmann density times operator $\hat{A}$, the propagator backward in time, the matrix element of operator $\hat{B}$ and the propagator forward in time. To pave the way for the so-called linearized approximation of the correlation function [13,15–18], the two propagators are expressed as path integrals in the hybrid coordinate-momenta representation, in which the resolution of the identity in the momentum basis (inserted, as usual, to evaluate the exponential of the kinetic energy contribution in the Trotter break up of the propagators) are not resolved analytically. The advantage of this representation is a closer analogy to the phase space representation of classical mechanics. The quantum expression of the correlation function thus obtained is then manipulated via a change of variables: the forward and backward paths are changed to semi-sum and difference paths (see the next section for a more precise definition of these paths), and the key approximation of the approach is introduced. A Taylor series expansion of the action to quadratic order in the difference path is performed, allowing all integrals on the difference variables, except the ones over the initial and endpoint in coordinate space, to be performed analytically. The integration results in a product of delta functions constraining the semi-sum path to be a classical (Hamiltonian) trajectory. The remaining integrals over the difference coordinates define Wigner transforms [19] of operators to give:

$$C_{A,B}^{l}(t;\beta) = \frac{1}{(2\pi\hbar)Z}\int dr dp \left[e^{-\beta\hat{H}}\hat{A}\right]_{w}(r,p)B_{w}(r(t),p(t)) \tag{4}$$

where, for example, $B_{w}(r(t),p(t)) = \int d\xi e^{\frac{i}{\hbar}p(t)\xi}\langle r(t) + \xi/2|\hat{B}|r(t) - \xi/2\rangle$. The superscript, l, indicates the linearization approximation. Compared to the semi-classical expression for the wave function, Equation (4) has the remarkable advantage of not containing unstable factors in the integrand, even though the approximation of the overall dynamics is accurate to the same order in $\hbar$. Indeed, the second order terms in the action expansion, which originated the $W(t)$ in Equation (1), cancel exactly when the difference of the action along the forward and backward paths is considered. The absence of the Ws suggests computing the approximate time correlation by combining Monte Carlo sampling of initial conditions and molecular dynamics. The serious difficulty with this idea lies in the sampling of the initial conditions. The probability density is, in fact, usually defined from the absolute value of $\left[e^{-\beta\hat{H}}\hat{A}\right]_{w}(r,p)/(2\pi\hbar)Z$, but computing this Wigner transform is far from trivial, and the available methods introduce further, uncontrolled, approximations. In addition to this practical difficulty, there is also a conceptual problem with linearized calculations: the classical evolution conserves the quantum probability density only for short times. The rapid decay time of correlations for standard condensed phase systems is usually invoked to mitigate the consequences of this pathology, but it is known that in some cases, e.g., low dimensional systems with a long time coherence, it may lead to unphysical results (this is the so-called zero energy leakage problem).

The problems and numerical cost of semi-classical calculations justify the development of the second, alternative approximation scheme mentioned at the beginning of this section: mixed quantum classical dynamics. In this approach, the degrees of freedom of the system are partitioned into two sets, usually based on their mass ratio. The first set (called the subsystem) is composed of a few degrees of freedom and is treated quantum mechanically; the second (called the environment or the bath) is often high dimensional and is treated classically. Existing quantum classical methods differ in the way in which the coupling among the classical evolution of the bath and the quantum propagation of the subsystem is taken into account. The first approach of this kind, still very popular due to its efficiency and ease of implementation, is Tully's surface hopping [20]. In this scheme, electrons and nuclei constitute the subsystem and the bath, respectively, and the coupling, designed to mimic dynamics beyond the Born–Oppenheimer approximation, is defined *ad hoc* based on heuristic arguments. In more recent, and more rigorous, developments, the coupling is derived starting from a fully quantum representation of the evolution equations for the system and then taking a partial classical limit on the bath's degrees of freedom. Examples of this type are schemes to propagate the full density matrix of the quantum subsystem, such as the Wigner Liouville mixed quantum dynamics [21,22], or the iterative linearized density propagation methods [23,24]. Both surface hopping and Wigner Liouville dynamics (with particular reference to its most recent developments aimed at computing correlation functions) are discussed (and/or criticized) in other contributions to this issue. Focusing on the latter, we quickly recall that it adopts a mixed representation in which the operators related to the bath's degrees of freedom are described using the Wigner representation, while for the subsystem an abstract operator representation is retained. The quantum evolution operator in this mixed representation is then expanded to first order in the ratio of the thermal De Broglie wavelength of subsystem to the bath to obtain the generator of the mixed quantum classical dynamics. This generator has the form of a generalized Lie bracket in which both a commutator (linked to the operators for the subsystem) and a Poisson parenthesis (acting on the bath's phase space) appear. Once a specific basis set is chosen for the subsystem (e.g., adiabatic electronic states [21,25] or, more recently, the so-called mapping representation [26,27]) the evolution equation for the density matrix, or any observable, becomes explicit, and several different algorithms, sharing the characteristic that the bath motion is obtained via classical evolution (possibly with generalized forces describing the influence of more than one electronic state), have been proposed to solve them. In spite of its merits, it has been shown that this mixed quantum classical dynamics lacks several properties that characterize fully quantum and classical dynamics [28]. In particular the mixed Lie bracket does not satisfy the Jacobi identity exactly, and, similar to linearized calculations, the quantum thermal density is not stationary under the mixed dynamics. The loss of formal properties with respect to classical and quantum mechanics arises, in different forms, in all current mixed quantum classical schemes (see also [29]).

While application driven calculations might not be paralyzed by the state of affairs described above, in particular, if and when it is possible to verify that these well-known pathologies have no uncontrolled effects on the results, it is important to pursue alternative approaches in an effort to derive more general schemes allowing for systematic improvement and/or assessment of the approximations employed. Indeed, a critical stumbling block common to semi-classical and mixed quantum classical methods is that it is essentially impossible to go beyond classical trajectories to approximate the quantum evolution of the full system (semi-classical) or of the bath (mixed). In the semi-classical case, including terms of higher order, the expansion of the action along the paths makes it impossible to obtain calculable expressions for the pre-factor in the expression of the wave function (already at third order, the integral corresponds to intractable Airy functions [2]), while in the linearized correlation function, it kills the emergence of delta functions that univocally determine the semi-sum path. In mixed quantum classical calculations, we refer to the Wigner Liouville formalism, but analogous problems appear, for example, in the iterative linearized propagation methods, including higher order terms in the mass ratio expansion of the propagator introduces terms in the phase space evolution of the bath that cannot be integrated numerically. In this paper, we summarize (in the spirit of an extended review) a recently developed method [30] attempting to overcome this problem. In this approach, the focus is not directly on the dynamics, but, rather, on statistical time-dependent averages, which are linked (via linear response theory) to experimental observables. In particular, we focus on time correlation functions expressed in the symmetrized form first introduced by Schofield [31]:

$$G_{A,B}(t;\beta) = \frac{1}{Z}\text{Tr}\left\{\hat{A}e^{\frac{i}{\hbar}\hat{H}t_c^*}\hat{B}e^{-\frac{i}{\hbar}\hat{H}t_c}\right\} \qquad (5)$$

where $t_c = t - \frac{i\beta\hbar}{2}$. The time Fourier transform of this complex time correlation function is related to the time Fourier transform of Equation (3) by a known multiplicative factor so both carry equivalent information. Furthermore, the symmetrized function shares some properties with its classical counterpart (e.g., it is a real function of time), which makes it a convenient starting point for developing approximations [32–37]. In the following, we summarize how the path integral formalism can be used to express the full complex time evolution in Equation (5) as a concatenation of segments alternating imaginary (*i.e.*, thermal sampling) and real-time propagation. The real-time propagation is then reduced to classical evolution via a linearization approximation. In our approach, the number of segments, L, plays a role analogous to that of the number of beads in standard thermal path integral calculations. Although the precise nature of the limit for $L \to \infty$ is still under investigation, this analogy and numerical calculations on relatively simple model systems indicate that increasing the number of segments systematically improves the results with respect to classical dynamics or to the previously mentioned linearization schemes. It may be worth stressing that, in this approach, the focus is on computing the correlation by defining an appropriate stochastic process inspired by the full quantum expression. Adopting this perspective, the dynamics does not have any meaning *per se* and is viewed simply as part of a sampling mechanism, which is implemented via a generalized Monte Carlo scheme. While this circumvents some of the inconsistency of standard semi-classical and mixed quantum classical schemes and justifies further investigation of the method, it remains to be seen whether the approach outlined in the following has practical value. In fact, due to the presence of an increasing number of phase factors in the Monte Carlo estimator of the correlation function, the numerical cost of the calculation scales very badly with the number of segments (and of degrees of freedom). In the second part of the paper, we then summarize (again reviewing published material) how, when only one segment is considered, it is possible to improve the situation via a cumulant expansion that tames the phase factor present already in this lowest order approximation of the result [38]. We then present a new formal development of our approach that generalizes the use of cumulants to the case of more than one propagation segment, and we give the explicit formal expression for the case $L = 2$. Future work will focus on testing the accuracy of this new result. We conclude by stating some of the open questions related to the approach and indicating possible further developments.

2. Theory

Let us begin by expressing the symmetrized correlation function, Equation (5), in the coordinate representation. Inserting resolutions of the identity, we have:

$$G_{A,B}(t;\beta) = \frac{1}{Z}\int dr_0 d\tilde{r}_0 dr_{t_c} d\tilde{r}_{t_c} \langle r_0|\hat{A}|\tilde{r}_0\rangle \langle \tilde{r}_0|e^{\frac{i}{\hbar}\hat{H}t_c^*}|\tilde{r}_{t_c}\rangle \langle \tilde{r}_{t_c}|\hat{B}|r_{t_c}\rangle \langle r_{t_c}|e^{-\frac{i}{\hbar}\hat{H}t_c}|r_0\rangle \qquad (6)$$

The structure of the integrand is represented in Figure 1 in which we show the sequence of matrix elements to be evaluated. Reading the figure from the bottom left corner up, we see the matrix element of operator $\hat{A}$, the backward complex time propagator (from $\tilde{r}_0$ to $\tilde{r}_c$), the matrix element of operator $\hat{B}$ and, finally, the forward propagator that closes the circuit representing the trace operation. The difficult task is the evaluation of the propagators in complex time. To set the stage for the approximation we intend to perform, we use the time composition property to divide the two propagations into L segments of duration $\tau_c = t_c/L$ (τ_c need not be infinitesimal) and rewrite, for example:

$$\langle r_{t_c}|e^{-\frac{i}{\hbar}\hat{H}t_c}|r_0\rangle = \int dr_1...dr_{L-1} \prod_{J=0}^{L-1} \langle r_{J+1}|e^{-\frac{i}{\hbar}\hat{H}\tau_c}|r_J\rangle \tag{7}$$

with $r_L = r_{t_c}$. Introducing an analogous expression for the backward propagator changes the scheme of the integrand as sketched in Figure 2, where each propagation lag (from J to $J+1$) is indicated by the segment with arrows. We can now pair corresponding segments of propagation along the forward and backward paths, as indicated by the red frame in the figure, and define the product $K(r_{J+1}, r_J; \tilde{r}_{J+1}, \tilde{r}_J) = \langle \tilde{r}_{J+1}|e^{\frac{i}{\hbar}\hat{H}\tau_c^*}|\tilde{r}_J\rangle\langle r_{J+1}|e^{-\frac{i}{\hbar}\hat{H}\tau_c}|r_J\rangle$ to rewrite the symmetrized correlation function as:

$$G_{A,B}(t;\beta) = \frac{1}{Z} \int d\tilde{r}_L dr_L \langle \tilde{r}_L|\hat{B}|r_L\rangle \left\{ \prod_{J=1}^{L-1} \int d\tilde{r}_J dr_J K(r_{J+1}, r_J; \tilde{r}_{J+1}, \tilde{r}_J) \right\}$$

$$\times \int dr_0 d\tilde{r}_0 K(r_1, r_0; \tilde{r}_1, \tilde{r}_0)\langle r_0|\hat{A}|\tilde{r}_0\rangle \tag{8}$$

Figure 1. Schematic representation of the integrand in the coordinate representation of the symmetrized time correlation function; see the text.

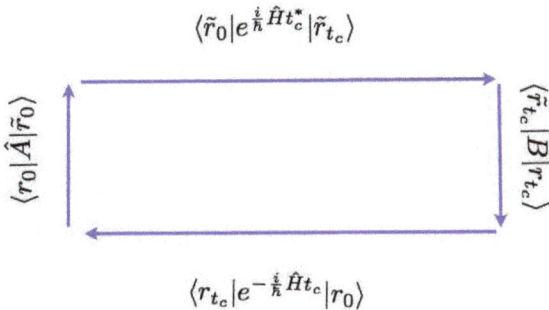

Figure 2. Schematic representation of the break up of the propagators in complex time: the short complex time propagators are represented as the segments with arrows along the forward and backward path, and the pairing mentioned in the text to obtain the K propagators is indicated by the red frame.

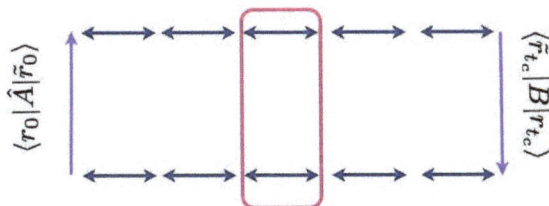

The expression above is an exact, incalculable, expression of the time correlation function. In the following, we will work on the generic K to obtain an approximate expression for it that has the key advantage of being analytically known as a product of functions calculable via an appropriate combination of Monte Carlo and molecular dynamics. To obtain this result, we first separate the real from the imaginary time part of the propagation by inserting one more resolution of the coordinate identity, thus:

$$K(r_{J+1}, r_J; \tilde{r}_{J+1}, \tilde{r}_J) = \langle \tilde{r}_{J+1}| e^{\frac{i}{\hbar}\hat{H}\tau_c^*}|\tilde{r}_J\rangle\langle r_{J+1}|e^{-\frac{i}{\hbar}\hat{H}\tau_c}|r_J\rangle \tag{9}$$

$$= \int d\tilde{r}_J^\nu dr_J^\nu \langle \tilde{r}_{J+1}|e^{\frac{i}{\hbar}\hat{H}\tau_t}|\tilde{r}_J^\nu\rangle\langle \tilde{r}_J^\nu|e^{-\tau_\beta \hat{H}}|\tilde{r}_J\rangle\langle r_{J+1}|e^{-\frac{i}{\hbar}\hat{H}\tau_t}|r_J^\nu\rangle\langle r_J^\nu|e^{-\tau_\beta \hat{H}}|r_J\rangle$$

The two (thermal) propagators in the integrand above, associated with the inverse temperature, τ_β, can be expressed relatively easily in the path integral formalism as positive definite functions and interpreted, via the so-called classical isomorphism, as probability densities associated with systems of polymers. Well established techniques allow one to sample these densities. The real-time propagators, on the other hand, are prohibitive, even in path integral form. Feynman's prescription requires, in fact, to generate "all possible paths" connecting the initial and final point of the propagation, but, in contrast to the thermal case in which the probability density provides us with a sampling mechanism for the paths, no rule is given to determine them. Furthermore, even if we had a recipe for generating the paths, we would have to sum a (potentially infinite) set of phase factors, the exponential weighting each path. Capturing accurately the interference among these factors is essentially impossible (this is the well-known dynamical sign problem). As shown in detail in [30], progress can be made by deriving an approximate form for the product of the two real-time propagators in Equation (9). In analogy with the standard linearization methods mentioned in the Introduction, this is done most conveniently using a hybrid coordinate momenta representation of the propagators. The propagation time is divided into n intervals of length $\delta_t = \tau_t/n$, and appropriate resolutions of the identity are introduced to isolate matrix elements of the propagator for each short time interval. As usual, after a Trotter break up of the exponential of the Hamiltonian is performed, the (diagonal) exponential of the potential can be trivially evaluated. The matrix element of the kinetic energy part of the propagator, on the other hand, is easily evaluated by inserting a resolution of the identity in the momenta. Contrary to what is done in standard path integrals, however, the resulting generalized Gaussian integral in the momenta is not performed analytically, but left in the expression. This sequence of operations results in:

$$\langle \tilde{r}_{J+1}|e^{\frac{i}{\hbar}\hat{H}\tau_t}|\tilde{r}_J^\nu\rangle\langle r_{J+1}|e^{-\frac{i}{\hbar}\hat{H}\tau_t}|r_J^\nu\rangle \approx \int \prod_{l=1}^{n}\frac{d\tilde{p}_J^l}{2\pi\hbar}\prod_{l=1}^{n-1}d\tilde{r}_J^{\nu+l}e^{-\frac{i}{\hbar}S(\{\tilde{r},\tilde{p}\})}\int\prod_{l=1}^{n}\frac{dp_J^l}{2\pi\hbar}\prod_{l=1}^{n-1}dr_J^{\nu+l}e^{\frac{i}{\hbar}S(\{r,p\})} \tag{10}$$

where $(\{r,p\})$ indicates the full set of path variables and $S(\{r,p\}) = \sum_{l=1}^{n}\left\{p_J^l(r_J^{(\nu+l)} - r_J^{(\nu+l-1)}) - \delta_t[(p_J^l)^2/2m - V(r_J^{(\nu+l-1)})]\right\}$ with analogous definitions for the tilde variables. The expression above becomes exact for $n \to \infty$. At this stage, the forward and backward path integrals above are independent. Proceeding in analogy with standard linearization methods, we combine them by introducing the semi-sum and difference variables:

$$\bar{r}_J^{\nu+l} = \frac{r_J^{\nu+l} + \tilde{r}_J^{\nu+l}}{2} \qquad \bar{p}_J^l = \frac{p_J^l + \tilde{p}_J^l}{2}$$

$$\Delta r_J^{\nu+l} \;=\; r_J^{\nu+l} - \tilde{r}_J^{\nu+l} \qquad\qquad \Delta p_J^l = p_J^l - \tilde{p}_J^l \tag{11}$$

with $l = 0, ..., n$. In these variables, the difference of actions in Equation (10) is a linear function in Δp_J^l. Integrals over the difference momenta can then be performed analytically and result in a set of delta functions (this originates the last set of deltas in Equation (12) below). The dependence of the difference of the actions on $\Delta r_J^{(\nu+l)}$ is more complicated: they appear in an explicit, linear term, but also in the argument of the potentials. This dependence can also be linearized via the expansion $V(\tilde{r}_J^{(\nu+l)} + \Delta r_J^{(\nu+l)}/2) - V(\tilde{r}_J^{(\nu+l)} - \Delta r_J^{(\nu+l)}/2) = \nabla V(\tilde{r}_J^{(\nu+l)})\Delta r_J^{(\nu+l)} + o[(\Delta r_J^{(\nu+l)})^3]$. This is the key approximation that we perform. An appropriate rescaling of the variables shows that the approximation is equivalent to a second order expansion in $\hbar$ of the phase, but a more precise analysis of its validity is required and under consideration (see, also, the discussion at the end of this section). Bearing this in mind, we observe that, once the expansion is performed, also the integrals on the $\Delta r_J^{(\nu+l)}$ variables can be analytically solved, producing a second set of delta functions. Thus:

$$\langle \bar{r}_J^{\nu} - \frac{\Delta r_J^{\nu}}{2}|e^{\frac{i}{\hbar}\hat{H}\tau_t}|\bar{r}_{J+1} - \frac{\Delta r_{J+1}}{2}\rangle\langle\bar{r}_{J+1} + \frac{\Delta r_{J+1}}{2}|e^{-\frac{i}{\hbar}\hat{H}\tau_t}|\bar{r}_J^{\nu} + \frac{\Delta r_J^{\nu}}{2}\rangle \approx$$

$$\int d\bar{r}_J^{\nu+1}...d\bar{r}_J^{\nu+n-1}\int d\bar{p}_J^1...d\bar{p}_J^n e^{\frac{i}{\hbar}\bar{p}_J^n \Delta r_J^{(\nu+n)}} e^{-\frac{i}{\hbar}\bar{p}_J^1 \Delta r_J^{\nu}} \tag{12}$$

$$\times \prod_{l=1}^{n-1}\delta\left[(\bar{p}_J^{(l+1)} - \bar{p}_J^l) + \delta_t\nabla V(\bar{r}_J^{(\nu+l)})\right]\prod_{l=1}^{n}\delta\left[\frac{\bar{p}_J^l}{m}\delta_t - (\bar{r}_J^{(\nu+l)} - \bar{r}_J^{(\nu+l-1)})\right]$$

The linearization approximation then has two crucial consequences: (1) by allowing the integration over the difference paths, it transforms the quantum expression of the correlation function, which, in the beginning, includes two propagators and, therefore, two paths, into a formula where only the semi-sum path appears, thus leading to a structure more similar to classical time correlations in which only one propagation is present; (2) (perhaps more importantly) it forces the semi-sum path to follow a, classical, Hamiltonian trajectory, as identified by the arguments of the delta functions.

The final step to obtain a suitable expression for Equation (9) does not introduce any further approximation. Let us consider again the product of the thermal propagators in the equation. As mentioned above, these can be written via standard coordinate path integrals. Once this is done, it is convenient to introduce also for these propagators semi-sum and difference path (this is important, in particular, to ensure that the common boundaries of the thermal and real-time propagations, r_J^{ν} and $\tilde{r}_J^{\nu}$, are represented coherently). In the semi-sum and difference variables, the product of the thermal propagators takes the form:

$$\langle \bar{r}_J - \frac{\Delta r_J}{2}|e^{-\hat{H}\tau_\beta}|\bar{r}_J^{\nu} - \frac{\Delta r_J^{\nu}}{2}\rangle\langle\bar{r}_J^{\nu} + \frac{\Delta r_J^{\nu}}{2}|e^{-\hat{H}\tau_\beta}|\bar{r}_J + \frac{\Delta r_J}{2}\rangle \approx \left[\frac{m}{2\pi\hbar\delta_\beta}\right]^{(\nu-1)}$$

$$\int d\bar{r}_J^1...d\bar{r}_J^{\nu-1}\int d\Delta r_J^1...d\Delta r_J^{\nu-1}e^{-\delta_\beta \sum_{\lambda=1}^{\nu}[V(\bar{r}_J^{(\lambda-1)}+\Delta r_J^{(\lambda-1)}/2)+V(\bar{r}_J^{(\lambda-1)}-\Delta r_J^{(\lambda-1)}/2)]} \tag{13}$$

$$\times\; e^{-\frac{\sigma_p^2}{2}\sum_{\lambda=1}^{\nu}(\Delta r_J^{\lambda}-\Delta r_J^{(\lambda-1)})^2}e^{-\frac{1}{\sigma_r^2}\sum_{\lambda=1}^{\nu}(\bar{r}_J^{\lambda}-\bar{r}_J^{(\lambda-1)})^2}$$

with $\sigma_p^2 = m/2\delta_\beta\hbar$ and $\sigma_r^2 = \hbar\delta_\beta/2m$. Substituting Equations (12) and (13) in Equation (9), it can be noted that the integral over Δr^{ν} is of a Gaussian form and can be performed analytically. Introducing the notation $\mathcal{X}_J = (\{\bar{r}_J^\lambda\}_{\lambda=0,...\nu}, \{\Delta r_J^\lambda\}_{\lambda=0,...\nu-1}, \{\bar{r}_J^{\nu+l}\}_{l=1,...n-1}, \{\bar{p}_J^l\}_{l=1,...,n})$, after integration over Δr^ν, the linearized short time propagator can be written as:

$$K^l(\bar{r}_{J+1}, \Delta r_{J+1}; \bar{r}_J, \Delta r_J) = \int d\mathcal{X}_J e^{\frac{i}{\hbar}\bar{p}_J^n \Delta r_{J+1}} \rho(\mathcal{X}_J, \bar{r}_{J+1}) e^{-\frac{i}{\hbar}\bar{p}_J^1 \Delta r_J^{(\nu-1)}} \qquad (14)$$

with:

$$
\begin{aligned}
\rho(\mathcal{X}_J, \bar{r}_{J+1}) &= \prod_{l=1}^{n-1} \delta\left[(\bar{p}_J^{(l+1)} - \bar{p}_J^l) + \delta_t \nabla V(\bar{r}_J^{(\nu+l)}) \right] \prod_{l=1}^{n} \delta\left[\frac{\bar{p}_J^l}{m}\delta_t - (\bar{r}_J^{(\nu+l)} - \bar{r}_J^{(\nu+l-1)}) \right] \\
&\times \quad e^{-\delta_\beta \sum_{\lambda=1}^{\nu}[V(\bar{r}_J^{(\lambda-1)} + \Delta r_J^{(\lambda-1)}/2) + V(\bar{r}_J^{(\lambda-1)} - \Delta r_J^{(\lambda-1)}/2)]} \\
&\times \quad e^{-\frac{(\bar{p}_J^1)^2}{2\sigma_p^2}} e^{-\frac{\sigma_r^2}{2}\sum_{\lambda=1}^{(\nu-1)}(\Delta r_J^\lambda - \Delta r_J^{(\lambda-1)})^2} e^{-\frac{1}{\sigma_r^2}\sum_{\lambda=1}^{\nu}(\bar{r}_J - \bar{r}_J^{(\lambda-1)})^2}
\end{aligned} \qquad (15)
$$

(The Gaussian in $\bar{p}_J^1$, the first factor in the last line of the expression above, is the result of the integration over Δr^ν.) Substituting the approximate form of the propagator between complex time slices J and $J+1$, we obtain for the symmetrized correlation function:

$$
\begin{aligned}
G_{A,B}^{(L)}(t,\beta) &= \frac{1}{Z} \int d\Delta r_{t_c} d\bar{r}_{t_c} \langle \bar{r}_{t_c} + \frac{\Delta r_{t_c}}{2} |\hat{B}| \bar{r}_{t_c} - \frac{\Delta r_{t_c}}{2}\rangle \\
&\times \prod_{J=1}^{L-1} \int d\mathcal{X}_J e^{\frac{i}{\hbar}\bar{p}_J^n \Delta r_{J+1}} \rho(\mathcal{X}_J, \bar{r}_{J+1}) e^{-\frac{i}{\hbar}\bar{p}_J^1 \Delta r_J^{(\nu-1)}} \\
&\times \int d\mathcal{X}_0 e^{\frac{i}{\hbar}\bar{p}_0^n \Delta r_1} \rho(\mathcal{X}_0, \bar{r}_1) e^{-\frac{i}{\hbar}\bar{p}_0^1 \Delta r_0^{(\nu-1)}} \langle \bar{r}_0 + \frac{\Delta r_0}{2} |\hat{A}| \bar{r}_0 - \frac{\Delta r_0}{2}\rangle
\end{aligned} \qquad (16)
$$

The expression above is interesting. First of all, assuming that the linearization approximation of each short time propagator improves when the propagation time goes to zero, there is potential for systematic improvement with increasing L, and indeed, numerical tests [30] indicate that this is the case. However, the limit for large L, and, in particular, the validity of the expansion in the difference path at the intermediate times of the propagation, is delicate. In fact, while it can be argued that the matrix elements of operators $\hat{A}$ and $\hat{B}$ (usually diagonal) force the forward and backward paths (the free and tilde variables in the upper panel of Figure 1) to start and end close to one another, and, therefore, that only small values of the difference among the paths will be relevant close to the initial and final time, truncating the expansion of the difference of the potentials along the whole pair of paths is considerably more delicate. This issue, and the nature of the dynamics when $L \to \infty$, are currently under investigation. In the meantime, note that the ρ functions are positive definite, so that they can be used to define a probability density for sampling the overall path variables (*i.e.*, the full set of $\{\mathcal{X}_J\}_{J=0,...,L-1}$ variables) as $\Pi = \frac{1}{\Omega}\prod_{J=0}^{L-1}\rho(\mathcal{X}_J, \bar{r}_{J+1})$, where Ω is the (unknown) normalization factor. The method to deal with this factor is illustrated in the next subsection for the case $L = 1$ and can be straightforwardly generalized to $L > 1$. The probability density, Π, corresponds to a stochastic process, which concatenates the thermal and time propagations within each short time propagator, K^l. The structure of the propagations in real and imaginary time is determined by the definition of ρ in Equation (15) and can be described as follows. For $L = 1$, there is only one real-time leg of duration $\tau_t = t$, while the imaginary time propagation corresponds to an inverse temperature $\beta/2$ for both the semi-sum and difference variables. The upper panel of Figure 3 illustrates these propagations with a sketch. In the figure, the horizontal axis is time and the vertical axis temperature. The vertical plane represents the space of configurations associated with the thermal path integral for both the semi-sum and difference variables; the thermal beads are represented with the red circles. The harmonic interactions in the thermal paths are indicated with zigzagged lines connecting adjacent beads, while the interactions among the two paths due to the potential are drawn as dashed lines. Note that the difference variables path, on the left in the vertical plane in the figure, has one less bead than the semi-sum variables path, due to the integration carried out to isolate a Gaussian probability density for the initial momenta. The propagation in real time is drawn as the curve on the horizontal plane, which represents the phase space of the system. The red and golden circle at $t = 0$ represents the initial conditions for the time evolution: the initial coordinate coincides with the last bead of the thermal path in the semi-sum variables, while the initial momentum is sampled from the Gaussian mentioned before. A phase factor is associated with the initial point of the classical propagation. The exponent of this phase couples the initial momentum of the trajectory with the last bead of the thermal path in the difference variables. A phase factor is also associated with the final point of the classical propagation, where, for $L = 1$, the exponent couples the momentum at time t with the variable, Δr_1. The integrals over $\Delta r_0^{(\nu-1)}$ and Δr_1 in the expression for $G_{AB}^1(t; \beta)$ involve products of these phase factors with the matrix elements of operators $\hat{A}$ and $\hat{B}$. The end-point integral reconstructs the Wigner transform of the operator, $\hat{B}$. To see this, consider Equation (16). For $L = 1$, the second line of the equation is absent, and boundary conditions impose $\Delta r_1 = \Delta r_{t_c}$ (with similar relationships for the sum variables). With this notation, the integral over Δr_{t_c} is recognizable as the Wigner transform of operator $\hat{B}$. The structure of the sequence of imaginary and real-time propagation for generic values of L can be inferred from the lower part of Figure 3, where we show what happens for $L = 2$. In this case, there are two segments of classical dynamics, each of duration $t/2$, and two propagations of semi-sum and difference variables in imaginary time, taking the system from zero inverse temperature to $\beta/4$ and from $\beta/4$ to $\beta/2$, respectively. As before, the first segment of dynamics starts, with a Gaussian initial momentum, from the last bead of the semi-sum variable thermal path at $t = 0$. The end-point of this leg of propagation is the initial configuration for the semi-sum variable thermal path at $t/2$, and the second segment of dynamics has as initial conditions the final coordinate of the semi-sum variable thermal path and a new momentum sampled from a Gaussian. The variances of the Gaussians associated with the momentum sampling are doubled with respect to the case $L = 1$. The integrand now contains four phase factors coupling the momenta at the beginning and end of each classical dynamics segment with the values of the difference path variables at the end and at the beginning of each thermal slice, respectively. The phase factor that depends on $\Delta r_2 \bar{p}_1^n$ (*i.e.*, the phase factor computed at time t) can again be combined with the matrix element of operator $\hat{B}$ to obtain the Wigner transform of this operator at the final time of the propagation, so that only three phases remain to contribute to the result. In general, $G_{AB}^{(L)}(t; \beta)$ involves L segments of classical propagation, each of duration t/L, interspersed with L pairs of thermal paths in the semi-sum and difference variables, each at an inverse temperature $\beta/2L$. The rules for connecting the coordinate and momenta at the initial and final time of the dynamics with the final and initial points

332

of the thermal paths and for constructing the $2L - 1$ phase factors contributing to the integrand (the phase factor at time t can always be absorbed in the Wigner transform of operator $\hat{B}$) are completely analogous to the $L = 2$ case.

Figure 3. Graphic representation of the propagators in real and imaginary times contributing to the approximate Schofield function for the case $L = 1$ (**upper panel**) and $L = 2$ (**lower panel**). The horizontal axis is real time, while the vertical axis is inverse temperature. The mean and difference coordinates in the thermal paths are represented as red dots on the vertical planes (in the upper panel, for example, $\nu = 6$, *i.e.*, we use six beads to represent the thermal path integrals at inverse temperature $\beta/2$). Segments of classical propagation in phase space are represented as continuous red curves in the horizontal planes. The golden circles indicate the connection between the coordinate-momentum representation of the dynamics in real time (horizontal planes) and the representation of the dynamics in imaginary time that takes place in coordinate space (vertical planes).

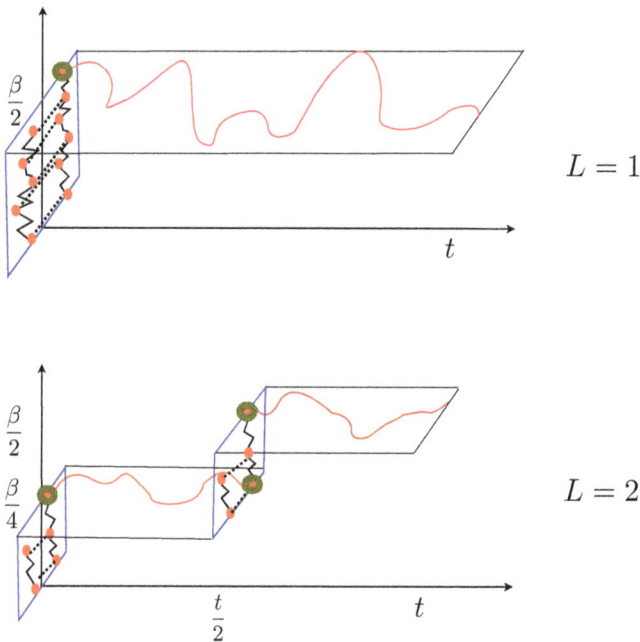

A Monte Carlo algorithm to sample Π for different values of L was illustrated in [30]. This Monte Carlo has several non-standard features, most notably the fact that the normalization of the probability density is unknown and that Π contains products of delta functions, *i.e.*, singular distributions. These difficulties can be circumvented as detailed in [30]. The first one is tackled by recasting (without further approximations) Equation (16) in the form of a ratio of expected values. The second is addressed via an appropriate choice of the trial moves and acceptance probabilities. The most serious numerical difficulty, unfortunately, comes from the estimator of the observable.

In fact, from Equation (16), it can be seen that, in addition to the matrix elements of the operators, the integrand contains a product of phase factors to be evaluated at the beginning and end of each real-time propagation segment. As mentioned above, the number of these phase factors for the "order L" approximation of the symmetrized correlation function is $2L - 1$, and their presence rapidly hinders the convergence of the calculation. Furthermore, for a system of n particles in three dimensions, the phases take the (generic) form $e^{\pm \frac{i}{\hbar} \mathbf{p} \cdot \mathbf{\Delta r}}$, where $\mathbf{p}$ and $\mathbf{\Delta r}$ are $3n$-dimensional vectors. The number of phase factors thus scales linearly with the number of degrees of freedom, so that, even for small values of L, convergence is problematic. The numerical tests performed so far on simple model systems confirm both the interest and the difficulties of Equation (16). In [30], we computed position autocorrelation functions for a set of one-dimensional systems (e.g., quartic potential) at temperatures low enough to ensure that the system was in the quantum regime. We observed that increasing the number of complex time slices did systematically improve the length of time for which we were able to get accurate results. However, the numerical effort to go beyond $L = 3$, though not exponential in time, became essentially prohibitive, even for these simple systems. To indicate possible means to reduce the numerical effort involved in these calculations, we now discuss a recent development of the method developed to address the problem of the phase in the simplest case, $L = 1$, in which it presents itself. We then illustrate how to formally extend this development, which, in its simplest form, has been successfully applied to realistic models of condensed phase systems, to the case $L > 1$.

3. L = 1: Fully Linearized Approximation

The expression for the $L = 1$ approximation of the symmetrized correlation function is given by (see Equation (16)):

$$
\begin{aligned}
G_{A,B}^{(1)}(t; \beta) &= \frac{1}{Z} \int d\Delta r_{t_c} d\bar{r}_{t_c} \int d\mathcal{X}_0 \langle \bar{r}_{t_c} + \frac{\Delta r_{t_c}}{2} |\hat{B}| \bar{r}_{t_c} - \frac{\Delta r_{t_c}}{2} \rangle e^{\frac{i}{\hbar} \bar{p}_0^n \Delta r_{t_c}} \\
&\times \rho(\mathcal{X}_0, \bar{r}_{t_c}) e^{-\frac{i}{\hbar} \bar{p}_0^1 \Delta r_0^{(\nu-1)}} \langle \bar{r}_0 + \frac{\Delta r_0}{2} |\hat{A}| \bar{r}_0 - \frac{\Delta r_0}{2} \rangle
\end{aligned}
\tag{17}
$$

We are now going to simplify the expression above using four steps: (1) observe that the integral over Δr_{t_c} in the first line of the equation above defines the Wigner transform of operator $\hat{B}$ (see, also, the definition below Equation (4)); (2) note that the product of δ functions in the definition of ρ (Equation (15)) forces $(\bar{r}_{t_c}, \bar{p}_{t_c}^n)$ to be endpoints of a classical trajectory of length t starting at $(\bar{r}_0^\nu, \bar{p}_0^1)$, so that, after integration over $\bar{r}^{(\nu+l)}$ ($l = 1, ..., n-1$) and $\bar{p}^l$ ($l = 1, ..., n$), $(\bar{r}_{t_c}, \bar{p}_{t_c}^n) = (r_t, p_t)$ (where (r_t, p_t) denote the classically evolved variables); (3) choose, for the sake of simplicity, to specialize the discussion to an operator, $\hat{A}$, which is diagonal in the coordinate representation (the case of generic operators is considered in the Appendix of [39]). This choice, producing a $\delta(\Delta r_0)$ in the evaluation of the matrix elements, allows one to integrate also over Δr_0. The surviving variables (i.e., the semi-sum and difference variables of the thermal path integral and the initial momentum of the classical trajectory) will be indicated collectively as $\Gamma = \{p^1, r^0, ..., r^\nu, \Delta r^1, ..., \Delta r^{\nu-1}\}$; (4) simplify the notation by dropping the bar from the semi-sum variables and the subscript, which

334

identifies the $J = 0$ propagation segment in Equation (16), since only one segment is now present. Once these operations are performed, the correlation function can be written as:

$$G_{A,B}^{(1)}(t;\beta) = \frac{\hat{Q}}{Z} \int d\Gamma \, B_w(r_t, p_t) P(\Gamma) e^{-\frac{i}{\hbar} p^1 \Delta r^{(\nu-1)}} A(r^0) \tag{18}$$

with:

$$\begin{aligned} P(\Gamma) = \; & \frac{1}{\hat{Q}} e^{-\frac{(p^1)^2}{2\sigma_p^2}} e^{-2\delta_\beta V(r^0)} e^{-\frac{1}{2\sigma_r^2}\sum_{\lambda=1}^{\nu}(r^\lambda - r^{\lambda-1})^2} e^{-\frac{\sigma_r^2}{2}\sum_{\lambda=2}^{\nu-1}(\Delta r^\lambda - \Delta r^{\lambda-1})^2} e^{-\frac{\sigma_r^2}{2}(\Delta r^1)^2} \\ & \times \; e^{-\delta_\beta \sum_{\lambda=2}^{\nu}\left[V\left(r^{(\lambda-1)}+\frac{\Delta r^{\lambda-1}}{2}\right)+V\left(r^{(\lambda-1)}-\frac{\Delta r^{\lambda-1}}{2}\right)\right]} \end{aligned} \tag{19}$$

and $\hat{Q}$ the normalization of $P(\Gamma)$. Note that the expression above for the probability is quite standard, being an explicit function of the semi-sum and difference variables, which can be sampled via Monte Carlo, multiplied by a Gaussian term for the momenta. As mentioned in the previous section, the ratio of the normalization over the partition function, which appears in Equation (18), is, in general, not known, and we estimate it via the autocorrelation of the identity, $G_{I,I}^{(1)} = 1 = \frac{\hat{Q}}{Z}\int d\Gamma P(\Gamma) e^{-\frac{i}{\hbar} p^1 \Delta r^{(\nu-1)}}$. Using this approach, the $L = 1$ estimator of the correlation function is given by the following ratio of expectation values over $P(\Gamma)$:

$$G_{A,B}^{(1)}(t;\beta) = \frac{\langle B_w(r_t, p_t) e^{-\frac{i}{\hbar} p^1 \Delta r^{(\nu-1)}} A(r^0)\rangle_P}{\langle e^{-\frac{i}{\hbar} p^1 \Delta r^{(\nu-1)}}\rangle_P} \tag{20}$$

As anticipated, both in the numerator and denominator of this expression, a phase factor appears, which, for high dimensional systems, hinders an efficient convergence of the calculation. To alleviate this problem, we proposed a method, described in detail in [39], which starts by obtaining an alternative expression for $G_{A,B}^{(1)}$. As will be shown in the following, the new expression does not introduce further (analytical) approximations, but it has the advantage of eliminating the phase factor from the observable. Let us consider in more detail the structure of the probability, $P(\Gamma)$. This probability is given by the product of a Gaussian for the momenta, $\rho_G(p) \propto e^{-\frac{p^2}{2\sigma_p^2}}$ (note that, with respect to Equation (19), we dropped the superscript, 1, on the momenta to simplify the notation), times a joint probability function for the semi-sum and difference thermal variables to be indicated in the following as $\tilde{\rho}(\mathbf{r}, \boldsymbol{\Delta r})$, where we have introduced the notation $\mathbf{r} = \{r^0, ..., r^\nu\}$ and $\boldsymbol{\Delta r} = \{\Delta r^1, ..., \Delta r^{(\nu-1)}\}$. This joint probability (whose form can be inferred from Equation (19) by taking out the momentum Gaussian) is most conveniently expressed as:

$$\tilde{\rho}(\mathbf{r}, \boldsymbol{\Delta r}) = \rho_c(\boldsymbol{\Delta r}|\mathbf{r})\rho_m(\mathbf{r}) \tag{21}$$

where:

$$\begin{aligned} \rho_m(\mathbf{r}) = \; & \frac{1}{\hat{Q}} e^{-2\delta_\beta V(r^0)} e^{-\frac{1}{2\sigma_r^2}\sum_{\lambda=1}^{\nu}(r^\lambda - r^{\lambda-1})^2} \int d\boldsymbol{\Delta r} \, e^{-\frac{\sigma_r^2}{2}\sum_{\lambda=2}^{\nu-1}(\Delta r^\lambda - \Delta r^{\lambda-1})^2} e^{-\frac{\sigma_r^2}{2}(\Delta r^1)^2} \\ & \times \; e^{-\delta_\beta \sum_{\lambda=2}^{\nu}\left[V\left(r^{(\lambda-1)}+\frac{\Delta r^{\lambda-1}}{2}\right)+V\left(r^{(\lambda-1)}-\frac{\Delta r^{\lambda-1}}{2}\right)\right]} \end{aligned} \tag{22}$$

is the marginal probability for the semi-sum variables and $\rho_c(\boldsymbol{\Delta r}|\mathbf{r}) \equiv \tilde{\rho}(\mathbf{r}, \boldsymbol{\Delta r})/\rho_m(\mathbf{r})$ is the conditional probability for the difference variables given the semi-sum variables. This rewriting of the probability density is convenient because the phase factors in Equation (20) depend only on the momenta and difference variables. We can use this observation to define:

$$F(p, \mathbf{r}) = \int d\Delta r e^{-\frac{i}{\hbar}p\Delta r^{(\nu-1)}} \rho_c(\Delta r|\mathbf{r}) \tag{23}$$

which is the average of the phase with respect to the conditional probability density, and investigate the properties of this function to see if we can use them to improve the convergence of our calculations. To that end, note that F is also, by definition, the cumulant generating function of the variable $\Delta r^{(\nu-1)}$ with respect to the conditional probability, ρ_c(see, for example, [40,41] for previous use of cumulants in this field). This means that the coefficients of the Taylor series expansion (with respect to $-ip/\hbar$):

$$\ln F(p, \mathbf{r}) = \sum_{n=1}^{\infty} \frac{(-ip/\hbar)^n}{n!} \langle (\Delta r^{(\nu-1)})^n \rangle^c_{\rho_c(\Delta r|\mathbf{r})} \tag{24}$$

(these coefficients are indicated above as $\langle (\Delta r^{(\nu-1)})^n \rangle^c_{\rho_c(\Delta r|\mathbf{r})}$) are the cumulant moments of $\Delta r^{(\nu-1)}$. Importantly, the conditional probability density is an even function of the difference variables, implying that only even order terms in the series above are non-zero and that the series corresponds to a real function that we will denote in the following with $E(p, \mathbf{r})$. We can then express the average of the phase as:

$$F(p, \mathbf{r}) = e^{-E(p,\mathbf{r})} \tag{25}$$

i.e., a positive definite function of the momenta and the semi-sum variables. We now use the function above to define a new probability density:

$$\mathcal{P}(p, \mathbf{r}) = \frac{\rho_g(p)e^{-E(p,\mathbf{r})}\rho_m(\mathbf{r})}{\int dp d\mathbf{r}\rho_g(p)e^{-E(p,\mathbf{r})}\rho_m(\mathbf{r})} \tag{26}$$

and note that, by direct substitution of this definition in the (explicit) expression of Equation (20), we obtain:

$$G^{(1)}_{A,B}(t; \beta) = \langle B_w(r_t, p_t)A(r^0) \rangle_{\mathcal{P}} \tag{27}$$

The key advantage of the expression above is that the observable does not contain phase factors anymore and is, therefore, well suited for a Monte Carlo estimate. Sampling the distribution, $\mathcal{P}$, however, is non-trivial, since this probability density contains two factors, $e^{-E(p,\mathbf{r})}$ and $\rho_m(\mathbf{r})$, that do not have an explicit analytic form, but, for each value of $\mathbf{r}$ and p, can only be estimated numerically. The numerical estimate of $E(\mathbf{r}, p)$, in particular, requires one to truncate the cumulant series at a given order. The convergence of the calculation with respect to truncation of the series can always be checked numerically, and, although the cost scales up where terms of higher order are included, it does not present any particular difficulty. (In all calculations performed so far, a second order cumulant expansion proved sufficient.) In the following subsection, we briefly describe how to combine two schemes, known as the Kennedy and Penalty methods, for Monte Carlo sampling of noisy probability densities and obtain $G^{(1)}_{A,B}(t; \beta)$. Our goal is to highlight the main differences among these schemes and standard Monte Carlo and to indicate where the algorithm is most affected by them. A detailed description of the algorithm can be found in [38,39].

3.1. Noisy Monte Carlo Algorithm

To simplify the discussion, we introduce some notation. Let us indicate the coordinate-dependent Gaussian terms in Equation (19) as:

$$
e^{-\frac{1}{2\sigma_r^2}\sum_{\lambda=1}^{\nu}(r^\lambda - r^{(\lambda-1)})^2} = e^{-\mathbf{V}_r(\mathbf{r})}
$$
$$
e^{-\frac{\sigma_p^2}{2}\sum_{\lambda=1}^{\nu-1}(\Delta r^\lambda - \Delta r^{(\lambda-1)})^2} = e^{-\mathbf{V}_\Delta(\Delta\mathbf{r})}
\tag{28}
$$

(above $\Delta r^0 = 0$) and write the potential term as:

$$
e^{-\delta_\beta \sum_{\lambda=2}^{\nu}\left[\mathbf{V}(r^{(\lambda-1)}+\frac{\Delta r}{2}^{(\lambda-1)})+\mathbf{V}(r^{(\lambda-1)}-\frac{\Delta r}{2}^{(\lambda-1)})\right]} e^{-2\delta_\beta \mathbf{V}(r^0)}
$$
$$
= e^{-\delta_\beta \bar{\mathbf{V}}(\mathbf{r},\Delta\mathbf{r})}
\tag{29}
$$

We also rewrite the marginal probability, $\rho_m(\mathbf{r})$, defined in Equation (22), isolating the terms that do not depend on $\Delta\mathbf{r}$ and have an explicit analytic expression, thus:

$$
\rho_m(\mathbf{r}) = \int d\Delta\mathbf{r}\tilde{\rho}(\mathbf{r},\Delta\mathbf{r}) = \frac{1}{Q}e^{-\mathbf{V}_r(\mathbf{r})}\int d\Delta\mathbf{r}\; e^{-\delta_\beta \bar{\mathbf{V}}(\mathbf{r},\Delta\mathbf{r})}e^{-\mathbf{V}_\Delta(\Delta\mathbf{r})}
$$
$$
= \frac{1}{Q}e^{-\mathbf{V}_r(\mathbf{r})}\rho'_m(\mathbf{r})
\tag{30}
$$

With the definitions above, $\mathcal{P}(p,\mathbf{r})$ takes the form:

$$
\mathcal{P}(p,\mathbf{r}) = \frac{1}{Q}\rho_G(p)e^{-E(\mathbf{r},p)}e^{-\mathbf{V}_r(\mathbf{r})}\rho'_m(\mathbf{r})
\tag{31}
$$

where Q is the normalization. The scheme that we use to perform the sampling is based on earlier work by Ceperley [42] and Kennedy [43]. Adapting their ideas to our case, we will introduce a Monte Carlo algorithm in which the definition of the probability to generate a new state of the system by changing either the coordinates or the momenta of the particle (unlike what happens in classical canonical densities, in our probability, the variables, p and $\mathbf{r}$, are not independent (with the momenta Gaussian and then integrable) and must be treated together) and/or to accept this new state is modified to guarantee that detailed balance is satisfied also when $\rho'_m(\mathbf{r})$ and $E(\mathbf{r},p)$ are estimated with significant noise. Both the Ceperley and Kennedy scheme require the introduction of appropriate numerical estimators of the unknown functions. These estimators will be indicated with calligraphic fonts.

The Monte Carlo scheme to sample Equation (31) is constructed as follows. Choose, with probability $1/2$, if the move will involve $\mathbf{r}$ or p.

(1) A move on p has been selected:

choose a new momentum according to $p' = p + \delta p$, where δp is a uniform random number centered on zero (the magnitude of the displacement is chosen so as to optimize the acceptance). Taking into account that the $\mathbf{r}$ variables are not being updated, detailed balance for this trial move takes the form:

$$
\rho_G(p)e^{-E(\mathbf{r},p)}A^p(p \to p') = \rho_G(p')e^{-E(\mathbf{r},p')}A^p(p' \to p)
\tag{32}
$$

where $A^p(p \to p')$ is the acceptance probability. The detailed balance relationship above has the same form as the one discussed by Ceperly *et al.* within the penalty method [42], a generalized Monte Carlo for sampling a density given by the exponential of a function, in our case $E(.,.)$, known with statistical errors. According to the penalty method, if a numerical estimate, $\Delta \mathcal{E}_p(p', p; \mathbf{r})$, of the difference $E(\mathbf{r}, p') - E(\mathbf{r}, p)$ has been obtained (for example, by averaging N_s values of a specific estimator) and an estimate of its variance, χ_p^2, is also known, detailed balance can be satisfied by defining the acceptance as:

$$A^p(p \to p') = \min \left[1, \frac{\rho_G(p')}{\rho_G(p)} e^{-\Delta \mathcal{E}_p(p', p; \mathbf{r}) - u_{\chi_p^2}} \right] \tag{33}$$

where:

$$u_{\chi_p^2} = \frac{\chi_p^2}{2} + \frac{\chi_p^4}{4(N_s + 1)} + \dots \tag{34}$$

The expression for the acceptance probability differs from the standard Metropolis prescription for the presence of $u_{\chi_p^2}$ and is valid when $\chi_p^2/n <$ $1/4$ (*here n is the size of the sample used to estimate the energy difference*) [42]. In the limit of an infinitely precise estimate of the difference, $u_{\chi_p^2} \to 0$ and the standard criterion is recovered; when non-zero, this function corrects, on average, for the effect of the noise.

(2) A move on $\mathbf{r}$ has been selected:

in this case, indicating with $T^r(\mathbf{r} \to \mathbf{r}')$ and $A^r(\mathbf{r} \to \mathbf{r}')$ the probability to generate and accept a new configuration, respectively, detailed balance is expressed, after simplifying $\rho_G(p)$, as:

$$e^{-E(\mathbf{r},p)} e^{-\mathbf{V}_r(\mathbf{r})} \rho'_m(r) T^r(\mathbf{r} \to \mathbf{r}') A^r(\mathbf{r} \to \mathbf{r}') =$$
$$e^{-E(\mathbf{r}',p)} e^{-\mathbf{V}_r(\mathbf{r}')} \rho'_m(\mathbf{r}') T^r(\mathbf{r}' \to \mathbf{r}) A^r(\mathbf{r}' \to \mathbf{r}) \tag{35}$$

The structure of this relationship is analogous to the one considered by Kennedy *et al.* [43], who adapted Monte Carlo sampling to probability densities given by an exponential term times a "noisy" (positive definite) function. They showed that detailed balance is satisfied if states are generated according to the probability:

$$T^r(\mathbf{r} \to \mathbf{r}') \propto e^{-E(\mathbf{r}',p)} e^{-\mathbf{V}_r(\mathbf{r}')} \tag{36}$$

(a method to sample $T^r(\mathbf{r} \to \mathbf{r}')$ is described after the next equation) and accepted with probability:

$$A^r(\mathbf{r} \to \mathbf{r}') = \begin{cases} c\,\mathcal{U}(\mathbf{r} \to \mathbf{r}') & \text{if } e^{-\delta_\beta \mathbf{V}(\mathbf{r},0)} > e^{-\delta_\beta \mathbf{V}(\mathbf{r}',0)} \\ c & \text{if } e^{-\delta_\beta \mathbf{V}(\mathbf{r},0)} \le e^{-\delta_\beta \mathbf{V}(\mathbf{r}',0)} \end{cases} \tag{37}$$

Above, $\mathcal{U}(\mathbf{r} \to \mathbf{r}')$ is an unbiased estimator of the ratio, $\rho'_m(\mathbf{r}')/\rho'_m(\mathbf{r})$, and $c < 1$ is a constant that ensures $A^r(\mathbf{r} \to \mathbf{r}') \in [0,1]$ (for details on the meaning and choice of c, see [43] and the discussion on page 8 of [39]). The conditions on the exponential of the potential enforce an ordering criterion on the states, whose optimal choice depends on the problem (here, we used the one adopted in our previous work [39]). In the usual implementation of the Kennedy method, the exponential part of the probability density is assumed to be known analytically, and the states are generated via a standard Monte Carlo method. In our case, the situation is more complicated, since $e^{-E(\mathbf{r}',\mathbf{p})}$ is only known with noise. To solve this problem, we employ the penalty method to obtain configurations distributed according to Equation (36). These configurations are generated using a Monte Carlo with transition probability $t(\mathbf{r} \rightarrow \mathbf{r}') \propto e^{-V_r(r')}$ and acceptance probability $a(r \rightarrow r') = \min[1, \exp(-\Delta\mathcal{E}_r(\mathbf{r}',\mathbf{r};p) - u_{\chi_r^2})]$, where $\Delta\mathcal{E}_r(\mathbf{r}',\mathbf{r};p)$ is an unbiased estimator of $E(\mathbf{r}',p) - E(\mathbf{r},p)$, and u_{χ^2} is defined in analogy with Equation (34).

This concludes the description of our Monte Carlo moves. The practical implementation of this algorithm requires the definition of the numerical estimators, $\mathcal{U}(\mathbf{r} \rightarrow \mathbf{r}')$, $\Delta\mathcal{E}_p(\mathbf{p}',\mathbf{p};\mathbf{r})$ and $\Delta\mathcal{E}_r(\mathbf{r}',\mathbf{r};\mathbf{p})$. While this is an important technical point, it only involves a set of calculations, each performed via an auxiliary Monte Carlo move, that are quite standard. To provide a typical example, we consider one of the estimators referring the reader to [39] for a detailed description of the others. Let us then consider $\mathcal{U}(\mathbf{r} \rightarrow \mathbf{r}')$. This quantity, necessary in the Kennedy acceptance test, see Equation (37), is obtained by writing the ratio of the marginal probabilities as:

$$
\begin{aligned}
\frac{\rho_m'(\mathbf{r}')}{\rho_m(\mathbf{r})} &= \frac{\int d\Delta\mathbf{r}\, e^{-\mathbf{V}_\Delta(\Delta\mathbf{r})} e^{-\delta_\beta \bar{\mathbf{V}}(\mathbf{r}',\Delta\mathbf{r})}}{\int d\Delta\mathbf{r}\, e^{-\mathbf{V}_\Delta(\Delta\mathbf{r})} e^{-\delta_\beta \bar{\mathbf{V}}(\mathbf{r},\Delta\mathbf{r})}} \\
&= \frac{\int d\Delta\mathbf{r}\, e^{-\mathbf{V}_\Delta(\Delta\mathbf{r})} e^{-\delta_\beta \bar{\mathbf{V}}(\mathbf{r},\Delta\mathbf{r})} e^{-\delta_\beta \left[\bar{\mathbf{V}}(\mathbf{r}',\Delta\mathbf{r}) - \bar{\mathbf{V}}(\mathbf{r},\Delta\mathbf{r})\right]}}{\int d\Delta\mathbf{r}\, e^{-\mathbf{V}_\Delta(\Delta\mathbf{r})} e^{-\delta_\beta \bar{\mathbf{V}}(\mathbf{r},\Delta\mathbf{r})}} \\
&= \left\langle e^{-\delta_\beta \left[\bar{\mathbf{V}}(\mathbf{r}',\Delta\mathbf{r}) - \bar{\mathbf{V}}(\mathbf{r},\Delta\mathbf{r})\right]} \right\rangle_{\rho_c(\Delta\mathbf{r}|\mathbf{r})}
\end{aligned}
\tag{38}
$$

whose unbiased estimator is:

$$
\mathcal{U}(\mathbf{r} \rightarrow \mathbf{r}') = \frac{1}{N_a} \sum_{i=1}^{N_a} e^{-\delta_\beta \left[\bar{\mathbf{V}}(\mathbf{r}',\Delta\mathbf{r}_i) - \bar{\mathbf{V}}(\mathbf{r},\Delta\mathbf{r}_i)\right]}
\tag{39}
$$

where $\{\Delta\mathbf{r}_i\}$ are a sample distributed according to $\rho_c(\Delta\mathbf{r}|\mathbf{r})$. This sample is obtained via an auxiliary (standard) Monte Carlo calculation over the conditional probability, $\rho_c(\Delta\mathbf{r}|\mathbf{r})$, in which new configurations are generated according to $T(\Delta\mathbf{r} \rightarrow \Delta\mathbf{r}') \propto \exp[-V_\Delta(\Delta\mathbf{r}')]$ (To do this, we use the staging method [44], which allows one to sample exactly a probability density containing Gaussian-like distributions, see Equation (28)), and moves are accepted or rejected based on $A(\Delta\mathbf{r} \rightarrow \Delta\mathbf{r}') = \min\left\{1, \exp[-\delta_\beta(\bar{V}(\mathbf{r},\Delta\mathbf{r}') - \bar{V}(\mathbf{r},\Delta\mathbf{r})]\right\}$. As can be seen from the expression above, calculating the estimator requires N_a steps in the auxiliary Monte Carlo calculation. A similar situation arises when the other estimators introduced above are considered, so that the total number of Monte Carlo moves in our scheme is given by $N_t = N_m \times N_a$, where we indicated with N_m the number of moves in the "main" Monte Carlo cycle (i.e., each choice of a move on r or p) and with N_a the number of auxiliary Monte Carlo steps per "main" move.

The computational overhead introduced by the auxiliary Monte Carlo calculation increases the cost of our calculation, but it is very small compared to the number of moves necessary to converge the estimate of Equation (20). The algorithm just described is, in fact, efficient enough to make possible calculations on realistic condensed phase systems with relatively little numerical effort. In particular, the algorithm was used to compute the dynamic structure factor of a model of liquid neon composed of 64 atoms [38]. Details of the calculation can be found in [38]. Here, we show, in

Figure 4, our results (red triangles with error bars in the figure) and compare them with experiments (green curve) and the results of a calculation (with the same empirical potential and simulation parameters) performed by Poulsen *et al.* [45] using the linearized approximation for quantum time correlation functions described in the Introduction (see Equation (4) and the discussion above it).

Figure 4. Dynamic structure factor for liquid neon (see the text). The solid green line shows the experimental curve, our results (with error bars) are the red triangles. We also report for comparison results obtained with the linearized IVRmethod by Polusen *et al.* (see the Introduction); blue circles.

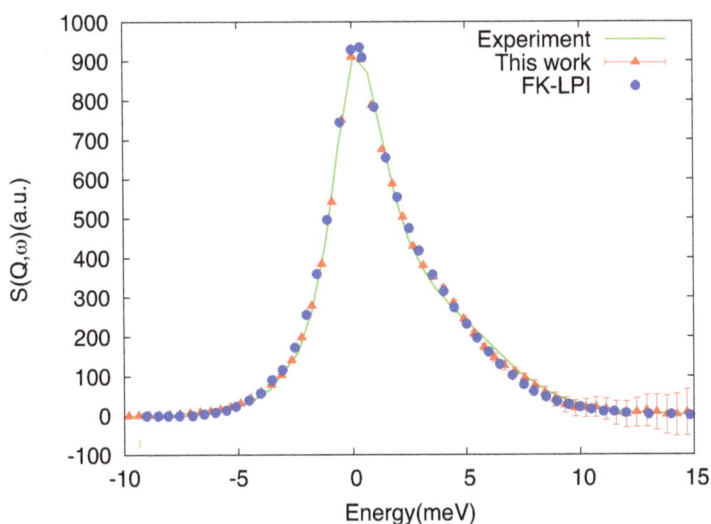

The results show a rather pronounced asymmetry around zero, due to detailed balance, that indicates the presence of relevant quantum effects in the system. The agreement between our calculations and experiments is very good, as it is the agreement with the standard linearized calculation by Poulsen (a state-of-the-art reference in the field). The numerical cost of the two calculations is very similar (about a million Monte Carlo steps in total for initial condition sampling), showing that the auxiliary steps, due to the noisy distributions in our approach, are essentially irrelevant. Indeed, other tests indicate that, depending on the system, the overall cost of our method can be less than that of alternative schemes with comparable or better accuracy. The approach described in this section, for example, has also been used to obtain the infrared spectra of simple models of molecules in the gas phase [46]. Although these systems are quite small, the calculations that we performed are known to pose a considerable challenge to alternative, less rigorous methods, such as Centroid Molecular Dynamics [47] and Ring Polymer Molecular Dynamics [48], which fail to capture the spectra and/or introduce spurious features. In contrast, even though obtaining the exact intensities is quite expensive, our method proved remarkably effective in identifying the positions of the peaks, which could be obtained with only about one hundred Monte Carlo moves.

3.2. L > 1

In this subsection, we present a new development of the approach summarized above that extends the use of cumulants to pre-average the phase factors in the expression of the symmetrized correlation function to the case $L > 1$ (This possibility came out in discussions with M. Montefermante.). For simplicity of notation, in this subsection, we describe how this can be done for $L = 2$, but the steps that we shall use can be generalized to a larger number of segments. In the following, we report the formal result, while the construction and test of an algorithm that generalizes the noisy Monte Carlo scheme described in the previous section will be the object of future work. Let us begin by rewriting the $L = 2$ approximation of the symmetrized correlation function, for diagonal $\hat{A}$, as follows (see Equations (16) and (20) for structure and notation):

$$G_{A,B}^{(2)}(t;\beta) = \frac{\int d\Gamma_1 d\Gamma_0 B_w(r_t^{(2)}, p_t^{(2)}) \left[e^{-\frac{i}{\hbar} p_1^1 \Delta r_1^{(\nu-1)}} P(\Gamma_1; r_0^n) e^{\frac{i}{\hbar} p_0^n \Delta r_1} \right] \left[e^{-\frac{i}{\hbar} p_0^1 \Delta r_0^{(\nu-1)}} P(\Gamma_0) \right] A(r_0)}{\int d\Gamma_1 d\Gamma_0 \left[e^{-\frac{i}{\hbar} p_1^1 \Delta r_1^{(\nu-1)}} P(\Gamma_1; r_0^n) e^{\frac{i}{\hbar} p_0^n \Delta r_1} \right] \left[e^{-\frac{i}{\hbar} p_0^1 \Delta r^{(\nu-1)}} P(\Gamma_0) \right]}$$

(40)

In the equation above, $(r_t^{(2)}, p_t^{(2)})$ is the endpoint of the propagation obtained by combining the two segments of classical dynamics described in the lower panel of Figure 3; $\Gamma_0 = \{p_0^1, r_0^0, ..., r_0^\nu, \Delta r_0^1, ..., \Delta r_0^{(\nu-1)}\}$ and $\Gamma_1 = \{p_1^1, r_1^1, ..., r_1^\nu, \Delta r_1^0, ..., \Delta r_1^{(\nu-1)}\}$ indicate the variables associated with the first and second set of thermal path integrals, respectively (the first set does not include Δr_0^0, since this variable can be integrated over for diagonal $\hat{A}$, and the second does not include $r_1^0 \equiv r_0^n$, since this is the endpoint of the, deterministic, classical propagation from zero to $t/2$ in Figure 3). $P(\Gamma_0)$ was defined in the previous section (see Equation (19)), and:

$$P(\Gamma_1; r_0^n) = \rho_G(p_1^1) \rho_m(\mathbf{r}_1; r_0^n) \rho_c(\Delta \mathbf{r}_1 | \mathbf{r}_1)$$

(41)

where $\mathbf{r}_1 = \{r_1^1, ..., r_1^\nu\}$ and $\Delta \mathbf{r}_1 = \{\Delta r_1^0, ..., \Delta r_1^{(\nu-1)}\}$. The Gaussian probability for the momenta, $\rho_G(p_1^1)$, and the marginal, $\rho_m(\mathbf{r}_1; r_0^n)$, and conditional, $\rho_c(\Delta \mathbf{r}_1 | \mathbf{r}_1)$, probabilities are defined in analogy with the expressions introduced in Section 3, with the caveat that for $J = 1$ (and, in general, for $J > 0$), the sum involving the potentials in the second line of Equation (22) runs from one to $\nu - 1$. In the marginal probability, we have also indicated the (parametric) dependence of the density on r_0^n, *i.e.*, the endpoint of the classical propagation of the first segment, which corresponds, due to the boundary condition mentioned above, to the first bead of the concatenated semi-sum thermal path. The square brackets in Equation (40) isolate the terms that play the most significant role in the following. The first bracket from the left (corresponding to the $J = 0$ term in the approximation of the correlation function) is the same as the one we encountered in the previous subsection, while the second shows the general structure of the terms involving the phase factors for $J > 0$. As summarized when discussing Figure 3, in this bracket the first phase factor is given by the product of the momentum at the endpoint of the first segment of classical dynamics (*i.e.*, a variable fixed by the classical evolution) and the first difference variable of the second thermal segment. The second phase factor is given by the product of the initial momentum of the second leg of classical dynamics (a variable to be sampled in analogy with p_0^1) and the final difference variable of the thermal path.

As in the $L = 1$ case, these phase factors do not depend on the semi-sum variables and can be pre-averaged with respect to the conditional probability density. Let us indicate this average as:

$$F(\pi_J, \mathbf{r}_J) = \int d\Delta \mathbf{r}_J e^{-\frac{i}{\hbar}\pi_J \cdot \delta \mathbf{r}_J} \rho_c(\Delta \mathbf{r}_J | \mathbf{r}_J) \tag{42}$$

where, for $J = 0$, $\pi_0 = p_0^1$ and $\delta \mathbf{r}_0 = \Delta r_0^{(\nu-1)}$, while for $J = 1$, and, more in general, for $J > 0$, $\pi_J = \{-p_{J-1}^n, p_J^1\}$ and $\delta \mathbf{r}_J = \{\Delta r_J^0, \Delta r_J^{(\nu-1)}\}$. The equation above is formally identical to Equation (23), with the important difference that, when $J > 0$, the phase is now given by the scalar product of two vectors and can be recognized as the definition of the *joint* cumulant generating function for the components of $\delta \mathbf{r}_J$ [40]. Although such joint cumulants are formally more complex, the cumulant moments of $\delta \mathbf{r}_J$ are still given by the coefficients of the expansion:

$$\ln F(\pi_J, \mathbf{r}_J) = \sum_{|\lambda| \geq 1}^{\infty} \frac{(-i)^{|\lambda|}}{\lambda! \hbar^{|\lambda|}} \pi_J^{\lambda} C_{\lambda}(\mathbf{r}_J) \tag{43}$$

For $J = 0$, the definition above is to be read as identical to Equation (24). For $J = 1$ (and, in general, $J > 0$), $\lambda = \{\lambda_1, \lambda_2\}$ is a vector of positive integers (including zero), $|\lambda|$ is their sum, $\lambda! = \lambda_1! \lambda_2!$ and:

$$C_{\lambda}(\mathbf{r}_J) = C_{\lambda_1, \lambda_2}(\mathbf{r}_J) = \left(\frac{\partial^{|\lambda|} \ln F(\pi_J, \mathbf{r}_J)}{\partial(-p_{J-1}^n)^{\lambda_1} \partial(p_J^1)^{\lambda_2}} \right)_{\pi_J = 0} \tag{44}$$

As in the previous subsection, the conditional distribution density is even with respect to the difference variables, implying that only even terms are non-zero in Equation (43). The function $F(\pi_J, \mathbf{r}_J)$ is then real and positive, so we can set $F(\pi_J, \mathbf{r}_J) = e^{-E(\pi_J, \mathbf{r}_J)}$ and define, in analogy with Equation (26), the probability density:

$$\mathcal{P}(\pi_J, \mathbf{r}_J; r_0^n) = \frac{\rho_G(p_J^1) \rho_m(\mathbf{r}_J; r_0^n) e^{-E(\pi_J, \mathbf{r}_J)}}{\int dp_J^1 d\mathbf{r}_J \rho_g(p_J^1) \rho_m(\mathbf{r}_J) e^{-E(\pi_J, \mathbf{r}_J)}} \tag{45}$$

Substitution of the definition above in the expression for the symmetrized correlation function shows that we can write the two-segment approximation as the following expectation value:

$$G_{A,B}^{(2)}(t; \beta) = \langle B_w(r_t^{(2)}, p_t^{(2)}) A(r_0) \rangle_{\mathcal{P}^{(2)}} \tag{46}$$

where $\mathcal{P}^{(2)} = \mathcal{P}(\pi_1, \mathbf{r}_1; r_0^n) \mathcal{P}(p_0^1, \mathbf{r}_0)$ (with a straightforward generalization of the notation adopted here, the L-segment approximation of the correlation function can be written as $G_{A,B}^{(L)}(t; \beta) = \langle B_w(r_t^{(L)}, p_t^{(L)}) A(r_0) \rangle_{\mathcal{P}^{(L)}}$, where $\mathcal{P}^{(L)} = \prod_{J=1}^{L-1} \mathcal{P}(\pi_J, \mathbf{r}_J; r_{J-1}^n) \mathcal{P}(p_0^1, \mathbf{r}_0))$. The average above presents the same immediate advantage of the $L = 1$ case in that the "observable" does not contain any explicit phase factors. It also presents the same numerical difficulties, given that the probability density contains analytically unknown quantities (the marginal probabilities, $\rho_m(\mathbf{r}_J)$, and the cumulants). Although it is possible to construct generalization of the noisy Monte Carlo scheme described in the previous section, it is not certain that the favorable convergence properties of the auxiliary sampling necessary, in particular, for computing the cumulants (the decisive ingredient in the $L = 1$ case) will be preserved in this more general situation. Developing and testing the most appropriate algorithm for this generalization will be the focus of future work.

4. Conclusions

In this paper, we summarized a recently developed method to approximate symmetrized quantum time correlation functions. The method recasts the problem as the calculation of averages over a stochastic process based on a linearized approximation of the complex time propagators in the correlation function. This approximation can be enforced either on the full length of the evolution (fully linearized approach) or in an iterative form obtained via the (complex) time composition property of the evolution operators. Thanks to the use of a cumulant expansion, which tames the phase factors present in the observable, the fully linearized approach has proven efficient and accurate in calculations on moderately quantum systems in the condensed phase. The iterative form offers, in principle, a way to improve the accuracy of the results with respect to the fully linearized case and may be useful when higher order quantum effects must be kept into account. While the potential for systematic improvement with respect to the fully classical limit for the dynamics is indeed the most interesting feature of the approach (and the one that distinguishes it from other available methods for which there is no way to improve upon the classical or semi-classical approximation), the practical use of the approach for $L > 1$ is currently hindered by numerical instabilities. In the final section of the paper, we have shown how to extend the use of the cumulant expansion to obtain a formal expression for this case that does not require one to average phase factors in the observable. This expression may be a promising starting point for considerable improvement of the algorithm for more than one segment, and future work will focus on developing and testing an appropriate algorithm. However, importantly, while numerical evidence on model systems supports the claim that systematic improvements can be obtained by higher order iterations, an exact statement on the convergence properties of the method is lacking, and further investigation is needed to formally assess the features of the scheme.

Acknowledgments

The authors are grateful to C. Pierleoni and M. Monteferrante for their substantial contributions to the earlier methods for symmetrized correlation functions summarized in this work. Funding from IIT-SEED grant No 259 "SIMBEDD" is also acknowledged.

Conflicts of Interest

The authors declare no conflict of interest.

References

1. Feynman, R. Space-time approach to non-relativistic quantum mechanics. *Rev. Mod. Phys.* **1948**, *20*, 367–387.
2. Schulman, L.S. *Techniques and Applications of Path Integration*; Dover Publications: Dover, UK, 2005; p. 448.

3. Van Vleck, V. The correspondence principle in the statistical interpretation of quantum mechanics. *Proc. Natl. Acad. Sci. USA* **1928**, *14*, 178–188.

4. Miller, W.H. Classical S matrix: Numerical application to inelastic collisions. *J. Chem. Phys.* **1970**, *53*, 3578.

5. Miller, W.H. The semi-classical initial value representation: a potentially practical way for adding quantum effects to classical molecular dynamics simulations. *J. Phys. Chem. A* **2001**, *105*, 2942–2955.

6. Kleinert, H. *Path Integrals in Quantum Mechanics, Statistics, and Polymer Physics, and Financial Markets*; World Scientific: River Edge, New Jersey, USA, 2004.

7. Glauber, R. Quantum theory of optical coherence. *Phys. Rev.* **1963**, *1449*, 2529–2539.

8. Kay, K.G. Integral expressions for the semiclassical time-dependent propagator. *J. Chem. Phys.* **1994**, *100*, 4377.

9. Caratzoulas, S.; Pechukas, P. Phase space path integrals in Monte Carlo quantum dynamics. *J. Chem. Phys.* **1996**, *104*, 6265.

10. Herman, M.F.; Kluk, E. A semiclassical justification for the use of non-spreading wavepackets in dynamics calculations. *Chem. Phys.* **1984**, *91*, 27–34.

11. Kluk, E.; Herman, M.F.; Davis, H.L. Comparison of the propagation of semiclassical frozen Gaussian wave functions with quantum propagation for a highly excited anharmonic oscillator. *J. Chem. Phys.* **1986**, *84*, 326.

12. Ankerhold, J.; Pollak, E.; Kay, K.G. The Herman Kluk approximation: Derivation and semiclassical corrections. *Chem. Phys.* **2006**, *322*, 3–12.

13. Shao, J.; Makri, N. Forward-Backward semiclassical dynamics with linear scaling. *J. Phys. Chem. A* **1999**, *103*, 9479.

14. Herman, M.F.; Coker, D.F. Classical mechanics and the spreading of localized wave packets in condensed phase molecular systems. *J. Chem. Phys.* **1999**, *111*, 1801.

15. Hernandez, R.; Voth, G.A. Quantum time correlation functions and classical coherence. *Chem. Phys.* **1998**, *233*, 243–255.

16. Poulsen, J.A.; Nyman, G.; Rossky, P.J. Practical evaluation of condensed phase quantum correlation functions: A Feynman–Kleinert variational linearized path integral method. *J. Chem. Phys.* **2003**, *119*, 12179.

17. Shi, Q.; Geva, E. A relationship between semiclassical and centroid correlation functions. *J. Chem. Phys.* **2003**, *118*, 8173.

18. Poulsen, J.A.; Nyman, G.; Rossky, P.J. Static and dynamic quantum effects in molecular liquids: A linearized path integral description of water. *Proc. Natl. Acad. Sci. USA* **2005**, *102*, 6709–6714.

19. Wigner, E. On the quantum correction for thermodynamic equilibrium. *Phys. Rev.* **1932**, *40*, 749–759.

20. Tully, J.C. Trajectory surface hopping approach to nonadiabatic molecular collisions: The reaction of H+ with D2. *J. Chem. Phys.* **1971**, *55*, 562.

21. Kapral, R.; Ciccotti, G. Mixed quantum classical dynamics. *J. Chem. Phys.* **1999**, *110*, 8919.

22. Kapral, R. Progress in the theory of mixed quantum classical dynamics. *Annu. Rev. Phys. Chem.* **2006**, *57*, 129–157.

23. Bonella, S.; Coker, D.F. LAND-map, a linearized approach to nonadiabatic dynamics using the mapping formalism. *J. Chem. Phys.* **2005**, *122*, 194102.

24. Bonella, S.; Montemayor, D.; Coker, D.F. Linearized path integral approach for calculating nonadiabatic time correlation functions. *Proc. Natl. Acad. Sci. USA* **2005**, *102*, 6715–6719.

25. MacKernan D.; Ciccotti, G.; Kapral, R. Trotter-based simulation of quantum classical dynamics. *J. Phys. Chem. B* **2008**, *112*, 424–432.

26. Kim, H.; Nassimi, A.; Kapral, R. Quantum classical Liouville dynamics in the mapping basis. *J. Chem. Phys.* **2008**, *129*, 084102.

27. Nassimi, A.; Kapral, R. Mapping Approach for Quantum Classical Time Correlation Functions. *Can. J. Chem.* **2009**, *87*, 880–890.

28. Nielsen, S.; Kapral, R.; Ciccotti, G. Statistical mechanics of quantum classical systems. *J. Chem. Phys.* **2001**, *115*, 5805.

29. Agostini, F.; Caprara, S.; Ciccotti, G. Do we have a consistent non adiabatic quantum classical mechanics? *Europhys. Lett.* **2007**, *78*, 30001.

30. Bonella, S.; Monteferrante, M.; Pierleoni, C.; Ciccotti, G. Path integral based calculations of symmetrized time correlation functions. II. *J. Chem. Phys.* **2010**, *133*, 164105.

31. Schofield, P. Space-time correlation function formalism for slow neutron scattering. *Phys. Rev. Lett.* **1960**, *4*, 239–240.

32. Filinov, V. Wigner approach to quantum statistical mechanics and quantum generalization molecular dynamics methods. Part 1. *Mol. Phys.* **1996**, *88*, 1517.

33. Filinov, V. Wigner approach to quantum statistical mechanics and quantum generalization molecular dynamics method. Part 2. *Mol. Phys.* **1996**, *88*, 1529.

34. Miller, W.H.; Schwartz, S.D.; Tromp, J.W. Quantum mechanical rate constants for bimolecular reactions. *J. Chem. Phys.* **1983**, *79*, 4889.

35. Frantsuzov, P.A.; Mandelshtam, V.A. Quantum statistical mechanics with Gaussians: Equilibrium properties of van der Waals clusters. *J. Chem. Phys.* **2004**, *121*, 9247–9256.

36. Jadhao, V.; Makri, N. Iterative Monte Carlo for quantum dynamics. *J. Chem. Phys.* **2008**, *129*, 161102.

37. Bonella, S.; Monteferrante, M.; Pierleoni, C.; Ciccotti, G. Path integral based calculations of symmetrized time correlation functions. I. *J. Chem. Phys.* **2010**, *133*, 164104.

38. Monteferrante, M.; Bonella, S.; Ciccotti, G. Quantum dynamical structure factor of liquid neon via a quasiclassical symmetrized method. *J. Chem. Phys.* **2013**, *138*, 054118.

39. Monteferrante, M.; Bonella, S.; Ciccotti, G. Linearized symmetrized quantum time correlation functions calculation via phase pre-averaging. *Mol. Phys.* **2011**, *109*, 3015–3027.

40. Lindley, D.V. *Introduction to Probability and Statistics from a Bayesian Viewpoint, Part 1, Probability*; Cambridge University Press: Cambridge, UK, 1965.

41. Causo, M.S.; Ciccotti, G.; Bonella, S.; Vuilleumier, R. An adiabatic linearized path integral approach for quantum time-correlation functions II: A cumulant expansion method for improving convergence. *J. Phys. Chem. B* **2006**, *110*, 16026–16034.

42. Ceperley, D.M.; Dewing, M. The penalty method for random walks with uncertain energies. *J. Chem. Phys.* **1999**, *110*, 9812.

43. Kennedy, A.; Kuti, J. Noise without noise: A new Monte Carlo method. *Phys. Rev. Lett.* **1985**, *54*, 2473–2476.

44. Sprik, M.; Klein, M.; Chandler, D. Staging: A sampling technique for the Monte Carlo evaluation of path integrals. *J. Chem. Phys.* **1985**, *31*, 4234–4244.

45. Poulsen, J.; Scheers, J.; Nyman, G.; Rossky, P. Quantum density fluctuations in liquid neon from linearized path-integral calculations. *Phys. Rev. B* **2007**, *75*, 1–5.

46. Beutier, J.; Bonella, S.; Monteferrante, M.; Vuilleumier, R.; Ciccotti, G. Gas Phase Infrared Spectra via the Phase Integration Quasi Classical Method. *Mol. Sim.* **2013**, 40, 196–207.

47. Cao, J.; Voth, G.A. A new perspective on quantum time correlation functions. *J. Chem. Phys.* **1993**, *99*, 10070.

48. Craig, I.R.; Manolopoulos, D.E. Quantum statistics and classical mechanics: Real time correlation functions from ring polymer molecular dynamics. *J. Chem. Phys.* **2004**, *121*, 3368–3373.

Reprinted from *Entropy*. Cite as: de Carvalho, F.F.; Bouduban, M.E.F.; Curchod, B.F.E.; Tavernelli, I. Nonadiabatic Molecular Dynamics Based on Trajectories. *Entropy* **2014**, *16*, 62–85.

Review

Nonadiabatic Molecular Dynamics Based on Trajectories

Felipe Franco de Carvalho, Marine E. F. Bouduban, Basile F. E. Curchod and Ivano Tavernelli *

Laboratory of Computational Chemistry and Biochemistry, Ecole Polytechnique Fédérale de Lausanne, CH-1015 Lausanne, Switzerland; E-Mails: felipe.francodecarvalho@epfl.ch (F.F.C.); marine.bouduban@epfl.ch (M.E.F.B.); basile.curchod@epfl.ch (B.F.E.C.)

* Author to whom correspondence should be addressed; E-Mail: ivano.tavernelli@epfl.ch; Tel.: +41-21-693-03-28.

Received: 18 September 2013; in revised form: 12 December 2013 / Accepted: 16 December 2013 / Published: 27 December 2013

Abstract: Performing molecular dynamics in electronically excited states requires the inclusion of nonadiabatic effects to properly describe phenomena beyond the Born-Oppenheimer approximation. This article provides a survey of selected nonadiabatic methods based on quantum or classical trajectories. Among these techniques, trajectory surface hopping constitutes an interesting compromise between accuracy and efficiency for the simulation of medium- to large-scale molecular systems. This approach is however, based on non-rigorous approximations that could compromise, in some cases, the correct description of the nonadiabatic effects under consideration and hamper a systematic improvement of the theory. With the help of an *in principle* exact description of nonadiabatic dynamics based on Bohmian quantum trajectories, we will investigate the origin of the main approximations in trajectory surface hopping and illustrate some of the limits of this approach by means of a few simple examples.

Keywords: nonadiabatic dynamics; trajectory surface hopping; Ehrenfest dynamics; Bohmian dynamics; Born-Oppenheimer approximation

1. Introduction

Traditionally, *ab-initio* molecular dynamics (AIMD) is described within the so-called Born-Oppenheimer approximation, which assumes that the electronic and nuclear dynamics can be adiabatically separated [1], due to a large difference in mass between nuclei and electrons. Within this approximation, one usually solves the time-independent electronic Schrödinger equation for a given nuclear configuration [2] and then computes the quantum mechanical forces acting on the nuclei from the gradient of the corresponding eigenvalues, which depend parametrically on the nuclear coordinates and form the so-called potential energy surfaces (PES). However, in the description of most photophysical and photochemical processes, the electronic and nuclear dynamics become entangled, and therefore, more accurate nonadiabatic molecular dynamics schemes that go beyond the Born-Oppenheimer (BO) approximation are required. The most commonly used *ab initio* nonadiabatic molecular dynamics schemes are those based on the mixed quantum/classical propagation of an ensemble of (quasi-) classical trajectories [3–6], which, to some extent, reproduce the quantum dynamics of the nuclei. These mixed quantum/classical methods are especially popular, because they only require that the necessary electronic structure properties be computed *on-the-fly*, *i.e.*, only at the points in the configuration space visited during the dynamics, therefore making the calculation of the full potential energy surfaces unnecessary. These approaches can be implemented numerically using electronic structure methods, such as Kohn-Sham Density Functional Theory (DFT) [7,8] and its time-dependent version (TDDFT) [9–12] or wavefunction-based approaches, such as Complete Active Space Self-Consistent Field (CASSCF), Multireference Configuration Interaction (MRCI) and Second-Order Approximate Coupled-Cluster (CC2) [13].

Among all nonadiabatic AIMD schemes, Tully's fewest switches trajectory surface hopping [14,15] (TSH) and its extensions to mixed quantum/classical dynamics [16] are probably the most widely used. In the framework of TSH, the nuclear wave packets on the different PESs are described as a swarm of *independent* classical trajectories, while the nonadiabatic couplings induce hops of the trajectories from one electronic state to another; the occurrence of a trajectory hop is governed by the evaluation of a hopping probability, which depends on the temporal evolution of state amplitudes (Tully's coefficients) and on the value of the nonadiabatic couplings.

Alternative schemes have been proposed for the description of the nonadiabatic dynamics of the nuclear degrees of freedom, among which we quote semiclassical approaches [17,18], extended surface hopping [19,20], quantum/classical Liouville approaches [21,22], hydrodynamic nonadiabatic dynamics [23], linearized nonadiabatic dynamics (LAND-map) [24] or correlated electron-ion dynamics methods [25].

Despite the success of the nonadiabatic trajectory-based approaches, there are many quantum mechanical phenomena that cannot be entirely captured within this framework, namely nuclear quantum effects, like wavepacket interference [22], decoherence [26–28] and tunneling. Quantum dynamics methods based on a quantum mechanical representation of both electronic and nuclear degrees of freedom have also become available (see, for example, [29]). However, their high computational cost and the need for a numerical fit of the relevant PESs prior to propagation have limited their application to just a few nuclear degrees of freedom, and they are therefore not yet suited for the simulation of complex molecular systems.

One possible way to account for quantum nuclear effects within a trajectory-based framework consists in the use of quantum (or Bohmian) trajectories [30–32]. This approach emerges from a transformation of the time-dependent Schrödinger equation using a polar representation of the complex nuclear wavefunction (see [33] and Section 2 below). Robert. E. Wyatt and coworkers have recently introduced a numerical formulation of Bohmian dynamics using a trajectory-based solution of the so-called quantum hydrodynamics equations [34], named the quantum trajectory method (QTM). In their approach, the spatial support of the nuclear wave packet is split into fluid elements (FEs) that represent volume elements in configuration space carrying quantum information (amplitude and phase). Each of them is propagated according to a Newton-like equation of motion augmented by a nonlocal quantum potential. The latter supplies correlation between the FEs and is, therefore responsible for most quantum nuclear effects. The QTM approach has been employed to address challenging quantum dynamics problems in low dimensional model systems (see [35–37] for an extended presentation of quantum trajectory methods). Generalizations of QTM for multiple electronic states have also been proposed [38–42]. These are, however, based on a diabatic representation of the PESs. In an attempt to extend this type of dynamics to the investigation of molecular systems, we have recently developed an *in principle* exact QTM approach, named NABDY (nonadiabatic Bohmian dynamics), which solves the non-relativistic quantum dynamics of nuclei and electrons within the framework of quantum hydrodynamics, using the adiabatic representation of the electronic states [43,44].

In this article, we review a number of trajectory-based nonadiabatic molecular dynamics schemes together with our recent work on nonadiabatic Bohmian dynamics. Our aim is to provide a unified picture of the field by trying to "derive" the different approaches starting from a common framework, namely the quantum hydrodynamics reformulation of the molecular time-dependent Schrödinger equation. In particular, we propose a classification of the different trajectory-based approaches based on the choice of the initial expansion of the molecular wavefunction (that depends on both the nuclear and the electronic degrees of freedom) into a sum or a single product of electronic and nuclear wavefunctions. Finally, we propose a rationalization of the TSH equation of motion based on our exact nonadiabatic Bohmian dynamics scheme, showing by means of tests on two simple model systems the origin of some typical failures of TSH.

2. Nonadiabatic Dynamics with Classical and Quantum Trajectories

In this Section, we briefly review the theoretical background of the different nonadiabatic molecular dynamics schemes that we have selected for this study. The selection is based on the fact that all these trajectory-based approaches can be classified according to the way the molecular wavefunction is represented in terms of the electronic and nuclear components.

Starting from the Born-Huang representation of the total molecular wavefunction, we first introduce the Born-Oppenheimer molecular dynamics (BO-MD), which is based on the adiabatic separation of the electronic and nuclear dynamics, the latter being described by a single classical trajectory. Nonadiabaticity is then reintroduced following different strategies. In trajectory surface hopping (TSH), when the classical trajectories enter a region of strong coupling between different PESs, they are allowed to *hop* from one surface to another according to a hopping algorithm designed by Tully [15]. An interesting improvement of this scheme consists in adding

Gaussian-expanded nuclear wavefunctions to the propagating trajectories; this approach is named Full Multiple Spawning [45–48] and is characterized by a balance between accuracy and numerical efficiency. Finally, we will describe a trajectory-based approach in which classical trajectories are replaced by Bohmian quantum trajectories that evolve under the influence of quantum adiabatic and nonadiabatic potentials. All these methods make use of the computationally advantageous adiabatic representation of all involved electronic states.

In the second part of this review, we discuss nonadiabatic AIMD approaches that can be derived from a single product ansatz for the total molecular wavefunction. Two of these methods will be investigated, namely the approximated Ehrenfest dynamics and the exact solution, named "Exact Factorization", which has recently been proposed by Gross and coworkers.

We begin by introducing the time-dependent Schrödinger equation (TDSE) for a molecular system, which, neglecting the nuclear and electronic spins, is given by

$$i\hbar\frac{\partial}{\partial t}\Psi(\boldsymbol{r},\boldsymbol{R},t) = \hat{H}(\boldsymbol{r},\boldsymbol{R})\Psi(\boldsymbol{r},\boldsymbol{R},t),\tag{1}$$

where $\Psi(\boldsymbol{r},\boldsymbol{R},t)$ is the total wavefunction of the system, $\boldsymbol{r} = (\boldsymbol{r}_1,\ldots,\boldsymbol{r}_k,\ldots,\boldsymbol{r}_{N_{el}})$ is the collective position vector for the N_{el} electrons and $\boldsymbol{R} = (\boldsymbol{R}_1,\ldots,\boldsymbol{R}_\gamma\ldots,\boldsymbol{R}_{N_n})$ the corresponding one for the N_n nuclei of mass M_γ. The molecular Hamiltonian can be expressed in the following form

$$\hat{H}(\boldsymbol{r},\boldsymbol{R}) = -\sum_\gamma\frac{\hbar^2}{2M_\gamma}\nabla_\gamma^2 - \frac{\hbar^2}{2}\sum_k\nabla_k^2 + \sum_{k<l}\frac{1}{|\boldsymbol{r}_k-\boldsymbol{r}_l|} - \sum_{\gamma,k}\frac{Z_\gamma}{|\boldsymbol{R}_\gamma-\boldsymbol{r}_k|} + \sum_{\gamma<\zeta}\frac{Z_\gamma Z_\zeta}{|\boldsymbol{R}_\gamma-\boldsymbol{R}_\zeta|}$$
$$= -\sum_\gamma\frac{\hbar^2}{2M_\gamma}\nabla_\gamma^2 + \hat{\mathcal{H}}_{el}(\boldsymbol{r};\boldsymbol{R}),\tag{2}$$

where $\hat{\mathcal{H}}_{el}(\boldsymbol{r};\boldsymbol{R})$ is the electronic Hamiltonian, which is parametrically dependent on the nuclear coordinates. In Equation (2) and in the ones that follow, atomic units are used, except for the reduced Planck's constant, $\hbar$, that will be kept for clarity.

2.1. Methods Based on the Born-Huang Expansion

The Born-Huang expansion gives an exact expression for the total wavefunction [49,50]

$$\Psi(\boldsymbol{r},\boldsymbol{R},t) = \sum_i^\infty \Omega_i(\boldsymbol{R},t)\Phi_i(\boldsymbol{r};\boldsymbol{R}).\tag{3}$$

The total wavefunction, $\Psi(\boldsymbol{r},\boldsymbol{R},t)$, is expanded in the complete basis set of of electronic eigenfunctions of $\hat{\mathcal{H}}_{el}(\boldsymbol{r};\boldsymbol{R})$, which depend parametrically on the nuclear positions, $\boldsymbol{R}$. The expansion "coefficients", $\Omega_i(\boldsymbol{R},t)$, are functions of the nuclear coordinates, $\boldsymbol{R}$, and are explicitly dependent on time. Inserting Equation (3) into the TDSE, multiplying by $\Phi_j^*(\boldsymbol{r};\boldsymbol{R})$ and then integrating over $\boldsymbol{r}$ gives the equation of motion for the amplitudes, $\Omega_j(\boldsymbol{R},t)$

$$i\hbar\frac{\partial}{\partial t}\Omega_j(\boldsymbol{R},t) = \left[-\sum_\gamma\frac{\hbar^2}{2M_\gamma}\nabla_\gamma^2 + E_j^{el}(\boldsymbol{R})\right]\Omega_j(\boldsymbol{R},t) + \sum_i^\infty \mathcal{F}_{ji}(\boldsymbol{R})\Omega_i(\boldsymbol{R},t),\tag{4}$$

where the $\mathcal{F}_{ji}(\boldsymbol{R})$ are the elements of the nonadiabatic coupling matrix

$$\mathcal{F}_{ji}(\boldsymbol{R}) = \int d\boldsymbol{r} \; \Phi_j^*(\boldsymbol{r}; \boldsymbol{R}) \left[-\sum_\gamma \frac{\hbar^2}{2M_\gamma} \nabla_\gamma^2 \right] \Phi_i(\boldsymbol{r}; \boldsymbol{R})$$

$$+ \sum_\gamma \frac{1}{M_\gamma} \left\{ \int d\boldsymbol{r} \; \Phi_j^*(\boldsymbol{r}; \boldsymbol{R}) \left[-i\hbar\nabla_\gamma \right] \Phi_i(\boldsymbol{r}; \boldsymbol{R}) \right\} \left[-i\hbar\nabla_\gamma \right]. \tag{5}$$

These terms induce nonadiabatic coupling between different electronic states (for $i \neq j$) due to nuclear motion. Equation (4) can be interpreted as a Schrödinger-like equation for "nuclear" wavefunctions $\Omega_j(\boldsymbol{R}, t)$ augmented by nonadiabatic coupling terms. In fact, the amplitudes $\Omega_j(\boldsymbol{R}, t)$ can be interpreted as nuclear wavefunctions in state j only when the coupling terms vanish.

2.1.1. The Born-Oppenheimer Approximation and Adiabatic Dynamics

The BO approximation consists in neglecting all off-diagonal terms $\mathcal{F}_{ji}(\boldsymbol{R})$ in Equation (4) (*i.e.*, neglecting inter-state couplings, but keeping intra-state electronic-nuclear couplings). The molecular wavefunction on each PES is therefore represented by the simple product $\Psi(\boldsymbol{r}, \boldsymbol{R}, t) = \Omega_j(\boldsymbol{R}, t)\Phi_j(\boldsymbol{r}; \boldsymbol{R})$. If the diagonal terms $\mathcal{F}_{jj}(\boldsymbol{R})$ are also neglected, then we obtain what is usually called the adiabatic BO approximation [51]. Introducing the polar representation of $\Omega_j(\boldsymbol{R}, t)$, we obtain

$$\Omega_j(\boldsymbol{R}, t) = A_j(\boldsymbol{R}, t) \exp\left[\frac{i}{\hbar} S_j(\boldsymbol{R}, t) \right], \tag{6}$$

where both the amplitude, $A_j(\boldsymbol{R}, t)$, and the phase, $S_j(\boldsymbol{R}, t)$, are real. By inserting Equation (6) into Equation (4) and separating the real and imaginary parts, we obtain, within the adiabatic BO approximation, two separate, but coupled, equations for the amplitude and the phases

$$\frac{\partial S_j(\boldsymbol{R}, t)}{\partial t} = \frac{\hbar^2}{2} \sum_\gamma M_\gamma^{-1} \frac{\nabla_\gamma^2 A_j(\boldsymbol{R}, t)}{A_j(\boldsymbol{R}, t)} - \frac{1}{2} \sum_\gamma M_\gamma^{-1} \left(\nabla_\gamma S_j(\boldsymbol{R}, t) \right)^2 - E_j^{el}(\boldsymbol{R}), \tag{7}$$

$$\frac{\partial A_j(\boldsymbol{R}, t)}{\partial t} = -\sum_\gamma M_\gamma^{-1} \nabla_\gamma A_j(\boldsymbol{R}, t) \cdot \nabla_\gamma S_j(\boldsymbol{R}, t) - \frac{1}{2} \sum_\gamma M_\gamma^{-1} A_j(\boldsymbol{R}, t) \nabla_\gamma^2 S_j(\boldsymbol{R}, t). \tag{8}$$

Taking the so-called classical limit $\hbar \to 0$ in Equation (7) leads to something akin to the classical Hamilton-Jacobi equation of motion

$$\frac{\partial S_j(\boldsymbol{R}, t)}{\partial t} = -\frac{1}{2} \sum_\gamma M_\gamma^{-1} \left(\nabla_\gamma S_j(\boldsymbol{R}, t) \right)^2 - E_j^{el}(\boldsymbol{R}), \tag{9}$$

where $S_j(\boldsymbol{R}, t)$ can now be interpreted as the classical Hamilton's principal function. In this case, $\nabla_\gamma S_j(\boldsymbol{R}, t)$ corresponds to the nuclear momentum $\boldsymbol{p}_j^\gamma(t)$. By rearranging Equation (9), we finally obtain the Newtonian equation of motion for the nuclei

$$M_\gamma \ddot{\boldsymbol{R}}_j^\gamma(t) = -\nabla_\gamma E_j^{el}(\boldsymbol{R}(t)). \tag{10}$$

The nuclei therefore evolve on a given potential energy surface, $E_j^{el}(\boldsymbol{R}(t))$ (selected by the initial conditions), while the electrons adiabatically follow the nuclei along their classical trajectories $\boldsymbol{R}(t)$.

Equation (8) represents a continuity equation for the nuclear amplitudes, $A_j(\boldsymbol{R},t)$, in an arbitrary state j, which, in the classical limit, is trivially fulfilled because of the conservation of the number of trajectories. The BO-MD method therefore consists in solving the time-independent electronic Schrödinger equation to get the potential and the forces acting on the nuclei; these are then used to propagate the nuclei for time step dt using Equation (10), and the process is iterated until the desired propagation time is reached.

2.1.2. Tully's Trajectory Surface Hopping

One of the most successful methods for nonadiabatic dynamics is Tully's trajectory surface hopping [14,15]. In this method, the nuclei are treated classically, and the only nuclear quantum effect that is accounted for is the nonadiabatic transfer of "amplitude" between electronic states. This is achieved classically through *hops* of trajectories from one electronic state to another according to a hopping probability determined by the strength of the nonadiabatic couplings and the values of the state amplitudes $C_i^{[\alpha]}(t)$ defined below. A swarm of trajectories needs to be propagated in order to reproduce the probability distribution associated with corresponding nuclear quantum wave packet.

In this section, we only give a brief introduction to TSH, while a more detailed description of the method is given in Section 3, where we attempt a "derivation" of TSH, starting from the nonadiabatic Bohmian dynamics equations of motion.

The main ansatz in TSH is given by the following description of the molecular wavefunction [5,15]

$$\Psi^{[\alpha]}(\boldsymbol{r},\boldsymbol{R},t) = \sum_i^{\infty} C_i^{[\alpha]}(t)\Phi_i^{[\alpha]}(\boldsymbol{r};\boldsymbol{R}), \tag{11}$$

which, in a way, constitutes a simplified version of the original Born-Huang expansion. When we introduce Equation (11) into the electronic time-dependent Schrödinger equation, we get a set of coupled equations of motion for the complex nuclear state amplitudes, $C_j^{[\alpha]}(t)$ (for trajectory α)

$$i\hbar\dot{C}_j^{[\alpha]}(t) = \sum_i^{\infty} C_i^{[\alpha]}(t)\left(E_i^{el}(\boldsymbol{R}^{[\alpha]})\delta_{ij} - i\hbar\sum_\gamma^{N_n} d_{ji}^\gamma(\boldsymbol{R}^{[\alpha]})\cdot\dot{\boldsymbol{R}}_\gamma^{[\alpha]}\right), \tag{12}$$

where $d_{ji}^\gamma(\boldsymbol{R}) = \int d\boldsymbol{r}\,\Phi_j^*(\boldsymbol{r};\boldsymbol{R})\nabla_\gamma\Phi_i(\boldsymbol{r};\boldsymbol{R})$ are the first-order nonadiabatic couplings (see Equation (5)). These coupled equations will be solved along a classical trajectory α, evolving *adiabatically* in a given electronic state j. The probability, $g_{ji}^{[\alpha]}(t,t+dt)$, for the trajectory α to jump from state j to state i during the time interval $[t,t+dt]$ is given by

$$g_{ji}^{[\alpha]}(t,t+dt) = 2\int_t^{t+dt} d\tau\frac{-\Re[C_i^{[\alpha]}(\tau)C_j^{[\alpha]*}(\tau)\Xi_{ij}^{[\alpha]}(\tau)]}{C_j^{[\alpha]}(\tau)C_j^{[\alpha]*}(\tau)}, \tag{13}$$

where $\Xi_{ij}^{[\alpha]}(\tau) = \sum_\gamma^{N_n} d_{ij}^\gamma(\boldsymbol{R}^{[\alpha]})\cdot\dot{\boldsymbol{R}}_\gamma^{[\alpha]}$.
A surface hop between two PESs, j and i ($j\to i$), occurs "stochastically" when, for a randomly generated number, $\zeta\in[0,1]$, we get:

$$\sum_{k\leq i-1} g_{jk}^{[\alpha]} < \zeta < \sum_{k\leq i} g_{jk}^{[\alpha]}. \tag{14}$$

This algorithm guarantees that a minimum number of hops is performed along each trajectory; for this reason, the method is also referred to as the "fewest switches algorithm".

2.1.3. Full Multiple Spawning

Full Multiple Spawning (FMS) [45–48] proposes an interesting compromise between accuracy and efficiency by representing nuclear wavefunctions as sums of time-dependent Gaussian basis functions, whose width is frozen and whose center evolves adiabatically according to classical mechanics. This ansatz on the classical evolution of the Gaussian centers is consistently applied throughout the full derivation of the FMS equations of motion.

In the FMS method, the nuclear wavefunction $\Omega_i(\boldsymbol{R}, t)$ in electronic state i, is represented by a linear combination of multidimensional Gaussian wave packets $\Omega_J^i(\boldsymbol{R}; \overline{\boldsymbol{R}}_J^i(t), \overline{\boldsymbol{p}}_J^i(t), \overline{\gamma}_J^i(t), \boldsymbol{\alpha}_J^i)$, products of one-dimensional Gaussian functions [52]

$$\Omega_i(\boldsymbol{R}, t) = \sum_{J=1}^{N_i(t)} C_J^i(t) \Omega_J^i(\boldsymbol{R}; \overline{\boldsymbol{R}}_J^i(t), \overline{\boldsymbol{p}}_J^i(t), \overline{\gamma}_J^i(t), \boldsymbol{\alpha}_J^i)$$

$$= \sum_{J=1}^{N_i(t)} C_J^i(t) \left[e^{i\overline{\gamma}_J^i(t)t} \prod_\rho^{3N_n} \left(\frac{2\alpha_{\rho J}^i}{\pi} \right)^{1/4} e^{-\alpha_{\rho J}^i \left(R_\rho - \overline{R}_{\rho J}^i(t) \right)^2 + i\overline{p}_{\rho J}^i(t) \left(R_\rho - \overline{R}_{\rho J}^i(t) \right)} \right]. \quad (15)$$

In Equation (15), the multidimensional Gaussian basis functions are labeled with index J, their time-independent width by $\boldsymbol{\alpha}_J^i$ and their time-dependent position, momentum and nuclear phase by $\overline{\boldsymbol{R}}_J^i(t)$, $\overline{\boldsymbol{p}}_J^i(t)$ and $\overline{\gamma}_J^i(t)$, respectively. $N_i(t)$ gives the number of Gaussian basis functions used in order to describe the nuclear wavefunction in electronic state i, and its time dependence comes from the possible "spawning" of new basis functions (as further discussed here below). The nuclear phases are propagated semi-classically, whereas the positions and momenta of the center of the Gaussians obey classical equations of motion in a given electronic state [52].

The time-evolution of the expansion coefficients $\boldsymbol{C}^i(t)$ is obtained through the solution of the following differential equation:

$$\frac{d\boldsymbol{C}^i(t)}{dt} = -i(\boldsymbol{S}_{ii}^{-1}) \left\{ \left[\boldsymbol{H}_{ii} - i\dot{\boldsymbol{S}}_{ii} \right] \boldsymbol{C}^i + \sum_{j \neq i} \boldsymbol{H}_{ij} \boldsymbol{C}^j \right\} \quad (16)$$

This equation is derived by plugging the Born-Huang expansion (Equation (3)) and the ansatz for the nuclear wavefunctions (Equation (15)) into the time-dependent Schrödinger equation. In Equation (16) the bold symbols emphasize that, for each electronic state i, there is a time-dependent coefficient per each Gaussian basis function, and $\boldsymbol{S}_{ii}$ and $\dot{\boldsymbol{S}}_{ii}$ represent different overlap matrices of the Gaussian functions (see [52] for the more details). The matrix elements of $\boldsymbol{H}_{ij}$ are given by:

$$(\boldsymbol{H}_{ij})_{KK'} = H_{iKjK'} = \langle \Omega_K^i \Phi_i | \hat{\mathcal{H}}_{el} + \hat{T}_R | \Omega_{K'}^j \Phi_j \rangle$$

$$= \langle \Omega_K^i | \mathcal{H}_{el}^{ij} | \Omega_{K'}^j \rangle_{\boldsymbol{R}} + 2D_{iKjK'} + G_{iKjK'} \quad (17)$$

where $\hat{\mathcal{H}}_{el}$ is the electronic Hamiltonian and $\hat{T}_R$ the kinetic energy operator for the nuclei. In Equation (17), $D_{iKjK'} = \langle \Omega_K^i | \sum_\gamma^{N_n} \frac{1}{2M_\gamma} \langle \Phi_i | \nabla_\gamma | \Phi_j \rangle_r \cdot \nabla_\gamma | \Omega_{K'}^j \rangle_{\boldsymbol{R}}$ and $G_{iKjK'} =$

$\langle \Omega_K^i | \sum_\gamma^{N_n} \frac{1}{2M_\gamma} \langle \Phi_i | \nabla_\gamma^2 | \Phi_j \rangle_r | \Omega_{K'}^j \rangle_R$ couple the electronic and nuclear motions ($\langle \cdots \rangle_R$ means integration over R and $\langle \cdots \rangle_r$ integration over r).

The spawning procedure takes place when a region of nonadiabaticity is detected along a trajectory (by monitoring the strength of nonadiabatic couplings in the adiabatic representation) and allows for the generation of new Gaussian basis functions (children), placed in the newly populated electronic state according to physical rules (like position or momentum conservation [52]) maximizing the coupling between the parent and children Gaussian basis functions [52] until the system leaves the region of strong nonadiabatic coupling [53]

The spawning procedure, therefore, limits the number of Gaussian basis function used in the calculation by defining precisely where and when they are needed. Moreover, the FMS method offers a numerically exact [48] solution when all matrix elements are computed exactly, and a complete Gaussian basis set is used.

While keeping a trajectory-based formalism, FMS fully incorporates nuclear quantum effects that are missing in methods like TSH. Furthermore, the nuclear propagation can be performed on-the-fly, by computing any electronic structure property needed, like electronic energies ($\mathcal{H}_{el}^{ii}$ in Equation (17)) or nonadiabatic couplings ($\langle \Phi_i | \nabla_R | \Phi_j \rangle_r$ and $\langle \Phi_i | \nabla_R^2 | \Phi_j \rangle_r$; note that the $G_{iKjK'}$ terms are normally small and usually neglected) with either an *ab initio* electronic structure or semiempirical methods (*Ab Initio* Multiple Spawning, AIMS [54]). AIMS, therefore, overcomes the limitations in accuracy of TSH, preserving efficiency all the while. For additional information about the derivation and numerical procedure of this method, the interested reader is referred to [52].

2.1.4. Nonadiabatic Bohmian Dynamics

Just as for the previous three methods, nonadiabatic Bohmian dynamics (NABDY) is also based on the propagation of trajectories. However, this time, the trajectories evolve under the action of additional quantum potentials (adiabatic and nonadiabatic), which make the dynamics exact in principle. In other words, this approach is able to capture all adiabatic and nonadiabatic nuclear quantum effects through the propagation of a sufficiently large (*i.e.*, converged) number of trajectories.

The derivation of the NABDY equations of motion starts from the insertion of the polar representation of the nuclear wavefunction in Equation (6) into Equation (4). After separation of the real and imaginary parts, we obtain

$$
\begin{aligned}
-\frac{\partial S_j(\boldsymbol{R}, t)}{\partial t} =& \sum_\gamma \frac{1}{2M_\gamma} \left(\nabla_\gamma S_j(\boldsymbol{R}, t) \right)^2 + E_j^{el}(\boldsymbol{R}) - \sum_\gamma \frac{\hbar^2}{2M_\gamma} \frac{\nabla_\gamma^2 A_j(\boldsymbol{R}, t)}{A_j(\boldsymbol{R}, t)} \\
&+ \sum_{\gamma i} \frac{\hbar^2}{2M_\gamma} D_{ji}^\gamma(\boldsymbol{R}) \frac{A_i(\boldsymbol{R}, t)}{A_j(\boldsymbol{R}, t)} \Re \left[e^{i\phi_{ij}(\boldsymbol{R}, t)} \right] - \sum_{\gamma, i \neq j} \frac{\hbar^2}{M_\gamma} \boldsymbol{d}_{ji}^\gamma(\boldsymbol{R}) \cdot \frac{\nabla_\gamma A_i(\boldsymbol{R}, t)}{A_j(\boldsymbol{R}, t)} \Re \left[e^{i\phi_{ij}(\boldsymbol{R}, t)} \right] \\
&+ \sum_{\gamma, i \neq j} \frac{\hbar}{M_\gamma} \frac{A_i(\boldsymbol{R}, t)}{A_j(\boldsymbol{R}, t)} \boldsymbol{d}_{ji}^\gamma(\boldsymbol{R}) \cdot \nabla_\gamma S_i(\boldsymbol{R}, t) \Im \left[e^{i\phi_{ij}(\boldsymbol{R}, t)} \right],
\end{aligned}
\tag{18}
$$

and

$$\hbar \frac{\partial A_j(\mathbf{R},t)}{\partial t} = -\sum_\gamma \frac{\hbar}{M_\gamma} \nabla_\gamma A_j(\mathbf{R},t) \cdot \nabla_\gamma S_j(\mathbf{R},t) - \sum_\gamma \frac{\hbar}{2M_\gamma} A_j(\mathbf{R},t) \nabla_\gamma^2 S_j(\mathbf{R},t)$$

$$+ \sum_{\gamma i} \frac{\hbar^2}{2M_\gamma} D_{ji}^\gamma(\mathbf{R}) A_i(\mathbf{R},t) \Im\left[e^{i\phi_{ij}(\mathbf{R},t)}\right] - \sum_{\gamma,i\neq j} \frac{\hbar^2}{M_\gamma} \mathbf{d}_{ji}^\gamma(\mathbf{R}) \cdot \nabla_\gamma A_i(\mathbf{R},t) \Im\left[e^{i\phi_{ij}(\mathbf{R},t)}\right]$$

$$- \sum_{\gamma,i\neq j} \frac{\hbar}{M_\gamma} A_i(\mathbf{R},t) \mathbf{d}_{ji}^\gamma(\mathbf{R}) \cdot \nabla_\gamma S_i(\mathbf{R},t) \Re\left[e^{i\phi_{ij}(\mathbf{R},t)}\right], \tag{19}$$

where $\phi_{ij}(\mathbf{R},t) = \frac{1}{\hbar}(S_i(\mathbf{R},t) - S_j(\mathbf{R},t))$ and $D_{ji}^\gamma(\mathbf{R})$ are the second-order nonadiabatic couplings. Equation (18) is equivalent to the classical Hamilton-Jacobi equation augmented by terms that are $\mathcal{O}(\hbar)$ and $\mathcal{O}(\hbar^2)$. The third term, $\mathcal{Q}(\mathbf{R},t)$, on the right-hand side of Equation (18) is called the quantum potential, and it includes all adiabatic quantum effects (adiabatic in the sense that the potential $\mathcal{Q}(\mathbf{R},t)$ acts on a single PES and does not include contributions from other surfaces). Unlike "classical" potentials, it is non-local in space, in the sense that it depends on the position of all particle in configuration space [55]. The last three terms on the right-hand side of Equation (18) describe inter-state nonadiabatic quantum effects and, like the quantum potential, $\mathcal{Q}(\mathbf{R},t)$, do not have a classical equivalent.

Equation (19) represents a continuity equation the for probability density, $|A_j(\mathbf{R},t)|^2$, with corresponding probability density flux $\mathbf{J}(\mathbf{R},t)$ [33,35,44]. The first two terms on the right-hand side describe the "adiabatic" probability density flow within a given state, j, while the remaining terms that depend on the first-order and second-order nonadiabatic couplings induce probability density exchanges across different electronic states. Of course, the overall nuclear amplitude (summed up over all states) is conserved.

The two equations for the phases and the amplitudes are coupled, and they therefore need to be solved simultaneously. Instead of solving complex differential equations for the two fields, ($A_j(\mathbf{R},t)$ and $S_j(\mathbf{R},t)$), we reintroduce trajectories in configuration space that drive the dynamics of "infinitesimal" volume elements called "fluid elements" [35]. The derivation of the equations of motion is similar to that described in Section 2.1.1 for the BO approach, with the important difference that in NABDY new fluid elements can be created at any time on any other PES according to the size of the nonadiabatic terms in Equation (19). The details of the numerical implementation of NABDY are given in [44], while a possible extension of NABDY to large dimensions (in the adiabatic case) is proposed in [56].

2.2. Methods Based on a Single Product Ansatz

2.2.1. Ehrenfest Dynamics

The equation of motion that drives Ehrenfest dynamics (EHD) is derived from a simpler ansatz for the total wavefunction than the Born-Huang expansion (Equation (3)) used for the methods presented in Section 2.1.

In EHD, the molecular wavefunction is described by the simple product

$$\Psi(r, R, t) = \Phi(r, t)\Omega(R, t) \exp\left[\frac{i}{\hbar}\int_{t_0}^{t} dt'\, E_{el}(t')\right] \tag{20}$$

where $\Phi(r, t)$ is the electronic wavefunction and $\Omega(R, t)$ is the nuclear wavefunction. Note that in this case, both amplitudes $\Phi(r, t)$ and $\Omega(R, t)$ depend explicitly on time. In addition, they also have a parametric dependence on the other set of coordinates ($\Phi(r, t)$ on R and $\Omega(R, t)$ on r), which is not explicitly shown, so as to simplify the notation.

The exponential in Equation (20) is named the phase term and is defined as

$$E_{el}(t) = \int\int dr\, dR\, \Phi^*(r, t)\Omega^*(R, t)\hat{\mathcal{H}}_{el}(r, R)\Phi(r, t)\Omega(R, t) \tag{21}$$

and it guarantees that the product wavefunction, $\Psi(r, R, t)$, fulfills the corresponding time-dependent Schrödinger equation.

Following the derivation proposed by Tully [57], we can substitute Equation (20) into the TDSE, multiply by $\Omega^*(R, t)$ and integrate over R, and assuming that $\Omega(R, t)$ is normalized, we finally obtain

$$i\hbar\frac{\partial\Phi(r, t)}{\partial t} = -\frac{\hbar^2}{2}\sum_k \nabla_k^2\Phi(r, t) + \tag{22}$$

$$\left\{\int dR\, \Omega^*(R, t)\left[-\frac{\hbar^2}{2}\sum_\gamma M_\gamma^{-1}\nabla_\gamma^2 + \hat{V}(r, R)\right]\Omega(R, t)\right\}\Phi(r, t) +$$

$$E_{el}(t)\Phi(r, t) - i\hbar\left[\int dR\, \Omega^*(R, t)\frac{\partial\Omega(R, t)}{\partial t}\right]\Phi(r, t)$$

where $\hat{V}(r, R)$ is the sum of all the potential energy terms in the molecular Hamiltonian.

Applying an analogous procedure, we can also derive the equation of motion for $\Omega(R, t)$

$$i\hbar\frac{\partial\Omega(R, t)}{\partial t} = -\frac{\hbar^2}{2}\sum_\gamma M_\gamma^{-1}\nabla_\gamma^2\Omega(R, t) + \tag{23}$$

$$\left\{\int dr\, \Phi^*(r, t)\left[\hat{\mathcal{H}}_{el}(r, R)\right]\Phi(r, t)\right\}\Omega(R, t)$$

$$+ E_{el}(t)\Omega(R, t) - i\hbar\left[\int dr\, \Phi^*(r, t)\frac{\partial\Phi(r, t)}{\partial t}\right]\Omega(R, t).$$

Using the relations [57]

$$\int dr\, \Phi^*(r, t)\frac{\partial\Phi(r, t)}{\partial t} = E_{el}(t) \tag{24}$$

and

$$\int dR\, \Omega^*(R, t)\frac{\partial\Omega(R, t)}{\partial t} = E \tag{25}$$

where E is the expectation value of the molecular Hamiltonian for the wavefunction appearing in Equation (20), and the fact that $E = E_{el}(t) + \langle T_R\rangle$ ($\langle T_R\rangle$ is the expectation value of the nuclear kinetic energy), we can further simplify Equations (22) and (23) and obtain the following differential equations for the two amplitudes

$$i\hbar\frac{\partial\Phi(\boldsymbol{r},t)}{\partial t} = -\frac{\hbar^2}{2}\sum_k \nabla_k^2\Phi(\boldsymbol{r},t) + \left[\int d\boldsymbol{R}\ \Omega^*(\boldsymbol{R},t)\hat{V}(\boldsymbol{r},\boldsymbol{R})\Omega(\boldsymbol{R},t)\right]\Phi(\boldsymbol{r},t) \qquad (26)$$

and

$$i\hbar\frac{\partial\Omega(\boldsymbol{R},t)}{\partial t} = -\frac{\hbar^2}{2}\sum_\gamma M_\gamma^{-1}\nabla_\gamma^2\Omega(\boldsymbol{R},t) + \left[\int d\boldsymbol{r}\ \Phi^*(\boldsymbol{r},t)\hat{\mathcal{H}}_{el}(\boldsymbol{r},\boldsymbol{R})\Phi(\boldsymbol{r},t)\right]\Omega(\boldsymbol{R},t) \qquad (27)$$

These are mean field coupled equations, in which the electrons move in a field generated by the nuclei (second term on the right hand side (r.h.s.) of Equation (26)) and the nuclei move in a field generated by the electrons (second term on the r.h.s. of Equation (27)). Strictly speaking, these are not yet the EHD equations of motion, but, rather, a version of the time-dependent self-consistent field equations. EHD implies the passage to the classical limit for the nuclear amplitudes, which is again accomplished through the use of the polar representation of the nuclear wavefunction (Equation (6)) in Equation (27).

Once again, we obtain a classical Hamilton-Jacobi equation, which can be transformed into a Newton equation of motion given by

$$M_\gamma\ddot{\boldsymbol{R}}_\gamma(t) = -\nabla_\gamma\langle\hat{\mathcal{H}}_{el}(\boldsymbol{r},\boldsymbol{R})\rangle_t = -\nabla_\gamma\left[\int d\boldsymbol{r}\ \Phi^*(\boldsymbol{r},t)\hat{\mathcal{H}}_{el}(\boldsymbol{r},\boldsymbol{R})\Phi(\boldsymbol{r},t)\right]. \qquad (28)$$

Notice that the potential acting on the nuclei is now given by the expectation value of the electronic Hamiltonian computed using the time-dependent electronic "wavefunction" $\Phi(\boldsymbol{r},t)$, which is not necessarily an eigenstate of $\hat{\mathcal{H}}_{el}(\boldsymbol{r},\boldsymbol{R}(t))$, but which can be expressed as a linear combination of the static solutions of the corresponding time-independent electronic Schrödinger equation for the same nuclear position, $\boldsymbol{R}$, at time t . For this reason, EHD is called a mean-field solution of the time-dependent molecular Schrödinger equation.

The equation of motion for the electronic amplitudes, Equation (26), also depends on the nuclear amplitudes, $\Omega(\boldsymbol{R},t)$. However, since the nuclei are treated as classical particles, we can set

$$|\Omega(\boldsymbol{R}(t))|^2 = \prod_\gamma \delta(\boldsymbol{R}_\gamma - \boldsymbol{R}_\gamma(t)) \qquad (29)$$

that is to say, we induce localization of the nuclear densities at a fixed position, $\boldsymbol{R}_\gamma(t)$. By plugging Equation (29) into Equation (26) we obtain a TDSE for the electronic amplitude

$$i\hbar\frac{\partial\Phi(\boldsymbol{r};\boldsymbol{R}(t),t)}{\partial t} = \hat{\mathcal{H}}_{el}(\boldsymbol{r};\boldsymbol{R}(t))\Phi(\boldsymbol{r};\boldsymbol{R}(t),t) \qquad (30)$$

where the Hamiltonian and the wavefunction both depend parametrically on the nuclear positions, which induces the coupling with the nuclear equation of motion (Equation (28)). As we mentioned before, in EHD the nuclei will evolve on a single time-dependent PES, which can be expressed at any instant of time as a linear combination of all adiabatic PESs. This implies that in EHD nonadiabatic effects are taken into account through the propagation of the electronic wavefunction [58]; a perspective that is indeed very different from what is observed in the approaches derived from the Born-Huang expansion (Section 2.1).

2.2.2. The Exact Factorization-Based Dynamics

Recently, Gross *et al.* [59,60] have shown that the (exact) solution of the molecular TDSE can be factorized into the product of an electronic and a nuclear wavefunction [61] (even when the Hamiltonian includes coupling to external electromagnetic fields)

$$\Psi(r, R, t) = \Phi(r; R, t)\Omega(R, t) .\tag{31}$$

Equation (31) might seem counter-intuitive at first, because the molecular Hamiltonian is not separable. In fact, while the factorization in Equation (31) can be made at any time, t, and at any position, r or R, the persistence of this kind of solution along the time propagation of the two wavefunctions is less obvious (as can be seen from the resulting evolution equations [59]). As discussed in [60], the factorization in Equation (31) can be justified using multivariate statistics, according to which any probability distribution (here, the square of the molecular wavefunction) can be factored into a marginal probability and a conditional probability. In this respect, it is also important to notice that $\Phi(r; R, t)$ depends (parametrically) on the nuclear coordinates, R, and there is, therefore, no loss of generality in applying Equation (31).

The factorization of $\Psi(r, R, t)$ does not simplify, *per se*, the task of solving the molecular TDSE. Nonetheless, this approach has a great interpretive power, since $\Phi(r; R, t)$ and $\Omega(R, t)$ have both a clear physical meaning: they are *the* exact electronic and nuclear time-dependent wavefunctions, respectively. One crucial requirement for this to be true is the so-called partial normalization condition

$$\int dr\, |\Phi(r; R, t)|^2 = 1.\tag{32}$$

This condition allows for the interpretation of $dr|\Phi(r; R, t)|^2$ as the conditional probability of finding an electron in volume element, dr, at position r given a nuclear configuration, R; that is to say, $|\Phi(r; R, t)|^2$ is an electronic probability density function. According to the standard interpretation of quantum mechanics, $\Phi(r; R, t)$ is then the corresponding electronic wavefunction. Similarly, $\Omega(R, t)$ is the marginal probability density for the nuclear position, R (marginal, and not conditional, because r is unknown), and $\Omega(R, t)$ is the corresponding nuclear wavefunction. Interestingly, just as in EHD, this factorization leads to the definition of a single time-dependent potential energy surface (because of the time-dependence of the electronic wavefunction), which this time is however, exact and unique. What is lost is the picture of a time-dependent nuclear wave packet (or corresponding trajectories) evolving on an ensemble of static PESs; a picture that has provided important insights for the understanding of many photophysical and photochemical processes.

The time evolution of the wavefunctions, $\Phi(r; R, t)$ and $\Omega(R, t)$, is described by two connected differential equations, which contain, besides the usual interaction terms, additional scalar and vector potential terms [59,60,62,63].

3. Trajectory Surface Hopping from the Nonadiabatic Bohmian Dynamics Equations

When it comes to nonadiabatic molecular dynamics, TSH is probably the most popular simulation scheme. As stated in Section 2, it relies on the description of nuclear wave packets by means of a swarm of classical trajectories. A complex coefficient, $C_j^{[\alpha]}$, for each electronic state, j, is propagated along a given classical trajectory, α, according to Equation (12). The classical trajectory may "hop" from its current electronic state, i, to another at any point in time, and the probability that a hop to state j occurs is given by Equation (13) [15,64–66].

In this Section, starting from Equations (18) and (19), we will present a "rationalization" of the TSH equations of motion based on the nonadiabatic Bohmian dynamics equations.

The following steps were reported in [44] and can be summarized as follows:

(a) The nuclear wave packet dynamics is discretized into a swarm of classical trajectories. Within the *independent trajectory approximation*, the quantum potential is set to zero, $Q_j(\boldsymbol{R}, t) = 0$, $\forall j$ and so is the divergence of the current, $\nabla_{\boldsymbol{R}} \cdot \boldsymbol{J}_j^{[\alpha]}(\boldsymbol{R}, t) = 0$, $\forall j, \forall \alpha$. The *independent trajectory approximation* arises from the assumption that all adiabatic non-local terms (involving a single electronic state, j) are set to zero. This will ensure that there is no adiabatic quantum transfer of amplitude among trajectories. In the adiabatic regime ($d_{ji}(\boldsymbol{R})$ and $D_{ji}(\boldsymbol{R}) \to 0$), the trajectories evolve independently from each other, and their equation of motion is obtained from the solution by characteristics of Equation (18) in the classical limit, which corresponds to the classical Newton equation of motion with classical forces $-\nabla_{\boldsymbol{R}} E_j^{el}(\boldsymbol{R})$.

(b) For the description of the nonadiabatic components of the dynamics (the three last terms in Equations (18) and (19)), we first move to a reference frame evolving along the classical trajectories for which

$$
\begin{aligned}
-\frac{\partial S_j(\boldsymbol{R}, t)}{\partial t} =& E_j^{el}(\boldsymbol{R}) + \sum_\gamma^{N_n} \sum_i^\infty \frac{\hbar^2}{2M_\gamma} D_{ji}^\gamma(\boldsymbol{R}) \frac{A_i(\boldsymbol{R}, t)}{A_j(\boldsymbol{R}, t)} \Re\left[e^{i\phi_{ij}(\boldsymbol{R}, t)}\right] \\
& - \sum_\gamma^{N_n} \sum_{i \neq j}^\infty \frac{\hbar^2}{M_\gamma} d_{ji}^\gamma(\boldsymbol{R}) \cdot \frac{\nabla_\gamma A_i(\boldsymbol{R}, t)}{A_j(\boldsymbol{R}, t)} \Re\left[e^{i\phi_{ij}(\boldsymbol{R}, t)}\right] \\
& + \sum_\gamma^{N_n} \sum_{i \neq j}^\infty \frac{\hbar}{M_\gamma} \frac{A_i(\boldsymbol{R}, t)}{A_j(\boldsymbol{R}, t)} d_{ji}^\gamma(\boldsymbol{R}) \cdot \nabla_\gamma S_i(\boldsymbol{R}, t) \Im\left[e^{i\phi_{ij}(\boldsymbol{R}, t)}\right]
\end{aligned}
\tag{33}
$$

and

$$
\begin{aligned}
\frac{\partial A_j(\boldsymbol{R}, t)}{\partial t} =& \sum_\gamma^{N_n} \sum_i^\infty \frac{\hbar}{2M_\gamma} D_{ji}^\gamma(\boldsymbol{R}) A_i(\boldsymbol{R}, t) \Im\left[e^{i\phi_{ij}(\boldsymbol{R}, t)}\right] \\
& - \sum_\gamma^{N_n} \sum_{i \neq j}^\infty \frac{\hbar}{M_\gamma} d_{ji}^\gamma(\boldsymbol{R}) \cdot \nabla_\gamma A_i(\boldsymbol{R}, t) \Im\left[e^{i\phi_{ij}(\boldsymbol{R}, t)}\right] \\
& - \sum_\gamma^{N_n} \sum_{i \neq j}^\infty \frac{1}{M_\gamma} A_i(\boldsymbol{R}, t) d_{ji}^\gamma(\boldsymbol{R}) \cdot \nabla_\gamma S_i(\boldsymbol{R}, t) \Re\left[e^{i\phi_{ij}(\boldsymbol{R}, t)}\right].
\end{aligned}
\tag{34}
$$

Note that due to the *independent trajectory approximation*, we assume that there is no amplitude exchange among the FEs propagated along the different trajectories.

Neglecting the second-order nonadiabatic couplings, $D_{ji}(\boldsymbol{R})$, due to their usually small size [67], we are left with an equation of motion for the phases and the amplitudes, which is equivalent to the following nuclear wavefunction time-evolution equation

$$i\hbar\frac{\partial\Omega_j(\boldsymbol{R},t)}{\partial t} = E_j^{el}(\boldsymbol{R})\Omega_j(\boldsymbol{R},t) - i\hbar\sum_{i\neq j}^{\infty}\sum_{\gamma}^{N_n}\frac{1}{M_\gamma}d_{ji}^\gamma(\boldsymbol{R})\cdot\hat{p}^\gamma\Omega_i(\boldsymbol{R},t) \tag{35}$$

where we have used the definition of the momentum operator $\hat{p}^\gamma = -i\hbar\nabla_\gamma$.

(c) In the derivation of the equation of motion for the nuclear amplitude coefficients, we start by assigning delta-like wave packets (denoted as the TSH wave packet in the following) to each trajectory, α, defined as

$$\tilde{\Omega}_j^{\lambda,[\alpha]}(\boldsymbol{R},t) = \tilde{A}_j^{[\alpha]}(t)\exp\left[\frac{i}{\hbar}\tilde{S}_j^{[\alpha]}(t)\right]g^\lambda(\boldsymbol{R}-\boldsymbol{R}^{[\alpha]}(t)) \tag{36}$$

where $\tilde{A}_j^{[\alpha]}(t)$ and $\tilde{S}_j^{[\alpha]}(t)/\hbar$ are real functions representing the amplitude and the phase of the TSH nuclear wave packet at $\boldsymbol{R}^{[\alpha]}(t)$ in electronic state j. The function $g^\lambda(\boldsymbol{R}-\boldsymbol{R}^{[\alpha]}(t)) = \frac{1}{\lambda\sqrt{\pi}}\exp\left(-(\boldsymbol{R}-\boldsymbol{R}^{[\alpha]}(t))^2/\lambda^2\right)$, localized at the position of the classical trajectory, α, is normalized to $\int d\boldsymbol{R}\, g^\lambda(\boldsymbol{R}-\boldsymbol{R}^{[\alpha]}(t)) = 1$ and becomes a δ-function in the limit $\lim_{\lambda\to\infty}g^\lambda(\boldsymbol{R}-\boldsymbol{R}^{[\alpha]}(t)) = \delta(\boldsymbol{R}-\boldsymbol{R}^{[\alpha]}(t))$. The total probability density of the nuclear wave packet in state j becomes $|\Omega_j(\boldsymbol{R},t)|^2 \sim \frac{1}{N_{traj}}\sum_{[\alpha]}\int_{t=0}^\infty dt'\,|\tilde{A}_j^{[\alpha]}(t')|^2 g^\lambda(\boldsymbol{R}-\boldsymbol{R}^{[\alpha]}(t'))\delta(t-t')$, where N_{traj} is the total number of trajectories. The *independent trajectory approximation* invoked in point (a) also has an important impact on the nonadiabatic component of the nuclear dynamics (due to their nonlocality; see Equations (18) and (19)). Indeed, it has the consequence that, for a given trajectory, α, the complex amplitudes, $\tilde{\Omega}_j^{\lambda,[\alpha]}(\boldsymbol{R},t)$, for each and every electronic state, j, share the same support (localized around the instantaneous nuclear position, $\boldsymbol{R}$, in configuration space). Said otherwise, the TSH nuclear wave packet component, $g^\lambda(\boldsymbol{R}-\boldsymbol{R}^{[\alpha]}(t))$, will be the same for all electronic states of a trajectory, α, at any time, t (this is why g^λ does not have an electronic state index). This is indubitably the strongest approximation made in the "derivation", since it induces "overcoherence" in the dynamics of the amplitudes, $C_j^{[\alpha]}(t)$, and suppresses all (nonadiabatic) decoherence effects that could occur, for example, at and after the branching of nuclear wave packets.

(d) Since we are working in the Lagrangian frame, we need only consider the explicit time-dependence of the amplitudes and phases. As a consequence, the TSH nuclear wave packet evolving in electronic state j follows the classical trajectory, α, on the support of the function, $g^\lambda(\boldsymbol{R}-\boldsymbol{R}^{[\alpha]})$ (where $\boldsymbol{R}^{[\alpha]}$ is the position vector in the Lagrangian frame), and is described by $\tilde{A}_j^{[\alpha]}(t)\exp\left[\frac{i}{\hbar}\tilde{S}_j^{[\alpha]}(t)\right]$.

If we substitute $\Omega_j(\boldsymbol{R},t)$ in Equation (35) by the form given in Equation (36) and then apply points (a), (b) and (d), we obtain [44]

$$-\dot{\tilde{S}}_j^{[\alpha]}(t) = E_j^{el}(\boldsymbol{R}^{[\alpha]}) + \hbar\sum_\gamma^{N_n}\sum_{i\neq j}^{\infty}\frac{\tilde{A}_i^{[\alpha]}(t)}{\tilde{A}_j^{[\alpha]}(t)}\left(d_{ji}^\gamma(\boldsymbol{R}^{[\alpha]})\cdot\dot{\boldsymbol{R}}_\gamma^{[\alpha]}\right)\Im\left[e^{i\tilde{\phi}_{ij}^{[\alpha]}(t)}\right] \tag{37}$$

$$\dot{\tilde{A}}_j^{[\alpha]}(t) = -\sum_\gamma^{N_n} \sum_{i\neq j}^\infty \tilde{A}_i^{[\alpha]}(t) \left(d_{ji}^\gamma(\boldsymbol{R}^{[\alpha]}) \cdot \dot{\boldsymbol{R}}_\gamma^{[\alpha]} \right) \Re \left[e^{i\tilde{\phi}_{ij}^{[\alpha]}(t)} \right]. \tag{38}$$

Here, $\tilde{\phi}_{ij}(t) = \frac{1}{\hbar}\left(\tilde{S}_i^{[\alpha]}(t) - \tilde{S}_j^{[\alpha]}(t) \right)$ and $\dot{\boldsymbol{R}}^{[\alpha]}$ are the nuclear velocities at time t along trajectory α.

Notice that Equations (37) and (38) are equivalent to the TSH equations for the complex coefficients, $C_j^{[\alpha]}(t)$ (Equation (12)), which is obtained using a polar representation of the complex coefficients, $C_j^{[\alpha]}(t) = \tilde{A}_j^{[\alpha]}(t) \exp\left[\frac{i}{\hbar} \tilde{S}_j^{[\alpha]}(t) \right]$.

We have described until now the dynamics of TSH nuclear wave packets following a single classical trajectory, α. At this point, we have to account for the fact that the nuclear dynamics in TSH is described by a "swarm" of classical trajectories that evolve according to the adiabatic and nonadiabatic components of the equation of motion (points (a) to (d)). In order to to this, we have to require that the following be maintained

$$(A_j^{TSH}(\boldsymbol{R},t))^2 d\boldsymbol{R} \approx (A_j(\boldsymbol{R},t))^2 d\boldsymbol{R} \tag{39}$$

at all times, for a sufficiently large number of trajectories. In Equation (39), $(A_j^{TSH}(\boldsymbol{R},t))^2$ is computed as the density (histogram) of configuration space points in the volume element, $d\boldsymbol{R}$, at $\boldsymbol{R}(t)$ in state j that are sampled by the ensemble of N_{traj} trajectories

$$(A_j^{TSH}(\boldsymbol{R},t))^2 d\boldsymbol{R} = \frac{N_j(\boldsymbol{R}, d\boldsymbol{R}, t)}{N_{traj}} \tag{40}$$

while the right-hand side is the corresponding nuclear density obtained from the corresponding quantum mechanically propagated nuclear wave packets. Note that Equation (39) is only valid when correlated quantum (Bohmian) trajectories are used [36].

In TSH, the balance described in Equation (39) is maintained (in an approximative way) through the use of the switching algorithm given in Equations (13) and (14), which can be motivated by the following considerations:

(e) In the *independent trajectory approximation*, the nonadiabatic terms in the time-evolution of the TSH nuclear wave packets (Equations (37) and (38)) induce trajectory surface hop transitions between different states according to the probability, $g_{ji}^{[\alpha]}(t, t + dt)$, which is a function of local variables computed along the propagation of a single trajectory.

(f) The switching probability is obtained from quantum mechanical arguments [15] under the assumption that

$$\int d\boldsymbol{R} \, (A_j^{TSH}(\boldsymbol{R},t))^2 = |C_j(t)|_{av}^2 \tag{41}$$

where the $|C_j(t)|_{av}$ are the norms of the coefficients defined in Equation (12) averaged over the ensemble of trajectories. Equation (41) is the internal consistency criterion described in [68]. However, in practice, one replaces the $|C_j(t)|_{av}^2$ with the corresponding amplitudes computed along a single trajectory, $|C_j^{[\alpha]}(t)|^2$, which are the coefficients that appear in Equation (13). The reason for this modification is that in the *independent trajectory approximation*, one computes single trajectories, and therefore, the average over the ensemble is not available during propagation. This replacement of $|C_j(t)|_{av}^2$ by $|C_j^{[\alpha]}(t)|^2$ is, in our opinion, more of an assumption than an approximation and remains without formal justification.

In summary, starting from the *exact* formulation of the nonadiabatic dynamics within the nonadiabatic Bohmian dynamics framework, we proposed a series of approximations/assumptions (points (*a*) to (*f*)) that help rationalize Tully's TSH equations of motion for the nuclear trajectories and amplitudes. In particular, the *independent trajectory approximation* (point (*a*)) implies that the amplitudes and phases associated with the classical trajectories are uncorrelated (which is also evident from the fact that the trajectories are propagated separately) and that quantum nonlocality is, therefore, lost. The assumption made in point (*f*) is particularly strong, as it states that the averaged TSH population amplitude (on a given electronic state, j) taken over the ensemble of trajectories can be replaced by the corresponding amplitude, $C_j^{[\alpha]}$, computed along a single trajectory, α. Furthermore, the nuclear amplitudes associated to each electronic state are evaluated strictly at the same position in space, at any time t, even though the different curvature of the PESs involved in the dynamics may drive the nuclear wave packets towards different regions in configuration space. This implies that TSH is strictly local in space and time or, equivalently, that equal-time corresponds to equal-space events, which leads to the loss of quantum mechanical nonlocality [37]. This is the case, even if we allow for retardation (causality), since the TSH equations have no memory. In other words, Equation (12) is obtained from

$$i\hbar \dot{C}_j^{[\alpha]}(t) = \sum_i \int_{t_0}^t dt' F(t - t') C_i^{[\alpha]}(t') \left(E_i^{el}(\boldsymbol{R}^{[\alpha]}) \delta_{ij} - i\hbar \sum_\gamma^{N_n} \boldsymbol{d}_{ji}^\gamma(\boldsymbol{R}^{[\alpha]}) \cdot \dot{\boldsymbol{R}}_\gamma^{[\alpha]} \right) \quad (42)$$

with the kernel, $F(t - t')$, replaced by a delta function, $\delta(t - t')$. Some implication of these approximations will be described in Section 4 for simple one-dimensional model systems.

4. Trajectory Surface Hopping at Work

While TSH is an elegant compromise between accuracy and efficiency for the simulation of nonadiabatic phenomena, its accuracy (either in its fewest-switches version or with additional corrections) has been challenged several times in the literature (see [22,68–72] for an non-exhaustive list). Recently, a series of simple one-dimensional model systems were used to highlight potential failures of the standard TSH, even with high initial momenta [28,73–76]. The "double arch" model is composed of a couple of potential energy curves, whose shapes strongly differ in the region where they are not degenerate ($-10 \leq x \leq 10$ a.u. in Figure 1).

In this model, a Gaussian wave packet launched from $x = -20$ a.u. with a positive initial momentum will first reach a region of strong nonadiabaticity (Figure 2, upper panel), leading to a population of both the ground state (GS) and the first excited state (S_1). Right after this nonadiabatic event occurs, the two potential energy curves will diverge, one exhibiting a strong positive slope (S_1 state), the other a negative one (GS). The wave packet contribution in each electronic state will therefore be spatially split and eventually recombined in a second nonadiabatic region at $x = 10$ a.u. (Figure 2, upper panel). The final population on S_1 after the second nonadiabatic region strongly depends on the spatial decoherence between the nuclear wave packets. However, such peculiar decoherence is hardly captured by TSH, due to the *independent trajectory approximation* (and other approximations discussed in Section 3). This is observed from its deviation with respect to an exact

362

nuclear wave packet propagation in the lower panel of Figure 2. TSH in general fails qualitatively for all different initial momenta tested here, which correspond in all cases to a propagation with no back reflections. Changing the initial conditions of TSH strongly alters the final population of S_1, but does not improve it substantially [75]. On the other hand, the correlated quantum trajectories (NABDY) provide an accurate description of the nuclear wave packet propagation with only minor deviations from the exact propagation (full numerical details can be found in [44]).

Figure 1. The double arch model in the adiabatic representation. The ground state (GS) (S_1) potential energy curve is represented with a red (dashed) line and nonadiabatic coupling with a blue dotted line. The initial nuclear wave packet is displayed in grey.

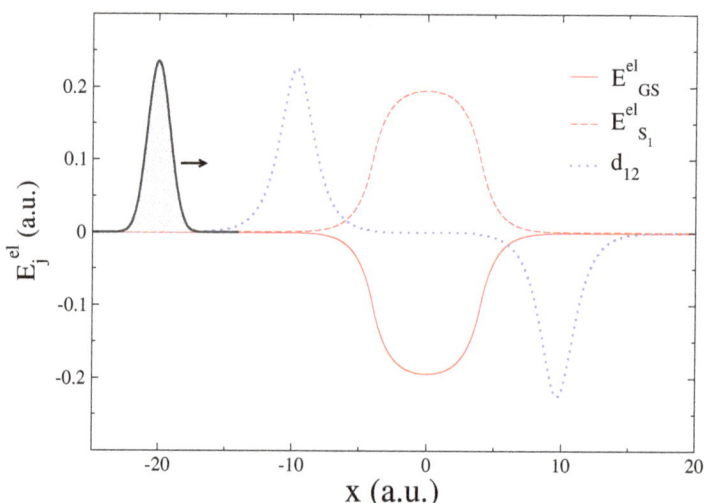

We further investigate the effects of overcoherence on the TSH dynamics by means of a second model system consisting of two coupled harmonic potentials, as depicted in the upper inset of Figure 3. A swarm of trajectories (and a corresponding Gaussian wave packet for the exact propagation) is initialized in the excited state (S_1) at $x = 0$ a.u., with a positive initial momentum $p_0 = 40$ a.u.. In this model system, a single nonadiabatic region is located at $x = 10$ a.u.; the initial conditions are chosen in such a way that the wave packets (and the classical trajectories) will reflect back shortly after the first transition through the nonadiabatic coupling region recrossing, and therefore the same coupling region a second time, with opposite velocity (for a total of two nonadiabatic events, see the lower inset of Figure 3).

The S_1 wave packet enters the strong coupling region at $x = 10$ a.u. (for $t < 1,000$ a.u.) and populates the GS (87%, Figure 3). This first nonadiabatic event is perfectly described by TSH (Figure 3, $1000 \leq t \leq 2000$ a.u.). Due to the difference between the potential energy curves (slope of $E_{S_1}^{el}$ larger than the one of E_{GS}^{el}), the wave packet component in the GS travels further towards positive x values, while the weak contribution in S_1 inverts the direction of its propagation and rapidly returns towards the nonadiabatic region at $x = 10$ a.u.. In the exact propagation, there is no interference with the wave packet evolving in the GS, since the two wave packets (GS and S_1) are spatially separated.

As for the first transition through the nonadiabatic region at $x \sim 10$ a.u., the S_1 wave packet is transferred almost entirely to the other electronic state (now, the GS, Figure 3, $t \geq 3000$ a.u.). On the other hand, in TSH, each independent trajectory carries a set of coherently coupled complex amplitudes (see point (c) of Section 3). When reaching the nonadiabatic coupling at $x = 10$ a.u. for the second time, the complex amplitudes, $C_{GS}^{[\alpha]}(t)$ and $C_{S_1}^{[\alpha]}(t)$, evolved along a given trajectory, α, in S_1 couple coherently because they share the same support (same position in space for any time t). This induces "overcoherence" in dynamics for the amplitudes, which leads to deviations from the exact propagation (Figure 3, $3000 \leq t \leq 5000$ a.u.). The total population in S_1 increases back to $\sim 78\%$ of the $t = 0$ value when the wave packet in the GS recrosses the nonadiabatic region at $t = 6000$ a.u.. Some additional deviations of TSH with respect to the exact propagation are observed, and they are linked to subsequent recrossings of the nuclear wave packets.

Figure 2. Nonadiabatic dynamics for the double arch system. (**Upper panel**) Time series (gray scale) of the nuclear wave packet probability density, $|A_j(x,t)|^2$, and trajectory surface hopping (TSH) histograms for $p_0 = 45$ a.u. (lower panel = GS; upper panel = S_1). The adiabatic potential energy curves are given in red, while the nonadiabatic coupling vectors are shown in blue. (**Lower panel**) Deviation of the final population in S_1 from an exact nuclear wave packet propagation obtained with TSH and nonadiabatic Bohmian dynamics (NABDY), for different initial momenta ("TSH": initial conditions sampled from a Gaussian distribution for positions and momenta, 1,500 trajectories; "TSH*": same initial conditions, momentum and position, for all 1,500 trajectories; "NABDY" is based on a maximum total number of 162 trajectories). The maximum total number of quantum trajectories used in NABDY is 162.

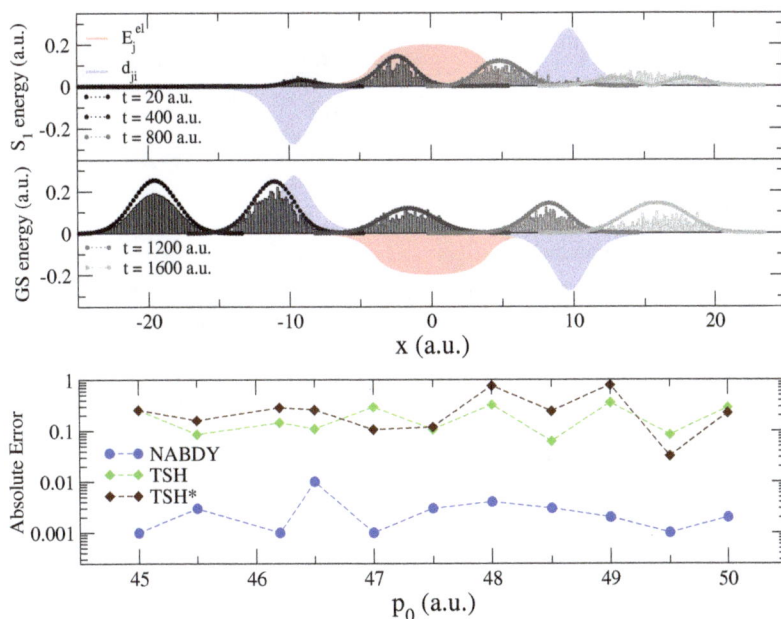

Figure 3. Nonadiabatic dynamics on two coupled harmonic potential energy curves. Population in the first excited state (S_1) along the dynamics for 3,444 TSH trajectories (green) and an exact propagation (red). (**Upper inset**) Schematic representation of the model. The GS (S_1) potential energy curve is represented with a continuous (dashed) black line and the nonadiabatic coupling with a blue dotted line. The initial nuclear wave packet is displayed in grey. (**Lower inset**) Time series of potential energies along a TSH trajectory. The trajectory is initially in S_1, then jumps to the GS after the first coupling and, finally, hops back to S_1 after it reaches back to the coupling region. This representation highlights that the model describes two nonadiabatic events with a single nonadiabatic region.

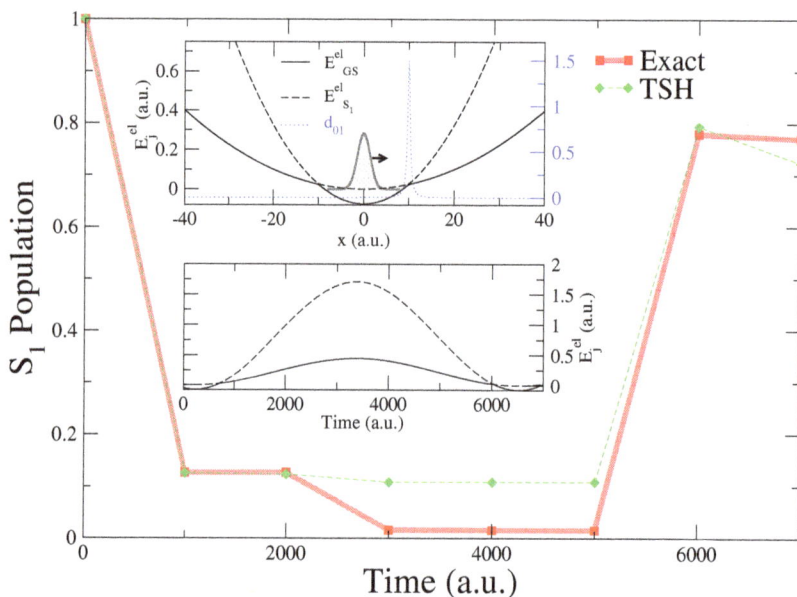

5. Conclusions

The description of the nonadiabatic dynamics of molecular systems is a challenging task for theory, due to the difficulty of providing both the electronic structure of a system *beyond* its electronic ground state and the corresponding nuclear dynamics *beyond* the Born-Oppenheimer approximation. Moreover, nonadiabatic phenomena require a description of the nuclear degrees of freedom that goes *beyond* the classical approximation and, finally a good compromise between accuracy and efficiency is also required when realistic molecular systems are investigated. In this article, we have summarized some of the main techniques for describing the nonadiabatic dynamics of molecular systems, namely Ehrenfest dynamics, nonadiabatic Bohmian dynamics, Multiple Spawning, the recently proposed Exact Factorization, and trajectory surface hopping. We have also shown how the latter method can be rationalized starting from the "exact" nonadiabatic Bohmian dynamics equations. Trajectory surface hopping is indeed one of the most commonly applied on-the-fly trajectory-based methods to describe the dynamics of molecular systems beyond the Born-Oppenheimer approximation in the (unconstrained) configuration space. This is possible at the cost of describing the nuclear wave packet dynamics with a swarm of uncorrelated classical trajectories with the consequent banishing of all quantum (de)coherence effects. Understanding the underlying limitations of trajectory surface hopping is of foremost importance for the future improvement of the theory and, in our opinion, quantum Bohmian dynamics can give valuable contributions in this direction.

Acknowledgments

We are grateful to Jiří Vaníček and Tomàš Zimmermann for providing us a version of their exact propagation code. COSTactions CM0702 and CM1204 and Swiss National Science Foundation grants 200021-137717 and 200021-146396 are acknowledged for funding and support.

Conflicts of Interest

The authors declare no conflicts of interest.

References

1. Born, M.; Oppenheimer, R. Zur quantentheorie der molekeln. *Annalen der Physik* **1927**, *389*, 457–484. (In German)
2. Ballhausen, C.; Hansen, A. Electronic spectra. *Annu. Rev. Phys. Chem.* **1972**, *23*, 15–38.
3. Marx, D.; Hutter, J. *Ab Initio Molecular Dynamics: Basic Theory and Advanced Methods*; Cambridge University Press: Cambridge, UK, 2009.
4. Kapral, R.; Ciccotti, G. Mixed quantum-classical dynamics. *J. Chem. Phys.* **1999**, *110*, 8919–8929.
5. Tully, J.C. Mixed quantum classical dynamics. *Faraday Discuss.* **1998**, *110*, 407–419.
6. Kapral, R. Progress in the theory of mixed quantum-classical dynimics. *Annu. Rev. Phys. Chem.* **2006**, *57*, 129–157.
7. Hohenberg, P.; Kohn, W. Inhomogeneous electron gas. *Phys. Rev. B* **1964**, *136*, B864–B871.
8. Kohn, W.; Sham, L.J. Self-consistent equations including exchange and correlation effects. *Phys. Rev.* **1965**, *140*, A1133–A1138.
9. Runge, E.; Gross, E.K.U. Density-functional theory for time-dependent systems. *Phys. Rev. Lett.* **1984**, *52*, 997–1000.
10. Casida, M.E. Time-dependent density-functional response theory for molecules. In *Recent Advances in Density Functional Methods*; Chong, D.P., Ed.; World Scientific: Singapore, Singapore, 1995; p. 155.
11. Petersilka, M.; Gossmann, U.J.; Gross, E.K.U. Excitation energies from time-dependent density-functional theory. *Phys. Rev. Lett.* **1996**, *76*, 1212–1215.
12. Appel, H.; Gross, E.K.U.; Burke, K. Excitations in time-dependent density-functional theory. *Phys. Rev. Lett.* **2003**, *90*, 043005.

13. Barbatti, M.; Granucci, G.; Ruckenbauer, M.; Plasser, F.; Pittner, J.; Persico, M.; Lischka, H. NEWTON-X, a package for Newtonian dynamics close to the crossing seam, version 1.2. Available online: http://www.newtonx.org (accessed on 20 December 2013).

14. Tully, J.C.; Preston, R.K. Trajectory surface hopping approach to nonadiabatic molecular collisions: The reaction of H^+ with D_2. *J. Chem. Phys.* **1971**, *55*, 562–572.

15. Tully, J.C. Molecular dynamics with electronic transitions. *J. Chem. Phys.* **1990**, *93*, 1061–1071.

16. Nielsen, S.; Kapral, R.; Ciccotti, G. Mixed quantum-classical surface hopping dynamics. *J. Chem. Phys.* **2000**, *112*, 6543–6553.

17. Herman, M.F. Nonadiabatic semiclassical scattering. I. Analysis of generalized surface hopping procedures. *J. Chem. Phys.* **1984**, *81*, 754–763.

18. Sun, X.; Miller, W.H. Semiclassical initial value representation for electronically nonadiabatic molecular dynamics. *J. Chem. Phys.* **1997**, *106*, 6346–6353.

19. Coker, D.; Xiao, L. Methods for molecular dynamics with nonadiabatic transitions. *J. Chem. Phys.* **1995**, *102*, 496–510.

20. Bittner, E.R.; Rossky, P.J. Quantum decoherence in mixed quantum-classical systems: Nonadiabatic processes. *J. Chem. Phys.* **1995**, *103*, 8130–8143.

21. Donoso, A.; Martens, C.C. Simulation of coherent nonadiabatic dynamics using classical trajectories. *J. Phys. Chem. A* **1998**, *102*, 4291–4300.

22. Horenko, I.; Salzmann, C.; Schmidt, B.; Schutte, C. Quantum-classical Liouville approach to molecular dynamics: Surface hopping Gaussian phase-space packets. *J. Chem. Phys.* **2002**, *117*, 11075–11088.

23. Burghardt, I.; Cederbaum, L. Hydrodynamic equations for mixed quantum states. II. Coupled electronic states. *J. Chem. Phys.* **2001**, *115*, 10312–10322.

24. Bonella, S.; Coker, D.F. LAND-map, a linearized approach to nonadiabatic dynamics using the mapping formalism. *J. Chem. Phys.* **2005**, *122*, 194102.

25. McEniry, E.J.; Wang, Y.; Dundas, D.; Todorov, T.N.; Stella, L.; Miranda, R.P.; Fisher, A.J.; Horsfield, A.P.; Race, C.P.; Mason, D.R.; *et al.* Modelling non-adiabatic processes using correlated electron-ion dynamics. *Eur. Phys. J. B* **2010**, *77*, 305–329.

26. Thachuk, M.; Ivanov, M.Y.; Wardlaw, D.M. A semiclassical approach to intense-field above-threshold dissociation in the long wavelength limit. II. Conservation principles and coherence in surface hopping. *J. Chem. Phys.* **1998**, *109*, 5747–5760.

27. Fang, J.Y.; Hammes-Schiffer, S. Improvement of the internal consistency in trajectory surface hopping. *J. Phys. Chem. A* **1999**, *103*, 9399–9407.

28. Subotnik, J.; Shenvi, N. A new approach to decoherence and momentum rescaling in the surface hopping algorithm. *J. Chem. Phys.* **2011**, *134*, 024105.

29. Meyer, H.D.; Manthe, U.; Cederbaum, L.S. The multi-configurational time-dependent hartree approach. *Chem. Phys. Lett.* **1990**, *165*, 73–78.

30. Bohm, D. A suggested interpretation of the quantum theory in terms of "hidden" variables. I. *Phys. Rev.* **1952**, *85*, 166–179.

31. Bohm, D. A suggested interpretation of the quantum theory in terms of "hidden" variables. II. *Phys. Rev.* **1952**, *85*, 180–193.

32. Takabayasi, T. On the formulation of quantum mechanics associated with classical pictures. *Prog. Theor. Phys.* **1952**, *8*, 143–182.

33. Holland, P.R. *The Quantum Theory of Motion—An Account of the de Broglie-Bohm Causal Interpretation of Quantum Mechanics*; Cambridge University Press: Cambridge, UK, 1993.

34. Lopreore, C.L.; Wyatt, R.E. Quantum wave packet dynamics with trajectories. *Phys. Rev. Lett.* **1999**, *82*, 5190–5193.

35. Wyatt, R.E. *Quantum Dynamics with Trajectories: Introduction to Quantum Hydrodynamics*; interdisciplinary applied mathematics; Springer: New York, NY, USA, 2005.

36. Chattaraj, P.K. (Ed.) *Quantum Trajectories*; Atoms, Molecules, and Clusters Series; CRC Press: Boca Raton, FL, USA, 2010.

37. Oriols, X., Mompart, J., Eds. *Applied Bohmian Mechanics, From Nanoscale Systems to Cosmology*; Pan Stanford Publishing Pte. Ltd.: Singapore, Singapore, 2012.

38. Wyatt, R.E.; Lopreore, C.L.; Parlant, G. Electronic transitions with quantum trajectories. *J. Chem. Phys.* **2001**, *114*, 5113–5116.

39. Lopreore, C.L.; Wyatt, R.E. Electronic transitions with quantum trajectories. II. *J. Chem. Phys.* **2002**, *116*, 1228–1238.

40. Gindensperger, E.; Meier, C.; Beswick, J.; Parlant, G. Combining fixed-and moving-grid methods to study direct dissociation processes involving nonadiabatic transitions. *J. Chem. Phys.* **2005**, *123*, 214107.

41. Poirier, B.; Parlant, G. Reconciling semiclassical and Bohmian mechanics: IV. Multisurface dynamics. *J. Phys. Chem. A* **2007**, *111*, 10400–10408.

42. Garashchuk, S.; Rassolov, V.A.; Schatz, G.C. Semiclassical nonadiabatic dynamics using a mixed wave-function representation. *J. Chem. Phys.* **2005**, *123*, 174108.

43. Curchod, B.F.E.; Tavernelli, I.; Rothlisberger, U. Trajectory-based solution of the nonadiabatic quantum dynamics equations: An on-the-fly approach for molecular dynamics simulations. *Phys. Chem. Chem. Phys.* **2011**, *13*, 3231–3236.

44. Curchod, B.F.E.; Tavernelli, I. On trajectory-based nonadiabatic dynamics: Bohmian dynamics versus trajectory surface hopping. *J. Chem. Phys.* **2013**, *138*, 184112.

45. Martínez, T.J.; Ben-Nun, M.; Levine, R.D. Multi-electronic-state molecular dynamics: A wave function approach with applications. *J. Phys. Chem.* **1996**, *100*, 7884–7895.

46. Martínez, T.J.; Ben-Nun, M.; Levine, R.D. Molecular collision dynamics on several electronic states. *J. Phys. Chem. A* **1997**, *101*, 6389–6402.

47. Martínez, T.J.; Levine, R.D. Non-adiabatic molecular dynamics: Split-operator multiple spawning with applications to photodissociation. *J. Chem. Soc. Faraday Trans.* **1997**, *93*, 941–947.

48. Ben-Nun, M.; Martínez, T.J. Nonadiabatic molecular dynamics: Validation of the multiple spawning method for a multidimensional problem. *J. Chem. Phys.* **1998**, *108*, 7244–7257.

49. Born, M. *Kopplung der Elektronen- und Kernbewegung in Molekeln und Kristallen* (in German); Vandenhoeck & Ruprecht: Göttingen, Germany, 1951.

50. Born, M.; Huang, K. *Dynamical Theory of Crystal Lattices*; Clarendon: Oxford, UK, 1954.

51. The term "Born-Oppenheimer approximation" is also used to name what should be referred to as the "adiabatic BO approximation".

52. Ben-Nun, M.; Martínez, T.J. *Ab Initio* Quantum Molecular Dynamics. In *Advances in Chemical Physics*; Volume 121; Wiley: New York, NY, USA, 2002; pp. 439–512.

53. The spawning process is rather involved, and the interested reader should refer to [52] for a very detailed discussion of the algorithm.

54. Ben-Nun, M.; Quenneville, J.; Martínez, T.J. *Ab initio* multiple spawning: Photochemistry from first principles quantum molecular dynamics. *J. Phys. Chem. A* **2000**, *104*, 5161–5175.

55. For an in-depth discussion on Bohmian mechanics and its physical meaning, see [33].

56. Tavernelli, I. *Ab initio*–driven trajectory-based nuclear quantum dynamics in phase space. *Phys. Rev. A* **2013**, *87*, 042501.

57. Tully, J.C. Nonadiabatic Dynamics. In *Modern Methods for Multidimensional Dynamics Computations in Chemistry*; Thompson, D.L., Ed.; World Scientific: Singapore, Singapore, 1998.

58. Tavernelli, I. Electronic density response of liquid water using time-dependent density functional theory. *Phys. Rev. B* **2006**, *73*, 094204.

59. Abedi, A.; Maitra, N.; Gross, E. Exact factorization of the time-dependent electron-nuclear wave function. *Phys. Rev. Lett.* **2010**, *105*, 123002.

60. Abedi, A.; Maitra, N.T.; Gross, E. Correlated electron-nuclear dynamics: Exact factorization of the molecular wavefunction. *J. Chem. Phys.* **2012**, *137*, 22A530.

61. Hunter, G. Conditional probability amplitudes in wave mechanics. *Int. J. Quantum Chem.* **1975**, *9*, 237–242.

62. Alonso, J.L.; Clemente-Gallardo, J.; Echenique-Robba, P.; Jover-Galtier, J.A. Comment on "Correlated electron-nuclear dynamics: Exact factorization of the molecular wavefunction" (J. Chem. Phys.137, 22A530 (2012)). *J. Chem. Phys.* **2013**, *139*, 087101.

63. Abedi, A.; Maitra, N.T.; Gross, E.K.U. Response to: Comment on "Correlated electron-nuclear dynamics: Exact factorization of the molecular wavefunction" (J. Chem. Phys. 139, 087101 (2013)). *J. Chem. Phys.* **2013**, *139*, 087102.

64. Barbatti, M.; Shepard, R.; Lischka, H. Computational and methodological elements for nonadiabatic trajectory dynamics simulations of molecules. In *Conical Intersections: Theory, Computation and Experiment*; Domcke, W., Yarkony, D.R., Koeppel, H., Eds.; World Scientific: Singapore, Singapore, 2011; p. 415.

65. Barbatti, M. Nonadiabatic dynamics with trajectory surface hopping method. *WIREs Comput. Mol. Sci.* **2011**, *1*, 620–633.

66. Curchod, B.F.E.; Rothlisberger, U.; Tavernelli, I. Trajectory-based nonadiabatic dynamics with time-dependent density functional theory. *Chem. Phys. Chem.* **2013**, *14*, 1314–1340.

67. Burant, J.C.; Tully, J.C. Nonadiabatic dynamics via the classical limit Schrödinger equation. *J. Chem. Phys.* **2000**, *112*, 6097.
68. Granucci, G.; Persico, M. Critical appraisal of the fewest switches algorithm for surface hopping. *J. Chem. Phys.* **2007**, *126*, 134114:1–134114:11.
69. Worth, G.A.; Hunt, P.; Robb, M.A. Nonadiabatic dynamics: A comparison of surface hopping direct dynamics with quantum wave packet calculations. *J. Phys. Chem. A* **2003**, *107*, 621–631.
70. Herman, M.F.; Moody, M.P. Numerical study of the accuracy and efficiency of various approaches for Monte Carlo surface hopping calculations. *J. Chem. Phys.* **2005**, *122*, 094104.
71. Granucci, G.; Persico, M.; Zoccante, A. Including quantum decoherence in surface hopping. *J. Chem. Phys.* **2010**, *133*, 134111.
72. Richter, M.; Marquetand, P.; González-Vázquez, J.; Sola, I.; González, L. SHARC: *Ab Initio* molecular dynamics with surface hopping in the adiabatic representation including arbitrary couplings. *J. Chem. Theory Comput.* **2011**, *7*, 1253–1258.
73. Shenvi, N.; Subotnik, J.; Yang, W. Phase-corrected surface hopping: Correcting the phase evolution of the electronic wavefunction. *J. Chem. Phys.* **2011**, *135*, 024101.
74. Subotnik, J.; Shenvi, N. Decoherence and surface hopping: When can averaging over initial conditions help capture the effects of wave packet separation? *J. Chem. Phys.* **2011**, *134*, 244114.
75. Shenvi, N.; Subotnik, J.; Yang, W. Simultaneous-trajectory surface hopping: A parameter-free algorithm for implementing decoherence in nonadiabatic dynamics. *J. Chem. Phys.* **2011**, *134*, 144102.
76. Shenvi, N.; Yang, W. Achieving partial decoherence in surface hopping through phase correction. *J. Chem. Phys.* **2012**, *137*, doi:10.1063/1.4746407.

Reprinted from *Entropy*. Cite as: Arnold, A.; Breitsprecher, K.; Fahrenberger, F.; Kesselheim, S.; Lenz, O.; Holm, C. Efficient Algorithms for Electrostatic Interactions Including Dielectric Contrasts. *Entropy* **2013**, *15*, 4569–4588.

Article

Efficient Algorithms for Electrostatic Interactions Including Dielectric Contrasts

Axel Arnold *, Konrad Breitsprecher, Florian Fahrenberger, Stefan Kesselheim, Olaf Lenz and Christian Holm *

Institute for Computational Physics, University of Stuttgart, Allmandring 3, Stuttgart 70569, Germany; E-Mails: konrad.breitsprecher@icp.uni-stuttgart.de (K.B.); florian.fahrenberger@icp.uni-stuttgart.de (F.F.); kessel@icp.uni-stuttgart.de (S.K.); olenz@icp.uni-stuttgart.de (O.L.)

* Authors to whom correspondence should be addressed;
  E-Mails: arnolda@icp.uni-stuttgart.de (A.A.); holm@icp.uni-stuttgart.de (C.H.);
  Tel.: +49-711-685-63593.

Received: 6 August 2013; in revised form: 15 October 2013 / Accepted: 18 October 2013 / Published: 24 October 2013

Abstract: Coarse-grained models of soft matter are usually combined with implicit solvent models that take the electrostatic polarizability into account via a dielectric background. In biophysical or nanoscale simulations that include water, this constant can vary greatly within the system. Performing molecular dynamics or other simulations that need to compute exact electrostatic interactions between charges in those systems is computationally demanding. We review here several algorithms developed by us that perform exactly this task. For planar dielectric surfaces in partial periodic boundary conditions, the arising image charges can be either treated with the MMM2D algorithm in a very efficient and accurate way or with the electrostatic layer correction term, which enables the user to use his favorite 3D periodic Coulomb solver. Arbitrarily-shaped interfaces can be dealt with using induced surface charges with the induced charge calculation (ICC*) algorithm. Finally, the local electrostatics algorithm, MEMD (Maxwell Equations Molecular Dynamics), even allows one to employ a smoothly varying dielectric constant in the systems. We introduce the concepts of these three algorithms and an extension for the inclusion of boundaries that are to be held fixed at a constant potential (metal conditions). For each method, we present a showcase application to highlight the importance of dielectric interfaces.

Keywords: computer simulation; electrostatics; implicit solvent; dielectric contrast

---

1. Introduction

Electrostatic interactions play an important role in many nano- or meso-scale systems. Almost every surface immersed in water develops a significant surface charge, due to the acid-base reactions of surface groups, and most biomolecules also carry charges. Therefore, it is often indispensable to include these long-ranged interactions in computer simulations. However, the system sizes that can be handled in simulations are orders of magnitude smaller than in real experiments, which drastically enhances the influence of boundary effects. To avoid artifacts due to an artificially small simulation volume, one typically uses periodic boundary conditions (PBC), which in simulations have to be taken into account by special electrostatics algorithms. These are often based on the idea of the Ewald summation [1–4], namely splitting the potential into a smooth, long-ranged and a singular, short-ranged contribution. In modern computer simulations, one usually uses mesh-based variants of the Ewald summation [5–8], which have a favorable $\mathcal{O}(N \log N)$ computing time scaling with respect to the number of charged particles, N.

When studying capacitors, membranes or thin films, one does not want PBC perpendicular to the surfaces of interest. For these systems, partially periodic boundary conditions with only two periodically replicated dimensions are desirable, while the third one has a finite extent h ($2D + h$ geometry). For these geometries, the Ewald summation becomes ineffective [9]. However, it has been shown that replicating the system artificially in the direction perpendicular to the surface results in reasonable accuracy when compared to the non-periodic system, provided a sufficient gap is left between the replicas, and a correction term for the summation order is included [10]. The ELC (electrostatic layer correction) approach [11,12] can, in principal, compute electrostatic interactions with the charges in the unwanted periodic dimension exactly. Practically, it allows one to tune the gap size to a desired accuracy, so that the $2D + h$ geometry is computationally tractable with any Coulomb solver for 3D PBC. It has to be stated that an algorithmic approach for partially periodic systems has been presented in [13,14] using a Monte Carlo extension to the local update scheme sketched in Section 5, but it is not implemented into the molecular dynamics implementation used for this article.

When studying large-scale problems, for example, the buckling of membranes, solutions containing charged polymers or colloids, DNA translocation or crystallization of charged colloids, an atomistic representation of the system under study is often unfeasible, even with periodic boundary conditions. This is related to the vast numbers of charges that would need to be treated, but also with the unfavorable scaling of the relaxation times of the system. One popular route to tackle such problems is to use implicit solvent models, which electrostatically represent the solvent as an effective dielectric medium of the representative dielectric constant at the investigated temperature. This works well if particles, for example, colloids or polymers, are in bulk solution, since, then, the dielectric medium is isotropic and, to a good approximation, homogeneous. However, if surfaces are involved, the dielectric constant of the solvent is usually drastically different from the material forming the surface. For example, water at room temperature has a dielectric constant of $\varepsilon_{rel} \approx 80$, while a typical membrane has a dielectric constant of $\varepsilon_{rel} \approx 2 - 5$.

In systems with spatial variations of the dielectric constant, the Poisson equation for electrostatics reads as:

$$\nabla \cdot (\varepsilon(\mathbf{r})\nabla\Phi) = -\rho \tag{1}$$

with the permittivity $\varepsilon(\mathbf{r}) = \varepsilon_0\varepsilon_{rel}(\mathbf{r})$, the electrostatic potential, Φ, and the charge distribution, ρ. Equation (1) leads to complex boundary conditions at dielectric interfaces, which need to be taken into account by the underlying electrostatics method. Conducting media can, in principle, be treated as the $\varepsilon \to \infty$ limit of the above equation and, thus, allow one to treat this case with very similar methods to those for dielectric contrast. Since these materials appear in important fields, such as energy storage and electrolyte capacitors, our described methods of treating dielectric contrast will also be useful there.

In the following, we will describe how to fulfill the dielectric boundary conditions for planar surfaces using the concept of image charges [15–17]. These approaches can even be extended to the special case of two *connected* conducting surfaces. As an alternative route, we present the ICC* (induced charge calculation) algorithm, which can handle arbitrarily-shaped surfaces [18,19]. Instead of computing image charge interactions, which is only feasible in some simple geometries, the method determines the induced charges at the surfaces self-consistently.

Both image charge and induced charge approaches can only handle boundaries between media of otherwise constant dielectric properties. In this article, we will also describe an extension of the Maxwell Equations Molecular Dynamics (MEMD) algorithm [20,21], which can also handle continuously varying dielectric constants. It solves a simplified version of the Maxwell electrodynamics equations on a discrete lattice.

The methods and results presented in this review were mostly published before in [16–19,22]. Implementations of all methods, including the features discussed here, can be found in ESPResSo, the Extensible Simulation Package for Research on Soft matter [22,23]. A review article on the general topic of long-range interactions in soft matter can be found in [24].

2. Planar Interfaces: Image Charges

We start by discussing the simplest case of dielectric interfaces, namely, two planar, parallel interfaces that enclose a set of charges. We assume a vertical orientation of the interfaces and refer to a left and a right (l/r) interface. The electric field between the two interfaces can be computed from these charges, plus additional image charges outside of the dielectric boundaries [25]. The positions and magnitude of these images charges are chosen to satisfy the boundary conditions for the electric field. If only one interface, the left or the right interface, were present, image charges would appear reflected at the respective interface, with the charge scaled down by a factor of:

$$\Delta_l = \frac{\varepsilon_m - \varepsilon_l}{\varepsilon_m + \varepsilon_l} \quad \text{and} \quad \Delta_r = \frac{\varepsilon_m - \varepsilon_r}{\varepsilon_m + \varepsilon_r} \tag{2}$$

where ε_m is the background permittivity in the main cell and ε_l and ε_r are the permittivities in the adjacent left and right media, respectively. Note that Δ_l and Δ_r can be positive, as well as negative, so that a charge can be attracted to the wall or repelled by it.

To construct the image charges in a situation with two interfaces, every image charge created by reflection on one interface also needs to be reflected again onto the other interface. This leads to an infinite set of images: A charge, q, is reflected at the right (left) interface and yields an image of the magnitude of $q\Delta_r$ (resp. $q\Delta_l$). The next reflection gives rise to another image charge, $q\Delta_l\Delta_r$ (resp. $q\Delta_r\Delta_l$), in the opposite dielectric domain, and so on. The infinite array of image charges is depicted in Figure 1: a charge, q, at position z will produce a series of mirror charges in the right dielectric domain (with ε_r) with charges:

$$q\Delta^{n+1} \text{ at positions } -2(n+1)L_z + z \quad \text{and} \quad q\Delta_r\Delta^n \text{ at positions } -2nL_z - z, \, n \geq 0 \quad (3)$$

where L_z denotes the distance between the two interfaces and $\Delta := \Delta_r\Delta_l$. In the left dielectric domain (with ε_l), the charges are:

$$q\Delta^{n+1} \text{ at positions } 2(n+1)L_z + z \quad \text{and} \quad q\Delta_l\Delta^n \text{ at positions } 2(n+1)L_z - z, \, n \geq 0 \quad (4)$$

Figure 1. Schematic summation scheme for Image Charge MMM2D (ICMMM2D): in order to take into account dielectric boundaries, image charges are introduced outside the dielectric boundaries, to the left and right of the original box. The dielectric contrasts, Δ_l and Δ_r, are computed from the dielectric jump at the left and right boundaries, respectively. Depending on the dielectric contrasts, charges will either be repelled by the surface (as in this sketch) or attracted to it. Note that in usual computer simulations, the system is periodically replicated in the dimensions parallel to the interfaces, as indicated in the figure.

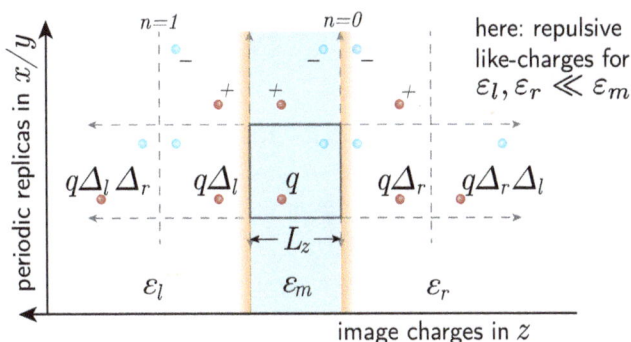

When computing the electrostatic interaction in a computer simulation with such parallel, infinite planar walls, it is often desired to employ periodic boundary conditions in the two directions parallel to the walls to minimize surface effects. The direct summation of periodic replicas is very costly, as the sum is only slowly convergent. Thus, special techniques to compute the electrostatic interactions are required. MMM2D is such an algorithm that computes the electrostatic interaction with two periodic dimensions [26,27] and is well suited for computing the interactions with the image charges. The key idea of the MMM2D algorithm is to use two different formulas for the interaction of two charges. One of the formulas, the near formula, is only used if the two particles are sufficiently close. In combination with image charges, it is only used for the closest of all images, and its discussion is beyond the scope of this article.

The second, far formula is accurate if a certain distance between the two charges is exceeded. We assume a simulation box of $L \times L \times L_z$ that is periodically replicated in the x and y dimensions. Then, the Coulomb potential of a unit charge placed at the origin evaluated at position (x, y, z) with $|z| > 0$, including periodic replicas in the x and y direction, can be written as:

$$\Phi(x, y, z) = \frac{1}{4\pi\varepsilon L^2} \sum_{p,q\neq0} \frac{e^{-2\pi f_{pq}|z|}}{f_{pq}} e^{(2\pi px + 2\pi qy)/L} + \frac{1}{2\varepsilon L^2}|z| \tag{5}$$

where $f_{pq} := \sqrt{p^2 + q^2}/L$. This formula follows from a Fourier transformation of the Poisson equation in the x and y direction and can be factorized into contributions. This makes it possible to compute the interactions between N separated charges in $\mathcal{O}(N)$ operations. Due to the unfavorable scaling of the near formula, MMM2D scales only like $\mathcal{O}(N^{5/3})$ overall. It, however, is still superior to Ewald-based methods [9] in partially periodic boundary conditions.

Coming back to our original problem of taking into account the infinite array of image charges, we note that all these charges, with the exception of those directly adjacent, i.e., with index $n \geq 1$, are far from the slab containing the real charges. Therefore, we can apply the fast far formula in order to compute the interaction with these charges. The infinite sums of image charges lead to geometric series of the form:

$$\sum_{n=0}^{\infty} \Delta^n \frac{1}{4\pi\varepsilon_m L^2} \sum_{p,q\neq0} \frac{e^{-2\pi f_{pq}(2nL_z \pm z)}}{f_{pq}} e^{(2\pi px + 2\pi qy)/L} \tag{6}$$

which is a geometric series that can be easily simplified to:

$$\frac{1}{4\pi\varepsilon_m L^2} \sum_{p,q\neq0} \frac{e^{\pm 2\pi f_{pq}z}}{f_{pq}(1 - \Delta e^{4\pi f_{pq}L_z})} e^{(2\pi px + 2\pi qy)/L} \tag{7}$$

In other words, an existing implementation of MMM2D can be easily enhanced in order to include dielectric interfaces, simply by modifying the prefactors of the p, q-Fourier sum. For the detailed expressions, see [16]. In the following, we denote such an MMM2D implementation for dielectric interfaces as ICMMM2D (Image Charge MMM2D).

Note that Equation (5) contains an additional term, $|z|/(2\varepsilon_m L^2)$, which represents the constant Fourier mode. The summation over the image charges leads to expressions of the form:

$$\frac{1}{2\varepsilon_m L^2} \sum_{n\geq0} \Delta^n (2nL_z \pm z) \tag{8}$$

since $2nL_z$ is larger than any possible particle distance z. Unlike Equation (6), this is not a simple geometric series. However, when computing the total potential in a charge neutral system, terms that do not depend on the positions cancel out, in particular, the $2nL_z$ terms. What remains from the four image charge sums are again geometric series that lead to a contribution of:

$$\frac{1}{2\varepsilon_m L^2}\frac{\Delta_r - \Delta_l}{1-\Delta}z \qquad (9)$$

If the two planar walls have the same dielectric contrast, this contribution to the potential vanishes. Furthermore, if one is only interested in energy or forces, this contribution vanishes, due to charge neutrality.

An alternative method for planar dielectric interfaces is based on an extension of our ELC method. It also uses the technique to sum up the image charges with the ICMM2D far formula, and we termed it ELCIC (ELC with Image Charges), see [17] for details of the method.

Note that it is also possible to consider systems that are not charge neutral. Formally, one assumes two equally charged plates at the positions of the dielectric interfaces that cancel the total charge [28]. The field of a charged plate is, however, exactly what the $|z|/(2\varepsilon_m L^2)$ term represents, so that the above considerations for this term still hold. Therefore, one can safely ignore this contribution.

Figure 2. (a) Sketch of our simulation setup, a 3:1 electrolyte, e.g., $AlCl_3$, between two walls with a dielectric constant different from that of water. The size difference between the ion types is neglected in this simulation. The slab is periodically replicated parallel to the walls, vertical in the sketch. (b) Density distribution of anions and cations of the trivalent electrolyte, near the dielectric interface. The dielectric interfaces is placed at $x = 0$, and a repulsive potential maintains a minimum distance of 0.5 nanometers for all ions. A good dielectric $\varepsilon_C = 800$, representing conducting material, strongly attracts cations, while anions are less attracted. A bad dielectric $\varepsilon_C = 2$, representing a typical biological membrane, repels cations. The univalent anions are much less repelled.

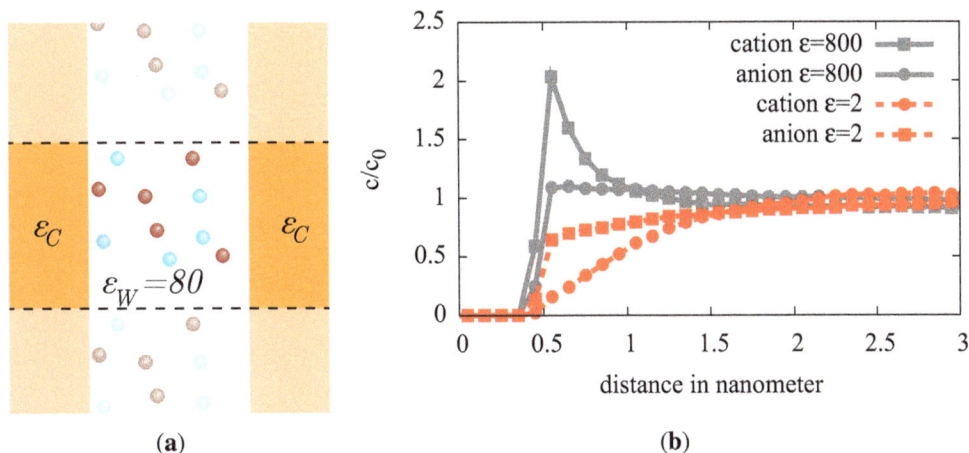

2.1. Example: Electrolyte between Dielectric Walls

As an example application of the ICMMM2D algorithm, we simulated a 3:1 electrolyte (e.g., $AlCl_3$) with a concentration of 0.01 mol/l confined by planar walls to a slab, as depicted in Figure 2a. The size difference between positive and negative ions has been neglected here. This is a good model for the narrow slit between the electrodes of a capacitor, as well as two biological membranes or glass plates. However, the dielectric properties of a metal electrode (here approximated with $\varepsilon_C \approx 800$) and a biological membrane ($\varepsilon_C \approx 2$) are very different and in both cases also differ strongly from the permittivity of the solvent; here, water ($\varepsilon_W = 80$). The resulting dielectric jumps at the surfaces have a pronounced effect on the distribution of ions near the walls.

Figure 2b shows this strong influence of the dielectric interfaces. Both cations and anions are attracted to the walls of high permittivity ($\varepsilon_C = 800$), but repelled by the low permittivity ($\varepsilon_C = 2$) walls. On a microscopic level, the electric field of a charge will give rise to a dielectric displacement in the wall. This displacement will weaken or pronounce the field within the dielectric medium, compared to the other side of the interface. It can be accounted for by imagining virtual mirror charges or surface charges directly on the interface. To correctly reproduce the field discontinuities at the boundary, these charges will be of an attractive nature in the region of lower permittivity and of a repulsive nature in the region of higher permittivity. Note also that the effect is more pronounced for multivalent ions. Ignoring the dielectric jumps would lead to a much more homogeneous charge distribution, so that one would strongly underestimate the effect of including multivalent ions.

3. Arbitrarily-Shaped Interfaces: Induced Charges

The concept of induced charges rather than image charges is a direct route to take into account dielectric interfaces of arbitrary shape. Conventionally, charge induction is considered to be the origin of Faraday's cage effect: applying an electric field to a conductor will trigger the mobile charges inside to move until the electric field vanishes inside and field lines end orthogonally to the surface of induced charges. The same concept, however, can also be applied to dielectrics, hence materials with immobile charges. This can be seen from the following mathematical consideration. The Poisson equation in an inhomogeneous dielectric medium (1) can be rewritten as:

$$\Delta\Phi = -\frac{1}{\varepsilon}\rho - \frac{1}{\varepsilon}\nabla\varepsilon \cdot \nabla\Phi \tag{10}$$

The term, $\nabla\varepsilon \cdot \nabla\Phi$, is identified as the *induced surface charge density*, σ. It is nonzero only on the dielectric interfaces, since $\nabla\varepsilon$ vanishes everywhere else.

Let us introduce the Green's function, G, for the Laplace operator. It can, e.g., be just $\frac{1}{4\pi|r-r'|}$, but may also include the desired periodicity. Then, it is possible to eliminate the Laplace operator from the equation above and express the potential by means of two integrals:

$$\Phi = \int_V G\left(r, r'\right)\rho\left(r'\right)/\varepsilon\,\mathrm{d}V' + \int_A G\left(r, r'\right)\sigma\left(r'\right)/\varepsilon\,\mathrm{d}A' \tag{11}$$

The volume integral extends over medium 1, and the surface integral extends over all dielectric interfaces. The potential is now expressed in terms of the Green's function of a homogeneous dielectric, yet the induced charge density, σ, is still unknown. We now assume that the charges are embedded in a medium with permittivity, ε_1, and for simplicity, only a second permittivity, ε_2. By taking the gradient and inserting this expression in the definition of the induced charge density, we obtain the following integral equation:

$$\sigma = 2\frac{\varepsilon_1 - \varepsilon_2}{\varepsilon_1 + \varepsilon_2}\left(\int_V \nabla_r G\left(r,r'\right)\rho\left(r'\right)/\varepsilon_1 dV' + \int_A \nabla_r G\left(r,r'\right)\sigma\left(r'\right)/\varepsilon_1 dA'\right) \tag{12}$$

This result is easily generalized to multiple regions with different permittivities The idea of the ICC* algorithm [18] is to determine this charge density self-consistently, after discretizing the surface.

In principle, this approach is a boundary element method, an approach that is very widely used, e.g., for a low Reynold number flow [29]. Different from other approaches is, however, that the efficient evaluation of the Green's function can be borrowed from standard Coulomb solvers. This can be seen from the following: assuming a discretized surface of m point charges on the dielectric interface, the equation above for discretization point k can be written as:

$$q_k = A_k\frac{\varepsilon_1 - \varepsilon_2}{\varepsilon_1 + \varepsilon_2}n_k \cdot \left[\sum_{i=1}^n q_i \nabla_{r_k} G\left(r_k, r_i\right) + \sum_{j=1,j\neq k}^m q_j \nabla_{r_k} G\left(r_k, r_j\right)\right]$$

where A_k is the surface area of the surface element, k. The term in square brackets is just the electric field acting at the position of point k, assuming a homogeneous dielectric constant, ε_1, in the system, created by conventional (not induced) charges. Any standard Coulomb solver can thus be used to perform the calculation. The desired solution of all q_k is the fix point of the following iteration:

$$q_k^{l+1} = (1-\omega)q_k^l + \omega A_k\frac{\varepsilon_1 - \varepsilon_2}{\varepsilon_1 + \varepsilon_2}n_k \cdot E\left([q_i],[q_j^l]\right)$$

It turned out that this iteration is very stable and with a choice of $\omega \approx 0.7$, no stability issues occur. In every MD step, only 1–3 iterations are necessary, as the particle positions change only slightly.

An important advantage of this algorithm is that the computationally most costly part, the evaluation of the electric field, can be done with *any* usual electrostatics solver without modifications. Thus, not only the computational efficiency, but also the periodicity is inherited from the Coulomb solver. The complexity of the algorithm remains unchanged by the presence of induced surface charges. However, the number of particles can increase considerably. We found it sufficient to discretize the surface with mutual particle distances equal to the distance of closest approach. For the system shown in Section 2, this would mean, in total, 1,600 surface charges per wall, compared to less than 100 ions in the system.

In our research, this algorithm was applied to investigate if dielectric effects can change the electrolytic conductance of very narrow pores, nanopores, through membranes or have an influence on the translocation of charged macromolecules through nanopores [19,30]. Here, dielectric boundary forces lead to a repulsion of (unpaired) ions. Taking into account dielectric effects in small pores will decrease the number of available ions and, thus, decrease the conductance.

The error in the obtained electric field depends on the applied resolution with which the surface is resolved. The method has been tested for planar and curved surfaces [18], and it was found that from a distance larger than one lattice spacing, the relative error remains smaller than 1%. Since the permissible error depends often on the desired application, we advise to determine the necessary accuracy specifically for each case. If interfaces with media with a high dielectric constant or even metallic boundaries are considered, charges are attracted to the surface and can come quite close to the interface, depending on the ion size. In this case, the necessary accuracy is clearly higher than for interfaces with lower dielectric media, from which particles are repelled.

3.1. Example: Ion Distribution in a Pore

As an example, we investigate the ion distribution in a cylindrical pore of radius 5 nm in a 40 nm-thick membrane in an aqueous electrolyte. This geometry resembles the so-called solid state nanopores [31–33] in silicon wafers. In several experiments (e.g., [34,35]), it was shown that single DNA molecules present in such a pore can be detected by the change of the electric conductance of such a system. To make the dielectric effect more pronounced, we again used a 3:1 electrolyte with a concentration of 10 mmol/L. Our setup is sketched in Figure 3a: the surface charges of the ICC* algorithm are displayed along with the ions, each as spheres. We assume the dielectric constant of the membrane to be $\varepsilon_P = 2$. In Figure 3b, the equilibrium distribution of ions near the center of the pore is shown. Ions, especially the trivalent ones, are repelled from the dielectric interface. This leads to an overall decrease of ions in the pore by around 20%. Thus, the conductance of the system can be expected to be reduced similarly, compared to a model that does not consider the dielectric contrast.

Figure 3. (a) The induced charge calculation (ICC*) example system: positive and negative ions are displayed as red and blue spheres, the ICC* discretization points by grey spheres; (b) ion density of both species in the pore measured close to the center of the pore.

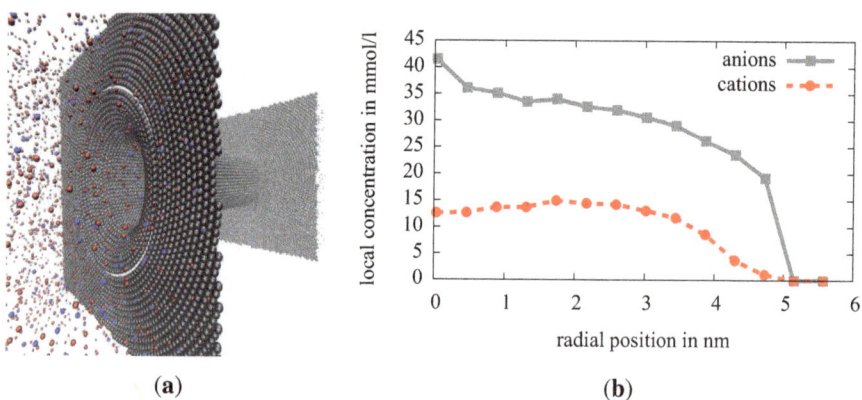

(a)   (b)

4. Metallic Interfaces: Corrections

Metallic boundary conditions are the $\varepsilon \to \infty$ limit of Equations (2) or (11), where ε denotes the dielectric constant of the surrounding medium. The corresponding dielectric contrasts become -1, so that the field automatically vanishes in the conductor. In the case of two dielectric interfaces, this simply leads to constant potentials on each of the two interfaces, but the potentials are not necessarily the same. It is, however, also possible to fix the electrostatic potential on surfaces in periodic systems. This has been done, for example, for the Monte Carlo implementation of the local algorithm presented in Section 5 [13,14]. The fixing of the surface potential in periodic systems can be achieved also for other algorithms, but the details are different, and we therefore briefly describe the necessary ingredients for the algorithms previously described.

Figure 4. (**a**) Illustration of the dielectric boundary problem of single charge q outside of a grounded conducting sphere. The problem can be solved by assuming an image charge, q', inside the sphere, leading to zero potential Φ on its surface. If the sphere is assumed to be conducting, but isolated, the excess charge, q', has to be canceled by adding a second charge, q'', in the center of the sphere, which leads to a constant surface potential, Φ. (adapted from [25]). (**b**) A more complex geometry with an upper and lower electrode (yellow). The electrodes are treated with the ICC* algorithm. If a Coulomb solver with periodic boundary conditions (BCs) in the vertical direction is applied, the potential difference between both electrodes is automatically zero. This is because the periodicity yields zero potential difference between an electrode and its periodic image, and the ICC* algorithm ensures that the two electrodes connected over periodic BCs are on equal potential.

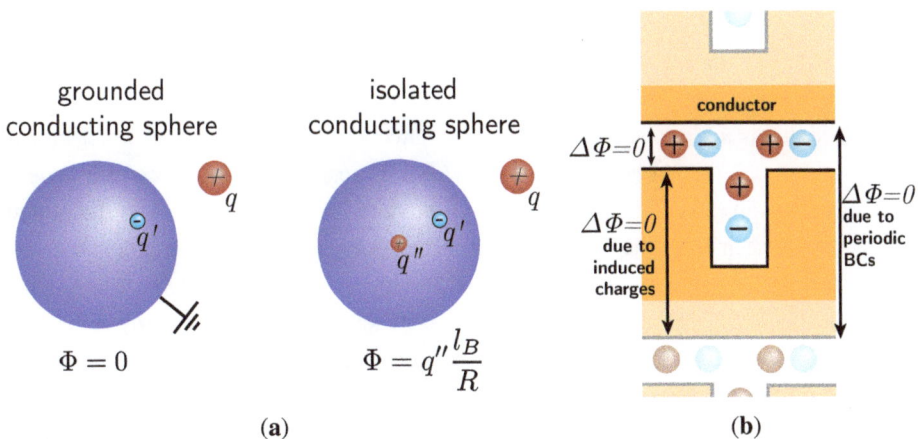

(**a**) (**b**)

The starting point is the textbook example of a single point charge outside of a conducting sphere, as depicted in Figure 4a. A metallic sphere brought into an electric field can either be isolated, *i.e.*, the charge on the sphere is constant, or on a constant electrostatic potential, typically grounded. Again, the boundary problem can be solved by adding an image charge opposite to the source charge in the sphere. This ensures that the surface potential of the sphere does not vary. A second image charge can be placed at the center, which dictates the electrostatic potential at the surface of the sphere: for an isolated sphere with zero net charge, the image charge at the center must be of the same magnitude as the other image charge, but with the opposite sign. For a grounded sphere, it is simply zero. In the following, we will show how conducting or grounded boundary conditions can be added to both the image charge and induced charge methods, in order to perform computer simulations with constant electrostatic metallic surface potentials.

Using the MMM2D far formula, the potential difference between the two plates appears as the $p = q = 0$ mode of the electrostatic potential, i.e., the $2\pi l_b|z|/L^2$ term in Equation (5). The higher modes, in the limit of $\varepsilon \to \infty$, serve to hold the potential constant on each conducting plate individually. Due to the absolute value, this term does not cancel when considering the interactions in the primary box. The potential difference, V, of the two plates is:

$$V = \frac{1}{2\varepsilon L^2} \int_0^L \int_0^L \int_0^{L_z} \rho\left(x, y, z\right) \underbrace{\left(|L_z - z| - |z|\right)}_{L_z - 2z} dz\, dy\, dx$$

where ρ denotes the charge density. In terms of the z-component of the dipole moment of the system, $P_z = \sum_i q_i z_i$, this equals:

$$V = -\frac{1}{\varepsilon L^2} P_z$$

since the L_z contribution vanishes once more due to charge neutrality. In order to cancel this potential difference, one has to apply a constant external field $\boldsymbol{E} = (0, 0, \frac{1}{\varepsilon L^2 L_z} P_z)$ in every MD step. This additional field corresponds to that created by the central charge in the spherical image charge picture, which puts the surfaces to equal potential. To obtain a different fixed potential $\Delta\Phi_0$, an additional field, $\Delta\Phi_0/L_z$, needs to be applied in the z direction.

Systems treated with ICC* require no special measures if the surface on the constant potential is connected within the simulation box. However, some attention is required when considering electrodes at the boundary of the simulation box, as depicted in Figure 4b. If an electrostatics method is applied that is not periodic in the respective direction, this will result in two electrically unconnected surfaces. To obtain electrically connected surfaces, such as two grounded plates, it is sufficient to use a solver periodic in the required direction and to leave a gap in the periodic images. This can be seen from the simple case depicted in Figure 4b. If a periodic solver with metallic boundary conditions is used, for example, the Ewald summation, the difference between the electrostatic potential at a given position and its nearest periodic image is necessarily zero, due to periodicity. However, since the induced charges create a constant potential in both sections of the conductor, these must be the same throughout the whole, periodically connected conductor.

To obtain a non-zero electrostatic surface potential, the solution of the Poisson equation with zero potential can be superimposed with a solution of the empty simulation box with nonzero potential on the surfaces. This requires a solution of the Laplace equation that is then applied as an external field, just as in the simple case of parallel plates. To do that, our simulation package, ESPResSo, supports reading in tabulated external potentials, which are applied to charged particles weighted with the according charge. It also takes care that the external potential is not applied to image charges. The solution of the Laplace equation has to be obtained externally. We use a finite element solution based on the DUNE framework [36,37]. Packages, like Matlab or Comsol, can, of course, also be applied.

4.1. Verification: Field and Potential in Metallic BCs

In order to illustrate the methods described above, we show its importance on a simple model system. We chose a planar geometry, because in this geometry, it is possible to use both the image charge method, as well as the induced charge method. Thus, our system consists of a set of charges confined by two parallel conducting planes. In the following, we will show that isolated plates can be simulated either by using the ICMMM2D algorithm without correction or by using the ICC* method with a Coulomb solver, which is not periodic in the direction of the planes' normal vectors. Connected plates at zero potential difference can either be simulated using the ICMMM2D method with the correction derived above or using the ICC* method with a Coulomb solver, which is periodic in the normal direction. As a solver for the fully periodic case, we use the P^{3}M algorithm [8,38]; as a solver for the partially periodic case, MMM2D [26,27].

Figure 5. (a) Sketch of the model system that was used to probe the influence of grounded and isolated metallic boundary conditions. The two possible setups are depicted by adding a switch to electrically connect the two plates. (b) The resulting electrostatic field, E, perpendicular to the plates and the electrostatic potential in the slab.

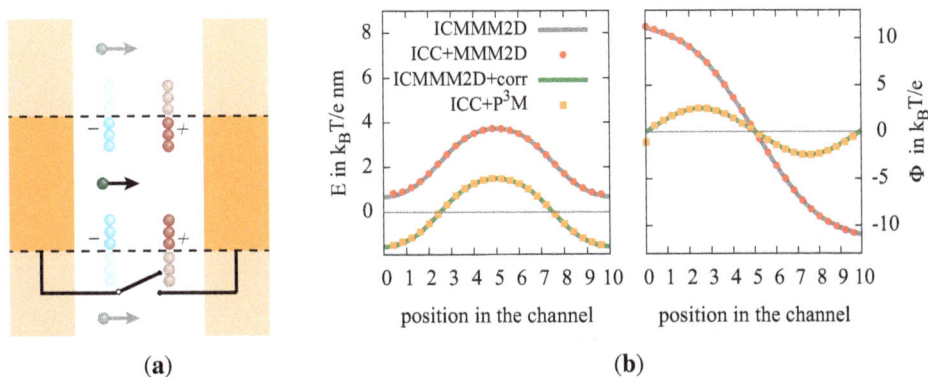

(a) (b)

For simplicity, we construct a constant charge distribution with a net dipole moment and probe the electrostatic field with a small test charge $q = 10^{-9}e$ that is moved through the system. Metallic boundary conditions are created at $z = 0$ and $z = L_z = 10$ nm by using the four algorithms described above. The dielectric permittivity between the surrounding metal plates is assumed to be $\varepsilon_W = 80$ for bulk water. The charge distribution we chose is depicted in Figure 5a: fixed, charged particles form two oppositely charged plates at $z = 0.25L_z$ and $0.75L_z$. In one of the two directions parallel to the surrounding metal plates, a void of a length of 0.5 L_x is left, so that the test charge can be moved through the gaps without getting close to the charged plates. We measure the electric field by performing a force calculation with the respective algorithm and dividing the obtained force by the small value of the test charge. In Figure 5b, we report the measured electric field and the potential obtained from the integration of the electric field. We observe the expected behavior: the shape of the electric field is identical in all cases up to a constant. In the cases where the the algorithms are supposed to simulate two connected metal electrodes, the electric field is shifted downwards, so that the integral is zero, and both electrodes have the same potential.

These two methods are, in our opinion, very interesting for investigating supercapacitors based on electrolytes or ionic liquids, (e.g., [39–41]). They are complimentary in the sense that the image charge method is computationally very cheap: the extra cost of image charges is typically negligible, and the only model parameter is the distance of closest approach between ions and the metallic interface. The induced charge methods, however, allow for arbitrarily-shaped surfaces, and one can investigate very complex geometries. The extra computational cost is feasible if the resolution of the surface can be relatively coarse or, in other words, a certain electrostatic "roughness" or inaccuracy is acceptable and can be considered as an adjustable model parameter.

5. Smooth Variations: Local Method

We have presented methods to deal with sharp dielectric interfaces of several types and shapes. Yet, none of these methods offer the possibility of a spatially smooth varying permittivity. More precisely, they are all restricted to single step-like changes and do not allow charges to pass through those regions of variation.

In the following, we sketch an algorithm that allows for an arbitrary distribution, $\varepsilon(r)$, of the permittivity that we call Maxwell Equations Molecular Dynamics (MEMD). The concept of diffusive field propagation was first presented by Maggs in 2002 [20,42] and adapted for molecular dynamics simulations simultaneously by Rottler and Maggs [43] and Dünweg and Pasichnyk [21]. The algorithm is not based on the static Poisson Equation (1), which is of a global nature. Instead, the time derivative of Gauss' law $\nabla D = \rho$ of electrodynamics, with $D = \varepsilon(r)E$, is extended to the following constraint of the most general form.

$$\dot{D} + j - \nabla \times \dot{\Theta} = 0 \tag{13}$$

where j denotes the local electric current and Θ is an arbitrary vector field representing an additional degree of freedom. If we apply this constraint to the system propagation via a Lagrange multiplier, A, fix the gauge degree of freedom from Equation (13) to:

$$\dot{A} = -D, \quad \text{define} \tag{14}$$
$$B := \nabla \times A \tag{15}$$

and introduce the magnetic field, B, then this so-called temporal (or Weyl) gauge will, via variational calculus, lead to the equations of motion for the charges and fields that are known as the Maxwell equations. The actual electrostatic potential, Φ, is never calculated in this algorithm, only the electric field, E, for Lorentz force calculation:

$$F_L = q\left(E + \frac{1}{c^2}v \times B\right) \tag{16}$$

It is remarkable that simply by applying constraint (13) and the Weyl gauge, (14), the complete equations of the electromagnetic formalism can be reproduced. It can even be shown [20] that the propagation speed of the magnetic field, an equivalent of the speed of light, c, can be reduced in a Car-Parrinello manner, and correct retarded solutions for statistic observables can still be maintained.

This reduces the elliptic partial differential Equation (1) to a set of hyperbolic differential equations for the propagation of magnetic fields and charges, requiring only local operations for the solution. It therefore opens up the possibility of arbitrary local dielectric permittivities within the system. If discretized on a lattice and coupled with a linear next neighbor interpolation scheme for the charges and electric currents, the permittivity can be set individually on every lattice link [44]. The discretization is carried out as seen in Figure 6a. This is in agreement with $\varepsilon(r)$ being a differential one-form, if we assume the tensor to only have identical diagonal entries (optically isotropic dielectric medium).

In this algorithm, the charges can move freely through a smoothly varying dielectric medium. At the current implementation state, the variations are only spatial, but temporal changes of the dielectric during the simulation are theoretically possible. It has also been shown that the field propagation within the system reproduces the classical Keesom potential interactions between two dielectric interfaces [45].

Figure 6. Maxwell Equations Molecular Dynamics (MEMD) interpolation of the charges onto the lattice. (**a**) The electric current, j, permittivity ε and electric field D are interpolated to the adjacent lattice links. The magnetic field component, B, is placed on the lattice plaquettes via a finite-differences curl $(\nabla \times)$ operator. (**b**) The numerical error of the algorithm is dependent on the lattice spacing. The error can be reduced by applying a coarser mesh, coming from the right in this graph, and, thereby, increasing the field propagation speed in the system. However, at large lattice spacings a, the interpolation error at small distances dominates and diverges. In a densely-populated system, like the examples seen here, a minimal relative error of 10^{-3} is achieved at mesh sizes comparable to the minimum distance of two charges, denoted here by σ. For reference, the errors compared to a high precision P3M force calculation are included for three example systems.

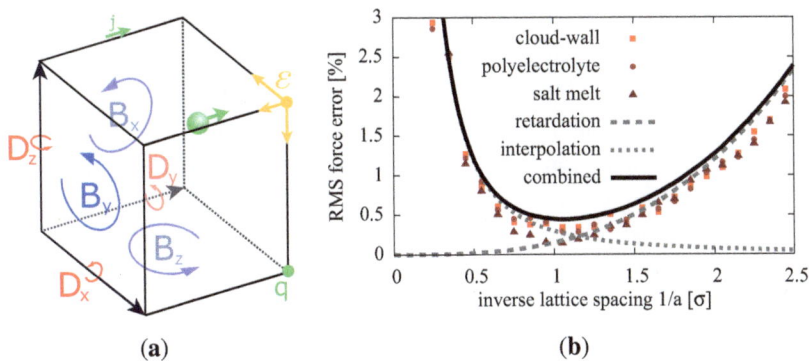

(a)   (b)

One source of numerical errors in this algorithm stems from the retarded solution of the Maxwell equations with the Lorentz force calculation (16) at the finite speed of light $c \ll \infty$. The second error is introduced via the linear lattice interpolation of the charges. The self-interaction of a charge with the lattice can be corrected, for example, using a lattice Greens function for constant dielectric backgrounds, but for locally varying dielectric properties, the implementation relies on a straight-forward direct subtraction scheme. Here, the local electric field created by the interpolated charges of one particle is calculated and subtracted from the resulting force. Still, an interpolation error remains for the coupling between two particles at close distances, since the charge is spread out on the lattice.

Since the field propagation speed within the system is proportional to the lattice spacing, a, the first of these errors depends on $(1/a^2)$ (see Equation (16)), whereas the second error for geometrical reasons scales like a^3. It can be seen in Figure 6b that the error can be reduced by making the mesh coarser, until the interpolation error at close distances dominates for a very coarse mesh. The plot shows a theoretical estimate of the error and simulation data for three example systems: an artificial system with a charged infinite plate (cloud-wall), a polyelectrolyte in aqueous solution and a silica salt melt. The second parameter, besides the lattice spacing, the artificial speed of light, c, was chosen to obey the Courant-Friedrichs-Lewy stability condition, $c \ll a/dt$, where dt is the time step of the MD simulation. For every lattice spacing, there exists a maximum speed of light parameter that keeps the algorithm stable. Here, we picked the speed of light $c = 0.1 \cdot a/dt$ as the parameter for all three setups. It turns out that a relative force error of 10^{-3} is achievable in sufficiently homogeneous systems, and the optimal lattice spacing is comparable to the minimal distance between charges. The interpolation error, and, therefore, the total error, can be reduced by splitting off a near field that spans across multiple mesh cells and applying a short-range calculation in this region. However, this is not possible for spatially varying permittivity and will not be discussed here.

5.1. Example: Colloid with Dielectric Jump and Continuous Dielectric Constant

An example where such smoothly varying changes in ε can play a substantial role is the simulation of a charged colloid in a salt solution (Figure 7a). The first approach at dielectric coarse graining is a sharp dielectric contrast between the colloid and the solvent. Such a system can be simulated by the ICC* method presented in Section 3 with a dielectric permittivity $\varepsilon_C = 2$ within the colloid and $\varepsilon_W = 80$ for the surrounding bulk water. However, in practice, the polarizability, and, therefore, bulk permittivity, of water close to charged surfaces and in regions of high ion concentrations is significantly reduced [46]. Many workarounds were proposed to address this behavior, including the introduction of an artificial Stern layer [47] to reproduce the desired Gouy-Chapman predictions. A more direct and more physical approach is to interpolate the bulk permittivity from the colloid to free water between the colloid surface and the solution (see the bottom of Figure 7b). A linear interpolation is sufficient to describe the local permittivity obtained from atomistic simulations [48].

In order to illustrate the difference between dielectric jump and continuous dielectric constant, we simulated a spherical setup with two different models of radial permittivity dependence $\varepsilon(|r|)$.

Model 1 includes a discontinuity of ε at the colloid surface, whereas model 2 uses a linear interpolation $\varepsilon(|r|) = \varepsilon_C + (\varepsilon_W - \varepsilon_C)/(4\sigma) \cdot (r - R)$ over four ion radii $\sigma = 0.425\,\mathrm{nm}$. Model 1 additionally has been simulated using the ICC* algorithm combined with the P³M Coulomb solver, for comparison. Figure 7b shows the radial charge density of the counterions around a colloid of charge $Z = 60$ with radius $R_c = 30\sigma = 12.75\,\mathrm{nm}$ in a monovalent electrolyte solvent of $c = 50\,\mathrm{mmol/L}$ concentration. The difference between the two models is drastic and can be explained by the stronger Coulomb attraction of counterions towards the colloid, due to the smaller dielectric permittivity close to the colloid surface. Another effect is the earlier occurrence of overcharging effects, because of increased ion correlations, due to their entering the region of lower permittivity. The comparison with the ICC* algorithm, on the other hand, shows that MEMD is also very well capable of simulating dielectric jumps, provided a sufficiently small mesh, comparable to the particle size, for the discretization of the electrodynamic equations can be realized. The computational overhead, compared to a simulation with MEMD at constant background permittivity, is negligible at less than 0.1% in this setup. MEMD ran 41% longer than the identical setup for the ICC* algorithm. This overhead could be reduced by further optimizing the mesh size, which was chosen to be well resolved here at a colloid diameter of 16 mesh spacings.

Figure 7. (a) A charged colloid (charge $Z = 60e$, radius $R = 12.75\,\mathrm{nm}$) is suspended in a salt solution (concentration $c = 50\,\mathrm{mmol/L}$). The dielectric constant is modeled as an abrupt jump to the bulk water permittivity with the MEMD and ICCP³M algorithms (gray points, green curve) or a linear radial increase within two ion diameters from the colloid surface between the two regimes (red). (b) The resulting radial charge density profiles, which exhibit a drastic difference between the dielectric jump in model 1 and a smooth interpolation in model 2.

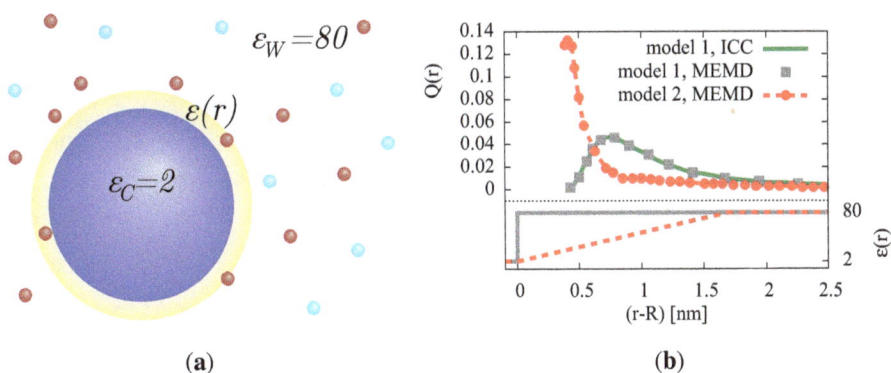

(a)   (b)

6. Conclusions

In this review, we have presented several methods to compute electrostatic interactions in the presence of dielectric interfaces in computer simulations and give examples that demonstrate the importance of taking the dielectric contrasts into account, like they appear in implicit solvent models.

In a planar slit pore, such as a plate capacitor, ions are attracted to the walls or are repelled by them, depending on whether the walls consist of an dielectric medium with high or low polarizability. The presented ICMMM2D [16] or the ELCIC [17] algorithms allow one to treat a slit pore with dielectric jumps at both confining interfaces, like capacitors or thin films.

In curved or more complex geometries, such as a nanopore, dielectric properties can also drastically alter the ion distribution or the translocation properties of charged macromolecules. The presented ICC* algorithm allows one to include arbitrarily-shaped dielectric interfaces in computer simulations, by computing the induced surface charges necessary to fulfill the dielectric boundary conditions on the fly.

As another interesting application, we presented the example of plates that are held electrically at a constant potential, such as one would use to study capacitors. This will lead to a different polarization than isolated grounded plates, and this effect has to be taken into account in simulations, with algorithms that are adjusted accordingly. Both the ICMMM2D and the ICC* algorithm can handle this special case, as we have demonstrated.

Finally, we have shown that smoothly varying dielectric properties again are very different from dielectric jumps as treated by ICMMM2D or ICC*. At present, only the sketched MEMD electrostatics algorithm [20,21,44] is able to treat such systems.

All algorithms presented here, including the extensions for varying dielectric constants, are implemented in the open source simulation package, ESPResSo [22,23]. MEMD is also part of the open source ScaFaCoS (Scalable Fast Coulomb Solvers) library for fast Coulomb solvers [49]. In conclusion, computational tools for the most common cases of varying dielectric properties are readily available for computer simulations. This will become more and more important with the increased interest of applying coarse-grained or implicit solvent models in nanopores, biological membranes, thin films and supercapacitors.

Acknowledgments

We would like to thank Anthony Maggs and Zhenli Xu for fruitful discussions.

This work was financially supported by the German Science Foundation (DFG) through the Collaborative Research Center (SFB) 716 and the funding programme Open Access Publishing. Furthermore funds were provided by the German Ministry of Science and Education (BMBF) under grant 01IH08001 and the cluster of excellence SimTech at the University of Stuttgart.

Conflicts of Interest

The authors declare no conflict of interest.

References

1. Ewald, P.P. Die Berechnung optischer und elektrostatischer Gitterpotentiale. *Ann. Phys.* **1921**, *369*, 253–287.

2. Heyes, D.M. Electrostatic potentials and fields in infinite point charge lattices. *J. Chem. Phys.* **1981**, *74*, 1924–1929.

3. De Leeuw, S.W.; Perram, J.W.; Smith, E.R. Simulation of electrostatic systems in periodic boundary conditions. I. Lattice sums and dielectric constants. *Proc. R. Soc. Lond. A* **1980**, *373*, 27–56.

4. De Leeuw, S.W.; Perram, J.W.; Smith, E.R. Simulation of electrostatic systems in periodic boundary conditions. II. Equivalence of boundary conditions. *Proc. R. Soc. Lond. A* **1980**, *373*, 57–66.

5. Hockney, R.W.; Eastwood, J.W. *Computer Simulation Using Particles*; IOP: London, UK, 1988.

6. Darden, T.; York, D.; Pedersen, L. Particle Mesh Ewald: An $N \log(N)$ method for Ewald sums in large systems. *J. Chem. Phys.* **1993**, *98*, 10089–10092.

7. Essmann, U.; Perera, L.; Berkowitz, M.L.; Darden, T.; Lee, H.; Pedersen, L. A smooth Particle Mesh Ewald method. *J. Chem. Phys.* **1995**, *103*, 8577–8593.

8. Deserno, M.; Holm, C. How to mesh up Ewald sums. I. A theoretical and numerical comparison of various particle mesh routines. *J. Chem. Phys.* **1998**, *109*, 7678–7693.

9. Widmann, A.H.; Adolf, D.B. A comparison of Ewald summation techniques for planar surfaces. *Comput. Phys. Commun.* **1997**, *107*, 167–186.

10. Yeh, I.C.; Berkowitz, M.L. Ewald summation for systems with slab geometry. *J. Chem. Phys.* **1999**, *111*, 3155–3162.

11. Arnold, A.; de Joannis, J.; Holm, C. Electrostatics in periodic slab geometries I. *J. Chem. Phys.* **2002**, *117*, 2496–2502.

12. De Joannis, J.; Arnold, A.; Holm, C. Electrostatics in periodic slab geometries II. *J. Chem. Phys.* **2002**, *117*, 2503–2512.

13. Levrel, L.; Maggs, A.C. Boundary conditions in local electrostatics algorithms. *J. Chem. Phys.* **2008**, *128*, 214103.

14. Thompson, D.; Rottler, J. Local monte carlo for electrostatics in anisotropic and nonperiodic geometries. *J. Chem. Phys.* **2008**, *128*, 214102 .

15. Smith, E.R. Electrostatic potentials for thin layers. *Mol. Phys.* **1988**, *65*, 1089–1104.

16. Tyagi, S.; Arnold, A.; Holm, C. ICMMM2D: An accurate method to include planar dielectric interfaces via image charge summation. *J. Chem. Phys.* **2007**, *127*, 154723.

17. Tyagi, S.; Arnold, A.; Holm, C. Electrostatic layer correction with image charges: A linear scaling method to treat slab 2D + h systems with dielectric interfaces. *J. Chem. Phys.* **2008**, *129*, 204102.

18. Tyagi, C.; Süzen, M.; Sega, M.; Barbosa, M.; Kantorovich, S.; Holm, C. An iterative, fast, linear-scaling method for computing induced charges on arbitrary dielectric boundaries. *J. Chem. Phys.* **2010**, *132*, 154112.

19. Kesselheim, S.; Sega, M.; Holm, C. Applying ICC* to DNA translocation. Effect of dielectric boundaries. *Comput. Phys. Commun.* **2011**, *182*, 33–35.

20. Maggs, A.C.; Rosseto, V. Local simulation algorithms for coulombic interactions. *Phys. Rev. Lett.* **2002**, *88*, 196402.

21. Pasichnyk, I.; Dünweg, B. Coulomb interactions via local dynamics: A molecular-dynamics algorithm. *J. Phys. Condens. Matter* **2004**, *16*, 3999–4020.

22. Arnold, A.; Lenz, O.; Kesselheim, S.; Weeber, R.; Fahrenberger, F.; Roehm, D.; Košovan, P.; Holm, C. ESPResSo 3.1—Molecular Dynamics Software for Coarse-Grained Models. In *Meshfree Methods for Partial Differential Equations VI*; Griebel, M., Schweitzer, M.A., Eds.; Springer: Berlin, Germany, 2013; Volume 89, *Lecture Notes in Computational Science and Engineering*, pp. 1–23.

23. Limbach, H.J.; Arnold, A.; Mann, B.A.; Holm, C. ESPResSo—An extensible simulation package for research on soft matter systems. *Comput. Phys. Commun.* **2006**, *174*, 704–727.

24. Arnold, A.; Holm, C. Efficient Methods to Compute Long Range Interactions for Soft Matter Systems. In *Advanced Computer Simulation Approaches for Soft Matter Sciences II*; Holm, C., Kremer, K., Eds.; Springer: Berlin, Germany, 2005; Volume II, *Advances in Polymer Sciences*, pp. 59–109.

25. Jackson, J.D. *Classical Electrodynamics*, 3rd ed.; Wiley: New York, NY, USA, 1999.

26. Arnold, A.; Holm, C. MMM2D: A fast and accurate summation method for electrostatic interactions in 2D slab geometries. *Comput. Phys. Commun.* **2002**, *148*, 327–348.

27. Arnold, A.; Holm, C. A novel method for calculating electrostatic interactions in 2D periodic slab geometries. *Chem. Phys. Lett.* **2002**, *354*, 324–330.

28. Ballenegger, V.; Arnold, A.; Cerda, J.J. Simulations of non-neutral slab systems with long-range electrostatic interactions in two-dimensional periodic boundary conditions. *J. Chem. Phys.* **2009**, *131*, 094107.

29. Katsikadelis, J.T. *Boundary Elements: Theory and Applications: Theory and Applications*; Elsevier: Oxford, UK, 2002.

30. Kesselheim, S.; Sega, M.; Holm, C. Effects of dielectric mismatch and chain flexibility on the translocation barriers of charged macromolecules through solid state nanopores. *Soft Matter* **2012**, *8*, 9480–9486.

31. Dekker, N.H.; Smeets, R.M.M.; Keyser, U.F.; Krapf, D.; Wu, M.Y.; Dekker, C. Salt dependence of ion transport and DNA translocation through solid-state nanopores. *Nano Lett.* **2006**, *6*, 89–95.

32. Dekker, C. Solid-state nanopores. *Nat. Nanotechnol.* **2007**, *2*, 209–215.

33. Siwy, Z.; Kosinska, I.D.; Fluinski, A.; Martin, C.R. Asymmetric diffusion through synthetic nanopores. *Phys. Rev. Lett.* **2005**, *94*, 048102.

34. Keyser, U.; van der Does, J.; Dekker, C.; Dekker, N. Optical tweezers for force measurements on DNA in nanopores. *Rev. Sci. Instrum.* **2006**, *77*, 105105.

35. Wanunu, M.; Morrison, W.; Rabin, Y.; Grosberg, A.; Meller, A. Electrostatic focusing of unlabelled DNA into nanoscale pores using a salt gradient. *Nat. Nanotechnol.* **2009**, *5*, 160–165.

36. Bastian, P.; Blatt, M.; Dedner, A.; Engwer, C.; Klöfkorn, R.; Kuttanikkad, S.; Ohlberger, M.; Sander, O. The Distributed and Unified Numerics Environment (DUNE). In Proceedings of the 19th Symposium on Simulation Technique, Hannover, Germany, 12–14 September 2006.

37. Bastian, P.; Heimann, F.; Marnach, S. Generic implementation of finite element methods in the Distributed and Unified Numerics Environment (DUNE). *Kybernetika* **2010**, *2*, 294–315.

38. Deserno, M.; Holm, C. How to mesh up Ewald sums. II. An accurate error estimate for the Particle-Particle-Particle-Mesh algorithm. *J. Chem. Phys.* **1998**, *109*, 7694.

39. Merlet, C.; Salanne, M.; Rotenberg, B. New coarse-grained models of imidazolium ionic liquids for bulk and interfacial molecular simulations. *J. Phys. Chem. C* **2012**, *116*, 7687–7693.

40. Kondrat, S.; Kornyshev, A. Superionic state in double-layer capacitors with nanoporous electrodes. *J. Phys. Condens. Matter* **2011**, *23*, 022201.

41. Feng, G.; Zhang, J.; Qiao, R. Microstructure and capacitance of the electrical double layers at the interface of ionic liquids and planar electrodes. *J. Phys. Chem. C* **2009**, *113*, 4549–4559.

42. Maggs, A.C. Auxilary field Monte Carlo for charged particles. *J. Chem. Phys.* **2004**, *120*, 3108–3118.

43. Rottler, J.; Maggs, A.C. Local molecular dynamics with Coulombic interactions. *Phys. Rev. Lett.* **2004**, *93*, 170201.

44. Fahrenberger, F.; Holm, C. Computing Coulomb Interaction in Inhomogeneous Dielectric Media via a Local Electrostatics Lattice Algorithm. 2013, arXiv:1309.7859. arXiv.org e-Print archive. Available online: http://arxiv.org/abs/1309.7859 (accessed on 14 October 2013).

45. Pasichnyk, I.; Everaers, R.; Maggs, A.C. Simulating van der Waals interactions in water/hydrocarbon-based complex fluids. *J. Phys. Chem. B* **2008**, *112*, 1761–1764.

46. Bonthuis, D.J.; Gekle, S.; Netz, R.R. Dielectric profile of interfacial water and its effect on double-layer capacitance. *Phys. Rev. Lett.* **2011**, *107*, 166102.

47. Drift, W.P.J.T.V.D.; Keizer, A.D.; Overbeek, J.T.G. Electrophoretic mobility of a cylinder with high surface charge density. *J. Colloid Interface Sci.* **1979**, *71*, 67–78.

48. Bonthuis, D.J.; Gekle, S.; Netz, R.R. Profile of the static permittivity tensor of water at interfaces: Consequences for capacitance, hydration interaction and ion adsorption. *Langmuir* **2012**, *28*, 7679–7694.

49. Arnold, A.; Bolten, M.; Dachsel, H.; Fahrenberger, F.; Gähler, F.; Halver, R.; Heber, F.; Hofmann, M.; Holm, C.; Iseringhausen, J.; *et al.* Comparison of scalable fast methods for long-range interactions. *Phys. Rev. E* **2013**, submitted.

Reprinted from *Entropy*. Cite as: Santiso, E.E.; Herdes, C.; Müller, E.A. On the Calculation of Solid-Fluid Contact Angles from Molecular Dynamics. *Entropy* **2013**, *15*, 3734–3745.

Article

On the Calculation of Solid-Fluid Contact Angles from Molecular Dynamics

Erik E. Santiso [1,2], **Carmelo Herdes** [1] **and Erich A. Müller** [1,*]

[1] Department of Chemical Engineering, Imperial College London, South Kensington Campus, London, SW7 2AZ, UK; E-Mail: eesantis@ncsu.edu (E.E.S.); c.herdes@imperial.ac.uk (C.H.)
[2] Department of Chemical and Biomolecular Engineering, North Carolina State University, Raleigh, NC 27695, USA

* Author to whom correspondence should be addressed; E-Mail: e.muller@imperial.ac.uk; Tel.: +44-207-594-1569.

Received: 21 July 2013; in revised form: 2 September 2013 / Accepted: 3 September 2013 / Published: 6 September 2013

Abstract: A methodology for the determination of the solid-fluid contact angle, to be employed within molecular dynamics (MD) simulations, is developed and systematically applied. The calculation of the contact angle of a fluid drop on a given surface, averaged over an equilibrated MD trajectory, is divided in three main steps: (i) the determination of the fluid molecules that constitute the interface, (ii) the treatment of the interfacial molecules as a point cloud data set to define a geometric surface, using surface meshing techniques to compute the surface normals from the mesh, (iii) the collection and averaging of the interface normals collected from the post-processing of the MD trajectory. The average vector thus found is used to calculate the Cassie contact angle (*i.e.*, the arccosine of the averaged normal *z*-component). As an example we explore the effect of the size of a drop of water on the observed solid-fluid contact angle. A single coarse-grained bead representing two water molecules and parameterized using the SAFT-γ Mie equation of state (EoS) is employed, meanwhile the solid surfaces are mimicked using integrated potentials. The contact angle is seen to be a strong function of the system size for small nano-droplets. The thermodynamic limit, corresponding to the infinite size (macroscopic) drop is only truly recovered when using an excess of half a million water coarse-grained beads and/or a drop radius of over 26 nm.

Keywords: cloud data set; interfacial tension; coarse-graining; water; line tension; graphene

1. Introduction

Wetting is the ability of a liquid to maintain contact with a solid surface, resulting from intermolecular interactions when the two are brought together. The presence of a liquid drop on a rigid surface is a reflection of the force balance between adhesive and cohesive forces and is commonly used to determine the wettability (the degree of wetting) of the solid-fluid system in terms of the solid-fluid contact angle, θ (see Figure 1). In this context, hydrophobicity is commonly referred to as the ability of a solid surface to repel water: if the water contact angle is smaller than 90°, the solid surface is considered hydrophilic and if the water contact angle is larger than 90°, the solid surface is considered hydrophobic.

Figure 1. Schematic of a liquid drop on a solid surface showing the contact angle.

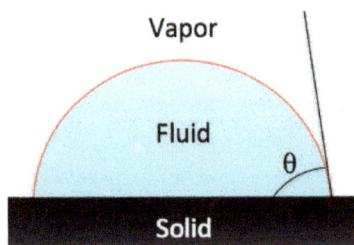

Despite the fact of being such a well-defined problem the amount of conflicting (both theoretical and experimental) reported values for a given system is intriguing (see Figure 2, data from reference [1]).

Figure 2. Frequency of contact angle values of water on graphite reported in literature; both from experimental results and numerical simulations [1].

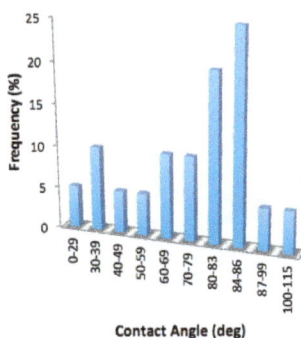

For instance, in the case of the graphite-water system, contact angles have been addressed extensively by experimental and theoretical approaches; however, a single general value has not been accepted [2–9].

A variety of causes for the discrepancies can be enumerated: heterogeneity and/or impurities at the surfaces or in the fluids, different methodological unstandardized calibration of equipment in experiments, possible system size effects and the distinct interaction potentials in simulations.

At the molecular scale, the main hurdle is that estimating contact angles for nanodroplets on surfaces is complicated by the fact that there are significant fluctuations in the shape of the droplet, and its geometry at a given step is often not axially symmetric. Furthermore, for very small nanoclusters, the fluid interfacial tension is a function of the curvature, and the planar limit is, in some cases, not recovered even after drop radii of 14 times the molecular diameter [10,11]. The change in the line tension with curvature (discussed in the latter part of this manuscript) is also an important factor affecting the result. It is seen that the contact angle will be, in the nanoscopic limit, a strong function of the system size [2]. In the analysis of simulations, contact angles [12,13] are commonly determined by using two-dimensional slices of the droplet and fitting its density profile to an empirical function, usually a circular section [14]. Such an approach, although appealing from the simplicity of the method, provides inconsistent results, particularly for small droplets [11]. Understandingly, different methods of increasing complexity have been devised for this purpose [15–17].

Figure 3 illustrates the difficulties associated with defining the contact angle using two-dimensional slices of molecular snapshots, especially for small droplets. Such droplets are often asymmetrical, and their shape changes substantially with time, due to capillary fluctuations. Using one or a few two-dimensional projections of a few molecular snapshots can thus potentially lead to large errors on the measured contact angle: for example, the values shown on Figure 3 vary over range of about 10°.

Figure 3. Two-dimensional projections of a given configuration of a water droplet on a surface. The contact angles, measured using the auxiliary lines depicted in black, are: (a) 63.77° (b) 60.52° (c) 64.56° (d) 54.93°.

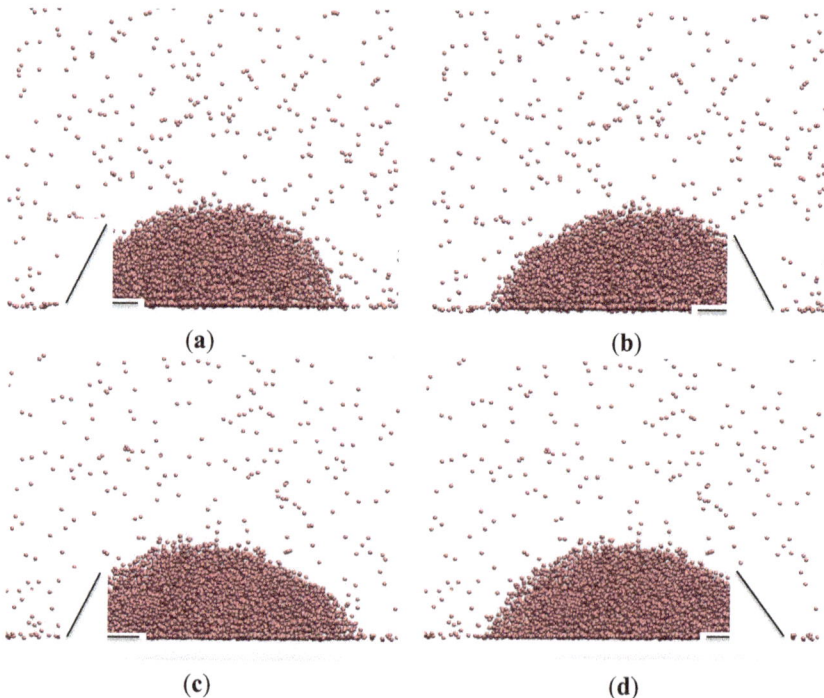

In addition to the inherent fluctuations of the drop's shape, there is a degree of arbitrariness in defining a single line separating the two phases in the two-dimensional projection. The interface layer is actually diffuse, as seen in Figure 3, spanning a width of several molecular diameters. This introduces another potential source of error, as the contact angle measured can change over a range of 5° or more depending on the choice of the contact line.

In this paper, we propose a method that does not make any *a priori* assumption about the shape of the drop, and uses the complete three-dimensional structure of the droplet near the surface to estimate the contact angle. For such analysis, we propose a geometrical estimation of the contact angle based on cloud point data breakdown of an equilibrated molecular dynamics (MD) trajectory of a drop on a given surface, this eliminates the effect of both spatial and temporal fluctuations on the estimated value of contact angle, yielding a value that better represents the average shape of the drop, as explained in detail in the next section.

2. Methodology

In order to estimate the average contact angle during an MD simulation, we first define a contact layer at each time step. This layer is defined as the set of molecules within the liquid-vapor interface that are close to (*i.e.*, within a maximum distance z_{max} from) the solid surface. We then estimate the normal to the interface for each molecule in the contact layer by finding the plane that best fits the local shape of the interface at that point. The average of all the local normal vectors for all the time steps, $\langle \mathbf{n} \rangle$, can be used to calculate the average contact angle as $\theta = \cos^{-1}\left(\langle \mathbf{n} \rangle \cdot \mathbf{n}_{surf}\right)$, where $\mathbf{n}_{surf}$ is the normal to the solid surface.

In practice, at each time step in the simulation, we carry out a two-step calculation: In the first step we use a discretized density profile to identify the molecules belonging to the liquid-vapor interface. In the second step, we estimate the local surface normal vectors for the molecules in the contact layer. The procedure is explained in detail below.

2.1. Identification of Interfacial Molecules

We identify the molecules belonging to the liquid-vapor interface using the following procedure:

(1) We divide the simulation box into subcells, and calculate the local number density in each subcell.
(2) We mark subcells as liquid if their number density is greater than a cutoff value ρ_c, otherwise we mark them as vapor cells. The cutoff density is chosen close to the average between the liquid and vapor number densities at the conditions of interest.
(3) Every subcell that is adjacent to at least one liquid cell and one vapor cell is marked as an interface cell.
(4) Finally, all molecules contained within the above-determined interface cells are marked as interfacial molecules.

The results obtained using the interface-sensing procedure are only weakly dependent on the choice of cutoff density, as long as this density is between the liquid and vapor densities. The number of subcells, on the other hand, should be chosen carefully: having more subcells identifies the interface with a higher resolution. However, if the number of subcells is too large, the density fluctuations within the droplet may cause the algorithm to incorrectly label too many cells as part of the interface. This can be easily detected by visualizing the interface molecules at one or a few simulation steps. Alternatively, one can use the average coordination number or the Minkowski-Bouligand dimension (obtained from a box-counting algorithm) [18,19] as a quantitative measure to select the appropriate subcell size. As the cells become smaller and the interface-sensing algorithm starts fitting the molecular-level fluctuations in the density rather than the actual interface, both the fractal dimension and the average coordination number will start deviating from the value expected for the larger subcells, indicating that the subcell size has become too small.

Figure 4. (a) A snapshot from a MD simulation showing a droplet on top of a given surface (not shown). **(b)** The discretized density profile from the same system obtained using cubic subcells of width 3 nm. Density values are in molecules/Å [3].

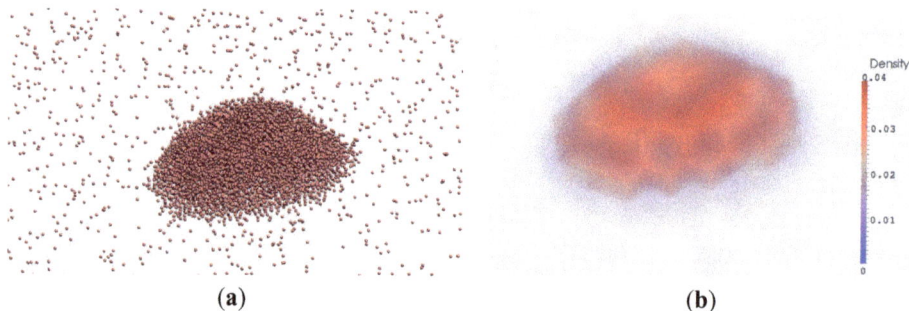

(a)   (b)

2.2. *Estimation of Local Contact Angles*

Having identified the molecules belonging to the interface, we choose a subset of the interface molecules as the interface contact layer. This layer contains the interface molecules that are within a given distance z_{max} from the solid surface. We then estimate the local contact angle at the position of each molecule i using the following procedure:

(a). Find all the interface molecules within a given cutoff radius, r_c, of molecule i.
(b). Find the average position $\mathbf{r}_{avg}$ of all the interface molecules found in step (a), including molecule i.
(c). Subtract the average position from the position of all the neighboring molecules, including i.
(d). Construct the covariance matrix Ω of the centered positions, $\mathbf{r}_j^{cent}$ found in step (c):

$$\Omega = \sum_j \mathbf{r}_j^{cent} \otimes \mathbf{r}_j^{cent}$$

(1)

where $\otimes$ denotes the outer (Kronecker) product.
(e). Find the eigenvector of Ω corresponding to its smallest eigenvalue. This is the normal to the plane that best fits the set of molecules in step (c). The sign of this normal is chosen to point away from the center of the droplet [20].

We average all the local normal vectors obtained at all the time steps to obtain the average normal. The component of this average normal perpendicular to the solid surface is the cosine of the average contact angle. The eigenvalues of the covariance matrix measure the variability of the positions along the three orthogonal directions defined by the eigenvectors. These can be used as a measure of how 1-, 2- or 3-dimensional the interface is. If one of the eigenvectors is close to zero, the interface is close to planar in that region.

The procedure described above to estimate the local contact angle requires arbitrarily setting a distance from the solid surface, z_{max}, to define the contact layer, and a cutoff radius r_c to estimate the tangent plane. In theory the contact angle would correspond to the limit when $z_{max} \to 0$. However, if z_{max} is chosen to be too small, the local density fluctuations within the contact layer cause the distribution of local contact angle values to be too noisy. Thus, one should choose the smallest value that gives a reasonable estimation error for the average contact angle. For the examples shown in this work, we have found $z_{max} = 0.5$ nm to give reasonable results.

For the cutoff radius, r_c, a compromise is also needed: if the value is too small, local density fluctuations cause the error in the estimated normal to be too large, whereas if the value is too large the orientation of the tangent plane will be affected by molecules far from the position of interest. The choice of r_c should be made by considerations similar to those used in choosing z_{max}. For the examples in this work, we have found $r_c = 1$ nm to be a reasonable value. A more detailed discussion of how to choose this cutoff can be found in reference [21]. Finally, we estimate the statistical error associated with the average contact angle computed by our method, we use the bootstrapping method [22].

2.3. Molecular Dynamics Details

Atomistic MD simulations are generally limited to nanometer-sized water droplets [2] Consequently, the apparent contact angle is usually drop-size dependent. To explore bigger systems, and aiming to find the optimal size for the MD contact angle calculation, we adopted a coarse-grained (CG) approach to describe the solid-water system in order to test the system size dependence beyond the atomistic limits, in a reasonable time [2].

We used the GROMACS simulation open source [23] suite to calculate the MD, which is well suited to implement Mie potentials [24]. Here, a single CG isotropic bead represents two water molecules [25]. Although several options are available for choosing the number of water molecules in a CG bead [26], this choice and the current parameterization produces sensible results, including a melting point, the surface tension, liquid densities and vapour pressures close to the experimental. The parameterization was carried out using SAFT-γ Mie approach, [27–29] where the water parameters were obtained by fitting to macroscopic properties, namely, the planar limit interfacial tension and liquid state density of water in a range from 0 °C to 40 °C. The SAFT EoS is a perturbation approach based on a well-defined Hamiltonian; here the CG beads are represented in the theory by a Mie potential, u:

$$u(r) = \left(\frac{\lambda_r}{\lambda_r - \lambda_a}\right)\left(\frac{\lambda_r}{\lambda_a}\right)^{\frac{\lambda_a}{\lambda_r - \lambda_a}} \varepsilon \left[\left(\frac{\sigma}{r}\right)^{\lambda_r} - \left(\frac{\sigma}{r}\right)^{\lambda_a}\right] = Ar^{-\lambda_r} - Cr^{-\lambda_a} \qquad (2)$$

where r is the intermolecular distance, and ε, and σ, are the adjustable parameters relating to the energy and distance scales, λ_a is the dispersion exponent and λ_r is the short-range repulsion. The potential is expressed in terms of two constants A and C for ease of tabulating in MD codes. Solid walls are modeled implicitly and described by an integrated potential of the same form in the z-dimension. Eight implicit solid surfaces of increasing $\varepsilon_{Wall\#\text{-}W}/k_B$ increments of 10 K, from 60 K to 130 K (labeled Wall01 to Wall08 respectively) are employed [30]. Table 1 summarizes the selected coarse-grained parameters.

Table 1. Coarse-grained parameters. W refers to a CG representation of water that accounts for two water molecules.

Interaction	σ [nm]	ε/k_B [K]	λ_r, λ_a	$C \times 10^2$ [kJ mol^{-1} nm$^{\lambda_a}$]	$A \times 10^4$ [kJ mol^{-1} nm$^{\lambda_r}$]
$\varepsilon_{Wall01\text{-}W}/k_B$		60		3.44107	1.15888
$\varepsilon_{Wall02\text{-}W}/k_B$		70		4.01458	1.35203
$\varepsilon_{Wall03\text{-}W}/k_B$		80		4.58810	1.54518
$\varepsilon_{Wall04\text{-}W}/k_B$	0.38716	90	10, 4	5.16161	1.73832
$\varepsilon_{Wall05\text{-}W}/k_B$		100		5.73512	1.93147
$\varepsilon_{Wall06\text{-}W}/k_B$		110		6.30863	2.12462
$\varepsilon_{Wall07\text{-}W}/k_B$		120		6.88214	2.31776
$\varepsilon_{Wall08\text{-}W}/k_B$		130		7.45565	2.51091
$\varepsilon_{W\text{-}W}/k_B$	0.37459	399.96	8, 6	8.71139	1.222380×10^2

The systems are run under a canonical (NVT) ensemble, where the total volume, concentration and temperature are kept constant. Periodic boundary conditions were applied in xy dimensions, meanwhile an attractive wall was placed at z = 0 and a repulsive one ($C = 5.73512 \times 10^{-4}$ kJ mol^{-1} nm^4, $A = 1.93147 \times 10^{-6}$ kJ mol^{-1} nm^{10} at the maximum height of the box. The number density of the atoms for each wall was set in 5 nm^{-2} (c.f. in a Si crystal the number of atoms per nm^2 on the (100), (111) and (110) planes are 6.78, 7.83 and 9.59 nm^{-2}, respectively [31]). The simulations are thermostated to 298.15 K every 1 ps by a Nose-Hoover algorithm, all non-bonded interactions were truncated at 2.0 nm. The trajectories were recorded every 1000 time-steps ($\Delta t = 0.01$ ps) for at least 2000 ps after equilibrium.

3. Results and Discussions

We simulate 16,000, 32,000, 64,000, 128,000, 256,000, 512,000 and 1,024,000 water molecules (8,000, 16,000, 32,000, 64,000, 128,000 and 512,000 beads) on the eight solid surfaces. The simulation boxes dimensions can be seen in Table 2.

Although arbitrary, the reason guiding the choice of the simulation box size is to prevent the interaction between the sample and its periodic images. The meshing was done by dividing the simulation domain with a $1.2 \times 1.2 \times 1.2$ nm^3 subcell. The water droplet density contour is obtained from the cloud point data set analysis explained in the methodology section.

To test the size dependence of the water contact angle on a given surface we chose the intermediate Wall05 (see Table 1). The contact angles obtained using a moderately hydrophilic substrate (Wall05) as a function of the drop size can be seen in the Figure 5. The water contact angle over Wall05 exhibits a marked system-size dependence even up to 256,000 water molecules (128,000 beads) Larger drops exhibit a less pronounced but nevertheless noticeable size dependence. It is interesting to note that the effect of this scale up is an increase in the hydrophobicity of this surface in correspondence with previous simulations [32] and experimental studies at the macroscale. For this substrate, a system with 1,024,000 water molecules shows an apparent limiting contact angle of 74.30° [8]. The Wall05 parameters are chosen to loosely relate to graphene, a hydrophilic substrate.

Table 2. Simulation box size for each system.

Water molecules	xy-dimensions [nm]	z-dimension [nm]
16,000		
32,000	60 × 60	18
64,000		
128,000		
256,000	80 × 80	37.2
512,000		
1,024,000	144 × 144	48

Figure 5. (a) Water contact angle as a function of the water molecules on Wall05, inset shows the correspondent drop diameter, dashed lines are guide to the eye. **(b)** Snapshots of the drop interfaces for the smallest and the biggest system studied.

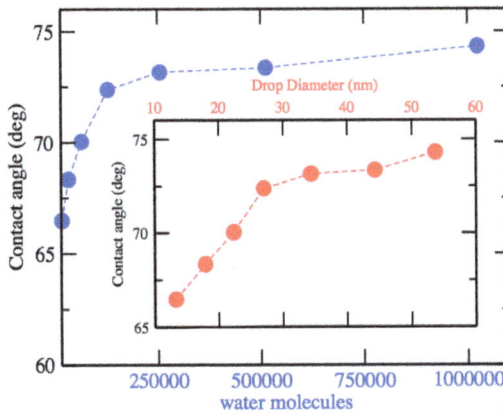

(a)

(b)

Figure 6. (a) Water contact angle as a function of the fluid-substrate interactions, solid circles are simulation results, dashed red line marks the hydrophobic-hydrophilic threshold. **(b)** Corresponding equilibrium interface snapshots depicting interfacial beads.

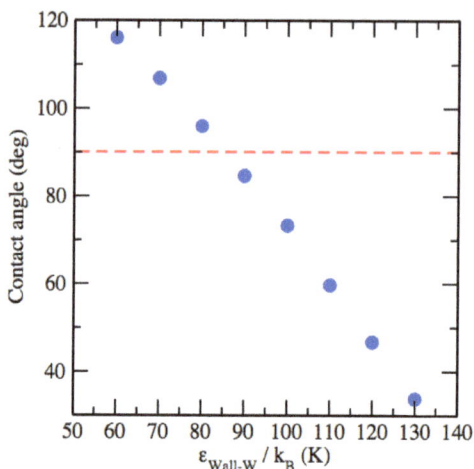

(a)

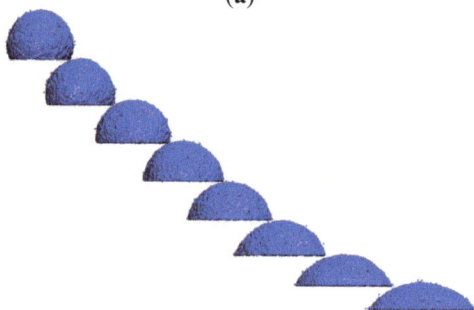

(b)

Following from the above, we use the system of 256,000 water molecules to study its contact angle as a function of the fluid-substrate interactions; the results can be seen in Figure 6. As expected, increasing the interaction energy between water molecules and the attractive wall diminishes the solid-fluid contact angle. For this particular coarse-grained surface model the functionality is linear, and a value of $\varepsilon_{Wall-W}/k_B \sim 85$ K can be taken as the boundary between hydrophilic and hydrophobic surface behavior.

An ancillary quantity of interest in the context of the drops on surfaces is the line tension. The line tension is the relation between the energy associated with the three phase contact line and the length of this line. This quantity is inherently dependent on the size of the drop [33] and is discussed as one of the reasons why the calculated tensions appear to be size-dependent. For large drops, one can obtain an estimate of the line tension using the approximation described by Weijs *et al.* [34]. For a typical case as shown above: the case of a drop of roughly 26 nm in diameter and considering the planar limit of the fluid-vapor tension (72 mN·m^{-1}) results in a contact tension line strength of $O(10^{-8})$ J·m^{-1}. While this number seems to be different from those estimated from micrometer drop experiments [35] or from molecular simulation of atomistic water models [36,37]; it is appropriate to point out that values as low as $\times 10^{-11}$ J·m^{-1} and as high as $\times 10^{-5}$ J·m^{-1} have been reported [33], which place our results in the correct context. Obviously, there is scope for much more detailed research into this topic.

4. Conclusions

We have proposed and validated a methodology for the unambiguous calculation of the solid-fluid contact angle from molecular dynamics simulations. We have tested model coarse-grained water-solid systems far beyond the limits commonly taken in atomistic simulation and showed, that for this particular model, more than 500,000 effective beads and/or drop diameters in excess of 50 nm would be required in order to obtain a result which is invariant of system size. So far the methodology has been applied over homogenous surfaces. However, is a well-known fact that surface roughness and energetic heterogeneities will have a profound effect on the contact angle calculations. The methodology presented is well suited to capture those effects.

Acknowledgments

This work was supported by the EPSRC through research grants (EP/I018212, EP/J014958 and EP/J010502).

Conflicts of Interest

The authors declare no conflict of interest.

References

1. Mattia, D. Templated growth and characterization of carbon nanotubes for nanofluidic applications. Ph.D. Thesis, Drexel University, Philadelphia, PA, USA, 2007.
2. Werder, T.; Walther, J.H.; Jaffe, R.L.; Halicioglu, T.; Koumoutsakos, P. On the water–carbon interaction for use in molecular dynamics simulations of graphite and carbon nanotubes. *J. Phys. Chem. B* **2003**, *107*, 1345–1352.
3. Fowkes, F.M.; Harkins, W.D. The state of monolayers adsorbed at the interface solid-aqueous solution. *J. Am. Chem. Soc.* **1940**, *62*, 3377–3386.
4. Schrader, M.E. Ultrahigh-vacuum techniques in the measurement of contact angles. LEED study of the effect of structure on the wettability of graphite. *J. Phys. Chem.* **1980**, *84*, 2774–2779.
5. Hirvi, J.T.; Pakkanee, T.A. Molecular dynamics simulations of water droplets on polymer surfaces. *J. Chem. Phys.* **2006**, *125*, 144712.
6. Hirvi, J.T.; Pakkanee, T.A. Enhanced hydrophobicity of rough polymer surfaces. *J. Phys. Chem. B* **2007**, *111*, 3336–3341.
7. Hirvi, J.T.; Pakkanee, T.A. Wetting of nanogrooved polymer surfaces. *Langmuir* **2007**, *23*, 7724–7729.
8. Ponter, A.B.; Yekta-Fard, M. The influence of environment on the drop size-contact angle relationship. *Colloid Polym. Sci.* **1985**, *263*, 673–681.

9. Li, H.; Zeng, X.C. Wetting and interfacial properties of water nanodroplets in contact with graphene and monolayer boron-nitride sheets. *ACS NANO* **2012**, *6*, 2401–2409.

10. Sampayo, J.G.; Malijevský, A.; Müller, E.A.; de Miguel, E.; Jackson, G. Communications: Evidence for the role of fluctuations in the thermodynamics of nanoscale drops and the implications in computations of the surface tension. *J. Chem. Phys.* **2010**, *132*, 141101.

11. Malijevský, A.; Jackson, G. A perspective on the interfacial properties of nanoscopic liquid drops. *J. Phys. Cond. Matt.* **2012**, *24*, 464121.

12. Ingebrigtsen, T.; Toxvaerd, S. Contact angles of Lennard-Jones liquids and droplets on planar surfaces. *J. Phys. Chem. C* **2007**, *111*, 8518–8523.

13. Shi, B.; Dhir, V.K. Molecular dynamics simulation of the contact angle of liquids on solid surfaces. *J. Chem. Phys.* **2009**, *130*, 034705.

14. De Ruijter, M.J.; Blake, T.D.; de Coninck, J. Dynamic wetting studied by molecular modeling simulations of droplet spreading. *Langmuir* **1999**, *15*, 7836–7847.

15. Hautman, J.; Klein, M.L. Microscopic wetting phenomena. *Phys Rev. Lett.* **1991**, *67*, 1763–1766.

16. Sergi, D.; Scocchi, G.; Ortona, A. Molecular dynamics simulations of the contact angle between water droplets and graphite surfaces. *Fluid Phase Equilib.* **2012**, *332*, 173–177.

17. Das, S.K.; Binder, K. Does Young's equation hold on the nanocale? A monte Carlo test for the binary Lennard-Jones fluid. *Europhys. Lett.* **2006**, *92*, 26006.

18. Falconer, K. *Fractal Geometry: Mathematical Foundations and Applications*; Wiley: Chichester, UK, 1990.

19. Mandelbrot, B. *The Fractal Geometry of Nature*; Macmillan: New York, NY, USA, 1983.

20. Berkmann, J., Caelli, T. Computation of surface geometry and segmentation using covariance techniques. *IEEE Trans. Pattern Anal.* **1994**, *16*, 1114–1116.

21. Mitra, N.J.; Nguyen, A.; Guibas, L. Estimating surface normals in noisy point cloud data. *Int. J. Comput. Geom. Appl.* **2004**, *14*, 261–276.

22. Efron, B.; Tibshirani, T. *An Introduction to The Bootstrap;* Chapman & Hall/CRC Monographs on Statistics and Applied Probability; Chapman and Hall/CRC: Boca Raton, FL, USA, 1993.

23. Van der Spoel, D.; Lindahl, E.; Hess, B.; Groenhof, G.; Mark, A.E.; Berendsen, H.J. GROMACS: Fast, flexible, and free. *J. Comput. Chem.* **2005**, *26*, 1701–1718.

24. Mie, G. Zur kinetischen Theorie der einatomigen Körper (In German). *Ann. Phys.* **1903**, *11*, 657–697.

25. Lobanova, O.; Avendaño, C.; Jackson, G.; Müller, E.A. SAFT-gamma force field for the simulation of molecular fluids: 5. A single-site coarse grained model of water. **2013**, In preparation.

26. Hadley, K.R.; McCabe, C. Coarse-grained molecular models of water: A review. *Mol. Simul.* **2012**, *38*, 671–681.

27. Avendaño, C.; Lafitte, T.; Galindo, A.; Adjiman, C.S.; Jackson, G.; Müller, E.A. SAFT-gamma force field for the simulation of molecular fluids: 1. A single-site coarse grained model of carbon dioxide. *J. Phys. Chem. B* **2011**, *115*, 11154–11169.

28. Avendaño, C.; Lafitte, T.; Adjiman, C.S.; Galindo, A.; Müller, E.A.; Jackson, G. SAFT-γ force field for the simulation of molecular fluids: 2. Coarse-grained models of greenhouse gases, refrigerants, and long alkanes. *J. Phys. Chem. B* **2013**, *117*, 2717–2733.

29. Lafitte, T.; Avendaño, C.; Papaioannou, V.; Galindo, A.; Adjiman, C.S.; Jackson, G.; Müller, E.A. SAFT-gamma force field for the simulation of molecular fluids: 3. Coarse-grained models of benzene and hetero-group models of n-decylbenzene. *Mol. Phys.* **2012**, *110*, 1189–1203.

30. Israelachvili, J.N. *Intermolecular and Surface Forces*, 3rd ed.; Elsevier: Amsterdam, The Netherlands, 2011; Chapter 11, pp. 208–211.

31. Shakouri, A. EE145 Spring 2002. Available online: http://classes.soe.ucsc.edu/ee145/Spring02/EE145hmwk1Sol.pdf (accessed on 10 July 2013).

32. Scocchi, G.; Sergi, D.; D'Angelo, C.; Ortona, A. Wetting and contact-line effects for spherical and cylindrical droplets on graphene layers: A comparative molecular-dynamics investigation. *Phys. Rev. E* **2011**, *84*, 061602.

33. Tadmor, R. Line energy, line tension and drop size. *Surf. Sci. Lett.* **2008**, *602*, L108–L111.

34. Weijs, J.H.; Marchand, A.; Andreotti, B.; Lohse, D.; Snoeijer, J.H. Origin of line tension for a Lennard-Jones nanodroplet. *Phys. Fluids* **2011**, *23*, 022001.

35. Pompe, T.; Herminghaus, S. Three-phase contact line energies from nanosclae surface topographies. *Phys. Rev. Lett.* **2000**, *85*, 1930–1933.

36. Ritchie, J.A.; Yazdi, J.S.; Bratko, D.; Luzar, A. Metastable sessile nanodroplets on nanopatterned surfaces. *J. Phys. Chem. C* **2012**, *116*, 8634−8641.

37. Daub, C.D.; Bartko, D.; Luzar, A. Electric control of wetting by salty nanodrops: Molecular dynamics simulations. *J. Phys. Chem. C* **2011**, *115*, 22393–22399.

Reprinted from *Entropy*. Cite as: Uline, M.J.; Corti, D.S. Molecular Dynamics at Constant Pressure: Allowing the System to Control Volume Fluctuations via a "Shell" Particle. *Entropy* **2013**, *15*, 3941–3969.

Review

Molecular Dynamics at Constant Pressure: Allowing the System to Control Volume Fluctuations via a "Shell" Particle

Mark J. Uline [1],* **and David S. Corti** [2]

[1] Department of Chemical Engineering, University of South Carolina, Columbia, SC 29208, USA
[2] School of Chemical Engineering, Purdue University, West Lafayette, IN 47907, USA;
   E-Mail: dscorti@purdue.edu

* Author to whom correspondence should be addressed; E-Mail: uline@cec.sc.edu;
   Tel.: +1-803-777-2030; Fax: +1-803-777-8265.

Received: 29 July 2013; in revised form: 6 September 2013 / Accepted: 16 September 2013 / Published: 23 September 2013

Abstract: Since most experimental observations are performed at constant temperature and pressure, the isothermal-isobaric (NPT) ensemble has been widely used in molecular simulations. Nevertheless, the NPT ensemble has only recently been placed on a rigorous foundation. The proper formulation of the NPT ensemble requires a "shell" particle to uniquely identify the volume of the system, thereby avoiding the redundant counting of configurations. Here, we review our recent work in incorporating a shell particle into molecular dynamics simulation algorithms to generate the correct NPT ensemble averages. Unlike previous methods, a piston of unknown mass is no longer needed to control the response time of the volume fluctuations. As the volume of the system is attached to the shell particle, the system itself now sets the time scales for volume and pressure fluctuations. Finally, we discuss a number of tests that ensure the equations of motion sample phase space correctly and consider the response time of the system to pressure changes with and without the shell particle. Overall, the shell particle algorithm is an effective simulation method for studying systems exposed to a constant external pressure and may provide an advantage over other existing constant pressure approaches when developing nonequilibrium molecular dynamics methods.

Keywords: isothermal-isobaric ensemble; molecular dynamics

1. Introduction

The molecular dynamics (MD) simulation method can be straightforwardly applied to the analysis of an isolated system or a system described by the microcanonical ensemble in which the energy, volume V and particle number N are held fixed. The equations of motion that describe the time evolution of the positions and momenta of the particles, *i.e.*, the resulting microcanonical ensemble phase space trajectory, follow directly from Newtonian mechanics. Energy, however, is not a variable of choice for experiments. Many experimental observations are carried out under conditions of constant pressure and temperature, such that the system is no longer isolated from its environment. Therefore, while the generation of dynamic information about these systems is of interest, how to modify the equations of motion to describe a system at constant temperature and/or constant pressure is arguably not an obvious task.

An extension of the MD method to systems not described by the microcanonical ensemble was presented by Andersen in 1980 [1]. Andersen showed, for example, that by modifying the Lagrangian of the system, a constant external pressure could be imposed within MD. Specifically, additional control variables were introduced into the Lagrangian, beyond the standard coordinate and momentum vectors needed to describe the classical N-particle system. The new variables served to drive the fluctuations of those variables no longer held fixed within the ensemble of interest. For a system in which a constant external pressure is imposed, the system volume is now introduced as a dynamic variable that serves to maintain, on average, mechanical equilibrium between the external and system pressure. Consequently, the system is exposed to a barostat, whereby a "piston" of arbitrary "mass" controls the dynamics of the volume. While ensemble averages are independent of the piston mass, the fictitious mass does affect the response time for volume fluctuations.

Andersen's extended Lagrangian approach was later adapted by Nosé [2,3] to simulate systems in contact with a thermostat using MD. Hoover [4,5] proposed another isothermal-isobaric (NPT) MD algorithm using a modification of Andersen's piston method for maintaining constant pressure and the thermostating method of Nosé. As Hoover was aware of, and as discussed in detail by Tuckerman *et al.* [6], this algorithm does not yield ensemble averages consistent with the then accepted form of the NPT ensemble partition function. Consequently, several new NPT MD algorithms have been introduced in the literature (a non-exhaustive list is given here [7–11]).

Yet, starting nearly 20 years ago, the foundation of the NPT ensemble (when the volume is considered to be a continuous variable) has been reconsidered [12–14]. What was noted was that the NPT partition function redundantly counts the configurations of the system. This problem of over-counting was removed by requiring that the volume, V, of the system be defined by a "shell" particle, where at least one particle resides in the volume, dV, encapsulating V. All of the NPT MD algorithms mentioned above are not, however, consistent with the proper shell-particle formulation of the NPT ensemble (we will show later that Hoover's algorithm does give the correct distribution of volumes if periodic boundary conditions are employed). As such, new NPT MD algorithms should be introduced in order to generate the correct NPT ensemble averages.

Corti [15] previously modified the Monte Carlo NPT algorithm to be consistent with the correct NPT partition function. The current authors performed a similar reformulation for the constant pressure MD algorithm for systems whose particles interact via continuous [16,17] and discontinuous [18] potentials. In these new MD algorithms, a shell particle is used to uniquely define the volume of a system exposed to a constant external pressure. Consequently, since the shell particle sets the volume of the system, no piston mass needs to be specified. In other words, the system itself controls the response time of volume fluctuations, as the mass of the shell particle is known, and not the user through the introduction of an arbitrary piston mass. Various benefits arise from the removal of this ambiguity in the NPT MD algorithm.

As a side note, Evans and Morriss [19,20] utilized constrained dynamics to develop an NPT MD algorithm. In this method, both the instantaneous pressure and kinetic energy are made strict constants of motion, and so, the Andersen piston is not employed. Nevertheless, this algorithm does not yield ensemble averages consistent with the NPT partition function (either with or without the shell particle), as the instantaneous pressure fluctuates within the NPT ensemble [15]. Even though the constraint dynamics also does not utilize a piston, the resulting equations of motion do not generate the proper NPT ensemble averages.

In this paper, we review our previous work on employing the shell particle to generate equations of motion that are consistent with the proper shell-particle formulation of the NPT ensemble. To begin, we provide in Section 2 an overview of the reformulation of the NPT ensemble partition function and the need to employ the shell particle to eliminate the redundant counting of configurations. In Section 3, the equations of motion required to properly generate a system within the NPT ensemble are presented, in which the piston of arbitrary mass is replaced with a shell particle of known mass. We include the previously derived equations in which an external temperature is imposed via the use of the Nosé-Hoover thermostat chains, as well as recently developed equations making use of a thermostat based on the configurational temperature. The Trotter expansion to the Liouville operator formalism [21–26] is used to factorize the classical propagator into analytically solvable operators. We also provide simulation results for the Lennard-Jones fluid, particularly for small system sizes, where interesting differences between the old and new NPT partition function appear for various ensemble averages. 'Nonequilibrium' simulations are presented in Section 4, in which the external pressure is changed after the system has equilibrated. As the system evolves to a new equilibrium state, we compare the dynamics of the volume as defined via the shell particle to that when the Hoover algorithm is used with different piston masses. Conclusions are provided in Section 5, as well as a discussion of some particular dynamic systems of interest that may benefit from the use of the shell-particle formalism.

2. The Volume Scale in Constant Pressure Ensembles

The original formulation of the isothermal-isobaric ensemble can be traced back to 1939, where Guggenheim [27] wrote the partition function, $\Delta(N, P, T)$, as:

$$\Delta(N, P, T) = \sum_{V} Q(N, V, T)e^{-PV/k_B T} \qquad (1)$$

where k_B is Boltzmann constant, $Q(N, V, T)$ is the canonical ensemble partition function of a system composed of N particles held in a volume, V, and at a temperature, T, and P is the external pressure to which the system is exposed as the volume is allowed to fluctuate. Although Equation (1) is formally correct, an ambiguity arises when dealing with systems in which the volume is a continuous variable. In the late 1950s, several authors [28–31] attempted to remove the conceptual difficulty associated with the sum over an unspecified set of discrete volumes by expressing $\Delta(N, P, T)$ as:

$$\Delta_0(N, P, T) = \frac{1}{V_0} \int_0^\infty Q(N, V, T)e^{-PV/k_B T} dV \tag{2}$$

The replacement of the sum in Equation (1) by an integral enables the inclusion of all volumes, but at the expense of generating a partition function that has the dimensions of volume. Consequently, this partition function must be rendered dimensionless through division by some constant with units of volume denoted by V_0 in Equation (2). Note that we wrote the partition function with a subscript ($\Delta_0(N, P, T)$) in Equation (2) to signify that this partition function uses V_0 as its volume scale. The constant, V_0, does cancels out when determining the ensemble average of a given variable and, so, need not be specified. Even so, Sack [30] showed in the thermodynamic limit that:

$$V_0 = \frac{k_B T}{P} \tag{3}$$

Hill [31] noted that in the thermodynamic limit, the choice of V_0 is arbitrary, due in part to the equivalency of the ensembles in the thermodynamic limit. Evaluation of ensemble averages of macroscopic systems using Equation (2) yields only a completely negligible error. Yet, the precise value of the volume scale is important when dealing with systems of sufficiently small size [12–14]. The volume scale must be chosen carefully, since it depends upon the properties of the boundary separating the system of interest from the surroundings [14]. The boundary serves to define the volume of the system and allows the system volume to fluctuate against the external pressure imposed by the surroundings. Hence, the boundary cannot be chosen arbitrarily, particularly when the system is not in the thermodynamic limit. In other words, the properties assigned to the boundary must conform to the actual physical situation in which the system is found.

As shown by Koper and Reiss [12] using the microcanonical ensemble, verified later by Corti and Soto-Campos [13] and Corti [14] using the canonical ensemble, when the boundary is not a physical object to which a mass or momentum can be assigned (*i.e.*, a mathematical construct to aid in the specification of the system volume), then the partition function in Equation (2) counts configurations of the system redundantly (whether or not V_0 is specified). The problem of over-counting is removed by requiring that the volume, V, of the system be specified by a "shell" molecule, where at least one molecule resides in the volume, dV, surrounding V. To illustrate this problem, turn to Figure 1, which demonstrates how several volumes may enclose the same configuration of n particles surrounded by $N - n$ particles. In the rigorous formulation of the NPT ensemble, each configuration of the system must correspond to only one specific volume state of the n particles. Otherwise, the same configuration will be counted more than once in Equation (2) [14]. The problem of over-counting, or redundancy, is resolved by defining a "shell" particle [12–14], in which at least one of the system particles resides in the shell that encapsulates the system volume. Defining the n-particle system with the shell particle means that a new and distinct state of the total N-particle system is necessarily created when the volume of the n-particle system is varied (whether or not the configuration of the surrounding $N - n$ particles changes), since the position of the shell particle changes, as well [14]. Consequently, the inclusion of configurations of the n particles common to larger values of the volume is explicitly avoided.

Figure 1. One particular configuration of N particles enclosed within a total volume, V, demonstrating how to uniquely define one specific volume state of n particles (shaded circles). The unshaded circles represent the surrounding $N - n$ particles that comprise the bath. Each particle center is marked by a dot and is surrounded by an effective diameter. The first step in determining the volume occupied by the n particles is to choose a particular reference point in V as the origin, r_c. Yet, several volumes (dashed circles) centered at r_c still enclose the n particles and, therefore, include common configurations. The exact volume, v (bold circle), of the n particles is defined by the presence of a shell particle that is farthest from r_c and resides in the shell, dv, encapsulating v. (Adapted from Figure 2 in reference [14].)

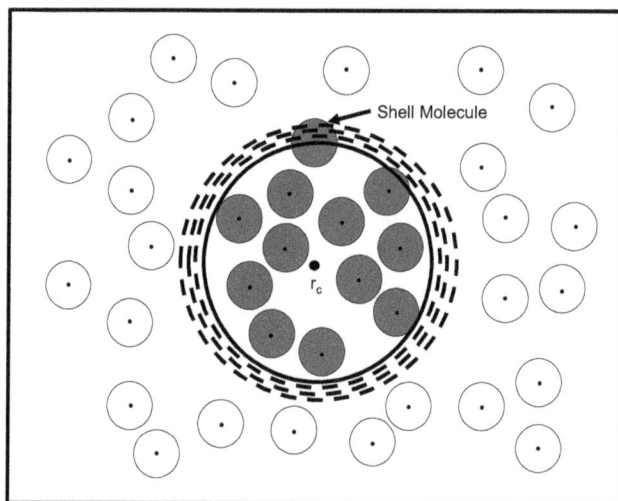

Therefore, the proper form of $\Delta(N, P, T)$, in which interactions between the system and surroundings are neglected, as in Equation (2), should be [12–14]:

$$\Delta(N, P, T) = \int_0^\infty Q^*(N, V, T)e^{-PV/k_BT}dV \qquad (4)$$

where $Q^*(N, V, T)dV$ represents the number of configurations in which at least one of the N particles resides in the shell, dV, surrounding V. Note that the above partition function is dimensionless, since $Q^*(N, V, T)dV$ is a pure number (or $Q^*(N, V, T)$ is a density of states). The shell particle is the correct volume scale when there is not a physical boundary to attach the volume of the system. Koper and Reiss [12] demonstrated that the states summed in Equation (4) do not contain common configurations, because the shell particle sets the volume. As shown above, all redundancies are eliminated by equating the system volume to the shell molecule.

2.1. Cubic System Volume

The application of the constant pressure ensemble to small systems also reveals the effects of additional variables, such as surface area and curvature, on the system's properties. In the thermodynamic limit, the shape of the container enclosing the system has no influence on its properties. As the size of the system is decreased, additional independent thermodynamic variables (e.g., surface area, curvature) must be introduced to ensure that the system's properties are described properly. These additional parameters are a function of the "shape" of the system volume. Therefore, the constant pressure ensemble partition function must be formulated differently in order to describe a system in which its volume is always either spherical (e.g., physical cluster) or cubic (the standard shape used to apply periodic boundary conditions). As a result, ensemble averages within the constant pressure ensemble will depend upon the shape of the system volume. This dependence upon shape, of course, becomes negligible in the thermodynamic limit. A reader interested in spherical systems is referred to [15]. However, since we are focusing on MD simulations with periodic boundary conditions, we are going to present results for a cubic volume [15].

The mathematical representation of $Q^*(N, V, T)dV$ for a cubic volume, $V = L^3$, whose length, L, lies between L and $L + dL$ with at least one particle in the shell, dL, is given by [15–18]:

$$Q^*_{cub}(N, L, T)dL = \frac{3dL}{(N-1)!\Lambda^{3N}} \int_A dy_1 dz_1 \int_{V^{N-1}} d\tau_{12}...d\tau_{1N} e^{-\beta U_N} \quad (5)$$

where $\beta = 1/k_B T$, Λ is the de Broglie wavelength, A represents the area of a face of the cube, dy_1 and dz_1 represent the differential change in the y and z coordinates of particle 1 (the shell particle), respectively, $\tau_{12}...\tau_{1N}$ are the coordinates of the remaining $N-1$ particles relative to the position of particle 1, and U_N is the interaction potential of all the N particles. The number three is required by the switch from dV to $3L^2 dL$, since volume changes occur with constant shape, and indicates the three sets of equivalent configurations generated if a particle is held fixed in the shell in either the x, y or z direction. Particle 1 cannot be integrated throughout the entire shell, but due to symmetry, can be integrated separately in the x direction (or the y or z direction). The above integral is therefore evaluated with the x coordinate of particle 1 held fixed in the plane that corresponds to one of the two faces of the cube perpendicular to the x-axis of the coordinate system.

With Equation (5), the isothermal-isobaric ensemble partition function for a cubic volume is now represented by [15–18]:

$$\Delta_{cub}(N, P, T) = \int Q^*_{cub}(N, L, T)e^{-PL^3/k_B T} dL \quad (6)$$

Within the NPT ensemble, the instantaneous pressure, P'', of the system fluctuates. The instantaneous pressure is defined as [15–18]:

$$P'' = \frac{k_B T}{3L^2}\left(\frac{\partial ln Q^*_{cub}}{\partial L}\right)_{T,N} \quad (7)$$

and is calculated during a simulation via the following equation [15]:

$$P'' = \frac{(N - 1/3)k_BT}{V} + \frac{\langle \sum_i \sum_{j>i} \vec{r}_{ij} \cdot \vec{f}_{ij} \rangle}{3V} \tag{8}$$

where the second term on the right side is the standard virial of the system, $\vec{r}_{ij}$ is the vector between the centers of particles i and j and $\vec{f}_{ij}$ is the corresponding force. The ideal, or kinetic, term now reflects the loss of one, out of $3N$, translational degree of freedom. By defining the system volume and, therefore, being directly coupled to the barostat via the changes in the volume, the shell particle does not translate freely, that is, independently of the volume, in the x direction and, so, does not impart any momentum in the x direction to the surface of the cube. The shell particle translates freely in only two directions (y and z). The virial term in Equation (8) remains unchanged, since the shell particle still interacts with the other particles in the system. The ensemble average of $\langle P'' \rangle$ is related to the externally imposed pressure, P, as follows [15]:

$$\langle P'' \rangle = P + \frac{2k_BT}{3N}\langle \rho \rangle \tag{9}$$

where $\langle \rho \rangle$ is the average density of the system.

While revising the Monte Carlo NPT algorithm to incorporate the shell particle, Corti [15] derived several relations that describe how ensemble averages obtained within the new NPT partition function, Equation (4) or (6), relate to ensemble averages obtained with the old no-shell NPT (Equation (2)) partition function, Δ_0, [15]. If $\langle V \rangle$ represents the ensemble-averaged volume defined via the shell particle and $\langle V \rangle_0$ represents the ensemble-averaged volume defined via the old definition, then [15]:

$$\langle V \rangle = \langle V \rangle_0 - \frac{k_BT}{P} \tag{10}$$

Consequently, $\langle V \rangle < \langle V \rangle_0$; the difference between these two average volumes is only apparent at small system sizes, since in the thermodynamic limit, $\langle V \rangle \to \langle V \rangle_0$ (k_BT/P is intensive).

2.2. Ideal Gas Results

The ideal gas offers a unique opportunity to obtain a closed form solution for the partition function. Using the shell molecule definition (Equation (4)), we obtain [14]:

$$\Delta(N, P, T) = \int_0^\infty Q_{id}^*(N, V, T)e^{-\beta PV}dV = \int_0^\infty \frac{V^{N-1}}{(N-1)!\Lambda^{3N}}e^{-\beta PV}dV = \left(\frac{1}{\beta P \Lambda^3}\right)^N \tag{11}$$

Using the following definition [31,32]:

$$\langle V \rangle = -k_BT\left(\frac{\partial ln\Delta}{\partial P}\right)_{\beta,N} \tag{12}$$

we get the following expression for the equation of state [14]:

$$P\langle V \rangle = Nk_BT \tag{13}$$

Now, we can use the older partition function (Equation (2)) and perform the same analysis. The partition function is [14]:

$$\Delta_0(N, P, T) = \frac{1}{V_0} \int_0^\infty Q_{id}(N, V, T)e^{-\beta PV}dV = \frac{1}{V_0} \int_0^\infty \frac{V^N}{(N)!\Lambda^{3N}}e^{-\beta PV}dV = \frac{1}{V_0}\frac{1}{\Lambda^{3N}(\beta P)^{N+1}} \quad (14)$$

We then get the following equation of state, noting that V_0 is a constant [14]:

$$P\langle V\rangle_0 = (N+1)k_B T \quad (15)$$

The use of $(N+1)$ or N is clearly inconsequential in the thermodynamic limit. Yet, the difference between Equations (13) and (15) is significant when the system is sufficiently small. In general, the ensemble averages calculated within different ensembles will not be the same for small systems. In contrast, ensemble averages are independent of the particular ensemble chosen to evaluate them when the system is in the thermodynamic limit. One exception, however, is the ideal gas. Due to the absence of inter-particle interactions, identical results should be obtained within all ensembles and for all system sizes. Hence, the small system NPT partition function of the ideal gas should yield Equation (13) and not Equation (15).

3. Shell Molecule Equations of Motion

In order to simulate the NPT ensemble, a technique for maintaining a constant temperature needs to be introduced into the equations of motion. As mentioned earlier, Nosé [2,3] and Hoover [4] proposed a completely dynamic method for maintaining constant temperature in an MD simulation. An additional variable, which serves to couple the system to a thermostat of fixed temperature T, is added to the Lagrangian of the N-body system. An effective mass is then associated with this new variable and controls the time scale for temperature fluctuations. While this scheme is usually effective, it does not always perform well [23]. In some cases, the resulting equations of motion do not generate phase-space trajectories that are ergodic [4]. To overcome this potential problem, the Nosé-Hoover chain method [23] was later developed. In this method, multiple thermostats are themselves successively coupled to adjacent thermostats, thereby forming a chain of thermostats.

The equations of motion for the NPT ensemble with the shell particle are straightforwardly obtained by employing an extended Lagrangian approach. The full derivation is presented in appendix A of reference [16]. For a cubic volume in which $V = L^3$, we let the $+x$ coordinate of particle 1, or the shell particle, define half the box length, $L/2$, of the simulation cell. We also choose q_i and p_i to represent the $3N$-generalized coordinates and conjugate momenta, respectively. Note that we always have that $q_1 = L/2$. We therefore get the following equations of motion for an isothermal-isobaric ensemble consistent with Equation (6) using a single thermostat chain for all of the particles [16]:

$$\dot{q}_i = \frac{p_i}{m_i} + \left(\frac{\dot{q}_1}{q_1}\right)q_i$$

$$\dot{p}_i = F_i - \left(\frac{\dot{q}_1}{q_1}\right)p_i - \dot{\xi}_1 p_i$$

$$\dot{q}_1 = \frac{p_1}{m_1}$$

$$\dot{p}_1 = 24q_1^2(P_{int} - P) - \dot{\xi}_1 p_1$$

$$\dot{\xi}_k = \frac{p_{\xi_k}}{Q_k}$$

$$\dot{p}_{\xi_1} = \sum_{j=1}^{3N} \frac{p_j^2}{m_j} - \dot{\xi}_2 p_{\xi_1} - g k_B T$$

$$\dot{p}_{\xi_k} = \frac{p_{\xi_{k-1}}^2}{Q_{k-1}} - (\dot{\xi}_{k+1}) p_{\xi_k} - k_B T$$

$$\dot{p}_{\xi_C} = \frac{p_{\xi_{C-1}}^2}{Q_{C-1}} - k_B T \tag{16}$$

where $i = 2, ..., 3N$, and the overdots signify time derivatives. F_i is the x, y or z component of the force acting on the particle represented by the ith-generalized coordinate, and m_i is the corresponding mass of the particle. Each ξ_k is a thermodynamic friction coefficient introduced to simplify the equations, and p_{ξ_k} is the corresponding momentum of ξ_k, whose effective mass is Q_k. C is the total number of coupled thermostats in the chain, so that $k = 1, ..., C$, and g denotes the total number of degrees of freedom of the momenta of the particles. The expression for the internal pressure, P_{int}, which follows from Equation (6), is:

$$P_{int} = \frac{1}{24q_1^3} \left[\sum_{i=2}^{3N} \frac{p_i^2}{m_i} + \sum_{i=1}^{3N} q_i F_i \right] \tag{17}$$

where the first summation runs from two to $3N$, indicating that the x momentum of the shell particle does not contribute to the internal pressure.

The extended Hamiltonian, H_{ext}, for this system is [16–18]:

$$H_{ext} = \sum_{i=1}^{3N} \frac{p_i^2}{2m_i} + \sum_i \sum_{j>i} U(r_{ij}) + 8q_1^3 P + \sum_{k=1}^{C} \frac{p_{\xi_k}^2}{2Q_k} + g k_B T \xi_1 + \sum_{k=2}^{C} k_B T \xi_k \tag{18}$$

where $V = 8q_1^3$, $U(r_{ij})$ is the interaction potential between particles i and j and $\sum_i \sum_{j>i}$ is a sum over all distinct pairs of particles. Although the equations of motion cannot be obtained directly from Equation (18), H_{ext} is a conserved quantity.

With the exception of those equations of motion that describe the velocity and acceleration of q_1, the proposed equations are the same as those of Andersen's method [1]. The expression for the acceleration of q_1 provides an interesting physical interpretation. Given that the area of a single face of the simulation cell is $4q_1^2$ (since $L = 2q_1$), the total surface area of the cube is $24q_1^2$. When multiplied by the difference between the internal and external pressures, we obtain the net force that drives the acceleration of q_1. This connection between the acceleration of q_1 and the pressure difference is more physically appealing than what appears in other methods [1,5,8]. Another benefit to the shell formulation is that the system itself sets the time scale for volume and pressure fluctuations, since the mass of the shell particle is known. In Andersen's method, there is an unknown piston mass that sets the response time of volume and pressure fluctuations.

In the simulations performed in this work, the forces acting in the y and z directions on each particle sum to zero when there are no external forces in the y and z directions. The sum of forces in the x direction will not be zero, since the x directional momentum of the shell particle is directly coupled to the barostat. Therefore, only the linear momenta in the y and z directions are conserved. Furthermore, to avoid particle drift during simulations when periodic boundary conditions are applied, the center-of-mass momentum in the y and z directions is set equal to zero. Again, the center-of-mass momentum in the x direction is driven by the external pressure and cannot be held fixed at a zero value, though volume fluctuations ensure that the total momentum in the x direction averages to zero. Consequently, it can be shown that $g = 3N - 2$ [16,33], indicating that the above equations of motion yield trajectories in phase space that are consistent with a $(3N - 2)PT$ partition function (there are $3N - 2$ momentum degrees of freedom) [16].

The new equations of motion that employ a shell particle to define the system volume provide another example of a non-Hamiltonian system, in that Equation (16) cannot be derived from the extended Hamiltonian in Equation (18). A systematic procedure for extending classical statistical mechanics to non-Hamiltonian systems was proposed by Tuckerman *et al.* [6,34]. The crux of their analysis relies on the notion that non-Hamiltonian phase space is compressible, as opposed to its Hamiltonian counterpart. For a non-Hamiltonian system, the Jacobian describing the transformation from an initial phase-space vector to a phase-space vector at time t is not equal to unity. The invariant phase-space metric for a non-Hamiltonian system is therefore not the same as the Hamiltonian system. Nevertheless, using the procedure of Tuckerman *et al.* [6,34], where the compressibility of the phase space is taken into account, the extended system partition function can still be derived from the equations of motion and the various constraints, or conservation relations, on the system. The detailed phase space analysis of Equation (16) presented in reference [16] shows that the proposed shell particle equations of motion are completely consistent with the shell particle partition function (Equation (6)) with and without periodic boundary conditions.

3.1. The Hoover Algorithm and Periodic Boundary Conditions

The NPT partition function in Equation (6) can be rewritten if the system is homogeneous and periodic boundary conditions are applied. Han and Son [35] showed that since periodic boundary conditions yield a transitionally symmetric system, particle 1 does not need to be held fixed inside the shell, dL. If all of the relative distances between the particles remain fixed, identical configurations will be generated if particle 1 is allowed to sample the entire instantaneous volume. Thus [15]:

$$Q_{cub}^*(N, L, T)dL = \frac{3dL}{(N-1)!\Lambda^{3N}} \int_A dy_1 dz_1 \int_{V^{N-1}} d\tau_{12}...d\tau_{1N} e^{-\beta U_N}$$

$$= \frac{3dL}{(N-1)!\Lambda^{3N}} \frac{1}{L} \int_V d\tau_1...d\tau_N e^{-\beta U_N} = \frac{N}{V} Q(N, V, T)dV \qquad (19)$$

where $Q(N, V, T)$ is the canonical partition function without the shell molecule. Using Equation (19), we can write the isothermal-isobaric partition function as:

$$\Delta_{PB}(N, P, T) = \int \frac{N}{V} Q(N, V, T) e^{-\beta PV} dV \qquad (20)$$

where the PB subscript signifies that the partition function is only valid under the symmetry imposed by periodic boundary conditions [15,16]. This volume scale was derived earlier using the information theory by Attard [36].

Let us now consider the following equations of motion proposed for the NPT ensemble by Hoover [4,5] with a single chained thermostat:

$$
\begin{aligned}
\dot{q}_i &= \frac{p_i}{m_i} + \left(\frac{\dot{V}}{3V}\right) q_i \\
\dot{p}_i &= F_i - \left(\frac{\dot{V}}{3V}\right) p_i - \dot{\xi}_1 p_i \\
\dot{V} &= \frac{3p_\epsilon}{M_p} V \\
\dot{p}_\epsilon &= 3V(P_{int} - P) - \dot{\xi}_1 p_\epsilon \\
\dot{\xi}_k &= \frac{p_{\xi_k}}{Q_k} \\
\dot{p}_{\xi_1} &= \sum_{j=1}^{3N} \frac{p_j^2}{m_j} + \frac{p_\epsilon^2}{M_p} - \dot{\xi}_2 p_{\xi_1} - g k_B T \\
\dot{p}_{\xi_k} &= \frac{p_{\xi_{k-1}}^2}{Q_{k-1}} - (\dot{\xi}_{k+1}) p_{\xi_k} - k_B T \\
\dot{p}_{\xi_C} &= \frac{p_{\xi_{C-1}}^2}{Q_{C-1}} - k_B T
\end{aligned}
\tag{21}
$$

where $i = 1, ..., 3N$ and:

$$
P_{int} = \frac{1}{3V}\left[\sum_{i=1}^{3N} \frac{p_i^2}{m_i} + \sum_{i=1}^{3N} q_i F_i\right]
\tag{22}
$$

The extended Hamiltonian for the Hoover NPT algorithm is:

$$
H_{ext} = \sum_{i=1}^{3N} \frac{p_i^2}{2m_i} + \sum_i \sum_{j>i} U(r_{ij}) + \frac{p_\epsilon^2}{2M_P} + PV + \sum_{k=1}^{C} \frac{p_{\xi_k}^2}{2Q_k} + g k_B T \xi_1 + \sum_{k=2}^{C} k_B T \xi_k
\tag{23}
$$

Tuckerman *et al.* [6] already performed the phase-space analysis on Hoover's equations of motion, in which they obtained a $1/V$ weighting in the volume distribution function when all three directional linear momenta are conserved, as well as the three center-of-mass momenta being set to zero. The appearance of the $1/V$ weighting of the volume distribution makes it completely consistent with the partition function introduced by Attard (Equation (20)). The Hoover algorithm does in fact lead to the correct sampling of volume states, but only for homogenous systems with periodic boundary conditions. In the absence of external forces, Hoover's algorithm yields a $(3N - 2)PT$ ensemble: there are a total of $(3N + 1)$ momentum degrees of freedom ($3N$ particles and one volume), but now, the total linear momentum in each of the three directions is conserved. Therefore, $g = 3N - 2$ in Equation (21).

Although the Hoover algorithm does sample phase space correctly (but only for periodic boundary conditions), there is still an unknown piston mass, which sets the response time of volume and pressure fluctuations, which must be specified. On the other hand, when the shell particle formulation is used, the system itself sets the time scale for volume and pressure fluctuations, since the mass of the shell particle is known. Furthermore, since the piston mass associated with the Hoover algorithm can have very different dynamics compared to the particles in the system, it is the suggested form of the equations of motion to use two separate chained thermostats, one coupled to the particles and the other to the volume [8]. The need to introduce another set of chained thermostats to drive the volume fluctuations in the Hoover algorithm requires that another set of unknown parameters, the additional thermostat masses, be specified [8]. This separate thermostat chain is not necessary with the shell particle algorithm, as the momentum of the shell particle is on the same scale as the rest of the particles of the system [16,17].

As a final point of interest and, again, to focus on the effects of the different barostats, we briefly consider the results of the phase-space analysis of the NPT equations of motion for the shell (Equation (16)) and the Hoover algorithm (Equation (21)). The explicit partition functions are derived in reference [16], where the influence of each barostat is clearly seen. By definition, the enthalpy, H, of the system is equal to $H = H(q, p) + PV$, where $H(p, q) = \sum_{i=1}^{3N} p_i^2/2m_i + U(q)$. Hoover's algorithm generates an extended Hamiltonian (Equation (23)) that contains an additional term associated with the kinetic energy of the volume $(p_\epsilon^2/2M_P)$, a quantity that should not appear in the enthalpy if the boundary used to describe the system volume is a mathematical construct to which a mass or momentum cannot be assigned [14]. In contrast, each configuration in the shell particle partition function corresponds to one and only one volume state, since the non-extended Hamiltonian is directly coupled to the volume (*i.e.*, there is a one to one correspondence with the non-extended Hamiltonian and the volume states) [16]. The redundant counting of volume states is not eliminated in the other algorithms, because those non-extended Hamiltonians are decoupled from the volume. Note that whenever we use the extended variables approach to thermostat systems in this manuscript that there is always a kinetic term associated with the thermostat variables in the extended Hamiltonian. We are focusing here on how the positions and momenta are sampling the correct distribution, and the preceding argument on the enthalpy is independent of this kinetic term, due to the thermostat variables.

We conclude this section by noting that the ensemble average pressure for a system whose partition function is described by Equation (20) obeys the following relation [15]:

$$\langle P'' \rangle = P + \frac{k_B T}{N} \langle \rho \rangle \tag{24}$$

The correction to the ensemble average volume is the same as is given in Equation (10).

3.2. Multicomponent Systems

In this section, we discuss the extension of the shell particle MD algorithm to multicomponent systems. In particular, we consider a binary mixture comprised of species A and B. In this case, the isothermal-isobaric partition function must include configurations in which the shell particle is of type A and configurations in which the shell particle is of type B. Therefore, the isothermal-isobaric partition function, Δ_{AB}, is given by (only the case of a cubical volume is considered) [15,17]:

$$\Delta_{AB}(N,P,T) \;=\; \int Q_{cub}^{*A}(N,V,T)e^{-\beta PL^3}\,dL + \int Q_{cub}^{*B}(N,V,T)e^{-\beta PL^3}\,dL \qquad (25)$$

where $Q_{cub}^{*A}(N,V,T)$, for example, is the total number of configurations of N_A particles of type A and N_B particles of type B contained in a volume $V = L^3$ in which at least one of the N_A particles resides in the shell, dL, encapsulating V.

When periodic boundary conditions are applied, one can show that the probability of a given configuration having a shell particle of type A is simply equal to the mole fraction of A. Begin with the partition function, Δ_A, that includes only those configurations in which the shell particle is of type A:

$$\Delta_A(N,P,T) \;=\; \int Q_{cub}^{*A}(N,V,T)e^{-\beta PL^3}\,dL \qquad (26)$$

For a homogeneous fluid in which periodic boundary conditions are employed, one can rewrite Δ_A, following the argument presented by Han and Son [35], as:

$$\Delta_A(N,P,T) \;=\; N_A \int \frac{Q(N,V,T)}{V} e^{-\beta PV}\,dV \qquad (27)$$

where $Q(N,V,T)$ is the canonical ensemble partition function for N_A and N_B particles without a shell particle used to define the volume. The fraction of configurations containing a shell particle of type A is therefore given by $N_A/(N_A + N_B) = x_A$ [17]. It was shown in reference [17] that the ensemble average of F (F being any given quantity) is given by:

$$\langle F \rangle = x_A \langle F \rangle_A + x_B \langle F \rangle_B \qquad (28)$$

where $\langle F \rangle_A$ is the ensemble average obtained with only A as the shell particle and $\langle F \rangle_A$ is the ensemble average obtained with only B as the shell particle. Hence, two separate simulations can be run, each with different identities of the shell particle, with the resulting ensemble averages simply weighted by the mole fractions of each component. Yet, one can proceed even further and demonstrate that only one simulation per state point is ultimately required, with the identity of the shell particle being completely arbitrary. With periodic boundary conditions, we showed in reference [17] that $\langle F \rangle = \langle F \rangle_A = \langle F \rangle_B$. Therefore, only one single simulation is required; the choice of which species to be the shell particle is solely a matter of convenience [17]. This conclusion also holds for mixtures with more than two components, again, only when periodic boundary conditions are employed.

3.3. Collision Dynamics for Discontinuous Potentials

In this section, we discuss the implementation of the shell particle formalism to simulate systems that have discontinuous intermolecular potentials [18]. Discontinuous molecular dynamics (DMD) have been widely used for quite some time, beginning with the initial work of Alder and Wainwright [37,38] in the microcanonical ensemble. Gruhn and Monson [39], following an analysis by de Smedt *et al.* [40], extended DMD for the hard-sphere potential to the NPT ensemble. Their method, however, was based on Andersen's constant pressure algorithm [1], which does not yield averages consistent with Equation (6). Gruhn and Monson [39] derived expressions for the discontinuous change of the momenta of two hard spheres upon collision, as well as the change of the velocity of the piston (or system volume) upon that same collision.

We use the shell particle equations of motion provided in Equation (16) to develop a constant pressure DMD algorithm for both the hard-sphere and square-well fluids that are consistent with the proper NPT ensemble partition function, Equation (6). Momentum changes upon the collision of any two particles, including those changes for the shell particle, whether or not it participates in the collision, were derived in reference [18] and presented below. Our method is based on that of Gruhn and Monson [39], though we utilize the conservation of the extended Hamiltonian to obtain the collision dynamics. We simply present the results below, so the reader interested in the detailed derivations are referred to [18].

In an additive hard-sphere system, the potential of interaction between two particles, i and j, with diameters, σ_i and σ_j, respectively, is represented by:

$$u(r) = \begin{cases} \infty, & r < \sigma \\ 0, & r \geq \sigma \end{cases} \tag{29}$$

where r is the distance between the particle centers and $\sigma = (\sigma_i + \sigma_j)/2$. In between collisions, the hard-sphere fluid evolves dynamically without any force interactions. When applying the shell particle equations of motion to a hard-sphere collision, one must consider two separate cases: (1) neither particle i nor particle j is the shell particle and (2) either i or j is the shell particle. Furthermore, even if the shell particle does not participate in a collision, its x momentum will still change, since the acceleration of the shell particle is proportional to the internal pressure, which varies upon any collision.

There are several variables that are present in all of the expressions for the collision dynamics. These are the reduced mass, μ, the center-to-center vector, $\vec{q}$, and $\dot{r}$. The reduced mass is

$$\mu = \frac{m_i m_j}{m_i + m_j} \tag{30}$$

where m_i and m_j are the masses of particles i and j, respectively. $\vec{q}$ is defined as $\vec{q} = \vec{q}_i - \vec{q}_j$, and $\dot{r}$ is $\dot{r} = (\vec{q} \cdot \dot{\vec{q}})/\sigma$, which is the time rate of change of $\vec{q}$ evaluated at $|\vec{q}| = \sigma$. When neither of the colliding particles are the shell particle, the collision dynamics are given by:

$$\begin{aligned} \bar{t}_d &= \frac{-2\mu\dot{r}}{1 + \mu\sigma^2/m_1 q_1^2} \\ \Delta\vec{p}_i &= \bar{t}_d \frac{\vec{q}}{\sigma} \\ \Delta\vec{p}_j &= -\bar{t}_d \frac{\vec{q}}{\sigma} \\ \Delta p_1 &= \bar{t}_d \frac{\sigma}{q_1} \end{aligned} \tag{31}$$

When one of the colliding particles is the shell particle, the collision dynamics are now given by:

$$\bar{t}_d = \frac{-2\mu\dot{r}}{1 + \mu\sigma^2/m_1q_1^2 - \mu q_x^2/m_1\sigma^2}$$

$$\Delta\vec{p}_i = \bar{t}_d\frac{\vec{q}}{\sigma}$$

$$\Delta p_{1,y} = -\bar{t}_d\frac{q_y}{\sigma}$$

$$\Delta p_{1,z} = -\bar{t}_d\frac{q_z}{\sigma}$$

$$\Delta p_1 = \bar{t}_d\frac{\sigma}{q_1} \tag{32}$$

where, for example, q_x is the x component of $\vec{q}$.

The square-well interaction potential is represented by:

$$u(r) = \begin{cases} \infty, & r < \sigma \\ -\epsilon, & \sigma \le r < \lambda\sigma \\ 0, & r \ge \lambda\sigma \end{cases} \tag{33}$$

where λ is the width and ϵ is the depth of the square-well (and may vary depending upon the interaction between any two particles). The interaction at $r \to \sigma$ is identical to the hard-sphere collision obtained above. There are three other types of collisions that occur in the square-well system at $r \to \lambda\sigma$. The capture interaction is the case where i and j start beyond $\lambda\sigma$. There are two types of collision that occur at $\lambda\sigma$ when the starting distance between i and j are within the attractive well. The dissociation collision occurs when the molecules have enough kinetic energy to overcome the attractive potential energy and the molecules no longer interact, and the bounce collision occurs when there is not enough kinetic energy to overcome the attractive energy and the particle centers stay within the attractive well. The bounce dynamics are analogous to the hard-sphere collision presented above with the only difference being setting $\sigma = \lambda\sigma$ in Equations (31) and (32). The mathematical condition for determining if the collision is a bounce or a dissociation collision is that if $\mu\dot{r}^2/2 \ge \epsilon(1 + \mu\sigma^2/m_1q_1^2)$ for the shell particle not taking part in the collision and $\mu\dot{r}^2/2 \ge \epsilon(1 + \mu\sigma^2/m_1q_1^2 - \mu q_x^2/m_1\lambda^2\sigma^2)$ when the shell particle is taking part in the collision, then the collision is a dissociation collision.

The collision dynamics for capture and dissociation differ only by a plus/minus sign, so we present them together. When neither of the colliding particles are the shell particle, the collision dynamics are given by:

$$\bar{t}_d = \frac{\mu\dot{r} \pm \mu[\dot{r}^2 \pm (2\epsilon/\mu)(1 + \mu\lambda^2\sigma^2/m_1q_1^2)]^{1/2}}{1 + \mu\lambda^2\sigma^2/m_1q_1^2}$$

$$\Delta\vec{p}_i = -\bar{t}_d\frac{\vec{q}}{\lambda\sigma}$$

$$\Delta\vec{p}_j = \bar{t}_d\frac{\vec{q}}{\lambda\sigma}$$

$$\Delta p_1 = -\bar{t}_d\frac{\lambda\sigma}{q_1} \tag{34}$$

with the "+" being for capture and the "−" being for dissociation. When one of the colliding particles is the shell particle, then the collision dynamics are given by:

$$\bar{t}_d = \frac{\mu \dot{r} \pm \mu[\dot{r}^2 \pm (2\epsilon/\mu)(1 + \mu\lambda^2\sigma^2/m_1q_1^2 - \mu q_x^2/m_1\lambda^2\sigma^2)]^{1/2}}{1 + \mu\lambda^2\sigma^2/m_1q_1^2 - \mu q_x^2/m_1\lambda^2\sigma^2}$$

$$\Delta\vec{p}_i = -\bar{t}_d\frac{\vec{q}}{\lambda\sigma}$$

$$\Delta p_{1,y} = \bar{t}_d\frac{q_y}{\lambda\sigma}$$

$$\Delta p_{1,z} = \bar{t}_d\frac{q_z}{\lambda\sigma}$$

$$\Delta p_1 = -\bar{t}_d\frac{\lambda\sigma}{q_1} \tag{35}$$

We integrate the NPT equations of motion in between collisions via the application of the generalized Trotter expansion formula to the extended phase space classical Liouville operator discussed in the appendix and [17,18,22,24,41]. Since the thermostat variables have no influence on a hard-sphere or square-well collision, the updates of the thermostats can be completely decoupled from the updates of the particle positions and the momentum changes upon a collision. The full integration scheme is presented in detail in [18].

3.4. Shell Particle Simulations Using the Configurational Temperature

The concept of a configurational temperature was introduced in 1997 in the seminal paper by Rugh [42], which provided a tractable statistical mechanical expression for the reciprocal of this temperature. The expression for the configurational temperature was later generalized by Jepps *et al.* [43]. Since then, several MD algorithms have been developed that make use of the configurational temperature, but the few that are most useful for the current discussion are by Braga and Travis [44,45]. They introduced NPT equations of motion that use the configurational temperature and showed the benefits of using this temperature, instead of the standard kinetic temperature, within nonequilibrium simulations [44,45].

The equations of motion that they derived are not consistent with the shell molecule partition function, however, and so, we reformulated their equations to account for this. The new equations of motion are:

$$\dot{q}_i = \frac{p_i}{m_i} + \left(\frac{\dot{q}_1}{q_1}\right)q_i + \left(\frac{\dot{q}_1}{q_1}\right)(3N-1)\frac{F_i}{\Delta'} + \dot{\xi}\frac{F_i}{\Delta'}$$

$$\dot{p}_i = F_i$$

$$\dot{q}_1 = \frac{p_1}{m_1}$$

$$\dot{p}_1 = 24q_1^2(P_{int} - P)$$

$$\dot{\xi} = \frac{p_\xi}{Q}$$

$$\dot{p}_\xi = \sum_{i=2}^{3N}\frac{F_i^2}{\Delta'} - k_B T \tag{36}$$

where $i = 2, ..., 3N$,

$$\Delta' = \sum_{i=2}^{3N} (\frac{\partial^2 U}{\partial q_i^2}) \tag{37}$$

where U is the total potential energy, and:

$$P_{int} = \frac{1}{24q_1^3} \left[(3N - 1) \sum_{i=2}^{3N} \frac{F_i^2}{\Delta'} + \sum_{j=1}^{3N} q_j F_j \right] \tag{38}$$

The extended Hamiltonian for the configurational temperature shell molecule system is:

$$H_{ext} = \sum_{i=1}^{3N} \frac{p_i^2}{2m_i} + \sum_i \sum_{j>i} U(r_{ij}) + 8q_1^3 P + \frac{p_\xi^2}{2Q} + k_B T \xi \tag{39}$$

The instantaneous configurational temperature, $k_B T_{conf}$, is:

$$k_B T_{conf} = \sum_{i=2}^{3N} \frac{F_i^2}{\Delta'} \tag{40}$$

The instantaneous configurational temperature appearing in the above shell particle equations of motion differs from that of Braga and Travis [44,45], whereby the sums appearing in Equations (37) and (40) run from two to $3N$, as compared to one to $3N$. The x-component of the shell particle is not included in these summations, although the shell particle does still contribute to the forces (and their derivatives) of the remaining particles. Since the configurational temperature thermostating appears to be preferred in nonequilibrium simulations, as known artifacts seen in some simulations with the Nosé-Hoover thermostat were not exhibited with the configurational temperature thermostat [44,45], we also include below new results for NPT MD simulations with the shell molecule using the configurational temperature.

4. Results and Discussion

Several tests of the new shell particle equations of motion (Equation (16)) have been published previously. The agreement between isobars predicted by the new shell molecule molecular dynamics, the shell molecule Monte Carlo algorithm [15] and the equation of state for the Lennard-Jones fluid introduced by Johnson et al. [46] is shown in Figure 1 of [17]. Similar agreement between the MD and MC results is presented in Figure 1 of [18] for the square-well potential, which also includes, for comparison, the predictions of an equation of state introduced by Patel et al. [47]. For both systems, the MD and MC simulation results agree with each other and with the appropriate equation of state to high accuracy over a very broad range of pressures. On the scale of the plots, the MD and MC results are nearly indistinguishable [17,18]. We also looked at the self-diffusion coefficients for various binary Lennard-Jones mixtures and compared them with results from MD simulations in the microcanonical ensemble [17]. The self-diffusion coefficients of each species were essentially identical within both ensembles.

Table 1. Comparison of various ensemble averages for the truncated and shifted Lennard-Jones fluid with a cuttoff of 1.5σ for $T^* = 1.5$ and $P^* = 0.5$. The top dataset was obtained with the shell particle Monte Carlo method [15]. The second dataset was generated from the shell particle molecular dynamics (MD) algorithm using the Nosé-Hoover chained thermostat. The third dataset was obtained from the shell particle MD algorithm using the configurational temperature thermostat. The fourth dataset was obtained from the Hoover algorithm. The bottom set are the results of constant pressure MC simulations without a shell particle. The numbers in parentheses indicate the error in the final significant digits.

N	$\langle P^*_{int}\rangle$	$\langle \rho^*\rangle$	$\langle V^*\rangle$	$\langle U^*\rangle$	$\langle T^*\rangle$
16	0.514(54)	0.227(43)	73(7)	−0.045(32)	1.5
32	0.507(49)	0.223(34)	145(11)	−0.045(18)	1.5
64	0.503(37)	0.222(16)	291(14)	−0.045(12)	1.5
108	0.502(24)	0.221(8)	490(18)	−0.045(10)	1.5
256	0.501(15)	0.221(2)	1161(32)	−0.045(9)	1.5
16	0.514(235)	0.229(31)	72(12)	−0.045(111)	1.49(31)
32	0.506(161)	0.224(27)	144(17)	−0.045(78)	1.50(22)
64	0.503(111)	0.223(19)	290(25)	−0.045(55)	1.50(15)
108	0.502(85)	0.221(14)	489(32)	−0.045(42)	1.50(12)
256	0.501(55)	0.221(9)	1161(49)	−0.045(28)	1.50(8)
16	—	—	—	—	—
32	0.506(346)	0.218(27)	148(18)	−0.038(86)	1.51(15)
64	0.503(289)	0.219(21)	294(27)	−0.041(55)	1.51(14)
108	0.502(203)	0.219(15)	493(36)	−0.043(42)	1.50(12)
256	0.501(133)	0.220(9)	1165(50)	−0.044(28)	1.50(5)
16	0.522(235)	0.230(40)	71(12)	−0.047(111)	1.50(31)
32	0.511(162)	0.226(27)	144(17)	−0.045(78)	1.50(22)
64	0.505(111)	0.223(19)	289(24)	−0.045(55)	1.50(15)
108	0.503(85)	0.222(14)	488(32)	−0.045(42)	1.50(12)
256	0.501(56)	0.221(9)	1163(48)	−0.045(28)	1.50(8)
16	0.500(62)	0.220(18)	75(8)	−0.045(26)	1.5
32	0.500(44)	0.220(15)	147(11)	−0.045(18)	1.5
64	0.500(31)	0.220(10)	293(14)	−0.045(13)	1.5
108	0.500(25)	0.220(8)	492(20)	−0.045(8)	1.5
256	0.500(16)	0.220(2)	1164(36)	−0.045(5)	1.5

The differences between systems that sample the rigorously correct volume distributions and those that do not can most readily be seen in small systems. Equations (9) and (10), for example, provide strict tests of the validity of the shell particle equations of motion when compared against simulation methodologies that don not employ the correct volume scale. Several state points for Lennard-Jones, hard-sphere and square-well fluids are compared in [17,18]. In each of the conditions studied, the relations derived earlier are found to be satisfied to a high accuracy. We also included results from the Hoover algorithm (Equation (21), where it is important to note that the internal pressure equation for the Hoover algorithm is given by Equation (24)).

As an additional test, we present here results for a pure component Lennard-Jones fluid with system sizes ranging from $N = 16$ to 256. To avoid the use of long-range corrections, as well as force profiles that would not sum to zero in the y and z directions, we utilized the truncated and shifted force Lennard-Jones potential [38]:

$$u(r) = 4\epsilon\left[\left(\frac{\sigma}{r}\right)^{12} - \left(\frac{\sigma}{r}\right)^{6}\right] - u_{LJ}(r_c) - (r - r_c)u'_{LJ}(r_c) \tag{41}$$

where r_c is the cutoff distance and $u'_{LJ}(r_c)$ is the derivative of the potential at the cutoff distance. At the chosen truncation distance, both the potential and force smoothly vanish. To prevent the truncation distance from exceeding half the box length at small system sizes, since periodic boundary conditions were employed, we chose $r_c = 1.5\sigma$. Up to 10^6 time steps of equilibration were performed, followed by up to 10^8 time steps at the smallest system sizes for the determination of ensemble averages. All simulations were run at $T^* = 1.5$. The results for a reduced external pressure of $P^* = 0.5$ are included in Table 1. Additionally, provided in the tables are the averages obtained from MC simulations both with and without the shell particle.

According to Equation (10), the average volume of the shell particle simulations should be three units lower than the no-shell simulations in Table 1. The simulation results agree quite well with these predictions, considering the large absolute volume fluctuations that are obtained and satisfy Equation (10) with similar accuracy as noted in [15]. As expected, the average volume, density and internal energy per particle obtained with the shell particle MD with the traditional Nosé-Hoover thermostat, the shell particle MD with the configurational thermostat and the Hoover algorithms are in agreement, at least within the error bars. Both sets of averages are nearly the same for $P^* = 0.5$. The average internal pressures differ, but each is seen to satisfy Equations (9) and (24) to a high degree of accuracy. Both of the shell particle results and Hoover results also agree, within the error bars, with the MC shell particle simulations. There is, however, a slight discrepancy between the MD and MC shell particle results at very small system sizes ($N \leq 64$), particularly for the values of P_{int} and the average density or volume. This difference can be attributed to the relatively large temperature fluctuations that develop within the MD simulations, as opposed to the strictly fixed temperature during the MC simulations. Statistical mechanics requires that the kinetic temperature of the system have a standard deviation of:

$$\sigma_T = \sqrt{\langle T^2 \rangle - \langle T \rangle^2} = T_{bath}\sqrt{\frac{2}{3N}} \tag{42}$$

where T_{bath} is the temperature of the surrounding temperature reservoir and N is the number of particles in the system. Equation (42) holds regardless of the usage of periodic boundary conditions [25,33]. In Table 1, we report the standard deviation by the number in parentheses indicating the error in the final significant digits. As an example, the number $1.50(31)$ means that the average is calculated to be 1.50 and the standard deviation is 0.31. The results for the temperature fluctuations in Table 1 agree very closely to Equation (42).

Furthermore, presented in Table 1 are the results for the shell molecule configurational temperature NPT. Note that the results for the configurational temperature are not provided for $N = 16$. At this small density and small number of particles, there is a chance that no pairs of particles reside within the cutoff distance. As a result, Δ' is equal to zero and the integration scheme breaks down for that time step. This problem only arose for the smallest system size ($N = 16$). Additionally, note in Table 1 that the configurational temperature yields the largest temperature fluctuations as compared to the other simulation methods. Again, at this relatively low density, the effects of adding or deleting one or two particle pairs within the potential cutoff for each time step are greatly enhanced for the configurational temperature (as compared to the kinetic temperature, which is based solely on the particle momenta). The average volumes are larger than they should be and the average potential energy is lower than it should be for the configurational temperature simulations. Interestingly, the results do seem to improve consistently as the number of particles increases. This may be due, in some small part, to the given expression for the configurational temperature, which as a measure of the system temperature is only accurate on the order of $(1/N)$ [44,45].

4.1. Discontinuous Pressure Jumps

Ultimately, the true benefits of the shell particle algorithm may become apparent for nonequilibrium simulations where the system itself sets the time scale for pressure/volume fluctuations. It is important to note that in a multicomponent system, there is freedom to choose the identity of the shell particle, although the masses of the various components comprising the mixture are still known. To gain some initial idea of how the shell particle equations of motion might behave in a nonequilibrium application, we ran a pressure-jump simulation in which the external pressure is abruptly changed after the system has equilibrated. For example, we first consider the response of the internal pressure to a sudden change in the external pressure from $P^* = 1.0$ to $P^* = 2.0$ and then back to $P^* = 1.0$ at $T^* = 2.0$ for the pure component Lennard-Jones fluid with $N = 500$ and long-range corrections applied after a potential cutoff of 3.0σ. The resulting time evolution of the internal pressure is shown in Figure 2.

The figure includes results for the shell molecule with the Nosé-Hoover thermostat, the shell molecule with the configurational temperature thermostat and results for the Hoover algorithm with the reduced piston mass, M_p^*, equal to $M_p^* = 10.0$ and $M_p^* = 5.0$, respectively. Both of the shell particle simulations and the Hoover simulation with $M_p^* = 10.0$ quickly adjust to the new external pressure, while the Hoover simulation with $M_p^* = 5.0$ requires a much longer time to re-equilibrate (again, the time scale obtained from the Hoover code is directly dependent upon the mass of the piston, whereas the time scale for the shell particle algorithm is automatically set by the system). This result is somewhat surprising, since the two piston masses are so close in their numerical values. The fluctuations of the internal pressure exhibited by both of the shell particle codes at the new external pressure of $P^* = 2.0$ and, then, again, at $P^* = 1.0$ are immediately identical to the fluctuations seen from regular equilibrium simulations at $P^* = 2.0$ and $P^* = 1.0$. The internal pressure for the Hoover simulations also adjusts to the new external pressure, but there does appear to be a considerable "decay" to the new set point after the pressure jump. This decay is dependent on the value of the piston mass. A smaller value of the piston mass yields a longer decay in the instantaneous pressure to the new equilibrium point.

Figure 2. Time response of the internal pressure of the pure component Lennard-Jones fluid to a sudden change of the external pressure from $P^* = 1.0$ to $P^* = 2.0$ and, then, back down to $P^* = 1.0$. For the given choice of the time origin, the pressure is increased after 2000 time steps and, then, reduced after another 4000 time steps. The solid line is the set external pressure, P. The dashed lines are the simulation results. In all cases, $T^* = 2.0$ and $N = 500$. The plot in the upper-left corner is the results for the shell molecule with the Nosé-Hoover thermostat. The plot in the upper-right corner is the results for the shell molecule with the configurational temperature thermostat; The plots in the lower-left and lower-right corner are the results for the Hoover algorithm with $M_p^* = 10.0$ and $M_p^* = 5.0$, respectively.

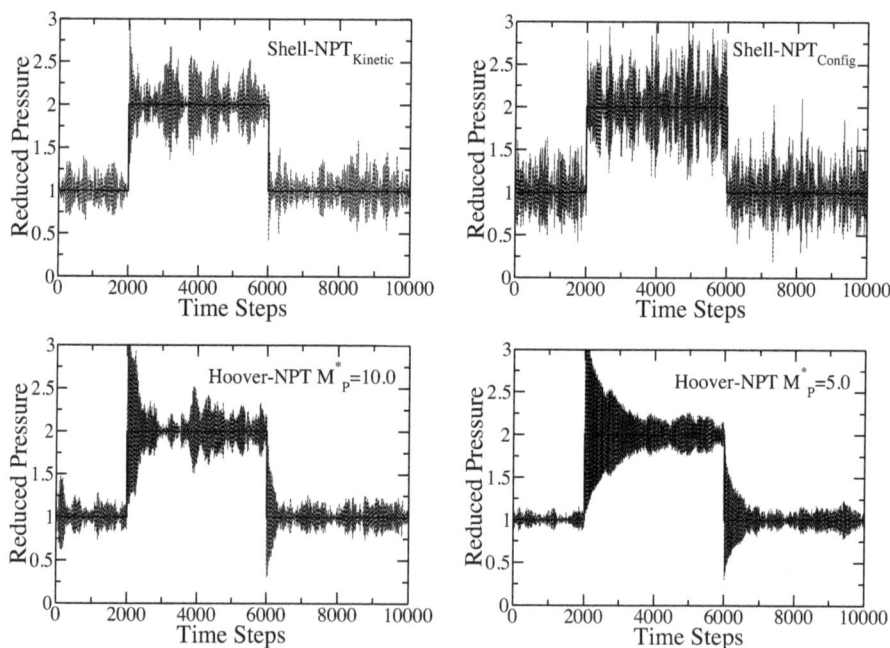

We also preformed an isothermal compression as a series of steps from $P^* = 1.0$ to $P^* = 4.0$ in increments of 1.0 unit of reduced pressure every 2000 time steps at $T^* = 2.0$ for the pure component Lennard-Jones fluid with $N = 500$ and long-range corrections applied after a potential cutoff of 3.0σ.

The results are presented in Figure 3. The value of the time steps, along with every other aspect of the simulations, are the same as those performed for Figure 2. As before, both of the shell particle simulations exhibit fluctuations of the internal pressure after the jumps to be almost immediately identical to the fluctuations seen from regular equilibrium simulations at the respective set external pressures at equilibrium. The internal pressure for the Hoover simulations also adjust to the new external pressure, but again, there appears to be a considerable decay to the new set point after the pressure jump. The simulation with $M_P^* = 5.0$ does not allow the internal pressure to equilibrate after an external pressure jump before the system takes the next jump. This shows that the value of the piston mass is critical to capturing the dynamics and fluctuations in nonequilibrium systems and that the results can be considerably different for values of the piston mass that are relatively close to one another.

Figure 3. Time response of the internal pressure of the pure component Lennard-Jones fluid to an isothermal compression from $P^* = 1.0$ to $P^* = 4.0$ in increments of 1.0 unit of reduced pressure every 2000 time steps. For the given choice of the time origin, the pressure is increased after 2000, 4000 and, again, after 6000 time steps. The solid line is the set external pressure, P. The dashed lines are the simulation results. In all cases, $T^* = 2.0$ and $N = 500$. The plot in the upper-left corner is the results for the shell molecule with the Nosé-Hoover thermostat. The plot in the upper-right corner is the results for the shell molecule with the configurational temperature thermostat. The plots in the lower-left and lower-right corner are the results for the Hoover algorithm with $M_p^* = 10.0$ and $M_p^* = 5.0$, respectively.

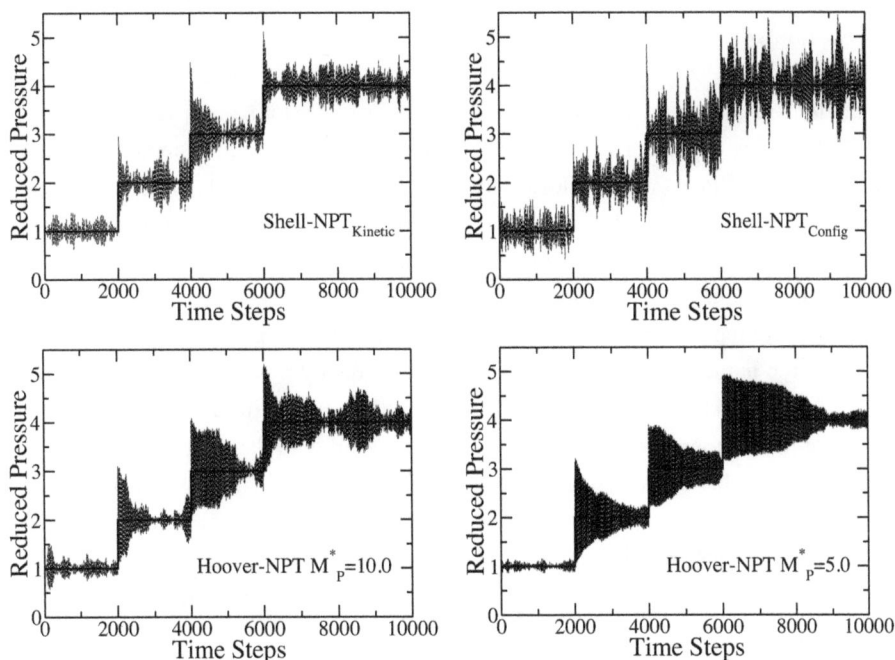

5. Conclusions

The MD NPT simulation method that employs the shell particle is based on equations of motion consistent with the proper statistical mechanical formulation of the NPT ensemble. Within other MD methods, a piston of arbitrary mass is introduced to control the response time of volume fluctuations. Now, the shell particle of known mass determines the time scale for volume and pressure fluctuations, in addition to performing the important function of eliminating the redundant counting of configurations through its unique definition of the volume of the system.

There are several benefits to using the shell particle algorithm for MD equilibrium simulations. For example, as the shell particle directly interacts with all other particles in the system, only a single Nosé-Hoover chained thermostat need be employed. Additionally, as noted above, the mass of the "volume", that is, the shell particle, is a known quantity.

Allowing the system itself to control the relaxation time of property fluctuations should ultimately provide a significant edge over piston-based methods, specifically for nonequilibrium systems (though that has yet to be shown). Adapting the shell particle approach to a simulation of isothermal-isobaric shear flow [44,45] or homogeneous nucleation in simple fluids [48–50] may provide another worthwhile test of the shell particle formalism. The piston mass is not known *a priori* and greatly affects the response time of the system. As such, for nonequilibrium simulations, the appropriate choice of a piston mass is unclear. On the other hand, at least for a pure component system, there is no ambiguity as to what should be the response time of the volume fluctuations; the mass of the shell particle is again a known input to the simulation. For mixtures, however, the identity of the shell particle is important (though not for equilibrium averages). Yet, the masses of the components comprising the mixture are still known. How the identity of the shell particle controls the response time of pressure and volume fluctuations is therefore, in the end, still a property of the system itself.

Future directions include incorporating the shell particle formalism into constant pressure molecular dynamics algorithms for both molecular systems and systems with long-range intermolecular interactions (such as electrostatic interactions). To date, we have not performed any simulations of these systems with the shell particle, so definitive statements would be inappropriate at this time. However, at first sight, it would appear that selecting the center of mass of one of the molecules as the volume scale is a logical choice for molecular systems. It also appears that the techniques already employed to deal with long-range interactions in previous algorithms (such as the Ewald summation in electrostatic systems) would apply equally to shell particle algorithms. There is nothing in the derivations given in this manuscript that suggests to us that intra- or long-range intermolecular interactions would have any effect on the validity of the shell molecule formulation.

Finally, various shape changing, or isotension, simulations [25] may also benefit from the use of the shell particle. The introduction of a shell particle into these codes would add some complexities, such as possibly allowing the shell particle to move along different faces of the system volume. Nevertheless, eliminating the need to specify a piston mass might be very helpful for these simulations.

Acknowledgments

This work was supported by the National Science Foundation/EPSCoR Grant(EPS-0903795) to Mark J. Uline.

Conflicts of Interest

The authors declare no conflict of interest.

Appendix

Integration Scheme Using the Liouville Operator

We integrate the NPT equations of motion via the application of the generalized Trotter expansion formula to the extended phase space classical Liouville operator [25,41]. What follows from this approach is an integration scheme that is time-reversible and volume-preserving in the appropriate extended phase space, yielding stable trajectories with no significant drift in the extended Hamiltonian. One begins with a unitary operator, called the classical propagator, that propagates the appropriate phase space vector, $\vec{\Gamma}$, from an initial state at time $t = 0$ to a final state at time t, *i.e.*, $\vec{\Gamma}(t) \to \vec{\Gamma}(0)$. The evolution as a function of time t can be formally written as the solution of the following equation:

$$\vec{\Gamma}(t) = exp(iLt)\vec{\Gamma}(0) \tag{43}$$

where iL is the Liouville operator, defined as $iL = \vec{\Gamma} \cdot \nabla_{\vec{\Gamma}}$, and $exp(iLt)$ is the classical propagator. The Liouville operator can be expressed as the sum of other operators, for example, $iL = iL_1 + iL_2$, such that the action of each separate classical propagator can be evaluated analytically. Since iL_1 and iL_2 do not, in general, commute, $exp(iLt)$ cannot be replaced by $exp(iL_1)exp(iL_2)$. The classical propagator can be rewritten, however, using the Trotter expansion theorem [21–26],

$$e^{iLt} = e^{(iL_1t+iL_2t)} = e^{(iL_1t/2)}e^{(iL_2t)}e^{(iL_1t/2)} + O(t^3) \tag{44}$$

Application of the above operator, along with the given equations of motion, provides a sequence of operations on the components of $\vec{\Gamma}$, thereby generating an integration scheme that updates $\vec{\Gamma}$ from time t to time $t + \Delta t$. Since we start from the Liouville formulation of classical mechanics and the classical propagator is a unitary operator, the resulting MD algorithm is guaranteed to be time-reversible and (phase space) volume-preserving (within finite machine precision and to the second order in the chosen time step, Δt).

The phase space vector appropriate for the shell particle equations of motion with a single chain of C thermostats is $\vec{\Gamma} = \vec{\Gamma}(q_1, p_1, q_i, p_i, \xi_k, p_{\xi_k})$, where $i = 2$ to $3N$ and $k = 1$ to C. In this case, the Liouville operator is given by [17]:

$$iL_{NPT} = \dot{q}_1\frac{\partial}{\partial q_1} + \dot{p}_1\frac{\partial}{\partial p_1} + \sum_{i=2}^{3N}\left[\dot{q}_i\frac{\partial}{\partial q_i} + \dot{p}_i\frac{\partial}{\partial p_i}\right] + \sum_{k=1}^{C}\left[\dot{\xi}_k\frac{\partial}{\partial \xi_k} + \dot{p}_{\xi_k}\frac{\partial}{\partial p_{\xi_k}}\right] \tag{45}$$

By changing the order of some of the terms and replacing the various time derivatives with their expressions given in Equation (16), we rewrite the full Liouville operator as a sum of the following five separate operators:

$$iL_{NPT} = iL_q + iL_{p_1} + iL_{p_i} + iL_{p_{1i}} + iL_{NH} \tag{46}$$

in which:

$$
\begin{aligned}
iL_q &= \frac{p_1}{m_1}\frac{\partial}{\partial q_1} + \sum_{i=2}^{3N}\left[\left(\frac{p_i}{m_i} + \left(\frac{\dot{q}_1}{q_1}\right)q_i\right)\frac{\partial}{\partial q_i}\right] \\[2mm]
iL_{p_1} &= G_\epsilon\frac{\partial}{\partial p_1} \\[2mm]
iL_{p_i} &= \sum_{i=2}^{3N}\left[F_i\frac{\partial}{\partial p_i}\right] \\[2mm]
iL_{p_{1i}} &= \sum_{i=2}^{3N}\left[\left(-\frac{\dot{q}_1}{q_1}\right)p_i\frac{\partial}{\partial p_i}\right] \\[2mm]
iL_{NH} &= \sum_{i=2}^{3N}\left[\left(-\frac{p_{\xi_1}}{Q_1}\right)p_i\frac{\partial}{\partial p_i}\right] - \frac{p_{\xi_1}}{Q_1}p_1\frac{\partial}{\partial p_1} + \sum_{k=1}^{C}\frac{p_{\xi_k}}{Q_k}\frac{\partial}{\partial \xi_k} + \sum_{k=1}^{C}G_{\xi_k}\frac{\partial}{\partial p_{\xi_k}} - \sum_{k=1}^{C-1}\frac{p_{\xi_{k+1}}}{Q_{k+1}}p_{\xi_k}\frac{\partial}{\partial p_{\xi_k}}
\end{aligned}
\tag{47}
$$

where:

$$
\begin{aligned}
G_\epsilon &= \left[\frac{1}{q_1}\left(\sum_{i=2}^{3N}\frac{p_i^2}{m_i} + \sum_{i=1}^{3N}q_i F_i\right) - 24q_1^2 P_{ext}\right] \\[2mm]
G_{\xi_1} &= \left[\sum_{i=1}^{3N}\frac{p_i^2}{m_i} - gk_B T\right] \\[2mm]
G_{\xi_k} &= \frac{p_{\xi_{k-1}}^2}{Q_{k-1}} - k_B T
\end{aligned}
$$

It is assumed in the above expressions that periodic boundary conditions are imposed, and we lose three of the momentum degrees of freedom in determining the instantaneous kinetic temperature (one from the x coordinate of the shell particle being coupled to the barostat and the center-of-mass momentum in the y and z directions being conserved and set equal to zero [33]). Under these circumstances, the instantaneous pressure is given in Equation (17), and the total number of the momentum degrees of freedom is $g = 3N - 2$.

The above decomposition of iL_{NPT} is slightly different from what was done previously for continuous potentials [17,24]. In particular, all particle positions have been removed from iL_{NH} and, so, are not altered by the action of this operator. What remains in iL_{NH} are the thermostat variables, as well as the influence of the thermostats on the particle momenta. Changes in the particle positions, along with further updates of the particle momenta due to the dilatation of the system volume, are now only generated by the action of the remaining four operators. We found slightly better conservation of the extended Hamiltonian by splitting up the operator in this way relative to our previous factorization [17]. This approach of completely separating the influence of the thermostat variables on the particle momentum was our approach to factorize the propagator when using discontinuous potentials [18]. The reason being that since the collision dynamics do not influence the thermostat variables, it is much more convenient to update the positions and momentum of the physical particles and their collision properties all at once separated from the thermostat variables [18].

To determine the effect of the full Liouville operator, or $exp(iL_{NPT}t)$, on $\vec{\Gamma}$, the Trotter expansion formula must be applied. Although there are several ways to do so, we follow a similar factorization proposed by Martyna *et al.* [24], whereby:

$$e^{iL_{NPT}t} = e^{iL_{NH}t/2}e^{iL_{P_1}t/2}e^{iL_{P_{1i}}t/2}e^{iL_{P_i}t/2}e^{iL_q t}e^{iL_{P_i}t/2}e^{iL_{P_{1i}}t/2}e^{iL_{P_1}t/2}e^{iL_{NH}t/2} + O(t^3) \tag{48}$$

The operator iL_{NH} has to be further divided, which we split in the following manner:

$$iL_{NH} = iL_{cv} + iL_{v\epsilon} + \sum_{k=1}^{C} iL_{G_{\xi_k}} + \sum_{k=1}^{C-1} iL_{v_{\xi_k}} + iL_\xi \tag{49}$$

where:

$$iL_{cv} = \sum_{i=2}^{3N} \left[\left(-\frac{p_{\xi_1}}{Q_1} \right) p_i \frac{\partial}{\partial p_i} \right]$$

$$iL_{v_\epsilon} = -\frac{p_{\xi_1}}{Q_1} p_1 \frac{\partial}{\partial p_1}$$

$$iL_{G_{\xi_k}} = G_{\xi_k} \frac{\partial}{\partial p_{\xi_k}}$$

$$iL_{v_{\xi_k}} = -\frac{p_{\xi_{k+1}}}{Q_{k+1}} p_{\xi_k} \frac{\partial}{\partial p_{\xi_k}}$$

$$iL_\xi = \sum_{k=1}^{C} \frac{p_{\xi_k}}{Q_k} \frac{\partial}{\partial \xi_k}$$

We again apply the Trotter expansion to factorize $exp(iL_{NH}t/2)$, the final form of which depends upon the number of thermostats in the chain. We present the results for $C = 3$ with $exp(iL_{NH}t/2)$ expanded as:

$$\begin{aligned}
e^{iL_{NH}t/2} = {}& e^{iL_{G_{\xi_3}}t/4} \left[e^{iL_{v_{\xi_2}}t/8} e^{iL_{G_{\xi_2}}t/4} e^{iL_{v_{\xi_2}}t/8} \right] \left[e^{iL_{v_{\xi_1}}t/8} e^{iL_{G_{\xi_1}}t/4} e^{iL_{v_{\xi_1}}t/8} \right] \\
& e^{iL_{v_\epsilon}t/4} \left[e^{iL_{cv}t/2} e^{iL_\xi t/2} \right] e^{iL_{v_\epsilon}t/4} \left[e^{iL_{v_{\xi_1}}t/8} e^{iL_{G_{\xi_1}}t/4} e^{iL_{v_{\xi_1}}t/8} \right] \\
& \left[e^{iL_{v_{\xi_2}}t/8} e^{iL_{G_{\xi_2}}t/4} e^{iL_{v_{\xi_2}}t/8} \right] e^{iL_{G_{\xi_3}}t/4}
\end{aligned} \tag{50}$$

Each of the above operators individually performs the following operations on the phase space vector [25]:

$$e^{iL_{G_{\xi_k}}t/4} \quad : \quad p_{\xi_k} \to p_{\xi_k} + G_{\xi_k}t/4$$

$$e^{iL_{v_{\xi_k}}t/8} \quad : \quad p_{\xi_k} \to p_{\xi_k} exp\left(-\frac{p_{\xi_{k+1}}}{Q_{k+1}}t/8 \right)$$

$$e^{iL_{v_e}t/4} \quad : \quad p_1 \to p_1 exp\left(-\frac{p_{\xi_1}}{Q_1}t/4\right)$$

$$e^{iL_{\xi}t/2} \quad : \quad \xi_k \to \xi_k + \frac{p_{\xi_k}}{Q_k}t/2; k = 1, ..., C$$

$$e^{iL_{cv}t/2} \quad : \quad p_i \to p_i exp\left(-\frac{p_{\xi_1}}{Q_1}t/2\right); i = 2, ..., 3N$$

$$e^{iL_{p_1}t/2} \quad : \quad p_1 \to p_1 + G_e t/2 \qquad\qquad (51)$$

$$e^{iL_{p_{1i}}t/2} \quad : \quad p_i \to p_i exp\left(-\frac{\dot{q}_1}{q_1}t/2\right); i = 2, ..., 3N$$

$$e^{iL_{p_i}t/2} \quad : \quad p_i \to p_i + F_i t/2; i = 2, ..., 3N$$

$$e^{iL_q t} \quad : \quad q_1 \to q_1 + \frac{p_1}{m_1}t$$

$$q_i \to q_i exp\left(\frac{\dot{q}_1}{q_1}t\right) + \frac{p_i}{m_i}t\frac{sinh\left(\frac{\dot{q}_1}{q_1}\frac{t}{2}\right)}{\left(\frac{\dot{q}_1}{q_1}\frac{t}{2}\right)}exp\left(\frac{\dot{q}_1}{q_1}t/2\right); i = 2, ..., 3N$$

where $sinh(x)/x$ can be expanded in a Maclaurin series to an arbitrarily high order [24] (we choose to truncate at the eighth order). In deriving the expression for q_i, we follow the literature and use a slightly different approach relative to propagating all of the other phase space variables forward in time [26]. Instead of using a Taylor series expansion (truncated at the second order) to express the action of the propagator on q_i [25], we instead rigorously solve the equation of motion (an ordinary first order differential equation with constant coefficients) for q_i, noting that all of the variables in the equation of motion are constant, except for q_i and t. Expanding $sinh(x)/x$ to an arbitrarily high order is equivalent to truncating the Taylor series expansion of the operator to an arbitrarily high order [21,24].

Now that the full operator $exp(iL_{NPT}t)$ has been factorized, we operate on the phase space vector by following the order of the expansion of $exp(iL_{NPT}t)$ from right to left, thereby sequentially propagating various components of $\vec{\Gamma}$ from time t to $t + \Delta t$.

References

1. Andersen, H.C. Molecular dynamics simulations at constant pressure and/or temperature. *J. Chem. Phys.* **1980**, *72*, 2384–2393.
2. Nosé, S. A unified formulation of the constant temperature molecular dynamics methods. *J. Chem. Phys.* **1984**, *81*, 511–518.
3. Nosé, S. A molecular dynamics method for simulations in the canonical ensemble. *Mol. Phys.* **1984**, *52*, 255–268.
4. Hoover, W.G. Canonical dynamics: Equilibrium phase-space distributions. *Phys. Rev. A* **1985**, *31*, 1695–1697.
5. Hoover, W.G. Constant-pressure equations of motion. *Phys. Rev. A* **1986**, *34*, 2499–2500.
6. Tuckerman, M.E.; Liu, Y.; Ciccotti, G.; Martyna, G.J. Non-Hamiltonian molecular dynamics: Generalizing Hamiltonian phase space principles to non-Hamiltonian systems. *J. Chem. Phys.* **2001**, *115*, 1678–1702.

7. Melchiomma, S.; Ciccotti, G.; Holian, B.L. Hoover NPT dynamics for systems varying in shape and size. *Mol. Phys.* **1993**, *78*, 533–544.

8. Martyna, G.J.; Tobias, D.J.; Klein, M.L. Constant pressure molecular dynamics algorithms. *J. Chem. Phys.* **1994**, *101*, 4177–4189.

9. Kalibaeva, G.; Ferrario, M.; Ciccotti, G. Constant pressure-constant temperature molecular dynamics: A correct constraint NPT ensemble using the molecular virial. *Mol. Phys.* **2003**, *101*, 765–778.

10. Keffer, D.J.; Baig, C.; Adhangale, P.; Edwards, B.J. A generalized Hamiltonian-based algorithm for rigorous equilibrium molecular dynamics simulation in the isobaric-isothermal ensemble. *Mol. Sim.* **2006**, *32*, 345–356.

11. Huang, C.; Li, C.; Choi, P.; Nandakumar, K.; Kostiuk, L. A novel method for molecular dynamics simulation in the isothermal-isobaric ensemble. *Mol. Phys.* **2011**, *109*, 191–202.

12. Koper, G.J.M.; Reiss, H. Length scale for the constant pressure ensemble: Application to small systems and relation to einstein fluctuation theory. *J. Phys. Chem.* **1996**, *100*, 422–432.

13. Corti, D.S.; Soto-Campos, G. Deriving the isothermal-isobaric ensemble: The requirement of a "shell" molecule and applicability to small systems. *J. Chem. Phys.* **1998**, *108*, 7959–7966.

14. Corti, D.S. Isothermal-isobaric ensemble for small systems. *Phys. Rev. E* **2001**, *64*, 016128.

15. Corti, D.S. Monte Carlo simulations in the isothermal-isobaric ensemble: The requirement of a "shell" molecule and simulations in small systems. *Mol. Phys.* **2002**, *100*, 1887–1904.

16. Uline, M.J.; Corti, D.S. Molecular dynamics in the isothermal-isobaric ensemble: The requirement of a "shell" molecule. I. Theory and phase-space analysis. *J. Chem. Phys.* **2005**, *123*, 164101.

17. Uline, M.J.; Corti, D.S. Molecular dynamics in the isothermal-isobaric ensemble: The requirement of a "shell" molecule. II. Simulation results. *J. Chem. Phys.* **2005**, *123*, 164102.

18. Uline, M.J.; Corti, D.S. Molecular dynamics in the isothermal-isobaric ensemble: The requirement of a "shell" molecule. III. Discontinuous potentials. *J. Chem. Phys.* **2008**, *129*, 014107.

19. Evans, D.J.; Morriss, G.P. Isothermal-isobaric molecular dynamics. *Chem. Phys.* **1983**, *77*, 63–66.

20. Evans, D.J.; Morriss, G.P. The isothermal-isobaric molecular dynamics ensemble. *Phys. Lett. A* **1983**, *98*, 433–436.

21. Raedt, H.D.; Raedt, B.D. Applications of the generalized Trotter formula. *Phys. Rev. A* **1983**, *28*, 3575–3580.

22. Tuckerman, M.E.; Berne, B.J.; Martyna, G.J. Reversible multiple time scale molecular dynamics. *J. Chem. Phys.* **1992**, *97*, 1990–2001.

23. Martyna, G.J.; Klein, M.L.; Tuckerman, M.E. Nosé-Hoover Chains—The canonical ensemble via continuous dynamics. *J. Chem. Phys.* **1992**, *97*, 2635–2643.

24. Martyna, G.J.; Tuckerman, M.E.; Tobias, D.J.; Klein, M.L. Explicit reversible integrators for extended system dynamics. *Mol. Phys.* **1996**, *87*, 1117–1157.

430

25. Frenkel, D.; Smit, B. *Understanding Molecular Simulation: From Algorithms to Applications*, 2nd ed.; Academic Press: San Diego, CA, USA, 2002.

26. Tuckerman, M.E. *Statistical Mechanics: Theory and Molecular Simulation*; Oxford Graduate Texts Series; Oxford University Press: Oxford, NY, USA, 2010.

27. Guggenheim, E.A. Grand potential functions and so-called "Thermodynamic Probability". *J. Chem. Phys.* **1939**, *7*, 103–107.

28. Byers Brown, W. Constant pressure ensembles in statistical mechanics. *Mol. Phys.* **1959**, *1*, 68–82.

29. Münster, A. Zur Theorie der Generalisierten Gesamtheiten. *Mol. Phys.* **1959**, *2*, 1–7. (In German)

30. Sack, R.A. Pressure dependent partition functions. *Mol. Phys.* **1959**, *2*, 8–22.

31. Hill, T.L. *Statistical Mechanics: Principles and Selected Applications*; Dover: Mineola, NY, USA, 1987.

32. McQuarrie, D.A. *Statistical Mechanics*; University Science Books: Sausalito, CA, USA, 2000.

33. Uline, M.J.; Siderius, D.W.; Corti, D.S. On the generalized equipartition theorem in molecular dynamics ensembles and the microcanonical thermodynamics of small systems. *J. Chem. Phys.* **2008**, *128*, 124301.

34. Tuckerman, M.E.; Mundy, C.J.; Martyna, G.J. On the classical statistical mechanics of non-Hamiltonian systems. *Europhys. Lett.* **1999**, *45*, 149–155.

35. Han, K.K.; San, H.S. On the isothermal-isobaric ensemble partition function. *J. Chem. Phys.* **2001**, *115*, 7793–7794.

36. Attard, P. On the density of states in the isobaric ensemble. *J. Chem. Phys.* **1995**, *103*, 9884–9885.

37. Alder, B.J.; Wainwright, T.E. Studies in molecular dynamics. I. General method. *J. Chem. Phys.* **1959**, *31*, 459–466.

38. Allen, M.P.; Tindesley, D.J. *Computer Simulation of Liquids*; Oxford University Press: Oxford, NY, USA, 1989.

39. Gruhn, T.; Monson, P.A. Isobaric molecular dynamics simulations of hard sphere systems. *Phys. Rev. E* **2001**, *63*, 061106.

40. De Smedt, Ph.; Talbot, J.; Lebowitz, J.L. Hard spheres in the isobaric-isoenthalpic ensemble. *Mol. Phys.* **1986**, *59*, 625–635.

41. Tuckerman, M.E.; Martyna, G.J. Understanding modern molecular dynamics: Techniques and applications. *J. Phys. Chem. B* **2000**, *104*, 159–178.

42. Rugh, H.H. Dynamical approach to temperature. *Phys. Rev. Lett.* **1997**, *78*, 772–774.

43. Jepps, O.G.; Ayton, G.; Evans, D.J. Microscopic expressions for the thermodynamic temperature. *Phys. Rev. E* **2000**, *62*, 4757–4763.

44. Braga, C.; Travis, K.P. A configurational temperature Nosé-Hoover thermostat. *J. Chem. Phys.* **2005**, *123*, 134101.

45. Travis, K.P.; Braga, C. Configurational temperature and pressure molecular dynamics: Review of current methodology and applications to the shear flow of a simple fluid. *Mol. Phys.* **2006**, *104*, 3735–3749.

46. Johnson, J.K.; Zollweg, J.A.; Gubbins, K.E. The Lennard-Jones equation of state revisited. *Mol. Phys.* **1993**, *78*, 591–618.

47. Patel, B.H.; Docherty, H.; Varga, S.; Galindo, A.; Maitland, G.C. Generalized equation of state for square-well potentials of variable ranges. *Mol. Phys.* **2005**, *103*, 129–139.

48. Wang, Z.-J.; Valeriani, C.; Frenkel, D. Homogeneous bubble nucleation driven by local hot spots: A molecular dynamics study. *J. Phys. Chem. B* **2009**, *113*, 3776–3784.

49. Torabi, K.; Corti, D.S. Molecular simulation study of cavity-generated instabilities in the superheated Lennard-Jones liquid. *J. Chem. Phys.* **2010**, *133*, 134505.

50. Meadley, S.L.; Escobedo, F.A. Thermodynamics and kinetics of bubble nucleation: Simulation methodology. *J. Chem. Phys.* **2012**, *137*, 074109.

Reprinted from *Entropy*. Cite as: Hülsmann, M.; Reith, D. SpaGrOW—A Derivative-Free Optimization Scheme for Intermolecular Force Field Parameters Based on Sparse Grid Methods. *Entropy* **2013**, *15*, 3640–3687.

Article

SpaGrOW—A Derivative-Free Optimization Scheme for Intermolecular Force Field Parameters Based on Sparse Grid Methods

Marco Hülsmann [1,*] and Dirk Reith [2]

[1] Fraunhofer Institute for Algorithms and Scientific Computing (SCAI), Schloss Birlinghoven, Sankt Augustin 53757, Germany

[2] Hochschule-Bonn Rhein-Sieg (HBRS), Grantham-Allee 20, Sankt Augustin 53757, Germany; E-Mail: dirk.reith@h-brs.de

* Author to whom correspondence should be addressed; E-Mail: marco.huelsmann@scai.fraunhofer.de; Tel.: +49-2241-14-2053; Fax: +49-2241-14-2656.

Received: 16 February 2013; in revised form: 15 July 2013 / Accepted: 28 August 2013 / Published: 6 September 2013

Abstract: Molecular modeling is an important subdomain in the field of computational modeling, regarding both scientific and industrial applications. This is because computer simulations on a molecular level are a virtuous instrument to study the impact of microscopic on macroscopic phenomena. Accurate molecular models are indispensable for such simulations in order to predict physical target observables, like density, pressure, diffusion coefficients or energetic properties, quantitatively over a wide range of temperatures. Thereby, molecular interactions are described mathematically by force fields. The mathematical description includes parameters for both intramolecular and intermolecular interactions. While intramolecular force field parameters can be determined by quantum mechanics, the parameterization of the intermolecular part is often tedious. Recently, an empirical procedure, based on the minimization of a loss function between simulated and experimental physical properties, was published by the authors. Thereby, efficient gradient-based numerical optimization algorithms were used. However, empirical force field optimization is inhibited by the two following central issues appearing in molecular simulations: firstly, they are extremely time-consuming, even on modern and high-performance computer clusters, and secondly, simulation data is affected by statistical noise. The latter provokes the fact that an accurate computation of gradients or Hessians is nearly impossible close to a local or global minimum, mainly because the loss function is flat. Therefore, the question arises of whether to apply a derivative-free method approximating the loss function by an appropriate model function. In this paper, a new Sparse Grid-based Optimization Workflow (SpaGrOW) is presented, which accomplishes this task robustly and, at the same time, keeps the number of time-consuming simulations relatively small. This is achieved by an efficient sampling procedure for the approximation based on sparse grids, which is described in full detail: in order to counteract the fact that sparse grids are fully occupied on their boundaries, a mathematical transformation is applied to generate homogeneous Dirichlet boundary conditions. As the main drawback of sparse grids methods is the assumption that the function to be modeled exhibits certain smoothness properties, it has to be approximated by smooth functions first. Radial basis functions turned out to be very suitable to solve this task. The smoothing procedure and the subsequent interpolation on sparse grids are performed within sufficiently large compact trust regions of the parameter space. It is shown and explained how the combination of the three ingredients leads to a new efficient derivative-free algorithm, which has the additional advantage that it is capable of reducing the overall number of simulations by a factor of about two in comparison to gradient-based optimization methods. At the same time, the robustness with respect to statistical noise is maintained. This assertion is proven by both theoretical considerations and practical evaluations for molecular simulations on chemical example substances.

Keywords: force field parameterization; molecular simulations; atomistic models; derivative-free optimization; sparse grids; smoothing procedures

Classification: PACS 34.20.Gj; 64.75.Gh

---

1. Introduction

In the last few decades, computer simulations have gained in importance for both science and industry, particularly due to the fast development of parallel high performance clusters. A denotative subarea of computer simulations are molecular simulations, which allow one to study the effects of modifications in microscopic states on macroscopic system properties. In contrast to simulations in process engineering, not the continuum, but the molecular level of a system is modeled. Thereby, the goal is to describe interatomic and intermolecular interactions, so that, on the one hand, certain accuracy demands are fulfilled, and on the other hand, the required computation time is as low as possible. The latter aspect is very important, because molecular simulations are numerically costly, also on modern computer clusters. The industrial relevance of molecular simulations originates from the fact that labor- and cost-intensive chemical experiments can be avoided. Hence, chemical systems can be simulated at temperatures and pressures that are very difficult to realize in a laboratory, the properties of toxic substances can be calculated without any risk and measure of precaution and processes on surfaces or within membranes are observable on a microscopic level. Another advantage of molecular simulations is that both the location and velocity of each particle are saved after certain time intervals, which results in a detailed observation of the behavior of the system. Altogether, molecular simulations have emerged as their own scientific discipline, and because of the continuous growth of computer resources, they will still become much more important in the coming years [1,2].

In principle, it is possible to describe interactions within a chemical system by quantum mechanics. Thereby, a partial differential equation, the so-called *Schrödinger equation* [3], has to be solved. As this turns out to be extremely difficult, especially for multi-particle systems, the problem is simplified by classical mechanical methods. Then, the system is considered on an atomistic level, *i.e.*, the smallest unit is an atom. Molecular simulation techniques are based on statistical mechanics, a subarea of classical mechanics. The most important simulation methods are molecular dynamics (MD) and Monte Carlo (MC). Thereby, both intra- and inter-molecular interactions are described by the foundation of a simulation, the force field. A force field consists of an analytic term and parameters to be adjusted. It is given by the potential energy, which has the following typical form [1,4]:

$$U_{pot}(r^M) := \sum_{Bonds} \frac{k_r}{2} (r - r_0)^2 + \sum_{Angles} \frac{k_\phi}{2} (\phi - \phi_0)^2 + \sum_{Dihedrals} \sum_{n=1}^{m} V_n \cos(n\omega)$$

$$+ \sum_{i<j}^{N} \left\{ 4\varepsilon_{ij} \left[\left(\frac{\sigma_{ij}}{r_{ij}} \right)^{12} - \left(\frac{\sigma_{ij}}{r_{ij}} \right)^6 \right] + \frac{q_i q_j}{4\pi\varepsilon_0 r_{ij}} \right\} \tag{1}$$

Thereby, $r^M \in \mathbb{R}^{3M}$ is a vector containing all three-dimensional coordinates of the interaction sites, where M is the number of particles in the system. The first row of Equation (1) describes the intramolecular and the second row the intermolecular part. The parameters of the intramolecular part, modeling the interactions caused by the modifications of bond lengths r, bond angles ϕ and dihedral angles ω can be computed by quantum mechanics. Please note that the index, $_0$, denotes the respective parameter in equilibrium and that k_r and k_ϕ are force constants. The factors, V_n, $n = 1, ..., m$, $m \in \mathbb{N}$, describe the rotation barriers around the molecular axes. The parameterization of the intermolecular part, modeling the interactions caused by dispersion—described by the Lennard-Jones (LJ) parameters, σ_{ij} and ε_{ij}, $i, j = 1, ..., M$, $i < j$, and electrostatic effects—described by partial atomic charges, q_i and q_j, $i, j = 1, ..., M$, $i < j$, is often tedious. The constant, $\varepsilon_0 = 8.854 \times 10^{-12}$ Fm^{-1}, is the dielectric constant.

1.1. Force Field Parameterization

It is known from the literature that many force fields describe molecular interactions accurately, qualitatively and quantitatively [5]. Intramolecular parameters can be determined by quantum mechanics, *i.e.*, by the minimization of a potential hyperplane. Partial atomic charges can also be computed from the position of nuclei and electrons. However, quantum mechanical methods require a high computational effort, especially for large molecules. This is why some simplifications were carried out, and the adjustment of the parameters was performed by fitting them to spectroscopic data. Some of the most famous force fields based on such semiempirical methods have been developed, e.g., by [6–8]. Quantum mechanical calculations of partial atomic charges were realized, e.g., by [9]. In related work [10,11], the automized optimization workflow, $WOLF_2PACK$, was created, which combines quantum mechanical algorithms with atomistic models and is capable of calculating both optimal intramolecular force field parameters and partial atomic charges. For the intermolecular part of Equation (1), the respective force field parameters were mostly adjusted to experimental target data, i.e., physical properties resulting from a molecular simulation were compared with experimental reference data. In particular, this empirical approach was realized by [12–14]. However, the parameters obtained cannot be considered as optimal, because they were not fitted to a large number of experimental data. Furthermore, in most cases, they ware adjusted manually, which is always time-consuming. They are transferable to other substances, but a subsequent readjustment is indispensable [15]. Many users take standard force fields from the literature for their simulations, which may lead to satisfactory, but not to optimal, target properties.

In the last decade, a few approaches were published realizing an automated force field parameterization procedure [16–20]. Thereby, physical target properties, like density, enthalpy of vaporization and vapor pressure, were fitted to their respective experimental reference data at different temperatures and pressures simultaneously. This was done via the minimization of a quadratic loss function between simulated and experimental data, i.e., by solving a mathematical optimization problem with numerical optimization algorithms. This approach is pursued in this paper, as well. The loss function to be minimized is given by:

$$F : \mathbb{R}^N \;\rightarrow\; \mathbb{R}_0^+ \qquad (2)$$

$$x \;\mapsto\; \sum_{i=1}^{n} w_i^2 \left(\frac{f_i^{\text{exp}} - f_i^{\text{sim}}(x)}{f_i^{\text{exp}}} \right)^2 \qquad (3)$$

where $x = (x_1, ..., x_N)^T$ is a force field parameter vector, N the dimension of the parameter space, n the number of considered physical properties, maybe at different temperatures, $f_i^{\text{sim}}(x)$, $i = 1, ..., n$ the simulated physical target properties as functions of the parameter vector, x, and f_i^{exp}, $i = 1, ..., n$ the respective experimental reference data. The weights, w_i^2, $i = 1, ..., n$, account for the fact that some properties are easier to reproduce or measured more accurately than others. The loss function, F, is minimized within a compact domain, $\Omega \subset \mathbb{R}^N$.

The optimization workflow is shown in Figure 1: the initial guess has to be reasonably close to the minimum. The target properties computed by a simulation tool are inserted into loss function in Equation (3) and compared with the experimental target properties. If a specified stopping criterion is fulfilled, the parameters are final, and the workflow terminates. Otherwise, the current parameter vector is passed on to the optimization procedure searching for new parameters with a lower loss function value.

There are two main requirements for the numerical optimization algorithms solving the minimization problem: Firstly, they have to be efficient, i.e., their convergency must be fast and the number of function evaluations has to be low, because for each function evaluation, time-consuming molecular simulations are needed, which can be parallelized for the different temperatures. Secondly, they have to be robust with respect to statistical noise, because simulation data is always affected by uncertainties. The reason for this is the fact that physical properties are computed by averaging over a certain time period in the case of MD simulations and over a certain number of system states in the case of MC simulations. In previous work [21–23], the software package, *Gradient-based Optimization Workflow (GROW)*, was developed. Thereby, efficient gradient-based numerical optimization algorithms were successfully applied to minimize loss function in Equation (3) in an efficient and robust way. Thereby, simple methods, like steepest descent and conjugate gradient algorithms, turned out to be most suitable. Algorithms requiring a Hessian were too time-consuming or not reliable whenever the Hessian was assumed to be positive definite, which it was not in most cases. Gradients and Hessians were approximated using first-order finite differences. The lengths of the descent directions were computed by an Armijo step length control mechanism.

Figure 1. Optimization workflow: The target properties are computed for an initial guess for the force field parameters. If they do not agree sufficiently well with the experimental target properties, the optimization procedure is performed searching for new parameters with a lower loss function value.

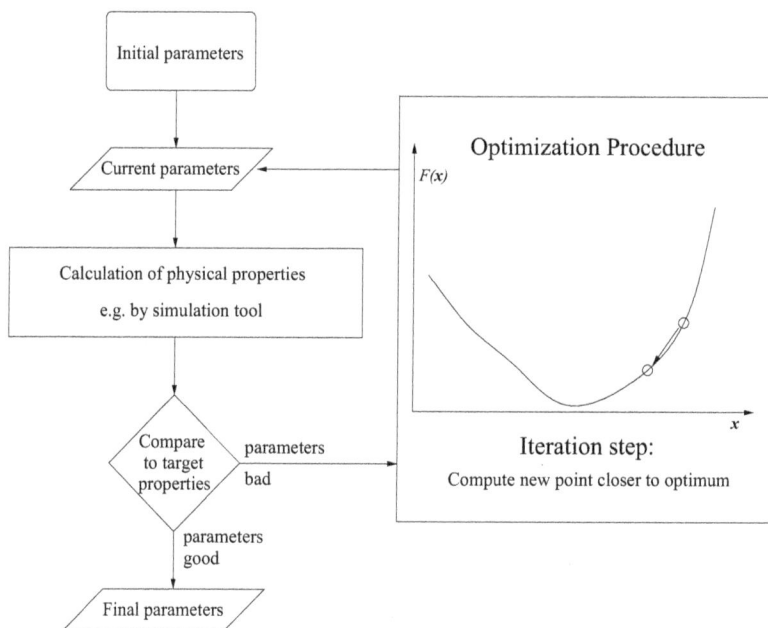

However, the accurate computation of a gradient or a Hessian is problematic close to a global or local minimum of F, due to the presence of statistical noise, *cf.* Figure 2. For the finite difference approximation of the gradient, two adjacent points are necessary. If these two points are situated too close to each other, their loss function values cannot be distinguished anymore, due to the error bars surrounding them. Therefore, the direction of the approximated gradient can be completely wrong.

This leads to the motivation to develop a new efficient derivative-free method counteracting this problem. The right side of Figure 2 shows that a possible solution is to approximate the loss function by an adequate model function and to determine the minimum of the model function. In this work, a new Sparse Grid-based Optimization Workflow (SpaGrOW) is presented in detail, implementing the aforesaid modeling approach. The main difficulty is to filter out the statistical noise during the approximation process, which is realized in SpaGrOW by regularization methods. As approximation is always based on sampling the parameter space and as molecular simulations are required for all selected points within the sampling process, sparse grids are involved in order to avoid high computational effort. The combination technique developed by [24] is applied in order to perform a piecewise multilinear interpolation from a sparse grid to a full grid. As sparse grids are fully occupied at their boundaries, the loss function is artificially set to zero at the boundary by a mathematical transformation. Then, the minimum on the full grid is determined, and a back-transformation is performed. As the combination technique requires certain smoothness properties, and as the loss function cannot be assumed to be smooth, it has to be preprocessed. Hence, it is approximated by a smooth model function before. The selection of a suitable model function is done by both theoretical considerations and practical evaluations.

Figure 2. Motivation of a derivative-free method in the case of noisy loss function values close to the minimum; the direction of a gradient can be completely wrong. Hence, an approximation of the loss function is necessary. This regression procedure has to filter out the statistical noise.

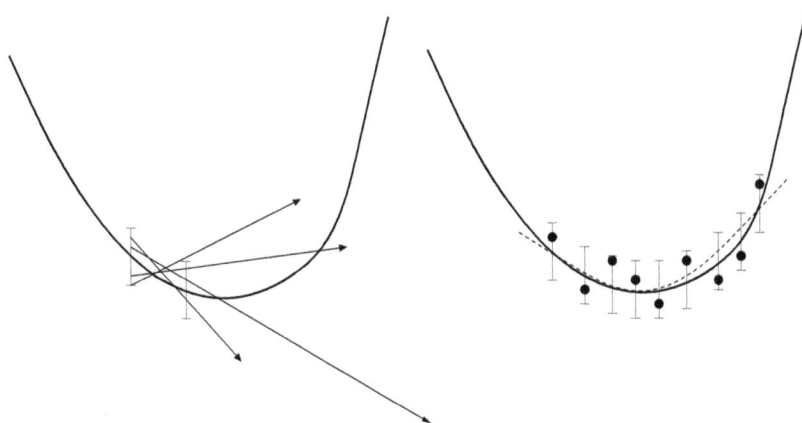

Another very important goal of SpaGrOW is to outperform gradient-based optimization algorithms with respect to computational effort, *i.e.*, the number of molecular simulations to be performed should be reduced by about a factor of two. At the same time, it should be at least as robust as gradient-based methods. Hence, SpaGrOW should also be applicable in domains that are further away from the minimum. This can only be realized efficiently by a local consideration. The minimization of SpaGrOW is performed in a small compact trust region. As the minimum can be situated at the boundary of the trust region, the method becomes an iterative procedure. The actual minimum is accepted as a new iteration, if the piecewise multilinear model predicts the decreasing trend of the loss function in a reliable way. Hence, SpaGrOW is combined with the approach of another category of optimization algorithms, the Trust Region methods, *cf.* [25,26]. The speed of convergency can be controlled by the size of the trust region. Altogether, SpaGrOW is shown to be an efficient combination of numerical methods, which has the following three very important issues: it is efficient, robust and gets very close to the minimum, if certain assumptions are fulfilled. However, these assumptions cannot be proven *a priori*, because the shape of the loss function and the amount of statistical noise is unknown in most cases. Therefore, a detailed practical evaluation is indispensable.

The methodology of SpaGrOW is presented in detail in Section 2, including complexity and convergency considerations. Section 4 shows the results of a detailed practical evaluation of the methodology, and finally, Section 5 concludes the paper.

1.2. Drawbacks of Gradient-Based Methods

Although gradient-based optimization methods turned out to be suitable and also efficient instruments for the parameterization of force fields in previous work [21–23], they unfortunately have the following disadvantages:

- Due to the statistical noise in the simulation data, the accuracy of a gradient is always limited. It can only predict the decreasing trend of the loss function. As motivated above, this becomes problematic close to the minimum. Theoretically, it is possible to counteract this problem by a suitable discretization for the calculation of the finite differences. However, optimizing the distance of two adjacent points would lead to much more computational effort. Moreover, even if the direction of the gradient can be calculated accurately, the step length control algorithm will only deliver insignificantly small improvements compared to the higher amount of computation time. This is a drawback of the Armijo step length control, but the application of more efficient step length control mechanisms would lead to more computational effort again.

- Whenever a descent direction is incorrect, a point with a smaller loss function value can only be found by chance, because there is no possibility to change the direction. In contrast to descent methods, Trust Region methods decrease the step length in order to find a new, maybe more suitable descent direction of the loss function. This is why these methods are able to get closer to the minimum. However, the Trust Region method applied in previous work requires a Hessian matrix, which leads to significantly more loss function evaluations.

- The amount of simulations is still quite high for gradient- and especially Hessian-based optimization algorithms. At each iteration, a gradient or, additionally, a Hessian have to be computed, which leads to N or, even, $N + N(N+1)/2$ loss function evaluations. Furthermore, the Armijo step length control algorithm requires additional functional evaluations.

These drawbacks of gradient-based methods motivate the development of an algorithm that is capable of both getting robustly closer to the minimum and solving the optimization problem with significantly less simulations. It should not assume the loss function to be smooth. Instead, it should use smoothing and regularization procedures in order to handle the statistical noise and jagged functions.

2. SpaGrOW Methodology: The Main Elements

In this work, a new derivative-free algorithm based on sparse grids and smoothing methods is presented. Figure 3 visualizes the gain in both efficiency and robustness by the combination of the Trust Region approach with the interpolation on sparse grids and smoothing procedures: the interpolation from a sparse grid to a full grid realizes a significant reduction of function evaluations, *i.e.*, simulations, without increasing the interpolation error significantly, on the other hand. At each iteration, the loss function is smoothed before the interpolation is carried out. The quality of the interpolation model is measured following the Trust Region idea. The actual minimum, which is determined on the full grid, is either accepted as a new iteration or the trust region is decreased. It is desisted from a continuous minimization within the trust region in order to avoid additional internal optimization iterations. There also exist approaches to determine the global minimum of a piecewise-linearly interpolated function via so-called subgradients [27]. However, only for the interpolation itself, at least 3^N function evaluations are required. In the present case, there are constraints, due to the minimization on a compact trust region. Hence, the sparse grid approach seems to be more reasonable. Another advantage of the Trust Region approach is the fact that it is able to be fast at the beginning of the optimization procedure and to jump over undesired intermediate local minima. This is due to the size of the step length, which is selected before a descent direction is calculated.

Figure 3. Overview of the Sparse Grid-based Optimization Workflow (SpaGrOW): the combination of the Trust Region approach with the interpolation on sparse grids requiring a smoothing procedure to be preceded leads to both increasing efficiency and robustness.

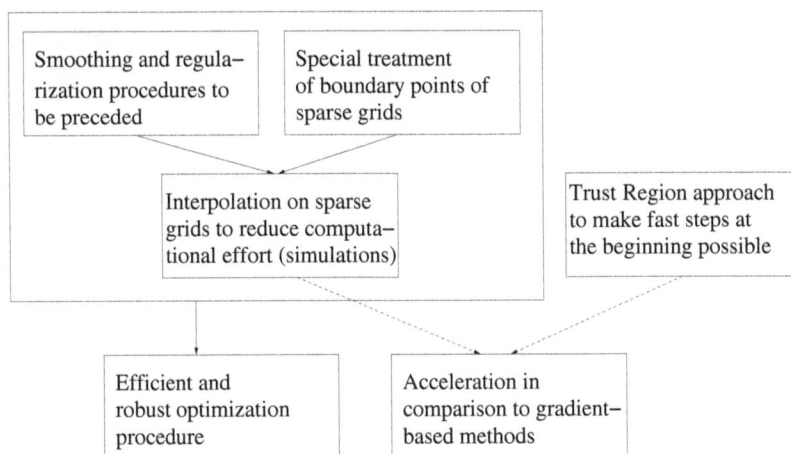

First, the two main elements of SpaGrOW are presented: In Section 2.1, the idea of sparse grids is introduced and the advantages with respect to reduction of computation time are presented. Suitable smoothing and regularization procedures are described in Section 2.2. The algorithm of SpaGrOW is introduced afterwards in Section 3.

2.1. Interpolation on Sparse Grids

The interpolation on sparse grids is a highly efficient discretization method in the field of Finite Element Methodology (FEM). In contrast to full grids, sparse grids possess significantly less grid points, especially in high dimensional spaces. The computational effort decreases from $\mathcal{O}(h^{-N})$ to $\mathcal{O}(h^{-1} \times (\log h^{-1})^{N-1})$, where h is the mesh size of the full grid. In the meantime, the interpolation error increases from $\mathcal{O}(h^2)$ to $\mathcal{O}(h^2 \times (\log h^{-1})^{N-1})$ only, with respect to the L_∞-norm [28]. The interpolation is founded on a tensor product approach for high dimensions and a linear combination of basis functions, e.g., of hierarchical hat functions [29]. Thereby, the basis functions with small coefficients within the linear combination are left out, and the remaining basis functions correspond to points of a sparse grid. Here, the combination technique by [24] is used, an efficient methodology interpolating a function on regular subgrids. The combination of these subgrids results in a sparse grid. The most important application of interpolation on sparse grids is the solution of partial differential equations, *cf.* [30].

2.1.1. Idea of Sparse Grids

The main idea of sparse grids is to reduce the computational effort without obtaining an intolerable increase of the interpolation error. Especially in high dimensions, the reduction of computational effort becomes notable; Table 1 shows the number of grid points of full and sparse grids of different resolutions and dimensions. In the following, full and sparse grids are introduced with the aid of basis functions for piecewise-bilinear functions, *i.e.*, for the two-dimensional case:

Table 1. Comparison of the number of grid points on full and sparse grids of different levels and dimensions. Especially for high levels and dimensions, the computational effort decreases significantly using sparse grids.

Level	$N = 1$	$N = 2$	$N = 3$	$N = 4$
1	$3 \to 3$	$9 \to 9$	$27 \to 27$	$81 \to 81$
2	$5 \to 5$	$25 \to 21$	$125 \to 81$	$625 \to 393$
3	$9 \to 9$	$81 \to 49$	$729 \to 225$	$6561 \to 1329$
4	$17 \to 17$	$289 \to 113$	$4913 \to 593$	$83521 \to 3921$
10	$1025 \to 1025$	$1.05 \times 10^6 \to 9217$	$1.07 \times 10^9 \to 47103$	$1.1 \times 10^{12} \to 1.78 \times 10^5$

Let $\Omega_{i,j}$ be the equidistant rectangular grid on the unit square, $\Omega := [0, 1]^2$, with mesh sizes $h_i := 2^{-i}$ in x- and $h_j := 2^{-j}$ in the y-direction. The vector, $\ell := (i, j) \in \mathbb{N}^2$, is called *level* of the grid, $\Omega_{i,j}$, with one-norm $|\ell|_1 := i + j$. Moreover, let $S_{i,j}$ be the space of piecewise-bilinear functions on the grid, $\Omega_{i,j}$. For reasons of simplicity, only those piecewise-bilinear functions are considered here, which fulfill homogeneous Dirichlet boundary conditions in $\Omega_{i,j}$. The respective space is denoted by $S_{i,j}^0$. It can be expressed as a tensor product of subspaces $T_{s,t}$, $s = 1, ..., i$, $t = 1, ..., j$, whose functions vanish on all grid points corresponding to $S_{s-1,t}^0$ and $S_{s,t-1}^0$:

$$S_{i,j}^0 = \bigotimes_{s=1}^{i} \bigotimes_{t=1}^{j} T_{s,t} \tag{4}$$

Uniquely determined piecewise-bilinear basis functions in $T_{s,t}$ with non-overlapping rectangular supports of size $1/2^{s-1} \times 1/2^{t-1}$ are introduced; the so-called *hierarchical basis* [29]. Thereby, each grid point corresponds to a specific hierarchical basis function. It is situated in the center of the support. Moreover, each grid point belongs to a grid of a certain level, ℓ. The full grid results from the combination of these grids, and the direct sums of the hierarchical basis functions are the standard basis functions of the full grid.

Each function, $u \in S_{i,j}^0$, can be expressed by a linear combination of hierarchical basis functions:

$$u = \sum_{s=1}^{i} \sum_{t=1}^{j} u_{s,t}, \; u_{s,t} \in T_{s,t}, \; s = 1, ..., i, \; t = 1, ..., j. \tag{5}$$

In [29], the inequality:

$$\|u_{s,t}\|_\infty \leq 4^{-s-t-1}|u| \tag{6}$$

was proven, where the seminorm, $|u|$, is given by:

$$|u| := \left\| \frac{\partial^4 u}{\partial x^2 \partial y^2} \right\|_\infty \tag{7}$$

In order to obtain a sparse grid, only the functions whose coefficients are greater than or equal to a certain tolerance value are chosen. In [29], this tolerance value is set to $4^{-\hat{\ell}-1}|u|$, where $\hat{\ell}$ is the level of the sparse grid, which is defined in the following. All remaining basis functions are neglected, and a sparse grid space can be defined as follows:

Definition 1 (Sparse grid space). *The space, $\hat{S}_{\hat{\ell}}^0$, spanned by the subspaces, $T_{i,j}$, with $i+j = \ell \leq \hat{\ell}+1$, i.e.,*

$$\hat{S}_{\hat{\ell}}^0 := \bigotimes_{s=1}^{\hat{\ell}} \bigotimes_{t=1}^{\hat{\ell}-s+1} T_{s,t} = \bigotimes_{s+t \leq \hat{\ell}+1} T_{s,t} \tag{8}$$

is called the sparse grid space. *The grid resulting from the combination of the subgrids corresponding to the subspaces, $T_{s,t}$, $s+t \leq \hat{\ell}+1$, is called the* sparse grid *and has the level, $\hat{\ell}$.*

The condition, $|\ell|_1 \leq \hat{\ell}+1$, leads to a triangular scheme of subspaces, $T_{i,j}$, which is depicted in Figure 4. Please note that $\ell_k = 0$, $k \in \{1,2\}$, is feasible, but it must hold $\forall_{k \in \{1,2\}} \ell_k < \hat{\ell}$. The transition to N-dimensional sparse grids of the level, $\hat{\ell}$, is trivial: all subgrids of the level, ℓ, with $|\ell|_1 \leq \hat{\ell}+N-1 \wedge \forall_{k=1,...,N} \ell_k < \hat{\ell}$ have to be combined, where $|\ell|_1 = \sum_{i=1}^{N} \ell_i$.

Figure 5 shows some examples of sparse grids of the levels 3, 4 and 5 in 2D and 3D, which were produced with an algorithm combining sparse grids of a given level from the respective subgrids.

Figure 4. Triangular scheme for the combination of a sparse grid of level 3 from two-dimensional subgrids meeting the condition, $|\ell|_1 \leq 3 + 2 - 1 = 4 \wedge \forall_{k \in \{1,2\}} \ell_k < 3$. If all eight subgrids are combined, a full grid of level 3 is obtained. If only the subgrids of levels (0,0), (1,0), (2,0), (3,0), (0,1), (0,2), (0,3), (1,1), (2,1), (1,2), (2,2), (3,1) and (1,3) are taken, *i.e.*, if the small triangle consisting of three grids at the bottom right is left out, the corresponding sparse grid is obtained, *cf.* Figure 5 top left.

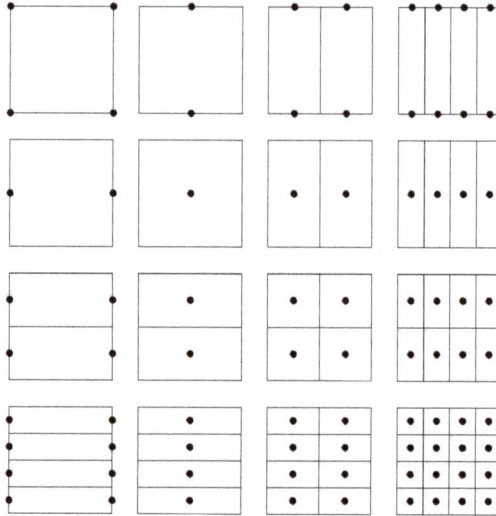

Figure 5. Sparse grids of the levels 3, 4, and 5 in 2D and 3D.

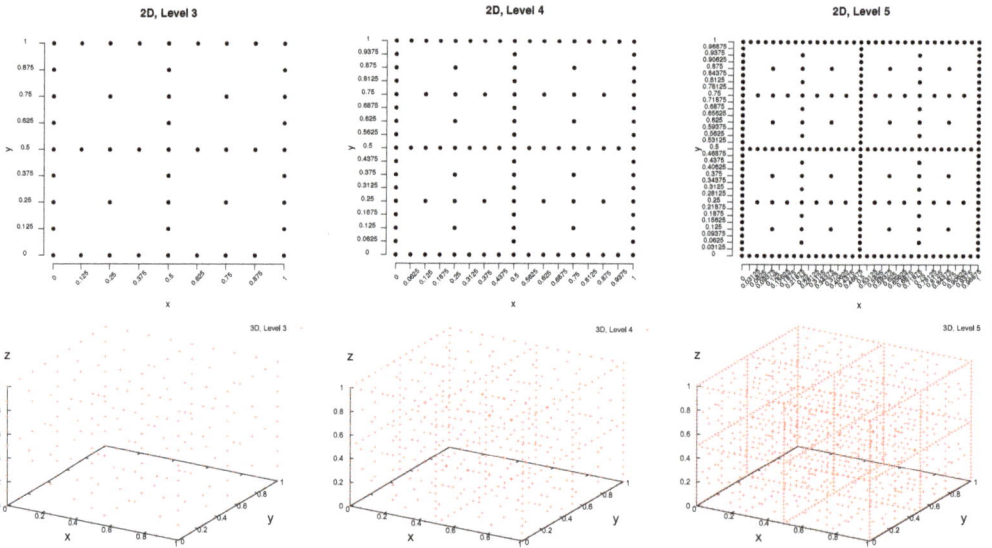

The coefficients corresponding to grid points of a grid with a high level do not contribute considerably to the interpolant. Hence, they can be neglected. However, a drawback of the sparse grid interpolation is the fact that this is only possible for functions that are sufficiently smooth. Because of inequality in Equation (6), the function that is interpolated must be at least four times continuously differentiable. Otherwise, the estimate cannot be made. As differentiability cannot be assumed in the present application, the combination technique developed by [24] is used here. A second problem concerning the computational effort is that sparse grids are still fully occupied at the boundary, *cf.* Figure 5. This problem can be handled using a transformation that sets the loss function to zero at the boundary. The methodology applied for SpaGrOW is described in the following.

2.1.2. Combination Technique

The combination technique developed by [24] combines problems on regular grids of the level, ℓ, with different mesh sizes, $\ell_1, ..., \ell_N$, to a sparse grid problem in a very efficient way. As this method has proven of value in practical applications and as it is also applicable to functions that are not necessarily smooth, it is taken as the sparse grid interpolation method for SpaGrOW.

As before, the two-dimensional case is considered first: Let $u : \mathbb{R}^2 \to \mathbb{R}$ be an arbitrary function. Every function defined on a sparse grid can be linearly combined from their interpolants on the regular subgrids, $\Omega_{i,j}$, $i + j \in \{\hat{\ell}, \hat{\ell} + 1\}$. If $u_{i,j} \in T_{i,j}$ is considered, a combination function, u_ℓ^c, can be defined as follows:

Definition 2 (Combination function (2D)). *The combination function of solutions $u_{i,j} \in T_{i,j}$ of FEM problems on regular two-dimensional grids of the level, (i, j), $i + j \in \{\hat{\ell}, \hat{\ell} + 1\}$, is given by:*

$$u_\ell^c := \sum_{i+j=\hat{\ell}+1} u_{i,j} - \sum_{i+j=\hat{\ell}} u_{i,j} \tag{9}$$

The error, $u - u_\ell^c$, is in $\mathcal{O}\left(h_{\hat{\ell}}^2 \log(h_{\hat{\ell}})^{-1}\right)$, *cf.* Section 3.3.1. Analogous to Definition 2, the following definition is given for the three-dimensional case:

Definition 3 (Combination function (3D)). *The combination function of solutions $u_{i,j,k} \in T_{i,j,k}$ of FEM problems on regular three-dimensional grids of the level, (i, j, k), $i + j + k \in \{\hat{\ell}, \hat{\ell} + 1, \hat{\ell} + 2\}$, is given by:*

$$u_\ell^c := \sum_{i+j+k=\hat{\ell}+2} u_{i,j,k} - 2 \sum_{i+j+k=\hat{\ell}+1} u_{i,j,k} + \sum_{i+j+k=\hat{\ell}} u_{i,j,k} \tag{10}$$

Via complete induction, this can directly be transferred to the N-dimensional case:

444

Definition 4 (Combination function (in general))**.** *The* combination function *of solutions $u_\ell \in T_\ell$ of FEM problems on regular N-dimensional grids of the level, ℓ, with $|\ell|_1 = \hat{\ell} + N - 1 - i$, $i = 0, ..., N - 1$, is given by:*

$$u_\ell^c := \sum_{i=0}^{N-1} (-1)^i \binom{N-1}{i} \sum_{|\ell|_1 = \hat{\ell}+N-1-i} u_\ell \qquad (11)$$

Remark 5. *Because of the condition, $|\ell|_1 = \hat{\ell}+N-1-i$, $i = 0, ..., N-1$, a multilinear interpolation on a full regular grid with mesh size $1/2^{\hat{\ell}}$ and level $\bar{\ell} = (\hat{\ell})$ can be executed as follows: All function values are computed on a sparse grid of the level, $(\hat{\ell}, ..., \hat{\ell}) \in \mathbb{N}^N$. Because of the hierarchical structure of the sparse grid, all function values of the regular subgrids of the level, ℓ, with $|\ell|_1 \le \hat{\ell} + N - 1$ are known, i.e., all function values required for the interpolation, cf. Equation (11).*

2.2. Smoothing Procedures

As already mentioned, the combination technique is applicable to functions that are not necessarily differentiable. However, an assumption of this method is the existence of an asymptotic error expansion for the discrete solution, *cf.* Section 3.3.1. Hence, the function to be interpolated should possess certain smoothness properties. At least, the statistical noise has to be filtered out as effectively as possible, and the quality of the continuity of the loss function should be high enough. As noise can be unfavorable for the combination technique, smoothing procedures have to be applied before the piecewise-linear interpolation is performed. In SpaGrOW, radial basis functions (RBFs) have turned out to be very suitable in order to approximate the loss function. In some cases, especially close to the minimum, a simple quadratic model is sufficient, as well. Additionally, the noise is filtered out via regularization procedures. In the approximation process, the respective nonlinear regression problem is not solved via least squares, but estimators with a lower variance. The regularization methods used in SpaGrOW are elastic nets and multi-adaptive regression splines. The applied smoothing and regularization algorithms are described in brief, and a theoretical selection is presented and discussed in the following. The final decision can only be made after a detailed practical evaluation, which is performed in Section 4.1.

2.2.1. Effects of Statistical Noise on Piecewise-linear Interpolation

Suppose u is linear and noisy. Then, the piecewise-linear interpolation should reproduce the function exactly. However, statistical noise can lead to a staggered function, which is shown in Figure 6. As this is not desired, a preprocessing, which approximates the function to be interpolated and eliminates the noise, is indispensable.

However, another problematic may appear whenever the sampled points used for the approximation are situated too close to each other: Figure 7 indicates that in this case, the trend of the function can be reproduced completely incorrectly. Therefore, SpaGrOW has to deal with this phenomenon, and the size of the trust region becomes too small has to be avoided.

Figure 6. Problematic in the case of piecewise-linear interpolation of a noisy function, the interpolation leads to a staggered function reproducing the noise.

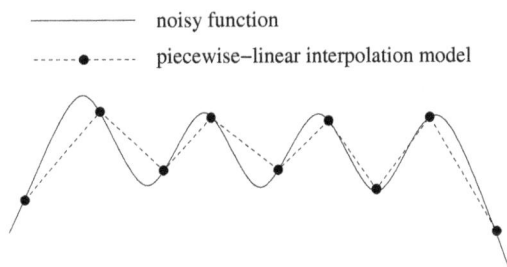

Figure 7. Problematic in the case of the approximation of a noisy function, when the points are situated too close to each other, the function values cannot be differentiated anymore, and the trend of the function can be reproduced completely incorrectly by the smoothing procedure.

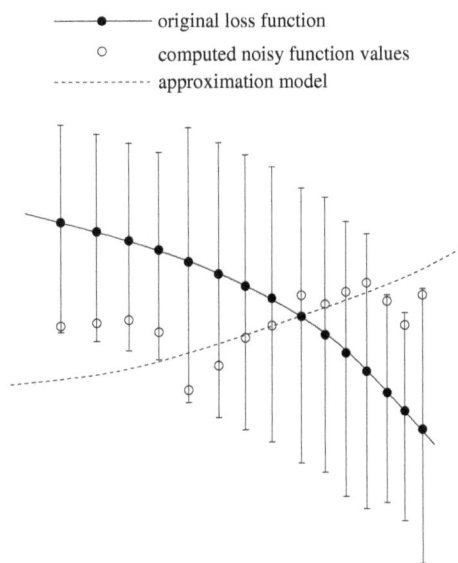

2.2.2. Smoothing Functions

In the following, the approximation of high-dimensional functions with radial basis functions (RBFs) is presented shortly.

RBFs are an efficient and common method for interpolation and approximation tasks [31]. The word *radial* indicates that they are functions of the distance of two adjacent points. Hence, points which are situated on a hyperball with radius r around a reference point have the same function value. Here are some examples:

$$\phi(r) = r \ \text{(linear)} \tag{12}$$

$$\phi(r) = \exp\left(-cr^2\right), \ c \in \mathbb{R}^+ \ \text{(Gaussian)} \tag{13}$$

$$\phi(r) = \sqrt{r^2 + c^2}, \ c \in \mathbb{R}^{\neq 0} \ \text{(multiquadric)} \tag{14}$$

$$\phi(r) = \frac{1}{\sqrt{r^2 + c^2}}, \ c \in \mathbb{R}^{\neq 0} \ \text{(inverse multiquadric)} \tag{15}$$

$$\phi(r) = r^3 \ \text{(cubic)} \tag{16}$$

$$\phi(r) = r^2 \log r \ \text{(thin plate splines)} \tag{17}$$

A function, $f : \mathbb{R}^N \to \mathbb{R}$, is approximated by:

$$f(X) \approx \sum_{i=1}^{K} \beta_i h_i \left(\|X - X_i\|\right) \tag{18}$$

where $X \in \mathbb{R}^N$ is an arbitrary test point and $K \leq m$, where $m \in \mathbb{N}$ is the total number of points. The points, $X_i \in \mathbb{R}^N$, $i = 1, ..., K$, are the K representative centers of the training set, on the basis of which the coefficients, $\beta_i \in \mathbb{R}$, $i = 1, ..., K$, are determined. The functions, $h_i \in \mathcal{H}$, $i = 1, ..., K$, are from the set of RBFs, $\mathcal{H}$. The computation of the coefficients is realized by so-called *RBF networks*, *cf.* Figure 8, which are directed neuronal networks [32] and proceed from the input to the output neurons. They possess one layer of hidden neurons only, which corresponds to the determination of the coefficients. In the case of an approximation, it holds $m \leq K$, and the following overdetermined linear equation system (LES) has to be solved:

$$H\beta = Y \tag{19}$$

where:

$$H = \begin{pmatrix} h_1\left(\|X_1 - X_1\|\right) & h_2\left(\|X_1 - X_2\|\right) & \cdots & h_K\left(\|X_1 - X_K\|\right) \\ h_1\left(\|X_2 - X_1\|\right) & h_2\left(\|X_2 - X_2\|\right) & \cdots & h_K\left(\|X_2 - X_K\|\right) \\ \vdots & \vdots & \vdots & \vdots \\ h_1\left(\|X_m - X_1\|\right) & h_2\left(\|X_m - X_2\|\right) & \cdots & h_K\left(\|X_m - X_K\|\right) \end{pmatrix} \in \mathbb{R}^{m \times K}$$

and:

$$Y = \begin{pmatrix} y_1 \\ y_2 \\ \vdots \\ y_m \end{pmatrix} = \begin{pmatrix} f(X_1) \\ f(X_2) \\ \vdots \\ f(X_m) \end{pmatrix} \in \mathbb{R}^m$$

The solution vector, $\beta := (\beta_1, ..., \beta_K)^T \in \mathbb{R}^K$, can be computed e.g., by using least squares.

In order to select K representative centers out of the m training data in an efficient way, an unsupervised learning method is used, namely the automated classification (*clustering*) procedure, *k-means* [33]. The K centers are also the centroids of the data classes obtained by the clustering algorithm.

A drawback of RBFs is the high computational effort required for the collocation of the matrix, H, in Equation (19). For $N = 5$, more than one thousand and for $N = 6$, over 72 million RBF evaluations are necessary. Hence, RBFs should not be used for $N > 4$.

For dimensions optimization problems, a simpler smoothing procedure is applied, whose complexity is only quadratic in the dimension. Instead of approximating the loss function itself, each of the physical target properties is approximated linearly, which results in a quadratic model of the loss function. This method is called *linear property approximation (Lipra)* and only requires the solution of an LES of the size N on a sparse grid. The loss function values can easily be obtained by inserting the approximated physical properties into the loss function. The complexity of the Lipra algorithm depends on the number of properties n and the dimension, N. It is $\mathcal{O}(nN^2)$.

Figure 8. Radial basis function (RBF) network: The network proceeds for a test point, X, from the input neurons, X^j, $j = 1, ..., N$, (components of X), over a layer of hidden neurons containing the RBF evaluations, $h_i(X - X_i)$, $i = 1, ..., K$, with the coefficients, β_i, $i = 1, ..., K$, to the output neuron, where the summation takes place. The result is the function value, $f(X)$.

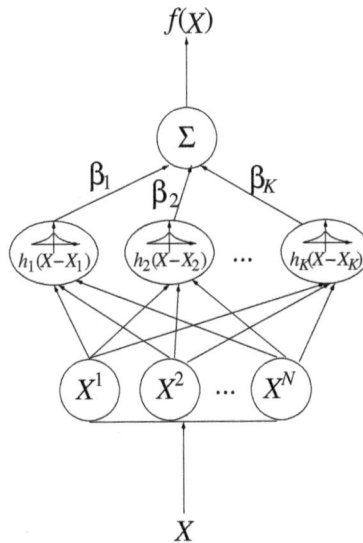

2.2.3. Regularization Algorithms

In order to avoid overfitting, the statistical noise the simulation data is affected with has to be eliminated. Therefore, the regression coefficients computed within the smoothing procedure may not be overestimated, so that not too much stress is laid on noisy data points. This can be achieved by a weighted regression, where a weighting factor is assigned to each data point that is small, whenever the noise on its function value is high, and *vice versa*. Another way is to constrain the coefficients in the optimization problem contained in the regression procedure. These so-called *regularization algorithms* do not solve the regression problem with the least square estimator, β_{LS}, but with low-variance estimators. The former is given by:

$$\beta_{LS} = \arg\min_{\beta} ||Y - H\beta||_2^2 \qquad (20)$$

and has the following drawbacks:

(1) Due to its high variance, the probability of overfitting is always high.
(2) The euclidean norm, $||\beta_{LS}||_2$, is, in general, high, as well.
(3) There is no variable selection, *i.e.*, none of the coefficients of β_{LS} can be zero. Hence, correlated variables have always an influence within the model, even if they do not have any impact on the loss function.
(4) If the dimension is greater than the number of data points, the least square method does not have a solution. However, due to $K << m$, this is not important in the present case.

In order to counteract these drawbacks, the regression coefficients have to be *shrinked* or set to zero in some cases (variable selection). This has the effect that especially correlated variables and outliers only have low influence within the model. Moreover, the variance of the estimator is reduced, so that the probability of overfitting decreases. In SpaGrOW, this is realized by a *Naive Elastic Net* [34]:

$$\beta_{NEN} = \arg\min_{\beta} ||Y - H\beta||_2^2, \text{ where } \alpha||\beta||_2^2 + (1-\alpha)|\beta|_1 \leq t, \ \alpha \in [0,1], \ t > 0 \qquad (21)$$

It contains two additional model parameters, $\alpha \in [0,1]$ and $t \in \mathbb{R}^{>0}$, which can be determined, e.g., by a ten-fold cross-validation. A naive elastic net possesses a so-called *grouping effect*, *i.e.*, correlated model variables have similar regression coefficients. A penalized formulation of Equation (21) is achieved by introducing Lagrangian multiplicators, $\lambda_1, \lambda_2 \geq 0$:

$$\beta_{NEN} = \arg\min_{\beta} \mathcal{L}(\beta, \lambda_1, \lambda_2), \ \mathcal{L}(\beta, \lambda_1, \lambda_2) := ||Y - H\beta||_2^2 + \lambda_2||\beta||_2^2 + \lambda_1|\beta|_1 \qquad (22)$$

It holds $\alpha = \frac{\lambda_2}{\lambda_1 + \lambda_2}$. For $\alpha \in \{0,1\}$, the two most famous methods in the field of biased regression are obtained: for $\alpha = 1$, it is the *Ridge Regression* [35], and for $\alpha = 0$, it is the *Least Absolute Shrinkage and Selection Operator (LASSO)* [36]. Both methods contain shrinkage, but only in the case of the LASSO, a variable selection is possible. For more details about Elastic Nets, Ridge Regression and the LASSO, *cf.* [34].

Another regularization method is based on the application of *Multivariate Adaptive Regression Splines (MARS)* [37]. The basis functions within this algorithm are products of so-called *hinge functions* of the order $q \in \mathbb{N}$:

$$S_q(x) = S_q(x - \nu) := [\pm(x - \nu)]_+^q, \ q \in \mathbb{N} \qquad (23)$$

where ν is a *node*. "+" means that for negative arguments, the functions becomes zero. Two hinge functions with opposite sign belong together and are mirrored at their common node. Hence, they divide the input space into two disjoint subsets. By the recursive selection of p nodes, the input space is divided into $p + 1$ disjoint subsets. A single hinge function or a product of hinge functions, which lead to even more divisions, is assigned to each subspace. This models the interaction of the input variables. The approximation of a function, $f : \mathbb{R}^N \to \mathbb{R}$, within the MARS algorithm is realized as follows:

$$f(x) \approx \sum_{i=1}^{p} (\beta_M)_i \prod_{j=1}^{s_i} S_q(x_{N(i,j)} - \nu_{i,j}) \tag{24}$$

Thereby, the s_i are the numbers of hinge functions considered in the partitioning, $i \in \{1, ..., p\}$, $\nu_{i,j}$, $i = 1, ..., p$, $j = 1, ..., s_i$ are the nodes of the recursive division and $N(i,j)$ is the dimension of the input vector, x, to which the division defined by the node $\nu_{i,j}$ refers. In total, f is approximated by a regression spline and, in the case of $q = 1$, by a piecewise-linear function. The lower probability of overfitting is reached by pruning the basis functions involved in the model. For more details about MARS, *cf.* [37]. A disadvantage of this regularization algorithm is the fact that it requires high computational effort for higher dimensions, due to the pruning procedure.

Please note that in the case of the Lipra algorithm, which is used for high-dimensional problems, no regularization procedure is applied, due to the high amount of computational effort.

2.2.4. Selection of Smoothing Procedures: Theoretical Considerations

The selection of the best smoothing procedure is made first from a theoretical perspective. A detailed practical evaluation is pointed out in Section 4.1. Theoretically, an approximation method can be selected, due to a reliable estimate of the approximation error within a domain, $\Omega \subset \mathbb{R}^N$. Such estimates exist for positive definite RBFs, *cf.* [38]. The Gaussian, the inverse multiquadric and so-called *Wendland functions* [39] are positive definite. The latter are RBFs with compact supports and are piecewise polynomial with a minimal degree dependent on differentiability and dimension.

An estimate of the approximation error can be realized via the introduction of *Native Spaces* [38]:

Definition 6 (Native Spaces). *The Native Space, $\mathcal{N}_{\phi(\Omega)}$, for a given positive definite RBF, ϕ, within the domain, Ω, is given by:*

$$\mathcal{N}_{\phi(\Omega)} := \overline{\{\phi(|| \cdot -x||), x \in \Omega\}} \tag{25}$$

where for $f_1 = \sum_{k=1}^{n_1} \alpha_k \phi(|| \cdot -x_k||) \in \mathcal{N}_{\phi(\Omega)}$ and $f_2 = \sum_{j=1}^{n_2} \beta_j \phi(|| \cdot -x_j||) \in \mathcal{N}_{\phi(\Omega)}$:

$$\langle f_1, f_2 \rangle_{\phi(\Omega)} := \sum_{k=1}^{n_1} \sum_{j=1}^{n_2} \alpha_k \beta_j \phi(||x_k - x_j||) \tag{26}$$

Thereby, $n_1, n_2 \in \mathbb{N}$, $\alpha_k, \beta_j \in \mathbb{R}$, $k = 1, ..., n_1$, $j = 1, ..., n_2$, and $x_k, x_j \in \Omega$, $k = 1, ..., n_1$, $j = 1, ..., n_2$.

The Hilbert space $\mathcal{N}_{\phi(\Omega)}$ is the completing of the pre-Hilbert space, $\{\phi(|| \cdot -x||), x \in \Omega\}$. As the approximation error depends on the size of the domain, Ω, the so-called *fill distance* is defined next:

Definition 7 (Fill distance). *For a given discrete training set, $\mathcal{X} \subset \Omega$, with training data, x_j, $j = 1, ..., m$, $m \in \mathbb{N}$, the fill distance, $\Delta_{\Omega,\mathcal{X}}$, is defined by:*

$$\Delta_{\Omega,\mathcal{X}} := \sup_{x \in \Omega} \min_{j=1,...,m} ||x - x_j|| \tag{27}$$

From Definitions 6 and 7, the following theorem for the estimation of the approximation error can be formulated:

Theorem 8 (Approximation error for positive definite RBFs). *Let Ω, $\mathcal{X}$, ϕ and f be defined as above. Let, furthermore, g be the approximating function for $f \in \mathcal{N}_{\phi(\Omega)}$, obtained by the positive definite RBF, ϕ. Then, the following estimate for the approximation error holds:*

$$||f - g||_{L_\infty(\Omega)} \leq h(\Delta_{\Omega,\mathcal{X}})||f||_{\mathcal{N}_{\phi(\Omega)}} \tag{28}$$

where $\lim_{\Delta_{\Omega,\mathcal{X}} \to 0} h(\Delta_{\Omega,\mathcal{X}}) = 0$.

Proof. [38] □

This theorem is very important for the convergency of SpaGrOW, *cf.* Section 3.3.2. However, for SpaGrOW, another condition, $\lim_{\Delta_{\Omega,\mathcal{X}} \to 0} h(\Delta_{\Omega,\mathcal{X}}) = 0$, has to be fulfilled. The following definition introduces the term *stability* of an approximation in the sense of SpaGrOW:

Definition 9 (Stable approximation). *Let Ω and $\mathcal{X}$ be defined as above. Let be $f \in \mathcal{H}$, where $\mathcal{H}$ is a Hilbert space of functions on Ω with the scalar product, $\langle \cdot, \cdot \rangle_{\mathcal{H}}$, and norm $||f||_{\mathcal{H}} = \sqrt{\langle f, f \rangle_{\mathcal{H}}}$ for $f \in \mathcal{H}$. An approximation within the domain, Ω, by a function, $g \in \mathcal{P}$, where $\mathcal{P}$ is a pre-Hilbert space with the same scalar product and $\bar{\mathcal{P}} = \mathcal{H}$, is called stable, if:*

$$||f - g||_{\mathcal{H}} \leq \kappa h(\Delta_{\Omega,\mathcal{X}}) \tag{29}$$

where $\kappa > 0$, $\lim_{\Delta_{\Omega,\mathcal{X}} \to 0} h(\Delta_{\Omega,\mathcal{X}}) = 0$ and $\lim_{\Delta_{\Omega,\mathcal{X}} \to 0} \frac{h(\Delta_{\Omega,\mathcal{X}})}{\Delta_{\Omega,\mathcal{X}}} = 0$.

The second important condition for the convergency of SpaGrOW is $\lim_{\Delta_{\Omega,\mathcal{X}} \to 0} \frac{h(\Delta_{\Omega,\mathcal{X}})}{\Delta_{\Omega,\mathcal{X}}} = 0$. The following corollary ascertains the stability of an approximation based on a Gaussian RBF:

Corollary 10 (Stability of an approximation based on a Gaussian RBF). *Let Ω, $\mathcal{X}$ and f be defined as in Theorem 8. Let, furthermore, $\phi(|| \cdot - x||) = \exp\left(-c|| \cdot - x||^2\right)$ for $x \in \Omega$ and $c \in \mathbb{R}^+$ a Gaussian RBF. Then, the approximation of f by $g(x) := \sum_{k=1}^{\nu} \alpha_k \phi(||x_k - x||)$, $\nu \in \mathbb{N}$, $\alpha_k \in \mathbb{R}$, $x_k \in \Omega$, $k = 1, ..., \nu$, is stable.*

Proof. For a Gaussian RBF, the following estimation holds (*cf.* [38]):

$$||f - g||_{L_\infty(\Omega)} \leq \exp\left(-d\left(\frac{\log(\Delta_{\Omega,\mathcal{X}})}{\Delta_{\Omega,\mathcal{X}}}\right)\right)||f||_{\mathcal{N}_{\phi(\Omega)}} \tag{30}$$

where $d > 0$ is a constant. With $h(\Delta_{\Omega,\mathcal{X}}) := \exp\left(-d\left(\frac{\log(\Delta_{\Omega,\mathcal{X}})}{\Delta_{\Omega,\mathcal{X}}}\right)\right)$, it holds $\lim_{\Delta_{\Omega,\mathcal{X}} \to 0} h(\Delta_{\Omega,\mathcal{X}}) = 0$ and $\lim_{\Delta_{\Omega,\mathcal{X}} \to 0} \frac{h(\Delta_{\Omega,\mathcal{X}})}{\Delta_{\Omega,\mathcal{X}}} = 0$. With $\kappa := ||f||_{\mathcal{N}_{\phi(\Omega)}}$ and the fact that $\mathcal{N}_{\phi(\Omega)}$ is a Hilbert space and g an element of the pre-Hilbert space, $\{\phi(|| \cdot - x||), x \in \Omega\}$, with $\overline{\{\phi(|| \cdot - x||), x \in \Omega\}} = \mathcal{N}_{\phi(\Omega)}$, it follows that the approximation based on a Gaussian RBF is stable with respect to Definition 9. □

Because of Corollary 10, the Gaussian RBF is selected for SpaGrOW for theoretical reasons. However, such estimates exist for other positive RBFs, as well. The final selection of the Gaussian RBF is made after a detailed practical evaluation, as mentioned above.

3. The SpaGrOW Algorithm

3.1. Ingredients of the Algorithm

Within SpaGrOW, the minimization of the interpolation model on a full grid is performed discretely. The corresponding complexity is $\mathcal{O}\left((n+1)^{2N}\right)$, where $n+1$ ($n \in \mathbb{N}$) is the number of points of the full grid in one dimension. The first question arising is whether the algorithm should be applied locally or globally. Globally, this means that the smoothing procedure and the sparse grid interpolation are performed on the complete feasible domain for the force field parameters. However, it results that a local consideration is the better way, *i.e.*, the combination with the Trust Region approach is highly important. This makes SpaGrOW an iterative local optimization method. The algorithm is described in detail in the following.

3.1.1. Local and Global Consideration

The combination technique delivers a piecewise-multilinear function, $q : \mathbb{R}^N \to \mathbb{R}$, which either interpolates the loss function, F, itself or an approximating function, G, from a sparse grid of the level, $\hat{\ell} \in \mathbb{N}$, on a full grid, $G_{\bar{\ell}}^N$, with the level, $\bar{\ell} = (\hat{\ell}, ..., \hat{\ell}) \in \mathbb{N}^N$. The total error resulting from the approximation and interpolation error can be measured via the L_2- or L_∞-distance between F and the interpolation model, q, on the unit square, $\Omega := [0,1]^N$:

$$||e||_{L_2,[0,1]^N} := ||F-q||_{L_2,[0,1]^N} = \left(\int_{[0,1]^N}(F(x)-q(x))^2\,dx\right)^{\frac{1}{2}} \tag{31}$$

$$||e||_{L_\infty,[0,1]^N} := ||F-q||_{L_\infty,[0,1]^N} = \max_{x\in[0,1]^N}|F(x)-q(x)| \tag{32}$$

If the function values, $F(x)$, are known for $x \in G_{\bar{\ell}}^N$, the error terms in Equations (31) and (32) can be approximated numerically.

The total error is important for the assessment of whether SpaGrOW can be applied globally or not. Hence, it has to be determined on a full grid for the local and the global variant. Moreover, the convergency behavior of SpaGrOW has to be analyzed. There are some arguments for a local consideration: Following the heuristics indicated in Section 2.2, it is indispensable to perform a smoothing procedure before interpolating. In the global case, this smoothing procedure would be executed on a sparse grid discretizing the complete feasible domain. As the loss function can be arbitrarily complex and jagged, it should be nearly impossible to reproduce them in an accurate way within a huge domain with only a small number of data points. Furthermore, the discretization error would be too high within a huge domain for the determination of the global minimum. However, it would be possible to apply the algorithm globally first and to make local refinements afterwards, but an inaccurate approximation of the loss function can lead far away from the minimum. Hence, a local consideration should be preferred, and SpaGrOW should be a local iterative procedure, like the gradient-based methods applied for the present task.

In order to verify these heuristic considerations, a practical analysis was performed, where the loss function was replaced by the simple two-dimensional function, $H(x_1, x_2) = -\exp\left(-(x_1 - 2)^2 - (x_2 + 1)^2\right)$. The global minimum is situated at $(x_1^*, x_2^*) = (2, -1)$, with corresponding function value, $H(2, -1) = -1$. Artificial statistical uncertainties, *i.e.*, uniformly distributed random numbers from the interval, $[-0.03H(x_1, x_2), +0.03H(x_1, x_2)]$, were added on the function values, $H(x_1, x_2)$. For reasons of brevity, the analysis is not reported here. The results were the following:

- **Comparison of the total errors:**
    - In the global consideration, the total errors were significantly higher than in the local consideration with respect to both the L_2- and the L_∞-norm.
    - The smoothing error is considerably higher in the global case than in the local one, *i.e.*, the function cannot be reproduced accurately. This was also the case when the function was not affected with statistical noise.

- **Convergency analysis:**
    - Without artificial uncertainties, the global variant of SpaGrOW only led to the minimum when the minimum was a grid point. Otherwise, the minimum could only be determined within the accuracy of the discretization.
    - With artificial uncertainties, both discretization and approximation error had a negative effect on the convergency. The resulting approximated minimum was far away from the real one.

To summarize, a local consideration is preferred to a global one.

3.1.2. Combination with the Trust Region Approach

For each iteration, a compact neighborhood, where the minimization problem is solved discretely, is determined. In most cases, the minimum is situated at the boundary of the compact domain. If certain assumptions are fulfilled, this boundary minimum becomes the new iteration and the center of a new compact neighborhood.

Following the idea of Trust Region methods [25,26], the compact neighborhood is a trust region of the size, $\Delta_k > 0$. Due to the sparse grid interpolation, it is not a ball, $B_{\Delta_k}(x^k)$, but a hyperdice, $\mathcal{W}^k$, of the form:

$$\mathcal{W}^k := \bigotimes_{i=1}^{N} \left[x_i^k - \Delta_k, x_i^k + \Delta_k\right] \qquad (33)$$

where x^k is the kth iteration of SpaGrOW. On the one hand, the size, Δ_k, has to be small enough, so that $\mathcal{W}^k$ is situated within the feasible domain of the force field parameters and the interpolation model is consistent with the original loss function, F, as already discussed in Section 3.1.1. On the other hand, Δ_k has to be large enough, so that the method converges as fast as possible to the global minimum within the feasible domain. Hence, the division of the domain into non-disjoint hyperdices should not be too fine.

The quality of the model, q, is estimated using the following ratio:

$$r^k := \frac{F(x^k) - F(x^*)}{q(x^k) - q(x^*)} = \frac{F(x^k) - F(x^*)}{G(x^k) - q(x^*)} \tag{34}$$

where x^* is the discrete minimum on a full grid within the hyperdice, $\mathcal{W}^k$. As x^k is a point of the sparse grid and the model, q, interpolates the approximating function, G, from the sparse grid on the full grid, it holds $q(x^k) = G(x^k)$.

In practice, two thresholds, $0 < \eta_1 < \eta_2$, and size parameters, $0 < \gamma_1 < 1 < \gamma_2$, are introduced, and the following three cases are considered:

- $\eta_2 > r^k \geq \eta_1$—in this case, the model is consistent with F, and the minimum, x^*, is taken as a new iteration: $x^{k+1} := x^*$.
- $r^k \geq \eta_2 > \eta_1$—then, x^* is taken as a new iteration, as well, and at the same time, the trust region is increased in order to accelerate the convergency: $\Delta^{k+1} := \gamma_2 \Delta_k$.
- $r^k < \eta_1$—in this case, the model is not consistent with F, and a better model has to be determined by decreasing the trust region: $\Delta^{k+1} := \gamma_1 \Delta_k$.

As for the Trust Region methods, a minimal $\Delta^{\min} > 0$ is defined *a priori*. The algorithm is stopped as soon as $\exists\, k\,:\, \Delta_k < \Delta^{\min}$.

3.1.3. Treatment of Boundary Points

After the application of a transformation, $\xi\,:\, \mathcal{W}^k \to [0,1]^N$, into the unit hyperdice, *cf.* Section 3.2.1, the loss function values are set to zero at its boundary by an appropriate modification. The reason of the realization of homogeneous Dirichlet boundary conditions is the fact that also sparse grids are fully occupied at their boundaries. In order to save a magnitude of simulations, the original loss function, F, is multiplied with a product of sine functions, *i.e.*, the modified function:

$$\bar{F}: [0,1]^N \to \mathbb{R}_0^+$$
$$y \mapsto \left(\prod_{i=1}^{N} \sin \pi y_i\right) F(y) \tag{35}$$

where $y = \xi(x)$ with $x \in \mathcal{W}^k$ is considered instead of F. Hence, the smoothing and interpolation procedures are applied for $\bar{F} \circ \xi\,:\, \mathcal{W}^k \to \mathbb{R}_0^+$. Due to Equation (35), it holds $\bar{F}|_{\partial[0,1]^N} = 0$. However, as F and not $\bar{F}$ has to be minimized and as F and $\bar{F}$ do not have the same minimum, the back-transformation:

$$F(x) = F(\xi^{-1}(y)) = \frac{\bar{F}(y)}{\prod_{i=1}^{N} \sin \pi y_i} \tag{36}$$

454

has to be applied before the discrete minimization is performed. Hence, the minimum of $F \circ \xi^{-1}$ has to be determined, which is not possible at the boundary of $\mathcal{W}^k$. As the minimum is expected to be situated at the boundary, only the grid:

$$\tilde{G}_{\hat{\ell}}^N := G_{\hat{\ell}}^N \backslash \partial G_{\hat{\ell}}^N \tag{37}$$

is considered for the minimization. This grid does not contain any boundary point of the original grid. However, this reduces the size of the trust region, but the loss in convergency speed is negligible, due to the number of simulations to be saved; for $N = 4$ and $\hat{\ell} = 2$, only nine instead of 393 simulations (*cf.* Table 1) per iteration are required.

Figure 9. Overview of the SpaGrOW algorithm, *i.e.*, the inner iteration of the optimization procedure visualized in Figure 1. The Trust Region size, Δ, is increased or decreased, depending on the quality of the approximation model on the sparse grid.

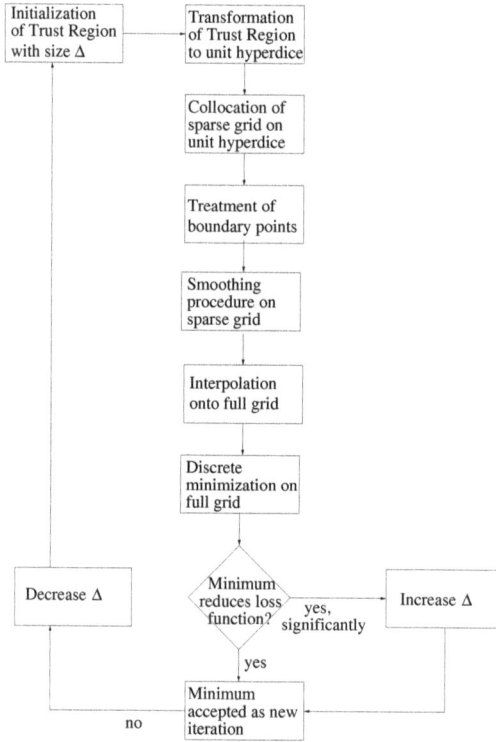

3.2. The Full Algorithm

3.2.1. Structure

The algorithm of SpaGrOW, *i.e.*, the inner iteration of the optimization procedure shown in Figure 1, is visualized in Figure 9 and has the following structure:

- **Initialization:** Choose an initial vector, x^0, and an initial step length, $D^0 > 0$, so that:

$$\forall_{i=1,...,N}\ c_i^0 \le x_i^0 \le C_i^0,\ \Delta_0 < C_i^0 - x_i^0,\ \Delta_0 < x_i^0 - c_i^0 \tag{38}$$

Thereby, $[c_i^0, C_i^0]$ is the feasible interval for the ith force field parameter. The maximal step length possible, $\Delta^{\max}$, is computed at the beginning, and Δ_0, as well as a minimal step length, $\Delta^{\min}$, are set in relation to it. Please note that on the one hand, Δ_0 must not be too small, due to the noise, and that, on the other hand, it must not be too large, so that the problematic described in Section 3.1.1 is not faced.

Let $k := 0$.

- **Transformation:** Via the transformation:

$$\xi : \mathcal{W}^k \rightarrow [0,1]^N$$
$$x \mapsto \frac{x - \left(\min_{i=1,...,N} x_i^k\right) \cdot e}{\left(\max_{i=1,...,N} x_i^k\right) \cdot e - \left(\min_{i=1,...,N} x_i^k\right) \cdot e}$$
$$= \frac{1}{2\Delta_k}\left(x - x^k + \Delta_k \cdot e\right) \tag{39}$$

the initial vector, x^0, is mapped from the hyperdice of the size, Δ_0, into the unit hyperdice. Thereby, $e = (1, ..., 1)^T \in \mathbb{R}^N$ and x_i^k, $i = 1, ..., N$ are the components of the vector, x^k. Please note that only the back-transformation obtained by the inverse function:

$$\xi^{-1} : [0,1]^N \rightarrow \mathcal{W}^k$$
$$y \mapsto 2\Delta_k y + x^k - \Delta_k \cdot e \tag{40}$$

is important, because, first, a sparse grid is simply collocated in $[0,1]^N$. Then, the grid points are back-transformed into the force field parameter space via ξ^{-1}, so that molecular simulations can be executed.

- **Sparse grid:** A sparse grid, $\hat{G}_\ell^N$, is determined within $[0,1]^N = \xi\left(\mathcal{W}^k\right)$. Note that the transformation, ξ, has not to be applied explicitly. As the combination technique is used, the grid points of the sparse grid are determined hierarchically from the required subgrids, cf. Equation (11). Let $y_{\ell,j}$ be a point of the subgrids of the level, $\ell = (\ell_1, ..., \ell_N)$, and position, $j = (j_1, ..., j_N)$, which is, at the same time, a non-boundary point of the sparse grid. For each of these points, the loss function value, $F(y_{\ell,j})$, is computed.

- **Boundary:** By multiplying F with a sine term:

$$\bar{F} : [0,1]^N \rightarrow \mathbb{R}_0^+ \tag{41}$$
$$y \mapsto \left(\prod_{i=1}^N \sin \pi y_i\right) F(y) \tag{42}$$

homogeneous Dirichlet boundary conditions are realized in order to reduce the computational effort significantly. The function, $\bar{F}$, is applied to each point, $y_{\ell,j}$, of the sparse grid.

- **Smoothing:** As the function to be minimized is affected with statistical noise, the function, $\bar{F}$, is smoothed and regularized by the methods indicated in Section 2.2.2. Hence, for each point, $y_{\ell,j}$, of the sparse grid, a value, $G(y_{\ell,j})$, of the approximating function is obtained.

- **Interpolation:** The function, G, is interpolated from the sparse grid, $\hat{G}_{\hat{\ell}}^N$, on the full grid, $G_{\tilde{\ell}}^N$, by the combination technique. Hence, an interpolation model, $\bar{q}$, is obtained for each point, $y \in G_{\tilde{\ell}}^N$.

  As the smoothing and interpolation procedures have been executed for $\bar{F}$, the following division has to be applied for each non-boundary point of the full grid:

  $$\forall_{y \in \tilde{G}_{\tilde{\ell}}^N} q(y) = \frac{\bar{q}(y)}{\prod_{i=1}^N \sin \pi y_i} \tag{43}$$

  The interpolation model, q, is only valid for all $y \in \tilde{G}_{\tilde{\ell}}^N$.

- **Discrete minimization:** Determine:

  $$y^* := \arg \min_{y \in \tilde{G}_{\tilde{\ell}}^N} q(y) \tag{44}$$

- **Iteration step $x^k \to x^{k+1}$:** The ratio:

  $$r^k := \frac{F(y^k) - F(y^*)}{q(y^k) - q(y^*)} = \frac{F(y^k) - F(y^*)}{G(y^k) - q(y^*)}, \ y^k = \xi(x^k) \tag{45}$$

  is determined, and the following three cases are differentiated:

  - $r^k \geq \eta_1 \Rightarrow x^{k+1} := \xi^{-1}(y^*) \wedge \Delta_{k+1} := \Delta_k$.
  - $r^k \geq \eta_2 > \eta_1 \Rightarrow x^{k+1} := \xi^{-1}(y^*) \wedge \Delta_{k+1} := \gamma_2 \Delta_k$.
  - $r^k < \eta_1 \Rightarrow x^{k+1} := x^k \wedge \Delta_{k+1} := \gamma_1 \Delta_k$.

  Thereby, $\eta_2 > \eta_1 > 0$ and $\gamma_2 > 1 > \gamma_1 > 0$ are global parameters.
  Let $k := k + 1$, and go to step 2.

- **Stopping criteria:** The general stopping criterion is:

  $$F(x^*) \leq \tau \tag{46}$$

where $\tau > 0$. However, an additional criterion is that the minimum is situated within the hyperdice and not at its boundary. In total, the following three stopping criteria are considered:

(i) $F(x^*) \leq \tau \wedge \xi(x^*) \notin U_0 \cup U_1$

(ii) $F(x^*) \leq \tau \wedge \xi(x^*) \notin U_0 \cup U_1 \wedge \Delta^* < \Delta^{\min}$

(iii) $\exists k \in \mathbb{N} : r^k < \eta_1 \wedge \Delta_k < \Delta^{\min}$.

Thereby:

$$U_0 := \{y \in [0,1]^N \mid \exists i \in 1, ..., N : y_i \in \{0, 1\}\},$$
$$U_1 := \{y \in [0,1]^N \mid \exists i \in 1, ..., N : y_i \in \{2^{-\hat{\ell}}, 1 - 2^{-\hat{\ell}}\}\}$$

Moreover, $\Delta^* := \Delta_k$, where $x^* = x^k$.

If the stopping criteria, (i) and (ii), are fulfilled, then SpaGrOW has converged successfully. Due to $\xi(x^*) \notin U_1$, it is excluded, as well, that the minimum is situated at the boundary of the grid, where the interpolation model, q, is valid. Stopping criterion (ii) contains the additional condition, $\Delta^* < \Delta^{\min}$, excluding that improvements can be achieved by local refinements. Hence, this is the ideal stopping criterion.

Stopping criterion (iii) means that SpaGrOW has not led to success, *i.e.*, even by decreasing the trust region, no accurate model, q, can be found. In particular, this is the case when the assumptions for the application of the combination technique are not fulfilled, which may be caused by an inaccurate smoothing procedure, wherein the noise has not been filtered out in a sufficient way.

3.2.2. Complexity

In the following, the complexity of SpaGrOW is discussed. The present section is organized like Section 3.2.1, but here, SpaGrOW is discussed with respect to complexity:

- **Initialization:** The effort is only caused by the allocation of the initial variables, like the initial force field parameters, their upper and lower bounds, as well as the maximal, initial and minimal step length.
- **Transformation:** The transformation is performed in an imaginary way only. The sparse grid is constructed directly in step 3. Only the execution of the back-transformation, ξ^{-1}, requires a small computational effort.
- **Sparse grid:** The complexity to obtain a sparse grid of the level, $\hat{\ell}$, is $\mathcal{O}\left(2^{\hat{\ell}} \cdot \hat{\ell}^{N-1}\right)$, which is still exponential in the dimension, but especially in the case of higher dimensions, it is significantly lower than for a full grid of the same level. The computational effort does not only concern the number of grid points to be computed, but in the first instance, the number of loss function evaluations, *i.e.*, simulations.
- **Boundary:** Setting the loss function values to zero at the boundary of the sparse grid does not mean any computational effort. For all other grid points, $2N + 1$ ($\mathcal{O}(N)$) multiplications have to be performed, *cf.* Equation (35). Furthermore, there are N sine evaluations. The number of boundary points of an N-dimensional sparse grid of the level, $\hat{\ell}$, is $\mathcal{O}\left(N2^{N-1}2^{\hat{\ell}}\right)$. Thereby, $N2^{N-1}$ is the number of edges of an N-dimensional hyperdice. Hence, the number of simulations to be executed is reduced to:

$$\mathcal{O}\left[2^{\hat{\ell}}\left(\hat{\ell}^{N-1} - N2^{N-1}\right)\right] \tag{47}$$

This number multiplied by $2N + 1$ is the number of required multiplications and multiplied by N, the number of sine evaluations required. The reduction of molecular simulations is achieved at the expense of

$$\mathcal{O}\left[N2^{\hat{\ell}}\left(\hat{\ell}^{N-1} - N2^{N-1}\right)\right] \tag{48}$$

multiplications and sine evaluations, a computational effort that can be neglected, if the high amount of computation time for a simulation is opposed.

- **Smoothing:** In the case of most approximation methods, a multivariate linear regression has to be performed with complexity $\mathcal{O}(mM^2 + M^3)$, where m is the number of data points and M, the number of basis functions (e.g., $M = K$ for RBFs). However, this complexity can often be reduced to $\mathcal{O}(mM^2 + M^2)$ or, even, to $\mathcal{O}(1)$ by smart numerical methods, e.g., by a Cholesky factorization, in the case of positive definiteness. In contrast to simulations, the computational effort required for a smoothing procedure is negligible, as well. However, one has to consider that m and M must be large enough, on the one hand, in order to achieve an approximation as accurate as possible, and also small enough, on the other hand, in order to keep the computational effort low and to avoid overfitting. Additional effort appears due to the selection of centroids by the *k-means* algorithm, whose convergency speed always depends on the random choice of the initial centroids. The evaluation of the model function is done by a summation of the centroids only.

Furthermore, the regularization methods require some amount of computational effort, in particular, due to the application of Newton-Lagrange algorithms for the constrained optimization. Please note that most of them have been parameterized, as well, e.g., by cross validation.

- **Interpolation:** In the multilinear interpolation, all adjacent points have to be considered for each grid point. Hence, the interpolation is in $\mathcal{O}\left[(2N+1) \cdot 2^{\ell} \cdot \hat{\ell}^{N-1}\right]$.

  After the multilinear interpolation, a division by a sine term has to be executed for each point situated inside the unit hyperdice. As the sine term has already been calculated for each point of the sparse grid, only:

  $$\mathcal{O}\left[(2N+1)\left((2^{\hat{\ell}}-1)^N - 2^{\hat{\ell}}\hat{\ell}^{N-1}\right)\right] \tag{49}$$

  multiplications and:

  $$\mathcal{O}\left[N\left((2^{\hat{\ell}}-1)^N - 2^{\hat{\ell}}\hat{\ell}^{N-1}\right)\right] \tag{50}$$

  sine evaluations have to be performed. The number of required divisions is equal to $(2^{\hat{\ell}}-1)^N$. Please note that the divisions have to be performed for each inner point of the sparse grid, as well, because only the approximating function, G, coincides with the interpolation model, but not $\bar{F}$.

- **Discrete minimization** For the minimization of q, a maximal function value of 10^6 is supposed. For each point on the full grid, a comparison has to be performed, in total, $(2^{\hat{\ell}}-1)^N$ comparisons.

- **Iteration step:** Only a few small operations are necessary in order to perform an iteration step, *i.e.*, simple subtractions, divisions, multiplications and comparisons.

- **Stopping criteria:** Only comparisons have to be performed in order to check the stopping criteria. The decision whether a vector is situated in $U_0 \cup U_1$ or not is made by comparing the components of y with the values zero, one, h and $1-h$. If one component coincides with one of these four values, the procedure is stopped, and $y \in U_0 \cup U_1$ is observed.

3.3. Convergency

In the following, some convergency aspects the SpaGrOW algorithm are considered. In [40], it was proven that the algorithm converges under certain assumptions. Thereby, both smooth and noisy objective functions were considered. In the smooth case, the interpolation error is relevant, and in the noisy case, the approximation error with respect to the original function, *i.e.*, the function without noise, has to be taken into account. Moreover, it is examined to what extent SpaGrOW manages with less simulations than gradient-based methods. In the case of the latter, simulations have to be performed for the gradient components, the entries of the Hessian matrix and the Armijo steps. In the case of SpaGrOW, they have to be executed for the sparse grid points and the Trust Region steps. The steepest descent algorithm and the conjugate gradient methods required significantly less simulations for the gradient than SpaGrOW for the sparse grid: for $N = 4$, four simulations are required for the gradient and nine for a sparse grid of the level 2. As for the step length control, it can be observed that both gradient-based methods and SpaGrOW mostly need one Armijo or, respectively, one Trust Region step at the beginning of the optimization, but many more close to the minimum. The reason for the reduction of computational effort in the case of SpaGrOW lies in the lower number of iterations, not in the lower number of function evaluations per iteration. The advantage is that the step length is determined before, so that, especially at the beginning of the optimization, large steps can be realized. For smooth functions, the combination technique delivers small interpolation errors in most cases, also when the trust region is quite large, but always a descent direction, mostly leading to the boundary of the actual trust region.

Close to the minimum, both approaches have the drawback that after a high number of step length control iterations, *i.e.*, after a large computational effort, only marginal improvements in the loss function values are observed. However, due to the statistical noise, the minimum can never be predicted exactly. At some point, the minimization has to be stopped, and the actual result has to be evaluated. However SpaGrOW is capable of searching for a smaller loss function value in more than one direction, due to the grid approach. Furthermore, its modeling approach increases the probability to get close to the minimum than gradient-based methods, as motivated in Section 1.

As the interpolation and smoothing errors are essential for the convergency of SpaGrOW, they are introduced and discussed in the following.

3.3.1. Interpolation Error

Under certain assumptions [24], the interpolation error for a sparse grid of the level, $\hat{\ell}$, is of the order $\mathcal{O}\left(h_{\hat{\ell}}^2 (\log(h_{\hat{\ell}})^{-1})^{\hat{\ell}-1}\right)$. First, the two-dimensional case is considered again. If the difference $u - u_{i,j}$, where $u_{i,j} \in T_{i,j}$, $i + j \in \{\hat{\ell}, \hat{\ell} + 1\}$, meets point-wise an asymptotic error expansion, *i.e.*,

$$u - u_{i,j} = C_1(h_i)h_i^2 + C_2(h_j)h_j^2 + D(h_i, h_j)h_i^2 h_j^2 \tag{51}$$

where $\forall_i \, |C_1(h_i)| \leq \kappa$, $\forall_j \, |C_2(h_j)| \leq \kappa$ and $\forall_{i,j} \, |D(h_i, h_j)| \leq \kappa$, $\kappa > 0$, then the interpolation error can be estimated as follows:

$$|u - \hat{u}_{\hat{\ell}}^c| \leq \left(1 + \frac{5}{4} \log\left(h_{\hat{\ell}}^{-1}\right)\right) \kappa h_{\hat{\ell}}^2 = \mathcal{O}\left(h_{\hat{\ell}}^2 \log(h_{\hat{\ell}})^{-1}\right) \tag{52}$$

For the N-dimensional case, an analogous asymptotic error expansion delivers the following formula:

$$|u - \hat{u}^c_{\hat{\ell}}| \leq \left(\sum_{l=0}^{N-1} \chi_l \left(\log \left(h_{\hat{\ell}}^{-1} \right) \right)^l \right) \kappa h_{\hat{\ell}}^2 = \mathcal{O} \left(h_{\hat{\ell}}^2 \left(\log(h_{\hat{\ell}})^{-1} \right)^{\hat{\ell}-1} \right) \tag{53}$$

Thereby, $u \in S^0_{\hat{\ell}}$, $\bar{\ell} = (\hat{\ell}, ..., \hat{\ell}) \in \mathbb{N}^N$ and $\chi_l \in \mathbb{N}$, $l = 0, ..., \hat{\ell} - 1$.

Please note that in order to guarantee the existence of such an asymptotic error expansion, the exact solution, u, must fulfill certain continuity and smoothness conditions. As u is supposed to reproduce F exactly on the sparse grid and as accurately as possible between the sparse grid points, these assumptions have to be transferred to F. This motivates again the need for a smoothing procedure in the case of statistical noise. In most cases, the existence of an asymptotic error expansion cannot be proven *a priori*. However, the combination technique was shown to deliver very good results in practice [24].

The interpolation error can be estimated above as follows:

Theorem 11 (Estimate for the interpolation error). *Let $\Delta > 0$ be the size of the hyperdice, $\mathcal{W}^\Delta(x)$, with center $x \in \mathbb{R}^N$, where a sparse grid of the level, $\hat{\ell}$, is defined. Let the approximated function, $G : \mathbb{R}^N \to \mathbb{R}^+_0$, be given on the sparse grid and interpolated by the model function, $q : B_\Delta(0) \to \mathbb{R}$, on a full grid of the level, $\ell := (\hat{\ell}, ..., \hat{\ell}) \in \mathbb{N}^N$, using the combination method. Then, for the interpolation error, $\varepsilon = |u - \hat{u}^c_{\hat{\ell}}|$, in Inequality (53), it holds for $u := G$ and $\hat{u}^c_{\hat{\ell}} := q$:*

$$\forall_{y \in \mathcal{W}^\Delta(x)} \; |\varepsilon(y)| \leq \kappa_\varepsilon(\hat{\ell}) f^\varepsilon_{\hat{\ell}}(\Delta) \tag{54}$$

where $\kappa_\varepsilon : \mathbb{N} \to \mathbb{R}^+$ and $f^\varepsilon_{\hat{\ell}} : \mathbb{R}^+ \to \mathbb{R}^+$ with $\lim_{\hat{\ell} \to \infty} \kappa_\varepsilon(\hat{\ell}) = 0$ and $\lim_{\Delta \to 0} f^\varepsilon_{\hat{\ell}}(\Delta) = 0$. The function, $f^\varepsilon_{\hat{\ell}}$, is continuous. Moreover, $\tilde{f}^\varepsilon_{\hat{\ell}} := \frac{f^\varepsilon_{\hat{\ell}}}{\Delta} : \mathbb{R}^+ \to \mathbb{R}^+$ is continuous, as well, with $\lim_{\Delta \to 0} \tilde{f}^\varepsilon_{\hat{\ell}}(\Delta) = 0$.

Proof. Let $y \in \mathcal{W}^\Delta(x)$ be an arbitrary point in the hyperdice. It holds $h_{\hat{\ell}} = 2^{1-\hat{\ell}}\Delta$. Following Inequality (53), there exist constants $\chi_l > 0$, $l = 0, ..., N - 1$ and a $\kappa > 0$, so that:

$$
\begin{aligned}
|\varepsilon(y)| &\leq \left(\sum_{l=0}^{N-1} \chi_l \left(\log \left(2^{\hat{\ell}-1}\Delta^{-1} \right) \right)^l \right) \kappa 2^{2-2\hat{\ell}} \Delta^2 \\
&= \left(\sum_{l=0}^{N-1} \chi_l \left(\hat{\ell} - 1 - \log \Delta \right)^l \right) \kappa 2^{2-2\hat{\ell}} \Delta^2 \\
&\leq \underbrace{N \cdot \max_{l=0}^{N-1} \chi_l \cdot \kappa 2^{2-2\hat{\ell}}}_{:=\kappa_\varepsilon(\hat{\ell})} \cdot \underbrace{\max_{l=0}^{N-1} \left(\hat{\ell} - 1 - \log \Delta \right)^l \cdot \Delta^2}_{:=f^\varepsilon_{\hat{\ell}}(\Delta)} \\
&= \kappa_\varepsilon(\hat{\ell}) f^\varepsilon_{\hat{\ell}}(\Delta)
\end{aligned}
\tag{55}
$$

It holds $\lim_{\hat{\ell} \to \infty} \kappa_\varepsilon(\hat{\ell}) = 0$, and also, because of l'Hospital's rule, $\lim_{\Delta \to 0} f^\varepsilon_{\hat{\ell}}(\Delta) = 0$. Due to $\tilde{f}^\varepsilon_{\hat{\ell}}(\Delta) = \max_{l=0}^{N-1} \left(\hat{\ell} - 1 - \log \Delta \right)^l \cdot \Delta$, it follows also from l'Hospital's rule that $\lim_{\Delta \to 0} \tilde{f}^\varepsilon_{\hat{\ell}}(\Delta) = 0$. $\square$

3.3.2. Smoothing Error

Let $||\cdot||$ be one of the two norms, $||\cdot||_{L_2}$ or $||\cdot||_{L_\infty}$. The smoothing error, μ, is the error with respect to $||\cdot||$ between the original function, $F = \tilde{F} - \Delta F$, without noise and the approximating function, G. Thereby, $\tilde{F}$ is the noisy function. It holds:

$$\mu := ||F - G|| = ||(F + \Delta F) - G - \Delta F|| = ||\tilde{F} - G - \Delta F|| \leq |\vartheta| + ||\Delta F|| \tag{56}$$

where ϑ denotes the training error. If Ω is a trust region, then define
$\forall_{x \in \Omega} \, \vartheta(x) := \tilde{F}(x) - G(x)$. In the ideal case, $\vartheta(x) = \Delta F_x$, it holds $\mu(x) = 0$. Otherwise, for $0 < \delta << |\Delta F_x|$, the following inequation must hold:

$$|\mu(x)| = |\Delta F_x - \vartheta(x)| \leq \delta \tag{57}$$

This means that the training error, ϑ, must not be too small. Figure 10a shows an overfitted model, G, which reproduces exactly the oscillations produced by the noise. In this case, $\vartheta = 0$, but $|\mu(x)| = |\Delta F_x| >> \delta$ for certain $x \in \Omega$. On the other hand, the training error must not be too high. In Figure 10b, it holds $\forall_{x \in \Omega} |\mu(x)| \approx |\Delta F_x| >> \delta$. Only Figure 10c depicts a feasible case; here, the training error is in the same order as the noise, and it holds $|\mu| \leq \delta$.

Figure 10. Approximation models with different training errors, ϑ, where the function, G, approximates the noisy function, $\tilde{F} = F + \Delta F$, overfitted model with $\vartheta = 0$ (**a**); model, where ϑ is too high (**b**); and feasible model with $|\Delta F - \vartheta| \leq \delta$ (**c**). In the ideal case, it holds $\vartheta(x) = \Delta F_x$.

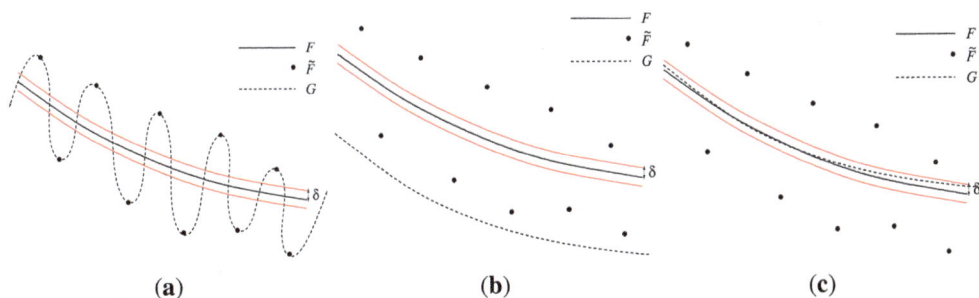

(**a**)　　　　(**b**)　　　　(**c**)

In order to keep the smoothing error low enough, a smoothing procedure has to be applied, which filters out the noise and reproduces the loss function, at least on the sparse grid, as exactly as possible. As a smoothing procedure can only be evaluated on a sparse grid, due to the high amount of computation time for molecular simulations, the condition, $|\mu| \leq \delta$, is considered only on the sparse grid. It does not matter if G has oscillations between the sparse grid points, because the piecewise-multilinear sparse grid interpolation is not capable of modeling them anyway.

The following theorem is essential for the convergency proof of SpaGrOW and gives an estimate for the smoothing error in the case of positive definite RBFs:

Theorem 12 (Estimate for the smoothing error). *Let $\Delta > 0$ be the size of the hyperdice, $\mathcal{W}^\Delta(x)$, with the center, $x \in \mathbb{R}^N$, in which a sparse grid of the level, $\hat{\ell}$, is constructed. Let the loss function, $F : \mathbb{R}^N \to \mathbb{R}_0^+$, be given on the sparse grid and approximated by the function, $G : \mathbb{R}^N \to \mathbb{R}$. Suppose that the approximation be stable. Then, the smoothing error μ from Equation (56) can be estimated as follows:*

$$\exists\, \kappa_\mu > 0 \,:\, \forall_{y \in \mathcal{W}^\Delta(x)} \; |\mu(y)| \leq \kappa_\mu f^\mu(\Delta) \tag{58}$$

where $f^\mu : \mathbb{R}^+ \to \mathbb{R}^+$ is continuous with $\lim_{\Delta \to 0} f^\mu(\Delta) = 0$. Furthermore, $\tilde{f}(\mu) := \frac{f^\mu}{\Delta} : \mathbb{R}^+ \to \mathbb{R}^+$ is continuous, as well, with $\lim_{\Delta \to 0} \tilde{f}^\mu(\Delta) = 0$.

Proof. Due to the stability of the smoothness, Inequality (29) holds, where $\Omega = \mathcal{W}^\Delta(x)$ and $\mathcal{X}$ is the sparse grid. The fill distance on the sparse grid and Δ only differ by a constant, $\omega > 0$, *i.e.*, $\Delta_{\Omega,\mathcal{X}} = \omega \Delta$. For $\kappa_\mu := \kappa$ and $f^\mu(\Delta) := h(\omega \Delta) = h(\Delta_{\Omega,\mathcal{X}})$, the estimate for the smoothing error in Equation (58) follows directly. $\square$

Remark 13. *Following Corollary 10, estimate Equation (58) is given in the case of a smoothing procedure based on Gaussian RBFs.*

Remark 14. *For Theorem 12 and the full convergency proof, it is irrelevant whether the original function, $F : \mathbb{R}^N \to \mathbb{R}_0^+$, or the transformed function, $\bar{F} : [0,1]^N \to \mathbb{R}_0^+$, is approximated by $G : \mathbb{R}^N \to \mathbb{R}$. The function to be smoothed only has to be continuous within the trust region. Please note that in the case of $\bar{F}$, the function, G, can have negative values, as $\bar{F}$ is equal to zero at the boundary of the trust region. For the original function, $G(x) \geq 0$ can be assumed, due to Theorem 12, when Δ is small enough. For $\bar{F}$, this can be assumed, as well. Otherwise, consider a translation that does not have any impact on either the approximation or the minimization.*

The convergency proof executed for SpaGrOW was related to a general convergency proof for derivative-free Trust Region methods [41]. However, the Trust Region method used in that paper is based on an interpolation with Newtonian fundamental polynomials. Hence, the partial proofs cannot be transferred, but have to be developed anew. Another crucial difference consists in the assumption for the loss function to be at least two times continuously differentiable with a bounded Hessian norm, which cannot be made in the case of SpaGrOW. The detailed convergency proof was performed in [40].

3.4. Speed of Convergency Compared to Gradient-Based Methods

The speed of convergency of SpaGrOW is discussed in the following, also with respect to statistical noise. As already mentioned, the trust region size, Δ, may not be too small, due to the noise. However, the convergency proof is based on the choice of a small Δ. This dilemma, which is also present in the case of gradient-based methods whenever two adjacent points are required for the computation of a partial derivative, leads to the need for an optimal parameterization of SpaGrOW. To achieve a high speed of convergency, primarily at the beginning of the optimization, the choice of a large Δ is required without hurting one of the assumptions for the convergency proof, *cf.* [40]. In the following, some heuristic considerations are made in this regard. Thereby, the index, $_g$, refers to gradient-based descent methods, the index, $_H$ to descent methods using a Hessian and the index, $_S$ to SpaGrOW. Furthermore, let $\bar{M}$ be the average number of function evaluations per iteration and $\bar{l}$, the average number of Armijo or Trust Region steps. Then, it holds:

$$\bar{M}_g = N + \bar{l}_g \tag{59}$$

$$\bar{M}_H = N + \frac{N(N+1)}{2} + \bar{l}_H \tag{60}$$

$$\bar{M}_S = 2N \cdot \bar{l}_S \tag{61}$$

The drawback of SpaGrOW lies in the multiplicative dependency of $\bar{M}_S$ on $\bar{l}_S$. This is due to the fact that a new sparse grid is used at each iteration step. However, at the beginning of the optimization, $\bar{l}_g = \bar{l}_H = \bar{l}_S = 1$ is assumed. Then, it holds $\bar{M}_g < \bar{M}_S < \bar{M}_H$. Hence, SpaGrOW requires less iterations on average than a method based on Hessians. However, in total, it has to manage with less iterations than a gradient-based method requiring less function evaluations per iteration: Let k be the general number of iterations; then it must hold: $k_S < k_g$. If:

$$k_S < \frac{N+1}{2N} k_g \tag{62}$$

SpaGrOW needs less iterations and function evaluations than a gradient-based method. Now, the question arises, how this can be steered. By choosing an initial Δ_0 (and also γ_1) that is large enough, a faster convergency can be achieved.

In the following, the initial phase of an optimization process is considered, and a short comparison between SpaGrOW and a gradient-based descent method is pointed out, *i.e.*, it is discussed under what conditions the speed of convergency is significantly higher in the case of SpaGrOW. Please note that *at the beginning of the optimization* means here that the number of trust region or Armijo steps is equal to one at each iterations and that k_g is chosen, so that the number of Armijo steps for x^{k_g} is still equal to one and for x^{k_g+1}, greater than one. Furthermore, choose k_S, so that $||x^{k_S} - x^0|| \le ||x^{k_g} - x^0||$, the size of the trust region is equal to Δ_0 and $x^{k_g} \in \Omega_k$. Hence, $x^{k_g} \notin \Omega_{k-1}$. This means that both x^{k_S} and x^{k_g} are reached by SpaGrOW with k_S steps of length Δ_0, which is depicted in Figure 11. The distance between x^0 and x^{k_S} is equal to $k_S \Delta_0$. It holds:

$$k_S \Delta_0 \le ||x^{k_g} - x^0|| \tag{63}$$

If:

$$\Delta_0 > \frac{||x^{k_g} - x^0||}{\underbrace{\frac{N+1}{2N} k_g}_{>1, \text{ if } k_g \ge 2}} \tag{64}$$

i.e., if:

$$\Delta_0 > \zeta ||x^{k_g} - x^0|| \tag{65}$$

for a maximal $\zeta \in (0, 1)$; then, Inequality (62) follows. In practice, a realistic case is $k_g = 4$. If:

$$\zeta := \frac{4N}{(N+1)k_g} = \frac{N}{N+1} < 1 \tag{66}$$

464

is chosen, SpaGrOW requires less than 50% of the iterations and function evaluations required by
the gradient-based descent method. Hence, a reduction of computation time by a factor of two is
plausible at the beginning of the optimization process.

Figure 11. Speed of convergency of SpaGrOW at the beginning of the optimization. For
an appropriate choice of the size of the initial trust region, Δ_0, the number of iterations
in the case of SpaGrOW (k_S) is significantly smaller than in the case of a gradient-based
method (k_g). It is realistic that the number of iterations and function evaluations can be
reduced by a factor of two.

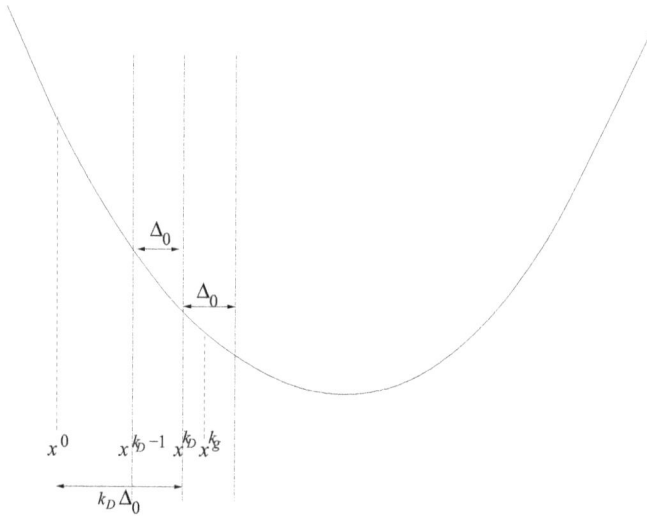

4. Practical Evaluation and Results

The methodology of SpaGrOW was implemented in *python* (version 2.4.3) and is modularly
constructed. The software consists of a main control script and secondary control scripts for specific
parts of the algorithm, e.g., for the sparse grid interpolations and the control of the smoothing
procedures, which were implemented in *S+* (version 2.1.10), scripting language related to the *R
Project for Statistical Computing* [42].

The implementation of SpaGrOW contains the full algorithm described in the previous section
and acts as an interface between optimization and simulation, providing all necessary control routines
for both tasks. On the optimization side, the method starts with an initial guess and evaluates the loss
function based on the results of a simulation. If one of the stopping criteria of SpaGrOW is fulfilled,
the optimization workflow terminates. Otherwise, the parameters are updated by SpaGrOW.

On the simulation side, a control script calls the simulation tool performing all preparation
routines and computing the trajectory, as well as the desired physical target properties. The latter
are passed on to the optimization workflow of SpaGrOW. In the case of a simultaneous optimization
of properties at different temperatures, the respective simulations are executed in parallel. In this

case, a script distributing K jobs at K temperatures is called. A script controlling the parallel environment and the simulation control script are called K times. If $m = n/K$ physical properties are fitted, the result of each job consists of m properties files. Figure 12 shows both the optimization and the simulation side and how they interact with each other in the case of parallel jobs at different temperatures.

Figure 12. Technical realization of optimization procedure for physical target properties at different temperatures. If the simulated properties do no coincide well with their experimental reference data, the optimization control script—depicted on the left—passes the current force field parameters on to a distribution control script, which submits parallel jobs at different temperatures. Then, a parallel environment control script is executed, and a simulation control script is called, which performs the following three tasks: preparation routines, the simulation itself and the computation of the simulated target properties. The properties are written into separate files, which are read by the optimization control script. Finally, the loss function is evaluated and the workflow continues.

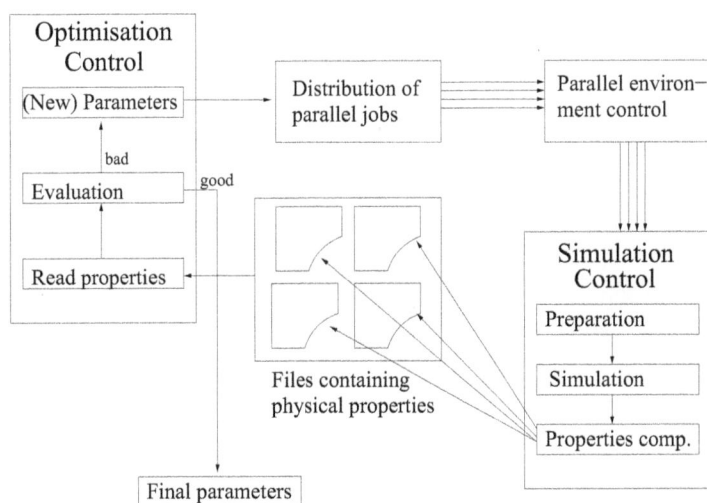

The simulations were performed on a parallel computer cluster with 215 available nodes, where each node is provided with two Intel-Nehalem-EP-Quadcore processors (Xeon X5550) with 24 GB of main storage, which are connected by a fast QDR infiniband interconnect with a 40-Gb/s Double Date Rate (DDR).

In this section, SpaGrOW is evaluated in practice and applied to molecular simulations as described above. The questions to be answered are the following:

- Which smoothing and regularization procedure are most suitable?
- Is SpaGrOW capable of saving simulations in comparison to gradient-based methods?
- How close does SpaGrOW get to the minimum?

4.1. Selection of Smoothing and Regularization Methods

As molecular simulations are extremely time-consuming, an analytical model replacing them was used for the selection. In previous work [22], a similar assessment has already been performed for gradient-based methods. Vapor-Liquid Equilibrium (VLE) data, like the saturated liquid density, ρ_l, the enthalpy of vaporization, $\Delta_v H$, and the vapor pressure, p_σ, can be evaluated directly as functions of certain force field parameters. These functions were determined by [43] by nonlinear regression using a high amount of simulation data for two-centered Lennard-Jones (LJ) fluids with a quadrupolar moment. Here, nitrogen was considered as an example for this kind of fluid. The model parameters to be optimized were the elongation, L, of the two LJ sites, the quadrupolar moment, Q^2, and the two LJ parameters, σ and ε. In order to mimic molecular simulations, uniformly distributed artificial uncertainties (0.5% for the density and 3% for the pressure) were added for the simultaneous optimization of ρ_l and p_σ at six different temperatures, $T/K \in \{65, 75, 85, 95, 105, 115\}$. As in [22], the weights of the properties in the loss function were all equal to one, because all properties were considered as homologous. As the simulation data were noisy, ten statistically independent random replicates were performed for each optimization run, whose results were averaged.

Due to theoretical considerations, the tendency consists in selecting positive definite RBFs, in particular, Gaussian RBFs. In the following, it is shown that this is also a good choice in practice.

In order to evaluate, whether a smoothing or regularization procedure is appropriate for SpaGrOW, the behavior of the algorithm combined with each preprocessing procedure is analyzed. Thereby, both efficiency and robustness with respect to noise are considered. Table 2 shows the candidates and their abbreviations.

Table 2. Candidates for smoothing and regularization procedures within SpaGrOW together with their abbreviations.

Smoothing procedures	Regularization procedures
Radial Basis Functions	Least squares
	Naive Elastic Nets (NENs)
	$\alpha = 0$: Least Absolute Shrinkage and Selection Operator (LASSO)
	$\alpha = 1$: Ridge Regression
Linear property approximation (Lipra)	Multivariate Adaptive Regression Splines (MARS)

Selection of the Best Smoothing Procedure It was already motivated that a smoothing procedure is indispensable, whenever noisy loss function values are present. A detailed analysis has shown that certain RBFs deliver better results by far than others. Suitable RBFs are the linear, cubic, Gaussian and thin plate spline RBF, *cf.* Section 2.2.2. The multiquadric functions were not reliable. Moreover, Wendland functions were considered, as well. Figure 13 shows box plots for the loss function values achieved by SpaGrOW combined with the different RBFs, which is a criterion for the quality of convergency. The results over ten statistically independent replicates are indicated. The lower the loss function achieved was, the closer got the algorithm to the minimum. The smallest loss function values were achieved robustly by the four RBFs mentioned above. A Gaussian RBF is selected here for the following reasons:

- No outliers were detected.
- A more accurate analysis of the approximation error has shown that the smoothing procedures based on cubic and thin plate spline RBFs reproduced the function, $\bar{F}$, cf. Section 3.2.1, worse than the one based on Gaussian RBFs. Figure 14 depicts the results for the Gaussian RBFs (Figure 14a) and the thin plate spline RBFs (Figure 14b) for the two-dimensional case: The approximating function and the original function values of $\bar{F}$, cf. Equation (35), on a sparse grid of the level 2 are plotted *versus* $\xi(Q^2)$ and $\xi(L)$, cf. Equation (39). The LJ parameters, σ and ε, were fixed. As can be seen, a smoothing procedure based on thin plate spline RBFs reproduced $\bar{F}$ at the boundary of the unit square in a very bad fashion, whereas Gaussian RBFs delivered a good approximation on the complete unit square.
- The Gaussian RBF is the only of the four RBFs mentioned above that is positive definite, *i.e.*, the selection of Gaussian RBFs is also founded theoretically according to the considerations in Section 2.2.4.

Please note that linear RBFs delivered good approximations at the beginning of the optimization, as well, which was not surprising, because the steepest descent method was also very successful [22]. For higher dimensions, the Lipra method, *i.e.*, a quadratic approximation of the loss function, could be convincing.

Figure 13. Box plots of the loss function values achieved by SpaGrOW in combination with a smoothing procedure based on Radial Basis Functions (RBFs). The RBFs were the linear, cubic, multiquadric (Multi), inverse multiquadric (Invers), Gaussian, thin plate spline RBF (TPS) and a Wendland function. Suitable RBFs were only the linear, cubic, Gaussian and thin plate spline RBF.

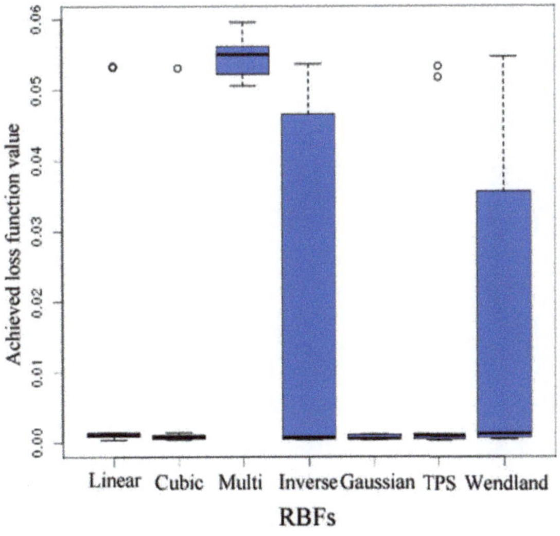

Figure 14. Approximations on the unit square, $[0,1]^2$, based on Gaussian RBFs (**a**) and thin plate spline RBFs (**b**). The blue points mark the original (noisy) function values of $\bar{F}$ on the sparse grid. It holds $x1 = \xi(Q^2)$, $x2 = \xi(L)$ and $y = \bar{F}(x1, x2)$. The smoothing procedure based on thin plate spline RBFs reproduces $\bar{F}$ at the boundary of the unit square in a very bad fashion.

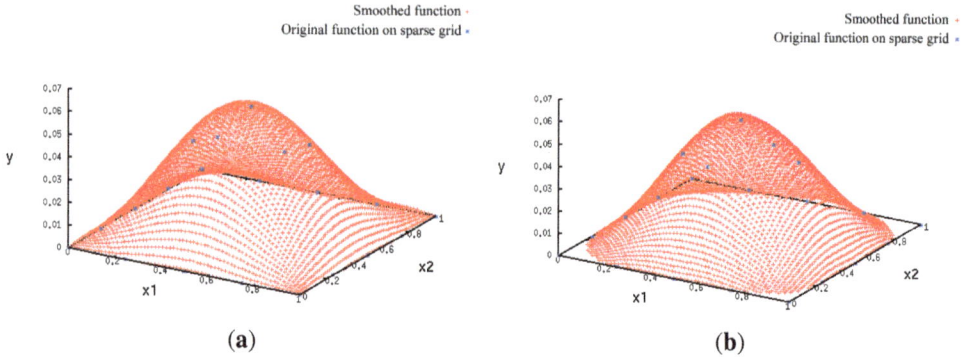

(a)   (b)

Selection of the Best Regularization Method As already mentioned, the selection of the best regularization method can only be achieved by practical evaluations. Candidates are least squares, NENs (LASSO for $\alpha = 0$ and Ridge Regression for $\alpha = 1$), as well as MARS.

Figure 15. Approximations of $\bar{F}$ on the unit square, $[0,1]^2$, based on Gaussian RBFs, combined with a LASSO regularization. The blue points mark the original (noisy) function values of $\bar{F}$ on the sparse grid. It holds $x1 = \xi(Q^2)$, $x2 = \xi(L)$ and $y = \bar{F}(x1, x2)$. At the boundary of the unit square, the function is reproduced in a bad fashion.

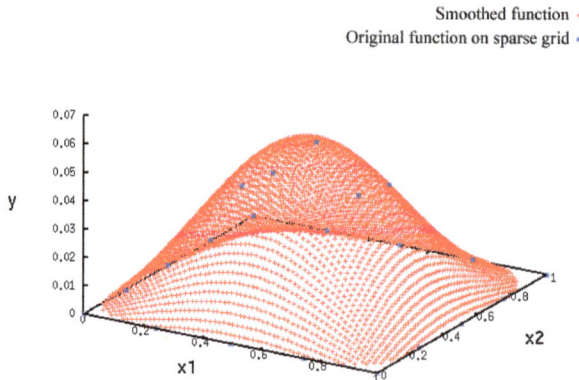

The regularization methods were evaluated in combination with the selected smoothing procedure based on Gaussian RBFs. The application of least squares is the same as not using any regularization method. All other regularization methods could improve the results achieved with least squares only. The LASSO algorithm performs a variable selection, *i.e.*, it tends to detect outliers by mistake. Hence, the model obtained by LASSO was often under-fitted. In contrast, the Ridge Regression estimator was more suitable for the present task, as it is a compromise between least squares, which often lead to overfitting, and the LASSO, which often leads to under-fitting. Figure 15 shows an approximation based on Gaussian RBFs in combination with LASSO. As can be seen, the function to be approximated was reproduced in a bad fashion at the boundary.

An NEN with $\alpha = 0.7$ delivered an even better quality of convergence; however, the computational effort to optimize α was too high compared to the benefit achieved. The application of an NEN with $\alpha \notin \{0, 1\}$ is not worthwhile for the present task.

Figure 16. Box plots of the loss function values (**a**) and of the number of function evaluations (**b**) resulting from the application of SpaGrOW combined with a smoothing procedure based on Gaussian RBFs and different regularization methods: Ridge Regression, LASSO, a weighted linear regression (rlm), an RBF approximation with an additional linear term (lt), an NEN (eln) with $\alpha = 0.7$ and MARS. It becomes clear that MARS is the algorithm to select for regularization. Ridge Regression is reliable, as well.

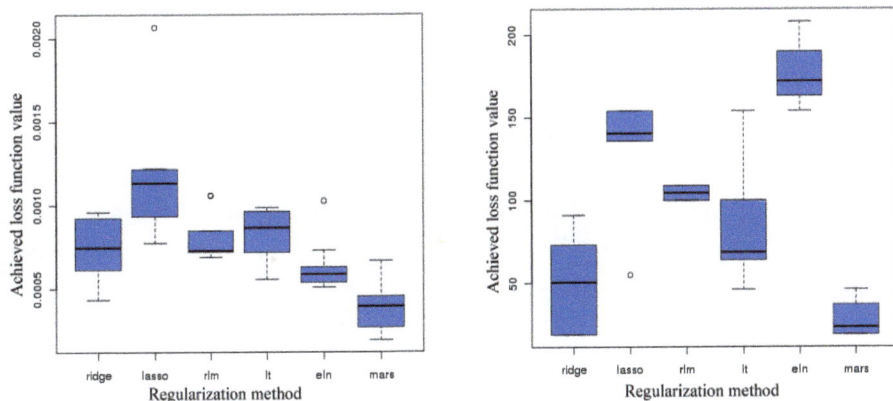

The best regularization method is the MARS algorithm, not only with respect to robustness and quality of convergency, but also with respect to the number of function evaluations: Figure 16 shows box plots of all regularization methods applied. Figure 16a shows the loss function values achieved and Figure 16b the number of function evaluations. Besides the methods considered here, two other ones were tried: a weighted linear regression based on M-estimators and an RBF approximation with an additional linear term. As can be seen, the MARS algorithm delivers the best results. Hence, it is selected as regularization method for SpaGrOW. However, Ridge Regression is reliable, as well, and suggested as the alternative. The NEN with $\alpha = 0.7$ achieved very low loss function values in a very robust way, but it always required more than a hundred function evaluations.

For Lipra, the least square estimator was the best regularization method. All other methods biased the quadratic approximation in a highly inappropriate way.

To summarize, Gaussian RBFs in combination with MARS and, in particular, for $N \geq 5$, the Lipra method together with the least square estimator have turned out to be most suitable for the present task and are implemented in SpaGrOW for this reason.

4.2. Application of SpaGROW to Molecular Simulations

Finally, SpaGrOW is applied to molecular simulations. Thereby, it is compared to gradient-based methods with respect to computational effort. Additionally, for an eight-dimensional problem, the Lipra method is evaluated, and it is analyzed how close SpaGrOW can get to the minimum.

4.2.1. Comparison to Gradient-Based Methods with Respect to Computation Time: Benzene and Ethylene Oxide

In the following, SpaGrOW is compared to GROW with respect to computational effort on the basis of two applications: benzene and ethylene oxide.

Benzene Benzene (C_6H_6) is a quite simple molecule, because of its symmetric structure and the fact that it does not possess a permanent dipolar moment. Furthermore, benzene has two chemically independent atom types only. Hence, the force field parameterization for benzene was deemed to be a relatively easy task. However, it is still challenging, because of the π interactions.

Figure 17 shows the comparison between SpaGrOW and GROW with respect to the computational effort required within the respective optimization procedures. The target observables were the enthalpy of vaporization (Figure 17a) and the saturated liquid density (Figure 17b), considered at three different temperatures. The values indicated on the y-axis are the Mean Absolute Percentage Errors (MAPE), *i.e.*, the absolute deviations from the respective experimental reference data in %, averaged over the range of temperatures. The experimental saturated liquid density was taken from [44] and the enthalpy of vaporization from the NISTdatabase [45]. The simulations performed were molecular dynamics simulations in the NpT ensemble executed with the software tool, *GROMACS* [46]. The non-bonded potential energy was computed by *Moscito* [47] using the trajectories collocated by GROMACS. Please note that the experimental target observables were VLE define data and that the simulated properties were only approximations, due to the lack of an explicit gas phase, which was assumed to be ideal. On the computer cluster mentioned above, three to four hours were required for the simulation of 1,000 benzene molecules for 2 ns using a time step of 2 fs. For SpaGrOW, the following variables were chosen: $\eta_1 = 0.2$, $\eta_2 = 0.7$, $\gamma_1 = 0.5$, $\gamma_2 = 1.1$, and $\Delta_0 = 0.3 \times \Delta_{\max}$. The force field parameters were $\sigma(H), \sigma(C), \varepsilon(H)$ and $\varepsilon(C)$. The initial parameters were taken from previous work [23]. Thereby, the saturated liquid density and the self-diffusion coefficient were optimized at the vapor-liquid coexistence curve. The feasible domain was defined as follows: σ was changed by no more than 30% and ε by no more than 80%. As it was a four-dimensional optimization problem, Gaussian RBFs were chosen for the smoothing and the MARS algorithm for the regularization procedure.

Figure 17. Mean Absolute Percentage Errors (MAPE) values with respect to $\Delta_v H$ (a) and ρ_l (b) for benzene during the SpaGrOW optimization in comparison to GROW. The smoothing procedure was based on Gaussian RBFs in combination with MARS. The force field parameters to be optimized were $\sigma(\mathrm{H}), \sigma(\mathrm{C}), \varepsilon(\mathrm{H})$ and $\varepsilon(\mathrm{C})$. A faster convergency of SpaGrOW could be confirmed.

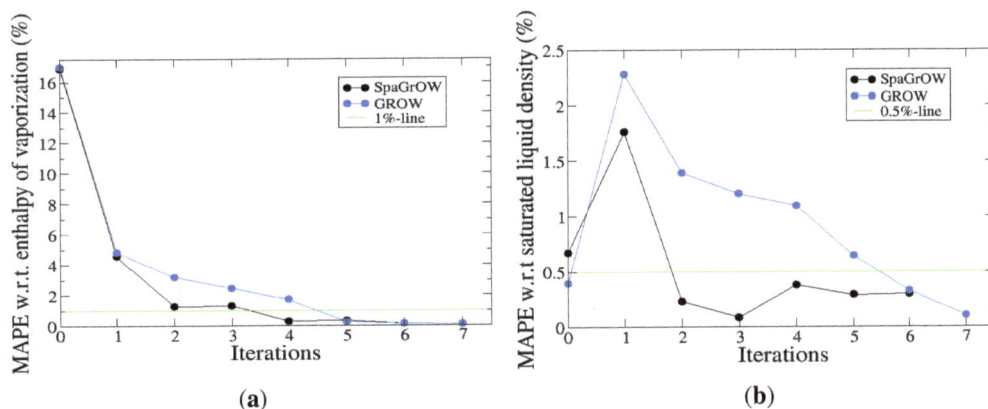

(a)   (b)

GROW needed seven iterations in total: five steepest descent and two conjugate gradient iterations. In contrast, SpaGrOW required six iterations only. Please note that an optimal force field with respect to $\Delta_v H$ and ρ_l was already achieved after four iterations, *i.e.*, both target observables were equal to their experimental reference data up to statistical uncertainties for all temperatures. Typical statistical uncertainties for $\Delta_v H$ and ρ_l are 1% and 0.5%, respectively, *cf.* e.g., [19]. For GROW, this was the case after seven iterations. The number of function evaluations, *i.e.*, simulations, for SpaGrOW and GROW was 37 and 62, respectively. However, in the latter case, seven simulations have to be subtracted for the comparison, because the partial atomic charged were optimized, as well. Hence, in the case of GROW, it was a five-dimensional optimization problem, and for the gradient calculation, one simulation more was required at each iteration. However, SpaGrOW was significantly faster than GROW.

Figure 18 depicts the development of the LJ Parameters for GROW and SpaGrOW (Figure 18a refers to σ and Figure 18b to ε). $\sigma(\mathrm{C})$ remained constant, and all other force field parameters were increased. At the fourth iteration of SpaGrOW, the parameters were very similar to the ones of GROW at the seventh iteration. However, the following interesting observation could be made. The parameter, $\sigma(\mathrm{C})$, remained constant, even during the whole optimization procedure. In the case of GROW, it was first decreased in order to obtain nearly the same value as before. As it is gradient-based, GROW tends to make detours. As it is grid-based, SpaGrOW is capable of keeping one or more parameters constant during the whole optimization procedure, because it can converge along a certain grid line or hyperplane. This is another reason for the faster convergency of SpaGrOW.

472

Only in the case of ε, some small detours were observed. The algorithm ran through the triangle indicated in Figure 18b. SpaGrOW delivered a set of force field parameters, which differed a little from the ones obtained by GROW. Hence, it achieved a different domain close to the minimum.

Figure 18. Development of the Lennard-Jones (LJ) parameters in the case of benzene ($\Delta_v H, \rho_l$) for GROW and SpaGrOW—$\sigma(H)$ and $\sigma(C)$ **(a)**—as well as $\varepsilon(H)$ and $\varepsilon(C)$ **(b)**. The unfilled circles indicate the optimal parameters. SpaGrOW led in a more direct way to the minimum than GROW. Only in the case of ε, some detours could be observed, due to the triangle.

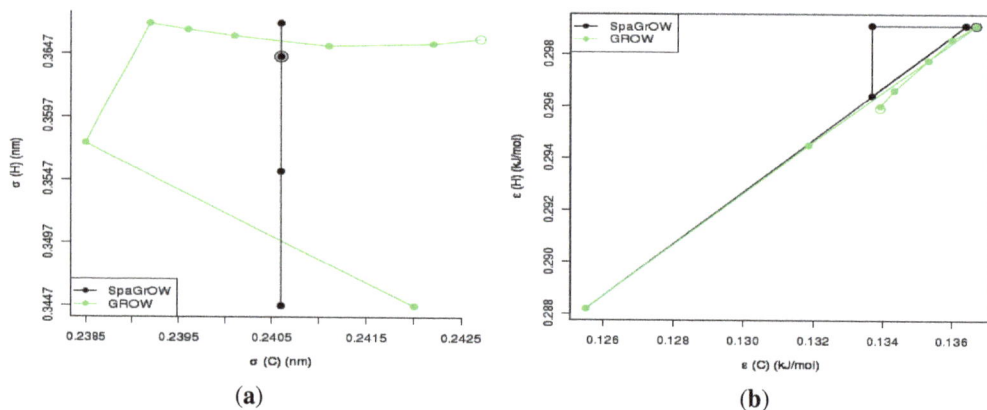

(a)   **(b)**

Ethylene oxide Ethylene oxide (C_2H_4O) is a highly toxic substance, but very relevant for industrial applications, because it is an educt for many industrially fabricated materials. It is very suitable for the evaluations of the optimization algorithms, because it has been characterized both experimentally [48,49] and by molecular simulations [14,50,51].

Figure 19 shows the MAPE values for ethylene oxide during the optimization procedures. The target observables were the saturated liquid density (Figure 19a), the enthalpy of vaporization (Figure 19b) and the vapor pressure (Figure 19c), considered at seven different temperatures. For pressures, the statistical uncertainty within molecular simulations is higher than for $\Delta_v H$ and ρ_l. Here, 5% were assumed. Experimental data was taken from [48]. The simulations performed were Grand-Equilibrium Monte-Carlo (GEMC) simulations based on a rigid united-atom model with a dipolar moment developed by [14]. They were executed with the software tool $ms2$ [52]. As GEMC simulations simulate the liquid and gas phase successively, the VLE data could be calculated correctly. The liquid simulation was performed in the NpT ensemble and the gas simulation in the μVT ensemble, where the chemical potential, μ, was kept constant. On the computer cluster mentioned above, eight to ten hours were required for the simulation of 500 liquid and 500 gaseous ethylene oxide molecules. Thereby, the total number of 345,000 Monte-Carlo steps was distributed on eight processors. The force field parameters to be optimized were $\varepsilon(CH_2), \varepsilon(O), \sigma(CH_2)$ and $\sigma(O)$. The initial parameters were the result of parameters obtained by a global pre-optimization based on random search [51]. Both parameters were changed by no more than 20%. The SpaGrOW variables were the same as for benzene. As the optimization problem was four-dimensional, Gaussian RBFs in combination with MARS were used again.

Figure 19. MAPE values with respect to ρ_l (**a**), $\Delta_v H$ (**b**) and p_σ (**c**) in the case of ethylene oxide during the Vapor-Liquid Equilibrium (VLE) optimization for SpaGrOW and GROW. The smoothing procedure was based on Gaussian RBFs combined with the MARS algorithm. The force field parameters were $\varepsilon(CH_2), \varepsilon(O), \sigma(CH_2)$ and $\sigma(O)$. The faster convergency of SpaGrOW compared to GROW could be confirmed again.

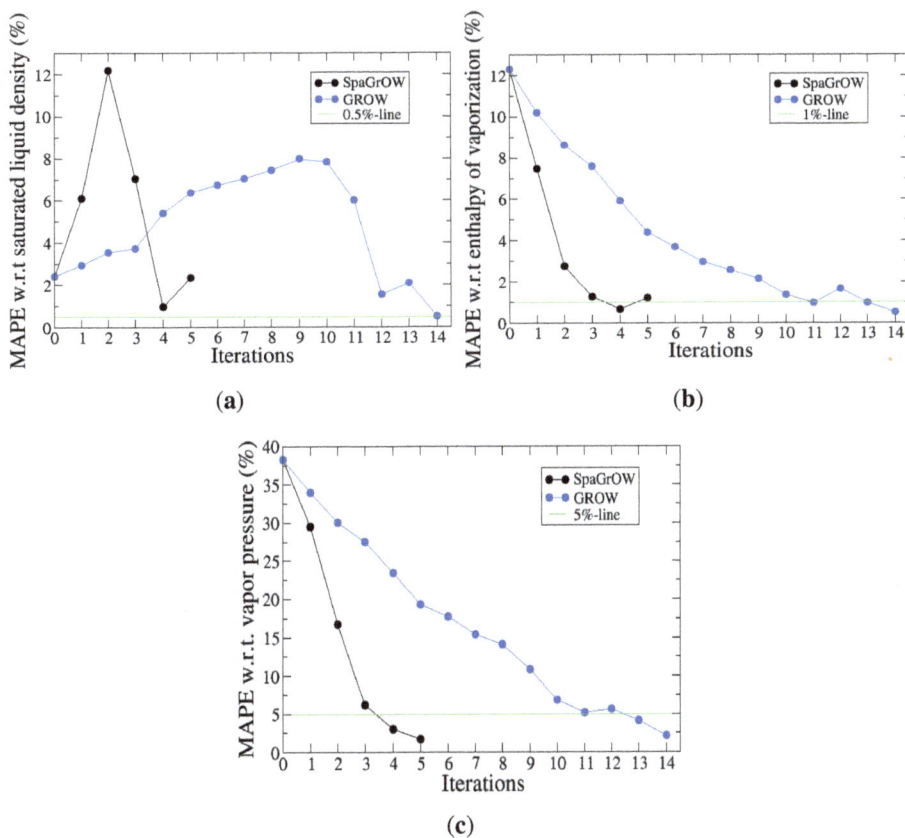

(a)

(b)

(c)

In total, GROW required 14 iterations for the optimization using the steepest descent method only. SpaGrOW only needed five to achieve a comparable loss function value in the same order of magnitude: for GROW, $F(x^{(14)}) = 2.4 \times 10^{-4}$ and, for SpaGrOW, $F(x^{(5)}) = 3.9 \times 10^{-4}$. Please note that the force field was not optimal and that other methods had to be applied in order to optimize it. For details, *cf.* [40]. SpaGrOW and GROW required 54 and 77 molecular simulations, respectively. Hence, SpaGrOW was significantly faster again.

Figure 20 shows the development of the LJ parameters during the optimization process in comparison to GROW (Figure 20a refers to ε and Figure 20b to σ). All parameters were decreased, and SpaGrOW delivered approximately the same parameters as GROW. Undesired detours were avoided again, except for σ. However, the detours made by GROW could be linearized.

Figure 20. Development of the LJ parameters in the case of ethylene oxide (VLE) for GROW and SpaGrOW: $\varepsilon(CH_2)$ and $\varepsilon(O)$ **(a)**, as well as $\sigma(CH_2)$ and $\sigma(O)$ **(b)**. The unfilled circles show the final parameters. SpaGrOW led to a more direct way to the minimum again.

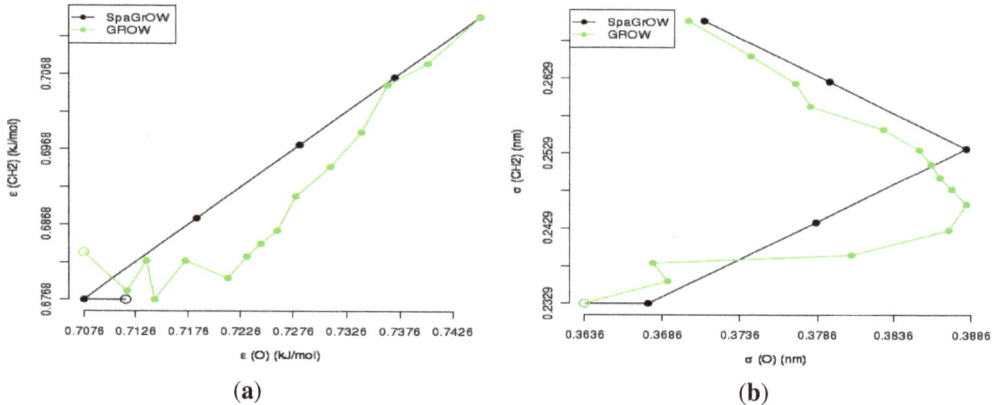

(a)   (b)

To summarize, SpaGrOW exhibits a significantly higher speed of convergency than GROW.

4.2.2. Comparison to Gradient-Based Methods Close to the Minimum: Dipropylene Glycol Dimethyl Ether

In the framework of the *Industrial Fluid Property Simulation Challenge (IFPSC) 2010* [53], a liquid-liquid equilibrium (LLE) between two liquid phases should be calculated with molecular models. One of the two component was water and the other, dipropylene glycol dimethyl ether ($C_8H_{18}O_3$), an inert, water-resistant, nontoxic and industrially highly relevant solvent.

The target observable was the liquid density, ρ, only, considered at four different temperatures and taken from [54]. The simulations performed were molecular dynamics simulations in the NpT ensemble executed with the software tool, *GROMACS* [46]. On the computer cluster mentioned above, three to four hours were required for the simulation of 512 ether molecules for 0.5 ns using a time step of 2 fs.

The LJ sites were located at the CH_3, the CH_2, the CH group and the oxygen. Hence, it was an eight-dimensional optimization problem. The initial force field parameters from [55]. As GROW could not achieve an optimal force field reproducing the liquid density of the ether [55], another gradient-based method was applied, based on a Taylor series up to the first member for the target observables, delivering a quadratic model for the loss function. It is a modified Gauss-Newton method combined with the Trust Region approach, *i.e.*, the quadratic model is minimized within a compact domain, as well. The algorithm has also been developed by the authors, *cf.* [40] for more details. After three iterations, the modified Gauss-Newton method could achieve optimal liquid densities.

Figure 21 indicates that SpaGrOW (with $\Delta_0 = 0.3 \times \Delta^{max}$) required two iterations only to do so. As the optimization problem was eight-dimensional, the Lipra method was applied as a smoothing procedure. However, SpaGrOW needed 38 simulations, three-times more than the modified Gauss–Newton method, which only had to execute twelve simulations. With an optimal Δ_0, the number of simulations could have been reduced to 19 in the case of SpaGrOW. The modified Gauss-Newton method is more reliable than SpaGrOW close to the minimum, but it could be shown that SpaGrOW is capable of getting closer to the minimum than the standard gradient-based algorithms used in GROW.

Figure 21. MAPE values with respect to ρ in the case of dipropylene glycol dimethyl ether during the optimization process of SpaGrOW in comparison to the modified Gauss-Newton method. The smoothing procedure was realized by the Lipra method. The optimization problem was eight-dimensional. SpaGrOW needed two iterations only, but significantly more simulations than the modified Gauss-Newton method in order to achieve optimal liquid densities.

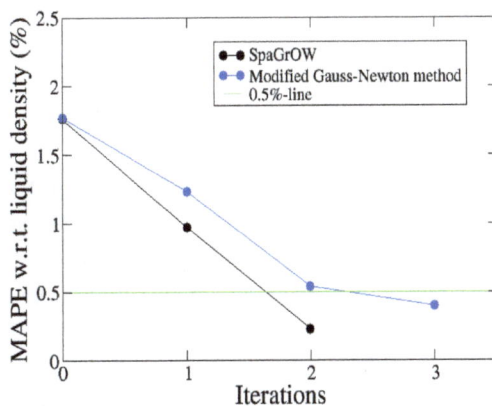

However, the Gauss-Newton method is only applicable close to the minimum, *i.e.*, when the loss function is nearly quadratic. In contrast, SpaGrOW is more generally applicable, and another drawback of the Gauss-Newton method is that it requires a gradient, which is not guaranteed to be computable, due to the reasons mentioned in Section 1. Please note that the density is not as noisy as other target properties, like the pressure or diffusion coefficient. Of course, the existence of an optimal Δ for SpaGrOW is not guaranteed either in practice.

Figure 22 shows the optimal liquid densities as a function of temperature for the modified Gauss-Newton method and SpaGrOW. Both methods could achieve an optimal force field with respect to ρ in contrast to GROW. However, SpaGrOW could reproduce the trend of the curve better than the modified Gauss-Newton method.

To summarize, SpaGrOW is capable of getting closer to the minimum than GROW, when the smoothing procedure and Δ_0 are chosen properly. However, the modified Gauss-Newton method needs significantly less simulations.

Figure 22. Optimization of the density ρ in the case of dipropylene glycol dimethyl ether using the modified Gauss-Newton method and SpaGrOW. Optimal densities could be achieved by both methods, but SpaGrOW needed many more simulations. However, SpaGrOW could reproduce the trend of the curve better.

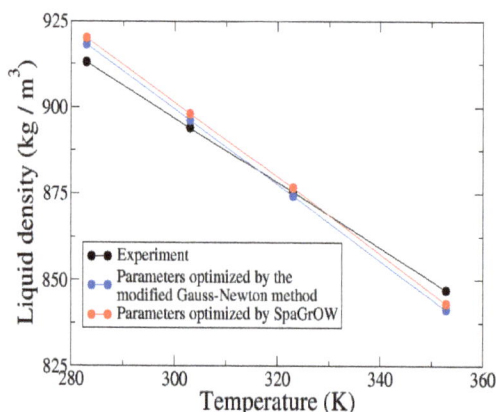

5. Conclusions

In this paper, the new derivative-free optimization method was presented in detail and applied to the parameterization of force fields in the field of molecular simulations. It is a combination of appropriate smoothing and regularization procedures, interpolation on sparse grids and the Trust Region approach. SpaGrOW turned out to be a highly efficient algorithm outperforming standard gradient-based methods with respect to the speed of convergency. Furthermore, it is capable of getting closer to the minimum. However, if a gradient can be calculated correctly, the gradient-based methods exhibit a slightly higher robustness than SpaGrOW. The new method is validated in the following with respect to the three criteria speed of convergency, local refinements and robustness:

- **Speed of convergency:** Whenever Δ_0 is chosen properly, SpaGrOW often requires only half of the number of simulations than gradient-based methods. The speed of convergency was also higher, *i.e.*, the number of iterations was significantly lower.
  The choice, $\Delta_0 = 0.3 \times \Delta^{max}$, was suitable in most cases, *i.e.*, the parameterization of SpaGrOW is not critical at the beginning of the optimization.
- **Local refinements:** SpaGrOW is capable of getting closer to the minimum than GROW. However, the choice of Δ becomes critical, which reduces the robustness of SpaGrOW: If Δ is too high, the smoothing and interpolation algorithms cannot deliver a reliable model for the loss functions. If it is too small, only noise can be reproduced. A modified Gauss-Newton method turned out to be more efficient and close to the minimum, if the associated gradients

can be computed correctly. An important advantage of SpaGrOW is the fact that the step length can be modified by the Trust Region steps, leading to different descent directions, which is not possible for gradient-based methods first determining the descent direction and, then, searching for a reliable step length.

- **Robustness:** SpaGrOW exhibits a slightly lower robustness than the gradient-based methods. Due to an inappropriate choice of Δ, the minimum of the model can be transferred under certain conditions, and the course of the algorithm can be modified. The variable must be chosen, so that a decreasing trend of the loss function is present within the actual trust region. However, this is not trivial here, because the shape of the loss function is not known *a priori*.

It is extremely difficult to find a method that is efficient and robust at the same time. These two properties often face each other: Stochastic global optimization methods, like simulated annealing or evolutionary algorithms, are very robust with respect to statistical noise, but require a high amount of computation time to determine the global minimum exactly. Gradient-based ones are fast-convergent, but are less robust and reliant on the differentiability of the function to be minimized. SpaGrOW is situated in between: The higher speed of convergency has to be compensated by a lower robustness. However, the robustness was not reduced significantly, and the assumption that the loss function is smooth does not have to be made. Furthermore, SpaGrOW may still be successful when gradient-based methods exist.

To summarize, SpaGrOW is a generic, efficient and, also, quite robust algorithm, which can be used for many optimization problems. For force field parameterization tasks, it is highly recommended and preferred to gradient-based algorithms.

Acknowledgments

We are grateful to Janina Hemmersbach for the detailed analysis and selection of appropriate smoothing procedures, as well as to Anton Schüller for fruitful discussions and advice.

Conflicts of Interest

The authors declare no conflict of interest.

References

1. Allen, M.P.; Tildesley, D.J. *Computer Simulation of Liquids*; Clarendon Press: Oxford, UK, 1987.
2. Frenkel, D.; Smit, B. *Understanding Molecular Simulation: From Algorithms to Applications*; Academic Press: Waltham, MA, USA, 2006.
3. Schrödinger, E. Quantisierung als eigenwertproblem. *Ann. Phys.* **1926**, *79*, 361–367.
4. Jensen, F. *Introduction to Computational Chemistry*; Wiley: Hoboken, NJ, USA, 1999.

5. Guevara-Carrion, G.; Hasse, H.; Vrabec, J. *Multiscale Molecular Methods in Applied Chemistry*; Kirschner, B., Vrabec, J., Eds.; Series: Topics in Current Chemistry, Volume 307; Springer: Berlin/Heidelberg, Germany, 2012; pp. 201–249.

6. Weiner, S.J.; Kollman, P.A.; Case, D.A.; Singh, U.C.; Ghio, C.; Alagona, G.; Profeta, S.; Weiner, P. A new force field for molecular mechanical simulation of nucleic acids and proteins. *J. Am. Chem. Soc.* **1984**, *106*, 765–784.

7. Berendsen, H.J.; van Gunsteren, W.F. *GROMOS87 Manual*, Library Manual; Biomos AG: Groningen, The Netherlands, 1987.

8. Jorgensen, W.L.; Maxwell, D.S.; Tirado-Rives, J. Development and testing of the OPLS all-atom force field on conformational energetics and properties of organic liquids. *J. Am. Chem. Soc.* **1996**, *118*, 11225–11236.

9. Duan, Y.; Wu, C.; Chowdhury, S.; Lee, M.C.; Xiong, G.; Zhang, W.; Yang, R.; Cieplak, P.; Luo, R.; Lee, T.; *et al.* A point-charge force field for molecular mechanics simulations of proteins based on condensed-phase quantum mechancical calculations. *J. Comput. Chem.* **2003**, *24*, 1999–2012.

10. Reith, D.; Kirschner, K.N. A modern workflow for force-field development—bridging quantum mechanics and atomistic computational models. *Comput. Phys. Commun.* **2011**, *182*, 2184–2191.

11. Kräemer-Fuhrmann, O.; Neisius, J.; Gehlen, N.; Reith, D.; Kirschner K.N. Wolf$_2$Pack—Portal based atomistic force-field development. *J. Chem. Inf. Mod.* **2013**, *53*, 802–808.

12. Jorgensen, W.L.; Madura, J.D.; Swensen, C.J. Optimized intermolecular potential functions for liquid hydrocarbons. *J. Am. Chem. Soc.* **1984**, *106*, 6638–6646.

13. Martin, M.G.; Siepmann, J.I. Transferable potentials for phase equilibria. 1. United-atom description of n-alkanes. *J. Phys. Chem. B* **1998**, *102*, 2567–2577.

14. Eckl, B.; Vrabec, J.; Hasse, H. On the application of force fields for predicting a wide variety of properties: Ethylene oxide as an example. *Fluid Phase Equilib.* **2008**, *274*, 16–26.

15. Peguin, R.P.S.; Kamath, G.; Potoff, J.J.; da Rocha, S.R.P. All-atom force field for the prediction of vapor–liquid equilibria and interfacial properties of HFA134a. *J. Phys. Chem. B* **2009**, *113*, 178–187.

16. Faller, R.; Schmitz, H.; Biermann, O.; Müller-Plathe, F. Automatic parameterization of force fields liquids by simplex optimization. *J. Comput. Chem.* **1999**, *20*, 1009–1017.

17. Ungerer, P.; Beauvais, C.; Delhommelle, J.; Boutin, A.; Rousseau, B.; Fuchs, A.H. Optimization of the anisotropic united atoms intermolecular potential for n-alkanes. *J. Comput. Phys.* **1999**, *112*, 5499–5510.

18. Bourasseau, E.; Haboudou, M.; Boutin, A.; Fuchs, A.H.; Ungerer, P. New optimization method for intermolecular potentials: Optimization of a new anisotropic united atoms potential for Olefins: Prediction of equilibrium properties. *J. Chem. Phys.* **2003**, *118*, 3020–3034.

19. Stoll, J.; Vrabec, J.; Hasse, H. A set of molecular models for carbon monoxide and halogenated hydrocarbons. *J. Chem. Phys.* **2003**, *119*, 11396–11407.

20. Sun, H. Prediction of fluid densities using automatically derived VDW parameters. *Fluid Phase Equilib.* **2004**, *217*, 59–76.

21. Hülsmann, M.; Köddermann, T.; Vrabec, J.; Reith, D. GROW: A gradient-based optimization workflow for the automated development of molecular models. *Comput. Phys. Commun.* **2010**, *181*, 499–513.

22. Hülsmann, M.; Vrabec, J.; Maaß, A.; Reith, D. Assessment of numerical optimization algorithms for the development of molecular models. *Comput. Phys. Commun.* **2010**, *181*, 887–905.

23. Hülsmann, M.; Müller, T.J.; Köddermann, T.; Reith, D. Automated force field optimisation of small molecules using a gradient-based workflow package. *Mol. Simul.* **2011**, *36*, 1182–1196.

24. Griebel, M.; Schneider, M.; Zenger, C. *A Combination Technique for the Solution of Sparse Grid Problems*; Technical Report; Institut für Informatik, Technische Universität München: Munich, Germany, 1990.

25. Moré, J.J. *Mathematical Programming: The State of the Art*; Bachem, A., Grötschel, M., Korte, B., Eds.; Springer: Berlin/Heidelberg, Germany, 1983; pp 258–287.

26. Nocedal, J.; Wright, S.J. *Numerical Optimization*; Springer: Berlin/Heidelberg, Germany, 1999.

27. Mangasarian, O.L.; Rosen, J.B.; Thompson, M.E. Global minimization via piecewise-linear underestimation. *J. Glob. Optim.* **2004**, *32*, 1–9.

28. Smolyak, S.A. Quadrature and interpolation formulas for tensor products of certain classes of functions. *Sov. Math. Doklady* **1963**, *4*, 240–243.

29. Zenger, C. *Parallel Algorithms for Partial Differential Equations, Notes on Numerical Fluid Mechanics*; Hackbusch, W., Ed.; Vieweg: Wiesbaden, Germany, 1991; Volume 31, pp. 241–251.

30. Bungartz, H.-J. Iterative Methods in Linear Algebra. In Proceedings of the IMACS International Symposium, Brüssel, Belgium, April 1991; de Groen, P., Beauwens, R., Eds.; North-Holland Publishing Co.: Amsterdam, The Netherlands, 1992; pp. 293–310.

31. Powell, M. *Algorithms for Approximation*; Mason, J.C., Cox, M.G., Eds.; Clarendon Press: Oxford, UK, 1987; pp. 143–167.

32. Bishop, C.M. *Pattern Recognition and Machine Learning*; Series: Information Science and Statistics; Springer: Berlin/Heidelberg, Germany, 2007.

33. MacQueen, J.B. Some Methods for Classification and Analysis of Multivariate Observations. In Proceedings of the 5th Berkeley Symposium on Mathematical Statistics and Probability, Berkeley, CA, USA, June/July 1965; University of California Press: Berkeley, CA, USA, 1967; pp. 281–297.

34. Zou, H.; Hastie, T. Regularization and variable selection via the elastic net. *J. Roy. Stat. Soc. Ser. B* **2005**, *67*, 301–320.

35. Hoerl, A.E.; Kennard, R.W. Ridge regression: Biased estimation for nonorthogonal problems. *Technometrics* **1970**, *12*, 55–67.

36. Tibshirani, R. Regression shrinkage and selection via the lasso. *J. Roy. Stat. Soc. Ser. B* **1996**, *58*, 267–288.

480

37. Friedman, J.H. Multivariate adaptive regression splines. *Ann. Stat.* **1991**, *19*, 1–67.
38. Wendland, H. *Scattered Data Approximation*; Cambridge University Press: Cambridge, UK, 2005.
39. Wendland, H. Konstruktion und Untersuchung radialer Basisfunktionen mit kompaktem Träger. Ph.D. Thesis, Universität Göttingen, Göttingen, Germany, 1996.
40. Hülsmann, M. Effiziente und neuartige Verfahren zur Optimierung von Kraftfeldparametern bei atomistischen Molekularen Simulationen kondensierter Materie. Ph.D. Thesis, Universität zu Köln, Cologne, Germany, 2012.
41. Conn, A.R.; Scheinberg, K.; Toint, P.L. *On the Convergence of Derivative-free Methods for Unconstrained Optimization*; Iserles, A., Buhmann, M., Eds.; Cambridge University Press: Cambridge, UK, 1997.
42. The R Project for Statistical Computing. Available online: http://www.r-project.org/ (accessed on 14 February 2013).
43. Stoll, J.; Vrabec, J.; Hasse, H.; Fischer, J. Comprehensive study of the vapour-liquid equilibria of the pure two-centre Lennard-Jones plus pointquadrupole fluid. *Fluid Phase Equilib.* **2001**, *179*, 339–362.
44. Yoshida, K.; Matubayasi, N.; Nakahara, M. Self-diffusion coefficients for water and organic solvents at high temperatures along the coexistence curve. *J. Chem. Phys.* **2008**, *129*, 214501–214509.
45. NIST Chemistry Webbook. Available online: http://webbook.nist.gov/chemistry/ (accessed on 14 February 2013).
46. GROMACS Molecular Simulation Tool. Available online: http://www.gromacs.org/ (accessed on 14 February 2013).
47. Moscito Molecular Simulation Tool. Available online: http://ganter.chemie.uni-dortmund.de/MOSCITO/ (accessed on 14 February 2013).
48. Buckles, C.; Chipman, P.; Cubillas, M.; Lakin, M.; Slezak, D.; Townsend, D.; Vogel, K.; Wagner, M. *Ethylene Oxide User's Guide*; Online Manual, Publisher: American Chemistry Council, 1999; Available online: http://www.ethyleneoxide.com (accessed on 14 February 2013).
49. Olson, J.D.; Wilson, L.C. Benchmarks for the fourth industrial fluid properties simulation Challenge. *Fluid Phase Equilib.* **2008**, *274*, 10–15.
50. Müller, T.J.; Roy, S.; Zhao, W.; Maaß, A.; Reith, D. Economic simplex optimization for broad range property prediction: Strengths and weaknesses of an automated approach for tailoring of parameters. *Fluid Phase Equilib.* **2008**, *274*, 27–35.
51. Maaß, A.; Nikitina, L.; Clees, T.; Kirschner, K.N.; Reith, D. Multiobjective optimisation on the basis of random models for ethylene oxide. *Mol. Simul.* **2010**, *36*, 1208–1218.
52. ms2 Molecular Simulation Tool. Available online: http://www.ms-2.de/ (accessed on 14 February 2013).
53. The Industrial Fluid Property Simulation Challenge, 2010. Available online: http://www.ifpsc.com (accessed on 14 February 2013).

54. Esteve, X.; Conesa, A.; Coronas, A. Liquid densities, kinematic viscosities, and heat capacities of some alkylene glycol dialkyl ethers. *J. Chem. Eng. Data* **2003**, *48*, 392–397.

55. Köddermann, T.; Kirschner, K.N.; Vrabec, J.; Hülsmann, M.; Reith, D. Liquid-liquid equilibria of dipropylene glycol dimethyl ether and water by molecular dynamics. *Fluid Phase Equilib.* **2011**, *310*, 25–31.

Reprinted from *Entropy*. Cite as: Takahashi, K.Z. Truncation Effects of Shift Function Methods in Bulk Water Systems. *Entropy* **2013**, *15*, 3249–3264.

Article

Truncation Effects of Shift Function Methods in Bulk Water Systems

Kazuaki Z. Takahashi

Department of Mechanical Engineering, Keio University, 3-14-1 Hiyoshi, Kohoku-ku, Yokohama, Kanagawa 223-8522, Japan; E-Mail: takahashi@mech.keio.ac.jp; Tel.: +81-45-566-1454 (ext. 47151); Fax: +81-45-566-1495

Received: 30 June 2013; in revised form: 5 August 2013 / Accepted: 7 August 2013 / Published: 13 August 2013

Abstract: A reduction of the cost for long-range interaction calculation is essential for large-scale molecular systems that contain a lot of point charges. Cutoff methods are often used to reduce the cost of long-range interaction calculations. Molecular dynamics (MD) simulations can be accelerated by using cutoff methods; however, simple truncation or approximation of long-range interactions often offers serious defects for various systems. For example, thermodynamical properties of polar molecular systems are strongly affected by the treatment of the Coulombic interactions and may lead to unphysical results. To assess the truncation effect of some cutoff methods that are categorized as the shift function method, MD simulations for bulk water systems were performed. The results reflect two main factors, *i.e.*, the treatment of cutoff boundary conditions and the presence/absence of the theoretical background for the long-range approximation.

Keywords: molecular-dynamics simulations; long-range interactions; liquid water; electrostatic interactions; reaction field

1. Introduction

In the calculation of thermodynamic, structural and dynamical properties by molecular dynamics (MD) simulations, the effect of long-range interactions is an important issue. Long-range interactions on the periodic boundary conditions (PBCs) can be calculated using the Ewald sum or cutoff methods. The Ewald sum [1] is the key standard method used in calculations involving long-range interactions with the periodic boundary condition. In this method, the total energy is split into real and reciprocal space contributions. Calculation of the Ewald sum involves three problems, the first being that the reciprocal part is computationally expensive. Particle mesh Ewald (PME) [2,3] reduces computational cost for the reciprocal part by using fast Fourier transform (FFT); however, FFT has problems, becoming a cause of a strong bottle neck in massively parallel computers [4]. The second is the inherent periodicity, which can develop artifacts [5–14]. The third is that the thermodynamic limit is unclear [15]. Notwithstanding the three problems, the Ewald sum and PME are methods of choice that most appropriately represent the long-range interactions.

Cutoff methods are often used to accelerate long-range interaction calculations. The interactions between molecular pairs only with a distance shorter than a given cutoff length are considered, and effects from more distant pairs are truncated or approximated. The plain cutoff, the cutoff with the switch/shift function and the reaction field (RF) method are some examples of typical cutoff methods for Coulombic systems. Simulations can be accelerated by using cutoff methods, but simple truncation or approximation of long-range interactions have serious defects for various systems. In Lennard-Jones (LJ) fluid systems, long-range interactions do not have a prominent effect on transport properties [16,17], but phase equilibria and interfacial properties change drastically [18–20]. In water systems, the electrostatic interactions dominate the physical properties, and truncation or continuum approximation may lead to unphysical results. A lot of cutoff methods applied to the Coulombic interaction offer insufficient accuracy [21–31], and all of the results are highly sensitive to the cutoff distance. Cutoff methods are also applied to macromolecular systems [12–14,32–43], and many results indicate that the aforementioned approximations have difficulties when estimating these systems. On the other hand, advanced cutoff-like methods have been developed to avoid the aforementioned difficulties and to accelerate long-range interaction calculations. Wolf *et al.* [44] developed a method to calculate electrostatic interactions, which is simpler than the Ewald sum. They took into account charge neutrality in the cutoff sphere and discovered that the electrostatic potential of condensed phases seems to have short range behavior. The modified method developed by Fennel and Gezelter [45] could reproduce some thermodynamic properties of homogeneous systems obtained by the Ewald sum. However, the method can hardly estimate heterogeneous systems [46,47]. Wu and Brooks developed the isotropic periodic sum (IPS) method [48–50]. The IPS method is a method that can calculate contributions from the infinite periodic structure without reciprocal space calculations. Some reports on the accuracy of the IPS method of homogeneous [48,50–54] and heterogeneous systems [47,50,55–57] show that the method yields estimates in good agreement with the results of the Ewald sum. Improved methods were developed to speed up calculations for large-scale systems [58] and to improve the accuracy for homogeneous and heterogeneous systems [59,60].

Some cutoff methods and the aforementioned two cutoff-like methods can be regarded as the shift function method, which produces the pairwise-potential shifted from the original potential function (e.g., Coulombic interaction) by any theoretical or other requirements. To discover the relation between truncation effects and approximation treatments of long-range interactions, in this work, we focused on shift function methods. MD simulations of bulk water systems were carried out for evaluating the truncation effects of the potential energy, self-diffusion, radial distribution function and the dipole-dipole correlation. The results reflect two main factors, *i.e.*, the treatment of cutoff boundary conditions and the presence/absence of the theoretical background for long-range approximation.

2. Experimental

MD simulations for bulk water systems were conducted to examine the truncation effects of shifted potentials, and physical properties were compared with those from the simulations of the Ewald sum. For shift function methods, CHARMm-shift [61], Ohmine-shift [62], the dumped-shifted-force potential of the Wolf method (Wolf-DSF) [45], the RFmethod with an infinite dielectric constant (RF-metal), the IPS method for non-polar systems (IPSn) [48], the IPS method for polar systems (IPSp) [50] and the linear-combination-based IPS (LIPS) method with a fifth-order cutoff boundary condition (LIPS-fifth) [59] were chosen. CHARMm-shift is used in CHARMm [61] for shifting Coulombic and LJ interactions. The cutoff boundary conditions are considered until the first-order differential of the interaction potential (first-order cutoff boundary condition). Ohmine-shift is originally one of the switching function methods. The method provides shifted pairwise-potential, if the switching point is set to zero. This potential has second-order cutoff boundary conditions. Wolf-DSF was developed by Fennel and Gezelter [45]. The charge neutrality assumption inside the cutoff sphere is the basic concept of the original Wolf method and Wolf-DSF. In the bulk water systems, the α-parameter of Wolf-DSF is set to $0.2\,\mathrm{nm}^{-1}$ [45]. It should be noted that better α-parameters for any other systems potentially exist [46,47]. RF-metal is the RF method with an infinite dielectric constant. In the RF theory, the Coulombic interaction can be modified for homogeneous systems, by assuming a constant dielectric environment beyond the cutoff sphere. Originally, the dielectric constant of the RF method should be set to realistic value. However, some results indicate that the RF method with an infinite dielectric constant is best for estimating bulk water systems [30]. Therefore, we set the dielectric constant of the RF method to the infinite value, like a bulk metal. IPSn and IPSp are two different versions of the IPS method. IPSn is applied to calculations for point charges, whereas IPSp calculates polar molecules. The IPS method assumes the isotropic periodic structure outside of the cutoff sphere. The contribution from this structure (periodic reaction field) determines the shape of the IPS potentials. LIPS-fifth is the potential produced by the improved IPS method, called the LIPS method. The LIPS method is based on the extended IPS theory that provides the design procedure of the periodic reaction fields.

In order to clarify the truncation effect of the Coulombic interaction for shift function methods, it was only applied for the Coulombic interaction, and a cutoff method was used for the LJ interaction. The cutoff radius of the LJ interaction was set as $1.2664\,\mathrm{nm}$, which is 4.0 in LJ length units. For the shift function methods, the cutoff radius, r_c, of the Coulombic interaction was changed from $1.2\,\mathrm{nm}$ to $2.0\,\mathrm{nm}$ by $0.2\,\mathrm{nm}$ increments. In this simulation, the extended simple point charge (SPC/E) model [63] was used for water molecules. The velocity Verlet algorithm [64] was used with three-dimensional periodic boundary conditions along with a time step of $2\,\mathrm{fs}$. The atoms in a water molecule were constrained by the RATTLEalgorithm [65]. The simulation was performed in a constant particle-number, volume and temperature ensemble with the Nosé-Hoover thermostat [66–68], where the number of water molecules was 6,192, the density was 0.997 cm^3 and the temperature was 298.15 K. After equilibrating the system, a total of 5×10^5 time steps (1 ns) were carried out for each cutoff radius of the shift function methods. The potential energy, U, the self-diffusion coefficient, D, the radial distribution function, $g(r)$, the distance dependence of the Kirkwood factor, $G_K(r)$, and the radial distribution of the dipole ordering, $s(r)$, were calculated. We calculated the self-diffusion coefficient for the transport coefficients. The self-diffusion coefficient can be determined either by the Einstein relation or the Green-Kubo formula, which are basically equivalent formulas. Here, we used the Einstein relation:

$$D = \lim_{t \to \infty} \frac{1}{6t} \langle |r_i(t) - r_i(0)|^2 \rangle_N \tag{1}$$

where t is the time, $r_i(t)$ is the position of particle i and $\langle \cdots \rangle_N$ denotes the particle average. The slope of the mean-squared displacement of the diffusing particle in the long-time limit is calculated for the diffusion coefficient. The radial distribution function, the distance dependence of the Kirkwood factor and the radial distribution of dipole ordering were calculated for the configuration of water. These properties are given as a function of the distance between two water molecules, denoted r. The conventional expressions give:

$$g(r) = \frac{V}{4\pi r^2 \Delta r N(N-1)} \left\langle \sum_i n_i(r) \right\rangle_e \tag{2}$$

$$G_K(r) = \frac{1}{N} \left\langle \sum_i \left(u_i \cdot \sum_{j, r_{ij} < r} u_j \right) \right\rangle_e \tag{3}$$

and:

$$s(r) = \frac{1}{N} \left\langle \sum_i \frac{1}{n_i(r)} \left(\sum_{j=1}^{n_i(r)} u_i \cdot u_j \right) \right\rangle_e \tag{4}$$

where $n_i(r)$ is the number of molecules that exist in the region between r and $r + \Delta r$ from molecule i. u_i and u_j are the normalized dipole moments of molecules i and j, respectively, while $\langle \cdots \rangle_e$ signifies an equilibrium ensemble average.

All of above properties calculated from the shift function methods were compared with that from the Ewald sum. For the Ewald sum, the cutoff radius for the real part was 2.8 nm. The α-parameter was determined by the following equation:

$$\text{erfc}(-\alpha r_c) \approx \exp(-\alpha^2 r_c^2) = \delta \tag{5}$$

where δ is a small number, which indicates the convergence of real space potentials in the Ewald sum. δ was 10^{-6}, so $\alpha r_c = (6 \ln 10)^{1/2}$.

3. Results and Discussions

3.1. Bulk Water

3.1.1. Potential Energy

The thermodynamic properties for the shift function methods and Ewald sum were calculated by potential energies. Figure 1 shows the potential energy per molecule with different cutoff radii, for the shift function methods and the Ewald sum. The results from CHARMm-shift, Ohmine-shift and Wolf-DSF are far from that of the Ewald sum. In contrast, RF-metal, IPSn, IPSp and LIPS-fifth are close to that of the Ewald sum.

> **Figure 1.** Potential energy for the shift function methods and the Ewald sum. The results from CHARMm-shift, Ohmine-shift and the Wolf method (Wolf-DSF) are far from that of the Ewald sum. In contrast, RF-metal, the isotropic periodic sum for non-polar systems (IPSn), for polar systems (IPSp) and the linear-combination-based IPS (LIPS)-fifth are close to that of the Ewald sum.

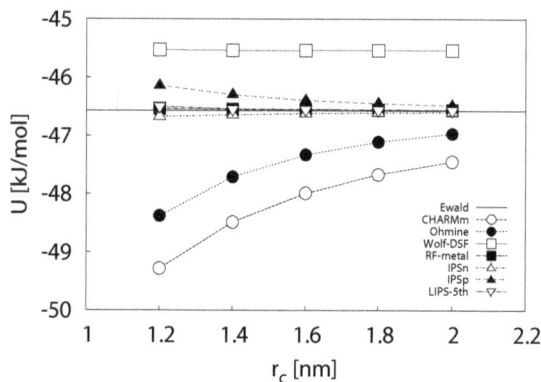

To examine the cutoff radius tendency of the potential energy thoroughly, we plotted the error of the potential energy calculated with the shift function methods against that determined with the Ewald sum, as shown in Figure 2. The error of the potential energy for each method decreases by an increment of the cutoff radius, except for the case of the Wolf-DSF. The fastest decline was observed in the case of LIPS-fifth; the error was roughly in proportion to r_c^{-4}. RF-metal and LIPS-fifth at $r_c = 2.0$ nm achieved the smallest error and had the same value as that of the Ewald sum within 0.02%. It is clearly shown that CHARMm-shift, Ohmine-shift and Wolf-DSF poorly estimated the potential energy for bulk water systems. The reason for this is related to the presence/absence of the theoretical background for contributions outside the cutoff sphere. CHARMm-shift and Ohmine-shift do not have the theory that justifies their shifting procedure. Wolf-DSF explains its own truncation treatment by the charge neutrality assumption inside the cutoff sphere, but contributions from outside are not considered. In contrast, RF-metal, IPSn and LIPS-fifth are, respectively, based on a definite theory that considers the contributions from long-range interactions. Therefore, RF-metal, IPSn and

LIPS-fifth have much better accuracy for estimating the potential energy. IPSp had intermediate values between the former and latter. This seems to be related to the counter-charge assumption of the IPSp. The counter-charge effect assumed at the cutoff boundary may partially interrupt long-range contributions.

Figure 2. The error of the potential energy calculated with the shift function methods against that determined with the Ewald sum. It is clearly shown that CHARMm-shift, Ohmine-shift and Wolf-DSF poorly estimated the potential energy for bulk water systems. In contrast, RF-metal, IPSn and LIPS-fifth have much better accuracy for estimating the potential energy. IPSp had intermediate values between the former and latter. The error of the potential energy for each method decreases by an increment of the cutoff radius, except for the case of the Wolf-DSF. The fastest decline was observed in the case of LIPS-fifth; the error was roughly in proportion to r_c^{-4}. RF-metal and LIPS-fifth at $r_c = 2.0$ nm achieved the smallest error and had the same value as that of the Ewald sum within 0.02%.

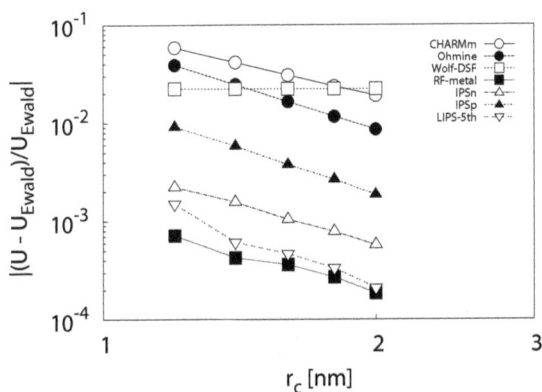

3.1.2. Self-Diffusion Coefficient

We calculated the self-diffusion coefficient for the Figure 3 shows the self-diffusion coefficient for shift function methods and the Ewald sum. The results from CHARMm-shift could not have a similar value to that of the Ewald sum at 1.2 nm $\leq r_c \leq 2.0$ nm. Other methods seem to estimate the self-diffusion coefficient with an adequate accuracy.

To examine the cutoff radius tendency of the self-diffusion coefficient thoroughly, we plotted the error of the self-diffusion coefficient calculated with the shift function methods against that determined with the Ewald sum, as shown in Figure 4. The convergence of the IPSp and LIPS-fifth is much faster than other methods. For IPSp, the self-diffusion coefficient is saturated at $r_c \geq 1.6$ nm, and the saturated value is almost the same as that of the Ewald sum (within 0.35%). For LIPS-fifth, the self-diffusion coefficient is saturated at $r_c \geq 1.4$ nm, and the saturated value is almost the same as that of the Ewald sum (within 0.36%). The difference of the cutoff radius tendency comes from the treatment of the cutoff boundary conditions and long-range interactions. The results show that the improvement of the cutoff boundary conditions or long-range interaction treatment strongly affects the accuracy of the self-diffusion coefficient. In CHARMm-shift, both treatments are insufficient. It has a first-order cutoff boundary condition and does not have any theoretical background for long-range interaction treatment. Ohmine-shift had improved accuracy in comparison with that of CHARMm-shift, even if it merely comes from an advantage on the cutoff boundary condition. RF-metal and IPSn consider the first-order cutoff boundary condition and the adequate treatment for long-range contributions. Therefore, these two methods have similar accuracy to that of Ohmine-shift. Wolf-DSF also had similar accuracy to the result of Ohmine-shift, despite the absence of the theoretical background for the long-range interaction treatment. Strictly, Wolf-DSF has a first-order cutoff boundary condition, but it can be regarded as an infinite-order cutoff boundary condition under the certain value of alpha. This is the reason for the results of Wolf-DSF. In IPSp, a faster convergence of errors were observed, because it has a third-order cutoff boundary condition and the long-range interaction treatment. LIPS-fifth achieved the fastest convergence. It has a fifth-order cutoff boundary condition and a reliable background for the long-range interaction treatment.

Figure 3. The self-diffusion coefficient for the shift function methods and the Ewald sum. The results from CHARMm-shift could not have a similar value to that of the Ewald sum at $1.2 \, \text{nm} \leq r_c \leq 2.0 \, \text{nm}$. Other methods seem to estimate the self-diffusion coefficient with an adequate accuracy.

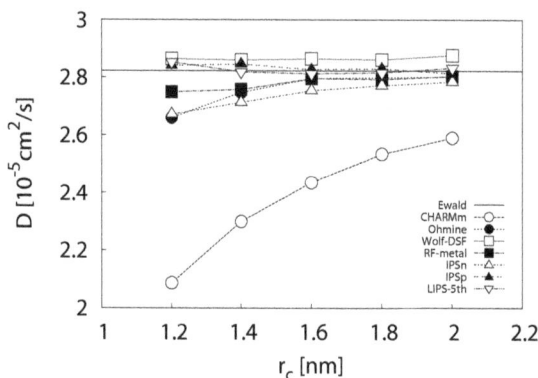

3.1.3. Radial Distribution Function

To examine the structure around a molecule for shift function methods, the radial distribution function, $g(r)$, was calculated. Figure 5 shows the oxygen-oxygen, $g(r)$, of the water molecule for shift function methods at $r_c = 2.0 \, \text{nm}$ and for the Ewald sum. In Figure 5, CHARMm-shift, Ohmine-shift, RF-metal and IPSn have notable deviations from the result of the Ewald sum. On the other hand, Wolf-DSF, IPSp and LIPS-fifth provided adequate accuracy. The oxygen-hydrogen and hydrogen-hydrogen, $g(r)$, have very similar behavior in comparison to oxygen-oxygen in Figure 5.

Figure 4. The error of the self-diffusion coefficient calculated with the shift function methods against that determined with the Ewald sum. The convergence of the IPSp and LIPS-fifth is much faster than other methods. For IPSp, the self-diffusion coefficient is saturated at $r_c \geq 1.6$ nm and the saturated value is almost the same as that of the Ewald sum (within 0.35%). For LIPS-fifth, the self-diffusion coefficient is saturated at $r_c \geq 1.4$ nm, and the saturated value is almost the same as that of the Ewald sum (within 0.36%).

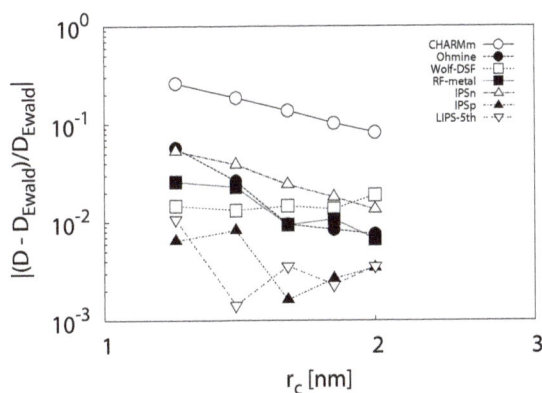

Figure 5. The oxygen-oxygen radial distribution function of the water molecule for the shift function methods at $r_c = 2.0$ nm and for the Ewald sum. CHARMm-shift, Ohmine-shift, RF-metal and IPSn have notable deviations from the result of the Ewald sum. On the other hand, Wolf-DSF, IPSp and LIPS-fifth provided adequate accuracy. The oxygen-hydrogen and hydrogen-hydrogen, $g(r)$, have very similar behavior in comparison to oxygen-oxygen (figures not shown).

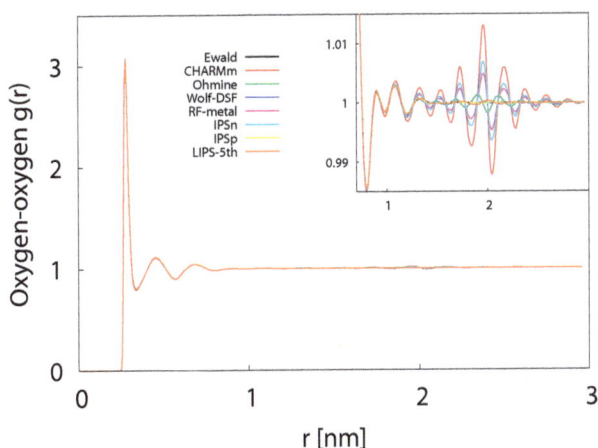

To examine the decrease of the deviation for r_c thoroughly, we plotted the root mean square deviation (RMSD) of the oxygen-oxygen, $g(r)$, for each shift function method against the Ewald sum at different cutoff radii in Figure 6a. The RMSDs of the oxygen-hydrogen and hydrogen-hydrogen,

$g(r)$, are also plotted in Figure 6b,c, respectively. The difference of the cutoff radius tendency is affected strongly by the treatment of the cutoff boundary conditions. In CHARMm-shift, RF-metal and IPSn that have a first-order cutoff boundary condition, the deviation decreases roughly in proportion to $r_c^{-2.5}$, r_c^{-3} and r_c^{-3}, respectively. Ohmine-shift has a second-order cutoff boundary condition, and the RMSD of $g(r)$ declines roughly in proportion to r_c^{-4}. The RMSD of LIPS-fifth has a similar tendency with these shift function methods for cutoff radii, but a faster decline is observed. The RMSD of LIPS-fifth decreases roughly in proportion to r_c^{-6}. Furthermore, LIPS-fifth gives accurate estimations of $g(r)$; the RMSD converges at $r_c \geq 2.0$ nm. Converged values of RMSD for LIPS-fifth are most accurate. On the other hand, the RMSDs of Wolf-DSF and IPSp have an adequate accuracy in any cutoff radius. The charge neutrality and counter-charge assumptions of Wolf-DSF and IPSp, respectively, seem to work better for bulk water systems.

Figure 6. The RMSDs of **(a)** the oxygen-oxygen, **(b)** the oxygen-hydrogen and **(c)** the hydrogen-hydrogen radial distribution function for the shift function method against the Ewald sum at different cutoff radii. In CHARMm-shift, RF-metal and IPSn that have a first-order cutoff boundary condition, the deviation decreases roughly in proportion to $r_c^{-2.5}$, r_c^{-3} and r_c^{-3}, respectively. Ohmine-shift has a second-order cutoff boundary condition, and the RMSD of $g(r)$ declines roughly in proportion to r_c^{-4}. The RMSD of LIPS-fifth has a similar tendency with these shift function methods for cutoff radii, but a faster decline is observed. The RMSD of LIPS-fifth decreases roughly in proportion to r_c^{-6}. Furthermore, LIPS-fifth gives accurate estimations of $g(r)$; the RMSD converges at $r_c \geq 2.0$ nm. Converged values of RMSD for LIPS-fifth are most accurate. The RMSDs of Wolf-DSF and IPSp have an adequate accuracy in any cutoff radius.

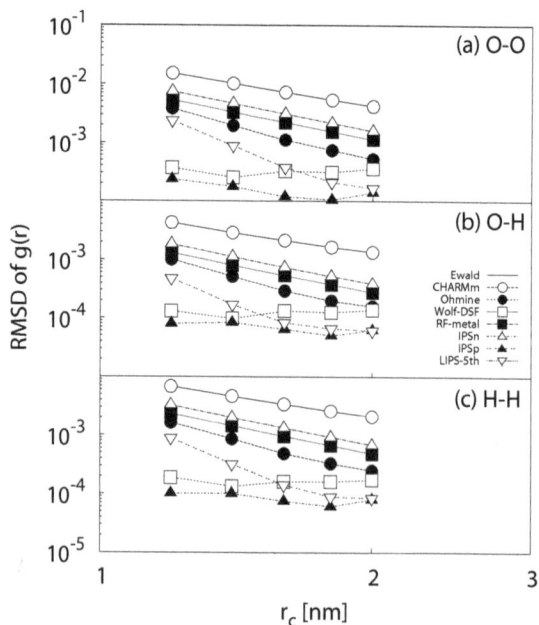

3.1.4. Dipole-Dipole Correlation

We focused on the distance dependence of the Kirkwood factor, $G_K(r)$, where one can see the dipole-dipole correlation of bulk water systems. $G_K(r)$ has a strong cutoff radius effect, and the influence of the interaction treatment is quantitatively-expressible by the shape of $G_K(r)$. An evident shortcoming of the cutoff-like method appears for the $G_K(r)$ value in bulk water systems. Thus, $G_K(r)$ of various cutoff radii were calculated using the shift function methods to evaluate the truncation effect of the dipole-dipole correlation.

Figure 7. Distance dependence of the Kirkwood factor for the shift function methods and the Ewald sum. It is clearly seen that $G_K(r)$ calculated with CHARMm-shift, Ohmine-shift, RF-metal and IPSn fluctuate near r_c as in $g(r)$, and this fluctuation still remains in spite of the increment of the cutoff radius. The artificial configuration of Ohmine-shift was smaller than that of the other three methods. The defect of $G_K(r)$ for these above shift function methods was not seen in Wolf-DSF, IPSp and LIPS-fifth. Wolf-DSF, IPSp and LIPS-fifth can estimate $G_K(r)$ more adequately than other shift function methods.

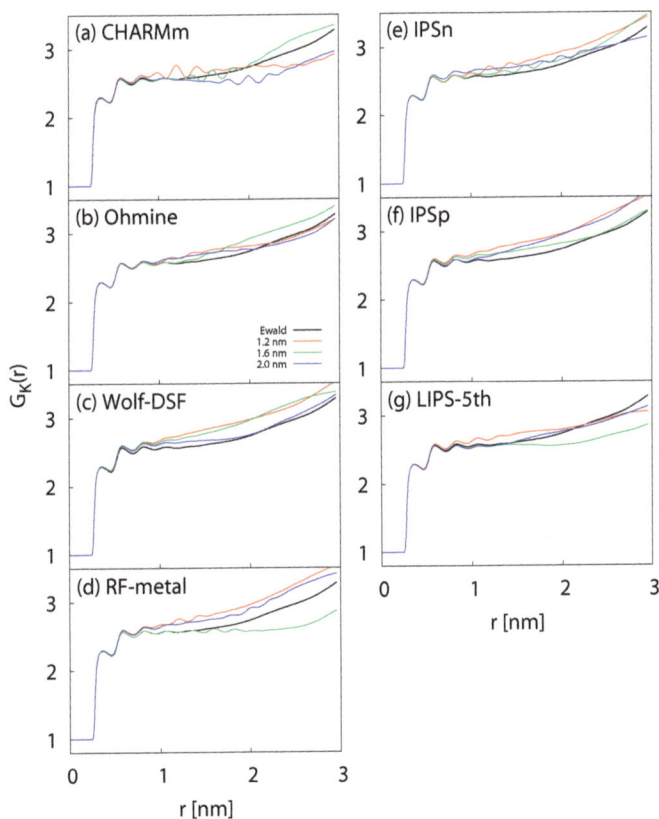

492

Figure 7 shows the shape of $G_K(r)$ determined using the shift function methods and the Ewald sum. It is clearly seen that $G_K(r)$ calculated with CHARMm-shift, Ohmine-shift, RF-metal and IPSn fluctuate near r_c as in $g(r)$, and this fluctuation still remains in spite of the increment of the cutoff radius. The artificial configuration of Ohmine-shift was smaller than that of the other three methods. The defect of $G_K(r)$ for these above shift function methods was not seen in Wolf-DSF, IPSp and LIPS-fifth. Wolf-DSF, IPSp and LIPS-fifth can estimate $G_K(r)$ more adequately than other shift function methods.

The result of $s(r)$ shows the radial distribution of the dipole ordering for water molecules. Figure 8 presents $s(r)$ calculated with the shift function methods at $r_c = 2.0\,\text{nm}$, along with that determined by the Ewald sum for comparison. $s(r)$ calculated with the CHARMm-shift, Ohmine-shift, RF-metal and IPSn fluctuate near r_c, like $g(r)$, despite the long cutoff radius. Wolf-DSF, IPSp and LIPS-fifth did not calculate any singular configurations of $s(r)$, like that for the other shift function methods. These three methods can estimate $s(r)$ with adequate accuracy.

The aforementioned characteristics of the truncation effect in the dipole-dipole correlation for the shift function methods are affected strongly by the treatment of the cutoff boundary condition, like the case of $g(r)$.

Figure 8. Radial distributions of dipole ordering calculated with the shift function methods at $r_c = 2.0\,\text{nm}$ and for the Ewald sum. $s(r)$ calculated with the CHARMm-shift, Ohmine-shift, RF-metal and IPSn fluctuate near r_c, like $g(r)$, despite the long cutoff radius. Wolf-DSF, IPSp and LIPS-fifth did not calculate any singular configurations of $s(r)$, like that for the other shift function methods. These three methods can estimate $s(r)$ with adequate accuracy.

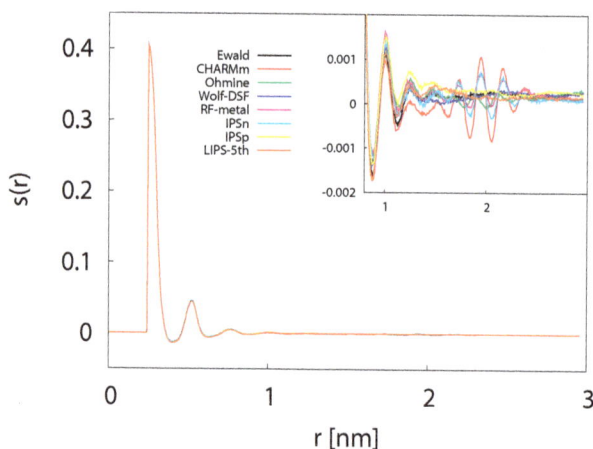

4. Conclusions

To assess the truncation effect of some cutoff methods that are categorized as the shift function method, MD simulations for bulk water systems were performed. The results reflect mainly two main factors, *i.e.*, the treatment of cutoff boundary conditions and the presence/absence of the theoretical background for long-range approximation.

The difference of estimated value of the potential energy is related to the presence/absence of the theoretical background for contributions outside the cutoff sphere. CHARMm-shift, Ohmine-shift and Wolf-DSF poorly estimated the potential energy, because these methods do not have a reliable theory that justifies their shifting procedure. In contrast, RF-metal, IPSn and LIPS-fifth are, respectively, based on a definite theory, which considers contributions from long-range interactions. RF-metal, IPSn and LIPS-fifth have much better accuracy for estimating the potential energy. The fastest decline was observed in the case of LIPS-fifth; the error was roughly in proportion to r_c^{-4}. RF-metal and LIPS-fifth at $r_c = 2.0$ nm achieved the smallest error and had the same value as that of the Ewald sum within 0.02%.

For estimating the self-diffusion coefficient, the difference of the cutoff radius tendency comes from the treatment of cutoff boundary conditions and long-range interactions. In IPSp, a faster convergence of errors was observed, because it has a third-order cutoff boundary condition and the long-range interaction treatment. For IPSp, the self-diffusion coefficient is saturated at $r_c \geq 1.6$ nm, and the saturated value is almost the same as that of the Ewald sum (within 0.35%). LIPS-fifth achieved the fastest convergence in this work. It has a fifth-order cutoff boundary condition and reliable treatment for long-range interactions. For LIPS-fifth, the self-diffusion coefficient is saturated at $r_c \geq 1.4$ nm, and the saturated value is almost the same as that of the Ewald sum (within 0.36%).

The truncation effect in the radial distribution function mainly reflects the treatment of cutoff boundary conditions. In CHARMm-shift, RF-metal and IPSn, which have a first-order cutoff boundary condition, the deviation decreases roughly in proportion to $r_c^{-2.5}$, r_c^{-3} and r_c^{-3}, respectively. Ohmine-shift has a second-order cutoff boundary condition, and the RMSD of $g(r)$ declines roughly in proportion to r_c^{-4}. The RMSD of LIPS-fifth has a similar tendency with these shift function methods for cutoff radii, but a faster decline is observed. The RMSD of LIPS-fifth decreases roughly in proportion to r_c^{-6}. Furthermore, LIPS-fifth gives accurate estimations of $g(r)$; the RMSD converges at $r_c = 2.0$ nm. Converged values of RMSD for LIPS-fifth are most accurate. On the other hand, the RMSDs of Wolf-DSF and IPSp have an adequate accuracy in any cutoff radius. The charge neutrality and counter-charge assumptions of Wolf-DSF and IPSp, respectively, seem to work better for bulk water systems.

The cutoff radius effect in the dipole-dipole correlation is very similar to that of $g(r)$. $G_K(r)$, and $s(r)$ calculated with CHARMm-shift, Ohmine-shift, RF-metal and IPSn fluctuate near r_c, as in $g(r)$. The artificial configuration of Ohmine-shift was smaller than that of the other three methods. The defect of $G_K(r)$ and $s(r)$ for these above shift function methods was not seen in Wolf-DSF, IPSp and LIPS-fifth. Wolf-DSF, IPSp and LIPS-fifth can estimate the dipole-dipole correlation more adequately than other shift function methods.

Overall, the shift function method that has a higher-order cutoff boundary condition and a reliable theoretical background for long-range interaction treatments achieves better accuracy. For estimating the potential energy and self-diffusion, LIPS-fifth is the most accurate shift function method. In the

494

estimation of the radial distribution function, Wolf-DSF and IPSp have good accuracy with relatively short cutoff radii, and LIPS-fifth becomes most accurate at $r_c = 2.0$ nm.

Acknowledgments

This work was supported by the Japan Society for the Promotion of Science (JSPS) Grants-in-Aid for Scientific Research (KAKENHI) Grant Number 25820065.

Conflicts of Interest

The authors declare no conflict of interest.

References

1. Ewald, P. The calculation of optical and electrostatic grid potential. *Ann. Phys.* **1921**, *64*, 253–287.
2. Darden, T.; York, D.; Pedersen, L. Particle mesh Ewald: An N log (N) method for Ewald sums in large systems. *J. Chem. Phys.* **1993**, *98*, 10089–10092.
3. Essmann, U.; Perera, L.; Berkowitz, M.; Darden, T.; Lee, H.; Pedersen, L. A smooth particle mesh Ewald method. *J. Chem. Phys.* **1995**, *103*, 8577–8593.
4. Yokota, R.; Barba, L.A.; Narumi, T.; Yasuoka, K. Petascale turbulence simulation using a highly parallel fast multipole method on GPUs. *Comput. Phys. Commun.* **2012**, *184*, 445–455.
5. Roberts, J.; Schnitker, J. How the unit cell surface charge distribution affects the energetics of ion–solvent interactions in simulations. *J. Chem. Phys.* **1994**, *101*, 5024–5031.
6. Roberts, J.; Schnitker, J. Boundary conditions in simulations of aqueous ionic solutions: A systematic study. *J. Chem. Phys.* **1995**, *99*, 1322–1331.
7. Luty, B.; van Gunsteren, W. Calculating electrostatic interactions using the particle-particle particle-mesh method with nonperiodic long-range interactions. *J. Chem. Phys.* **1996**, *100*, 2581–2587.
8. Hünenberger, P.; McCammon, J. Ewald artifacts in computer simulations of ionic solvation and ion–ion interaction: A continuum electrostatics study. *J. Chem. Phys.* **1999**, *110*, 1856–1873.
9. Hünenberger, P.; McCammon, J. Effect of artificial periodicity in simulations of biomolecules under Ewald boundary conditions: A continuum electrostatics study. *Biophys. Chem.* **1999**, *78*, 69–88.
10. Weber, W.; Hünenberger, P.; McCammon, J. Molecular dynamics simulations of a polyalanine octapeptide under Ewald boundary conditions: Influence of artificial periodicity on peptide conformation. *J. Phys. Chem. B* **2000**, *104*, 3668–3675.
11. Patra, M.; Karttunen, M.; Hyvönen, M.; Falck, E.; Vattulainen, I. Lipid bilayers driven to a wrong lane in molecular dynamics simulations by subtle changes in long-range electrostatic interactions. *J. Phys. Chem. B* **2004**, *108*, 4485–4494.
12. Beck, D.; Armen, R.; Daggett, V. Cutoff size need not strongly influence molecular dynamics results for solvated polypeptides. *Biochemistry* **2005**, *44*, 609–616.

13. Monticelli, L.; Simões, C.; Belvisi, L.; Colombo, G. Assessing the influence of electrostatic schemes on molecular dynamics simulations of secondary structure forming peptides. *J. Phys. Condens. Matter* **2006**, doi:10.1088/0953-8984/18/14/S15.

14. Reif, M.; Krautler, V.; Kastenholz, M.; Daura, X.; Hunenberger, P. Molecular dynamics simulations of a reversibly folding β-heptapeptide in methanol: Influence of the treatment of long-range electrostatic interactions. *J. Phys. Chem. B* **2009**, *113*, 3112–3128.

15. Linse, P.; Andersen, H.C. Truncation of Coulombic interactions in computer simulations of liquids. *J. Chem. Phys.* **1986**, *85*, 3027–3041.

16. Hoheisel, C. Bulk viscosity of model fluids. A comparison of equilibrium and nonequilibrium molecular dynamics results. *J. Chem. Phys.* **1987**, *86*, 2328–2334.

17. Hoheisel, C.; Vogelsang, R.; Schoen, M. Bulk viscosity of the Lennard-Jones fluid for a wide range of states computed by equilibrium molecular dynamics. *J. Chem. Phys.* **1987**, *87*, 7195–7198.

18. Smit, B. Phase diagrams of Lennard-Jones fluids. *J. Chem. Phys.* **1992**, *96*, 8639–8640.

19. Trokhymchuk, A.; Alejandre, J. Computer simulations of liquid/vapor interface in Lennard-Jones fluids: Some questions and answers. *J. Chem. Phys.* **1999**, *111*, 8510–8524.

20. Lopez-Lemus, J.; Alejandre, J. Thermodynamic and transport properties of simple fluids using lattice sums: Bulk phases and liquid-vapour interface. *Mol. Phys.* **2002**, *100*, 2983–2992.

21. Neumann, M.; Steinhauser, O. The influence of boundary conditions used in machine simulations on the structure of polar systems. *Mol. Phys.* **1980**, *39*, 437–454.

22. Alper, H.; Levy, R. Computer simulations of the dielectric properties of water: Studies of the simple point charge and transferrable intermolecular potential models. *J. Chem. Phys.* **1989**, *91*, 1242–1251.

23. Kitchen, D.; Hirata, F.; Westbrook, J.; Levy, R.; Kofke, D.; Yarmush, M. Conserving energy during molecular dynamics simulations of water, proteins, and proteins in water. *J. Comput. Chem.* **1990**, *11*, 1169–1180.

24. Tasaki, K.; McDonald, S.; Brady, J. Observations concerning the treatment of long-range interactions in molecular dynamics simulations. *J. Comput. Chem.* **1993**, *14*, 278–284.

25. Smith, P.; van Gunsteren, W. Consistent dielectric properties of the simple point charge and extended simple point charge water models at 277 and 300 K. *J. Chem. Phys.* **1994**, *100*, 3169–3175.

26. Feller, S.; Pastor, R.; Rojnuckarin, A.; Bogusz, S.; Brooks, B. Effect of electrostatic force truncation on interfacial and transport properties of water. *J. Phys. Chem.* **1996**, *100*, 17011–17020.

27. Van der Spoel, D.; van Maaren, P.; Berendsen, H. A systematic study of water models for molecular simulation: Derivation of water models optimized for use with a reaction field. *J. Chem. Phys.* **1998**, *108*, 10220–10231.

28. Mark, P.; Nilsson, L. Structure and dynamics of liquid water with different long-range interaction truncation and temperature control methods in molecular dynamics simulations. *J. Comput. Chem.* **2002**, *23*, 1211–1219.

29. Yonetani, Y. A severe artifact in simulation of liquid water using a long cut-off length: Appearance of a strange layer structure. *Chem. Phys. Lett.* **2005**, *406*, 49–53.

30. Van der Spoel, D.; van Maaren, P. The origin of layer structure artifacts in simulations of liquid water. *J. Chem. Theory Comput.* **2006**, *2*, 1–11.

31. Yonetani, Y. Liquid water simulation: A critical examination of cutoff length. *J. Chem. Phys.* **2006**, *124*, 204501–204512.

32. Loncharich, R.; Brooks, B. The effects of truncating long-range forces on protein dynamics. *Proteins Struct. Funct. Bioinform.* **1989**, *6*, 32–45.

33. Schreiber, H.; Steinhauser, O. Cutoff size does strongly influence molecular dynamics results on solvated polypeptides. *Biochemistry* **1992**, *31*, 5856–5860.

34. Schreiber, H.; Steinhauser, O. Molecular dynamics studies of solvated polypeptides: Why the cut-off scheme does not work. *Chem. Phys.* **1992**, *168*, 75–89.

35. Schreiber, H.; Steinhauser, O. Taming cut-off induced artifacts in molecular dynamics studies of solvated polypeptides. The reaction field method. *J. Mol. Biol.* **1992**, *228*, 909–923.

36. Saito, M. Molecular dynamics simulations of proteins in water without the truncation of long-range Coulomb interactions. *Mol. Simul.* **1992**, *8*, 321–333.

37. Guenot, J.; Kollman, P. Conformational and energetic effects of truncating nonbonded interactions in an aqueous protein dynamics simulation. *J. Comput. Chem.* **1993**, *14*, 295–311.

38. Saito, M. Molecular dynamics simulations of proteins in solution: Artifacts caused by the cutoff approximation. *J. Chem. Phys.* **1994**, *101*, 4055–4062.

39. Oda, K.; Miyagawa, H.; Kitamura, K. How does the electrostatic force cut-off generate non-uniform temperature distributions in proteins? *Mol. Simul.* **1996**, *16*, 167–177.

40. Norberg, J.; Nilsson, L. On the truncation of long-range electrostatic interactions in DNA. *Biophys. J.* **2000**, *79*, 1537–1553.

41. Patra, M.; Karttunen, M.; Hyvönen, M.; Falck, E.; Lindqvist, P.; Vattulainen, I. Molecular dynamics simulations of lipid bilayers: Major artifacts due to truncating electrostatic interactions. *Biophys. J.* **2003**, *84*, 3636–3645.

42. Mazars, M. Long ranged interactions in computer simulations and for quasi-2D systems. *Phys. Rep.* **2011**, *500*, 43–116.

43. Piana, S.; Lindorff-Larsen, K.; Dirks, R.M.; Salmon, J.K.; Dror, R.O.; Shaw, D.E. Evaluating the effects of cutoffs and treatment of long-range electrostatics in protein folding simulations. *PLoS One* **2012**, *7*, e39918.

44. Wolf, D.; Keblinski, P.; Phillpot, S.; Eggebrecht, J. Exact method for the simulation of Coulombic systems by spherically truncated, pairwise r summation. *J. Chem. Phys.* **1999**, *110*, 8254–8283.

45. Fennell, C.J.; Gezelter, J.D. Is the Ewald summation still necessary? Pairwise alternatives to the accepted standard for long-range electrostatics. *J. Chem. Phys.* **2006**, *124*, e234104.

46. Mendoza, F.; López-Lemus, J.; Chapela, G.; Alejandre, J. The wolf method applied to the liquid-vapor interface of water. *J. Chem. Phys.* **2008**, *129*, e024706.

47. Takahashi, K.Z.; Narumi, T.; Yasuoka, K. Cutoff radius effect of the isotropic periodic sum and Wolf method in liquid-vapor interfaces of water. *J. Chem. Phys.* **2011**, *134*, e174112.

48. Wu, X.; Brooks, B. Isotropic periodic sum: A method for the calculation of long-range interactions. *J. Chem. Phys.* **2005**, *122*, e44107.

49. Wu, X.; Brooks, B. Using the isotropic periodic sum method to calculate long-range interactions of heterogeneous systems. *J. Chem. Phys.* **2008**, *129*, e154115.

50. Wu, X.; Brooks, B. Isotropic periodic sum of electrostatic interactions for polar systems. *J. Chem. Phys.* **2009**, *131*, e024107.

51. Takahashi, K.; Yasuoka, K.; Narumi, T. Cutoff radius effect of isotropic periodic sum method for transport coefficients of Lennard-Jones liquid. *J. Chem. Phys.* **2007**, *127*, e114511.

52. Takahashi, K.; Narumi, T.; Yasuoka, K. Cutoff radius effect of the isotropic periodic sum method in homogeneous system. II. Water. *J. Chem. Phys.* **2010**, *133*, e014109.

53. Takahashi, K.; Narumi, T.; Yasuoka, K. Cut-off radius effect of the isotropic periodic sum method for polar molecules in a bulk water system. *Mol. Simul.* **2012**, *38*, 397–403.

54. Kameoka, S.; Takahashi, K.Z.; Nozawa, T.; Narumi, T.; Yasuoka, K. Application of isotropic periodic sum method for macromolecular systems. I. 4-pentyl-4'-cyanobiphenyl liquid crystal. **2013**, submitted.

55. Klauda, J.; Wu, X.; Pastor, R.; Brooks, B. Long-range Lennard-Jones and electrostatic interactions in interfaces: Application of the isotropic periodic sum method. *J. Chem. Phys. B* **2007**, *111*, 4393–4400.

56. Venable, R.; Chen, L.; Pastor, R. Comparison of the extended isotropic periodic sum and particle mesh ewald methods for simulations of lipid bilayers and monolayers. *J. Chem. Phys. B* **2009**, *113*, 5855–5862.

57. Nakamura, H.; Ohto, T.; Nagata, Y. Polarizable site charge model at liquid/solid interfaces for describing surface polarity: Application to structure and molecular dynamics of water/rutile TiO$_2$ (110) interface. *J. Chem. Theory Comput.* **2013**, *9*, 1193–1201.

58. Takahashi, K.Z.; Narumi, T.; Yasuoka, K. A combination of the tree-code and IPS method to simulate large scale systems by molecular dynamics. *J. Chem. Phys.* **2011**, *135*, e174108.

59. Takahashi, K.Z.; Narumi, T.; Suh, D.; Yasuoka, K. An improved isotropic periodic sum method using linear combinations of basis potentials. *J. Chem. Theory Comput.* **2012**, *8*, 4503–4516.

60. Takahashi, K.Z. Design of a reaction field using a linear-combination-based isotropic periodic sum method. *J. Chem. Phys.* **2013**, submitted.

61. Brooks, B.R.; Bruccoleri, R.E.; Olafson, B.D.; States, D.J.; Swaminathan, S.; Karplus, M. CHARMM: A program for macromolecular energy, minimization and dynamic calculations. *J. Comput. Chem.* **1983**, *4*, 187–217.

62. Matsumoto, M.; Saito, S.; Ohmine, I. Molecular dynamics simulation of the ice nucleation and growth process leading to water freezing. *Nature* **2002**, *416*, 409–413.

63. Berendsen, H.; Grigera, J.; Straatsma, T. The missing term in effective pair potentials. *J. Phys. Chem.* **1987**, *91*, 6269–6271.

64. Swope, W.; Andersen, H.; Berens, P.; Wilson, K. A computer simulation method for the calculation of equilibrium constants for the formation of physical clusters of molecules: Application to small water clusters. *J. Phys. Chem.* **1982**, *76*, 637–650.

65. Andersen, H. Rattle: A "velocity" version of the shake algorithm for molecular dynamics calculations. *J. Comput. Phys.* **1983**, *52*, 24–34.

66. Nosé, S. A molecular dynamics method for simulations in the canonical ensemble. *Mol. Phys.* **1984**, *52*, 255–268.

67. Nosé, S. A unified formulation of the constant temperature molecular dynamics methods. *J. Chem. Phys.* **1984**, *81*, 511–520.

68. Hoover, W. Canonical dynamics: Equilibrium phase-space distributions. *Phys. Rev. A* **1985**, *31*, 1695–1697.

Reprinted from *Entropy*. Cite as: Ono, S. Elastic Properties of CaSiO₃ Perovskite from *ab initio* Molecular Dynamics. *Entropy* **2013**, *15*, 4300–4309.

Article

Elastic Properties of CaSiO₃ Perovskite from *ab initio* Molecular Dynamics

Shigeaki Ono [1,2]

[1] Institute for Research on Earth Evolution, Japan Agency for Marine-Earth Science and Technology, 2-15 Natsushima-cho, Yokosuka-shi, Kanagawa 237-0061, Japan; E-Mail: sono@jamstec.go.jp; Tel.: +81-46-867-9631; Fax: +81-46-867-9625

[2] Earthquake Research Institute, University of Tokyo, 1-1-1 Yayoi, Bunkyo-ku, Tokyo 113-0032, Japan

Received: 26 June 2013 / Accepted: 6 October 2013 / Published: 10 October 2013

Abstract: *Ab initio* molecular dynamics simulations were performed to investigate the elasticity of cubic CaSiO₃ perovskite at high pressure and temperature. All three independent elastic constants for cubic CaSiO₃ perovskite, C_{11}, C_{12}, and C_{44}, were calculated from the computation of stress generated by small strains. The elastic constants were used to estimate the moduli and seismic wave velocities at the high pressure and high temperature characteristic of the Earth's interior. The dependence of temperature for sound wave velocities decreased as the pressure increased. There was little difference between the estimated compressional sound wave velocity (V_P) in cubic CaSiO₃ perovskite and that in the Earth's mantle, determined by seismological data. By contrast, a significant difference between the estimated shear sound wave velocity (V_S) and that in the Earth's mantle was confirmed. The elastic properties of cubic CaSiO₃ perovskite cannot explain the properties of the Earth's lower mantle, indicating that the cubic CaSiO₃ perovskite phase is a minor mineral in the Earth's lower mantle.

Keywords: perovskite; first-principles calculation; seismic wave velocity

PACS Codes: 62.20.de; 91.60.Gf

1. Introduction

Mineral physics constraints on the composition of the Earth's lower mantle rely on knowledge of the equations of state (EOSs) and sound wave velocities in candidate minerals. According to reliable estimates of the composition of the Earth, an MgO-FeO-SiO₂-CaO-Al₂O₃ system could comprise about 99% of the mantle volume [1]. Three minerals have been proposed to be possible hosts of the MgO-FeO-SiO_2-CaO-Al_2O_3 system in the Earth's lower mantle. A recent phase equilibrium study using a more representative composition of the mantle shows that Mg, Fe, and Al are mostly accommodated in orthorhombic $(Mg,Fe)SiO_3$ perovskite and ferropericlase, $(Mg,Fe)O$. On the other hand, a number of other experimental studies indicate that the most likely Ca-bearing phase is $CaSiO_3$ perovskite [2,3]. Thus, the Earth's lower mantle may be composed mainly of aluminous $(Mg,Fe)SiO_3$ perovskite, $CaSiO_3$ perovskite, and ferropericlase. To gain an understanding of the structure and dynamics of the Earth's lower mantle, it is important to investigate the elastic properties of these minerals under the pressure and temperature conditions found in this region. It is easy to investigate the physical properties of orthorhombic $(Mg,Fe)SiO_3$ perovskite and ferropericlase, because both minerals can be recovered under ambient conditions. By contrast, $CaSiO_3$ perovskite is unstable under ambient conditions, and it readily transforms to glass on the release of pressure. Therefore, it is difficult to measure some of its physical properties.

The structure of $CaSiO_3$ perovskite has tetragonal or orthorhombic symmetry at high pressures and room temperature, e.g., [4,5]. The structure of low-symmetry $CaSiO_3$ perovskite is still an open question [6–8]. A phase transformation of $CaSiO_3$ perovskite from this low symmetry into cubic symmetry with an increase in temperature was found in a previous study [5], indicating that the cubic structure of $CaSiO_3$ perovskite is stable under the conditions of the Earth's lower mantle, and that its physical properties are important for understanding the dynamics and evolution of the Earth's interior. The elastic properties of cubic $CaSiO_3$ perovskite were calculated at 0 K [9,10] and high temperatures [8]. These data relating to the elastic properties of cubic $CaSiO_3$ perovskite are insufficient to discuss the composition of the Earth's lower mantle, because an internally consistent data set describing the density-V_P-V_S relationship in cubic $CaSiO_3$ perovskite is needed in order to compare the density-V_P-V_S relationship of a preliminary reference earth model (PREM) [11] with that estimated based on cubic $CaSiO_3$ perovskite. Therefore, it is necessary to reevaluate the density-V_P-V_S relationship using *ab initio* calculations.

We employed the *ab initio* molecular dynamics (AIMD) method using density functional theory (DFT) to determine the density values and sound wave velocities for cubic $CaSiO_3$ perovskite at pressures typical of the Earth's lower mantle. We also used the experimental data to correct the calculated values of the density and the sound wave velocities for cubic $CaSiO_3$ perovskite.

2. Method

We performed the AIMD calculations based on DFT using the VASP code [12]. The interactions between the electrons and the ionic cores were described using the projector augmented wave (PAW) method [13] with generalized gradient approximations, known as PBE [14]. The advantage of this code is that the *ab initio* energy of the system can be combined with the molecular dynamics method to simulate the properties of cubic $CaSiO_3$ perovskite at high pressure and high temperature simultaneously. The PAW potentials of Ca, Si, and O had core radii of 2.3, 1.5, and 1.1 a.u., respectively. Single particle orbitals were expanded in plane waves, with a plane-wave cut-off of 900 eV. The calculations were performed on the basis of a self-consistency convergence on the total energy of 10^{-4} eV per simulation cell. We used a 135-atom supercell, with Γ-point Brillouin zone sampling and a time step of 1 fs at a constant volume. The simulations were run in the constant *NVT* ensemble with a Nosé thermostat [15] for at least 5 ps after equilibration. The computation time required to reach equilibrium varied between configurations, and depended on the starting atomic position, velocity, temperature, and pressure. In previous studies, we have confirmed that useful data for the elastic properties of solids in high pressure and temperature conditions can be acquired using the previous AIMD calculations, e.g., [16,17]. In this study, AIMD calculations were performed under 27 selected pressure and temperature conditions up to 175 GPa and 4000 K. A comprehensive description of our method as applied to the modeling of condensed matter has been described previously [18].

The elastic constants can be determined from the computation of the second derivatives of the free energy as a function of small strains [19]. For a cubic crystal, the three elastic moduli, C_{11}, C_{12}, and C_{44}, fully describe its elastic behavior. The values of C_{11} and C_{12} can be determined from the bulk modulus K and shear constant C_S:

$$K = (C_{11} + 2C_{12})/3 \tag{1}$$

$$C_S = (C_{11} - C_{12})/2 \tag{2}$$

The following tetragonal strains were applied to obtain C_S:

$$\varepsilon = \begin{pmatrix} e & 0 & 0 \\ 0 & e & 0 \\ 0 & 0 & (1+e)^{-2} - 1 \end{pmatrix} \tag{2}$$

where e is the strain magnitude. The change in the free energy of the strained structure, $\Delta E(e)$, is related to e as follows:

$$\Delta E(e) = 3V(C_{11} - C_{12})e^2 + O(e^3) \tag{3}$$

where V is the volume of the cell. C_{44} was calculated by applying the volume-conserving orthorhombic strain:

$$\varepsilon = \begin{pmatrix} 0 & e & 0 \\ e & 0 & 0 \\ 0 & 0 & e^2/(1-e^2) \end{pmatrix} \tag{4}$$

The energy associated with this strain is:

$$\Delta E(e) = 2C_{44}Ve^2 + O(e^4) \tag{5}$$

According to the calculations for unstrained and strained structures, the elasticity of cubic CaSiO$_3$ perovskite at high pressure and high temperature can be determined.

3. Results

The EOS of cubic CaSiO$_3$ perovskite has been investigated in a previous experimental study [20]. Recent theoretical studies have investigated the physical properties of materials under high pressure and high temperature using first-principles calculations. We noticed that the scatter of the EOS determined by experimental study was smaller than that obtained from first-principles calculations [18], indicating that the EOS determined in experiments was more accurate than that determined from first-principles calculations. By contrast, the AIMD calculations present significant advantages for investigating the elastic properties of materials under high temperatures and high temperatures. Therefore, we used the experimental data to determine the EOS for cubic $CaSiO_3$ perovskite. The pressures estimated by the AIMD calculations were corrected based on the EOS determined by the experimental data. The combination of first-principles molecular dynamics calculations and high pressure experimental data led us to determine reliable physical properties over a wide range of pressures and temperatures. The EOS for a solid can be described in a general form as a functional relationship between pressure, volume, and temperature:

$$P_{Total}(V,T) = P_{st}(V,300) + P_{th}(V,T)$$
(7)

A fit of the volume-pressure data yielded volume and bulk modulus values of $V_0 = 45.58$ Å^3, $K_{T0} = 236$ GPa and $K'_{T0} = 3.9$ [20] for a third-order Birch-Murnaghan EOS [21]:

$$P_{st} = \frac{3}{2}K_{T0}\left[\left(\frac{V_0}{V}\right)^{\frac{7}{3}} - \left(\frac{V_0}{V}\right)^{\frac{5}{3}}\right]\left\{1 - \frac{3}{4}(4 - K'_{T0})\left[\left(\frac{V_0}{V}\right)^{\frac{2}{3}} - 1\right]\right\}$$
(8)

where V_0 and K_{T0}, and K'_{T0} are the volume, isothermal bulk modulus, and first pressure derivative of the isothermal bulk modulus, respectively. The thermal pressure, P_{th}, of the thermal pressure EOS can be written as follows:

$$P_{th} = \left[\alpha K_T(V_0,T) + \left(\frac{\partial K_T}{\partial T}\right)_V \ln\left(\frac{V_0}{V}\right)\right](T - T_0)$$
(9)

A least squares fit of the high temperature data from the AIMD calculations yields $\alpha K_T(V_0,T) = 0.0083$ and $(\delta K_T/\delta T)_V = -0.0031$. The value of P_{th} of cubic $CaSiO_3$ perovskite was not sensitive to changes in volume at the pressures investigated in this study, because the values of $(\delta K_T/\delta T)_V$ were very small. The fitting parameters of the third-order Birch-Murnaghan EOS combined with the thermal pressure EOS were V_0, K_{T0}, K_{T0}', $\alpha K_T(V_0,T)$, and $(\delta K_T/\delta T)_V$. The results of the fit of our P-V-T data to the thermal pressure equation of state are summarized in Table 1.

Table 1. The thermoelastic parameters of cubic $CaSiO_3$ perovskite. The third-order Birch–Murnaghan EOS was used to calculate the parameters of cubic $CaSiO_3$ perovskite. Key: K_{T0}, isothermal bulk modulus at 0 GPa and 300 K; K'_{T0}, first pressure derivation of the bulk modulus; V_0, volume at 0 GPa and 300 K. The terms $\alpha K_T(V_0,T)$ and $(\partial K_T/\partial T)_V$ are parameters of the thermal pressure.

Parameter	Value
V_0 (Å^3)	45.58 [a]
K_{T0} (GPa)	236 [a]
K'_{T0}	3.9 [a]
$\alpha K_T(V_0,T)$ (GPa/K)	0.0083(3)
$(\partial K_T/\partial T)_V$ (GPa/K)	−0.0031(31)

[a] The parameters are from Shim et al. [20].

We determined the elastic constant by computing the stress generated by small deformations of the equilibrium cell. Figure 1 shows three elastic constants of cubic $CaSiO_3$ perovskite (C_{11}, C_{12}

and C_{44}) at 2000 K as a function of pressure up to 160 GPa. The bulk modulus of an isotropic aggregate cubic crystal is well defined, whereas the shear modulus can be constrained by the Voigt-Reuss-Hill scheme [22]:

$$G^V = \frac{1}{5}(2C_S + 3C_{44})$$

(10)

$$G^R = \left[\frac{1}{5}\left(\frac{2}{C_S} + \frac{3}{C_{44}}\right)\right]^{-1}$$

(11)

$$G^H = \frac{1}{2}(G^V + G^R)$$

(12)

We also show the bulk modulus K and Hill's average G at 2000 K as a function of pressure in Figure 1. Karki and Crain [9] calculated elastic constants and moduli of cubic CaSiO₃ perovskite at 0 K up to 140 GPa. Our results for elastic parameters calculated at 2000 K were in general agreement with those at 0 K reported by previous studies.

Figure 1. Pressure dependence of three elastic constants, C_{11}, C_{12}, and C_{44}, and the isotropic bulk (K) and shear (G) moduli of cubic CaSiO₃ perovskite at 2000 K. The solid circles and diamonds represent the elastic constants and the elastic moduli, respectively. The solid and dashed lines are the fits of each parameter.

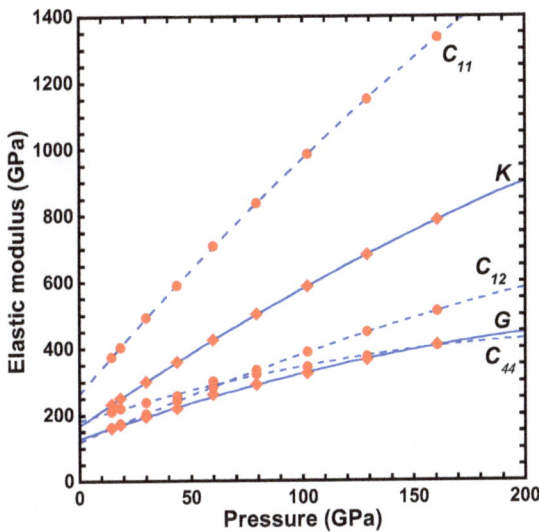

The three isotropically averaged aggregate sound velocities could be derived from the bulk modulus K and shear modulus G:

$$V_P = \left[\left(K + \frac{4}{3}G\right)/\rho\right]^{\frac{1}{2}}$$

(13)

$$V_B = \left(\frac{K}{\rho}\right)^{\frac{1}{2}}$$

(14)

$$V_S = \left(\frac{G}{\rho}\right)^{\frac{1}{2}} \tag{15}$$

where V_P, V_B, and V_S are the compressional, bulk, and shear sound wave velocities, respectively, and ρ is the density. The three sound wave velocities, V_P, V_B, and V_S, increased with increasing pressure at 2000 K in Figure 2. Our results for sound wave velocities were in good agreement with those reported by Li *et al.* [8].

> **Figure 2.** Sound wave velocities in cubic CaSiO₃ perovskite at 2000 K calculated from the elastic constants. The solid circles, squares, and diamonds represent the compressional, bulk, and shear velocities, respectively.

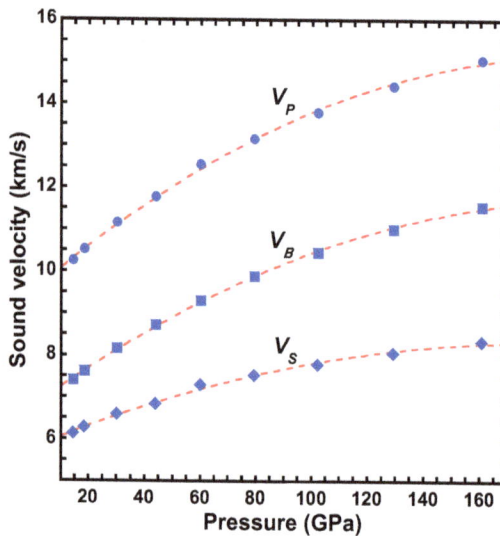

The effect of temperature on the sound wave velocities was investigated at high temperatures corresponding to conditions in the Earth's mantle. In Figure 3, the results of the AIMD simulations at high temperatures of $2000 \leq T(K) \leq 4000$ showed that sound wave velocities decreased with increasing temperature. However, there were only small dependencies on temperature. As the pressure increased, these dependencies on temperature became small. The sound wave velocities were fitted to the following equation as functions of temperature and pressure:

$$v = a + (b + cP)T + d \ln(P) \tag{16}$$

where a, b, c and d are fitted parameters, and T and P are given in K and GPa, respectively. The results of the fitted parameters are summarized in Table 2.

Figure 3. Sound wave velocities in cubic CaSiO$_3$ perovskite at high temperatures. The diamonds represent the compressional and shear wave velocities calculated by the AIMD simulations. The dashed lines represent the fitted velocities for 2000, 3000, and 4000 K.

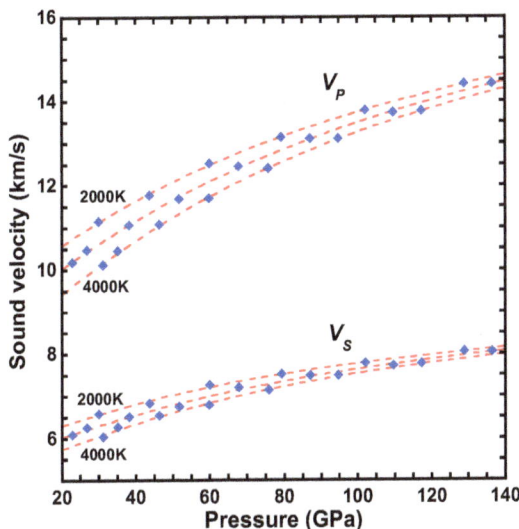

Table 2. Parameters of the compressional and shear sound velocities. The parameters are given by $v = a + (b + cP)T + d\ln(P)$, where T and P are the temperature (K) and the pressure (GPa), respectively. The conditions for applying these parameters to Equation (16) are 15 GPa $< P <$ 140 GPa and 1500 K $< T <$ 4500 K.

	a	b	c	d
V_P	7.12(0.27)	$-6.96(0.33) \times 10^{-4}$	$4.07(0.30) \times 10^{-7}$	1.56(0.06)
V_S	4.65(0.15)	$-3.20(0.02) \times 10^{-4}$	$1.75(0.17) \times 10^{-6}$	0.74(0.04)

4. Discussion

It is important to assess the uncertainties of the *ab initio* calculations to understand the implications of the calculated results. In general, different types of approximation have led to different values in previous *ab initio* studies, e.g., [18]. The difference between Local Density Approximation (LDA) and Generalized Gradient Approximation (GGA) leads to a change in cell volume of a few percent. This uncertainty is non-negligible in the context of discussing the behavior of the Earth's mantle. Although GGA was used in the present study, we corrected the calculated values according to the experimental EOS to minimize the uncertainties related to approximations used in *ab initio* simulations. Therefore, our discussion of the comparison between estimated elastic properties and PREM values is more reliable than those of previous studies.

It is believed that the lower mantle contains three minerals; (Fe,Al)-bearing Mg-perovskite, ferropericlase, and Ca-perovskite [2,3]. We calculated the density, and the compressional and shear sound wave velocities for cubic CaSiO$_3$ perovskite in order to compare them with the values from the PREM [11]. The values for cubic CaSiO$_3$ perovskite were calculated using the EOS defined in

Table 1 and Equation (16) defined in Table 2. The estimated values for cubic CaSiO$_3$ perovskite and the PREM are compared in Figure 4. The adiabatic temperature profile (geotherm) was used as the temperature profile in the Earth's lower mantle [3]. The calculated density of cubic CaSiO$_3$ perovskite was in good agreement with that estimated by seismological data (PREM). Li *et al.* [8] estimated the density of cubic CaSiO$_3$ perovskite, and the estimated density was higher than that in the PREM data. Although our AIMD method was similar to that used by Li *et al.* [8], the discrepancy between this and the previous study was confirmed. As the pressure was corrected to estimate density accurately in this study, the difference between our estimated density of cubic CaSiO$_3$ perovskite and that from the PREM data should be small. For the compressional sound wave velocity, the discrepancy between the calculated values for cubic CaSiO$_3$ perovskite and the observed data was small. The difference increased as the depth increased. By contrast, the shear sound wave velocity in cubic CaSiO$_3$ perovskite was much higher than that from the PREM data. If cubic CaSiO$_3$ perovskite is a major mineral in the Earth's lower mantle, the sound wave velocity profiles cannot be explained. Therefore, our study implies that cubic CaSiO$_3$ perovskite is a minor mineral in the Earth's lower mantle. In a previous study, the shear sound wave velocity in orthorhombic (Mg,Fe)SiO$_3$ perovskite was reported to be lower than that in cubic CaSiO$_3$ perovskite at 0 K [23]. As the shear sound wave velocity in ferropericlase, (Mg,Fe)O, is much lower than that of orthorhombic (Mg,Fe)SiO$_3$ perovskite, sound wave velocities in orthorhombic (Mg,Fe)SiO$_3$ perovskite at higher temperatures are therefore in general agreement with those from PREM, indicating that orthorhombic (Mg,Fe)SiO$_3$ perovskite might be a major mineral in the Earth's lower mantle.

Figure 4. Density and sound wave velocities for cubic CaSiO$_3$ perovskite compared with PREM data. The solid circles represent the values from PREM [11]. The solid lines represent the calculated values under the conditions of the Earth's lower mantle. The values for cubic CaSiO$_3$ perovskite were calculated using the equations defined in Tables 1 and 2 and the adiabatic temperature profile (geotherm) in the Earth's interior [3]; a: compressional sound velocity; b: shear sound velocity; c: density.

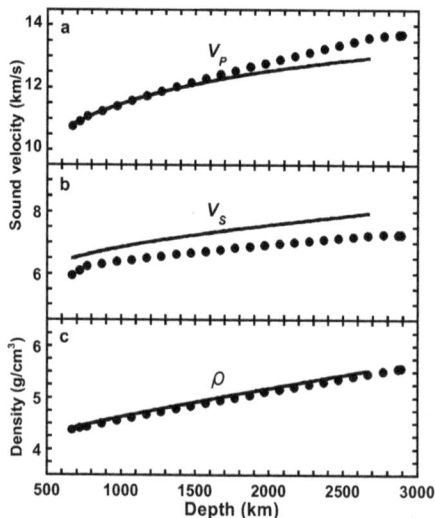

In general, a certain quantity of most elements dissolves in three minerals, namely (Fe,Al)-bearing Mg-perovskite, ferropericlase, and Ca-perovskite, and the partition coefficients of minor elements between the three minerals change with temperature and pressure. In this study, the physical properties of a pure cubic $CaSiO_3$ perovskite host were calculated, because the complicated chemical composition would need a very large simulation time, and it would have been difficult to perform a reliable AIMD study. The effects of the minor elements on the density and the sound wave velocities therefore require further investigation.

5. Conclusions

We have predicted from the first-principles theory the sound velocities of cubic $CaSiO_3$ perovskite at high pressures and temperatures corresponding to the Earth's lower mantle. Comparison of the elastic properties of cubic $CaSiO_3$ perovskite with the lower mantle properties estimated from the seismic observations supports the prevailing hypothesis that the lower mantle consists primarily of $(Mg,Fe)SiO_3$ perovskite.

Acknowledgments

This work made use of the super computer system of JAMSTEC and of the computer systems of the Earthquake Information Center of the Earthquake Research Institute. This work was supported by Grant-in-Aid for Scientific Research from JSPS and the Earthquake Research Institute cooperative research program, Japan.

Conflicts of Interest

The author declares no conflict of interest.

References

1. Anderson, D.L. Composition of the Earth. *Science* **1989**, *243*, 367–370.
2. Irifune, T.; Ringwood, A.E. Phase Transitions in Primitive MORB and Pyrolite Compositions to 25 GPa and Some Geophysical Implications. In *High-Pressure Research in Mineral Physics*; Manghnanii, M.H., Syono, Y., Eds.; Terra Scientific Publishing Company, Tokyo/American Geophysical Union: Washington, DC, USA, 1987; pp. 231–242.
3. Ono, S. Experimental constraints on the temperature profile in the lower mantle. *Phys. Earth Planet. Inter.* **2008**, *170*, 267–273.
4. Shim, S.H.; Jeanloz, R.; Duffy, T.S. Tetragonal structure of $CaSiO_3$ perovskite above 20 GPa. *Geophys. Res. Lett.* **2002**, *29*, doi:10.1029/2002GL016148.
5. Ono, S.; Ohishi, Y.; Mibe, K. Phase transition of Ca-perovskite and stability of Al-bearing Mg-perovskite in the lower mantle. *Am. Mineral.* **2004**, *89*, 1480–1485.
6. Caracas, R.; Wentzcovitch, R.; Price, G.D.; Brodholt, J. $CaSiO_3$ perovskite at lower mantle pressures. *Geophys. Res. Lett.* **2005**, *32*, L06306.
7. Adams, D.J.; Oganov, A.R. *Ab inito* molecular dynamics study CaSiO3 perovskite at *P-T* conditions of Earth's lower mantle. *Phys. Rev. B* **2006**, *73*, 184106.

8. Li, L.; Weidner, D.J.; Brodholt, J.; Alfè, D.; Price, G.D.; Caracas, R.; Wentzcovitch, R. Phase stability of CaSiO$_3$ perovskite at high pressure and temperature: Insights from *ab initio* molecular dynamics. *Phys. Earth Planet. Inter.* **2006**, *155*, 260–268.

9. Karki, B.B.; Crain, J. First-principles determination of elastic properties of CaSiO$_3$ perovskite at lower mantle pressures. *Geophys. Res. Lett.* **1998**, *25*, 2741–2744.

10. Stixrude, L.; Lithgow-Bertelloni, C.; Kiefer, B.; Fumagalli, P. Phase stability and shear softening in CaSiO$_3$ perovskite at high pressure. *Phys. Rev. B* **2007**, *75*, 024108.

11. Dziewonski, A.M.; Anderson, D.L. Preliminary reference earth model. *Phys. Earth Planet. Inter.* **1981**, *25*, 297–356.

12. Kresse, G.; Furthmuller, J. Efficient iterative schemes for *ab initio* total-energy calculations using a plane-wave basis set. *Phys. Rev. B* **1996**, *54*, 11169–11186.

13. Blöchl, P.E. Projector augmented-wave method. *Phys. Rev. B* **1994**, *50*, 17953–17979.

14. Perdew, J.P.; Burke, K.; Ernzerhof, M. Generalized gradient approximation made simple. *Phys. Rev. Lett.* **1996**, *77*, 3865–3868.

15. Nosé, S. A molecular dynamics method for simulations in the canonical ensemble. *Mol. Phys.* **1984**, *52*, 255–268.

16. Ono, S.; Brodholt, J.P.; Alfè, D.; Alfredsson, M.; Price, G.D. *Ab initio* molecular dynamics simulations for thermal equation of state of B2-type NaCl. *J. Appl. Phys.* **2008**, *86*, 5801–5808.

17. Ono, S. First-principles molecular dynamics calculations of the equation of state for tantalum. *Int. J. Mol. Sci.* **2009**, *10*, 4342–4351.

18. Ono, S. Synergy between First-Principles Computation and Experiment in Study of Earth Science. In *Some Applications of Quantum Mechanics*; Pahlavani, M.R., Ed.; InTech: Vienna, Austria, 2012; pp. 91–108.

19. Mehl, M.J.; Osburn, J.E.; Papaconstantopoulos, D.A.; Klein, B.M. Structural properties of ordered high-melting-temperature intermetallic alloys from first-principles total-energy calculations. *Phys. Rev. B* **1990**, *41*, 10311–10323.

20. Shim, S.H.; Duffy, T.S. The stability and P-V-T equation of state of CaSiO$_3$ perovskite in the Earth's lower mantle. *J. Geophys. Res.* **2000**, *105*, 25955–25968.

21. Poirier, J.P. *Introduction to the Physics of the Earth's Interior*; Cambridge University Press: Cambridge, UK, 2000.

22. Anderson, O.L. *Equations of State of Solids for Geophysics and Ceramic Science*; Oxford University Press: New York, NY, USA, 1995.

23. Karki, B.B.; Stixrude, L. Seismic velocities of major silicate and oxide phases of the lower mantle. *J. Geophys. Res.* **1999**, *104*, 13025–13033.

Reprinted from *Entropy*. Cite as: Wang, B.-B.; Wang, X.-D.; Chen, M.; Xu, J.-L. Molecular Dynamics Simulations on Evaporation of Droplets with Dissolved Salts. *Entropy* **2013**, *15*, 1232–1246.

Article

Molecular Dynamics Simulations on Evaporation of Droplets with Dissolved Salts

Bing-Bing Wang [1,2], **Xiao-Dong Wang** [1,2,*], **Min Chen** [3,*] and **Jin-Liang Xu** [1,2]

[1] State Key Laboratory of Alternate Electrical Power System with Renewable Energy Sources, North China Electric Power University, Beijing 102206, China
[2] Beijing Key Laboratory of Multiphase Flow and Heat Transfer for Low Grade Energy, North China Electric Power University, Beijing 102206, China
[3] Department of Engineering Mechanics, Tsinghua University, Beijing 100084, China

* Authors to whom correspondence should be addressed;
  E-Mails: wangxd99@gmail.com (X.-D.W); mchen@mail.tsinghua.edu.cn (M.C.);
  Tel.: +86-10-62321277 (X.-D.W.); +86-10-62773776 (M.C.);
  Fax: +86-10-62321277 (X.-D.W); +86-10-62795832 (M.C.).

Received: 6 January 2013; in revised form: 18 March 2013 / Accepted: 18 March 2013 / Published: 8 April 2013

Abstract: Molecular dynamics simulations are used to study the evaporation of water droplets containing either dissolved LiCl, NaCl or KCl salt in a gaseous surrounding (nitrogen) with a constant high temperature of 600 K. The initial droplet has 298 K temperature and contains 1,120 water molecules, 0, 40, 80 or 120 salt molecules. The effects of the salt type and concentration on the evaporation rate are examined. Three stages with different evaporation rates are observed for all cases. In the initial stage of evaporation, the droplet evaporates slowly due to low droplet temperature and high evaporation latent heat for water, and pure water and aqueous solution have almost the same evaporation rates. In the second stage, evaporation rate is increased significantly, and evaporation is somewhat slower for the aqueous salt-containing droplet than the pure water droplet due to the attracted ion-water interaction and hydration effect. The Li$^+$-water has the strongest interaction and hydration effect, so LiCl aqueous droplets evaporate the slowest, then NaCl and KCl. Higher salt concentration also enhances the ion-water interaction and hydration effect, and hence corresponds to a slower evaporation. In the last stage of evaporation, only a small amount of water molecules are left in the droplet, leading to a significant increase in ion-water interactions, so that the evaporation becomes slower compared to that in the second stage.

Keywords: molecular dynamics simulations; evaporation; aqueous droplet; salts

1. Introduction

The physics of droplet evaporation in an infinite space has attracted much interest due to the crucial role played in energy engineering, in chemical engineering as well as in environmental processes [1], and has been under investigation for many years by different methods: hydrodynamics [2], kinetic theory [3], and molecular simulations [4–11]. Molecular dynamics attempts to simulate the real behavior of Nature by identifying each atom and following their motion in time through the basic laws of classical mechanics [7]. The system behavior and temporal evolution of its thermodynamic and transport properties can be obtained by statistically averaging the results of all molecular motions. Molecular dynamics simulating an evaporation process has no need of some assumptions made by CFD (computational fluid dynamics), so this method was adopted to study the droplet evaporation [4–11] and the evaporation of flat thin liquid films on solid surfaces [12–14]. These studies focused on evaporation of droplets consisting of one component under various conditions, and compared the evaporation rates simulated by molecular dynamics and predicted by classical kinetic theory, such as the D^2 law [5,6,9]. The D^2 law was derived based on the droplet evaporation in an infinite space [15], and predicts that the derivative of the square of the droplet diameter with respect to time is constant, or $dD^2/dt = -K$, where K is the evaporation constant. However, Semenov et al. [16] presented that for the evaporation of sessile drops on hydrophobic substrates, the evaporation rate is proportional to the radius of the three phase line instead of being proportional to the area of the surface of the droplet. Semenov et al. [17] also investigated the effect of the influence of kinetic effects on evaporation of pinned sessile water droplets of submicrometer size placed on a heat conductive substrate. Their computer simulation model took into account the following phenomena: influence of curvature of the droplet's surface on the saturated vapor pressure above the surface (Kelvin's equation), the effect of latent heat of vaporization, thermal Marangoni convection, and Stefan flow inside an air domain above the droplet.

The evaporation of droplets of dissolved salts (aqueous droplets) has extensive applications in many industrial processes such as crystallization [2,18], electrospraying [19], electrospinning [20], and atmospheric science [21–25]. Starov and Churaev [2] investigated the crystallization process of aqueous solutions in a thin capillary using hydrodynamics, and they presented that the evaporation flux differs significantly from that predicted by the classical solution with a one-component liquid. Recently, some studies have used molecular dynamics simulations to investigate the evaporation of a small water cluster dissolving a single kind of charged ion [19,26–28]. Caleman and Spoel [26] simulated the evaporation from a water cluster ($N = 216$ and 512) containing either Cl^-, $H_2PO_4^-$, Na^+ or NH_4^+ ($N = 0$, 4 and 8) under a vacuum, and their results showed a somewhat slower evaporation rate for clusters with Cl^- and Na^+ than those with $H_2PO_4^-$ and NH_4^+. Daub and Cann [27] studied evaporation and condensation of a small cluster ($N = 10$, 20, 30 and 40) of water or methanol containing one single Ca^{2+}, Na^+ or Cl^- ion in either a vacuum or under argon gas. Daub and Cann [19] also studied the evaporation of a water cluster ($N = 10$, 15 or 20) containing one Na$^+$ ion or one Ca^{2+} ion under the action of an electric field. Daub and Cann's results demonstrated that the interaction between ions and water molecules affects the evaporation of the cluster [19,27]. Köhler [21] investigated the process of formation of liquid cloud drops based on equilibrium thermodynamics, where water vapor condensed with existence of nucleus (solutes), and proposed the well-known Köhler theory expressed as:

$$\ln\left(\frac{p_w(D)}{p^0}\right) = \frac{4M_w\sigma_w}{RT\rho_w D} - \frac{6n_s M_w}{\pi\rho_w D^3} \tag{1}$$

where p_w is the water vapor pressure outside the droplet, p^0 is the corresponding saturation vapor pressure over a flat surface, σ_w is the droplet surface tension, ρ_w is the density of pure water, n_s is the moles of solute, M_w is the molecular weight of water, and D is the droplet diameter. According to Köhler's theory, the droplet diameter, water surface tension, and molar concentration of the solute significantly affect the water vapor pressure and hence the droplet evaporation rate.

Generally, an aqueous solution is electrically neutral and it includes the same number of cations and anions for monovalent salts, the interaction between cations and anions may influence the evaporation properties of the aqueous droplet, however, few molecular dynamics simulations have been carried out considering this process. The salt crystallization process from an evaporating NaCl aqueous solution has been studied by Mucha and Jungwirth [18] using molecular dynamics simulations, but they did not focus on evaporation rates. This work uses molecular dynamics simulations to investigate the evaporation of water droplets containing either dissolved LiCl, NaCl or KCl salt. The droplet is surrounded and heated by a nitrogen gas atmosphere at a constant temperature. The effects of the salt concentration and salt type on the evaporation rate are examined. By analyzing the spatial position and the interaction energy of ions and water molecules, the differences between evaporation rates for various cases are explained in detail.

2. Molecular Dynamics Simulations

2.1. Interatomic Potential and Initial Configuration

For molecular dynamics simulations, selecting a proper intermolecular potential function and constructing a correct initial configuration of system are important to correctly describe the physical process concerned. This work simulates the evaporation of water droplets with or without dissolved salts. Three salts, LiCl, NaCl or KCl, at various concentrations are added to the water droplet. The droplet is surrounded and heated by nitrogen gas at a constant temperature. Since the system includes nitrogen molecules, water molecules, Li$^+$, Na$^+$, K$^+$, and Cl$^-$ ions, the long-range Coulombic force between ions must be considered. Thus, a combined potential model of Lennard-Jones 12-6 potential and Coulombic potential is adopted here, which can be expressed as [29,30]:

$$U_{ij} = \frac{q_i q_j}{r_{ij}} + 4\varepsilon_{ij}\left[\left(\frac{\sigma_{ij}}{r_{ij}}\right)^{12} - \left(\frac{\sigma_{ij}}{r_{ij}}\right)^6\right] \tag{2}$$

where the subscripts i and j denote ith and jth particles (atoms or ions), q is the charge of particle, r is the distance between particles, σ and ε are the minimum energy and the zero energy separation distance. The water molecules are characterized by the SPC/E model [30]. The values of the potential parameters q, σ and ε for the same particles are summarized in Table 1 [29,30]. The following mixing rules are adopted to describe the potential parameters between different particles, or:

$$\sigma_{ij} = \left(\sigma_i + \sigma_j\right)/2 \tag{3}$$

$$\varepsilon_{ij} = \sqrt{\varepsilon_i \varepsilon_j} \tag{4}$$

Table 1. Values of potential parameters.

Particle	σ (Å)	ε (KJ mol^{-1})	q (e)
O	3.169	0.6502	−0.8476
H	0.000	0.0000	+0.4238
Na$^+$	2.583	0.4184	+1.0000
Li$^+$	1.505	0.6904	+1.0000
K$^+$	3.331	0.4184	+1.0000
Cl$^-$	4.400	0.4184	−1.0000
N	3.710	0.6990	+0.0000

The truncated distances for short-range and long-range forces are taken as 10 Å and the PPPM summation technique [31] is used to modify the long-range Coulomb interaction. The equations of motion are integrated using the Velocity-Verlet algorithm [5]. The method of constraints [29] is applied for nitrogen molecules and water molecules to maintain their bond lengths and angles.

The initial configuration of the system is shown in Figure 1, where the nitrogen molecules, water molecules, and ions are distinguished with different colors. The droplet is placed in the center of a cubic box with a side length of 12 nm, and nitrogen molecules surround the droplet. The number of water molecules in the droplet is 1,120, and the number of nitrogen molecules is 600, corresponding to a 16.47 kg·m^{-3} gas density. The number of salt molecules is assumed to be 0, 40, 80, and 120, respectively, to analyze the effect of salt concentration on droplet evaporation. It is noted that the maximum salt mole concentration is 9.7%, which is less than its saturation concentration, and hence salt crystallization cannot occur. The droplet radius is fixed to 2 nm for all cases. This value corresponds to a density of 1 g·cm^{-3} as the droplet is composed of pure water.

Figure 1. Initial configuration of system: green balls are N, white balls are H, red balls are O, blue balls are positive ions, and purple balls are chloride ions.

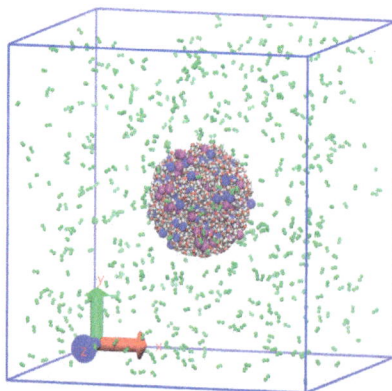

2.2. Preparation of Initial Equilibrium State

Before the onset of evaporation a well-defined system has to be prepared. Initial velocities of particles in both the gaseous phase (nitrogen) and the droplet are generated by assuming a Maxwell-Boltzmann distribution based on the initial temperature of 298 K. Periodic boundary condition is applied to the three coordinates of the box. A time step length $\delta t = 1$ fs is used for all cases. The system with the initial configuration is simulated in an NVT ensemble and it reaches an equilibrium state after 100,000 time steps. For each time step, the velocities of the gaseous phase and the droplet are separately rescaled to maintain a constant temperature of 298 K. It is noted that a very small amount of water molecules (less than 10 water molecules for the case in Figure 2a) escape from the droplet and occur in the surrounding vapor when the initial equilibrium state is reached, as shown Figure 2a.

To analyze the temporal evolution of the evaporation rate, one must define whether a water molecule belongs to the droplet or to the vapor. The method originally proposed by Shigeo *et al.* [32] is adopted here, which is based on counting the number of neighboring water molecules around each water molecule. Neighbor molecules are determined as the molecules within a distance of 4.34 Å from the molecule of interest. A molecule is considered to be in the droplet if its neighbor molecule number $n_{neighbor} \geq 9$, in the vapor phase if $n_{neighbor} \leq 1$, or in interface region if $2 \leq n_{neighbor} \leq 8$. The interface is ignored in the present work, so $n_{neighbor} \geq 4$ is used as a threshold value to determine the droplets.

Figure 2. Snapshots of the simulation boxes and corresponding number densities of water molecules, Na$^+$ and Cl$^-$ ions at different evaporation instants: (**a**) $t = 0$ ps; (**b**) $t = 500$ ps; (**c**) $t = 1,000$ ps; (**d**) $t = 1,600$ ps.

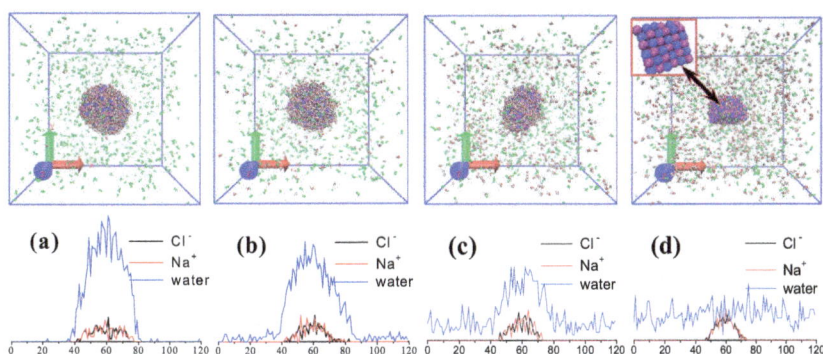

2.3. Droplet Evaporation

The temperature of the gaseous phase is abruptly increased to 600 K to trigger the droplet evaporation and this temperature is kept by the velocity-rescaling method [4] for nitrogen molecules for every time step throughout the evaporation process for all cases. Total evaporation time is set as 1,600 ps. The instantaneous positions and velocities of particles are recorded every 1 ps and all quantities of interest are calculated statistically for 10 recorded results to reduce statistical fluctuations. Figure 2 presents snapshots for the water droplets with 120 dissolved NaCl molecules and the number densities of water molecules, Na^+ and Cl^- ions at different evaporation instants. It can be seen that the droplets deviate from the initial spherical shape and their volume gradually decreases with time, however, Na^+ and Cl^- ions cannot escape from the droplet and finally crystallize as the droplet evaporates completely.

3. Results and Discussion

3.1. Effect of Salt Concentration

To analyze the effect of the salt concentration on the droplet evaporation, 0, 40, 80, and 120 LiCl molecules are added into 1,120 water molecules, respectively, to prepare aqueous droplets. The initial temperature of the droplet is 298 K, the droplet is heated by 600 K nitrogen gas and evaporates. The temporal evolution of the water molecule number in the droplet is shown in Figure 3.

Figure 3 shows that the evaporation process can be divided into three stages. At the beginning of evaporation the evaporation rates for pure water and three aqueous solutions are low, since only a small amount of heat is transferred to the droplet and water has high latent heat of evaporation, and no visible difference is observed for four droplets which means that Li^+ and Cl^- have not yet affected the evaporation. Later, the evaporation rates increase because more heat is transferred to the droplet and the difference between four droplets occurs, the water droplet evaporates faster than three aqueous droplets and vanishes about at $t = 1,150$ ps. The aqueous droplet with high LiCl concentration has a lower evaporation rate than that with low LiCl concentration. In the last stage of evaporation (at about $t > 1,200$ ps), the evaporation rates for aqueous droplets decrease compared to that in the second stage.

Figure 3. Temporal evolution of the water molecule number in the droplet with various salt concentrations.

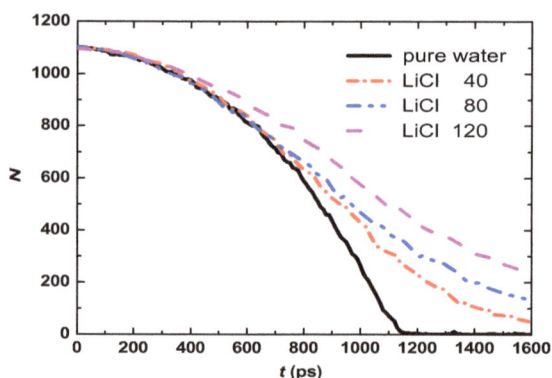

The radial distribution functions and their integrals of Li^+-O, Cl^--O for the aqueous droplet with 80 LiCl molecules at different evaporation instants are shown in Figure 4. The hydration number in the present work is defined as the average number of water molecules around an ion in the first solvation shell, and can be expressed as:

$$N_{\text{ion-O}}^{r_{\text{sol}}} = \rho_{\text{ion}} 4\pi \int_0^{r_{\text{sol}}} g_{\text{ion-O}}(r) r^2 dr \qquad (5)$$

where, ρ_{ion} is the number density of ions (Li$^+$ or Cl$^-$), $g_{\text{ion-O}}(r)$ is the radial distribution function, and r_{sol} is the radius of the first solvation shell. Figure 4 shows that the first peak values of $g_{\text{Li}^+\text{-O}}(r)$ and $g_{\text{Cl}^-\text{-O}}(r)$ occur at $r = 1.95$ Å and 3.25 Å, and the first valley values at 2.85 Å and 4.15 Å at $t < 600$ ps, which implies that the water molecules located at a distance $r < 2.85$ from Li$^+$ and 4.15 Å from Cl$^-$ are attracted strongly by the ions, thus, 2.85 Å and 4.15 Å can be regarded as the radius of the first solvation shell for Li$^+$ and Cl$^-$, respectively.

Figure 4. Radial distribution functions and hydration numbers at various evaporation instants for aqueous droplet with 80 LiCl molecules: (a) $g_{\text{Li}^+\text{-O}}(r)$ and $N_{\text{Li}^+\text{-O}}(r)$; (b) $g_{\text{Cl}^-\text{-O}}(r)$ and $N_{\text{Cl}^-\text{-O}}(r)$.

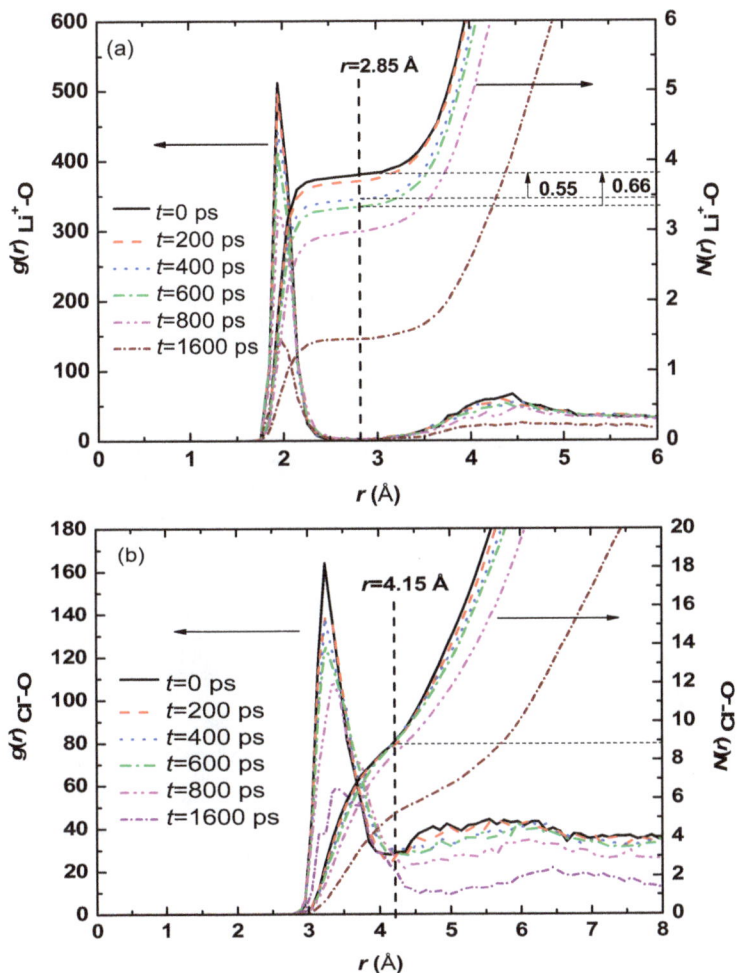

Figure 4 shows that hydration number of Li$^+$ is 3.80 at $t = 0$ ps, 3.71 at $t = 200$ ps, 3.45 at $t = 400$ ps, as well as 3.34 at $t = 600$ ps, with only 9.2% decrease from $t = 0$ ps to $t = 600$ ps; however, hydration number of Cl$^-$ is 8.70 at $t = 0$ ps, 8.68 at $t = 200$ ps, 8.67 at $t = 400$ ps, and 8.65

at $t = 600$ ps. Therefore, only 52 and four water molecules escape the confinement of Li^+ and Cl^-, respectively. At the same period, about 270 water molecules escape from the droplet due to evaporation (Figure 3). The results above demonstrates that the free water molecules with a weak interaction with ions made the biggest contribution to evaporation rate at the beginning of evaporation, and hence no visible difference is observed for pure water and aqueous solution with various LiCl concentrations.

The hydration numbers of Li^+ and Cl^- at $t = 600$ ps for aqueous droplets with 40, 80, and 120 LiCl molecules are listed in Table 2. Although high LiCl concentration leads to a small hydration number, the hydration effect is enhanced because the total number of water molecules bounded by Li^+ and Cl^- is increased. The addition of Li^+ and Cl^- into the water droplet also affects the interaction between water molecules. Table 3 lists the coordination number of water molecular at $t = 0$ ps, which is defined as the average number of water molecules in a sphere with 0.35 nm radius around a water molecule. The value of 0.35 nm chosen here is based on the fact that it is a standard length to determine the formation of hydrogen bonds between water molecules [26]. Table 3 shows that the coordination number of water molecular is reduced for high LiCl concentration, thus, the interaction between water molecules becomes less with increased LiCl concentration. The average interaction energies between water molecules and ions (Li^+ and Cl^-) for various LiCl concentrations are calculated by Equation (2) and shown in Figure 5. The negative value means that water molecules are attracted by ions. The interaction energy is stronger for high LiCl concentration at $t < 1,200$ ps. Based on results above, the low evaporation rate of the droplet with high LiCl concentration can be attributed to stronger hydration effect and stronger attractive force to water imposed by Li^+ and Cl^- as compared to that with low LiCl concentration.

Table 2. hydration number of Li^+ and Cl^- for different cases at $t = 600$ ps.

Case	Hydration number of Li^+	Hydration number of Cl^-
40 LiCl	3.76	9.02
80 LiCl	3.34	8.65
120 LiCl	3.06	8.16

Table 3. Coordination number of water molecule at $t = 0$ ps.

Case	0 LiCl	40 LiCl	80 LiCl	120 LiCl
Hydration number	4.97	4.79	4.67	4.42

Figure 5 also shows that the interaction energy is significantly elevated at $t > 1,200$ ps, because less and less water molecules are left in the droplet, hence, the evaporation becomes slower at $t > 1,200$ ps. The vapor pressure is low at the beginning stage of evaporation, and it gradually increases as water molecules escape from the droplet. High vapor pressure means larger evaporation resistance, which is another important factor for the slower evaporation rate in the last stage of evaporation than that in the second stage.

The aqueous droplet with 40 dissolved LiCl molecules has the highest evaporation rate at $t < 1,200$ ps (Figure 3), thus, less water molecules are left in the droplet compared to the droplets with 80 and 120 LiCl, so that each water molecule in the droplet is surrounded by more ions at $t > 1,200$ ps, which leads to a stronger ion-water interaction for the droplet with 40 LiCl.

Therefore, the crossover of curves of ion-water interaction for various LiCl concentrations is observed at $t = 1,200$ s.

Figure 5. Average interaction energies between water molecules and ions (Li^+ and Cl^-) for various LiCl concentrations.

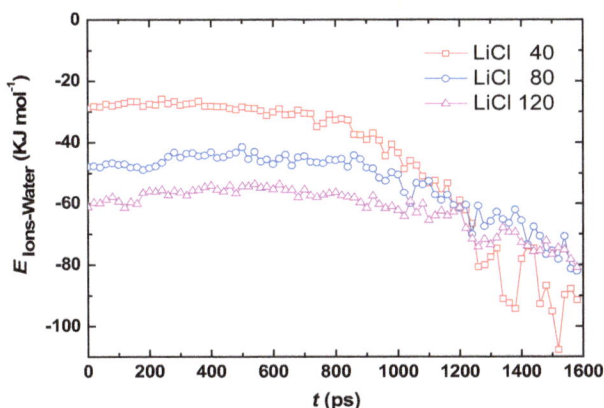

3.2. Effect of Salt Category

Due to the difference of interactions between water molecules and various ions, the evaporation rates of droplets with dissolved different salts may be different. Three common salts LiCl, NaCl and KCl are used to analyze this effect. Figure 6 shows temporal evolution of the number of water molecules in the droplets with 120 LiCl, NaCl or KCl molecules. Again, three stages are observed during evaporation for all the three aqueous droplets. The evaporation rates of aqueous droplets are lower than that of pure water droplet, and the slowest is LiCl aqueous droplet, then NaCl and KCl.

Figure 6. Temporal evolution of the water molecule number in the droplets with various salts.

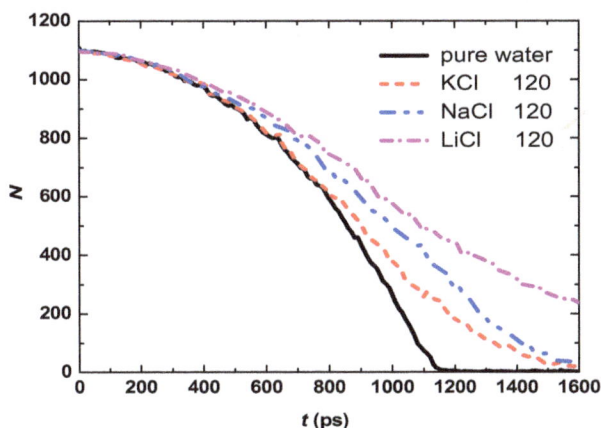

Figure 7 shows the distribution of water molecules in a sphere with 0.45 nm radius around Li^+, Na^+, and K^+ at evaporation instants of 0, 300, and 1,000 ps. Oxygen atoms are closer to cations than hydrogen atoms due to attracted Coulombic interaction. The number of water molecules around Li^+, Na^+, and K^+ differs significantly, more water molecules occur around Li^+, then Na^+ and K^+. As the droplet evaporates, the water molecules around cations are gradually reduced.

Figure 7. Snapshots of local distribution of water molecules around cations at different evaporation instants for various salts: (**a1**), (**a2**) and (**a3**) KCl at 0, 300, and 1,000 ps; (**b1**), (**b2**) and (**b3**) NaCl at 0, 300, and 1,000 ps; (**c1**), (**c2**) and (**c3**) LiCl at 0, 300, and 1,000 ps. (White balls: H, red balls: O, purple balls: Cl^- blue ball: K^+, Na^+ or Li^+).

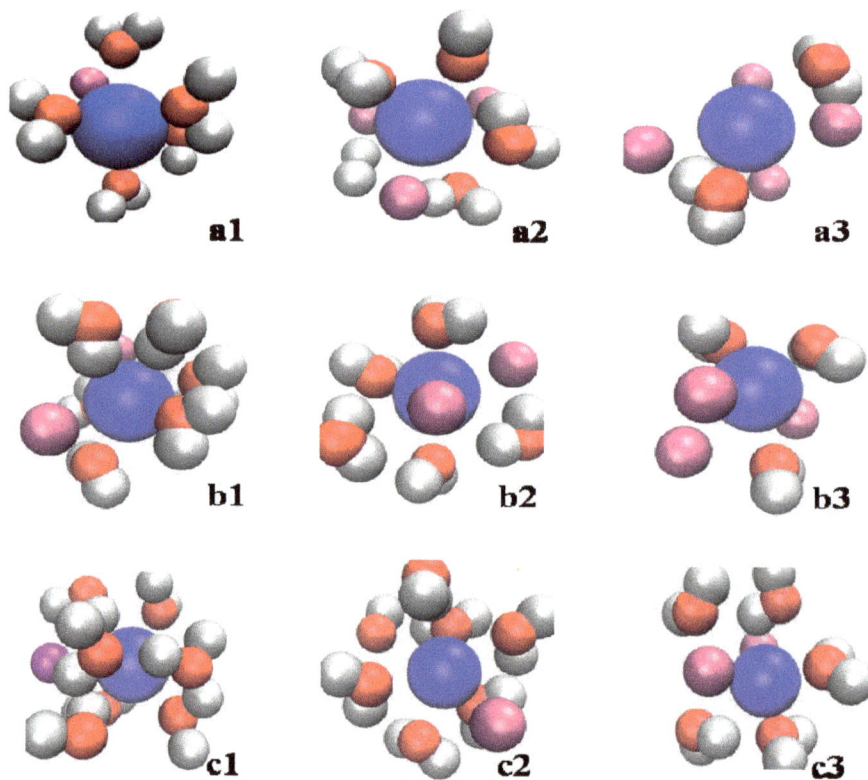

The radial distribution functions $g(r)_{Cation-Cl^-}$, $g(r)_{Cation-O}$, $g(r)_{Cl^--O}$ for LiCl, NaCl, and KCl aqueous droplets at various evaporation instants of 0, 600, 1,300, and 1,600 ps are show in Figure 8, where the subscript "Cation" denotes K^+, Na^+, or Li^+, respectively. The positions of first peak of $g(r)_{Cation-Cl^-}$ are 0.24 nm, 0.27 nm and 0.33 nm for LiCl, NaCl, and KCl aqueous droplets (Figure 8a), and the positions are almost unchanged throughout the evaporation process. However, the peak values of $g(r)_{Cation-Cl^-}$ are elevated with the time, which means that more and more cations and chloride ions aggregate together. Eventually, a crystal will form when all water molecules in the droplet evaporate completely. The peak values of $g(r)_{Cation-O}$ (Figure 8b) and $g(r)_{Cl^--O}$ (Figure 8c) for LiCl aqueous droplet are the largest throughout the evaporation process, then for NaCl and the smallest for KCl. Thus, the strongest hydration effect occurs for LiCl aqueous droplet according to Equation (4).

Comparison of Figure 8b,c indicates that the difference of $g(r)_{Cation-O}$ for three aqueous droplets is more significant than that of $g(r)_{Cl^--O}$. Therefore, only the average interaction energy between water molecules and cations (K^+, Na^+, or Li^+) is calculated by Equation (2) and is plotted in Figure 9. The attractive force between water molecules and Li^+ is the strongest, while the weakest is for K^+. The results confirm again that the strong hydration effect and attractive force are responsible for the slow evaporation. The results can also be connected to the Hoffmeister series effect [33] in term of structure breakers or structure enhancer cations. Hofmeister series is a classification of ions in order of their ability to salt out. The order of cations is usually given as: $K^+ > Na^+ > Li^+$ in Hofmeister series, therefore, the present results are in good agreement with the Hofmeister series effects.

Figure 8. The radial distribution functions at different evaporation instants: **(a)** $g(r)_{Cation-Cl^-}$; **(b)** $g(r)_{Cation-O}$; **(c)** $g(r)_{Cl^--O}$. (cation denotes K^+, Na^+, or Li^+).

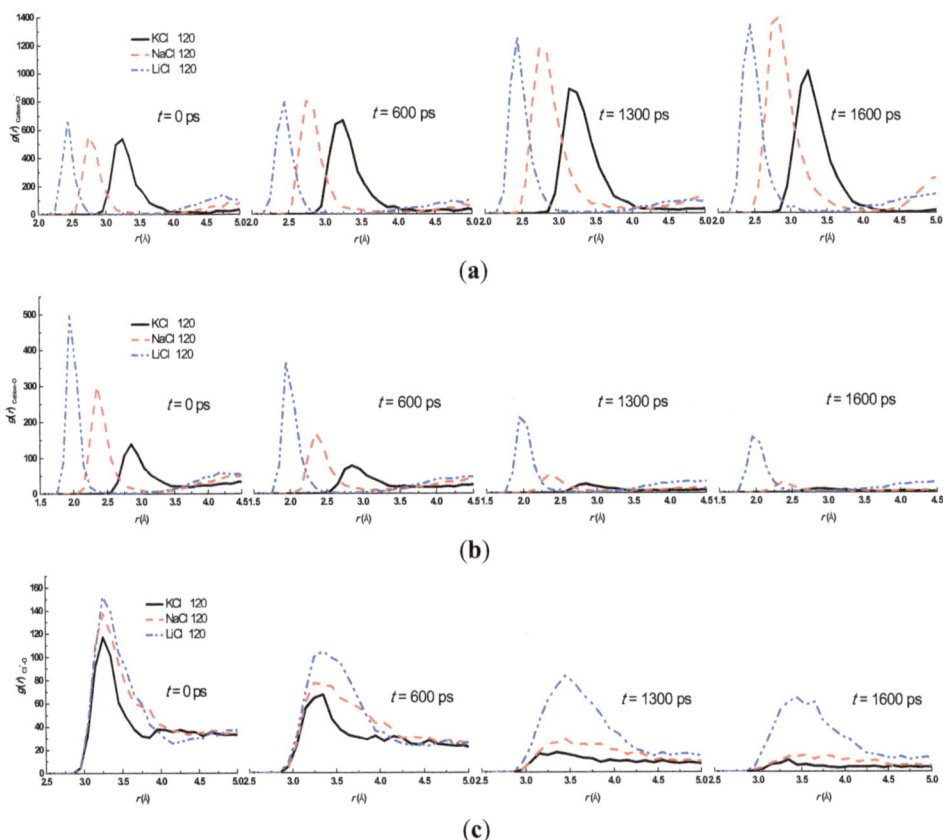

In Nature, many aqueous solutions include two or more solutes, so it is necessary to discuss the evaporation of water droplet simultaneously dissolved various kinds of salts. The evaporation of KCl + LiCl aqueous droplet is simulated and the results are shown in Figure 10, where KCl20+LiCl60 means that droplet dissolves simultaneously 20 KCl molecules and 60 LiCl molecules. With the same salt concentration, the evaporation rates of KCl20 + LiCl60 and

KCl40 + LiCl40 aqueous droplets are between the ones of KCl and LiCl aqueous droplets, and faster evaporation occurs at KCl40 + LiCl40 since K^+ has weaker hydration effect and smaller attractive force towards water molecules than Li^+.

Figure 9. Average interaction energies between water molecules and ions (Li^+ and Cl^-) for various salts.

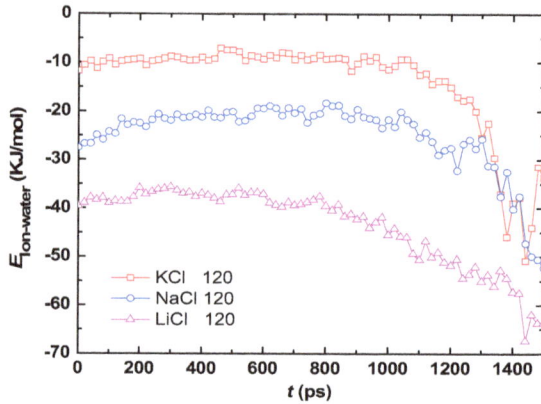

Figure 10. Temporal evolution of the water molecule number in the droplets with various salts.

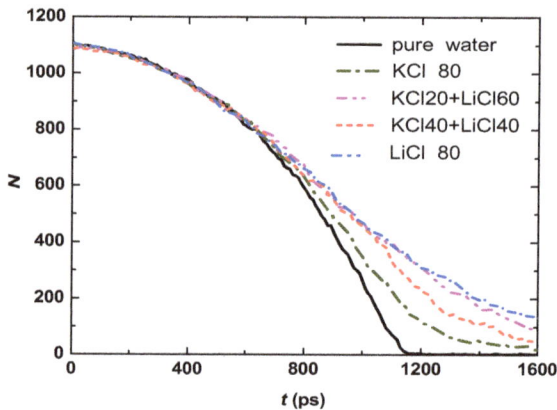

4. Conclusions

The evaporations of pure water droplets as well as NaCl, KCl and LiCl aqueous droplets are studied by molecular dynamics simulations. The droplets are placed in a gaseous nitrogen surrounding and heated by the surrounding with a constant high temperature of 600 K. The evaporation of aqueous droplet can be divided into three stages with different evaporation rates. The rate is slow at the beginning of evaporation, because only a small amount of heat is transferred to the droplet and water has high latent heat of evaporation. The rate is increased in the second stage as more and more heat is transferred to the droplet, however, the rate is again decreased at the last stage of evaporation due to the much stronger ion-water interaction.

The addition of salts into water droplet results in a slower evaporation rate compared to pure water droplets, which can be attributed to the strong hydration effect and strong attractive force on the water imposed by cations and anions. The evaporation rates for various aqueous droplets are LiCl < NaCl < KCl, and the evaporation becomes slower for high salt concentrations, due to the stronger hydration effect and attractive force occur in LiCl aqueous droplets and at high salt concentration.

The interaction potential model of particles is important for MD simulation results, however, the present study focuses on the effect of the salts concentration and category on the evaporation of the aqueous droplets, the conclusions are expected to be still applicable when a different interaction potential model is adopted. Furthermore, the previous MD studies [5,6,9] showed that because there was no "bulk" liquid for nano-scale droplets, its evaporation rate deviated from the classical D^2 law, which was regarded as a good description of evaporation of micro- and milli-scale droplets. Therefore, it is worth studying further that whether the present results may extend to micro- and milli-scale droplets.

Acknowledgments

This study was partially supported by the National Natural Science Foundation of China (No. 51076009), by the 111 Project (No. B12034), and by the National Natural Science Foundation of China (Nos. 50876049 and 51210011).

References

1. Semenov, S.; Starov, V.M.; Velarde, M.G.; Rubio, R.G. Droplets evaporation: Problems and solutions. *Eur. Phys. J.-Spec. Top.* **2011**, *197*, 265–278.
2. Starov, V.M.; Churaev, N.V. Crystal groth at the end of a capillary on solution evaporation. *J. Eng. Phys.* **1988**, *54*, 443–446.
3. Sone, Y.; Sugimoto, H. Kinetic theory analysis of steady evaporating flows from a spherical condensed phase into a vacuum. *Phys. Fluids* **1993**, *5*, 1491–1511.
4. Sumardiono, S.; Fischer, J. Molecular simulations of droplet evaporation by heat transfer. *Microfluid. Nanofluid.* **2007**, *3*, 127–140.
5. Landry, E.S.; Mikkilineni, S.; Paharia, M.; McGaughey, A.J.H. Droplet evaporation: A molecular dynamics investigation. *J. Appl. Phys.* **2007**, *102*, 124301.
6. Long, L.N.; Micci, M.M.; Wong, B.C. Molecular dynamics investigation of droplet evaporation. *Comput. Phys. Commun.* **1996**, *96*, 167–172.
7. Mason, P.E. Molecular dynamics study on the microscopic details of the evaporation of water. *J. Phys. Chem. A* **2011**, *115*, 6054–6058.
8. Consolini, L.; Aggarwal, S.K.; Murad, S. A molecular dynamics simulation of droplet evaporation. *Int. J. Heat Mass Tran.* **2003**, *46*, 3179–3188.
9. Walther, J.H.; Koumoutsakos, P. Molecular dynamics simulation of nanodroplet evaporation. *J. Heat Tran.* **2001**, *123*, 741–748.
10. Kaltz, T.L.; Long, L.N.; Micci, M.M. Supercritical vaporization of liquid oxygen droplets using molecular dynamics. *Combust. Sci. Technol.* **1998**, *136*, 279–301.

11. Bhansali, A.P.; Bayazitoglu, Y.; Maruyama, S.; Little, J.K. Molecular dynamics simulation of an evaporating sodium droplet. *Int. J. Therm. Sci.* **1999**, *38*, 66–74.

12. Wang, Z.J.; Chen, M.; Guo, Z.Y.; Yang, C. Molecular dynamics study on the liquid vapor interfacial profiles. *Fluid Phase Equilibr.* **2001**, *183*, 321–329.

13. Carey, V.P.; Wemhoff, A.P. Disjoining pressure effects in ultra-thin liquid films in micropassages-comparison of thermodynamic theory with predictions of molecular dynamics simulations. *J. Heat Transfer* **2006**, *128*, 1276–1284.

14. Yu, J.P.; Wang, H. A molecular dynamics investigation on evaporation of thin liquid films. *Int. J. Heat Mass Transfer* **2012**, *55*, 1218–1225.

15. Faeth, G.M. Current status of droplet and liquid combustion. *Prog. Energy Combust. Sci.* **1977**, *3*, 191–224.

16. Semenov, S.; Starov, V.M.; Rubio, R.G.; Agogo, H.; Velarde, M.G. Evaporation of sessile water droplets: Universal behaviour in presence of contact angle hysteresis. *Colloid. Surface. A* **2011**, *391*, 135–144.

17. Semenov, S.; Starov, V.M.; Rubio, R.G.; Velarde, M.G. Computer simulations of evaporation of pinned sessile droplets: influence of kinetic effects. *Langmuir* **2012**, *28*, 15203–15211.

18. Mucha, M.; Jungwirth, P. Salt crystallization from an evaporating aqueous solution by molecular dynamics simulations. *J. Phys. Chem. B* **2003**, *107*, 8271–8274.

19. Daub, C.D.; Cann, N.M. How are completely desolvated ions produced in electrospray ionization: Insights from molecular dynamics simulations. *Anal. Chem.* **2011**, *83*, 8372–8376.

20. Theron, S.A.; Zussman, E.; Yarin, A.L. Experimental investigation of the governing parameters in the electrospinning of polymer solution. *Polymer* **2004**, *45*, 2017–2030.

21. Köhler, H. The nucleus in and the crowth of hygroscopic droplets. *Trans. Faraday Soc.* **1936**, *32*, 1152–1161.

22. Jungwirth, P.; Tobias, D.J. Molecular stucture of salt solution: A new view of the interface with implications for heterogeneous atmospheric chemistry. *J. Phys. Chem. B* **2001**, *105*, 10468–10472.

23. Jungwirth, P.; Tobias, D.J. Ions at the air/water interface. *J. Phys. Chem. B* **2002**, *106*, 6361–6373.

24. Li, X.; Hede, T.; Tu Y.Q.; Leck, C.; Ågren, H. Surface-active cis-pinonic acid in atmopheric droplet: A molecular dynamics study. *J. Phys. Chem. Lett.* **2010**, *1*, 769–773.

25. Sun, L.; Li, X.; Hede, T.; Tu, Y.Q.; Leck, C.; Ågren, H. Molecular dynamics simulations of the surface tension and structure of salt solution and clusters. *J. Phys. Chem. B* **2012**, *116*, 3198–3204.

26. Caleman, C.; Spoel, D.V.D. Evaporation from water clusters containing singly charged ions. *Phys. Chem. Chem. Phys.* **2007**, *9*, 5105–5111.

27. Daub, C.D.; Cann, N.M. Molecular dynamics simulations to examine structure, energetics, and evaporation/condensation dynamics in small charged clusters of water or methanol containing a single monatomic ion. *J. Phys. Chem. A* **2012**, *116*, 10488–10495.

28. Znamenskiy, V.; Marginean, I.; Vertes, A. Solvated ion evaporation from charged water nanodroplets. *J. Phys. Chem. A* **2003**, *107*, 7406–7412.

29. Bouazizi, S.; Nasr, S. Local order in aqueous lithium chloride solutions as studied by X-ray scattering and molecular dynamics simulations. *J. Mol. Struct.* **2007**, *837*, 206–213.
30. Chowdhuri, S.; Chandra, A. Molecular dynamics simulations of aqueous NaCl and KCl solutions: Effects of ion concentration on the single-particle, pair, and collective dynamical properties of ions and water molecules. *J. Phys. Chem.* **2001**, *115*, 3732–3741.
31. Beckers, J.V.L.; Lowe, C.P.; Leeuw, S.W.D. An iterative PPPM method for simulating coulombic systems on distributed memory parallel computers. *Mol. Simulat.* **1998**, *3*, 369–383.
32. Shigeo, M.A.; Sohei, M.; Akihiro, O. Surface phenomena of molecular clusters by molecular dynamics method. *Therm. Sci. Eng.* **1994**, *2*, 77–84.
33. Zhang, Y.J.; Cremer, P.S. Interactions between macromolecules and ions: the Hofmeister series. *Curr. Opin. Chem. Biol.* **2006**, *10*, 658–663.

524

Reprinted from *Entropy*. Cite as: Akhter, T.; Rohlf, K. Quantifying Compressibility and Slip in Multiparticle Collision (MPC) Flow through a Local Constriction. *Entropy* **2014**, *16*, 418–442.

Article

Quantifying Compressibility and Slip in Multiparticle Collision (MPC) Flow through a Local Constriction

Tahmina Akhter [1] **and Katrin Rohlf** [2,*]

[1] Department of Applied Mathematics, University of Waterloo, Waterloo, ON N2L 3G1, Canada;
E-Mail: takhter@uwaterloo.ca
[2] Department of Mathematics, Ryerson University, 350 Victoria Street, Toronto,
ON M5B 2K3, Canada

* Author to whom correspondence should be addressed; E-Mail: krohlf@ryerson.ca;
Tel.: +1-416-979-5000, Fax: +1-416-598-5917.

Received: 27 October 2013; in revised form: 13 December 2013 / Accepted: 16 December 2013 / Published: 2 January 2014

Abstract: The flow of a compressible fluid with slip through a cylinder with an asymmetric local constriction has been considered both numerically, as well as analytically. For the numerical work, a particle-based method whose dynamics is governed by the multiparticle collision (MPC) rule has been used together with a generalized boundary condition that allows for slip at the wall. Since it is well known that an MPC system corresponds to an ideal gas and behaves like a compressible, viscous flow on average, an approximate analytical solution has been derived from the compressible Navier–Stokes equations of motion coupled to an ideal gas equation of state using the Karman–Pohlhausen method. The constriction is assumed to have a polynomial form, and the location of maximum constriction is varied throughout the constricted portion of the cylinder. Results for centerline densities and centerline velocities have been compared for various Reynolds numbers, Mach numbers, wall slip values and flow geometries.

Keywords: multiparticle collision (MPC) dynamics; constriction; slip; Karman–Pohlhausen method; compressible; ideal gas

1. Introduction

Flows through microchannels and microtubes have become recent areas of interest due to new developments in the fabrication technology of microfluidic devices. Examples of applications include micro-gas turbine generators and bio-analytical devices. In order to implement flow control measures or to optimize the design of bio-analytical devices, for example, a proper understanding of the flow through the device has to be developed. On the other hand, in gas microflows, compressibility effects can be important, and wall slip can be measurable, requiring incorporation of these in any numerical or analytical studies in this field. Particle-based methods, such as multiparticle collision dynamics (MPCD), are a means to simulate flows of a Newtonian, compressible, ideal gas, and slip effects can be incorporated very easily. Additionally, a constricted geometry is an ideal flow domain where compressibility effects can be important, for which an analytical solution is feasible. Our goal in this paper is to develop a better understanding, both theoretically and numerically, of the effects of compressibility and wall slip in a flow through a local constriction.

Flows through constrictions are popular in blood flow studies, and the analytical method used in this paper is an extension of the pioneering analysis carried out in [1–3]. The method used is called the Karman–Pohlhausen method, which essentially leads to the determination of the axial velocity profile. In [1–3], the fluid is considered to be Newtonian and incompressible, and the no-slip assumption is used, as would be common for blood flow applications. A more accurate pressure distribution was later developed for the same flow problem and presented in [4]. The same method was also used in [5], where the flow of an incompressible couple-stress fluid through a constriction was developed. In [6], a modified Karman–Pohlhausen method was proposed, and a general $(2M)$-degree polynomial was used for the flow field rather than a fourth degree polynomial, as per the original Karman–Pohlhausen method. In [7], slip was incorporated for incompressible, Newtonian flow through a local constriction. Weakly compressible flow with slip was later considered by [8,9], who also allowed for a flow geometry that is not necessarily symmetric about the location of maximum constriction. The results presented here are extensions of the results given in [8,9], giving more accurate expressions for the axial velocity profile.

Numerical works for flow through constrictions are two-fold. Discretization of the Navier–Stokes equations of motion for steady flow through stenoses was carried out by a number of authors for a Newtonian fluid [10–16]. Non-Newtonian models were considered numerically in [17–19] to name a few. All but [19] used the no-slip boundary condition. All of these works are for incompressible flows as they are applied to blood flow studies. Particle-based numerical methods, such as the Lattice-Boltzmann method [20], dissipative particle dynamics (DPD) [21,22] and multiparticle collision (MPC) dynamics [8,9,16], have more recently led to numerical solutions for flow through a local constriction. The Lattice-Boltzmann method has also recently been used for blood flow studies in complex flow geometries for realistic cardiovascular flow domains [23–25]. The method has been reviewed recently in [26], and its use for complex flows has been reviewed in [27]. Except for [8,9], the results are numerical. Since compressible flows through constrictions can exhibit significant compressibility effects and since particle-based methods have compressibility built-in, such methods are ideal numerical means for simulating compressible flow through local constrictions.

Additional particle-based methods applied to blood flow studies in microvessels, for which deformable particles are modeled separately from the fluid in which they are suspended, include simulations with MPC [28,29] and DPD [30–33]. In [32], a Y-shaped bifurcation is considered, and [29] consider a complex flow domain. The simulations in this paper differ from these references in that the MPC fluid in this paper has point particles that neither deform nor aggregate, and there is only one type of particle in the system.

In this paper, the Karman–Pohlhausen method is used to develop the axial velocity distribution for steady, Newtonian flow through a stenosed vessel, allowing for slip at the wall, as well as compressibility. The analysis is a natural extension of [1] and an improvement to the results given in [8,9]. The flow geometry considered is axisymmetric, but asymmetric about the location of maximum constriction. Effects of compressibility, slip and flow geometry are assessed. Numerical results for flow through the same geometry using multiparticle collision (MPC) dynamics are also obtained and compared to the analytical solution.

2. Multiparticle Collision Dynamics

The particle system contains N identical point particles of unit mass that are distributed uniformly over cells on a regular three-dimensional lattice. Each cell, ξ, contains n particles on average. At discrete time intervals, Δt, the continuous positions, $\mathbf{r}_i$, and velocities, $\mathbf{v}_i$ $(i = 1, \dots N)$, are updated according to the multiparticle collision (MPC) dynamics originally developed in [34]. So as to ensure Galilean invariance, a random grid shift is implemented prior to each collision step as first introduced in [35]. The idealized collisions of the MPC algorithm then update the velocity of particle i according to:

$$\mathbf{v} \rightarrow \mathbf{V}_\xi + \hat{\omega}_\xi (\mathbf{v}_i - \mathbf{V}_\xi) \tag{1}$$

where $\hat{\omega}_\xi$ is a stochastic rotation matrix that rotates the velocities by either $+\pi/2$ or $-\pi/2$ about a randomly chosen axis that varies from cell to cell and in time, and $\mathbf{V}_\xi$ is the average velocity of all particles in cell ξ in the pre-collision state [34].

Next, a constant external force accelerates the post-collision velocity of particle i in the z-direction according to:

$$v_z^i \rightarrow v_z^i + g\Delta t \tag{2}$$

where v_z^i is the z-component of the velocity of particle i and g is the acceleration value.

To simulate isothermal flow conditions, a thermostat is applied to the system, so as to remove the energy that the external force pumps into the system. The velocity of each particle is rescaled according to a profile-unbiased Galilean invariant thermostat first introduced by [36], the details of which can be found in [8,9,16].

Finally, free-streaming of the particles updates the positions according to:

$$\mathbf{r}_i \rightarrow \mathbf{r}_i + \mathbf{v}_i \Delta t \tag{3}$$

where the velocity here is the velocity after the collision, acceleration and thermostatting steps have taken place.

2.1. Boundary Conditions

Periodic boundary conditions are applied in the z-direction, and collisions with the cylinder walls follow the generalized boundary condition [8,9,16,37,38]:

$$\mathbf{v}_n \to -\mathbf{v}_n \tag{4}$$

$$\mathbf{v}_t \to (2\lambda - 1)\mathbf{v}_t \tag{5}$$

which is capable of incorporating macroscopic slip by means of changing the value of $\lambda \in [0,1]$. No-slip flow is obtained with the $\lambda = 0$ bounce-back rule, while elastic collisions ($\lambda = 1$) would result in uniform flow through the pipe. For our simulations, we use $\lambda \in [0, 0.5]$.

In order to compare the particle-based method with the analytical results, the particle-system is subjected to a cumulative averaging procedure as outlined in [16], where it was found that the averaging method is ideal for determining the macroscopic velocity profile for MPC flows.

Theoretical expressions for the viscosity coefficient of an MPC flow have been developed, and it has been shown that for our choice in $\hat{\omega}$:

$$\mu = \mu_{kin} + \mu_{coll} \tag{6}$$

where:

$$\mu_{kin} = \left(\frac{nk_BT}{m(\Delta x)^3}\right)\Delta t\left[\frac{5n}{6(n-1+e^{-n})} - \frac{1}{2}\right] \tag{7}$$

$$\mu_{coll} = \frac{m}{18\Delta x\Delta t}(n-1+e^{-n}) \tag{8}$$

and k_B is the Boltzmann constant, T the system temperature, Δx the length of a cubic cell in the lattice and n the average number of particles in a cell [34,35,39–43].

3. Theoretical Analysis

The governing equations of motion for a compressible, isothermal, viscous flow of an ideal gas are given by:

$$\frac{\partial\rho}{\partial t} + \nabla \cdot (\rho\mathbf{u}) = 0, \quad \text{(conservation of mass)} \tag{9}$$

$$\rho\frac{D}{Dt}\mathbf{u} = -\nabla P + \rho\mathbf{f} + \mu\nabla^2\mathbf{u} + (\kappa - \frac{2}{3}\mu)\nabla(\nabla \cdot \mathbf{u})$$
$$\text{(conservation of momentum)} \tag{10}$$

$$P = \frac{k_BT}{m}\rho, \quad \text{(equation of state)} \tag{11}$$

where ρ is the density, t is time, $D/Dt = \partial/\partial t + \mathbf{u} \cdot \nabla$ is the material derivative, $\mathbf{u}$ is the velocity vector, P is the pressure, $\mathbf{f}$ corresponds to an external force, μ is the viscosity, κ is the bulk viscosity, m is the mass of the fluid particle, k_B is the Boltzmann constant and T is the constant fluid temperature.

Assuming steady-state and axisymmetry, the velocity vector in cylindrical coordinates is assumed to have the form:

$$\mathbf{u} = (u_r, u_\theta, u_z) = (u(r, z), 0, w(r, z)) \tag{12}$$

together with $\rho = \rho(r, z)$. Under the Stokes assumption ($\kappa = 0$), the governing equations, with an external force in the form $\mathbf{f} = (f_r, f_\theta, f_z) = (0, 0, \rho g)$ become:

$$\frac{\partial}{\partial r}(\rho u) + \frac{\partial}{\partial z}(\rho w) + \frac{\rho u}{r} = 0, \quad \text{(mass)} \tag{13}$$

$$\rho\left(u\frac{\partial u}{\partial r} + w\frac{\partial u}{\partial z}\right) = -\frac{\partial P}{\partial r} + \mu\left(\frac{\partial^2 u}{\partial r^2} + \frac{1}{r}\frac{\partial u}{\partial r} + \frac{\partial^2 u}{\partial z^2} - \frac{u}{r^2}\right)$$
$$+ \frac{\mu}{3}\frac{\partial}{\partial r}(\nabla \cdot \mathbf{v}), \quad \text{(r-momentum)} \tag{14}$$

$$\rho\left(u\frac{\partial w}{\partial r} + w\frac{\partial w}{\partial z}\right) = \rho g - \frac{\partial P}{\partial z} + \mu\left(\frac{\partial^2 w}{\partial r^2} + \frac{1}{r}\frac{\partial w}{\partial r} + \frac{\partial^2 w}{\partial z^2}\right)$$
$$+ \frac{\mu}{3}\frac{\partial}{\partial z}(\nabla \cdot \mathbf{v}), \quad \text{(z-momentum)} \tag{15}$$

$$P(r, z) = \frac{k_B T}{m}\rho(r, z), \quad \text{(equation of state)} \tag{16}$$

where:

$$\nabla \cdot \mathbf{v} = \frac{u}{r} + \frac{\partial u}{\partial r} + \frac{\partial w}{\partial z} \tag{17}$$

and the θ-momentum equation is identically satisfied.

As per [1], for a mild stenosis geometry, the r-momentum Equation (14) can be approximated as $\frac{\partial P}{\partial r} = 0$, in which case, Equation (16) implies $\rho = \rho(z)$, which can be used in Equation (13) to give:

$$\frac{u}{r} + \frac{\partial u}{\partial r} = -\frac{1}{\rho}\frac{\partial}{\partial z}(\rho w) \tag{18}$$

Using this in the last term of Equation (15), together with the assumption that $u\frac{\partial w}{\partial r} \ll w\frac{\partial w}{\partial z}$ allows us to write the system for determining $w(r, z)$ and $P(z)$ as:

$$\rho w\frac{\partial w}{\partial z} = \rho g - \frac{dP}{dz} + \mu\left(\frac{\partial^2 w}{\partial r^2} + \frac{1}{r}\frac{\partial w}{\partial r} + \frac{4}{3}\frac{\partial^2 w}{\partial z^2}\right)$$
$$- \frac{\mu}{3}\frac{\partial}{\partial z}\left(\frac{1}{\rho}\frac{\partial}{\partial z}(\rho w)\right) \tag{19}$$

$$P(z) = \frac{k_B T}{m}\rho(z) \tag{20}$$

Following [1], we now assume that the radial dependence of the axial velocity, w, is a fourth-order polynomial in the form:

$$\frac{w}{W} = A\eta + B\eta^2 + C\eta^3 + D\eta^4 + E \tag{21}$$

where $\eta = \frac{R-r}{R}$, and $W = W(z)$ is the as yet unknown centerline velocity. Constants A to E are determined by imposing:

  (i) $w = \frac{w_s}{\sqrt{1+R'^2}}$ at $r = R$ (slip boundary condition),
  (ii) $\frac{\partial w}{\partial r} = 0$ at $r = 0$ (axisymmetric flow),
  (iii) $w = W$ at $r = 0$ (by definition of centerline velocity W),
  (iv) $\frac{\partial^2 w}{\partial r^2} = -\frac{2(W-w_s)}{R^2}$ at $r = 0$ (nearly parabolic flow with slip),
  (v) $\frac{dP}{dz} \approx \rho g + \mu \left(\frac{\partial^2 w}{\partial r^2} + \frac{1}{r}\frac{\partial w}{\partial r} \right)$ at $r = R$ (using (19)).

Condition (i) follows from solving $\mathbf{u} \cdot \mathbf{n} = 0$ (the vanishing normal component of velocity) and $\mathbf{u} \cdot \mathbf{t} = w_s$ (the tangential component of velocity is w_s) for w, while (iv) comes from the velocity profile:

$$w^{poi}(r) = (W - w_s)\left[1 - \frac{r^2}{R^2} \right] + w_s \tag{22}$$

which is Poiseuille flow in an unconstricted tube with slip, w_s, at the wall ($r = R$) and W is centerline velocity.

Imposing (i)–(v) and solving for the unknown constants gives:

$$A = \frac{1}{7}\left(-\lambda + 10 - 12E + T + 2\frac{w_s}{W} \right) \tag{23}$$

$$B = \frac{1}{7}\left(3\lambda + 5 - 6E - 3T + \frac{w_s}{W} \right) \tag{24}$$

$$C = \frac{1}{7}\left(-3\lambda - 12 + 20E + 3T - 8\frac{w_s}{W} \right) \tag{25}$$

$$D = \frac{1}{7}\left(\lambda + 4 - 9E - T + 5\frac{w_s}{W} \right) \tag{26}$$

$$E = \frac{w_s}{W\sqrt{1 + R'^2}} \tag{27}$$

where:

$$\lambda = \frac{R^2}{\mu W}\frac{dP}{dz} \tag{28}$$

and:

$$T = \frac{\rho g R^2}{\mu W} \tag{29}$$

By definition, the flow rate is given by:

$$Q = \pi \rho R^2 \overline{W} = \int_0^R 2\pi \rho w r \, dr \tag{30}$$

Substituting Equation (21) for w, using Equations (23)–(27) and solving for W in terms of $\overline{W}$, gives the relationship:

$$W = \frac{210}{97}\overline{W} + \frac{2}{97}\frac{R^2}{\mu}\frac{dP}{dz} - \frac{2}{97}\frac{R^2 \rho g}{\mu} - \frac{11}{97}w_s - \frac{102}{97}\frac{w_s}{\sqrt{1 + R'^2}} \tag{31}$$

The details pertaining to the next step involving the derivation of the equation for $\frac{dP}{dz}$ are outlined in Appendix A. The result is:

$$
\begin{aligned}
\frac{R^2}{\mu \overline{W}} \frac{dP}{dz} &\left(1 - \frac{388}{225} Ma^2 + \frac{97}{225} \frac{w_s}{\overline{W}} Ma^2 - \frac{194}{225} \frac{w_s}{\overline{W}} \frac{R'}{\sqrt{1+R'^2}} \frac{Ma^2}{Re}\right) = \frac{388}{225} R' Re + \frac{gR}{\overline{W}^2} Re \\
&-8 - \frac{8}{25} \frac{w_s}{\overline{W}} - \frac{97}{225} \frac{w_s^2}{\overline{W}^2} R' Re + \frac{w_s}{\overline{W}} \frac{1}{\sqrt{1+R'^2}} \left(\frac{208}{25} - \frac{194}{75} \frac{d}{dz}(RR')\right) \\
&+\frac{97}{75} \frac{R'}{1+R'^2} \frac{w_s^2}{\overline{W}^2} Re - \frac{388}{75} RR' \frac{w_s}{\overline{W}} \frac{d}{dz}(1+R'^2)^{-1/2}
\end{aligned}
\tag{32}
$$

where we have defined the local Reynolds and Mach numbers as:

$$
Re = \frac{\rho \overline{W} R}{\mu}
\tag{33}
$$

and:

$$
Ma = \frac{\overline{W}}{\sqrt{\frac{k_B T}{m}}}
\tag{34}
$$

respectively.

Finally, substitution of Equations (31) and (32) in Equation (21) and subsequent simplification gives the axial velocity as:

$$
\frac{w(\eta, z)}{\overline{W}} = \frac{G\eta + H\eta^2 + I\eta^3 + J\eta^4}{\left(1 - \frac{388}{225} Ma^2 + \frac{97}{225} \frac{w_s}{\overline{W}} Ma^2 - \frac{194}{225} \frac{w_s}{\overline{W}} \frac{R'}{\sqrt{1+R'^2}} \frac{Ma^2}{Re}\right)} + K
\tag{35}
$$

where G, H, I, J and K are given in Appendix B.

Substituting $\eta = 1$ and simplifying gives the centerline velocity as:

$$
\begin{aligned}
&\left[\frac{w(\eta = 1, z)}{\overline{W}} - \frac{w_s}{\overline{W}\sqrt{1+R'^2}}\right] \left(1 - \frac{388}{225} Ma^2 + \frac{97}{225} \frac{w_s}{\overline{W}} Ma^2 - \frac{194}{225} \frac{w_s}{\overline{W}} \frac{R'}{\sqrt{1+R'^2}} \frac{Ma^2}{Re}\right) \\
&= 2 + Re \frac{dR}{dz} \left[\frac{8}{225} - \frac{2}{225} \frac{w_s^2}{\overline{W}^2} + \frac{6}{225} \frac{1}{1+R'^2} \frac{w_s^2}{\overline{W}^2}\right] \\
&+ \frac{1}{75} \frac{w_s}{\overline{W}} \left[-9 - \frac{141}{\sqrt{1+R'^2}} - \frac{4}{\sqrt{1+R'^2}}(R'^2 + RR'') + 8\frac{RR'^2 R''}{(1+R'^2)^{3/2}}\right] \\
&+ Ma^2 \left[-\frac{56}{15} + \frac{2}{225} \frac{gR}{\overline{W}^2} \left(4Re - \frac{w_s}{\overline{W}} Re + 2\frac{w_s}{\overline{W}} \frac{R'}{\sqrt{1+R'^2}}\right)\right. \\
&+ \frac{1}{225} \frac{w_s}{\overline{W}} \left(254 - 11\frac{w_s}{\overline{W}} + \frac{796}{\sqrt{1+R'^2}} - \frac{199}{\sqrt{1+R'^2}} \frac{w_s}{\overline{W}}\right) \\
&\left.+ \frac{2}{225} \frac{1}{Re} \frac{w_s}{\overline{W}} \left(-210\frac{R'}{\sqrt{1+R'^2}} + 11\frac{w_s}{\overline{W}} \frac{R'}{\sqrt{1+R'^2}} + 199\frac{w_s}{\overline{W}} \frac{R'}{1+R'^2}\right)\right]
\end{aligned}
\tag{36}
$$

Note that substituting $Ma = 0$ and $\frac{dR}{dz} = 0$ leads to $\frac{W}{\overline{W}} = 2 - \frac{w_s}{\overline{W}}$, which agrees with Equation (A9) for $w = w^{poi}$, as it should, and that substitution of $Ma = 0$ and $w_s = 0$ for $\frac{dR}{dz} \neq 0$ in the above solution gives Forrester and Young's [1] result for incompressible no-slip flow.

4. Equation for Density

In order to plot the velocity profile obtained in the previous section, the explicit solution for $\rho(z)$ has to be found, since Re and Ma depend on $\rho(z)$, due to their local nature. To achieve this, the ideal gas equation of state Equation (20) can be used to replace pressure terms with density in Equation (A2), while constant flow rate can be used to replace local Re and Ma numbers with upstream values and $\rho(z)$ terms. Specifically, constant flow rate implies (see Equation (30)):

$$\overline{W} = \frac{\overline{W}_0 \rho_0 R_0^2}{\rho R^2} \tag{37}$$

where the zero subscript indicates constant upstream values in the unconstricted portion of the cylinder. Thus:

$$
\begin{aligned}
Re &= \frac{\rho \overline{W} R}{\mu} = \frac{\rho \overline{W}_0 \rho_0 R_0^2 R}{\rho R^2 \mu} \\
&= Re_0 \frac{R_0}{R}
\end{aligned} \tag{38}
$$

and:

$$
\begin{aligned}
Ma &= \frac{\overline{W}}{\sqrt{\frac{k_B T}{m}}} = \frac{\overline{W}_0 \rho_0 R_0^2}{\rho R^2 \sqrt{\frac{k_B T}{m}}} \\
&= Ma_0 \frac{\rho_0}{\rho} \left(\frac{R_0}{R} \right)^2
\end{aligned} \tag{39}
$$

Lastly, the dimensionless slip velocity can be written as:

$$\frac{w_s}{\overline{W}} = \frac{w_s}{\overline{W}_0} \left(\frac{R}{R_0} \right)^2 \frac{\rho}{\rho_0} \tag{40}$$

It follows that the pressure equation can be written in terms of $\rho(z)$ using the equation of state, Equation (20), giving:

$$
\begin{aligned}
&-\frac{R_0^2}{\mu \overline{W}_0} \left(\frac{R}{R_0} \right)^4 \frac{\rho}{\rho_0} \frac{k_B T}{m} \frac{d\rho}{dz} \left[1 - \frac{388}{225} Ma_0^2 \left(\frac{\rho_0}{\rho} \right)^2 \left(\frac{R_0}{R} \right)^4 + \frac{97}{225} \frac{w_s}{\overline{W}_0} Ma_0^2 \frac{\rho_0}{\rho} \left(\frac{R_0}{R} \right)^2 \right. \\
&\left. -\frac{194}{225} \frac{w_s}{\overline{W}_0} \frac{R'}{\sqrt{1+R'^2}} \frac{Ma_0^2}{Re_0} \frac{\rho_0}{\rho} \frac{R_0}{R} \right] = \frac{388}{225} R' Re + \frac{gR}{\overline{W}^2} Re \\
&-8 - \frac{8}{25} \frac{w_s}{\overline{W}} - \frac{97}{225} \frac{w_s^2}{\overline{W}^2} R' Re + \frac{w_s}{\overline{W}} \frac{1}{\sqrt{1+R'^2}} \left(\frac{208}{25} - \frac{194}{75} \frac{d}{dz} (RR') \right) \\
&+ \frac{97}{75} \frac{R'}{1+R'^2} \frac{w_s^2}{\overline{W}^2} Re - \frac{388}{75} RR' \frac{w_s}{\overline{W}} \frac{d}{dz} (1+R'^2)^{-1/2}
\end{aligned} \tag{41}
$$

where Re and Ma must be written in terms of Re_0, Ma_0 and ρ as given by Equations (38)–(40).

5. Flow Geometry

In order to be able to consider an asymmetric stenosis, the radius is taken to have the idealized polynomial form:

$$R(z) = \begin{cases} R_0, & \text{for } z \leq z_1 \\ az^3 + bz^2 + cz + d, & \text{for } z_1 \leq z \leq z_2 \\ ez^3 + fz^2 + gz + h, & \text{for } z_2 \leq z \leq z_3 \\ R_0 & \text{for } z \geq z_3 \end{cases} \tag{42}$$

where $z_2 = z_1 + l_1$ and $z_3 = z_2 + l_2$.

Imposing that $R(z)$ be continuously differentiable and that $R(z_2) = R_0 - \delta$ and $R'(z_2) = 0$ requires:

$$a = \frac{2\delta}{l_1^3} \tag{43}$$

$$b = -\frac{3\delta(2z_1 + l_1)}{l_1^3} \tag{44}$$

$$c = \frac{6\delta z_1(z_1 + l_1)}{l_1^3} \tag{45}$$

$$d = -\frac{2\delta z_1^3 + 3\delta z_1^2 l_1 - R_0 l_1^3}{l_1^3} \tag{46}$$

$$e = -\frac{2\delta}{l_2^3} \tag{47}$$

$$f = \frac{3\delta(2z_1 + 2l_1 + l_2)}{l_2^3} \tag{48}$$

$$g = -\frac{6\delta(z_1 + l_1)(z_1 + l_1 + l_2)}{l_2^3} \tag{49}$$

$$h = \frac{3\delta l_2(z_1^2 + l_1^2) + 6\delta z_1 l_1(z_1 + l_1 + l_2) + 2\delta(z_1^3 + l_1^3) + (R_0 - \delta)l_2^3}{l_2^3} \tag{50}$$

The resulting axisymmetric flow domain is shown in Figure 1. As can be seen from the figure, by construction, δ controls the severity of the constriction, while l_1 can be used to create the asymmetry about the z_2 location.

Figure 1. Flow geometry.

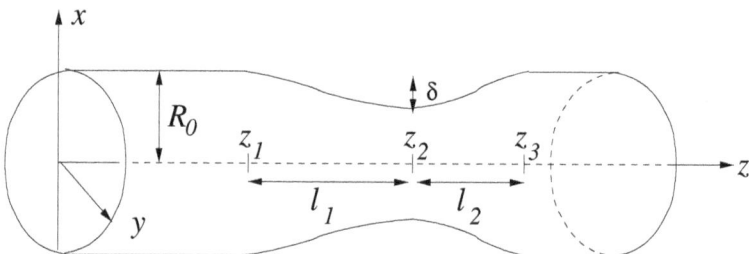

For all results that follow, $R_0 = 10.5$, $z_1 = 600.5$ and $l_1 + l_2 = 30$.

6. Numerical Results and Discussion

For all (dimensionless) MPC simulations that follow, there were approximately $N = 8.5$ million particles of unit mass $m = 1$ in the system, $\Delta x = 1 = \Delta y = \Delta z$; there were $1,200$ cells in the z direction and 25 cells in the x and y directions, respectively. The time step was taken to be $\Delta t = 1$ and $k_B T = 1$, together with $n = 20$. For the cumulative average, the averaging started after $5,000$ time steps and was performed for $35,000$ time steps thereafter. The initial system was set up with x and y velocities drawn from a Maxwellian velocity distribution, and z velocity drawn from the steady velocity profile of flow through a cylinder of fixed radius R_0.

A length of $1,200$ cells in the z-direction was chosen so as to ensure that periodic boundary conditions are valid. For this cylinder length, the velocity settled back to the expected parabolic profile in an unconstricted cylinder prior to reaching the exit for all constrictions considered here. In addition, since the velocity and density were found to be affected upstream in some simulations, starting the constriction at $z = 600.5$ ensured that there was a region upstream for which this effect was not present. Although some constrictions did not require a length of 1,200, this length was fixed for all simulations, so as to ensure that the most severe constriction with the highest Reynolds number would satisfy the periodic boundary condition.

The initial velocity distribution in the z direction was chosen, so as to reduce the simulation time. Test simulations (not reported here) were performed using a Maxwellian velocity distribution in all three directions as the initial state. The system maintained the Maxwellian velocity distribution in the x and y directions, and on average, the expected z velocity distribution that was later chosen as the initial state. In this way, the system reached equilibrium earlier, and the cumulative averaging could start after 5,000 time steps in all cases considered.

Table 1. Parameter values used in the analytical solution in Figure 2 for comparison with the particle-based method for compressible no-slip flow ($\lambda = 0$, $w_s = 0$).

g	ρ_0	ρ^{equil}	$\frac{\rho_0 - \rho^{equil}}{\rho_0}$	$\overline{W}_0$	Re_0
0.005	20.025321	20.55929025	−0.0267	0.168938215102975	4.126
0.01	20.187479	20.73938113	−0.0273	0.338610045766591	8.277
0.02	20.9408427	21.57604487	−0.0303	0.683610724092841	16.770

Table 2. Parameter values used in the analytical solution in Figure 3 for comparison with the particle-based method for compressible flow with slip ($\lambda = 0.5$).

g	ρ_0	ρ^{equil}	$\frac{\rho_0 - \rho^{equil}}{\rho_0}$	w_s	$\overline{W}_0$	Re_0
0.005	12.3602273	12.70014982	−0.0275	0.0358641	0.164985484616928	3.784
0.01	12.4565002	12.57105274	−0.0092	0.0733492	0.331580364858540	7.616
0.02	12.8636085	12.53377480	+0.0256	0.145642	0.670483591369728	15.479

Simulations were done using serial code on an Intel Xeon X5482 3.2 GHz machine with 8 GB RAM. Typical run times were 3–4 days.

To obtain the required upstream values for ρ_0, the particle-based numerical results were averaged over the centerline density values for $z \in [0, 100]$, and a best parabolic fit to the cross-section at $z = 100.5$ gave rise to the values for $\overline{W}_0$ and w_s, as provided in Tables 1 and 2. These values were then used to determine the density from numerical integration of Equation (41).

Figure 2. Comparison of analytical results with the particle-based method for variation in the Reynolds number in a constriction for which $\delta = 0.5$, $l_1 = 20$, $\lambda = 0$ (no slip) and $w_s = 0$. **(a)** Numerical and theoretically-predicted scaled centerline densities; and **(b)** corresponding numerical and analytical scaled centerline velocities. See also Table 1.

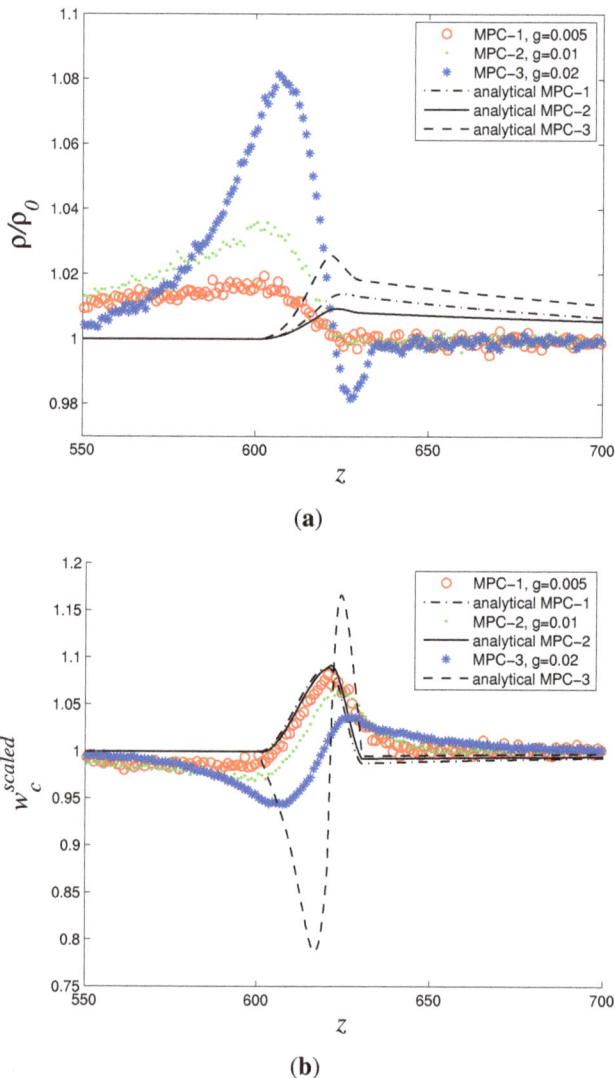

(a)

(b)

Figure 3. Comparison of analytical results with the particle-based method for variation in the Reynolds number in a constriction for which $\delta = 0.5$, $l_1 = 20$, $\lambda = 0.5$ (slip). (**a**) Numerical and approximate scaled centerline densities; and (**b**) corresponding numerical and analytical scaled centerline velocities. See also Table 2.

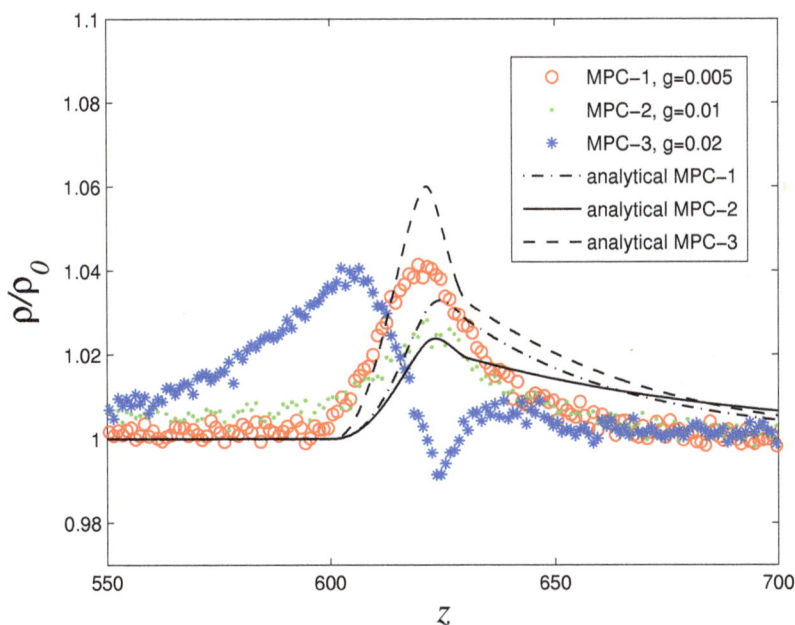

(**a**)

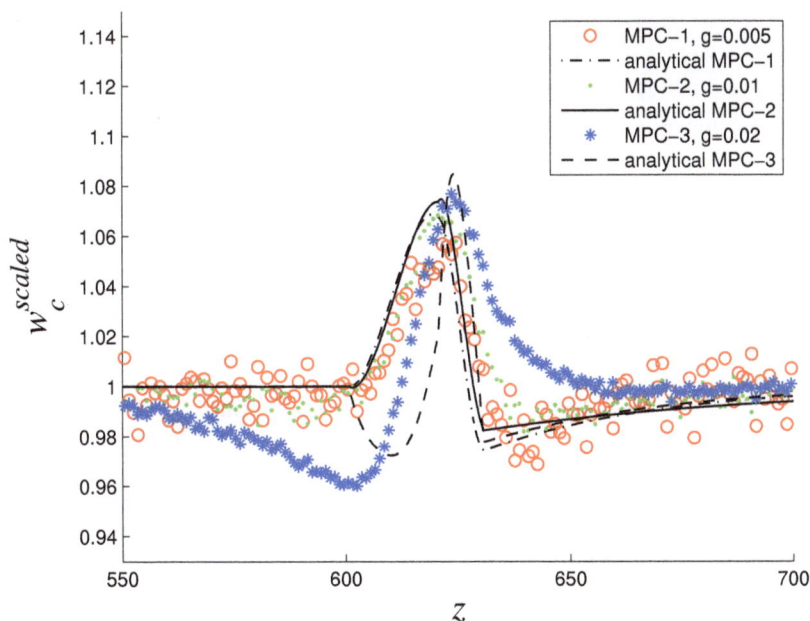

(**b**)

536

It can be seen in Figure 4 that the bounce-back rule (MPC-BB, $\lambda = 0$) correctly leads to the expected zero velocity at the wall, while slip is clearly present in the MPC-LIT($\lambda = 0.5$) case.

Figure 4. Cross-section velocity profile at various z locations far upstream of the constriction for $\lambda = 0$ (bounce-back, multiparticle collision (MPC)-BB) and for $\lambda = 0.5$ (loss-in-tangential, MPC-LIT) together with a best parabolic fit. MPC-BB correctly leads to the no-slip boundary condition, while MPC-LIT clearly has finite slip at the wall.

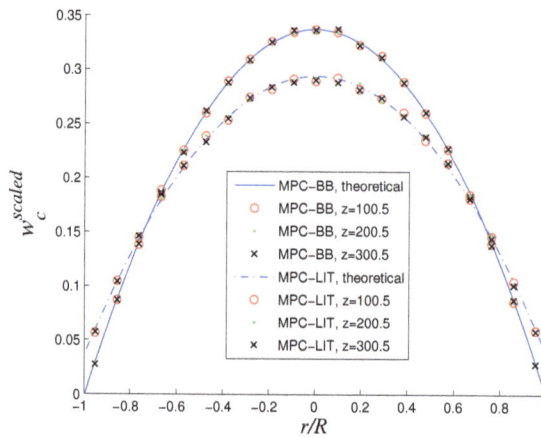

The differential equation for density was found to have a stable positive steady state, ρ^{equil}, that differed slightly from the ρ_0 determined from the MPC results. The values have been added to the Tables, as well as the relative errors from ρ_0. The density equation was solved numerically using the fourth-order Runge–Kutta scheme with $\Delta z = 0.001$ using MAPLE. Since the geometry is a piecewise defined function, the equation was solved one piece at a time, and instead of imposing ρ_0 as an initial condition at $z = 0$, ρ^{equil} was used. The differential equation was then solved on $[0, z_1]$ with the value at z_1 becoming the initial condition for the differential equation on $[z_1, z_2]$, and so on. In this way, the numerical solution was found for $z \in [0, 1200]$. Since the system has a steady state, the density settled back to the equilibrium value downstream of the constriction, thus ensuring that periodic boundary conditions are obtained in the analysis allowing comparison with the MPC results.

6.1. Compressible No-Slip Flow

In Figure 2a, a comparison of the theoretically-predicted centerline density arising from the numerical solution of Equation (41) is made with the particle-based MPC density results in the no-slip case. It can be seen that although there are some discrepancies between the predicted density curves and those obtained from the MPC simulations, both predict a density increase through the constriction, and the best agreement is found for the lowest Reynolds number considered ($g = 0.005$ curves). Worth noting in Table 1 is the increase in ρ_0 as Re_0 increases, which is consisted with the increase in ρ^{equil}.

Using the theoretically-predicted density curves in the centerline velocity expression (36) gives rise to the theoretically-predicted centerline velocity curves in Figure 2b. It can be seen that the theoretically-predicted centerline velocity agrees fairly well with the MPC result for $g = 0.005$, but as the Reynolds number increases, the agreement worsens. Worth noting is the appearance of a dip in the centerline velocity in both the theoretically-predicted and MPC results as a result of the constriction for the largest Reynolds number considered ($g = 0.02$).

6.2. Compressible Flow with Slip

For compressible flow with slip at the wall, relevant parameter values arising from the theoretical and numerical results are shown in Table 2. Theoretical scaled centerline densities and centerline velocities are compared to MPC results in Figure 3. It can be seen in (a) of the figure that there is some discrepancy between the theoretically predicted and MPC density results, but that the agreement is somewhat better than in the no-slip case. Likely due to the better agreement between the density curves, the scaled centerline velocities agree better, as well, and the dip for the largest Reynolds number ($g = 0.02$) is slightly overestimated by the theoretical predictions, contrary to the no-slip case.

Worth noting here is that, although the density curves seem to match better in the slip case, glancing at Table 2, ρ_0 is found to increase as the Reynolds number increases, while the reverse is predicted with ρ^{equil}.

Table 3. Parameter values used in the analytical solution in Figure 5 for comparison with particle-based method for compressible flow through constrictions of varying degrees.

δ	λ	ρ_0	ρ^{equil}	w_s	$\overline{W}_0$	Re_0
0.5	0	20.025321	20.55929025	0	0.168938215102975	4.126
0.5	0.5	12.3602273	12.70014982	0.0358642	0.164985484616928	3.784
1.5	0	19.9307477	20.43003857	0	0.168324883949003	4.109
1.5	0.5	12.3319318	12.67373252	0.0379518	0.165379390657969	3.792
2	0	19.8767493	20.32769045	0	0.167506930369402	4.088
2	0.5	12.3103817	12.61661932	0.0356837	0.163995726106273	3.759

6.3. Effect of the Severity of the Constriction

For the smallest Reynolds number considered ($g = 0.005$), the severity of the constriction is varied for both slip ($\lambda = 0.5$) and no-slip ($\lambda = 0$) flow. Corresponding parameter values are given in Table 3, and resulting scaled centerline velocity plots are shown in Figure 5. It can be seen that there is relatively good agreement between the theoretically-predicted curves and those from the MPC results for the mildest constriction ($\delta = 0.5$) and that there is some discrepancy as the constriction becomes more severe. The appearance of a dip in the scaled centerline velocity for the more severe constrictions is captured in the slip case, while the decrease in scaled centerline velocity

538

upstream of the constriction is found in the MPC no-slip results, but not in the theoretical predictions. On these same no-slip plots, the theoretical results predict a lower scaled centerline velocity in the post-constriction region, while MPC results do not show this feature.

Figure 5. Comparison of analytical results with the particle-based method as the severity of the constriction varies with $g = 0.005$, $l_1 = 20$. (a) Scaled centerline velocities for no-slip flow ($\lambda = 0$); (b) scaled centerline velocities for flow with slip ($\lambda = 0.5$). See also Table 3.

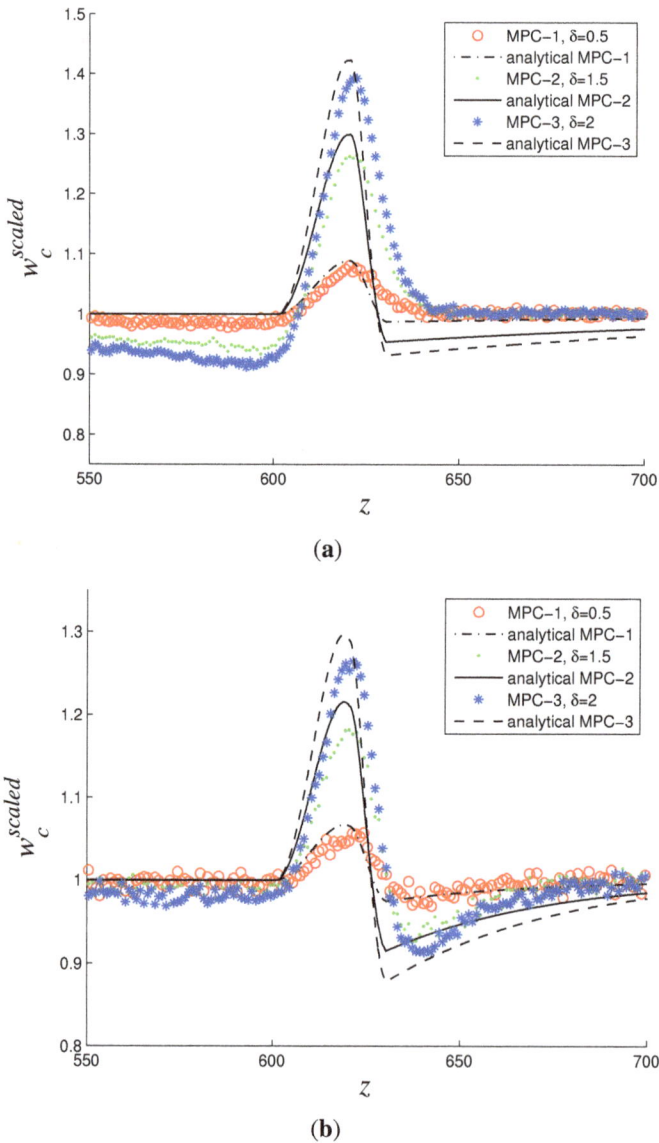

(a)

(b)

6.4. Effect of Increasing Slip

Increasing the slip parameter, λ, and, thus, the wall slip, w_s, leads to Figure 6. Parameter values used in the analytical velocity profiles are provided in Table 4. There is very good agreement between the analytical and the numerical results as the slip is varied, and the equilibrium density values from the theoretical predictions agree well with the centerline densities obtained in the MPC results.

Table 4. Parameter values used in the analytical solution in Figure 6 for comparison with the particle-based method for compressible flow through a constriction with $\delta = 0.5$, $g = 0.005$, $l_1 = 20$ and variable slip parameter values.

λ	ρ_0	ρ^{equil}	w_s	$\overline{W}_0$	Re_0
0	20.025321	20.55929025	0	0.168938215102975	4.126
0.2	14.9570845	14.90941499	0.00852200	0.155209851730363	3.663
0.4	12.6323301	12.66736550	0.0242417	0.155669931290677	3.583
0.5	12.3602273	12.70014982	0.0358642	0.164985484616928	3.784

Figure 6. Comparison of analytical and numerical scaled centerline velocities for varying values of the wall slip through a constriction with $g = 0.005$, $\delta = 0.5$ and $l_1 = 20$. See also Table 4.

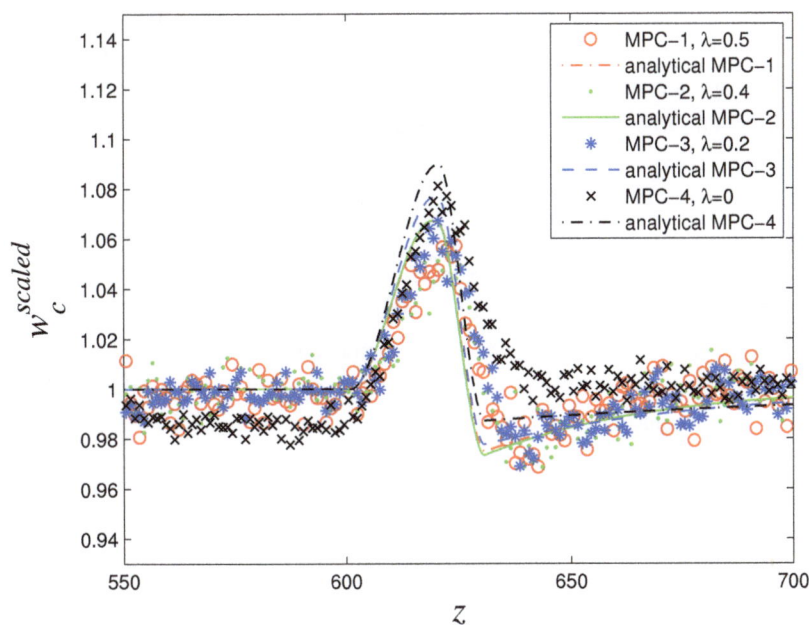

6.5. Contour Plot Comparison

Figure 7 shows contour plots for the scaled centerline velocity for both the analytical and numerical particle-based method results for a constriction with $\delta = 2$, $g = 0.005$, $l_1 = 20$ and $\lambda = 0.5$. For the analytical results, the values of the last row of Table 3 were used.

Figure 7. (Color online) Contour plots for the scaled velocity with $\delta = 2$, $\lambda = 0.5$, $l_1 = 20$ and $g = 0.005$ for (**a**) the analytical results and (**b**) the particle-based method.

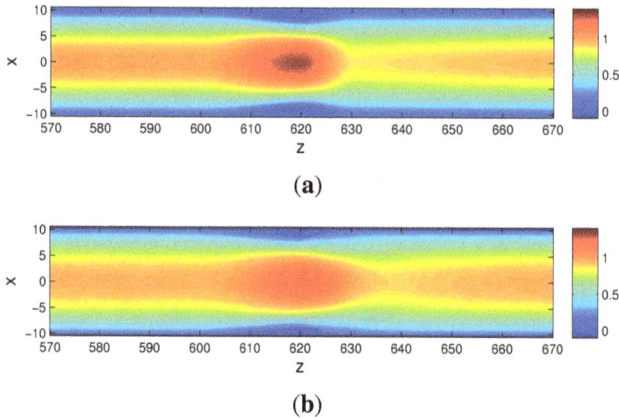

(a)

(b)

7. Discussion and Conclusions

An approximate analytical solution for the density, and for the axial velocity distribution, in an asymmetric constriction have been developed and compared to the numerical solution of a particle-based system governed by the multiparticle collision (MPC) dynamics. The solutions in all cases correspond to compressible flow with slip at the cylinder wall. Reynolds numbers varied from approximately four to 17.

Analysis of results revealed that increasing the Reynolds number in a fixed geometry leads to the appearance of a dip in the scaled centerline velocity in the entry region of the constriction, together with more pronounced flow acceleration following the location of maximum constriction. This is true with and without slip. In addition, as the Reynolds number increases, there is an increase in scaled centerline density, ρ_0, in all cases considered, except in the analytical results with slip that predict a decrease in centerline density (ρ^{equil}) instead. As the severity of the constriction increases, both slip and no-slip results show acceleration through the constriction, although the analytical and MPC results agree best for the mildest constriction $(\delta = 0.5)$ considered. Consistent with theory and MPC is the appearance of a dip in the scaled centerline velocity in the post-constriction region that is more pronounced as the severity of the constriction increases. This dip is, however, missing from the no-slip MPC results, which, instead, show a dip in the upstream section that is not captured in the theoretical predictions. Lastly, increasing slip has the effect of leading to faster flow through the constriction with the appearance of a dip in the post-constriction region that is consistent with the MPC results.

Figure 8. Comparison of W *versus* $W_{approx} = 2\overline{W} - w_s$ for both (**a**) no-slip; and (**b**) slip. The best agreement is found for $g = 0.005$.

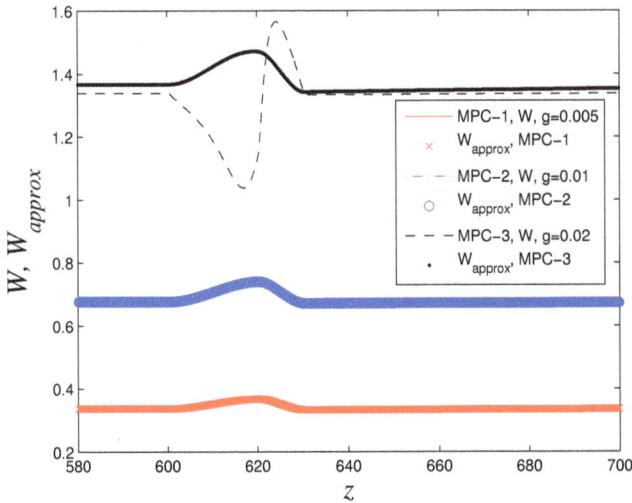

(a)

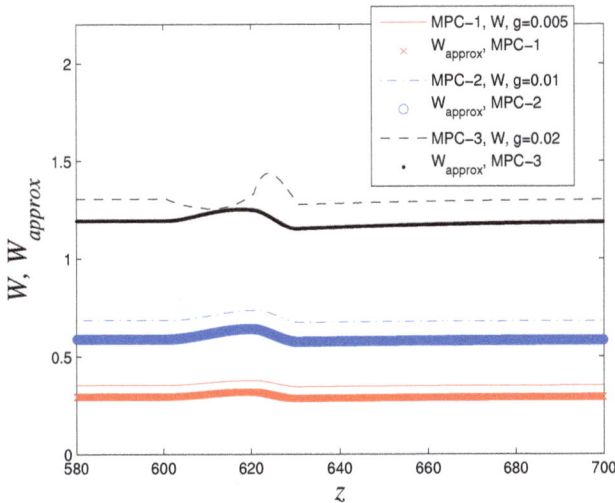

(b)

Since many key features compare well between the theoretically predicted results and those obtained by MPC, it is expected that improvements in the theory will lead to even better agreement in the constriction region and thereafter. In particular, an approximation was made for $\int_0^R rw^2 dr$, which led to some errors in the pressure equation and all equations in the subsequent analysis. In Figure 8, plots of W and $W_{approx} = 2\overline{W} - w_s$ can be found for the constrictions considered in Figures 2 and 3. It can be seen that relationship Equation (A9) is true for the smallest constriction considered and fails to hold for the higher Reynolds numbers, more so for the no-slip case in (a). This is likely a

542

key reason as to why the agreement between MPC and theory is worse for larger Reynolds numbers. Furthermore, all quadratic $(dP/dz)^2$ and second-order d^2P/dz^2 terms were dropped in the analysis, which likely led to some errors, as well. It would be interesting to explore whether or not keeping such terms in the analysis leads to significant improvements over what was found here, and this is currently under investigation. An additional source of discrepancy between the results could be the use of a thermostat in the MPC simulations that is applied uniformly, rather than locally, and whether or not using a local thermostat leads to better agreement is currently under investigation. A discussion on the use of thermostats in MPC simulations has been given in [43,44], and it would be interesting to see whether or not changing the thermostat in the simulations can lead to better agreement with the theoretical results.

In summary, an analytical solution for the flow of a compressible Newtonian fluid with slip at the wall was developed and found to compare fairly well to a numerical solution for a particle-based fluid governed by MPC in mild constrictions with low Reynolds numbers. Various Reynolds numbers, Mach numbers, wall slip values and flow geometries were considered in the analysis for asymmetric flow domains.

Acknowledgments

This work was supported by an equipment grant from the Natural Sciences and Engineering Research Council of Canada, as well as grants from the same agency for graduate student support. Additionally, T. Akhter gratefully acknowledges funding support from her Ontario Graduate Scholarship.

Conflicts of Interest

The authors declare no conflict of interest.

References

1. Forrester, J.H.; Young, D.F. Flow through a converging-diverging tube and its implications in occlusive vascular disease—I: Theoretical development. *J. Biomech.* **1970**, *3*, 297–305.
2. Forrester, J.H.; Young, D.F. Flow through a converging-diverging tube and its implications in occlusive vascular disease—II: Theoretical and experimental results and their implications. *J. Biomech.* **1970**, *3*, 307–316.
3. Morgan, B.E.; Young, D.F. Integral method for analysis of flow in arterial stenoses. *Bull. Math. Biol.* **1974**, *36*, 39–53.
4. Yao, L.; Li, D.Z. Pressure and pressure gradient in an axisymmetric rigid vessel with stenosis. *Appl. Math. Mech. Engl.* **2006**, *27*, 347–351.
5. Pralhad, R.N.; Schultz, D.H. Modeling of arterial stenosis and its applications to blood diseases. *Math. Biosci.* **2004**, *190*, 203–220.
6. Najeme, A.; Zagzoule, M.; Mauss, J. Numerical analysis of flow in arterial stenoses. *Mech. Res. Commun.* **1992**, *19*, 379–384.

7. Verma, V.K.; Singh, M.P.; Katiyar, V.K. Mathematical modeling of blood flow through stenosed tube. *J. Mech. Med. Biol.* **2008**, *8*, 27–32.

8. Akhter, T. Role of Compressibility and Slip in Blood Flow through a Local Constriction. Master Thesis, Ryerson University, Toronto, ON, Canada, 2012.

9. Akhter, T.; Rohlf, K. Weakly Compressible Flow with Slip through a Local Constriction. In Proceedings of the 24th CANCAM (Canadian Congress of Applied Mechanics), Saskatoon, Saskatchewan, Canada, 2–10 June 2013; pp. FM54–FM58.

10. Lee, J.S.; Fung, Y.C. Flow in nonuniform small blood vessels. *Microvasc. Res.* **1971**, *3*, 272–287.

11. Wille, S.O. Pressure and flow in arterial stenoses simulated in mathematical models. *Appl. Math. Model.* **1980**, *4*, 483–488.

12. O'Brien, V.; Ehrlich, L.W. I. Simple Pulsatile flow in an artery with a constriction. *J. Biomech.* **1985**, *18*, 117–127.

13. Wong, P.K.C.; Johnston, K.W.; Ethier, C.R.; Cobbold, R.S.C. Computer simulation of blood flow patterns in arteries of various geometries. *J. Vasc. Surg.* **1991**, *14*, 658–667.

14. Varghese, S.S.; Frankel, S.H.; Fischer, P.F. Direct numerical simulation of stenotic flows. Part 1. Steady flow. *J. Fluid Mech.* **2007**, *582*, 253–280.

15. Deshpande, M.D.; Giddens, D.P.; Mabon, R.F. Steady laminar flow through modelled vascular stenoses. *J. Biomech.* **1976**, *9*, 165–174.

16. Bedkihal, S.; Kumaradas, J.C.; Rohlf, K. Steady flow through a constricted cylinder by multiparticle collision dynamics. *Biomech. Model. Mechan.* **2013**, *12*, 929–939.

17. Pontrelli, G. Blood flow through an axisymmetric stenosis. *P. I. Mech. Eng. H* **2001**, *215*, 1–10.

18. Tandon, P.N.; Rana, U.V.S. A new model for blood flow through an artery with axisymmetric stenosis. *Int. J. Biomed. Comput.* **1995**, *38*, 257–267.

19. Misra, J.C.; Shit, G.C. Role of slip velocity in blood flow through stenosed arteries: A non-Newtonian model. *J. Mech. Med. Biol.* **2007**, *7*, 337–353.

20. Zhou, J.G. Axisymmetric lattice Boltzmann method. *Phys. Rev. E* **2008**, *78*, 036701:1–036701:7.

21. Darias, J.R.; Quiroga, M.; Medina, E.; Colmenares, P.J.; Paredes, V.R. Simulation of suspensions in constricted geometries by dissipative particle dynamics. *Mol. Simul.* **2003**, *29*, 443–449.

22. Feng, R.; Xenos, M.; Girdhar, G.; Kang, W.; Davenport, J.W.; Deng, Y.; Bluestein, D. Viscous flow simulation in a stenosis model using discrete particle dynamics: A comparison between DPD and CFD. *Biomech. Model. Mech.* **2012**, *11*, 119–129.

23. Melchionna, S.; Bernaschi, M.; Succi, S.; Kaxiras, E.; Rybicki, F.J.; Mitsouras, D.; Coskun, A.U.; Feldman, C.L. Hydrokinetic approach to large-scale cardiovascular blood flow. *Comput. Phys. Commun.* **2010**, *181*, 462–472.

24. Bernaschi, M.; Melchionna, S.; Succi, S.; Fyta, M.; Kaxiras, E.; Sircar, J.K. MUPHY: A parallel MUlti PHYsics/scale code for high performance bio-fluidic simulations. *Comput. Phys. Commun.* **2009**, *180*, 1495–1502.

544

25. Bernaschi, M.; Bisson, M.; Fatica, M.; Melchionna, S.; Succi, S. Petaflop hydrokinetic simulations of complex flows on massive GPU clusters. *Comput. Phys. Commun.* **2013**, *184*, 329–341.

26. Benzi, R.; Succi, S.; Vergassola, M. The lattice Boltzmann equation: Theory and applications. *Phys. Rep.* **1992**, *222*, 145–197.

27. Aidun, C.K.; Clausen, J.R. Lattice-Boltzmann method for complex flows. *Ann. Rev. Fluid Mech.* **2010**, *42*, 439–472.

28. Noguchi, H.; Gompper, G. Shape transitions of fluid vesicles and red blood cells in capillary flows. *Proc. Natl. Acad. Sci. USA* **2005**, *102*, 14159–14164.

29. Noguchi, H.; Gompper, G.; Schmid, L.; Wixforth, A.; Franke, T. Dynamics of fluid vesicles in flow through structured microchannels. *EPL* **2010**, *89*, 28002:1–28002:6.

30. Steiner, T.; Cupelli, C.; Zengerle, R.; Santer, M. Simulation of advanced microfluidic systems with dissipative particle dynamics. *Microfluid. Nanofluid.* **2009**, *7*, 307–323.

31. McWhirter, J.L.; Noguchi, H.; Gompper, G. Flow-induced clustering and alignment of vesicles and red blood cells in microcapillaries. *Proc. Natl. Acad. Sci. USA* **2009**, *106*, 6039–6043.

32. Li, X.; Popel, A.S.; Karniadakis, G.E. Blood-plasma separation in Y-shaped bifurcating microfluidic channels: A dissipative particle dynamics simulation study. *Phys. Biol.* **2012**, *9*, 026010:1–026010:12.

33. Lei, H.; Fedosov, D.A.; Caswell, B.; Karniadakis, G.E. Blood flow in small tubes: Quantifying the transition to the non-continuum regime. *J. Fluid Mech.* **2013**, *722*, 214–239.

34. Malevanets, A.; Kapral, R. Mesoscopic model for solvent dynamics. *J. Chem. Phys.* **1999**, *110*, 8605–8613.

35. Ihle, T.; Kroll, D.M. Stochastic rotation dynamics: A Galilean-invariant mesoscopic model for fluid flow. *Phys. Rev. E* **2001**, *63*, 020201:1–020201:4.

36. Padding, J.T.; Louis, A.A. Hydrodynamic interactions and Brownian forces in colloidal suspensions: Coarse-graining over time and length scales. *Phys. Rev. E* **2006**, *74*, 031402:1–031402:29.

37. Chikkadi, V.; Alam, M. Slip velocity and stresses in granular Poiseuille flow via event-driven simulation. *Phys. Rev. E* **2009**, *80*, 021303:1–021303:16.

38. Whitmer, J.K.; Luijten, E. Fluid-solid boundary conditions for multiparticle collision dynamics. *J. Phys.: Condens. Matter* **2010**, *22*, 104106:1–104106:14.

39. Kapral, R. Multiparticle collision dynamics: Simulation of complex systems on mesoscales. *Adv. Chem. Phys.* **2008**, *140*, 89–146.

40. Noguchi, H.; Gompper, G. Transport coefficients of off-lattice mesoscale-hydrodynamics simulation techniques. *Phys. Rev. E* **2008**, *78*, 016706:1–016706:12.

41. Kikuchi, N.; Pooley, C.M.; Ryder, J.F.; Yeomans, J.M. Transport coefficients of a mesoscopic fluid dynamics model. *J. Chem. Phys.* **2003**, *119*, 6388–6395.

42. Ihle, T.; Tüzel, E.; Kroll, D.M. Resummed Green-Kubo relations for a fluctuating fluid-particle model. *Phys. Rev. E* **2004**, *70*, 035701:1–035701:4.

43. Gompper, G.; Ihle, T.; Kroll, D.M.; Winkler, R.G. Multi-particle collision dynamics: A particle-based mesoscale simulation approach to the hydrodynamics of complex fluids. *Adv. Polym. Sci.* **2009**, *221*, 1–87.
44. Huang, C.C.; Chatterji, A.; Sutmann, G.; Gompper, G.; Winkler, R.G. Cell-level canonical sampling by velocity scaling for multiparticle collision dynamics simulations. *J. Comput. Phys.* **2010**, *229*, 168–177.

Appendix A

In this Appendix, the details of obtaining pressure Equation (A19) are shown.

To obtain an expression for $\frac{dP}{dz}$, we first integrate Equation (19) across the cylinder to get:

$$\frac{1}{2}\int_0^R \rho r \frac{\partial}{\partial z}(w^2)dr \;=\; \rho g \frac{R^2}{2} - \frac{dP}{dz}\frac{R^2}{2} + \mu R \left(\frac{\partial w}{\partial r}\right)\bigg|_{r=R} + \frac{4}{3}\mu \int_0^R \frac{\partial^2 w}{\partial z^2}dr \tag{A1}$$

$$-\frac{\mu}{3}\int_0^R \frac{\partial}{\partial z}\left(\frac{1}{\rho}\frac{\partial}{\partial z}(\rho w)\right) \tag{A2}$$

Next, we divide by $\rho = \rho(z)$ and take all z-derivatives outside of the integral using:

$$\frac{d}{dz}\int_0^R h(r,z)dr \;=\; h(R,z)\frac{dR}{dz} + \int_0^R \frac{d}{dz}h(r,z)dr \tag{A3}$$

to get:

$$\frac{1}{2}\frac{d}{dz}\int_0^R rw^2 dr - \frac{1}{2}\frac{RR'w_s^2}{1+R'^2} = g\frac{R^2}{2} - \frac{1}{\rho}\frac{dP}{dz}\frac{R^2}{2} + \nu R\left(\frac{\partial w}{\partial r}\right)\bigg|_{r=R}$$

$$+\frac{4}{3}\nu\left[\frac{d^2}{dz^2}\int_0^R rw\,dr - \frac{w_s}{\sqrt{1+R'^2}}\frac{d}{dz}(RR') - 2RR'w_s\frac{d}{dz}(1+R'^2)^{-1/2}\right]$$

$$+\frac{\nu}{3\rho^2}\frac{d\rho}{dz}\frac{d}{dz}\int_0^R r\rho w\,dr - \frac{\nu}{3\rho}\frac{d\rho}{dz}\frac{RR'w_s}{\sqrt{1+R'^2}} - \frac{\nu}{3\rho}\frac{d^2}{dz^2}\int_0^R r\rho w\,dr \tag{A4}$$

$$+\frac{\nu}{3}\frac{w_s}{\sqrt{1+R'^2}}\frac{d}{dz}(RR') + \frac{2}{3}\nu RR'w_s\frac{d}{dz}(1+R'^2)^{-1/2} + \frac{2\nu}{3\rho}\frac{d\rho}{dz}\frac{RR'w_s}{\sqrt{1+R'^2}}$$

where $\nu = \frac{\mu}{\rho}$. Taking $w \approx w^{poi}$ in the integral on the left-hand side gives,

$$\frac{1}{2}\frac{d}{dz}\int_0^R rw^2\,dr \;\approx\; \frac{1}{2}\frac{d}{dz}\int_0^R r(w^{poi})^2\,dr \tag{A5}$$

$$=\; \frac{1}{2}\frac{d}{dz}\left[\frac{R^2}{6}(W^2 + Ww_s + w_s^2)\right] \tag{A6}$$

$$= \frac{d}{dz}\left[\frac{1}{3}R^2\overline{W}^2 - \frac{1}{6}R^2\overline{W}w_s + \frac{1}{12}R^2w_s^2\right] \tag{A7}$$

$$= \frac{d}{dz}\left[\frac{1}{3}\frac{Q^2}{\pi^2\rho^2R^2} - \frac{1}{6}\frac{Qw_s}{\pi\rho} + \frac{1}{12}R^2w_s^2\right] \tag{A8}$$

where:

$$\overline{W} = \frac{1}{2}(W + w_s) \tag{A9}$$

has been used in Equation (A6) to replace W in terms of $\overline{W}$, and Equation (30) has been used in Equation (A7) to replace $\overline{W}$ in terms of Q. The relationship in Equation (A9) follows from using w^{poi} as given in Equation (22), in flow rate Equation (30). Although this relationship is exact for $w = w^{poi}$ in an unconstricted portion of the cylinder, it is also assumed to hold throughout the constriction, thus potentially giving rise to some error in the analysis.

Now:

$$\frac{dQ}{dz} = \frac{d}{dz}\int_0^R 2\pi\rho w r\, dr \quad \text{(from (30))} \tag{A10}$$

$$= 2\pi\rho R\, w|_{r=R}\frac{dR}{dz} + 2\pi\int_0^R r\frac{d(\rho w)}{dz}\, dr \tag{A11}$$

$$= 2\pi\rho R\, w|_{r=R}\frac{dR}{dz} - 2\pi\int_0^R r\rho\left(\frac{u}{r} + \frac{\partial u}{\partial r}\right)\, dr \quad \text{(using (13))} \tag{A12}$$

$$= 2\pi\rho R\, w|_{r=R}\frac{dR}{dz} - 2\pi\rho\int_0^R \frac{\partial}{\partial r}(ur)\, dr \tag{A13}$$

$$= 2\pi\rho R\, w|_{r=R}\frac{dR}{dz} - 2\pi\rho R\, u|_{r=R} \tag{A14}$$

$$= 2\pi\rho R\frac{w_s}{\sqrt{1+R'^2}}R' - 2\pi\rho R\frac{w_s R'}{\sqrt{1+R'^2}} \tag{A15}$$

$$= 0. \tag{A16}$$

Thus, Equation (A8) gives:

$$\frac{1}{2}\frac{d}{dz}\int_0^R rw^2\, dr \approx \frac{2Q}{3\pi^2\rho^2R^2}\left(\frac{dQ}{dz} - \frac{Q}{\rho}\frac{d\rho}{dz} - \frac{Q}{R}\frac{dR}{dz}\right)$$

$$- \frac{w_s}{6\pi\rho}\left(\frac{dQ}{dz} - \frac{Q}{\rho}\frac{d\rho}{dz}\right) + \frac{1}{6}w_s^2 R\frac{dR}{dz} \tag{A17}$$

$$= -\frac{4}{3}R\overline{W}\, u|_{r=R} - \frac{2}{3}R^2\overline{W}^2\frac{m}{\rho k_B T}\frac{dP}{dz} - \frac{2}{3}\overline{W}^2 R\frac{dR}{dz}$$

$$+ \frac{1}{3}Rw_s\, u|_{r=R} + \frac{1}{6}R^2\overline{W}w_s\frac{m}{\rho k_B T}\frac{dP}{dz}$$

$$+ \frac{1}{6}R\frac{dR}{dz}w_s^2 \tag{A18}$$

where we have also used equation of state Equation (16) to write $\frac{d\rho}{dz}$ in terms of $\frac{dP}{dz}$ and flow rate Equation (30) to write Q in terms of $\overline{W}$.

Substituting Equation (A18) in Equation (A4), writing all integrals in terms of Q and differentiating, noting that $dQ/dz = 0$, using $\left(\frac{\partial w}{\partial r}\right)\big|_{r=R} = -\frac{AW}{R}$ from Equation (21) together with Equations (23), (28), (29) and (31), and $u|_{r=R} = \frac{w_s dR/dz}{\sqrt{1+R'^2}}$, gives:

$$
\frac{R^2}{\mu \overline{W}} \frac{dP}{dz} \left(1 - \frac{388}{225}Ma^2 + \frac{97}{225}\frac{w_s}{\overline{W}}Ma^2 - \frac{194}{225}\frac{w_s}{\overline{W}}\frac{R'}{\sqrt{1+R'^2}}\frac{Ma^2}{Re}\right) = \frac{388}{225}R'Re + \frac{gR}{\overline{W}^2}Re
$$

$$
-8 - \frac{8}{25}\frac{w_s}{\overline{W}} - \frac{97}{225}\frac{w_s^2}{\overline{W}^2}R'Re + \frac{w_s}{\overline{W}}\frac{1}{\sqrt{1+R'^2}}\left(\frac{208}{25} - \frac{194}{75}\frac{d}{dz}(RR')\right) \tag{A19}
$$

$$
+\frac{97}{75}\frac{R'}{1+R'^2}\frac{w_s^2}{\overline{W}^2}Re - \frac{388}{75}RR'\frac{w_s}{\overline{W}}\frac{d}{dz}(1+R'^2)^{-1/2}
$$

where we have defined the local Reynolds and Mach numbers as:

$$
Re = \frac{\rho \overline{W} R}{\mu} \tag{A20}
$$

and

$$
Ma = \frac{\overline{W}}{\sqrt{\frac{k_B T}{m}}} \tag{A21}
$$

respectively.

This is the pressure equation provided in Equation (32).

Appendix B In this Appendix, we provide the coefficients of η in axial velocity Equation (35):

$$
G = 4 + Re\frac{dR}{dz}\left[-\frac{44}{225} + \frac{11}{225}\frac{w_s^2}{\overline{W}^2} - \frac{33}{225}\frac{1}{1+R'^2}\frac{w_s^2}{\overline{W}^2}\right]
$$

$$
+\frac{2}{75}\frac{w_s}{\overline{W}}\left[6 - \frac{156}{\sqrt{1+R'^2}} + \frac{11}{\sqrt{1+R'^2}}(R'^2 + RR'') - 22\frac{RR'^2 R''}{(1+R'^2)^{3/2}}\right]
$$

$$
+Ma^2\left[-\frac{16}{3} + \frac{11}{225}\frac{gR}{\overline{W}^2}\left(-4Re + \frac{w_s}{\overline{W}}Re - 2\frac{w_s}{\overline{W}}\frac{R'}{\sqrt{1+R'^2}}\right)\right. \tag{A22}
$$

$$
+\frac{4}{75}\frac{w_s}{\overline{W}}\left(21 + \frac{w_s}{\overline{W}} + \frac{104}{\sqrt{1+R'^2}} - \frac{26}{\sqrt{1+R'^2}}\frac{w_s}{\overline{W}}\right)
$$

$$
\left.+\frac{8}{75}\frac{1}{Re}\frac{w_s}{\overline{W}}\left(-25\frac{R'}{\sqrt{1+R'^2}} - \frac{w_s}{\overline{W}}\frac{R'}{\sqrt{1+R'^2}} + 26\frac{w_s}{\overline{W}}\frac{R'}{1+R'^2}\right)\right]
$$

$$
H = -2 + \frac{43}{225}Re\frac{dR}{dz}\left[4 - \frac{w_s^2}{\overline{W}^2} + \frac{3}{1+R'^2}\frac{w_s^2}{\overline{W}^2}\right]
$$

$$
+\frac{2}{75}\frac{w_s}{\overline{W}}\left[-3 + \frac{78}{\sqrt{1+R'^2}} - \frac{43}{\sqrt{1+R'^2}}(R'^2 + RR'') + 86\frac{RR'^2 R''}{(1+R'^2)^{3/2}}\right]
$$

$$
+Ma^2\left[-\frac{8}{3} + \frac{43}{225}\frac{gR}{\overline{W}^2}\left(4Re - \frac{w_s}{\overline{W}}Re + 2\frac{w_s}{\overline{W}}\frac{R'}{\sqrt{1+R'^2}}\right)\right. \tag{A23}
$$

$$
\left.+\frac{2}{75}\frac{w_s}{\overline{W}}\left(21 + \frac{w_s}{\overline{W}} + \frac{104}{\sqrt{1+R'^2}} - \frac{26}{\sqrt{1+R'^2}}\frac{w_s}{\overline{W}}\right)\right]
$$

$$+\frac{4}{75}\frac{1}{Re}\frac{w_s}{\overline{W}}\left(-25\frac{R'}{\sqrt{1+R'^2}}-\frac{w_s}{\overline{W}}\frac{R'}{\sqrt{1+R'^2}}+26\frac{w_s}{\overline{W}}\frac{R'}{1+R'^2}\right)\Bigg]$$

(A24)

$$
\begin{aligned}
I \;=\;& Re\frac{dR}{dz}\left[-\frac{4}{5}+\frac{1}{5}\frac{w_s^2}{\overline{W}^2}-\frac{3}{5}\frac{1}{1+R'^2}\frac{w_s^2}{\overline{W}^2}\right]\\
&+\frac{2}{5}\frac{w_s}{\overline{W}}\left[-2+\frac{2}{\sqrt{1+R'^2}}+\frac{3}{\sqrt{1+R'^2}}(R'^2+RR'')-6\frac{RR'^2R''}{(1+R'^2)^{3/2}}\right]\\
&+Ma^2\left[\frac{32}{5}+\frac{1}{5}\frac{gR}{\overline{W}^2}\left(-4Re+\frac{w_s}{\overline{W}}Re-2\frac{w_s}{\overline{W}}\frac{R'}{\sqrt{1+R'^2}}\right)\right.\\
&\quad+\frac{4}{225}\frac{w_s}{\overline{W}}\left(2-23\frac{w_s}{\overline{W}}-\frac{452}{\sqrt{1+R'^2}}+\frac{113}{\sqrt{1+R'^2}}\frac{w_s}{\overline{W}}\right)\\
&\quad\left.+\frac{8}{225}\frac{1}{Re}\frac{w_s}{\overline{W}}\left(90\frac{R'}{\sqrt{1+R'^2}}+23\frac{w_s}{\overline{W}}\frac{R'}{\sqrt{1+R'^2}}-113\frac{w_s}{\overline{W}}\frac{R'}{1+R'^2}\right)\right]
\end{aligned}
$$

(A25)

(A26)

$$
\begin{aligned}
J \;=\;& Re\frac{dR}{dz}\left[\frac{4}{15}-\frac{1}{15}\frac{w_s^2}{\overline{W}^2}+\frac{3}{15}\frac{1}{1+R'^2}\frac{w_s^2}{\overline{W}^2}\right]\\
&+\frac{1}{5}\frac{w_s}{\overline{W}}\left[3-\frac{3}{\sqrt{1+R'^2}}+\frac{2}{\sqrt{1+R'^2}}(R'^2+RR'')-4\frac{RR'^2R''}{(1+R'^2)^{3/2}}\right]\\
&+Ma^2\left[-\frac{32}{15}+\frac{1}{15}\frac{gR}{\overline{W}^2}\left(4Re-\frac{w_s}{\overline{W}}Re+2\frac{w_s}{\overline{W}}\frac{R'}{\sqrt{1+R'^2}}\right)\right.\\
&\quad+\frac{1}{75}\frac{w_s}{\overline{W}}\left(-44+21\frac{w_s}{\overline{W}}+\frac{244}{\sqrt{1+R'^2}}-\frac{61}{\sqrt{1+R'^2}}\frac{w_s}{\overline{W}}\right)\\
&\quad\left.+\frac{2}{75}\frac{1}{Re}\frac{w_s}{\overline{W}}\left(-40\frac{R'}{\sqrt{1+R'^2}}-21\frac{w_s}{\overline{W}}\frac{R'}{\sqrt{1+R'^2}}+61\frac{w_s}{\overline{W}}\frac{R'}{1+R'^2}\right)\right]
\end{aligned}
$$

(A27)

$$K \;=\; \frac{w_s}{\overline{W}\sqrt{1+R'^2}}$$

(A28)

Reprinted from *Entropy*. Cite as: Herman, A. Shear-Jamming in Two-Dimensional Granular Materials with Power-Law Grain-Size Distributio. *Entropy* **2013**, *15*, 4802–4821.

Article

Shear-Jamming in Two-Dimensional Granular Materials with Power-Law Grain-Size Distribution

Agnieszka Herman

Institute of Oceanography, University of Gdansk, Pilsudskiego 46, Gdynia 81-378, Poland;
E-Mail: oceagah@ug.edu.pl; Tel.: +48-58-5236887; Fax: +48-58-5236678

Received: 13 August 2013; in revised form: 28 October 2013 / Accepted: 31 October 2013 / Published: 5 November 2013

Abstract: Although substantial progress has been made in recent years in research on sheared granular matter, relatively few studies concentrate on the behavior of materials with very strong polydispersity. In this paper, shear deformation of a two-dimensional granular material composed of frictional disk-shaped grains with power-law size distribution is analyzed numerically with a finite-difference model. The analysis of the results concentrates on those aspects of the behavior of the modeled system that are related to its polydispersity. It is demonstrated that many important global material properties are dependent on the behavior of the largest grains from the tail of the size distribution. In particular, they are responsible for global correlation of velocity anomalies emerging at the jamming transition. They also build a skeleton of the global contact and force networks in shear-jammed systems, leading to the very open, "sparse" structure of those networks, consisting of only ∼35% of all grains. The details of the model are formulated so that it represents fragmented sea ice moving on a two-dimensional sea surface; however, the results are relevant for other types of strongly polydisperse granular materials, as well.

Keywords: granular materials; finite-element simulation; shear deformation; jamming phase transition; polydispersity; force networks

1. Introduction

Granular materials are an example of systems in which relatively simple interactions between similar discrete objects (grains, or particles) produce very complex emergent behavior. Extensive experimental and numerical research on granular materials in recent years produced many important insights into the dynamics of those systems. One group of studies has concentrated on the jamming phase transition, revealing new details of the (relatively well understood) isotropic jamming (e.g., [1,2]), as well as the existence of previously unexplored jammed states in systems subject to shear strain [3–10]. However, the behavior of very strongly polydisperse materials in those settings remains very poorly understood. Most works, including those cited above, concentrate on materials with narrow grain-size distributions (GSD). How polydispersity influences the system dynamics close to and at the jamming phase transition remains an open question.

An example of a granular material with a very wide GSD is sea ice, especially close to the ice edge (the so-called marginal ice zone) or, more generally, in regions where, due to the action of wind, ocean surface waves and currents, the ice cover is fragmented into separate floes. A typical example of this ice cover type is shown in Figure 1. Because the vertical dimension (thickness) of the floes is much smaller than their horizontal dimension (diameter), sea ice can be regarded as two-dimensional (2D). The shape of the ice floes may vary from very irregular through polygonal to nearly circular, depending on the external forcing (especially waves) and the ice age and thickness. However, in most situations, the geometrical properties of the floes, like, e.g., the aspect ratio, remain within a relatively narrow range independently of the area and conditions of observation [11–13]. More importantly, the observed floe-size distributions (FSDs) are very wide and have power-law tails with an exponent $\alpha < 2$ [11–16]. Although it is generally acknowledged that the granular nature of fragmented sea ice influences its dynamics (see, e.g., [17]), most large-scale sea ice models treat ice as a viscous-plastic continuum; our knowledge of how and when the processes taking place at a floe level influence the large-scale behavior of sea ice is very limited.

Figure 1. Fragment of a satellite image of fragmented sea ice in the marginal ice zone off the Antarctic Peninsula (source: Landsat [18]).

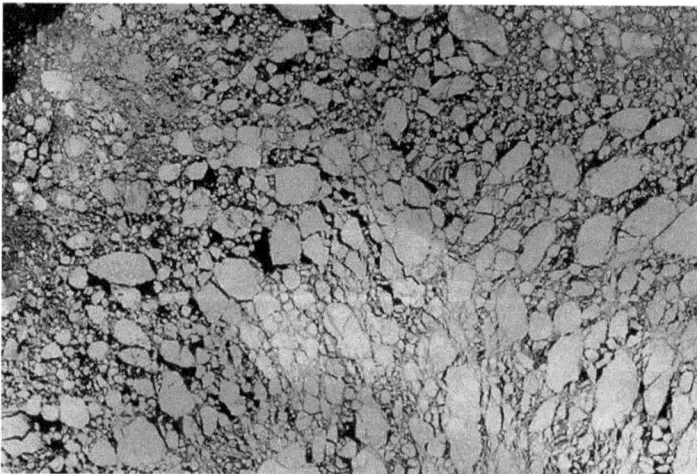

This work is a continuation of previous numerical studies on sea ice composed of disk-shaped floes with power-law size distribution [19–21]. It examines the behavior of a 2D polydisperse granular material composed of frictional grains under pure-shear deformation (constant packing fraction, or, in the sea-ice nomenclature, ice concentration A). The grains are placed on a frictional substrate (representing the ocean) and interact with each other by means of Hertzian contact forces. Although the details of the model are formulated so that it can represent sea ice moving on the sea surface, the results are relevant in a more general context of sheared, strongly polydisperse granular materials. Therefore, the specific sea-ice terminology is generally avoided in the rest of this paper, with an exception yo some parts of the discussion in the last section.

The paper is structured as follows: the next section contains the description of the model—its assumptions, governing equations and numerical formulation. The results are presented and discussed in Section 3, with an emphasis on those aspects of the model behavior that are related to the polydispersity of the material. In particular, it is demonstrated that grains from the tail of the GSD play a crucial role in the development of the force and contact networks during the jamming phase transition and are responsible for the emergence of domain-wide correlations between velocity anomalies of individual grains. Finally, conclusions are formulated in Section 4.

2. Model Description

2.1. Model Equations

The modeled system consists of $i = 1, \ldots, N$ disk-shaped grains with radii r_i, thickness h_i and density ρ_i, occupying a certain two-dimensional region, $\mathcal{S}$. Let us denote the surface area, volume and mass of the i-th grain with S_i, V_i and m_i, respectively. Obviously, $m_i = \rho_i V_i = \pi \rho_i h_i r_i^2$. The grains move within $\mathcal{S}$, due to both external forcing (e.g., friction against the underlying material) and interactions with neighboring grains. The external forcing acting on the individual grains can be expressed in terms of the density of the surface and body forces, denoted with $\check{\mathbf{f}}_{s,i}$ and $\check{\mathbf{f}}_{b,i}$, respectively. The net interaction force acting on grain i at a given time instance, t, is a sum of all pairwise interaction forces with grains that are in contact with i at time t. The set of those grains will be denoted with $\mathcal{C}_i(t)$. For $j \in \mathcal{C}_i$, $\hat{\mathbf{F}}_{ij,n}$ and $\hat{\mathbf{F}}_{ij,t}$, denote the normal and tangential components, respectively, of the grain-grain interaction force. Under these assumptions, the general form of the equations for the linear and angular momentum of the i-th grain is:

$$m_i \frac{d\mathbf{u}_i}{dt} = \int_{S_i} \check{\mathbf{f}}_{s,i} \mathrm{d}s + \int_{V_i} \check{\mathbf{f}}_{b,i} \mathrm{d}v + \sum_{j \in \mathcal{C}_i(t)} \hat{\mathbf{F}}_{ij,n} \tag{1}$$

and:

$$m_i \frac{r_i^2}{2} \frac{d\omega_i}{dt} = \mathbf{k} \cdot \left[\int_{S_i} \mathbf{r} \times \check{\mathbf{f}}_{s,i} \mathrm{d}s + \int_{V_i} \mathbf{r} \times \check{\mathbf{f}}_{b,i} \mathrm{d}v + \sum_{j \in \mathcal{C}_i(t)} \mathbf{r}_{ij} \times \hat{\mathbf{F}}_{ij,t} \right] \tag{2}$$

where $\mathbf{u}_i$ denotes the velocity of the grain's mass center, ω_i (its angular velocity), $\mathbf{k}$ (a unit vector pointing vertically upward), $\mathbf{r}$ (the horizontal distance from the grain's center) and $\mathbf{r}_{ij}$ (a vector pointing from the center of grain i to the contact point with grain j). The interaction forces are calculated based on the Hertzian contact model. The normal force, $\hat{\mathbf{F}}_{ij,n}$, has two components, a contact force (dependent on the overlap between grains) and a damping force (dependent on the relative normal velocity between grains). The tangential force, $\hat{\mathbf{F}}_{ij,t}$, has two component,s as well, namely the shear force (the so-called "history effect" that accounts for the tangential displacement of the interacting grains during contact) and the damping force (dependent on the relative tangential velocity between grains). All four of those forces depend on the effective radius of grains i and j, $r_{ij} = r_i r_j / (r_i + r_j)$, and on their material properties, *i.e.*, the elastic modulus, E, and Poisson's ratio, ν, assumed constant for all grains. Details concerning the formulation of $\hat{\mathbf{F}}_{ij,n}$ and $\hat{\mathbf{F}}_{ij,t}$ can be found, e.g., in [22,23] and in the documentation of the numerical model (see below).

Further details concerning the model formulation are given in [21]. The previous works [19–21] stressed the importance of the size-dependent response of individual grains (ice floes) to the forcing acting on them, relevant at low and medium packing fractions ($A \ll 1$). However, the focus of this paper is on a slow deformation of a compact material in or close to the jammed state. Therefore, the simulations described further were performed with a simplified set of equations, without the form-drag terms responsible for the size-dependent response (see [21] for comparison). Furthermore, it is assumed that the frictional substrate is at rest (*i.e.*, both the wind speed and the current speed are zero); the inertial effects (Coriolis term) are omitted, as well. Linearized formulae are used for the grain-substrate (ice-ocean) friction term. Thus, Equations (1) and (2) simplify to:

$$m_i \frac{d\mathbf{u}_i}{dt} = -\pi r_i^2 C_f \mathbf{u}_i + \sum_{j \in \mathcal{C}_i(t)} \hat{\mathbf{F}}_{ij,n} \tag{3}$$

and:

$$m_i \frac{r_i^2}{2} \frac{d\omega_i}{dt} = -\pi \frac{r_i^4}{2} C_f \omega_i + \mathbf{k} \cdot \sum_{j \in \mathcal{C}_i(t)} \mathbf{r}_{ij} \times \hat{\mathbf{F}}_{ij,t} \tag{4}$$

where C_f denotes the friction coefficient (in the context of sea ice, $C_f = \rho_w C_{hw}$, where ρ_w is the water density and C_{hw}, the water-ice drag coefficient).

As described in [21], the model is based on the LAMMPS (Large-scale Atomic/Molecular Massively Parallel Simulator) library [24,25], designed for simulating large systems of interacting objects (particles, molecules, *etc.*). For the purpose of sea ice modeling, LAMMPS has been extended to disk-shaped particles moving within two-dimensional domains. The Hertzian contact model, available in the official version of LAMMPS, but only for spherical particles, has been modified in order to account for non-spherical grain shape. The modification concerns the relationship between the overlap between the grains and the shape and size of the contact area between them, which, in turn, determines the resulting interaction force. In the case of spherical particles, the contact area is circular, in the case of cylindrical particles, rectangular.

For all N grains, Newton equations of motion (1) and (2) (or, in this particular case, (3) and (4)) are solved by means of the velocity-Verlet integrator.

2.2. Model Configuration and Simulations

In the simulations described in this work, the model domain, S, was rectangular, with length L_x, width $L_y = L_x/2$ and surface area $S = L_x L_y = \pi \sum_{i=1}^{N} r_i^2/A$. In isotropic-compression simulations, periodic boundary conditions were used in both x and y directions; in pure-shear simulations, only along the x-axis, with the grains along the lower model boundary defined as "frozen" (velocity set to zero throughout the simulation) and the grains along the upper model boundary moving with a prescribed velocity $\mathbf{u}_i = [u_b, 0]$.

A complete list of the model parameters can be found in Table 1. The simulations were conducted for grains with a power-law (PL) GSD, with the mean grain radius $\bar{r} = 4.0$ m and the slope of the distribution $\alpha = 1.8$, a typical value observed in sea ice (see, e.g., [15,16]). The sample of $N = 2 \cdot 10^4$ grain radii was generated with a maximum-likelihood method (e.g., [26], chapter 6.5), which provides an estimate of the most probable value of the i-th element in the rank-ordered sample of finite size N from a given distribution. Thus, deviations from a power law in the tail of the GSD, resulting from the finite sample size, are properly accounted for.

Table 1. Physical and numerical model parameters used in the simulations. GSD, grain-size distributions.

Parameter	Symbol	Value	Units
Grain density	ρ_i	910	kg/m^3
Friction coefficient	C_f	1.025	kg/m^2/s
Disk thickness	h_i	1.5	m
Mean grain radius	$\bar{r}$	4.0	m
Exponent of the power-law GSD	α	1.8	—
Elastic modulus	E	$9.0 \cdot 10^9$	Pa
Poisson's ratio	ν	0.33	—
Static yield criterion	μ	0.7	—
No. of grains	N	20,000	—
Speed at the upper boundary	u_b	0.2–1.0	m/s
Time step	Δt	$5 \cdot 10^{-4}$	s

In order to better illustrate the role of the extreme polydispersity in systems with a PL GSD, additional simulations were performed with a narrow, bidisperse (BD) GSD, corresponding to that used by Bi and colleagues [7,8], i.e., with the ratio of the radii of the coarse and fine fraction $r_1/r_2 = 1.16$ and the respective numbers of grains $n_1 = 0.2N$ and $n_2 = 0.8N$. The total number, N, and mean grain radius, $\bar{r}$, were the same as in the model setup with the PL GSD, resulting in $r_1 = 4.503$ m, $r_2 = 3.874$ m. In the remaining parts of the paper, all results and comments relate to the PL-GSD simulations unless clearly stated otherwise. Experiments with PL GSD with other values of α from the range $[1.5, 2.0]$ produced very similar results and will not be discussed here.

The simulations were performed in two stages: (i) uniform, biaxial compression up to the jamming phase transition; and (ii) pure shear for a set of combinations of packing fraction A and strain rate ϵ values (with $\epsilon = u_b/L_y$). The simulations of the second stage were initialized by sampling the results of the first stage at selected values of A and letting the system relax before applying shear strain.

3. Results and Discussion

The general model behavior in uniform-compression simulations is described in [21]. The jamming phase transition in the analyzed case occurs at $A_J \approx 0.918$. It is accompanied by a rapid increase of the internal pressure, p, the fraction of non-rattler grains (defined here as grains with at least two contacts) and the mean contact number, η_c, i.e., changes indicative of the percolation of the contact and force network. Additional simulations performed with different values of the GSD exponent, α, showed that, not surprisingly, the jamming packing fraction, A_J, increases with decreasing α (the wider the GSD, the denser the packing fraction attainable), but the course of the jamming transition (for example, the shape of the $p(A - A_J)$ curve) remains almost unaffected. This suggests that the results presented here are relevant for a wider range of model parameters than those actually used in the simulations.

The analysis below concentrates on the pure-shear model runs, with an emphasis on the role of polydispersity in the model behavior close to and at the jamming phase transition. All results have been obtained for packing fractions $A < A_J$, i.e., below the isotropic jamming point. Hence, the "jammed states" in the discussion below refer to regions of shear-jammed and fragile states on the jamming phase diagram proposed by Bi and colleagues [8]. Anticipating the further analysis of the results, the term "jammed state" used throughout the rest of the paper refers to states in which the largest contact network has percolated the whole system in both directions (and which are accompanied by certain global characteristics described further).

3.1. Shear Jamming: General Characteristics

The general behavior of the modeled system under pure shear deformation depends on the packing fraction, A, and the strain rate, ϵ [8]. At low A, the system remains in an unjammed state, in which the internal stress is generated via short, binary collisions between neighboring grains. Regions of jammed, more densely packed grains develop only locally (Figure 2); they are short-lived and disperse, due to interactions with the surrounding, more loosely packed regions. Hence, the internal stress level at the system scale remains very low, a few orders of magnitude lower than in the jammed states (Figure 3), when the force network between grains percolates the whole system (Figure 2) and the neighboring grains remain in contact for many seconds or even minutes (see further Section 3.2), i.e., periods of time up to a few orders of magnitude longer than the duration of a typical binary collision.

Figure 2. Snapshots of contact networks in the modeled system in the unjammed (**left**; $A = 0.890$, $u_b = 0.5$ m/s) and jammed (**right**; $A = 0.905$, $u_b = 1.0$ m/s) state. For each grain, i, a line is drawn from its center to the center of the neighboring grain, j, if $j \in C_i(t)$. Grains belonging to the 'frozen' and moving boundaries are not shown.

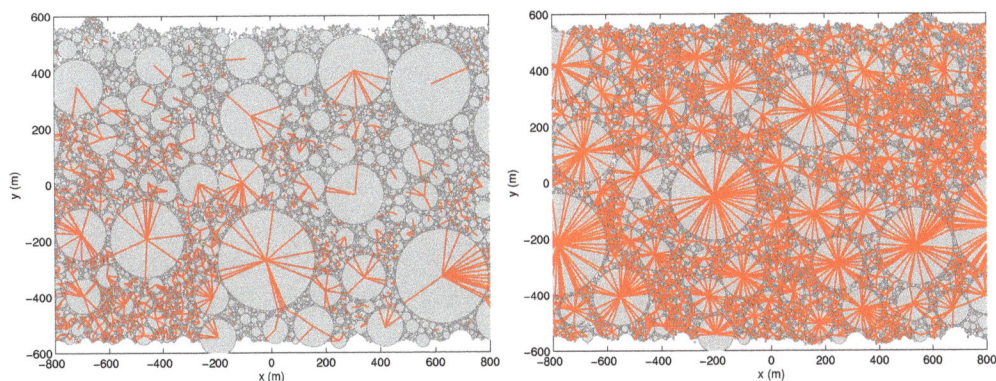

Figure 3. Time series of the average contact number, η_c (**a**), force-network anisotropy η_a (**b**), pressure p (**c**), shear stress τ (**d**) and the principal angle, θ_p (**e**), during simulations with: $A = 0.908$ and $u_b = 1.0$ m/s (blue); $A = 0.905$ and $u_b = 0.5$ m/s (black); $A = 0.905$ and $u_b = 0.2$ m/s (magenta).

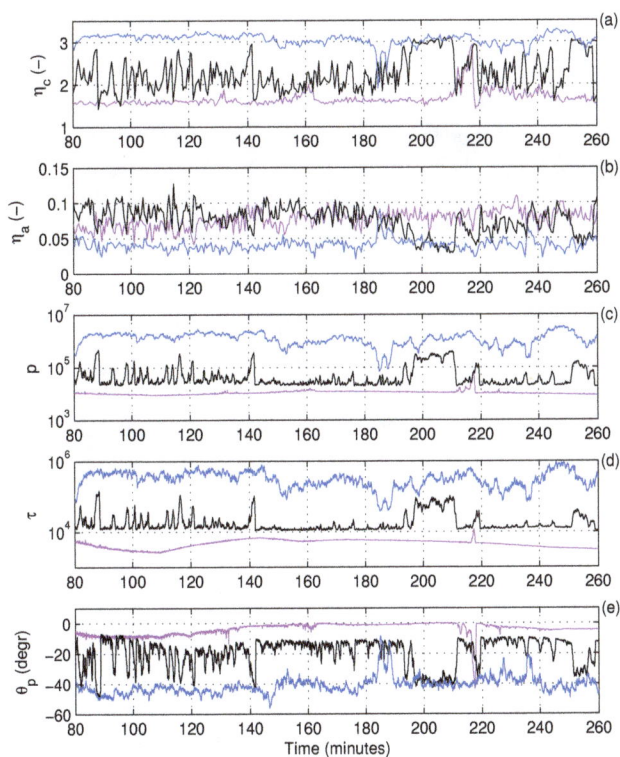

556

The large-scale system behavior can be described by means of the properties of the stress and fabric tensors: the pressure $p = (\sigma_1 + \sigma_2)/2$ and the shear stress $\tau = (\sigma_2 - \sigma_1)/2$ are calculated from the principal stresses, σ_1 and σ_2; the mean contact number $\eta_c = \lambda_1 + \lambda_2$ and the contact-network anisotropy $\eta_a = (\lambda_2 - \lambda_1)/\eta_c$ are calculated from the eigenvalues of the fabric tensor, λ_1 and λ_2 (see [8,21] for details). Both in the jammed and unjammed states, far from the jamming-transition point, those four large-scale system characteristics—p, τ, η_c and η_a—remain relatively stable in time, and the system recovers fast from short rearrangement events that sporadically take place (Figure 3). In between those two extremes, the system undergoes rapid changes and shifts from unjammed to jammed states and *vice versa* (black lines in Figure 3). Between those two extremes, the force networks often have a fragile, "openwork" structure, with relatively large unjammed areas where the stress remains very low and with forces transmitted via long "strands" of approximately linearly aligned grains. As in the case of fragile states observed recently [8], those force networks may span the whole model domain in only one (compressive) direction, giving the material anisotropic strength in response to deformation, which manifests itself in high values of η_a (see, also, Section 3.2). The present results suggest that, even in constant strain conditions, the fragile states are short-lived, at least in the range of A and ϵ combinations analyzed here.

Apart from the properties of the stress and contact-fabric tensors, a signature of jamming is also present in the grains' velocity, both means and their anomalies. Let us define $\mathbf{u}_m(y,t)$ and $\sigma_m(y,t)$, as the mean and standard deviation, respectively, of the velocity of all grains that at time t have their y-coordinate within a certain small distance, δ, from y (*i.e.*, that lie inside a stripe of length L_x and width 2δ). Further, let $\langle \mathbf{u}_m(y) \rangle$ and $\langle \sigma_m(y) \rangle$ denote the time mean of $\mathbf{u}_m(y,t)$ and $\sigma_m(y,t)$ over the whole simulation time and $\mathbf{u}'_i(\mathbf{x},t)$—the velocity anomaly of grain i, *i.e.*, $\mathbf{u}'_i(\mathbf{x},t) = \mathbf{u}_i(\mathbf{x},t) - \langle \mathbf{u}_m(y) \rangle$.

The profiles of $\langle \mathbf{u}_m(y) \rangle$ and $\langle \sigma_m(y) \rangle$ are shown in Figure 4 for a range of A values corresponding to unjammed and jammed states. At low packing fractions, the motion of the grains is confined to the region close to the moving boundary, and a narrow zone of strong shear separates this region from the rest of the model domain, remaining almost at rest. To the contrary, jammed states are characterized by an almost constant velocity gradient $d\langle \mathbf{u}_m(y) \rangle /dy$ and constant standard deviation of velocity $\langle \sigma_m(y) \rangle$, independently on the distance from the moving boundary, *i.e.*, the strain is distributed over the whole system.

In order to characterize the variability of velocity anomalies, it is convenient to define a measure analogous to entropy (as used in statistical mechanics), characterizing the spread of velocity anomalies of individual grains at a given time, t:

$$E(t) = -c \sum_{i=1}^{n} (p_i \log_2 p_i) \tag{5}$$

where n denotes the number of bins of the discrete pdfof $\mathbf{u}'_i(\mathbf{x},t)$, p_i is the probability density of the i-th bin and $c = 1/\log_2 n$—a normalization constant, introduced so that the maximum value of $E = 1$. In order to account for different ranges of $\mathbf{u}'_i(\mathbf{x},t)$ in different model runs, the pdfs were estimated by dividing the range, $[q_{0.01}, q_{0.99}]$, into $n = 100$ bins of equal width, where $q_{0.01}$, $q_{0.99}$ denote the 1% and 99% quantiles of the data, respectively. Thus, E analyzed here reflects the shape of the pdfs within the respective inter-quantile range, and not their widths, which, as shown in Figure 4b, is much larger in the jammed than in the unjammed states.

Figure 4. Profiles of the average (**a**) and standard deviation (**b**) of the x-component of grain velocity in the function of the normalized y-distance ($y = 0$ at the "frozen" boundary and $y = 1$ at the moving boundary). Results obtained with $u_b = 1.0$ m/s.

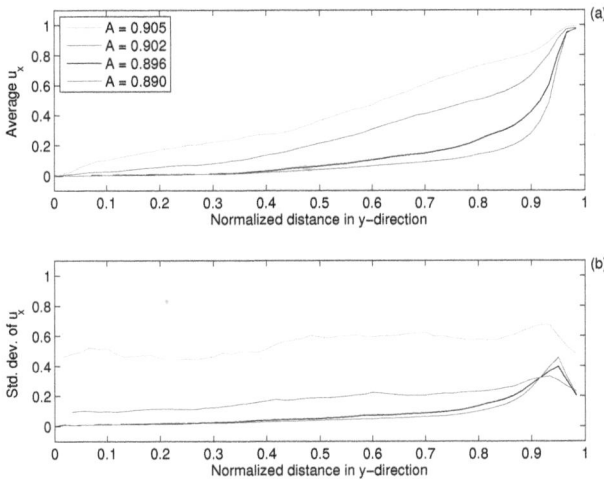

Figure 5. Normalized entropy E of the anomalies, $\mathbf{u}'_i(\mathbf{x}, t)$, in the function of the grain packing fraction, A (**a**), and grain size (**b**). On each box, the central mark is the median, the edges of the box are the 25th and 75th percentiles and the whiskers extend to the most extreme data points not considered outliers. In (b), for the two selected values of A (0.890 and 0.905), the statistics are calculated three times: for all N grains and for the subsets of the 10% largest and 10% smallest grains, respectively. Results obtained with $u_b = 1.0$ m/s.

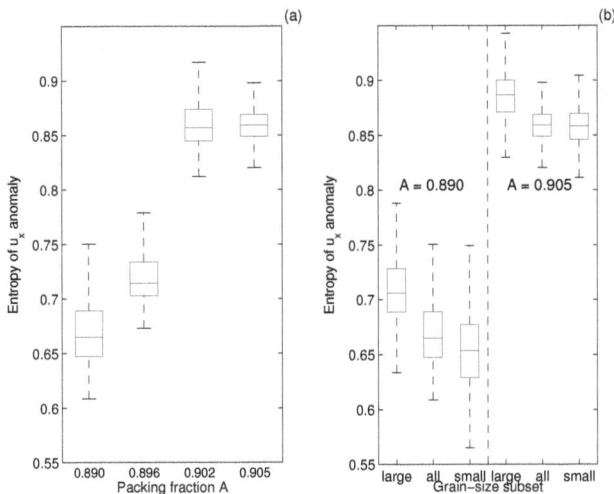

558

As can be seen in Figure 5a, E increases with increasing packing fraction A. It has highest values, exceeding 0.85, and lowest time variability (see the boxes and whiskers in Figure 5) in shear-jammed states. In unjammed states, E, most of the time remains within the 0.65–0.7 range. Thus, the range of instantaneous velocity anomalies in jammed systems is significantly larger, even corrected for the width of the respective pdfs. On the other hand, jamming is associated with a transition from local to global correlations of $\mathbf{u}'_i(\mathbf{x}, t)$, as illustrated in Figures 6 and 7, showing the linear correlation coefficient, C, between pairs of grains in two selected model runs (for two grains, i and j, C is a Pearson correlation coefficient between the x-components of $\mathbf{u}'_i$ and $\mathbf{u}'_j$ over time $t_c = 100$ min). At low A, statistically significant correlation of velocity anomalies is observed only between grains within a small spatial distance from each other. At high A, the correlation remains high within the whole model domain. Those two facts—velocity anomalies correlated on the system-scale and high values of E—indicate that in a jammed state, the grains tend to have large velocity anomalies that are of the same sign.

Figure 6. Snapshots of the modeled system for $u_b = 1.0$ m/s and the packing fraction $A = 0.809$ (**a**) and $A = 0.905$ (**b**), showing the linear correlation coefficient, C, between the velocity anomalies, $\mathbf{u}'_i(\mathbf{x}, t)$, of a selected grain (dark brown, $C = 1$) and all other grains in the system. C was calculated for a period of time equal to 100 minutes. Grains belonging to the "frozen" and moving boundaries are not shown.

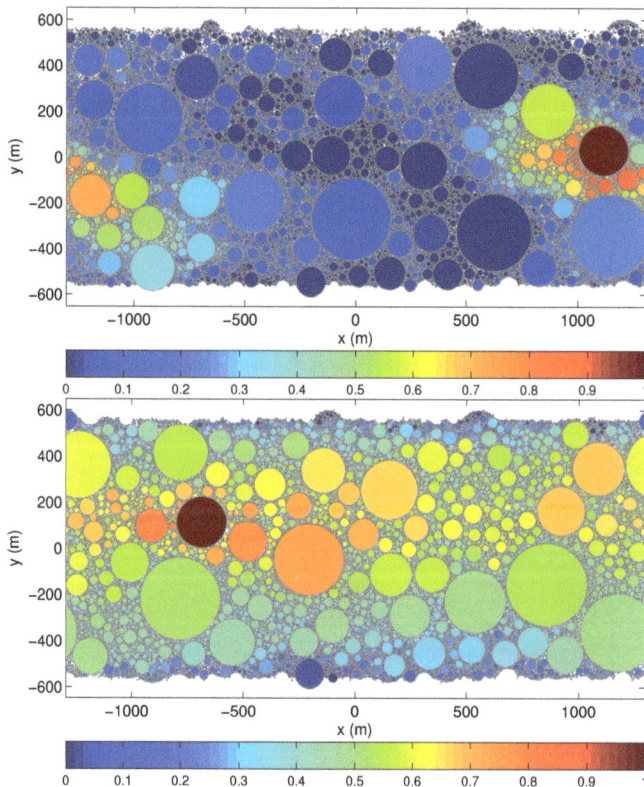

Figure 7. The correlation coefficient, C, between the velocity anomalies, $\mathbf{u}_i'(\mathbf{x}, t)$, calculated for pairs of grains from a subset of the 10% largest (continuous lines) and 10% smallest (dashed lines) grains in the whole ensemble, in the function of the grain-grain distance. Results of simulations with $u_b = 1.0$ m/s and $A = 0.890$ (blue), $A = 0.905$ (red).

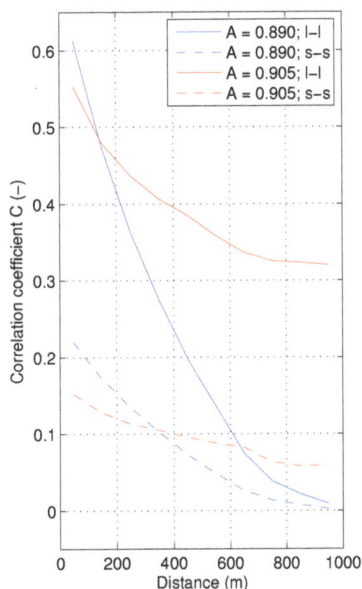

3.2. The Role of Polydispersity

In systems with power-law GSD, the largest grains occupy a substantial part of the model domain (even with increasing system size N), and it is their locations and relative movement that have a deciding influence on the system as a whole. Sub-regions of the model domain that at a given time instance are filled with small grains can change their shape (and thus react to strain deformation) more easily than assemblies of large grains. In many respects, assemblies of very small grains act as a plastic, easily deformable 'filler' occupying empty spaces between very large grains. An analysis of animations illustrating the time evolution of the modeled system reveal that the rapid jamming and un-jamming events mentioned earlier (black curves in Figure 3) tend to be associated with reorganization of the positions of the largest grains. This observation seems confirmed by the fact that, at high packing fractions, the analyzed measures of the grains' velocity anomalies, like the entropy, E, are strongly correlated to the global instantaneous pressure, p, and shear stress τ and that this correlation is higher for a subset of the largest grains than for the whole system. For example, in the model run with $A = 0.905$ and $u_b = 1$ m/s, the correlation of E with $\log(p)$ equals 0.83 and 0.95 for, respectively, all and the subset of 10% of the largest grains.

Previous experiments with an earlier version of the model demonstrated that polydispersity plays an important role in many aspects of the dynamics of sea ice composed of floes with power-law size distribution, including the formation of clusters in response to wind [19,20]. Not surprisingly, polydispersity also influences the behavior of the sheared systems studied here. Many global characteristics of the system, including those analyzed above, have different values when they are calculated for a subset of the largest or smallest grains, revealing their different response to the forcing and interactions with neighboring grains. In particular, the entropy, E, of velocity anomalies of the largest grains in an ensemble is higher than the system average at all packing fractions analyzed, *i.e.*, both in jammed and unjammed states (Figure 5b). The emergence of long-range correlations between velocity anomalies at the jamming transition, described in the previous section, takes place almost exclusively due to correlations between the largest grains in the system (Figures 7 and 8). Similarly, at low A, the high values of C within clusters (0.5–0.6 on average) are observed only for pairs of the largest grains. Furthermore, whereas at low A, those values drop rapidly with increasing grain-grain distance, the rate of that decrease is much slower in jammed states (compare the continuous curves in Figure 7), resulting in a shift of the pdf of C towards larger values, representing statistically significant correlation (Figure 8b). To the contrary, the pdfs of C of the smallest grains (in this case, $r_i < 1.87$ m) hardly change at the jamming transition, with most values of C remaining at a very low, statistically insignificant level.

Figure 8. pdfsof the correlation coefficient, C, between the velocity anomalies, $u_i'(\mathbf{x}, t)$, calculated for pairs of grains from a subset of the 10% largest (blue) and 10% smallest (red) grains in the whole ensemble. Results of simulations with $u_b = 1.0$ m/s and $A = 0.890$ (**a**), $A = 0.905$ (**b**).

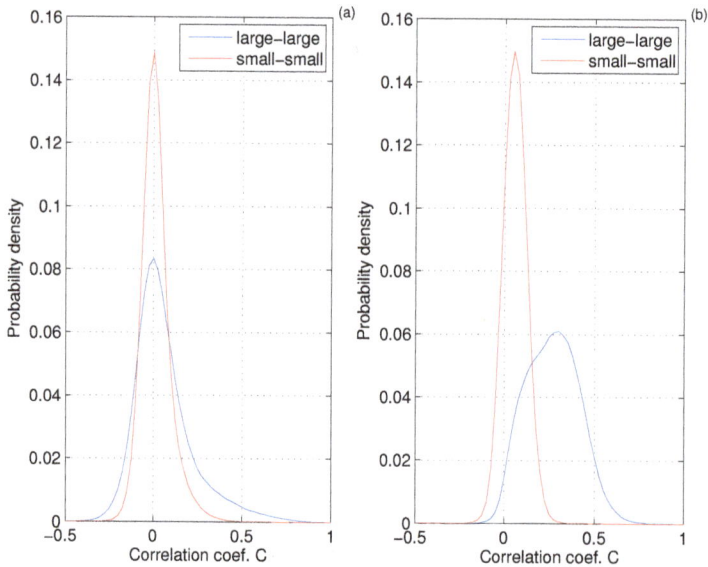

Thus, the increase of the packing fraction, A, towards jamming is accompanied by the growth of clusters of coordinated motion of the relatively small subset of the largest grains (notably, the range of sizes of those grains is still very wide, between 6.5 and 180 m). It is worth noting that similar behavior has been described recently for sheared bidisperse granular systems, in which the dominant dynamical modes were associated with reorganization of grains within localized clusters [10]. Similarly, Weeks and colleagues [27] observed cooperative motion of particles within clusters in colloidal supercooled fluids, with the size of clusters rapidly increasing when the system approached the glass transition.

Figure 9. Selected properties of the contact networks in the modeled system for a number of packing fractions, A: number of contacts of individual non-rattler grains, $n_{c,i}$ (**a**); $n_{c,i}$ scaled with grain perimeter $2\pi r_i$ (**b**); percentage of the simulation time when individual grains were non-rattler grains (**c**); percentage of grains with at least three contacts (**d**); average contact number η_c (**e**); and contact-network anisotropy η_a (**f**). In (**d–f**), the elements of the box symbols are the same as in Figure 5; they reflect the time variability of the analyzed variables during the simulation.

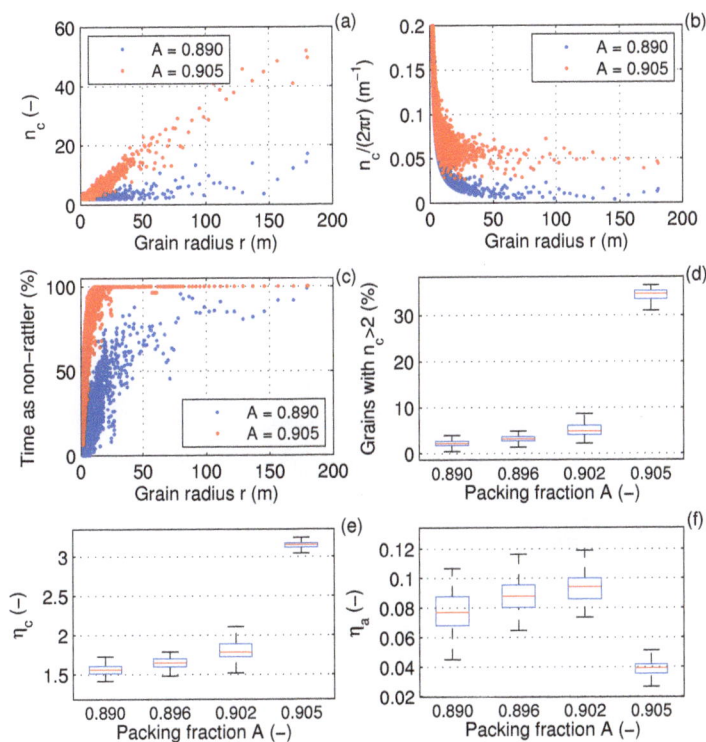

Due to obvious geometrical reasons, in strongly polydisperse materials, the number of contacts of individual grains strongly varies. Interestingly, the jamming transition (inferred from the size of the largest connected cluster) in the analyzed cases still takes place when the average contact number, η_c, exceeds the value of three (packing fraction $A = 0.905$ in Figure 9e), similarly as in monodisperse and weakly polydisperse systems ([8] and Figure 10). For the large grains from the tail of the GSD, the contact number of individual grains, $n_{c,i}$, is an approximately linear function of their radius (and perimeter), with $n_{c,i}/(2\pi r_i) \approx 0.05$ in the jammed state (Figure 9a,b), e.g., $n_{c,i} \sim 30$ for $r_i = 100$ m.

A linear relationship for $n_c(r)$ has been obtained recently by Shaebani and colleagues [28] in simulations of 2D uniformly compressed, weakly polydisperse systems, in agreement with their mean-field solution. In our simulations, smaller grains often have just one or two neighbors (hence the points on the left side of Figure 9b tend to lie on the r^{-1} curve), and importantly, it is their incorporation into the system-wide contact network that leads to its consolidation at the jamming transition: whereas the largest grains are non-rattler grains most of the time, even in unjammed states, the smallest grains switch at jamming from predominantly freely moving to predominantly non-rattler (Figure 9c; see also Figure 2). Consequently, jamming is associated with a decrease of the mean radius of grains forming the main contact network (not shown). On the other hand, it is the largest grains that build the stable skeleton of the global contact and force network, in the sense that the great majority of grains, predominantly from the left part of the GSD, does not participate in its formation. Even in a jammed state, only ~35% of grains have three or more contacts (Figure 9d), and even though $\eta_c > 3$, the mean contact number for the 90% smaller grains (*i.e.*, excluding the 10% largest) is smaller than three. The "sparse" character of the grain-grain contacts is clearly seen in Figure 11. In the jammed state at $A = 0.905$, the force network percolates the whole model domain (see also Figure 2), but it has an "openwork" structure, with the largest grains incorporated into long force chains and irregular "cells", surrounding unjammed regions, usually filled with very small grains. As already mentioned, the assemblies of the smallest grains act as a semi-plastic "filler", adjusting its shape to the deforming cells of the main force network.

All those properties of the analyzed system are directly related to its extreme polydispersity. In the bidisperse reference case (Figure 10), the jamming transition is much more rapid in terms of the amount of grains that are incorporated into the global contact network: as soon as η_c exceeds the value of three, roughly 80% of grains become non-rattler grains (Figure 10d,e). Moreover, the coarser and finer fractions contribute similarly to the contact network (compare Figure 10b,c), and the (very small) difference between the average number of neighbors of individual grains from the two fractions (Figure 10a) results simply from the difference between their perimeters.

Back to the PL-GSD simulation, it is also worth noticing that although the contact numbers of the largest grains are high independent of the packing fraction (Figures 9c and 11), in unjammed states, most of those contacts do not form part of stable force strains, but reflect individual collisions as they 'fight their way' among smaller neighbors (in Figure 11 at $A = 0.890$, most of the lines outgoing from the centers of the largest grains are black, *i.e.*, they lead to grains with a number of contacts lower than three). Such contacts rarely survive more than a few seconds. Indeed, the exceedance probability of contact lifetime is at low A similar for all grain sizes (Figure 12) and only ~10% of contacts survive for longer than one minute, as compared to ~25% and 40% of contacts between the smallest and largest grains, respectively, observed in the jammed state.

Moreover, it must be remembered that the condition, $n_{c,i} \geq 3$, alone is not sufficient to stabilize the position of individual grains within the contact network. The arrangement of the contacts around the grain's perimeter is important, as well. For a given grain, i, this arrangement is determined by a set of vectors, $\mathbf{r}_{ij}$, for all $j \in \mathcal{C}_i$ (see Section 2.1), which divide the grain into $n_{c,i}$ sectors. The central angle of the widest of those sectors (for brevity, we will call it the maximum contact angle,

α_{max}) provides a useful measure of the above-mentioned stability of the i-th grain within the force network. Obviously, $\alpha_{max} < 180°$, possible only if $n_{c,i} > 2$, is necessary for stability; $\alpha_{max} < 120°$ is possible only if $n_{c,i} > 3$.

Figure 10. Selected properties of the contact networks in the bidisperse (BD) model case for a number of packing fractions, A: mean number of contacts of individual non-rattler grains, $n_{c,i}$, from the finer and coarser fractions (**a**); percentage of the simulation time when individual grains from the finer (**b**) and coarser (**c**) fraction were non-rattler grains; percentage of grains with at least three contacts (**d**); average contact number η_c (**e**); and contact-network anisotropy η_a (**f**). In (**b–f**), the elements of the box symbols are the same as in Figure 5; they reflect the time variability of the analyzed variables during the simulation.

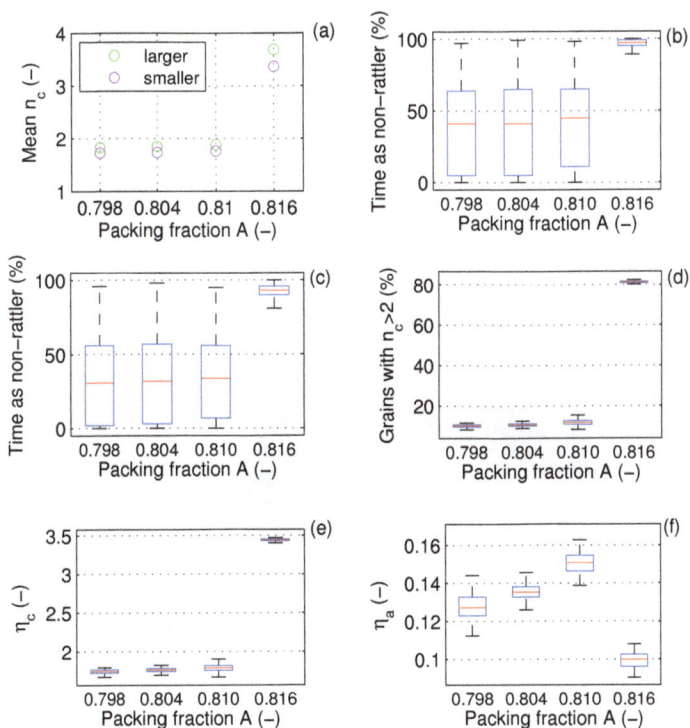

As can be seen in Figure 11b, many non-rattler grains, *i.e.*, those with $n_{c,i} > 2$, have $\alpha_{max} > 180°$. In fact, ~14% out of the ~35% of grains building the percolated contact network (*i.e.*, ~5% of the total) have $\alpha_{max} > 180°$. For comparison, out of 82%–83% of grains building the global network in the BD case, only ~1% are unstable. Figure 13 shows the pdfs of α_{max} for the PL and BD cases in shear-jammed states. In the BD simulations, the pdfs have high peaks close to the 120° value, indicating the prevailing—very stable—contact arrangement with $n_{c,i} = 3$ and roughly uniform distribution of neighbors around the grain's perimeter. Notably, the pdfs for the coarser and finer fractions have very similar shapes, with an exception of a small second peak close to 180° for the finer fraction. To the contrary, in the PL simulations, the pdf of α_{max} of the smallest grains has a wide maximum shifted towards the unstable region and a long tail corresponding to individual, instantaneous collisions.

Figure 11. Zoomed fragments of the modeled system (top: $A = 0.890$; bottom: $A = 0.905$) corresponding to the situations shown in Figure 2. Color scale: number of contacts, $n_{c,i}$, of individual grains (dark blue: zero; light blue: one; yellow: two; brown: three or more). Red lines: forces between grains with $n_c \geq 3$; black lines: the remaining forces.

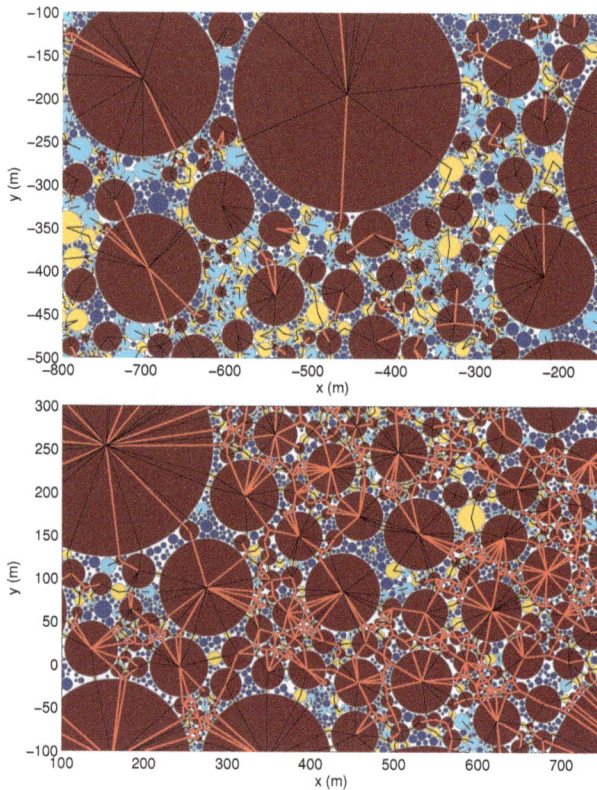

Finally, another very important property of force networks in sheared systems is their anisotropy, η_a. It has been identified as an order parameter for shear-jammed states, in that it is non-zero in such states and zero in isotropically jammed systems [8]. The values of η_a obtained in this work, in the BD and, especially, PL simulations, are lower than those reported in [8], which may be related to the details of the contact-model formulation. Characteristically, for the PL case, the η_a values are roughly 50% higher for the 10% largest grains than the ensemble mean (not shown), again underlying the special role of those grains in shaping the force network structure. The overall variability of $\eta_a(A)$ is, however, similar in both PL and BD simulations, *i.e.*, a rapid drop of η_a is observed at jamming, preceded by a slight increase when the jamming is approached from below (Figures 9f and 10f). The strongest anisotropy was obtained in those model runs, in which the system underwent strong shifts between unjammed and jammed states (black curves in Figure 3).

Figure 12. Exceedance probability of the contact lifetime (in seconds) for two values of the packing fraction ($A = 0.890$, dashed lines; $A = 0.905$, continuous lines), calculated for contacts between the 10% largest (blue) and 10% smallest (red) grains.

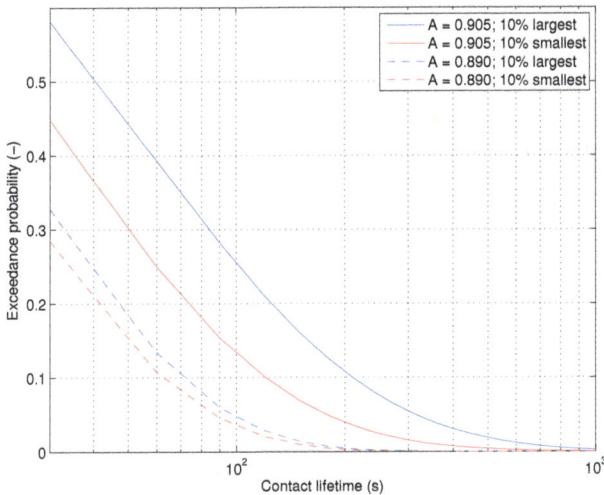

Figure 13. pdfs of the maximum contact angle, α_{max}, for two model runs: with bidisperse (BD; $A = 0.816$) and power-law (PL; $A = 0.905$) grain-size distribution. For the BD case, the pdfs are calculated separately for the coarser and finer grain fraction; for the PL case—for the subsets of 10% largest and 10% smallest grains. The vertical dotted and dashed lines mark the values $\alpha_{max} = 120°$ and $\alpha_{max} = 180°$, respectively (see the text for a description).

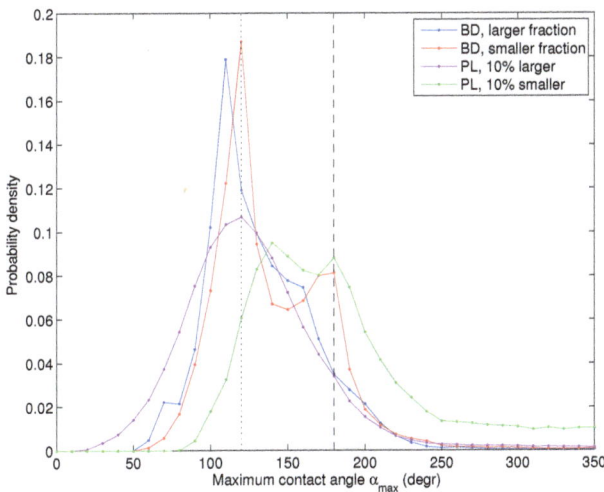

4. Conclusions

The results of the present study suggest that many global characteristics of granular materials with very wide GSD, including those indicative of the jamming phase transition, are determined by the behavior of and interactions between a relatively small subset of the largest grains from the tail of the GSD. They build the "core" of the contact and force networks in the material and, consequently, react in a coordinated manner to strain deformation.

To summarize, the percolation of the contact networks in the analyzed system with PL GSD is associated with the following changes of the global system properties: (i) rapid increase of the entropy of grain velocity anomalies; (ii) emergence of large-scale correlation between the velocity anomalies of the largest grains; (iii) rapid increase of the mean contact number and the fraction of non-rattler grains, accompanied by a rapid decrease of the contact network anisotropy; and (iv) rapid increase of the contact lifetimes, especially between the largest grains. In comparison to less strongly polydisperse systems, the percolated contact networks are built of a significantly smaller subset of grains in stable positions, capable of sustaining non-zero strain.

In the context of sea ice (and presumably other real-world polydisperse granular material, as well), the behavior of the largest grains is relevant from a practical point of view: very likely, it is they that are subject to observation. Measuring equipment is usually mounted on the largest and thickest ice floes, providing stable and relatively safe observational platforms. Similarly, in the analysis of remote sensing (satellite or airborne) images of sea ice, it is the largest floes that can be easily identified and tracked. Thus, it is relevant to understand how the behavior of floes (or, more generally, grains) from the tail of the GSD is different from the behavior of the remaining grains and how it is related to the properties of the granular material as a whole. Even though the simulations described in this paper are highly idealized (no wind, ocean currents, *etc.*), in view of the fact of how little is known about the influence of granular effects on sea ice dynamics, they provide a starting point for further, more advanced studies.

Acknowledgments

The calculations described in this paper were carried out at the Academic Computer Center (TASK) in Gdansk, Poland [29]. I would like to thank the anonymous reviewers for their insightful and very valuable comments and suggestions that helped to improve the quality of this paper.

Conflicts of Interest

The author declares no conflict of interest.

References

1. Zhang, H.; Makse, H. Jamming transition in emulsions and granular materials. *Phys. Rev. E* **2005**, *72*, 011301.

2. Bandi, M.; Rivera, M.; Krzakala, F.; Ecke, R. Fragility and hysteretic creep in frictional granular jamming. *Phys. Rev. E* **2013**, *87*, 042205.

3. Cates, M.; Wittmer, J.; Bouchaud, J.P.; Claudin, P. Jamming, force chains, and fragile matter. *Phys. Rev. Lett.* **1998**, *81*, 1841–1844.

4. Howell, D.; Behringer, R.; Veje, C. Stress fluctuations in a 2D granular Couette experiment: A continuous transition. *Phys. Rev. Lett.* **1999**, *82*, 5241–5244.

5. Schöllmann, S. Simulation of a two-dimensional shear cell. *Phys. Rev. E* **1999**, *59*, 889–899.

6. Veje, C.; Howell, D.; Behringer, R. Kinematics of a two-dimensional granular Couette experiment at the transition to shearing. *Phys. Rev. E* **1999**, *59*, 739–745.

7. Zhang, J.; Majmudar, T.; Tordesillas, A.; Behringer, R. Statistical properties of a 2D granular material subjected to cyclic shear. *Granul. Matter* **2010**, *12*, 159–172.

8. Bi, D.; Zhang, J.; Chakraborty, B.; Behringer, R. Jamming by shear. *Nature* **2011**, *480*, 355–358.

9. Pastore, R.; Ciamarra, M.; Coniglio, A. 'Flow and jam' of frictional athermal systems under shear stress. *Phil. Mag.* **2011**, *91*, 2006–2013.

10. Banigan, E.; Illich, M.; Stace-Naughton, D.; Egolf, D. The chaotic dynamics of jamming. *Nat. Phys.* **2013**, *9*, 288–292.

11. Toyota, T.; Takatsuji, S.; Nakayama, M. Characteristics of sea ice floe size distribution in the seasonal ice zone. *Geophys. Res. Lett.* **2006**, *33*, doi:10.1029/2005GL024556.

12. Lu, P.; Li, Z.; Zhang, Z.; Dong, X. Aerial observations of floe size distribution in the marginal ice zone of summer Prydz Bay. *J. Geophys. Res.* **2008**, *113*, doi:10.1029/2006JC003965.

13. Toyota, T.; Haas, C.; Tamura, T. Size distribution and shape properties of relatively small sea-ice floes in the Antarctic marginal ice zone in late winter. *Deep Sea Res. II* **2011**, *9–10*, 1182–1193.

14. Rothrock, D.; Thorndike, A. Measuring the sea-ice floe size distribution. *J. Geophys. Res.* **1984**, *89*, 6477–6486.

15. Steer, A.; Worby, A.; Heil, P. Observed changes in sea-ice floe size distribution during early summer in the western Weddell Sea. *Deep-Sea Res. II* **2008**, *55*, 933–942.

16. Herman, A. Sea-ice floe-size distribution in the context of spontaneous scaling emergence in stochastic systems. *Phys. Rev. E* **2010**, *81*, 066123.

17. Feltham, D. Granular flow in the marginal ice zone. *Phyl. Trans. R. Soc. A* **2005**, *363*, 1677–1700.

18. Landsat. Available online: http://earthexplorer.usgs.gov/ (accessed on 31 October 2013).

19. Herman, A. Molecular-dynamics simulation of clustering processes in sea-ice floes. *Phys. Rev. E* **2011**, *84*, 056104.

20. Herman, A. Influence of ice concentration and floe-size distribution on cluster formation in sea ice floes. *Cent. Eur. J. Phys.* **2012**, *10*, 715–722.

21. Herman, A. Numerical modeling of force and contact networks in fragmented sea ice. *Ann. Glaciol.* **2013**, *54*, 114–120.

22. Brilliantov, N.; Spahn, F.; Hertzsch, J.M.; Pöschel, T. Model for collisions in granular gases. *Phys. Rev. E* **1996**, *53*, 5382–5392.

23. Schwager, T. Coefficient of restitution for viscoelastic disks. *Phys. Rev. E* **2007**, *75*, 051305.

24. Plimpton, S. Fast parallel algorithms for short-range molecular dynamics. *J. Comp. Phys.* **1995**, *117*, 1–19.

25. LAMMPS. Available online: http://lammps.sandia.gov/ (accessed on 31 October 2013).

26. Sornette, D. *Critical Phenomena in Natural Sciences*, 2nd ed.; Springer: Berlin/Heidelberg, Germany, 2006.

27. Weeks, E.; Crocker, J.; Levitt, A.; Schofield, A.; Weitz, D. Three-dimensional direct imaging of structural relaxation near the colloidal glass transition. *Science* **2000**, *287*, 627–631.

28. Shaebani, M.; Madadi, M.; Luding, S.; Wolf, D. Influence of polydispersity on micromechanics of granular materials. *Phys. Rev. E* **2012**, *85*, 011301.

29. TASK. Available online: http://www.task.gda.pl/ (accessed on 31 October 2013).

Reprinted from *Entropy*. Cite as: Potestio, R.; Peter, C.; Kremer, K. Computer Simulations of Soft Matter: Linking the Scales. *Entropy* **2014**, *16*, 4199–4245.

Review

Computer Simulations of Soft Matter: Linking the Scales

Raffaello Potestio [1,*]**, Christine Peter** [2] **and Kurt Kremer** [1]

[1] Max-Planck-Institut für Polymerforschung, Ackermannweg 10, 55128 Mainz, Germany;
  E-Mail: kremer@mpip-mainz.mpg.de
[2] Department of Chemistry, Universität Konstanz, Universitätsstr.10, 78464 Konstanz, Germany;
  E-Mail: christine.peter@uni-konstanz.de

* Author to whom correspondence should be addressed; E-Mail:potestio@mpip-mainz.mpg.de;
  Tel.: +49-6131-379-201.

Received: 3 June 2014; in revised form: 10 July 2014 / Accepted: 11 July 2014 / Published: 28 July 2014

Abstract: In the last few decades, computer simulations have become a fundamental tool in the field of soft matter science, allowing researchers to investigate the properties of a large variety of systems. Nonetheless, even the most powerful computational resources presently available are, in general, sufficient to simulate complex biomolecules only for a few nanoseconds. This limitation is often circumvented by using coarse-grained models, in which only a subset of the system's degrees of freedom is retained; for an effective and insightful use of these simplified models; however, an appropriate parametrization of the interactions is of fundamental importance. Additionally, in many cases the removal of fine-grained details in a specific, small region of the system would destroy relevant features; such cases can be treated using dual-resolution simulation methods, where a subregion of the system is described with high resolution, and a coarse-grained representation is employed in the rest of the simulation domain. In this review we discuss the basic notions of coarse-graining theory, presenting the most common methodologies employed to build low-resolution descriptions of a system and putting particular emphasis on their similarities and differences. The AdResS and H-AdResS adaptive resolution simulation schemes are reported as examples of dual-resolution approaches, especially focusing in particular on their theoretical background.

Keywords: soft matter; coarse-graining; adaptive resolution simulations

1. Introduction

Since the pioneering work carried out by Berni Alder [1] in the 1950s, *in silico* experiments, such as Molecular Dynamics (MD) or Monte Carlo (MC) simulations, allowed researchers to obtain major advancements in the understanding of systems with many degrees of freedom. In particular, during the last few decades, the increasing accuracy of the force-fields, the improvement of the algorithms, and the steady boost of computer power made it possible to perform insightful simulations of a broad variety of systems of increasing size and complexity, ranging from simple liquids -composed of idealized, point-like molecules interacting via simple potentials- to biomolecules. Nonetheless, the amount of available computational resources can be insufficient to simulate, for a physically meaningful time, even the simplest nontrivial macromolecule. It is often the case, in fact, that "interesting" phenomena in these systems occur on very long time-scales: a simple example of this is provided by the diffusion of a polymer in a melt [2,3]; the same behavior can observed in conformational changes of proteins [4–9], at least in those cases in which the force field provides a good approximation to the real atomistic interactions.

At the same time, in many cases the massive amount of data that are produced in a simulation is composed mostly of non-useful information. A prototypical example is given by the solvent: the water molecules that solvate a protein or a membrane are typically discarded from the analysis that follows the simulation, with the possible exception of a few solvation shells around the molecule itself. In this case a large fraction of the computational power is employed in the integration of the equations of motion of degrees of freedom which are extremely relevant *during* the simulations, but are completely neglected *afterwards*.

In order to overcome this limitation, *coarse-grained models* [10–15] have been developed, where the structure and interactions of the original system are replaced with simpler ones, which are easier to describe, model, simulate and understand. The assumption underlying the coarse-graining of a system is that above a given length scale the low-level, chemistry-specific detail of the model affects some properties of the system only in a simple, functionally trivial way - often through prefactors. Examples of systems for which this approach proved to be extremely successful are molecular fluids, polymers [2,3,16,17], elastic network models of proteins [18–23], lipid membranes and other biomolecular systems, just to mention a few.

In recent years, systematic coarse graining approaches have gained importance, where the interactions in the coarse-grained (CG) model are derived systematically from atomistic reference simulations in a bottom-up fashion. These models are often used in a multiscale simulation framework, where the closeness of higher and lower levels of resolution allows a switching back and forth between them. Below, we will review several systematic coarse graining approaches and address some of the most important methodological issues and challenges.

The smaller amount of degrees of freedom that are retained in coarse-grained models and the simpler force-fields employed allow the characterization of relevant properties of a system at a cheaper computational cost compared to the high-resolution atomistic models; on the other hand, there are cases in which the chemical detail *in a small region of the system* plays a crucial role, such that no simplification of the description is possible: think, for example, of the active site of a large enzyme, where fine-grained chemical processes take place. A high-resolution modeling of each part of the system would not be necessary, but at the same time a coarse-graining approach would delete important information.

This last observation naturally leads us to identify a particular class of soft matter systems among those that are studied with the help of computer simulations. Specifically, we can consider those systems where the focus is on a small, well-defined subregion of the simulation box. To this class belong, for example, certain solvated (macro)molecules, active sites of enzymes, the interaction of specific polymer ends at a surface, or simply a small spherical region in a homogeneous fluid whose radius is of the length scale of the property we are interested in.

For such systems the remaining, "non-interesting" region consists of the volume containing all those degrees of freedom which will be eventually neglected and/or discarded once the simulation is done, such as the solvent or large parts of a macromolecule which do not play an active role in the process of interest (e.g., all atoms sufficiently far from the active site of an enzyme). Usually, detailed knowledge about structural, energetic and thermodynamical properties of these large sections of the system is not required; nonetheless these "non-interesting" degrees of freedom have to be explicitly present and integrated, inasmuch as they "scaffold" the target object of the simulation and represent a reservoir of energy and molecules.

A method is thus required that allows one to perform a simulation where the largest part of the computational resources is concentrated on that region of the system that will be subsequently analyzed. *Adaptive resolution simulations methods* [24–34] were developed to solve the contradiction between the necessity of simulating all parts of the system and the fact that, eventually, the detailed information from a large subgroup of them will be neglected. The underlying idea is to replace these "non-interesting" degrees of freedom of the system with a simpler, coarse-grained representation, such that a sensibly smaller number of computations (e.g., force calculations) is required, while the "interesting" region is treated at a higher resolution.

This approach gives rise to at least two important conceptual problems that have to be solved:

(1) what is the smallest number of properties of the original system that have to be retained in the coarser model, and which are they;
(2) how to interface the low-resolution, "non-interesting" region and the high-resolution region to preserve the correct physics at least in the latter.

These two problems are obviously interconnected, since the way the high- and low-resolution regions interact at the interface naturally depends on the specific properties of the models used in each of them; a thorough discussion of these aspects will be carried out in the context of the Adaptive Resolution Simulation (AdResS) [24–32] and Hamiltonian AdResS [33,34] (H-AdResS) methods.

The present review is composed of two principal parts: in Section 2 the basics of coarse-graining theory are presented together with a few examples of the most commonly used techniques, e.g., Force Matching, Boltzmann Inversion and Relative Entropy; in Section 3 we discuss two strategies, the adaptive resolution simulation (AdResS) scheme and the Hamiltonian AdResS

(H-AdResS) to perform simulations in which different regions of the same system are modeled with different resolution.

Large parts of the present review are based on course material that was compiled for two workshops at the Forschungszentrum Jülich ("Hierarchical Methods for Dynamics in Complex Molecular Systems, 2012" [35], and "Workshop on Hybrid Particle-Continuum Methods in Computational Materials Physics", 2013 [36]), as well as on original publications on the respective methodologies [33,34,37–40].

2. Coarse-Graining

As was mentioned in the Introduction, there are many interesting physical problems for which a detailed description of the system at the all-atom (AA) level is not necessary to obtain the relevant information. In these cases a simpler model might be used, where a given high-resolution, computationally expensive model is replaced with a simpler one.

These Coarse-Grained models possess a number of features that make them particularly appealing. For example, a smaller amount of computational resources is required to perform a simulation: this is due to both the reduced number of degrees of freedom and the simpler form of the interactions. Another important characteristic is that since many interaction centers are replaced with a single one, the fluctuations of the force experienced by a molecule are generally much smaller; this results in smoother free energy profiles and, as a consequence, in faster diffusive processes, allowing the system to reach larger time-scales with less computations. This last aspect implies that one typically has to determine a rescaling factor between the simulation timescale (usually given in Lennard Jones units) and the corresponding real world time (or the corresponding timescale in a higher resolution system). A detailed discussion of these dynamic aspects with further references can be found in Reference [41]. Finally, coarse-grained models are designed to reproduce large length-scale properties of the system, such as the global, collective conformational changes of a protein or the diffusive process of a polymer in a melt, that can be strongly insensitive to the fine-grained, chemistry-specific details; as a consequence, the parametrization of the coarse-grained interactions is also advantageously simpler.

Many CG models are generic, i.e., they were not developed to model a specific chemical system but rather with the aim of studying a physical phenomenon such as folding or aggregation in general. One example is generic CG lipid models, which have been successfully employed to study the self assembly of micelles, bilayers and other structures [42–46]. Generic CG models have also been employed to study folding and aggregation of peptides and proteins [47–59]. For polymers, such generic models were especially successful. Following the so called 1/N theorem of de Gennes [60–62] it was shown that properties such as the overall chain extension as a function of the polymerisation index follow the same power law with the same exponent for all polymers, independent of the chemical species. The results of these scaling theories were instrumental in the development of generic and thus very efficient models, as well as in the interpretation of experiments. For dynamical properties generic models simulations provided the first direct evidence of the reptation/tube concept put forward by Edwards and de Gennes [63,64]. The reptation model is based on the fact that the dynamics of long polymer chains is dominated by the constraint that polymer chains cannot simply cut through each other.

A wide range of approaches have been developed that aim for consistency between a CG model and either experimental data or simulations of accurate high resolution models. Typically, these approaches are divided into thermodynamics-based and so-called *structure-based* ones. In thermodynamic coarse graining approaches, individual elements of the CG interaction function are separately parameterized based on thermodynamic reference data such as solvation free energies and partitioning data, liquid densities, surface tension, *etc.* [65–76]. (These are usually experimental reference data, but in a multiscale simulation approach the reference data can of course also be obtained from an atomistic simulation, to keep the CG and atomistic level thermodynamically consistent). In another group of approaches, one numerically generates CG interaction functions with the aim of reproducing the configurational phase space sampled in an atomistic reference simulation. These approaches may rely on different types of reference properties such as structure functions [77–89], mean forces [90–95] or relative entropies [96–98]. In the following subsection, a few basic notions of coarse-graining theory will be introduced, together with examples of the strategies that can be employed to perform the coarse-graining in practice.

2.1. The Mapping Function and the Potential of Mean Force

In a multiscale approach, one first needs to define the relationship between the two levels of resolution. This is typically done via mapping functions which determine the CG Cartesian coordinates of each site as a linear combination of coordinates for the atoms that are involved in the site (that could be via a center-of-mass or a center-of-geometry mapping or some other geometric construction). This means the CG coordinates $\mathbf{R}$ are constructed from the atomistic coordinates $\mathbf{r}$ via

$$\mathbf{R} = \mathbf{M}\mathbf{r} \tag{1}$$

where $\mathbf{M}$ is an $n \times N$ matrix (n and N being the number of particles in the atomistic and CG system, respectively). In the (canonical) sampling of the atomistic and CG systems with respective interaction potentials $V^{AA}(\mathbf{r})$ and $V^{CG}(\mathbf{R})$ the corresponding configuration functions $P^{AA}(\mathbf{r})$ and $P^{CG}(\mathbf{R})$ are given by

$$P^{AA}(\mathbf{r}) = Z_{AA}^{-1} \exp[-\beta V^{AA}(\mathbf{r})] \tag{2}$$

and

$$P^{CG}(\mathbf{R}) = Z_{CG}^{-1} \exp[-\beta V^{CG}(\mathbf{R})] \tag{3}$$

with $Z_{AA} = \int \exp[-\beta V^{AA}(\mathbf{r})]\mathrm{d}\mathbf{r}$ and $Z_{CG} = \int \exp[-\beta V^{CG}(\mathbf{R})]\mathrm{d}\mathbf{R}$ being the respective partition functions and $\beta = 1/k_B T$. If one analyses the atomistically sampled system in CG coordinates one can determine the probability distribution of sampling atomistic coordinates that map to a given CG coordinate $\mathbf{r}$)

$$P^{AA}(\mathbf{R}) = \langle \delta(\mathbf{M}\mathbf{r} - \mathbf{R}) \rangle \tag{4}$$

(Here, we follow the notation used by Noid and collaborators, e.g., in References [99,100]). The angular brackets indicate canonical sampling of the atomistic system (*i.e.*, according to $P^{AA}(\mathbf{r})$).

One can formulate the aim of many systematic coarse graining approaches in the following way: To sample the part of phase space which is sampled by the atomistic system with the same probability distribution. Following this, one possible definition of consistency between atomistic and CG level of resolution is that the two models are consistent if the canonical configurational distribution sampled by the CG model $P^{CG}(\mathbf{R})$ is equal to the probability distribution $P^{AA}(\mathbf{R})$ obtained after mapping the atomistic system to CG coordinates. In a canonical ensemble, independent degrees of freedom q are Boltzmann distributed and the Boltzmann inverse of $P(q)$

$$V(q) = -k_B T \ln P(q) \tag{5}$$

is a many-dimensional potential of mean force (PMF), which, when used for example as an interaction potential in a CG simulation, reproduces the distribution $P(q)$. This means that Boltzmann inversion of $P^{AA}(\mathbf{R})$ defines, uniquely up to an additive constant, a high-dimensional CG potential

$$V_{PMF}^{CG}(\mathbf{R}) = -k_B T \ln P^{AA}(\mathbf{R}) + const \tag{6}$$

which will result in a sampling of CG configurations consistent with the atomistic reference simulation. This high-dimensional, many-body CG potential contains both energetic and entropic contributions from the configurational sampling in the high-resolution model and the mapping between high-resolution and CG model (Equation (4)). Therefore, the resulting CG model is state point dependent and not necessarily readily transferable. While it is conceptually easy to formulate the PMF as a solution of the systematic coarse graining task, it is practically unfeasible. In most cases the PMF cannot be easily determined, and even if it were possible, the resulting high-dimensional potentials are computationally prohibitive. In addition, $V_{PMF}^{CG}(\mathbf{R})$ is a function of $\mathbf{R}$, i.e., this PMF as is can in principle only be applied to a system which is identical in size to the atomistic reference system; if this limitation cannot be overcome, e.g., by breaking it down to short-range interactions, it would defeat the purpose of coarse graining. Therefore, one has to decompose the PMF into simpler independent terms and approximate it by simpler interaction functions, ideally ones that resemble interaction functions typically used in molecular mechanics forcefields, i.e., short range bonded contributions and pair potentials or similar. Conceptually, one can decompose the PMF into a series of many-body terms up to an N-body term, where N is the number of particles on the system. However, this itself does not solve the problem since these multi-body interactions are again computationally unfeasible.

$$
\begin{aligned}
V_{PMF}^{CG}(\mathbf{R}) &= \sum_{i,j} V_2(r_{ij}) + \sum_{i,j,k} V_3(r_{ij}, r_{jk}, r_{ik}) + \cdots + const \\
&\approx \sum_{i,j} V_{\text{eff}}(r_{ij}) + const
\end{aligned} \tag{7}
$$

In Equation (7) one approximates the series by an effective pair interaction which also contains contributions from the higher order terms in Equation (7) (some approaches also include three-body terms for systems where this is necessary [101]). There are many approaches to this task of determining effective CG interactions, and all the resulting CG models are (only) approximations to $V_{PMF}^{CG}(\mathbf{R})$.

2.2. Multi-Scale Coarse-Graining

Probably the most painful limitation in the use of the many-body PMF is the fact that, in general, it cannot be decomposed into a sum of local contributions depending on the interactions between two to a few particles. A simple strategy would therefore be to decide a simple functional form of the potential, e.g., a sum of pairwise, radial interactions, which depend on a set of parameters; the values of the latter are then chosen so that the CG potential is as close as possible to the true PMF. This approach was pioneered by Ercolessi and Adams in 1994 [102] and Tschöp and coworkers in 1998 [103]. Later, Izvekov and Voth [104,105] made use of the force-matching concept of Ercolessi and Adams in the development of the Multi-Scale Coarse-Graining (MS-CG) method. These approaches have been successfully applied to a multitude of biomolecular and other soft matter systems, in particular to biomolecules [90–95].

The central idea of Force Matching is to use a variational (*i.e.*, non-iterative) approach for constructing the CG potential based on the atomistic reference simulation (the recorded forces from the atomistic simulation). The numerical implementation of this variational principle works in such a way that the exact many-body PMF (Equation (6)) is represented by a linear combination of basis functions that are functions of the CG site coordinates [14,15]. For a given configuration of the CG coordinates, in fact, the average of the total atomistic force $\mathbf{f}_\alpha$ acting on a CG site α is equal to the derivative of the many-body PMF:

$$\langle \mathbf{f}_\alpha \rangle_\mathbf{R} \equiv -\frac{\partial U[\mathbf{R}]}{\partial \mathbf{R}_\alpha} \tag{8}$$

where the subscript $\mathbf{R}$ on the averages indicates that the sampling is constrained to those configurations of the AA system having the CG sites in a fixed configuration. The CG force field depends on M parameters $g_1, \cdots, g_M$, that can be prefactors of analytical functions, tabulated values of the interaction potentials, or coefficients of splines used to describe these potentials. These parameters have to be optimized so that the CG force field reproduces the forces in the atomistic system (after mapping) as close as possible. To this end, one minimizes the difference between the average AA force $\langle \mathbf{f}_\alpha \rangle_\mathbf{R}$ and the force $\mathbf{F}_\alpha$ due to the CG potential by minimizing the following quadratic function:

$$\chi^2[\mathbf{F}] = \frac{1}{3N} \left\langle \sum_{\alpha=1}^{N} |\mathbf{f}_\alpha - \mathbf{F}_\alpha|^2 \right\rangle \tag{9}$$

Equation (9) can be rephrased in terms of generalized scalar products of elements in a multi-dimensional vector space; these elements are the $3N$-dimensional force-fields $\mathbf{f}$ and $\mathbf{F}$ acting on the CG sites, with the scalar product and the corresponding norm given by:

$$\mathbf{F}^a \cdot \mathbf{F}^b \equiv \left\langle \sum_{\alpha=1}^{N} \mathbf{F}_\alpha^a \cdot \mathbf{F}_\alpha^b \right\rangle \tag{10}$$

$$\|\mathbf{F}\| \equiv \sqrt{\mathbf{F} \cdot \mathbf{F}}$$

Given the definitions in Equation (10), it can be shown that minimizing the function χ^2 in the MS-CG method is equivalent to minimizing the 'distance' between the many-body PMF and the CG potential:

$$\chi^2[\mathbf{F}] = \chi^2[\mathbf{F}^{\mathbf{PMF}}] + \frac{1}{3N}||\mathbf{F}^{\mathbf{PMF}} - \mathbf{F}||^2 \qquad (11)$$

The force-matching strategy thus *projects* the true many-body PMF onto the basis of functions that are used to define the CG force-field; a thorough formal explanation of this interpretation can be found in Reference [14,15].

It should be noted, however, that the CG force field is still an approximation to the high dimensional PMF within the limitations of the types of CG forces chosen (for example pair forces that can be derived either from analytical or from numerical tabulated potentials). This also implies that a CG model obtained from force matching does not by construction reproduce the pair correlation functions in the system, and the reproduction of local structural properties such as pair distributions may (or may not) be imperfect depending on the importance of cross-correlations between degrees of freedom. An exact reproduction of the underlying atomistic problem by matching mean forces therefore potentially requires the introduction of higher order (e.g., three-body) interactions. Noid and coworkers have extended the force matching method and demonstrated that the CG force field can be directly determined from structural correlation functions obtained from the atomistic system instead of the forces [99]. Their theoretical approach also allows an assessment of the correlations between different interactions that are neglected by straightforward Boltzmann inversion and allows the quantification of the importance of many-body correlations in CG models. In a recent study, Rudzinski and Noid explore these aspects in detail [106]. They demonstrate how the balance between accurately reproducing individual correlation functions (such as pair correlation functions or angle distributions) and also reproducing cross correlations between the respective degrees of freedom is affected by the mapping scheme and the coarse graining method (or more accurately its targets, namely the mean forces versus the individual correlation functions).

2.3. Boltzmann-Inversion Based Methods

In contrast to the Force Matching or Multi-scale coarse graining scheme, other structure-based methods provide CG interactions that reproduce pre-defined target structure properties—often a set of radial distribution functions [77–89]. This means that the many-body PMF (Equations (6) and (7)) is replaced as a target by a set of simpler structural correlation functions. If the interactions in the CG model are statistically independent or only weakly coupled then direct Boltzmann inversion determines each term in the potential immediately from the corresponding distribution function [77,107–109]; for non-bonded interactions in dense systems, though, this is typically not the case. This means that the individual distribution functions and their corresponding potentials of mean force, e.g., a radial distribution function of a simple liquid $g_{target}(r)$ and its Boltzmann inverse, the pair PMF, $V_0^{CG}(r) = -k_B T \ln g_{target}(r)$, cannot be directly used as an interaction function since they correspond not only to the interaction potential but also to the correlated contributions from the surroundings. These multi-body effects of the environment need to be removed from the PMF in order to generate an effective pair potential that reproduces the target structure, for example the pair correlation function in the liquid. It can be shown that such a pair potential is unique up to an arbitrary constant [110] and exists [96,111–113]. There are several numerical methods to generate this pair potential (tabulated interaction function).

Iterative Boltzmann inversion (IBI) [81,114,115] is a natural extension of the Boltzmann inversion method. Here, a numerical CG potential is iteratively refined until the target structure is reproduced within a predefined error. Each step in the iteration procedure is a CG simulation with potential $V_i^{CG}(r)$ which yields an RDF $g_i(r)$ that differs from the target $g_{target}(r)$. The potential is then modified by a correction term $\Delta V(r)$ according to

$$V_{i+1}^{CG}(r) = V_i^{CG}(r) + \Delta V_i(r) = V_i^{CG}(r) + k_B T \ln \frac{g_i(r)}{g_{target}(r)}$$

Sometimes the potential correction $\Delta V_i(r)$ is multiplied by a prefactor $0 < \lambda \leq 1$ to avoid overshooting in the numerical procedure. The iterative procedure is often initiated with the pair potential of mean force $V_0^{CG}(r) = -k_B T \ln g_{target}(r)$, but that is not mandatory. Different starting potentials can be useful, in particular for more complex mixed systems where the iterative procedure may be unstable because intermediate CG models lead to phase separation. This is for example observed in the case of hydrophobic molecules in aqueous solution where both above-mentioned precautions have found to be useful to prevent strong oscillations or even instability of the IBI procedure.

IBI is by no means the only numerical method that solves the above task. Another numerical scheme is the so called inverse Monte Carlo (or more recently renamed Newton inversion) method [78,79,83,84] which, according to Henderson's theorem, should lead to the same numerical solution for the pair potential corresponding to a given pair correlation function. While in IBI the potential update ΔV_i is ad hoc, in IMC it is computed using rigorous statistical mechanical arguments (for details see Reference [78]). In the case of multicomponent systems, where several pair potentials need to be updated, IMC accounts for correlations between observables, *i.e.*, the updates for the different potentials are interdependent. In contrast, for IBI each potential is updated independently, which might lead to oscillations and convergence problems in the iteration procedure. The disadvantage of IMC on the other hand is a high computational cost and problems with numerical stability; for a detailed comparison see Reference [116]. Related to IMC, there are several other recent developments, e.g., a molecular renormalization group approach [85–87] or an approach that relies on relative entropies [96–98] (which will be discussed in more detail below). While the above structure-based methods by construction reproduce *exactly*, within the error of the numerical procedure, the local pair structures and thus are well-suited to the reinsertion of atomistic coordinates, it can be expected *a priori* that they will not be equally well suited to the reproduction of thermodynamic properties (pressure, phase behavior, *etc.*) of the reference system; in this respect, water provides a prototypical case and a reference for testing. Note also that CG models based on pair correlation functions do not necessarily reproduce higher-order (e.g., three-body) structural correlations [116] since the pair correlation functions as structural targets are just an approximation to the total conformational distribution function obtained from the atomistic sampling, $P^{AA}(\mathbf{R})$ (Equation (4)). This means that if higher order correlations are a crucial part of the many-body PMF, models based on pair structures may fail to represent these, and it may even be possible that models which are limited to pair potentials may fail to reproduce these correlations irrespective of the parametrization methodology. One example where this is studied in detail is liquid water [101,116–119]. Recently Noid and coworkers have analyzed these aspects using concepts from liquid state theory [100,120].

One more note concerning Henderson's theorem: even though there is in principle one *exact* solution for the effective pair potential that reproduces a given pair correlation function, different potentials might give a reasonably close representation of the structure, *i.e.*, the above inverse problem is mathematically ill-posed [116,121]. This effect becomes even more pronounced in complex systems where several interaction functions corresponding to several RDFs need to be numerically determined. This can to some extent be turned into an advantage since it allows one to impose thermodynamic constraints in the parametrization procedure. This will result in interaction functions which do *not exactly* reproduce the target structure but give a very close representation while at the same time producing the desired thermodynamic behavior. One example of this is pressure correction terms [81,117]. Here, an additional linear pressure correction is applied during the iterative Boltzmann inversion procedure with

$$\Delta V_{i,P}^{CG}(r) = A_i \left(1 - \frac{r}{r_{cut}}\right) \tag{12}$$

where r_{cut} is the radial cutoff distance of the non-bonded interaction and the constant A is determined via the virial expression for the pressure to

$$-\left[\frac{2\pi N \rho}{3 r_{cut}} \int_0^{r_{cut}} r^3 g_i(r)\mathrm{d}r\right] A_i \approx (P_i - P_{target}) V \tag{13}$$

V is the volume of the system, P_i the pressure of the CG model in the i-th iteration, and P_{target} the target pressure. The price to pay for this adjustment, however, is the loss of the perfect compressibility match. This phenomenon is of course a direct consequence of the state point dependency of coarse grained interactions. Further details on this topic can be found in Reference [117]. Recently, different functional forms of pressure correction terms and the influence of the cutoff length have been explored by Fu *et al.* [122].

It is to be expected that there will be more development in this direction (using other types of thermodynamic constraints) since in particular for complex soft matter system the balancing of structural and thermodynamic behavior in CG models is an ongoing field of research [88,89].

The IBI method is in its original form designed and best suited for systems with uniform density distributions. Recently, Jochum *et al.* have shown how it can be generalized for non-bonded potentials for inhomogeneous systems [123]. For a system with a slab geometry (such as systems of solvent slabs in vacuum or phase-separated systems consisting of two liquid slabs in contact with each other), the method is analogous to IBI but the iterative update of the interaction potential consists of two terms, one based on the radial distribution function calculated in a slab geometry and one that accounts for the slab and interfacial widths. These latter geometric features are very sensitive to the thermodynamic properties (surface tension) of the interface. Therefore the two update terms allow for a balance between the local liquid structure and the thermodynamic properties of the liquid/vapor or liquid/liquid interface. In addition to water/vapor and methanol/vapor interfaces, the method has also been successfully applied to a solute-solvent system of a single benzene molecule at the vacuum-water interface, *i.e.*, it is possible to account to some extent for the partitioning behavior of a solute between bulk and interface, an aspect that makes this method promising in the context of designing transferable CG models for phase separation processes (see below).

Last but not least, one should mention the particular case of Boltzmann-inversion based approaches for mixed systems where (at least) one component is very dilute (from now on termed solute), e.g., biomolecules in aqueous solution. In this case, iterative Boltzmann inversion and similar methods are problematic. While one can easily compute the solvent-solvent and the solute-solvent radial distribution functions, and therefore determine the corresponding CG potentials with for example IBI, this is not so straightforward for the interactions between the low concentration component (solute). (Note that for simplicity only solutes that are represented by a single CG bead will be discussed here.) In these cases, obtaining the PMF through brute force sampling of a radial distribution function is not advisable. One should rather compute the solute-solute pair PMF (between two solute particles) with an advanced sampling method such as umbrella sampling or thermodynamic integration (using distance constraints) [124,125].

When solvent degrees of freedom are not explicitly present in the CG system, this solute-solute PMF can be used directly as an effective solute-solute non-bonded interaction since the environmental (solvent) effects within the PMF are not explicitly represented through solvent degrees of freedom in the CG model. For many types of solutes the solute-solute PMF has been used as an interaction potential in implicit solvent models [126,127]. One prominent example is the use of the solute-solute pair PMF for implicit solvent models of aqueous electrolyte solutions, *i.e.*, implicit solvent ion models [37,79,85,128,129].

The case is somewhat different if some sort of explicit solvent representation, for example in the form of a CG water model, is present in the CG system. In this case, effective solute-solute non-bonded pair interactions are needed from which the solvent contributions are removed in the same way they are removed by IBI in other systems. However, due to the sampling problem of the PMF between dilute components, an iterative procedure is prohibitive for solute-solute interactions. To solve this problem, an approximate method has been developed by Villa *et al.* [38,130]. Here, the CG solvent-solvent and solute-solvent interactions are first determined, for example through normal IBI. Now the pair PMF between the solutes $V_{PMF}^{AA}(r)$ is computed (from atomistic umbrella sampling or thermodynamic integration) and used as a target, in other words the resulting CG model is parameterized to reproduce the solute-solute association strength observed in the atomistic system. In order to remove the solvent contribution from $V_{PMF}^{AA}(r)$, a subtraction procedure is employed. One conducts a separate PMF calculation (again with umbrella sampling or thermodynamic integration), this time in a CG system, where the (previously determined) CG solvent-solvent and solute-solvent interactions are present but no direct interaction between the solute particles is turned on. The resulting PMF $V_{PMF,excl}^{CG}(r)$ *only* consists of the environmental contributions (in the CG environment). By subtracting $V_{PMF,excl}^{CG}(r)$ from the target PMF one obtains the missing direct pair interaction

$$V^{CG}(r) = V_{PMF}^{AA}(r) - V_{PMF,excl}^{CG}(r) \tag{14}$$

which by construction reproduces the target PMF. Note that this subtraction procedure is not necessarily limited to CG solvent-solvent or solute-solvent interactions determined by IBI. In principle other types of CG solvent-solvent or solute-solvent interactions could also be used to determine $V_{PMF,excl}^{CG}(r)$. If one then applies Equation (14), one obtains an effective solute-solute interaction $V^{CG}(r)$ which reproduces the atomistically observed solute-solute association strength (*i.e.*, $V_{PMF}^{AA}(r)$) in the particular CG solvent that was chosen.

2.4. Relative Entropy

Aiming at reproducing different properties or objective functions of the reference, atomistic system, IBI and Force Matching have manifestly different algorithms and produce qualitatively different results. Recently a different coarse-graining strategy has been developed, namely the Relative Entropy method [131–133], which relies on a quantitative measure of the loss of information that follows from the description of a system in terms of different interaction potentials and/or different resolution. Remarkably, it is possible to demonstrate that the information function employed in this strategy connects Relative Entropy, IBI and Force Matching together. The functional form of this measure function is given by the relative entropy, or Kullback–Leibler distance:

$$S_{rel} = \sum_{\nu} \mathcal{P}_{AA}(\nu) \cdot \ln\left(\frac{\mathcal{P}_{AA}(\nu)}{\mathcal{P}_{CG}(\nu)}\right) \tag{15}$$

In Equation (15), ν labels a given microstate or atomistic configuration, $\mathcal{P}_{AA}(\nu)$ is the probability of sampling a configuration ν in the fully atomistic system, and $\mathcal{P}_{CG}(\nu)$ is the probability of sampling the same (atomistic) configuration in the system with coarse-grained interactions, but still described by a high-resolution structure. This latter probability is degenerate with respect to the atomistic-potential configurations, as many of them correspond to the same coarse-grained configuration $\mathcal{V}$. It is therefore advantageous to write the probability to sample a given atomistic configuration in the CG system in terms of the function that maps the fine-grained configurations onto the coarse-grained ones:

$$\mathcal{P}_{CG}(\nu) \equiv \frac{\mathcal{P}'_{CG}(\mathcal{V})}{\Omega(\mathcal{V})}$$
$$\mathcal{V} \equiv \mathbf{M}(\nu) \tag{16}$$

Here, $\mathcal{P}'_{CG}(\mathcal{V})$ is the probability of sampling the CG configuration $\mathcal{V}$ in the low-resolution system and $\Omega(\mathcal{V}) = \sum_{\nu} \delta(\mathbf{M}(\nu) - \mathcal{V})$ is a measure of the degeneracy of the configuration $\mathcal{V}$ in the atomistic system. It should be noted that this last quantity depends only on the mapping function M and not on the coarse-grained interactions; this term can therefore be separated out in the definition of the relative entropy to obtain:

$$S_{rel} = S_{map} + \langle \phi \rangle$$

with:

$$\langle \mathcal{Q} \rangle \equiv \sum_{\nu} \mathcal{P}_{AA}(\nu) \cdot \mathcal{Q}(\nu) \tag{17}$$

$$S_{map} = \langle \ln \left(\Omega(\mathbf{M}(\nu)) \right) \rangle$$

$$\phi(\nu) = \ln \left(\frac{\mathcal{P}_{AA}(\nu)}{\mathcal{P}_{CG}(\mathbf{M}(\nu))} \right) \tag{18}$$

The quantity $\phi(\nu)$ can be interpreted as the amount of information in the configuration ν which discriminates between the atomistic and the coarse-grained probability. The definition in Equation (17) is particularly appealing because it shows that the relative entropy can be computed as the sum of operator averages. In the special, but quite common case of systems in thermal equilibrium, the probability distributions $\mathcal{P}$ are simply given by the Boltzmann weights, and the relative entropy reduces to the form:

$$S_{rel} = S_{map} + \beta \left[(A_{AA} - A_{CG}) - \langle U_{AA} - U_{CG} \rangle_{AA} \right]$$

with A_{AA} (resp. A_{AA}) being the free energy of the atomistic (resp. CG) system. For a given choice of the mapping function $\mathbf{M}$, the optimal coarse-grained potential is obtained by minimizing the relative entropy functional with respect to the parameters in terms of which the aforementioned potential is defined: common choices for non-bonded, two-body interactions are the coefficients of a Lennard-Jones potential or the nodes of a spline.

As anticipated at the beginning of this section, IBI and Force Matching can be connected using the concept of relative entropy. In fact, a straightforward minimization of S_{rel} making use of two-body coarse-grained potentials can be shown to be equivalent to the IBI algorithm; on the other hand, the Force Matching scheme is retrieved if the average of the function $|\nabla \phi|^2$ is minimized instead of the average of ϕ [134]: the squared gradient of the ϕ function with respect to the Cartesian coordinates, in fact, is proportional to the squared difference of the forces obtained from the AA and the CG descriptions, so that:

$$\chi^2[\mathbf{F}] = \chi^2[\mathbf{F}^{\mathbf{PMF}}] + \frac{(k_B T)^2}{3N} \langle |\nabla \phi|^2 \rangle \tag{19}$$

In conclusion, it therefore appears evident that different coarse-graining schemes are obtained through the minimization of different functionals of the same information function ϕ, which represents the unifying element between various approaches.

2.5. Transferability of Coarse-Grained Models

From the preceding sections we have seen that there are different approaches to the systematical parameterization of CG models which by construction will not be equally well suited to the reproduction of thermodynamic and structural properties of the system. It is not *a priori* clear whether structure-based potentials reproduce macroscopic thermodynamic properties and, vice versa, if thermodynamics-based potentials reproduce microscopic structural properties. However, the interplay of structure and thermodynamics is crucial for the investigation of structure formation processes, in particular for biomolecular aggregation in aqueous solution where partitioning and phase separation play a decisive role. All CG models (in fact also all classical atomistic forcefields) are state-point dependent and cannot necessarily be—without reparametrization—transferred to different thermodynamic conditions or a different chemical environment compared to the one where they had been derived. This means "transferability" can refer to a change in temperature, density, concentration, system composition, phase, *etc.*, but also a change in chemical environment, e.g., the change of length or sequence of an amino acid chain. Structure-motivated CG models which approximate the high dimensional PMF obtained from an atomistic reference are by construction heavily state point dependent, and several studies have addressed questions regarding their ability to reproduce thermodynamic properties. One system that has been of particular interest in this context is liquid water [112,117,135]. The reason is on the one hand of course its immense importance in all questions regarding biomolecular systems. In addition, it is of particular methodological interest because for single bead models of water it is known that three-body correlations play a decisive role and the potential compromise between reproducing pair- or higher order structural correlations is particularly relevant for the properties of the model [101,116,117]. Different studies have been carried out that compare structure-motivated and thermodynamics-based CG models [121,136,137]. While CG models where the parametrization targets had been solvation and partitioning properties are particularly well suited to reproduce processes where for example hydrophilicity/hydrophobicity arguments play a decisive role, they do not per se reproduce the structure of the system [121,136]. Related to their ability to reproduce the thermodynamic properties of certain chemical units, these models exhibit considerable transferability and can often be applied to a variety of molecular systems and a range of thermodynamic conditions. Motivated by these observations, intensive research is currently being carried out to derive CG potentials that are both thermodynamically as well as structurally consistent with the underlying higher-resolution description, thus ensuring for example a certain state point transferability [38,88,89,94,138].

One possibility to improve transferability in this context is to exploit—similar to the case of the pressure correction described above—the fact that the derivation of a CG model based on the reproduction of structural properties (potentials of mean force) is an ill-posed problem which allows a reproduction of the original target property within a given error while at the same time including certain thermodynamic target properties during parametrization. One approach developed by Ganguly *et al.* for multicomponent systems that follows this idea combines the IBI method with Kirkwood–Buff integrals as additional targets which are related to the activity coefficients of the components [139]. With this approach transferability over a certain concentration range can be achieved.

Yet another non-structure-based method that produces CG pair potentials with remarkable state point transferability is the conditional reversible work method by Brini *et al.* [140–142]. Here, several calculations of pair potentials of mean force on the atomistic level are used to assess and correctly account for the indirect contribution by the environment to the effective CG pair forces. The observed transferability of this method can be ascribed to the fact that the method relies on direct pairwise interactions in the atomistic reference system. In other words, the method does *not* rely on reproducing a structural property such as a pair PMF or multi-body PMF, *i.e.*, on properties that are extremely dependent on the precise thermodynamic state of the reference system.

It has been mentioned before that effective pair potentials account for multi-body effects, for example, three body interactions. For this reason, they are only to a limited extent additive, which limits the transferability of the potentials [38,143]. Understanding the physical nature of non-additivity in the system of interest can help to make a CG model transferable. In principle, there are various possibilities to approach the question of transferability of effective pair potentials: (*i*) One applies a model derived at/optimized for a given state point unaltered to a range of state points nearby; in that case, one has to carefully investigate the range in which this is permitted [144–146]; (*ii*) One creates a new set of potentials for each state point one wants to investigate [144]; (*iii*) One specifically designs a single CG model with the aim of transferability (for example specific density dependent potentials [94,147,148], CG models that are designed to be applicable for a range of mixture compositions [71,138], or CG models that are capable of capturing a liquid crystalline phase transition [88,89]); (*iv*) One uses a model derived at one state point and (analytically) modifies it to be applicable to different conditions (one example being the rescaling of potentials in order to apply them to a different temperature [149]).

The approach of using a model at a specific state point and then testing its transferability over a reasonably wide range of different physical conditions has traditionally been applied in the case of classical polymer melts. In this field, structure-based models have been very successfully applied, and decent temperature [77,150–152] and pressure transferability [153] have been found. In fact in the first papers by Tschöp *et al.* [77,103] the temperature transferability already allowed the semi-quantitative prediction of shifts in Vogel Fulcher temperature for different polycarbonate modifications. This observation appears to hold for classical isotropic polymer melt systems where the behavior is largely dominated by the correct representation of the chain conformations and the excluded volume of the chain. As soon as more specific chemical interactions play a role, the case of transferability becomes more delicate.

In the following, we discuss three examples which illustrate that understanding the physical basis behind the limitations in transferability can help to design transferable models.

Binary mixtures have in general been widely used as model systems to explore various aspects of the transferability of CG models [37,38,71,128,129,138,143,147,148]. The transferability to different concentrations of liquid mixtures or solutions is of vital importance for simulation of processes such as (bio)molecular aggregation which are characterized by spatially varying structure and fluctuating concentrations.

Following the above-described method to apply Boltzmann-inversion derived methods in dilute solute solvent systems, a CG model for mixed systems of benzene in water had been derived [38]. This means that the CG benzene-benzene potential had been parameterized on the basis of the benzene-benzene PMF of two benzene molecules in aqueous solution, *i.e.*, at "infinite" dilution. Benzene-water mixtures of different composition have been studied with this CG model and analyzed using the Kirkwood-Buff theory of solutions [154]. Kirkwood-Buff theory provides a link between local structural information and thermodynamic properties of the solution. This CG model, parametrized at infinite dilution of benzene, reproduces the Kirkwood-Buff integrals of mixtures at various concentrations obtained with the detailed-atomistic model. It reproduces the changes in the benzene chemical potential and the activity coefficients of the mixtures over a range of mixture compositions (up to concentrations where benzene and water demix in the atomistic reference simulation). A possible explanation is that hydrophobic interactions between benzene solutes are short-ranged, and the multi-body correlations involved in hydrophobic association can be described by pairwise additive effective potentials (category (*i*) of the above list). The observed transferability of the potential supports the idea that hydrophobic interactions between small molecules are pairwise additive. Villa *et al.* also found that a different CG model for benzene-benzene interactions that had been derived for pure benzene (via IBI) is neither suited to describe benzene-benzene interactions in aqueous solution at different concentrations nor a phase-separated benzene/water system with a bulk benzene layer [38].

To reproduce the actual phase separation process as well as the behavior of the mixed (or dilute) systems is much more complicated (yet it is of vital importance in the parameterization of bottom-up CG models that are able to reproduce biological partitioning and self aggregation phenomena). Here, a combination of a wise choice of one or possibly several reference state points is promising, in particular combining the reference of infinite dilution with the phase separated one. For the latter, application of the IBI extension by Jochum *et al.* for inhomogeneous systems with an interface/phase boundary can be utilized [123].

In this context it should also be mentioned that similar transferability problems exist in other areas, for example in the simulations of solids (e.g., with embedded atom potentials). As soon as one encounters surfaces or interfaces the local environment of an atom differs substantially from the bulk (crystalline) phase, which was used to parameterize the interaction potentials. Consequently the transferability of the potentials will affect the ability to model processes such as crack formation or the relocation of grain boundaries [155,156].

In the second example, the situation is different. Here, the transferability of CG (in this case implicit-solvent) ion models in aqueous solution had been investigated. Due to long-range electrostatic interactions, the ions affect the behavior of water increasingly strongly with increasing ion concentration. More specifically, the presence of many ions reduces the orientational fluctuations of the water molecules and thus the dielectric permittivity of the solvent. Therefore, effective ion-ion potentials parametrized at infinite dilution are not directly transferable to higher salt concentrations. Hess *et al.* developed a reduced-resolution (in this case implicit-solvent) potential for aqueous electrolyte solutions where an ion-concentration-dependent Coulomb term was added to the (ion-specific) pair interaction. Thus, by using a concentration-dependent dielectric permittivity for water, part of the multi-body effects in the system were accounted for in the ion-ion pairwise interaction in the implicit solvent model [128,129]. This approach reproduced the NaCl solution osmotic properties and the ion coordination up to a concentration of 2.8 M (mol/L). While in the case of the CG model of benzene/water mixtures [38] the short-range hydrophobic interactions parameterized at infinite dilution were directly transferable to higher benzene concentrations, the ion-ion interactions determined at infinite dilution had to be split into a short-range ion-specific and a long-range electrostatic part. The interactions were then made transferable by keeping the short-ranged part constant and analytically modifying the long-ranged electrostatic part (category (iv) of the above list). Shen et al. have further investigated the structure and osmotic properties of electrolyte solutions over a wide range of concentrations [37]. Using a concentration-dependent dielectric constant one also obtains very good structural properties of the electrolyte solution at low and intermediate salt concentrations while for larger salt concentrations multi-body ion-ion correlations limit straightforward transferability. Guided by this structural analysis, the transferability of the implicit-solvent model could also be improved for high ion concentrations. One obtains transferable implicit-solvent effective pair potentials which are both structurally and thermodynamically consistent with an explicit solvent reference model.

The third example again stresses the immense importance of a good reference state point. It also shows how the reference choice can be guided by understanding the underlying physics.

One highly relevant case of a transferability problem is the ability of a CG model to correctly describe a phase transition while being (reasonably) faithful at both phases below and above, a prominent example being liquid crystalline systems. For such systems, coarse graining can gain access to large system sizes with local disorder, domains etc., and a bottom-up, non-generic CG model has the power to include molecular flexibility and other chemistry specific details. This means that the model should on the one hand faithfully represent the structure in the LC ordered state and on the other hand reproduce the LC/isotropic phase transition.

For an azobenzene-based liquid crystalline compound (8AB8) it was found that state point transferability could be achieved by choosing as an appropriate state point for the reference simulation the supercooled liquid just below the smectic-isotropic phase transition. This reference state is characterized by a high degree of local nematic order while being overall isotropic. The primary idea behind this choice of reference state is the observation that—in the spirit of arguments from classical density functional theories of liquids [157]—the short ranged correlations in the ordered phase are not very different from the local correlations present in a disordered phase at suitable thermodynamic state (density, temperature, etc.) (as one approaches the transition from the high-temperature side). If one captures these local correlations and builds them into the (structure-based) potentials, then these potentials should be able to describe phases on both sides of the transition. For 8AB8, indeed an excellent structural correspondence with the atomistic reference in the smectic state has been found. With the resulting CG model it is possible to switch between the atomistic and the CG levels (and vice versa) in a seamless manner maintaining values of all the relevant order parameters which describe the LC ordered state (see Figure 1). At the same time, this CG model shows remarkable state point transferability and reproduces the LC-isotropic phase transition upon heating and cooling [39]. Such a CG LC model—since it is on the one hand sufficiently coarse grained to study a variety of processes in the LC phase while being at the same time still very closely related to an underlying chemically realistic atomistic description, e.g., allowing for realistic molecular flexibility—is able to give new insights into for example microscopic dynamics in LC phases [40]

Figure 1. A transferable coarse-grained (CG) model for a liquid crystalline molecule that reproduces the ordered/disordered phase transition while at the same time being highly consistent with an atomistic level of resolution. This is achieved by the choice of reference state point, namely the supercooled liquid just below the smectic-isotropic phase transition which is characterized by a high degree of local nematic order while being overall isotropic, for details see Reference [39]. **Left** panel: snapshot of a CG simulation in the LC state with a backmapped atomistic structure superimposed; **Right** panel: This model allows mechanistic studies of dynamic processes in smectic systems, where the influence of the intrinsic flexibility of the molecules on the free energy of different permeation pathways can be elucidated (reprinted from [40]).

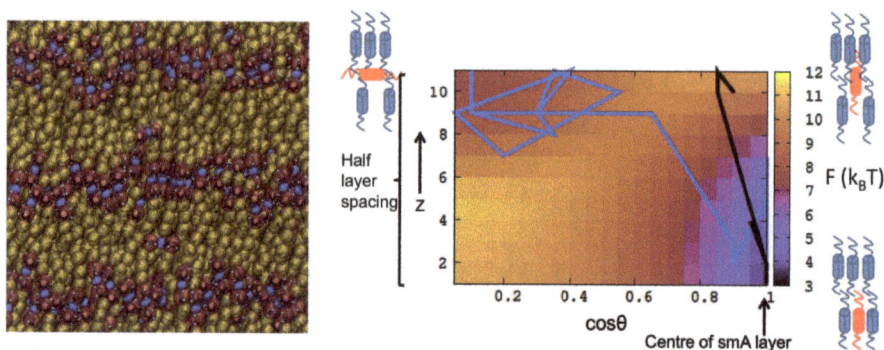

3. Adaptive Resolution Simulations

In the introduction we defined a class of systems for which the focus of the researcher's interest is on a (possibly small) subregion of the simulated system: this is the case, for example, of the hydrogen bond network at the surface of a solvated molecule in water. The bulk of water molecules has to be simulated in order to sustain the thermodynamical properties of the subsystem of interest—the interfacial water—and to provide the correct exchange of molecules. Nonetheless, the fine-grained detail of molecules far from the interface is not relevant; it would be therefore desirable to replace the atomistic, expensive interactions of hydrogen and oxygen atoms with a coarser model.

We can then introduce a geometrical separation between an "inside" and an "outside", *i.e.*, an all-atom and a coarse-grained region, and assign different types of representations and interactions to the molecules according to their position in the simulation domain.

This idea has a long and successful history: to investigate crack propagation in hard matter, for example, several authors [158–162] made use of a hybrid description of the system, where a "high resolution" description is employed only for the area in the proximity of the crack, and the material far from the crack is treated with a simpler model. Another important example of hybrid resolution simulation is provided by Quantum Mechanics/Molecular Mechanics (QM/MM) methods [163–167]. In this case the structure of the system is described at the same (atomistic) level everywhere; however, the interactions are obtained from a classical force-field in the bulk of the system, but in a small region *ab initio* methods -such as Density Functional Theory, DFT- are employed to calculate the forces. Many different "flavors" of this approach have been developed; in all of them, though, one of the crucial aspects is how to interface the two domains where interactions are different, and in most of the established methods the identity or resolution of the particle is not allowed to change. In general, one has to answer the two following questions:

(1) how should two atoms/molecules in different domains interact?
(2) how should the properties of an atom/molecule *change* in crossing the interface?

The last question is of particular importance for all systems whose components can diffuse on large length scales (at last of the order of the molecules' size) in the simulation time. It appears natural to introduce a *transition region* (often called hybrid region, or healing region) that allows for a smooth interpolation from a given representation of the molecule's structure/interaction to another; a schematic representation of this setup is provided in Figure 2. The choice of the specific way this interpolation is implemented depends, as we mentioned earlier, on the properties that have to be preserved in the CG region.

Figure 2. Typical scheme of an adaptive resolution simulation: a high-resolution region, where molecules are described at the atomistic level, is coupled to a low-resolution region where a simpler, coarse-grained model is employed. These two sub-parts of the system are interfaced via a hybrid region, in which the molecule's representation smoothly changes from one to the other, depending on their positions. It is on this last region and its properties (*i.e.*, the way molecules change resolution) that the complexity of adaptive resolution schemes concentrates.

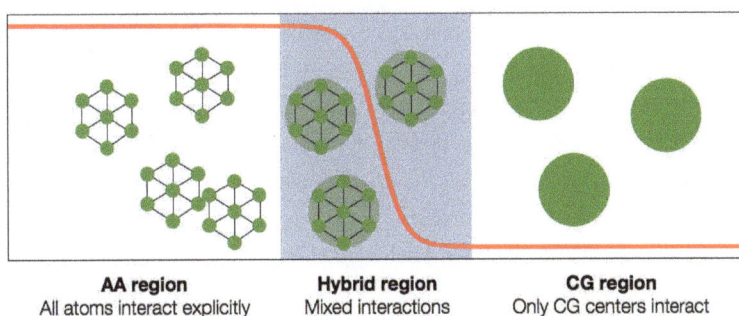

AA region	Hybrid region	CG region
All atoms interact explicitly	Mixed interactions	Only CG centers interact

Irrespective of the chosen method to interface the two regions of the system, however, it is natural to expect that the equilibrium state that will be reached in the absence of external driving forces will not be the desired one. A further crucial point is then to find the simplest way to impose the desired thermodynamics.

The central, strong requirement that has to be satisfied is that molecules should be free to diffuse from any region of the simulation box to any other. Additionally, in a hybrid resolution model thermal equilibrium should be preserved, *i.e.*, the temperature of the system has to be constant during the simulation. Another possible constraint is to impose a uniform density across the box, irrespective of the specific resolution; nonetheless, we'll see that there are cases where this is neither strictly necessary nor desirable.

These are the fundamental constraints that can be imposed on the system as a whole. Other, more specific ones can be introduced on the properties of the CG region as well as the transition region, which will "drive" us towards a specific formulation of a double-resolution simulation method.

3.1. The Adaptive Resolution Simulation Scheme

The Adaptive Resolution Scheme (AdResS) represents the first effective and computationally efficient method to simulate a system where two different models, e.g., an all-atom one and a coarse-grained one, are *simultaneously* employed in different subregions of the simulation domain, interfaced in such a way to allow molecules to freely diffuse from one region to the other.

The basic constraint that was enforced in the original version of this scheme is that Newton's 3rd law has to be exactly satisfied everywhere in the simulation domain. This requirement rules out any form of potential energy interpolation: it can in fact be formally demonstrated [168] that no method exists to smoothly "blend" the interaction between two molecules from a given potential energy to another without generating forces that cannot be recast in a form that satisfies Newton's Third Law. In order to preserve the latter, then, a *force-interpolation scheme* is required, such that the force that a given molecule receives due to the interaction with a second one is antisymmetric under exchange of the molecules' labels:

$$\mathbf{F}_{\alpha|\beta} = -\mathbf{F}_{\beta|\alpha} \tag{20}$$

A second, less strict requirement is that CG molecules possess CG degrees of freedom only; this determines the specific way the force mixing is performed: a molecule in the CG region loses completely its atomistic detail (thus retaining, for example, the center of mass coordinates only), and interacts with a molecule in the AA or even the transition region only via its CG degrees of freedom. Formally, this constraint imposes that the atomistic forces vanish when at least one of the two interacting molecules is in the CG domain.

These two constraints are sufficient to define the force-field interpolation; the force acting between molecules α and β is given by:

$$\mathbf{F}_{\alpha\beta} = \lambda(\mathbf{R}_\alpha)\lambda(\mathbf{R}_\beta)\mathbf{F}_{\alpha\beta}^{AA} + (1 - \lambda(\mathbf{R}_\alpha)\lambda(\mathbf{R}_\beta))\,\mathbf{F}_{\alpha\beta}^{CG} \tag{21}$$

In Equation (21) $\lambda(x)$ is any smooth function that goes from 1 in the AA region to 0 in the CG region. $\mathbf{R}_\alpha$ (resp. $\mathbf{R}_\beta$) is the CoM coordinate of molecule α (resp. β). $\mathbf{F}_{\alpha\beta}^{AA}$ and $\mathbf{F}_{\alpha\beta}^{CG}$ are, respectively, the atomistic and the coarse-grained forces acting on molecule α due to the interaction with molecule β.

The CG force is computed between the coarse grained centers of the molecules and then redistributed to the atoms weighted by the ratio of the atom's mass to the mass of molecule [169]; in the transition region this operation is required by the fact that molecules interact at both the AA and the CG level. AA degrees of freedom thus *have* to be explicitly integrated, at least into the hybrid region. In the CG region, on the other hand, it is in principle not necessary to conserve the atomistic detail of the molecules, so that the CG force could be applied directly to the CoM coordinate; a molecule's internal structure can thus be removed when it enters the CG region, and reintroduced (e.g., taking it from a reservoir/repertoire of equilibrated atomistic molecules) as soon as it approaches the hybrid region. In all AdResS versions implemented so far, though, the atomistic DoFs are retained for simplicity of implementation [24]; the CoM of the molecule is nonetheless decoupled from the internal atomistic structure, and it evolves only subject to the CG force.

It was previously mentioned that no energy interpolation is possible, that is compatible with the requirement of having Newton's 3rd law preserved everywhere in the system [168]; as a consequence, a force interpolation had to be chosen. It is evident, then, that the AdResS scheme cannot be formulated in terms of a Hamiltonian, thus making it impossible to perform microcanonical, *i.e.*, energy-conserving simulations. The force-field used in this adaptive resolution simulation framework is not conservative in the transition region, and when crossing it a molecule receives a surplus of energy that has to be removed in order to prevent the system from artificially heating up. This excess energy can be removed with a *local* thermostat, such as Langevin thermostat: in this way, the temperature of the system is kept constant everywhere. The equilibrium state of the system is then *dynamical*: the thermostat takes care of absorbing the extra heat produced in the transition region by non-conservative forces, and the system samples equilibrium configurations according to Boltzmann's distribution [24–32].

The pressure difference between an AA system and a low-resolution model typically resulting from coarse graining procedures determines the onset of a non-uniform density profile. For example, a one-site CG model of water obtained with IBI can have a pressure $\sim$6000 times the atomistic reference value [117]. Therefore, the densities in the two subregions will change in order to equate the pressures. A possible solution to this density imbalance is to parametrize the CG potential to the target pressure. In the IBI framework this can be achieved by introducing a "pressure correction" [81]. This approach can provide a CG potential that has the target pressure, but this would also result in a modified compressibility [117].

Another option to preserve a uniform density across the simulation domain without modifying the CG potential is to introduce an external force which counterbalances the high pressure of the CG model. This *thermodynamic force* can be obtained with an iterative procedure via the following expression [169–171]:

$$\mathbf{f}_{th}^{i+1} = \mathbf{f}_{th}^{i} - \frac{1}{\rho^{\star}\kappa_T}\nabla\rho^{i}(r) \tag{22}$$

where $\rho^{\star}$ is the reference molecular density, κ_T is the system's isothermal compressibility and $\rho^{i}(r)$ is the molecular density profile as a function of the position in the direction perpendicular to the CG-AA interface. The thermodynamic force is initialized to zero, $\mathbf{f}_{th}^{0} = 0$, while the initial density profile is the one calculated from an AdResS simulation with $\mathbf{f}_{th} = 0$. As can be easily seen, the iterative procedure converges once the density profile is flat ($\nabla \rho(r) = 0$).

This approach guarantees a flat density profile without having to modify the CG potential: because of its very definition, the thermodynamic force only acts on those molecules that cross the hybrid region, leaving the others unaffected. It can also be shown [24,169] that the integral of the thermodynamic force across the interface, *i.e.*, the work due to this force performed by a molecule while crossing the hybrid region, is proportional to the local pressure profile, the proportionality factor being the reference density ρ^*.

In summary, the thermodynamic force allows us to couple a system at atomistic resolution to a coarse-grained counterpart whose pressure, for given values of density and temperature, is significantly different. The global properties of the force, whose direct effect is restricted to the hybrid region, only depend on the pressure difference between the two coupled subsystems; the detailed profile of the force, on the other hand, can be obtained via a system-specific iterative procedure. This method not only allows one to preserve the desired structure of the system in the CG region; in principle, in fact, an arbitrary CG force-field, with pressure *and structure* completely different from the target atomistic ones, can be used. Consequently, the AA region behaves as an open system [169] that exchanges energy and molecules with a reservoir: the molecule number fluctuations, the pressure and all other thermodynamically relevant quantities are the same as if the AA region were simply 'cut' from a large all-atom simulations. It is relevant to stress here that because of the thermodynamic force this condition can be established *irrespective of the specific model used in the CG region*.

3.2. Applications

The possibility of treating a system with a reduced number of degrees of freedom except where it is strictly necessary was explored, making use of the AdResS method, in several applications [24–30,172]. From the numerical/computational point of view it clearly represents an advantage, since a much smaller number of force calculations are required in the coarse-grained region: this is particularly true for parallel MD codes such as GROMACS [173], where a dynamical decomposition of the simulation box allows one to subdivide the box with a finer grid in the AA and hybrid region, while a smaller number of processors is assigned to the CG region. For example, for a water system with an AA region covering $1/6$ of the total simulation box, simulated with GROMACS on a 16-cores processor, the speed-up is about a factor three. This factor is nonetheless small compared to what can be achieved with other simulation packages, such as ESPRESSO++ [174]: in fact, water simulation in GROMACS is extremely optimized, and any hacking of the standard code can introduce a bottleneck.

A major strength of the AdResS method is the fact that it introduces a decoupling between a given region of the system and the rest while keeping the thermodynamic properties of both regions under control: as a consequence, it is possible to conceive numerical experiments in which the spatial extension of correlations in the system is investigated. More specifically, one can study the structural properties of the high-resolution region as a function of its size, in order to determine their dependency on the interaction with molecules in the bulk region. This kind of experiments is different from the study of finite-size effects: in the latter, in fact, the system has the same resolution and interaction type everywhere, and the change of a property with the box size depends on the asymptotic approach to the thermodynamic limit. In the AdResS setup, on the other hand, finite-size effect can be neglected for sufficiently large boxes, thus allowing one to characterize the response of the system's properties in a small subregion when atomistic interactions with the bulk are switched off, but the thermodynamics is the same as in a fully-atomistic simulation. An example of this applications is provided by the work in Reference [175]: here a molecule with both hydrophilic and hydrophobic interactions was solvated in water and put at the center of the high-resolution region, while the water molecules far from the surface were treated at the coarse-grained level. The ordering degree of the hydrogen bond network on the molecule's surface was measured as a function of the size of the all-atom region: the results showed a dependency of the ordering for water molecules close to the surface of the repulsive solute, while no relevant effect was observed for the attractive case.

The same strategy has been applied to investigate the extent of spatial correlations in a quantum fluid, namely low-temperature para-hydrogen [30,176]. The latter is the spin-zero singlet state of molecular hydrogen. Because of the spherical symmetry of the global wave function, para-hydrogen in the solid and gas phase can be modeled as a classical, point-like particle interacting via a simple radial potential, such as Lennard-Jones or the more accurate Silvera-Goldman potential [177,178]. The same classical potential has been shown to correctly reproduce the experimental results both in the solid and the gas phase [178]. In the fluid phase, however, nuclear delocalization effects become important, and a quantum mechanical treatment of the problem is necessary. This can be achieved through the path integral formalism [179,180], which allows for the explicit inclusion of nuclear quantum effects in a "classical" description; unfortunately, this also implies a significant increase in the number of degrees of freedom that have to be simulated, since each molecule becomes a collection of P beads connected by springs. The possibility to simulate a quantum system in a classical framework such as classical MD makes it possible to couple quantum a classical descriptions with the AdResS scheme. In particular, a low-temperature para-hydrogen system was simulated making use of the explicit path integral representation only in a small spherical subregion of the domain, while the molecules in the outer region were treated at the purely classical level, *i.e.*, point-like particles interacting through a coarse-grained potential [30]; in Figure 3 a snapshot of the simulated system is provided. This study showed that a few molecules in a small ($\sim$0.6 nm radius) region of the system are sufficient to reproduce the quantum pair correlation function obtained from a full path integral simulation, but treating the molecules in the outer region at the CG level; this result opens the way to simulate large systems of low-temperature para-hydrogen taking advantage of a double resolution without disrupting the thermodynamical and structural properties of the small, purely quantum region, thus saving computational time in the CG region.

More recently the AdResS scheme has been successfully employed to perform simulations of biologically relevant systems such as methanol-water mixtures [181] and triglycine in aqueous urea [171], and to study the coil-globule transition of a PNIPAm molecule in aqueous methanol [182]. In all these cases a crucial necessity is to correctly reproduce the solvation free energies of the system, a condition that is verified only when the particle number fluctuations are compatible with those observed in the Grand Canonical ensemble. The large system sizes necessary to fulfill this requirement in a standard, all atom simulation often make the latter unfeasible; the employment of dual-resolution simulation methods, possibly coupled to a Monte Carlo scheme [182] to enforce fluctuations in the total number of molecules, see Figure 4, allows one to keep the computational cost low and obtain results that would otherwise require a significantly longer time.

Figure 3. Set-up of the Adaptive Resolution Simulation (AdResS) para-hydrogen simulation performed in Reference [30] (figure adapted from therein). A small sphere in the center of the box, having radius as small as 0.6 nm, is treated at the path integral level (**red rings**), while the rest is described by point-like molecules (the **white spheres**); the hybrid region (**blue**) interfaces these two representations.

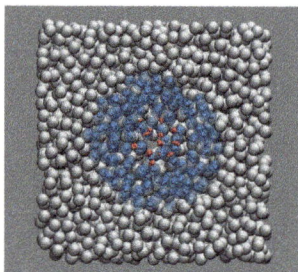

Figure 4. Schematic representation of the schemes used for the simulations of a PNIPAm molecule solvated in aqueous methanol: (**a**) Conventional AdResS scheme, where a small all-atom (AA) region is coupled to a large "closed boundary" coarse-grained reservoir; (**b**) Particle exchange adaptive resolution scheme (PE-AdResS), where an AA region is coupled to a much smaller open boundary coarse-grained reservoir, where particle exchange is performed at the eight corners of the simulation domain to avoid depletion effects; (**c**) Mapping scheme representing the smooth coupling between AA and CG particle representations. Figure from [182].

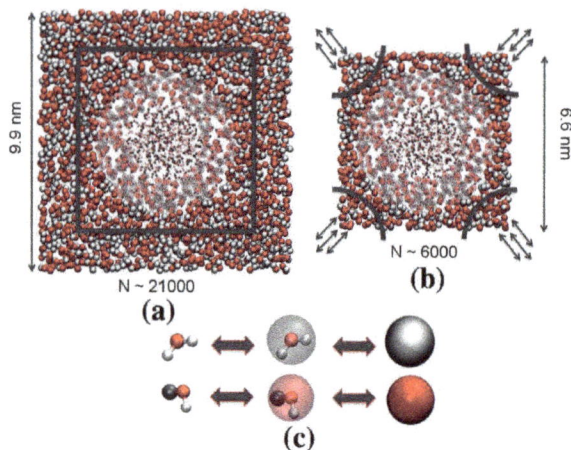

3.3. The Limitations of the Force-Based Approach

The AdResS method discussed so far represents a simple, effective way to perform double-resolution simulations, *i.e.*, simulations where the model used to represent a molecule and its interactions with the others changes according to the molecule position. Assigning the lowest-resolution model to the largest region allows one to save computational resources and characterize the bulk dependence of structural properties of the high-resolution subsystem. A majorly important point is given by the possibility to keep the thermodynamics of the system under control: this can be achieved by direct intervention on the CG model's properties, or by introducing an external field -the thermodynamic force- in the hybrid region to compensate for density imbalances. This second streategy is crucial, since it allows one to *couple arbitrarily different systems* while keeping locally well-defined temperature, pressure and energy.

The AdResS method was conceived based on the requirement that Newton's Third Law has to be exactly satisfied everywhere. This constraint poses a strict limitation to the possible ways to interface the two representations of the system: specifically, no potential energy interpolation is possible, via a position-dependent switching function, that preserves Newton's Third Law [168]; as a consequence, the only acceptable interpolation can be performed on the forces.

A posteriori, the lack of a global energy function proves not to be a major problem: equilibrium and canonical sampling can be enforced making use of a local Langevin thermostat. A theoretical analysis of the AdResS dual resolution scheme has been recently carried out in Reference [183], where the presence of a local thermostat and the thermodynamic force have been shown to be necessary and sufficient conditions to guarantee the equivalence of the atomistic region to an open region of a fully atomistic simulation up to second order correlation functions (density profile and radial distribution function). These results have been obtained from a completely general model of a dual resolution setup under the assumption of the thermodynamic limit; the generality of this approach makes it thus applicable to different types of adaptive resolution schemes, independently of the detailed form of the method chosen to interpolate the resolutions.

Nonetheless, the lack of a Hamiltonian has negative consequences on the usage of the AdResS method; the four major ones are: microcanonical, *i.e.*, energy-conserving simulation are not possible; no partition function can be written for the system as a whole; no Monte Carlo scheme can be implemented. Finally, due to the non-conservative nature of the forces in the hybrid region the system necessarily has to be locally thermostatted to compensate for the heat that is produced in the hybrid region, so that an AdResS simulation is found to be in a state of *dynamical equilibrium* [32], with a constant flux of heat between the system and the thermostat.

In the next section a method is discussed, named H-AdResS [33] (for Hamiltonian Adaptive Resolution Simulation scheme), that provides a solution to the aforementioned problems; clearly, as no free lunch is usually available, there is a price to pay: the Hamiltonian formulation requires a local breakdown of Newton's Third Law.

3.4. The Hamiltonian Adaptive Resolution Scheme

As was discussed in the previous section, the force-based AdResS method was developed on the basis of a central requirement, namely that Newton's 3rd law has to be exactly satisfied everywhere. A consequence of this constraint is that no Hamiltonian formulation is possible [168]: if a position-dependent interpolation of the potential energies is done, in fact, the resulting forces include a term proportional to the derivatives of the switching function λ that cannot be recast in a form that satisfies Newton's Third Law. The only method developed in the past that allows one to explicitly conserve the energy in an adaptive resolution simulation is that proposed by Heyden and Truhlar, [184,185], where a sum of the Lagrangians of all possible groupings of atomistic and coarse-grained molecules is done. Due to its combinatoric nature, this approach is extremely difficult to implement efficiently; moreover, the resulting Lagrangian includes a position-dependent kinetic energy term for which a specific, non-symplectic integrator is required.

In the H-AdResS method [33], which we now describe, the aforementioned constraints are relaxed in order to develop an energy-based, Hamiltonian adaptive resolution simulation scheme. As will be clear in a few lines, the particular choice of energy "mixing" gives rise to forces that do not comply with the first constraint; nevertheless, the physical interpretation of these terms is immediate and naturally points towards the solution -though approximate- of the Newton's Third Law breakdown.

The core idea of the energy-based approach is to weight the *total energy* of each molecule with a position-dependent function:

$$H = \mathcal{K} + V^{int} + \sum_{\alpha} \left\{ \lambda_\alpha V_\alpha^{AA} + (1 - \lambda_\alpha) V_\alpha^{CG} \right\} \tag{23}$$

where $\mathcal{K}$ is the (all-atom) kinetic energy of the molecules, V^{int} is the interaction internal to the molecules, and:

$$\begin{cases} V_\alpha^{AA} \equiv \dfrac{1}{2} \sum_{\beta,\beta\neq\alpha}^{N} \sum_{ij} V^{AA}(|\mathbf{r}_{\alpha i} - \mathbf{r}_{\beta j}|) \\[2ex] V_\alpha^{CG} \equiv \dfrac{1}{2} \sum_{\beta,\beta\neq\alpha}^{N} V^{CG}(|\mathbf{R}_\alpha - \mathbf{R}_\beta|) \\[2ex] \lambda_\alpha = \lambda(\mathbf{R}_\alpha) \end{cases}$$

The switching function λ goes from 0 (purely CG) to 1 (purely AA). The force acting on atom i in molecule α is obtained through differentiation of the Hamiltonian in Equation (23):

$$\mathbf{F}_{\alpha i} = \mathbf{F}_{\alpha i}^{int} + \sum_{\beta,\beta\neq\alpha} \left\{ \frac{\lambda_\alpha + \lambda_\beta}{2} \mathbf{F}_{\alpha i|\beta}^{AA} + \left(1 - \frac{\lambda_\alpha + \lambda_\beta}{2} \right) \mathbf{F}_{\alpha i|\beta}^{CG} \right\} - \left[V_\alpha^{AA} - V_\alpha^{CG} \right] \nabla_{\alpha i} \lambda_\alpha \tag{24}$$

The forces $\mathbf{F}_{\alpha i|\beta}^{AA}$ and $\mathbf{F}_{\alpha i|\beta}^{CG}$ are defined as:

$$\mathbf{F}_{\alpha i|\beta}^{AA} \equiv \sum_{j=1}^{n_\beta} -\frac{\partial}{\partial \mathbf{r}_{\alpha i}} V(|\mathbf{r}_{\alpha i} - \mathbf{r}_{\beta j}|)$$

$$\mathbf{F}_{\alpha i|\beta}^{CG} \equiv -\frac{m_{\alpha i}}{M_\alpha} \frac{\partial}{\partial \mathbf{R}_\alpha} V^{CG}(|\mathbf{R}_\alpha - \mathbf{R}_\beta|) \tag{25}$$

The redistribution of the CG force on the atomistic degrees of freedom follows the same rules as applied in the case of the force-based AdResS method. It's worth noting that in this energy-based scheme the atomistic degrees of freedom are retained and integrated everywhere in the system, a necessary requirement in order to perform a *microcanonical* simulation making use of a Hamiltonian.

We now detail the various components of the force, Equation (24). The first term, $\mathbf{F}_{\alpha i}^{int}$, is due to the interactions *internal* to the molecule; as such, it automatically satisfies Newton's Third Law. The second term is a sum of pairwise forces obtained from all-atom and coarse-grained Hamiltonians, weighted by a function that is symmetric under molecule label exchange, that is $\alpha \leftrightarrow \beta$; this force also complies with Newton's Law. Up to this point we modified only one aspect of the original AdResS scheme, that is, the force weights are not given by the product of the two molecules' switching function, rather by the average; consequently, the molecules in the coarse-grained region are also allowed to interact through their atomistic degrees of freedom.

The third term of the forces in Equation (24) is the part that breaks down Newton's Third Law: in fact, it cannot be written as a sum of terms antisymmetric under molecule label exchange. This force, which is nonzero *only in the hybrid region*, is proportional to the difference between the potential energies of a given molecule in the AA and the CG representation; if a systematic difference exists between the AA and the CG potentials, the effect of this term is to push molecules into one of the two bulk regions. The hybrid region thus behaves as an *active membrane*, inducing a density imbalance and a non-flat pressure profile. One is then naturally led to ask how strong is the drift term $\mathbf{F}_{\alpha}^{dr} = -\left[V_{\alpha}^{AA} - V_{\alpha}^{CG}\right]\nabla_{\alpha i}\lambda_{\alpha}$; if it is negligible in some cases; which these cases are; and if there is a general way at least to minimize its effect without giving up the Hamiltonian character of these forces. We shall now address these questions.

The optimal case in which this term is minimized is *when the CG potential perfectly reproduces the many-body PMF*. If this is true, in fact, the drift term vanishes on average:

$$V_{\alpha\beta}^{CG} \equiv \left\langle V_{\alpha\beta}^{AA}\right\rangle \;\Rightarrow\; \left\langle\mathbf{F}_{\alpha}^{dr}\right\rangle \propto \left\langle\left[V_{\alpha}^{AA} - V_{\alpha}^{CG}\right]\right\rangle \to 0$$

This can be numerically verified with a simple toy model, for which a pairwise CG potential represents an excellent approximation to the PMF. Such a model is provided by a low-density fluid of purely repulsive tetrahedral molecules [24], whose CG potential has been obtained from IBI. This model was used in an energy-conserving H-AdResS simulation, and the resulting density profile is plotted in Figure 5. The molecular density attains the same value in both the AA and CG regions; in the hybrid region a small depletion is present, because the free energy of the mixed potential is different from the free energy of the "pure" (*i.e.*, purely AA or purely CG) potentials. The same behavior has been systematically observed in AdResS simulations [24].

Needless to say, this particular case is very fortunate: as we discussed in the previous sections, the CG potentials almost never reproduce the many-body potential of mean force [117,186]. The difference between an atomistic model and its coarse-grained representation therefore results in a thermodynamic imbalance, that is, both pressure and density of the two bulk (AA and CG) regions are different [13]. The solution to this problem is again to introduce a compensation term in the

Hamiltonian, as was done in the AdResS scheme with the thermodynamic force. More specifically, we modify the Hamiltonian as follows:

$$H_\Delta = H - \sum_{\alpha=1}^{N} \Delta H(\lambda(\mathbf{R}_\alpha)) \tag{26}$$

where $\Delta H(\lambda)$ is a function to be defined. It's worth noting that this term preserves the conservative nature of the Hamiltonian.

Figure 5. H-AdResS simulation of a system of tetrahedral molecules coupled to point-like molecules interacting through an Iterative Boltzmann inversion (IBI)-CG potential (reprinted from the Supporting Information of Reference [33]). Top: density profile; bottom: radial distribution functions of the atomistic (red lines) and coarse-grained (blue lines) degrees of freedom in the all-atom region; the solid lines are the reference RDFs calculated in the all-atom system, while the dashed lines are obtained from a H-AdResS simulation.

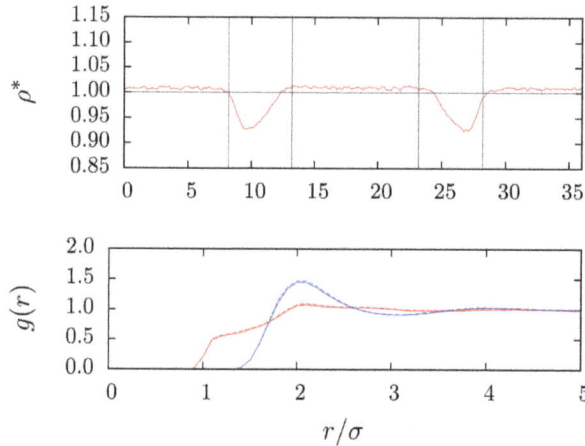

In order to determine the specific form of ΔH we impose that the drift force cancels on average:

$$\left.\frac{d\Delta H(\lambda)}{d\lambda}\right|_{\lambda_\alpha} \nabla_\alpha \lambda_\alpha + \langle \mathbf{F}_\alpha^{dr} \rangle \equiv 0 \tag{27}$$

or equivalently:

$$\left.\frac{d\Delta H(\lambda)}{d\lambda}\right|_{\lambda=\lambda_\alpha} = \left\langle \left[V_\alpha^{AA} - V_\alpha^{CG}\right]\right\rangle_{\mathbf{R}_\alpha} \tag{28}$$

where the subscript in the average indicates that the latter has to be performed constraining the CG site of molecule α in the position $\mathbf{R}_\alpha$.

In principle, Equation (28) provides us with the way to compute the compensating function—or, more precisely, its derivative; nonetheless, an approximation to ΔH might be sufficient. A way to do this is the following:

$$\langle [V_\alpha^{AA} - V_\alpha^{CG}] \rangle_{\mathbf{R}_\alpha} \simeq \frac{1}{N} \langle [V^{AA} - V^{CG}] \rangle_{\lambda'} \tag{29}$$

where $\lambda' \equiv \lambda(\mathbf{R}_\alpha)$ is the same for all molecules. The approximate function ΔH is obtained by integration:

$$\Delta H(\lambda) = \int_0^\lambda d\lambda' \frac{d\Delta H(\lambda')}{d\lambda'} \simeq \frac{1}{N} \int_0^\lambda d\lambda' \langle [V^{AA} - V^{CG}] \rangle_{\lambda'} = \frac{\Delta F(\lambda)}{N}$$

Most interestingly, we see from Equation (30) that the compensation needed to cancel $\langle \mathbf{F}_\alpha^{dr} \rangle$ is related to the *Helmholtz free energy* difference between AA and CG system [187]. Therefore, it is possible to calculate the compensating function needed to restore, on average, Newton's Third Law by performing a Kirkwood thermodynamic integration.

The "Helmholtz free energy compensation" thus cancels the active effect of the hybrid region, restoring a flat pressure profile. Nonetheless, coarse-grained models have, in general, a substantially different pressure with respect to their atomistic reference [117], thus inducing a further density imbalance (usually larger than the one due to the different Helmholtz free energy). In order to restore a flat density profile a second term has then to be added to the compensating function, that counterbalances the pressure difference.

The right way to introduce the pressure into the compensating function is to balance, rather than Helmholtz free energy, the *Gibbs free energy* difference per particle, that is, the chemical potential $\Delta\mu = \Delta G/N$:

$$\Delta H(\lambda) \equiv \Delta\mu(\lambda) = \frac{\Delta F(\lambda)}{N} + \frac{\Delta p(\lambda)}{\rho^\star} \tag{30}$$

Figure 6 shows the density and pressure profiles for the three possible cases we discussed: the previously mentioned system of tetrahedral molecules was coupled to a coarse-grained fluid of purely repulsive point-like molecules; the pressure of this fluid has a larger pressure then the reference all-atom one for the same temperature and density. In the plot, the red lines correspond to the case in which no free energy compensation is introduced: the density is higher in the AA region, due to the molecules in the CG region that "push" with a higher pressure. The profile of the pressure is also not flat: the Helmholtz free energy of two systems differs, therefore an active force exists in the hybrid region. When the Helmholtz free energy compensation is applied we have the situation shown by the green lines: the density is still higher in the AA region, but the pressure profile is now flat: the forces that break Newton's Third Law in the hybrid region are cancelled on average, and the density imbalance decreases. Finally, when the Gibbs free energy compensation is applied the densities of the AA and CG regions attain the same value, but for a small deviation due to the fluctuations present in the hybrid region (that the compensation function ΔH, computed in a homogeneous system, cannot remove). The pressure, on the other hand, is different: in fact, in each region it reaches the value that corresponds to the reference state of density and temperature. Analogous results are obtained in a thermostatted simulation of a water box, as shown in Figure 7: here the system is composed of a slab of water molecules described at atomistic resolution, coupled to a CG bulk where particles interact via a purely repulsive WCA potential. As in the previous case, the CG interaction was parametrized to induce an increase of the density in the atomistic region, as can be seen in Figure 8 (upper panel). The Free Energy Compensation restores the correct density profile, and guarantees that in the AA region the pairwise correlations, *i.e.*, the radial distribution functions, are the same that one would measure in a fully atomistic simulation, as shown in Figure 8 (bottom panel). We notice that Gibbs free energy compensation, even though it equates the densities in the bulk regions, is not sufficient to remove small fluctuations (of the order of ∼3%) in the hybrid region: these deviations from the reference value are due to the fact that the compensation ΔH is computed in a homogeneous system, where all molecules have the same value of λ—that is, a regular Kirkwood thermodynamic integration Hamiltonian. The molecules in the hybrid region, on the other hand, interact with other molecules having different λ values. The resulting fluctuations are expected to decrease with increasing size of the hybrid region, in which case the environment of a given molecule approaches the condition of homogeneous λ. Another strategy to flatten the density profile is clearly provided by the iterative approach of the thermodynamic force (Equation (22)), a few iterations of which would be sufficient to modify the ΔH function by the small amount necessary to remove the fluctuations.

Figure 6. Plots showing the effect of the free energy compensations on the density profile (upper panel) and pressure profile (lower panel) in a H-AdResS simulation with CG potential having larger pressure, for identical temperature and density values, than the all-atom one (reprinted from Reference [33]). The red line corresponds to the case where no compensating function was employed; the green line to the Helmholtz free energy compensation; and the blue line to the Gibbs free energy compensation. All densities are normalized to the value of the fully atomistic simulation (dotted line at $\rho = 1$). All pressures are normalized to the value of the fully atomistic simulation (dash-dot line); the dotted line indicates the normalized pressure of the fully coarse-grained simulation.

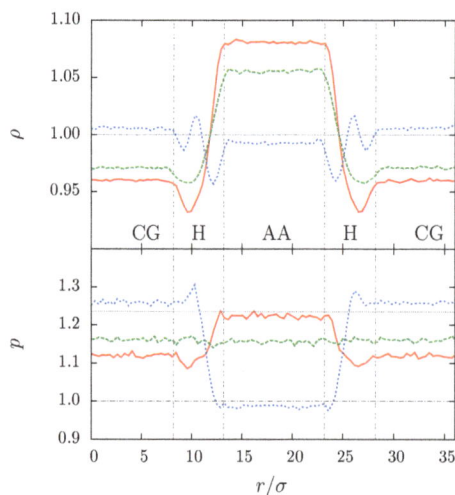

Figure 7. Schematic view of a dual-resolution simulation of water: the central slab of the box is described at atomistic resolution, while in the bulk the molecules are point-like particles interacting via a purely repulsive WCA potential.

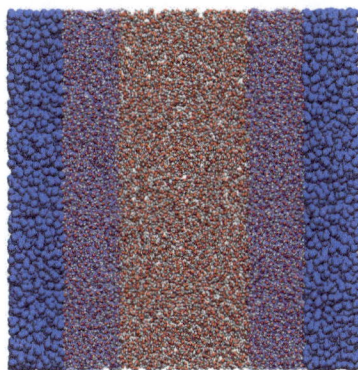

Figure 8. Top panel: density profile of the water system along the x coordinate. The red dotted line corresponds to the H-AdResS simulation without FEC, while the solid back line has been obtained using the FEC. Bottom panel: radial distribution functions of the water atoms in the central (AT) slab of the box, as obtained from a fully atomistic simulation (solid lines) and a H-AdResS simulation with FEC (dots).

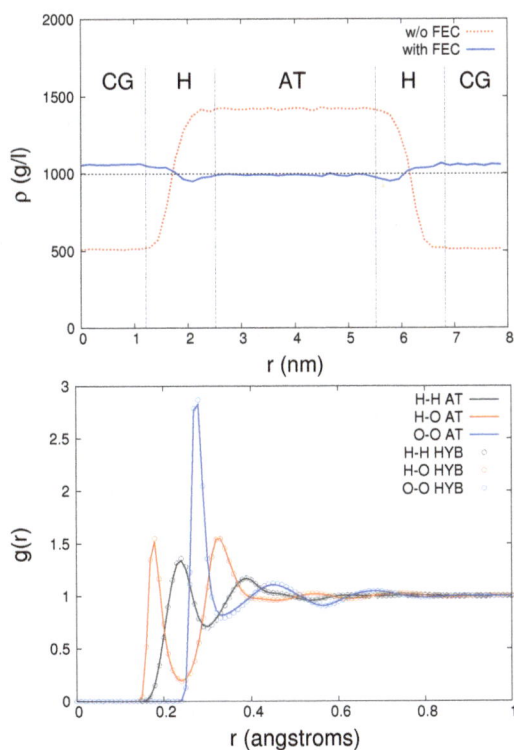

The Free Energy Compensation (FEC) strategy, defined by Equation (26), can be extended to multi-component systems. To illustrate this idea we consider a molecular liquid composed by two types of molecules, A and B, indexed with a and b, respectively. The corresponding H-AdResS Hamiltonian for this system reads:

$$H^{MIX} = K + V^{int} + \sum_{a \in A} \left[\lambda_a V_a^{AA} + (1 - \lambda_a) V_a^{CG} \right] + \sum_{b \in B} \left[\lambda_b V_b^{AA} + (1 - \lambda_b) V_b^{CG} \right] \tag{31}$$

with $\lambda_a = \lambda(\mathbf{R}_a)$ and $\lambda_b = \lambda(\mathbf{R}_b)$. The intermolecular potential energy terms are given by the following expressions:

$$V_a^{AA} = \frac{1}{2} \left[\sum_{\substack{a' \in A \\ a' \neq a}} \sum_{ij} V[AA]_{ai;a'j}^{AA} + \sum_{b \in B} \sum_{ij} V[AB]_{ai;bj}^{AA} \right]$$

$$V_a^{CG} = \frac{1}{2} \left[\sum_{\substack{a' \in A \\ a' \neq a}} V[AA]_{aa'}^{CG} + \sum_{b \in B} V[AB]_{ab}^{CG} \right]$$

$$\tag{32}$$

$$V_b^{AA} = \frac{1}{2} \left[\sum_{\substack{b' \in B \\ b' \neq b}} \sum_{ij} V[BB]_{bi;b'j}^{AA} + \sum_{a \in A} \sum_{ij} V[AB]_{bi;aj}^{AA} \right]$$

$$V_b^{CG} = \frac{1}{2} \left[\sum_{\substack{b' \in B \\ b' \neq b}} V[BB]_{bb'}^{CG} + \sum_{a \in A} V[AB]_{ba}^{CG} \right]$$

where $V[XY]$ is the non-bonded interaction between a molecule of type X and a molecule of type Y, with $X, Y = A, B$, and the indices i, j labeling the atoms.

In analogy with one-component systems we introduce a FEC term for each species to compensate for the free energy difference between the AA and the CG regions:

$$H_\Delta^{MIX} = H^{MIX} - \sum_{a \in A} \Delta H_A(\lambda_a) - \sum_{b \in B} \Delta H_B(\lambda_b) \tag{33}$$

An *Ansatz* for the compensation term of a given species $k = a, b$ can be obtained from TI as follows:

$$\Delta H_k(\lambda) = \frac{\Delta F_k(\lambda)}{N_k} + \frac{\Delta p_k(\lambda)}{\rho_k^\star}$$

$$\Delta F_k(\lambda) = \int_0^\lambda d\lambda' \left\langle \left[V_k^{AA} - V_k^{CG} \right] \right\rangle_{\lambda'}$$

$$\Delta p_k(\lambda) = p_k(\lambda) - p_k(0) \tag{34}$$

where the N_k, $\rho_k^\star \equiv N_k/V$ and p_k are, respectively, the number of molecules, the reference partial density and the partial virial pressure of species k. We stress that all the quantities in Equation (34) can be computed in a single TI of the mixture from AA to CG at the concentration of interest, irrespective of the number of species. All the cross-interactions between different types of molecules are automatically included in the free energy contribution of each species. Additionally, the Free Energy Compensation $\Delta H_k(\lambda)$ is an *intensive* quantity and does not depend on the specific geometry of the H-AdResS setup. It is therefore possible to perform the TI in a relatively small system, provided that it is statistically representative, *i.e.*, finite size effects are negligible.

The effectiveness of this strategy has been proven by the Monte Carlo simulations of binary mixtures performed in Reference [34]. Here we report one of these simulations, specifically the mixture of 70% A-type molecules and 30% B-type molecules, both made of four identical atoms; the A–A and B–B interactions are identical WCA potentials, while the A–B interaction is a Lennard-Jones potential. In the CG region both molecules are represented as spherical particles with identical, purely repulsive WCA A–A, B–B and A–B interactions, resulting in a particularly large thermodynamic mismatch between AA and CG domains. This can be directly observed in the snapshot of the simulation reported in Figure 9 (top) as well as in the density profiles (dotted lines in Figure 10): the chemical potential imbalance between the two resolutions leads to a large accumulation of B-molecules in the AA zone. As a consequence, neither the total density nor the relative concentrations in the AA zone obtained using the uncompensated adaptive resolution Hamiltonian in Equation (31) correspond to the reference atomistic system.

Figure 9. Snapshots of a H-AdResS Monte Carlo simulation (reprinted from [34]). **Top** panel: Equilibrated configuration, without FEC. **Bottom** panel: Equilibrated configuration, with FEC. The A-type atoms are represented in gray, the B-type atoms in orange. Molecules in the coarse-grained (CG) region are represented as large spheres. White vertical lines mark the boundaries of the CG-hybrid and hybrid-atomistic regions.

According to Equation (34), a thermodynamic integration was performed to determine the thermodynamic mismatch between the AA and the CG zone. The Helmholtz and Gibbs free energy differences per molecule between the CG and AA models as a function of the coupling parameter

λ, computed for both species *simultaneously* in a single TI, are shown in Figure 11. In spite of the same interaction between molecules of the same type ($V[AA] \equiv V[BB]$), the uneven relative concentration of the two species determines a much larger free energy difference between the AA and CG models for the B-type. In fact, the latter shows a Gibbs free energy difference per particle $|\Delta G_B/N_B| > 2 |\Delta G_A/N_A|$. This is mainly due to the fact that the interaction between A and B types is attractive only in the AA representation, thus determining a lower chemical potential for the minority type (B) in the AA region. In addition, in both cases the sign of ΔG favors the densification of particles in the AA region, as can be seen in Figure 10. To counterbalance the mismatch in chemical potentials a FEC was introduced in the H-AdResS Hamiltonian according to Equation (33), using the free energy functions shown in Figure 11. The resulting density profiles (solid lines in Figure 10) demonstrate the success of the procedure.

Figure 10. Density profiles along the direction of resolution change (reprinted from [34]). Dotted lines: H-AdResS simulations without FEC; solid lines: With FEC. Vertical dashed lines indicate the boundaries between the AT, hybrid and CG regions; horizontal dashed lines mark the reference value of the density (normalized to the total density) as expected in a fully atomistic simulation of the system.

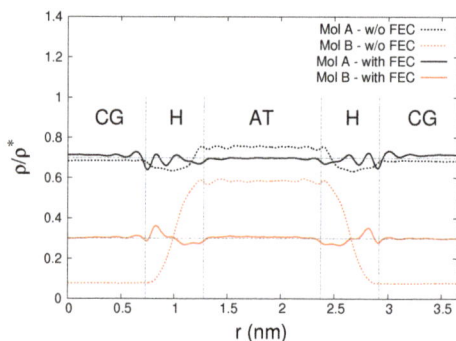

Figure 11. Free energy differences per molecule between the AA and CG models as a function of the mixing parameter λ (reprinted from [34]). The Helmholtz free energy is represented by the dotted lines, the Gibbs free energy by the solid lines. Molecular species A corresponds to the black curves, species B to the orange curves.

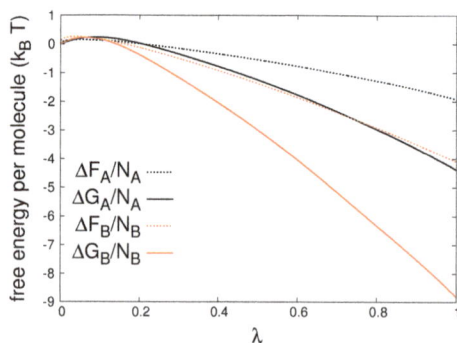

In this section we discussed the H-AdResS method, which allows for a seamless coupling of two models of the same system with different resolution within a Hamiltonian framework. In order to define an energy-based mixing rule for the two models, the requirement to preserve Newton's Third Law everywhere in the system had to be relaxed. Nonetheless, the "undesired" term that appears in the forces due to the differentiation of the switching function λ is non-zero only in the hybrid region, and its particular form naturally indicates how to introduce, in a physically sound manner, a compensation function that cancels the average effect of the drift force without disrupting the Hamiltonian character of the model. The computational cost of the H-AdResS simulations is comparable to that of the AdResS method, the only difference being the need to calculate the drift force $\mathbf{F}_\alpha^{dr}$ in the hybrid region: nonetheless, the number of molecules that are affected by this force is typically small (both the AA and the CG regions are expected to be much larger than the hybrid region), and the quantities involved, namely interaction energies and molecules' CoM coordinates, are normally computed in a MD simulation.

In spite of its simple formulation and relatively small difference with respect to the force-based method, H-AdResS represents a major step forward in terms of understanding and practical advantages. In fact, the existence of a Hamiltonian allows one to precisely formulate a statistical physics theory of double-resolution systems, providing a deep insight into the properties of a given all-atom model, its coarse-grained counterpart and the relation between them. In particular, the free energy compensations provide a simple and effective way to modulate the thermodynamic balance of AA and CG regions, thus leaving to the user the choice of the environment for the AA region most appropriate for the specific problem under examination. Last but not least, this scheme broadens the spectrum of physical ensembles that can be simulated to the microcanonical ensemble, and allows the use use of simulations techniques—e.g., Monte Carlo—that were not accessible in the force-based AdResS framework, with the *a priori* guarantee that real equilibrium configurations are sampled.

4. Conclusions

The characterization of the properties of new materials, as well as the investigation of biological macromolecular machineries, have largely benefited from *in silico* experiments. In spite of a steady increase in available computational power for very large systems and long timescales of the processes involved these resources turn out to be insufficient, due to the extraordinarily large amount of data that has to be stored and force/energy calculations that have to be performed. To overcome these limitations, the field of multiscale simulations has vastly expanded over recent years, and in the present review we have covered two aspects that are central to many multiscale approaches.

In the first part of the review we have addressed methodological questions associated with the development of coarse grained models, where atoms are grouped into super-atoms to reduce the number of degrees of freedom in the system. We have summarized the current approaches to bottom-up coarse graining and addressed some of the ongoing coarse graining issues such as the choice of parametrization targets and the choice of interaction functions used for the coarse grained model. These choices lead to several possibilities (*i.e.*, coarse graining methodologies) for solving the inverse problem of finding parameters for the coarse grained interaction functions given the selected target properties. We have briefly discussed these (statistical-mechanically interrelated) methods in context with each other. An inevitable question that arises from having to choose coarse graining target properties and approximations to solve the parametrization problem is the question of representability of different thermodynamic and structural properties. These representability challenges go hand in hand with the question of transferability, *i.e.*, to which extent a reduced-resolution model is applicable to a state-point that is different from the one where it was parametrized. In general it can be said that transferability problems increase with decreasing level of resolution, *i.e.*, the coarser a model the more limited is its applicability range, which then needs to be very carefully assessed. However, as a positive aspect one should mention that the investigation of transferability issues can help to gain insight into physical-chemical principles that drive the behavior of the system. We have illustrated transferability-related questions with the help of a few examples. In conclusion, one should mention that transferability problems are *not* specific to coarse grained models. Such problems are well known for classical atomistic forcefield models as well. A good example is simulations of mineral systems in contact with electrolyte or polyelectrolyte solutions. Here, forcefields for ions in solution and in the mineral solid have to be combined. This combination leads to transferability issues since electronic polarizability is not represented in a classical atomistic forcefield, and the compromises that are made to approximately account for its effects in a classical parametrization are different in the different phases. As a consequence, the typical "recipes" to combine parameters for different components cannot be straightforwardly applied, resulting in a significant parametrization effort for such problems [188–190]. The increasing awareness of transferability as a modeling challenge and the solution strategies developed in the context of coarse grained models may therefore very well benefit other areas of model development such as classical atomistic force-fieds for multicomponent materials systems.

In the second part of the review we have discussed the recent advances in the field of adaptive resolution approaches. The above mentioned limitation in system size comes together with the disappointing fact that a considerable fraction of the simulated data is often discarded afterwards: the solvent, for example, is usually not involved in the analysis of the system, but it is nonetheless required by the simulation. Adaptive resolution methods try to reduce the amount of resources dedicated to the simulation of large, non-interesting regions of the system by replacing them with a simpler, coarse-grained representation of their content. Such "dual-resolution" schemes are built with the constraint that the thermodynamical properties of the region of interest (*i.e.*, the one with the higher resolution) do not differ from those that an equivalent subdomain of the system would have in a fully high-resolution simulation.

In the present work we discussed two methods to achieve this goal: the Adaptive Resolution Simulation (AdResS) scheme, based on the interpolation of two different force-fields, and its Hamiltonian formulation, H-AdResS, where the all-atom and coarse-grained potential energies are interpolated. These methods have been successfully used to interface different molecular fluids, treated at the atomistic level, with their coarse-grained models; the different properties of the AA and the CG potentials naturally induce thermodynamical imbalances in the corresponding sub-regions, but simple and effective ways to overcome this problem have been described.

The possibility of replacing vast regions of the simulated system with a crude, cheap-to-compute representation and concentrating the computational resources on smaller parts while keeping the relative thermodynamics under control makes it possible to sensibly reduce the amount of calculations required to perform a simulation, and opens the way to a broad spectrum of applications, such as large-scale simulations of complex biomolecules in solution and efficient open-boundary simulations with varying number of particles.

Acknowledgments

We thank all present and past members of the multiscale modeling group at the Max Planck Institute for Polymer Research, the theoretical chemistry group at the University of Konstanz as well many further colleagues for fruitful and enjoyable collaborations, in particular Luigi Delle Site, Rafael Delgado Buscalioni, Davide Donadio, Pep Español, Ralf Everaers, Sebastian Fritsch, Mara Jochum, Christoph Junghans, Biswaroop Mukherjee, Simon Poblete, Matej Praprotnik, and Nico van der Vegt. We would also like to thank the Kavli Institute for Theoretical Physics for sponsoring and hosting the workshop "Physical Principles of Multiscale Modeling, Analysis and Simulation in Soft Condensed Matter" and the participants of this workshop for many stimulating discussions. CP acknowledges financial support by the German Science Foundation within the Emmy Noether Programme (grant PE 1625/1-1) and by the Volkswagen Foundation within the call "New Conceptual Approaches to Modeling and Simulation of Complex Systems". Kurt Kremer acknowledges research funding through the European Research Council under the European Union's Seventh Framework Programme (FP7/2007-2013) / ERC grant agreement n. 340906-MOLPROCOMP. All authors express their deep gratitude to Aoife Fogarty for a careful proofreading of the manuscript.

Author Contributions

All authors contributed equally to the composing as well as the writing of this paper. All authors have read and approved the final published manuscript.

Conflicts of Interest

The authors declare no conflict of interest.

References

1. Alder, B.; Wainwright, T. Phase transition for a hard sphere system. *J. Chem. Phys.* **1957**, *5*, 1208–1209.
2. Grest, G.; Kremer, K. Molecular dynamics simulation for polymers in the presence of a heat bath. *Phys. Rev. A* **1986**, *33*, 3628–3631.
3. Kremer, K.; Grest, G.; Carmesin, I. Crossover from Rouse to Reptation Dynamics: A Molecular-Dynamics Simulation. *Phys. Rev. Lett.* **1988**, *61*, 566–569.
4. McCammon, J.; Karplus, M. Internal motions of antibody molecules. *Nature* **1977**, *268*, 765–766.

5. Karplus, M.; McCammon, J. Protein structural fluctuations during a period of 100 ps. *Nature* **1979**, *277*, 578–578.

6. Raiteri, P.; Laio, A.; Gervasio, F.L.; Micheletti, C.; Parrinello, M. Efficient reconstruction of complex free energy landscapes by multiple walkers metadynamics. *J. Phys. Chem. B* **2006**, *110*, 3533–3539.

7. Lou, H.; Cukier, R.I. Molecular dynamics of apo-adenylate kinase: A distance replica exchange method for the free energy of conformational fluctuations. *J. Phys. Chem. B* **2006**, *110*, 12796–12808.

8. Arora, K.; Brooks, C.L. Large-scale allosteric conformational transitions of adenylate kinase appear to involve a population-shift mechanism. *Proc. Natl. Acad. Sci. USA* **2007**, *104*, 18496–18501.

9. Pontiggia, F.; Zen, A.; Micheletti, C. Small and large scale conformational changes of adenylate kinase: A molecular dynamics study of the subdomain motion and mechanics. *Biophys. J.* **2008**, *95*, 5901–5912.

10. Kremer, K. Computer simulations in soft matter science. In *Soft and Fragile Matter: Non Equilibrium Dynamics, Metastability And Flow*; IOP Publishing Ltd.: Bristol, UK, 2000; SUSSP Proceedings, Volume 53, pp. 145–184.

11. Kremer, K.; Müuller-Plathe, F. Multiscale problems in polymer science: Simulation approaches. *MRS Bull.* **2001**, *26*, 205–210.

12. Van der Vegt, N.A.; Peter, C.; Kremer, K. *Structure-Based Coarse- and Fine-Graining in Soft Matter Simulations*; CRC Press—Taylor and Francis Group: Boca Raton, FL, USA, 2009; pp. 379–397.

13. Hijón, C.; Vanden-Eijnden, E.; Delgado-Buscalioni, R.; Español, P. Mori-Zwanzig formalism as a practical computational tool. *Faraday Discuss.* **2010**, *144*, 301–322.

14. Noid, W. Systematic Methods for Structurally Consistent Coarse-Grained Models. In *Biomolecular Simulations*; Humana Press: New York, NY, USA, 2013; Volume 924, pp. 487–531.

15. Noid, W.G. Perspective: Coarse-grained models for biomolecular systems. *J. Chem. Phys.* **2013**, *139*, 090901.

16. Yelash, L.; Mueller, M.; Paul, W.; Binder, K. How well can coarse-grained models of real polymers describe their structure? The case of polybutadiene. *J. Chem. Theory Comput.* **2006**, *2*, 588–597.

17. Spyriouni, T.; Tzoumanekas, C.; Theodorou, D.; Müller-Plathe, F.; Milano, G. Coarse-Grained and Reverse-Mapped United-Atom Simulations of Long-Chain Atactic Polystyrene Melts: Structure, Thermodynamic Properties, Chain Conformation, and Entanglements. *Macromolecules* **2007**, *40*, 3876–3885.

18. Tirion, M.M.; ben Avraham, D. Normal mode analysis of G-actin. *JMB* **1993**, *230*, 186–195.

19. Tirion, M.M. Large amplitude elastic motions in proteins from a single–parameter, atomic analysis. *Phys. Rev. Lett.* **1996**, *77*, 1905–1908.

20. Bahar, I.; Atilgan, A.R.; Erman, B. Direct evaluation of thermal fluctuations in proteins using a single parameter harmonic potential. *Fold. Des.* **1997**, *2*, 173–181.

21. Micheletti, C.; Carloni, P.; Maritan, A. Accurate and efficient description of protein vibrational dynamics: comparing molecular dynamics and Gaussian models. *Proteins* **2004**, *55*, 635–645.

22. Potestio, R.; Pontiggia, F.; Micheletti, C. Coarse-grained description of proteins' internal dynamics: An optimal strategy for decomposing proteins in rigid subunits. *Biophys. J.* **2009**, *96*, 4993–5002.

23. Globisch, C.; Krishnamani, V.; Deserno, M.; Peter, C. Optimization of an Elastic Network Augmented Coarse Grained Model to Study CCMV Capsid Deformation. *PLoS One* **2013**, *8*, e60582.

24. Praprotnik, M.; Delle Site, L.; Kremer, K. Adaptive resolution molecular-dynamics simulation: Changing the degrees of freedom on the fly. *J. Chem. Phys.* **2005**, *123*, 224106–224114.

25. Praprotnik, M.; Delle Site, L.; Kremer, K. Adaptive resolution scheme for efficient hybrid atomistic-mesoscale molecular dynamics simulations of dense liquids. *Phys. Rev. E* **2006**, *73*, 066701.

26. Praprotnik, M.; Delle Site, L.; Kremer, K. A macromolecule in a solvent: Adaptive resolution molecular dynamics simulation. *J. Chem. Phys.* **2007**, *126*, 134902.

27. Praprotnik, M.; Delle Site, L.; Kremer, K. Multiscale Simulation of Soft Matter: From Scale Bridging to Adaptive Resolution. *Ann. Rev. Phys. Chem.* **2008**, *59*, 545–571.

28. Fritsch, S.; Junghans, C.; Kremer, K. Structure Formation of Toluene around C60: Implementation of the Adaptive Resolution Scheme (AdResS) into GROMACS. *J. Chem. Theory Comput.* **2012**, *8*, 398–403.

29. Poma, A.B.; Site, L.D. Classical to Path-Integral Adaptive Resolution in Molecular Simulation: Towards a Smooth Quantum-Classical Coupling. *Phys. Rev. Lett.* **2010**, *104*, 250201.

30. Potestio, R.; Delle Site, L. Quantum locality and equilibrium properties in low-temperature parahydrogen: A multiscale simulation study. *J. Chem. Phys.* **2012**, doi:10.1063/1.3678587.

31. Ensing, B.; Nielsen, S.; Moore, P.; Klein, M.; Parrinello, M. Energy Conservation in Adaptive Hybrid Atomistic/Coarse-Grain Molecular Dynamics. *J. Chem. Theory Comput.* **2007**, *3*, 1100–1105.

32. Praprotnik, M.; Poblete, S.; Delle Site, L.; Kremer, K. Comment on "Adaptive Multiscale Molecular Dynamics of Macromolecular Fluids". *Phys. Rev. Lett.* **2011**, *107*, 099801.

33. Potestio, R.; Fritsch, S.; Español, P.; Delgado-Buscalioni, R.; Kremer, K.; Everaers, R.; Donadio, D. Hamiltonian Adaptive Resolution Simulation for Molecular Liquids. *Phys. Rev. Lett.* **2013**, *110*, 108301.

34. Potestio, R.; Español, P.; Delgado-Buscalioni, R.; Everaers, R.; Kremer, K.; Donadio, D. Monte Carlo Adaptive Resolution Simulation of Multicomponent Molecular Liquids. *Phys. Rev. Lett.* **2013**, *111*, 060601.

608

35. Marx, D.; Sutmann, G.; Grotendorst, J.; Gompper, G. *Hierarchical Methods for Dynamics in Complex Molecular Systems*; Forschungszentrum Jülich: Jülich, Germany, 2012; Volume 10.

36. Müser, M.; Sutmann, G.; Winkler, R. *Hybrid Particle-Continuum Methods in Computational Material Physics*; NIC Series; Forschungszentrum Jülich, John von Neumann Institute: Jülich, Germany, 2013; Volume 46.

37. Shen, J.W.; Li, C.; van der Vegt, N.F.A.; Peter, C. Transferability of Coarse Grained Potentials: Implicit Solvent Models for Hydrated Ions. *J. Chem. Theory Comput.* **2011**, *7*, 1916–1927.

38. Villa, A.; Peter, C.; van der Vegt, N.F.A. Transferability of Nonbonded Interaction Potentials for Coarse-Grained Simulations: Benzene in Water. *J. Chem. Theory Comput.* **2010**, *6*, 2434–2444.

39. Mukherjee, B.; Delle Site, L.; Kremer, K.; Peter, C. Derivation of Coarse Grained Models for Multiscale Simulation of Liquid Crystalline Phase Transitions. *J. Phys. Chem. B* **2012**, *116*, 8474–8484.

40. Mukherjee, B.; Peter, C.; Kremer, K. Dual translocation pathways in smectic liquid crystals facilitated by molecular flexibility. *Phys. Rev. E* **2013**, *88*, 010502.

41. Fritz, D.; Koschke, K.; Harmandaris, V.A.; van der Vegt, N.F.A.; Kremer, K. Multiscale modeling of soft matter: scaling of dynamics. *Phys. Chem. Chem. Phys.* **2011**, *13*, 10412–10420.

42. Lopez, C.; Nielsen, S.; Moore, P.; Shelley, J.; Klein, M. Self-assembly of a phospholipid Langmuir monolayer using a coarse-grained molecular dynamics simulations. *J. Phys.: Condens. Matter* **2002**, *14*, 431–9444.

43. Cooke, I.R.; Kremer, K.; Deserno, M. Tunable generic model for fluid bilayer membranes. *Phys. Rev. E* **2005**, *72*, 011506.

44. Müller, M.; Katsov, K.; Schick, M. Biological and synthetic membranes: What can be learned from a coarse-grained description? *Phys. Rep.* **2006**, *434*, 113–176.

45. Reynwar, B.J.; Illya, G.; Harmandaris, V.A.; Müller, M.M.; Kremer, K.; Deserno, M. Aggregation and vesiculation of membrane proteins by curvature-mediated interactions. *Nature* **2007**, *447*, 461–464.

46. Klein, M.L.; Shinoda, W. Large-scale molecular dynamics simulations of self-assembling systems. *Science* **2008**, *321*, 798–800.

47. Go, N. Theoretical-studies of protein folding. *Annu. Rev. Biophys. Bioeng.* **1983**, *12*, 183–210.

48. Thirumalai, D.; Klimov, D.K. Deciphering the timescales and mechanisms of protein folding using minimal off-lattice models. *Curr. Opin. Struct. Biol.* **1999**, *9*, 197–207.

49. Liwo, A.; Arlukowicz, P.; Czaplewski, C.; Oldziej, S.; Pillardy, J.; Scheraga, H.A. A method for optimizing potential-energy functions by a hierarchical design of the potential-energy landscape: Application to the UNRES force field. *Proc. Natl. Acad. Sci. USA* **2002**, *99*, 1937–1942.

50. Favrin, G.; Irback, A.; Wallin, S. Folding of a small helical protein using hydrogen bonds and hydrophobicity forces. *Proteins* **2002**, *47*, 99–105.

51. Head-Gordon, T.; Brown, S. Minimalist models for protein folding and design. *Curr. Opin. Struct. Biol.* **2003**, *13*, 160–167.

52. Nguyen, H.D.; Hall, C.K. Molecular dynamics simulations of spontaneous fibril formation by random-coil peptides. *Proc. Natl. Acad. Sci. USA* **2004**, *101*, 16180 – 16185.

53. Buchete, N.V.; Straub, J.E.; Thirumalai, D. Development of novel statistical potentials for protein fold recognition. *Curr. Opin. Struct. Biol.* **2004**, *14*, 225–232.

54. Clementi, C. Coarse-grained models of protein folding: Toy models or predictive tools? *Curr. Opin. Struc. Biol.* **2008**, *18*, 10–15.

55. Derreumaux, P.; Mousseau, N. Coarse-grained protein molecular dynamics simulations. *J. Chem. Phys.* **2007**, *126*, 025101.

56. Bellesia, G.; Shea, J.E. Self-assembly of beta-sheet forming peptides into chiral fibrillar aggregates. *J. Chem. Phys.* **2007**, *126*, 245104.

57. Bereau, T.; Deserno, M. Generic coarse-grained model for protein folding and aggregation. *J. Chem. Phys.* **2009**, *130*, 235106.

58. Tozzini, V. Minimalist models for proteins: A comparative analysis. *Q. Rev. Biophys.* **2010**, *43*, 333–371.

59. Wu, C.; Shea, J.E. Coarse-grained models for protein aggregation. *Curr. Opin. Struc. Biol.* **2011**, *21*, 209–220.

60. De Gennes, P. *Scaling Concepts in Polymer Physics*; Cornell University Press: Ithaca, NY, USA, 1979; p. 324.

61. De Gennes, P.G. Some conformation problems for long macromolecules. *Rep. Prog. Phys.* **1969**, *32*, doi:10.1088/0034-4885/32/1/304.

62. De Gennes, P.G. Exponents for the excluded volume problem as derived by Wilson method. *Phys. Lett. A* **1972**, *A 38*, 339–340.

63. Kremer, K.; Grest, G.S. Dynamics of entangled linear polymer melts: A molecular-dynamics simulation. *J. Chem. Phys.* **1990**, *92*, 5057–5086.

64. Doi, M.; Edwards, S.F. *The Theory of Polymer Dynamics*; Oxford University Press: Oxford, UK, 1986.

65. Nielsen, S.O.; Lopez, C.F.; Srinivas, G.; Klein, M.L. A coarse grain model for n-alkanes parameterized from surface tension data. *J. Chem. Phys.* **2003**, *119*, 7043–7049.

66. Marrink, S.J.; deVries, A.H.; Mark, A.E. Coarse Grained Model for Semiquantitative Lipid Simulations. *J. Phys. Chem. B* **2004**, *108*, 750–760.

67. Marrink, S.J.; Risselada, H.J.; Yefimov, S.; Tieleman, D.P.; de Vries, A.H. The MARTINI Force Field: Coarse Grained Model for Biomolecular Simulations. *J. Phys. Chem. B* **2007**, *111*, 7812–7824.

68. Shinoda, W.; DeVane, R.; Klein, M.L. Multi-property fitting and parameterization of a coarse grained model for aqueous surfactants. *Mol. Simul.* **2007**, *33*, 27–36.

69. Monticelli, L.; Kandasamy, S.K.; Periole, X.; Larson, R.G.; Tieleman, D.P.; Marrink, S.J. The MARTINI Coarse-Grained Force Field: Extension to Proteins. *J. Chem. Theory Comput.* **2008**, *4*, 819–834.

70. Mognetti, B.M.; Yelash, L.; Virnau, P.; Paul, W.; Binder, K.; Mueller, M.; Macdowell, L.G. Efficient prediction of thermodynamic properties of quadrupolar fluids from simulation of a coarse-grained model: The case of carbon dioxide. *J. Chem. Phys.* **2008**, *128*, 104501.

71. Mognetti, B.M.; Virnau, P.; Yelash, L.; Paul, W.; Binder, K.; Müller, M.; Macdowell, L.G. Coarse-grained models for fluids and their mixtures: Comparison of Monte Carlo studies of their phase behavior with perturbation theory and experiment. *J. Chem. Phys.* **2009**, *130*, 044101.

72. López, C.A.; Rzepiela, A.J.; de Vries, A.H.; Dijkhuizen, L.; Hünenberger, P.H.; Marrink, S.J. Martini Coarse-Grained Force Field: Extension to Carbohydrates. *J. Chem. Theory Comput.* **2009**, *5*, 3195–3210.

73. DeVane, R.; Shinoda, W.; Moore, P.B.; Klein, M.L. Transferable Coarse Grain Nonbonded Interaction Model for Amino Acids. *J. Chem. Theory Comput.* **2009**, *5*, 2115–2124.

74. DeVane, R.; Klein, M.L.; Chiu, C.C.; Nielsen, S.O.; Shinoda, W.; Moore, P.B. Coarse-Grained Potential Models for Phenyl-Based Molecules: I. Parametrization Using Experimental Data. *J. Phys. Chem. B* **2010**, *114*, 6386–6393.

75. He, X.; Shinoda, W.; DeVane, R.; Klein, M.L. Exploring the utility of coarse-grained water models for computational studies of interfacial systems. *Mol. Phys.* **2010**, *108*, 2007–2020.

76. Yesylevskyy, S.O.; Schafer, L.V.; Sengupta, D.; Marrink, S.J. Polarizable Water Model for the Coarse-Grained MARTINI Force Field. *PLoS Comput. Biol.* **2010**, *6*, e1000810.

77. Tschöp, W.; Kremer, K.; Batoulis, J.; Burger, T.; Hahn, O. Simulation of polymer melts. I. Coarse-graining procedure for polycarbonates. *Acta Polym.* **1998**, *49*, 61–74.

78. Lyubartsev, A.P.; Laaksonen, A. Calculation of effective interaction potentials from radial-distribution functions—A reverse Monte-Carlo approach. *Phys. Rev. E* **1995**, *52*, 3730–3737.

79. Lyubartsev, A.P.; Laaksonen, A. Osmotic and activity coefficients from effective potentials for hydrated ions. *Phys. Rev. E* **1997**, *55*, 5689–5696.

80. Müller-Plathe, F. Coarse-graining in polymer simulation: From the atomistic to the mesoscopic scale and back. *ChemPhysChem* **2002**, *3*, 754–769.

81. Reith, D.; Pütz, M.; Müller-Plathe, F. Deriving effective mesoscale potentials from atomistic simulations. *J. Comput. Chem.* **2003**, *24*, 1624–1636.

82. Peter, C.; Delle Site, L.; Kremer, K. Classical simulations from the atomistic to the mesoscale: Coarse graining an azobenzene liquid crystal. *Soft Matter* **2008**, *4*, 859–869.

83. Murtola, T.; Karttunen, M.; Vattulainen, I. Systematic coarse graining from structure using internal states: Application to phospholipid/cholesterol bilayer. *J. Chem. Phys.* **2009**, *131*, 055101.

84. Lyubartsev, A.; Mirzoev, A.; Chen, L.J.; Laaksonen, A. Systematic coarse-graining of molecular models by the Newton inversion method. *Faraday Discuss.* **2010**, *144*, 43–56.

85. Savelyev, A.; Papoian, G.A. Molecular renormalization group coarse-graining of electrolyte solutions: Application to aqueous NaCl and KCl. *J. Phys. Chem. B* **2009**, *113*, 7785–7793.

86. Savelyev, A.; Papoian, G.A. Molecular Renormalization Group Coarse-Graining of Polymer Chains: Application to Double-Stranded DNA. *Biophys. J.* **2009**, *96*, 4044–4052.

87. Savelyev, A.; Papoian, G.A. Chemically accurate coarse graining of double-stranded DNA. *Proc. Natl. Acad. Sci. USA* **2010**, *107*, 20340–20345.

88. Megariotis, G.; Vyrkou, A.; Leygue, A.; Theodorou, D.N. Systematic Coarse Graining of 4-Cyano-4 '-pentylbiphenyl. *Ind. Eng. Chem. Res.* **2011**, *50*, 546–556.

89. Mukherje, B.; Delle Site, L.; Kremer, K.; Peter, C. Derivation of a Coarse Grained model for Multiscale Simulation of Liquid Crystalline Phase Transitions. *J. Phys. Chem. B* **2012**, 116, 8474–8484.

90. Izvekov, S.; Voth, G.A. A multiscale coarse-graining method for biomolecular systems. *J. Phys. Chem. B* **2005**, *109*, 2469–2473.

91. Ayton, G.S.; Noid, W.G.; Voth, G.A. Multiscale modeling of biomolecular systems: In serial and in parallel. *Curr. Opin. Struct. Biol.* **2007**, *17*, 192–198.

92. Zhou, J.; Thorpe, I.F.; Izvekov, S.; Voth, G.A. Coarse-grained peptide modeling using a systematic multiscale approach. *Biophys. J.* **2007**, *92*, 4289–4303.

93. Hills, R.D.; Lu, L.; Voth, G.A. Multiscale Coarse-Graining of the Protein Energy Landscape. *PLoS Comput. Biol.* **2010**, *6*, e1000827.

94. Izvekov, S.; Chung, P.W.; Rice, B.M. The multiscale coarse-graining method: Assessing its accuracy and introducing density dependent coarse-grain potentials. *J. Chem. Phys.* **2010**, *133*, 064109.

95. Mullinax, J.W.; Noid, W.G. Recovering physical potentials from a model protein databank. *Proc. Natl. Acad. Sci. USA* **2010**, *107*, 19867–19872.

96. Shell, M.S. The relative entropy is fundamental to multiscale and inverse thermodynamic problems. *J. Chem. Phys.* **2008**, *129*, 144108.

97. Chaimovich, A.; Shell, M.S. Relative entropy as a universal metric for multiscale errors. *Phys. Rev. E* **2010**, *81*, 060104.

98. Chaimovich, A.; Shell, M.S. Coarse-graining errors and numerical optimization using a relative entropy framework. *J .Chem. Phys.* **2011**, *134*, 094112.

99. Mullinax, J.W.; Noid, W.G. A Generalized-Yvon-Born-Green Theory for Determining Coarse-Grained Interaction Potentials. *J. Phys. Chem. C* **2010**, *114*, 5661–5674.

100. Ellis, C.R.; Rudzinski, J.F.; Noid, W.G. Generalized-Yvon-Born-Green Model of Toluene. *Macromol. Theory Simul.* **2011**, *20*, 478–495.

101. Larini, L.; Lu, L.; Voth, G.A. The multiscale coarse-graining method. VI. Implementation of three-body coarse-grained potentials. *J. Chem. Phys.* **2010**, *132*, 164107.

102. Ercolessi, F.; Adams, J. Interatomic Potentials from First-Principles Calculations: The Force-Matching Method. *Europhys. Lett.* **1994**, *26*, doi:10.1209/0295-5075/26/8/005.

103. Tschöp, W.; Kremer, K.; Hahn, O.; Batoulis, J.; Burger, T. Simulation of polymer melts. II. From coarse-grained models back to atomistic description. *Acta Polym.* **1998**, *49*, 75–79.

104. Izvekov, S.; Voth, G. Effective Force Field for Liquid Hydrogen Fluoride from Ab Initio Molecular Dynamics Simulation Using the Force-Matching Method. *J. Phys. Chem. B* **2005**, *109*, 6573–6586.

105. Izvekov, S.; Voth, G. Multiscale coarse graining of liquid-state systems. *J. Chem. Phys.* **2005**, *123*, 134105.

106. Rudzinski, J.F.; Noid, W.G. Investigation of Coarse-grained Mappings via an Iterative Generalized Yvon-Born-Green Method. *J. Phys. Chem. B* **2014**, in press.

107. Jernigan, R.L.; Bahar, I. Structure-derived potentials and protein simulations. *Curr. Opin. Struct. Biol.* **1996**, *6*, 195–209.

108. Bahar, I.; Jernigan, R.L. Inter-residue potentials in globular proteins and the dominance of highly specific hydrophilic interactions at close separation. *J. Mol. Biol.* **1997**, *266*, 195–214.

109. Akkermans, R.L.C.; Briels, W.J. A structure-based coarse-grained model for polymer melts. *J. Chem. Phys.* **2001**, *114*, 1020–1031.

110. Henderson, R.L. Uniqueness Theorem for Fluid Pair Correlation-Functions. *Phys. Lett. A* **1974**, *A49*, 197–198.

111. Chayes, J.T.; Chayes, L.; Lieb, E.H. The Inverse Problem in Classical Statistical-Mechanics. *Commun. Math. Phys.* **1984**, *93*, 57–121.

112. Johnson, M.E.; Head-Gordon, T.; Louis, A.A. Representability problems for coarse-grained water potentials. *J. Chem. Phys.* **2007**, *126*, 144509.

113. D'Alessandro, M.; Cilloco, F. Information-theory-based solution of the inverse problem in classical statistical mechanics. *Phys. Rev. E* **2010**, *82*, 021128.

114. Schommers, W. A pair potential for liquid rubidium from the pair correlation function. *Phys. Lett.* **1973**, *43*, 157–158.

115. Soper, A.K. Empirical potential Monte Carlo simulation of fluid structure. *Chem. Phys.* **1996**, *202*, 295–306.

116. Rühle, V.; Junghans, C.; Lukyanov, A.; Kremer, K.; Andrienko, D. Versatile Object-Oriented Toolkit for Coarse-Graining Applications. *J. Chem. Theory Comput.* **2009**, *5*, 3211–3223.

117. Wang, H.; Junghans, C.; Kremer, K. Comparative atomistic and coarse-grained study of water: What do we lose by coarse-graining? *Eur. Phys. J. E* **2009**, *28*, 221–229.

118. Molinero, V.; Moore, E.B. Water Modeled As an Intermediate Element between Carbon and Silicon. *J. Phys. Chem. B* **2009**, *113*, 4008–4016.

119. Moore, E.B.; Molinero, V. Structural transformation in supercooled water controls the crystallization rate of ice. *Nature* **2011**, *479*, 506–508.

120. Rudzinski, J.F.; Noid, W.G. The Role of Many-Body Correlations in Determining Potentials for Coarse-Grained Models of Equilibrium Structure. *J. Phys. Chem. B* **2012**, *116*, 8621–8635.

121. Rzepiela, A.J.; Louhivuori, M.; Peter, C.; Marrink, S.J. Hybrid simulations: Combining atomistic and coarse-grained force fields using virtual sites. *Phys. Chem. Chem. Phys.* **2011**, *13*, 10437–10448.

122. Fu, C.C.; Kulkarni, P.M.; Scott Shell, M.; Gary Leal, L. A test of systematic coarse-graining of molecular dynamics simulations: Thermodynamic properties. *J. Chem. Phys.* **2012**, *137*, 164106.

123. Jochum, M.; Andrienko, D.; Kremer, K.; Peter, C. Structure-based coarse-graining in liquid slabs. *J. Chem. Phys.* **2012**, *137*, 064102.

124. Torrie, G.M.; Valleau, J.P. Non-Physical Sampling Distributions in Monte-Carlo Free-Energy Estimation: Umbrella Sampling. *J. Comput. Phys.* **1977**, *23*, 187–199.

125. Den Otter, W.K.; Briels, W.J. The calculation of free-energy differences by constrained molecular-dynamics simulations. *J. Chem. Phys.* **1998**, *109*, 4139–4146.

126. Villa, A.; Peter, C.; van der Vegt, N.F.A. Self-assembling dipeptides: Conformational sampling in solvent-free coarse-grained simulation. *Phys. Chem. Chem. Phys.* **2009**, *11*, 2077–2086.

127. Carr, R.; Comer, J.; Ginsberg, M.D.; Aksimentiev, A. Atoms-to-microns model for small solute transport through sticky nanochannels. *Lab Chip* **2011**, *11*, 3766–3773.

128. Hess, B.; Holm, C.; van der Vegt, N.F.A. Osmotic coefficients of atomistic NaCl (aq) force fields. *J. Chem. Phys.* **2006**, *124*, 164509.

129. Hess, B.; Holm, C.; van der Vegt, N.F.A. Modeling multibody effects in ionic solutions with a concentration dependent dielectric permittivity. *Phys. Rev. Lett.* **2006**, *96*, 147801.

130. Villa, A.; van der Vegt, N.F.A.; Peter, C. Self-assembling dipeptides: Including solvent degrees of freedom in a coarse-grained model. *Phys. Chem. Chem. Phys.* **2009**, *11*, 2068–2076.

131. Shell, M.S. The relative entropy is fundamental to multiscale and inverse thermodynamic problems. *J. Chem. Phys.* **2008**, *129*, 144108.

132. Chaimovich, A.; Shell, M.S. Anomalous waterlike behavior in spherically-symmetric water models optimized with the relative entropy. *Phys. Chem. Chem. Phys.* **2009**, *11*, 1901–1915.

133. Chaimovich, A.; Shell, M.S. Relative entropy as a universal metric for multiscale errors. *Phys. Rev. E* **2010**, *81*, 060104.

134. Rudzinski, J.F.; Noid, W.G. Coarse-graining entropy, forces, and structures. *J. Chem. Phys.* **2011**, *135*, 214101.

135. Chaimovich, A.; Shell, M.S. Anomalous waterlike behavior in spherically-symmetric water models optimized with the relative entropy. *Phys. Chem. Chem. Phys.* **2009**, *11*, 1901–1915.

136. Baron, R.; Trzesniak, D.; de Vries, A.H.; Elsener, A.; Marrink, S.J.; van Gunsteren, W.F. Comparison of thermodynamic properties of coarse-grained and atomic-level simulation models. *ChemPhysChem* **2007**, *8*, 452–461.

137. Betancourt, M.R.; Omovie, S.J. Pairwise energies for polypeptide coarse-grained models derived from atomic force fields. *J. Chem. Phys.* **2009**, *130*, 195103.

138. Mullinax, J.W.; Noid, W.G. Extended ensemble approach for deriving transferable coarse-grained potentials. *J. Chem. Phys.* **2009**, *131*, 104110.

139. Ganguly, P.; Mukherji, D.; Junghans, C.; van der Vegt, N.F.A. Kirkwood–Buff Coarse-Grained Force Fields for Aqueous Solutions. *J. Chem. Theory Comput.* **2012**, *8*, 1802–1807.

140. Brini, E.; Marcon, V.; van der Vegt, N.F.A. Conditional reversible work method for molecular coarse graining applications. *Phys. Chem. Chem. Phys.* **2011**, *13*, 10468–10474.

141. Brini, E.; van der Vegt, N.F.A. Chemically transferable coarse-grained potentials from conditional reversible work calculations. *J. Chem. Phys.* **2012**, *137*, 154113.

142. Brini, E.; Algaer, E.A.; Ganguly, P.; Li, C.; Rodríguez-Ropero, F.; van der Vegt, N.F.A. Systematic coarse-graining methods for soft matter simulations—A review. *Soft Matter* **2013**, *9*, 2108–2119.

143. Silbermann, J.R.; Klapp, S.H.L.; Schoen, M.; Chennamsetty, N.; Bock, H.; Gubbins, K.E. Mesoscale modeling of complex binary fluid mixtures: Towards an atomistic foundation of effective potentials. *J. Chem. Phys.* **2006**, *124*, 074105.

144. Ghosh, J.; Faller, R. State point dependence of systematically coarse–grained potentials. *Mol. Simul.* **2007**, *33*, 759–767.

145. Fritz, D.; Harmandaris, V.A.; Kremer, K.; van der Vegt, N.F.A. Coarse-Grained Polymer Melts Based on Isolated Atomistic Chains: Simulation of Polystyrene of Different Tacticities. *Macromolecules* **2009**, *42*, 7579–7588.

146. Wang, Y.L.; Lyubartsev, A.; Lu, Z.Y.; Laaksonen, A. Multiscale coarse-grained simulations of ionic liquids: Comparison of three approaches to derive effective potentials. *Phys. Chem. Chem. Phys.* **2013**, *15*, 7701–7712.

147. Allen, E.C.; Rutledge, G.C. A novel algorithm for creating coarse-grained, density dependent implicit solvent models. *J. Chem. Phys.* **2008**, *128*, 154115.

148. Allen, E.C.; Rutledge, G.C. Evaluating the transferability of coarse-grained, density-dependent implicit solvent models to mixtures and chains. *J. Chem. Phys.* **2009**, *130*, 034904.

149. Krishna, V.; Noid, W.G.; Voth, G.A. The multiscale coarse-graining method. IV. Transferring coarse-grained potentials between temperatures. *J. Chem. Phys.* **2009**, *131*, 024103.

150. Harmandaris, V.A.; Adhikari, N.P.; van der Vegt, N.F.A.; Kremer, K.; Mann, B.A.; Voelkel, R.; Weiss, H.; Liew, C. Ethylbenzene Diffusion in Polystyrene: United Atom Atomistic/Coarse Grained Simulations and Experiments. *Macromolecules* **2007**, *40*, 7026–7035.

151. Carbone, P.; Varzaneh, H.A.K.; Chen, X.; Müller-Plathe, F. Transferability of coarse-grained force fields: The polymer case. *J. Chem. Phys.* **2008**, *128*, 064904.

152. Fritz, D.; Harmandaris, V.A.; Kremer, K.; van der Vegt, N.F.A. Coarse-Grained Polymer Melts Based on Isolated Atomistic Chains: Simulation of Polystyrene of Different Tacticities. *Macromolecules* **2009**, *42*, 7579–7588.

153. Harmandaris, V.A.; Floudas, G.; Kremer, K. Temperature and Pressure Dependence of Polystyrene Dynamics through Molecular Dynamics Simulations and Experiments. *Macromolecules* **2011**, *44*, 393–402.

154. Ben-Naim, A. *Solvation Thermodynamics*; Plenum Press: New York, NY, USA, 1987.

155. Pastewka, L.; Pou, P.; Pérez, R.; Gumbsch, P.; Moseler, M. Describing bond-breaking processes by reactive potentials: Importance of an environment-dependent interaction range. *Phys. Rev. B* **2008**, *78*, 161402.

156. Pizzagalli, L.; Godet, J.; Guénolé, J.; Brochard, S.; Holmstrom, E.; Nordlund, K.; Albaret, T. A new parametrization of the Stillinger–Weber potential for an improved description of defects and plasticity of silicon. *J. Phys.-Condens. Matter* **2013**, *25*, 055801.

157. Curtin, W.; Ashcroft, N. Density-functional theory and freezing of simple liquids. *Phys. Rev. Lett.* **1986**, *56*, 2775–2778.

158. Rudd, R.; Broughton, J. Concurrent coupling of length scales in solid state systems. *Phys. Status Solidi B-Basic Res.* **2000**, *217*, 251–291.

159. Rottler, J.; Barsky, S.; Robbins, M. Cracks and Crazes: On Calculating the Macroscopic Fracture Energy of Glassy Polymers from Molecular Simulations. *Phys. Rev. Lett.* **2002**, *89*, 148304.

160. Csanyi, G.; Albaret, T.; Payne, M.C.; Vita, A.D. "Learn on the Fly": A Hybrid Classical and Quantum-Mechanical Molecular Dynamics Simulation. *Phys. Rev. Lett.* **2004**, *93*, 175503.

161. Jiang, D.; Carter, E. First principles assessment of ideal fracture energies of materials with mobile impurities: Implications for hydrogen embrittlement of metals. *Acta Mater.* **2004**, *52*, 4801–4807.

162. Lu, G.; Tadmor, E.; Kaxiras, E. From electrons to finite elements: A concurrent multiscale approach for metals. *Phys. Rev. B* **2006**, *73*, 024108.

163. Warshel, A.; Levitt, M. Theoretical Studies of Enzymic Reactions—Dielectric, Electrostatic and Steric Stabilization of Carbonium-Ion in Reaction of Lysozyme. *J. Mol. Biol.* **1976**, *103*, 227–249.

164. Gao, J.; Lipkowitz, K.; Boyd, D. *Methods and Applications of Combined Quantum Mechanical and Molecular Mechanical Potentials*; Wiley: Hoboken, NJ, USA 1995; pp. 119–185.

165. Svensson, M.; Humbel, S.; Froese, R.; Matsubara, T.; Sieber, S.; Morokuma, K. ONIOM: A multilayered integrated MO+MM method for geometry optimizations and single point energy predictions. A test for Diels-Alder reactions and Pt(P(t-Bu)(3))(2)+H-2 oxidative addition. *J. Phys. Chem.* **1996**, *100*, 19357–19363.

166. Carloni, P.; Rothlisberger, U.; Parrinello, M. The role and perspective of a initio molecular dynamics in the study of biological systems. *Acc. Chem. Res.* **2002**, *35*, 455–464.

167. Bulo, R.; Ensing, B.; Sikkema, J.; Visscher, L. Toward a Practical Method for Adaptive QM/MM Simulations. *J. Chem. Theory Comput.* **2009**, *5*, 2212–2221.

168. Delle Site, L. Some fundamental problems for an energy-conserving adaptive-resolution molecular dynamics scheme. *Phys. Rev. E* **2007**, *76*, 047701.

169. Fritsch, S.; Poblete, S.; Junghans, C.; Ciccotti, G.; Delle Site, L.; Kremer, K. Adaptive resolution molecular dynamics simulation through coupling to an internal particle reservoir. *Phys. Rev. Lett.* **2012**, *108*, 170602.

170. Poblete, S.; Praprotnik, M.; Kremer, K.; Delle Site, L. Coupling different levels of resolution in molecular simulations. *J. Chem. Phys.* **2010**, *132*, 114101.

616

171. Mukherji, D.; van der Vegt, N.F.A.; Kremer, K. Preferential Solvation of Triglycine in Aqueous Urea: An Open Boundary Simulation Approach. *J. Chem. Theory Comput.* **2012**, *8*, 3536–3541.

172. Delle Site, L.; Leon, S.; Kremer, K. BPA-PC on a Ni(111) Surface: The Interplay between Adsorption Energy and Conformational Entropy for Different Chain-End Modifications. *J. Am. Chem. Soc.* **2004**, *126*, 2944–2955.

173. Hess, B.; Kutzner, C.; van der Spoel, D.; Lindahl, E. Gromacs 4: Algorithms for highly efficient, load-balanced, and scalable molecular simulation. *J. Chem. Theory Comput.* **2008**, *4*, 435–447.

174. Halverson, J.D.; Brandes, T.; Lenz, O.; Arnold, A.; Bevc, S.; Starchenko, V.; Kremer, K.; Stuehn, T.; Reith, D. ESPResSo++: A modern multiscale simulation package for soft matter systems. *Comput. Phys. Commun.* **2013**, *184*, 1129–1149.

175. Lambeth, B.J.; Junghans, C.; Kremer, K.; Clementi, C.; Delle Site, L. Communication: On the locality of Hydrogen bond networks at hydrophobic interfaces. *J. Chem. Phys.* **2010**, *133*, 221101.

176. Poma, A.; Delle Site, L. Adaptive resolution simulation of liquid para-hydrogen: Testing the robustness of the quantum-classical adaptive coupling. *Phys. Chem. Chem. Phys.* **2011**, *13*, 10510–10519.

177. Silvera, I.; Goldman, V. The isotropic intermolecular potential for H_2 and D_2 in the solid and gas phases. *J. Chem. Phys.* **1978**, *69*, 4209–4213.

178. Silvera, I. The solid molecular hydrogens in the condensed phase: Fundamentals and static properties. *Rev. Mod. Phys.* **1980**, *52*, 393–452.

179. Feynman, R.P. Atomic Theory of the Two-Fluid Model of Liquid Helium. *Phys. Rev.* **1954**, *94*, 262–277.

180. Tuckermann, M.E. *Statistical Mechanics: Theory and Molecular Simulation*; Oxford University Press: Oxford, UK 2010.

181. Mukherji, D.; van der Vegt, N.F.A.; Kremer, K.; Delle Site, L. Kirkwood-Buff Analysis of Liquid Mixtures in an Open Boundary Simulation. *J. Chem. Theory Comput.* **2012**, *8*, 375–379.

182. Mukherji, D.; Kremer, K. Coil-Globule-Coil Transition of PNIPAm in Aqueous Methanol: Coupling All-Atom Simulations to Semi-Grand Canonical Coarse-Grained Reservoir. *Macromolecules* **2013**, *46*, 9158–9163.

183. Wang, H.; Hartmann, C.; Schütte, C.; Delle Site, L. Grand-Canonical-like Molecular-Dynamics Simulations by Using an Adaptive-Resolution Technique. *Phys. Rev. X* **2013**, *3*, 011018.

184. Heyden, A.; Truhlar, D.G. Conservative Algorithm for an Adaptive Change of Resolution in Mixed Atomistic/Coarse-Grained Multiscale Simulations. *J. Chem. Theory Comput.* **2008**, *4*, 217–221.

185. Park, J.H.; Heyden, A. Solving the equations of motion for mixed atomistic and coarse-grained systems. *Mol. Simul.* **2009**, *35*, 962–973.

186. Johnson, M.E.; Head-Gordon, T.; Louis, A.A. Representability problems for coarse-grained water potentials. *J. Chem. Phys.* **2007**, *126*, 144509.

187. Kirkwood, J. Statistical Mechanics of Fluid Mixtures. *J. Chem. Phys.* **1935**, *3*, 300–313.

188. Raiteri, P.; Demichelis, R.; Gale, J.D.; Kellermeier, M.; Gebauer, D.; Quigley, D.; Wright, L.B.; Walsh, T.R. Exploring the influence of organic species on pre-and post-nucleation calcium carbonate. *Faraday Discuss.* **2012**, 159, 61–85.

189. Shen, J.W.; Li, C.; van der Vegt, N.F.A.; Peter, C. Understanding the Control of Mineralization by Polyelectrolyte Additives: Simulation of Preferential Binding to Calcite Surfaces. *J. Phys. Chem. C* **2013**, *117*, 6904–6913.

190. Kahlen, J.; Salimi, L.; Sulpizi, M.; Peter, C.; Donadio, D. Interaction of Charged Amino-Acid Side Chains with Ions: An Optimization Strategy for Classical Force Fields. *J. Phys. Chem. B* **2014**, *118*, 3960–3972.

MDPI
St. Alban-Anlage 66
4052 Basel
Switzerland
Tel. +41 61 683 77 34
Fax +41 61 302 89 18
www.mdpi.com

Entropy Editorial Office
E-mail: entropy@mdpi.com
www.mdpi.com/journal/entropy

www.ingramcontent.com/pod-product-compliance
Lightning Source LLC
Chambersburg PA
CBHW050346230326
41458CB00102B/6428